THE CITY READER
Fourth edition

The fourth edition of the highly successful *The City Reader* brings together the very best of publications on the city. Classic writings by such authors as Lewis Mumford, Ernest W. Burgess, Le Corbusier, Lewis Wirth, Jane Jacobs, and Kevin Lynch meet the best contemporary writings of, among others, Sir Peter Hall, Richard Florida, Mike Davis, Michael Porter, Robert Putnam, Andrés Duany, Saskia Sassen, and Manuel Castells. New to the fourth edition are important classic writings on urban economics by Wilbur Thomson and on bosses and machines by James Bryce, Jane Addams, and William L. Riordan, and new contemporary material on sustainable urban development, the creative class, metropolitics, occidentalism, Asian megacities, and urban futurism by The Bruntland Commission, Richard Florida, Myron Orfield, Ian Buruma and Avishai Margalit, Aprodicio Laquian, and Joel Kotkin.

Fifty-seven generous selections are included: a combination of forty-six readings from the third edition and eleven entirely new selections. Structured to aid student understanding, the anthology features main and part introductions, as well as individual introductions to the selected articles. Each selection is introduced with a brief intellectual biography and a review of the author's writings and related literature, an explanation of how the piece fits into the broader context of urban history and practice, competing ideological perspectives on the city, and the major current debates concerning race and gender, globalization, terrorism, the impact of information technology on cities, civic engagement, and postmodernism.

The City Reader provides the comprehensive mapping of the terrain of Urban Studies, old and new. It is illustrated with over forty photographs and is essential reading for anyone interested in the city.

The City Reader, fourth edition, is the anchor text in Routledge's new Urban Reader Series described on the next two pages. It provides an interdisciplinary introduction to the study of the city for use by itself or in conjunction with more focused disciplinary and topical readers in the series.

Richard T. LeGates is Professor of Urban Studies at San Francisco State University.

Frederic Stout is Lecturer in Urban Studies at Stanford University.

THE ROUTLEDGE URBAN READER SERIES

Series editors

Richard T. LeGates
Professor of Urban Studies, San Francisco State University

Frederic Stout
Lecturer in Urban Studies, Stanford University

The Routledge Urban Reader Series responds to the need for comprehensive coverage of the classic and essential texts that form the basis of intellectual work in the various academic disciplines and professional fields concerned with cities.

The readers focus on the key topics encountered by undergraduates, graduates students, and scholars in urban studies and allied fields. They discuss the contributions of major theoreticians and practitioners and other individuals, groups, and organizations that study the city or practice in a field that directly affects the city.

As well as drawing together the best of classic and contemporary writings on the city, each reader features extensive general, section and selection introductions prepared by the volume editors to place the selections in context, illustrate relations among topics, provide information on the author and point readers towards additional related bibliographic material.

Each reader contains:

- Approximately thirty-six *selections* divided into approximately six sections (eight sections and fifty-seven readings in *The City Reader*). Almost all of the selections are previously published works that have appeared as journal articles or portions of books.
- A *general introduction* describing the nature and purpose of the reader.
- Two- to three-page *section introductions* for each section of the reader to place the readings in context.
- A one-page *selection introduction* for each selection, describing the author, the intellectual background of the selection, competing views of the subject matter of the selection, and bibliographic references to other readings by the same author and other readings related to the topic.
- A plate section with twelve to fifteen plates (forty-two in *The City Reader*) and illustrations at the beginning of each section.
- An index.

The series consists of the following titles:

THE CITY READER

The City Reader: Fourth edition – an interdisciplinary urban reader aimed at urban studies, urban planning, urban geography and urban sociology courses – is the *anchor urban reader*. Routledge published a first edition of *The City Reader* in 1996, a second edition in 2000, and a third edition in 2003. *The City Reader* has become one of the most widely used anthologies in urban studies, urban geography, urban sociology and urban planning courses in the world.

URBAN DISCIPLINARY READERS

The series contains *urban disciplinary readers* organized around social science disciplines. The urban disciplinary readers include both classic writings and recent, cutting-edge contributions to the respective disciplines. They are lively, high-quality, competitively priced readers which faculty can adopt as course texts and which also appeal to a wider audience.

TOPICAL URBAN ANTHOLOGIES

The urban series includes *topical urban readers* intended both as primary and supplemental course texts and for the trade and professional market.

INTERDISCIPLINARY ANCHOR TITLE

The City Reader: Fourth edition
Richard T. LeGates and Frederic Stout (eds)

URBAN DISCIPLINARY READERS

The Urban Geography Reader
Nicholas Fyfe and Judith Kenny (eds)

The Urban Sociology Reader
Jan Lin and Christopher Mele (eds)

The Urban Politics Reader
Elizabeth Strom and John Mollenkopf (eds)

The Urban and Regional Planning Reader
Eugenie Birch (ed.)

The Urban Design Reader
Michael Larice and Elizabeth Macdonald (eds)

TOPICAL URBAN READERS

The City Cultures Reader: Second edition
Malcolm Miles, Tim Hall with Iain Borden (eds)

The Cybercities Reader
Stephen Graham (ed.)

The Sustainable Urban Development Reader
Stephen M. Wheeler and Timothy Beatley (eds)

The Global Cities Reader
Neil Brenner and Roger Keil (eds)

■ ■ ■ ■ ■ ■

For further Information on The Routledge Urban Reader Series please visit our website:
www.geographyarena.com/geographyarena/urbanreaderseries
or contact

Andrew Mould
Routledge
2 Park Square
Milton Park
Abingdon
Oxon OX14 4RN
England
Andrew.mould@tandf.co.uk

Richard T. LeGates
Urban Studies Program
San Francisco State University
1600 Holloway Avenue
San Francisco, California 94132
(415) 338-2875
dlegates@sfsu.edu

Frederic Stout
Urban Studies Program
Stanford University
Stanford, California 94305–6050
(650) 725–6321
fstout@stanford.edu

The City Reader
Fourth edition

Edited by

Richard T. LeGates

and

Frederic Stout

Routledge
Taylor & Francis Group

LONDON AND NEW YORK

First published 1996
by Routledge
2 Park Square, Milton Park, Abingdon, Oxon OX14 4RN

Simultaneously published in the USA and Canada
by Routledge
270 Madison Ave, New York, NY 10016

Second edition 2000

Third edition 2003

Fourth edition 2007

Routledge is an imprint of the Taylor & Francis Group, an informa business

© 1996, 2000, 2003, 2007 selection and editorial matter Richard T. LeGates and Frederic Stout

Designed and typeset in Amasis MT Lt and Akzidenz Grotesk
by Keystroke, 28 High Street, Tettenhall, Wolverhampton

Printed and bound in Spain by Graphos S.A. Arte Sobre Papel

British Library Cataloguing in Publication Data
A catalogue record for this book is available from the British Library

Library of Congress Cataloging in Publication Data
The city reader / [edited by] Richard T. LeGates & Frederic Stout. – 4th ed.
    p. cm.– (Urban reader series)
Includes bibliographical references and index.
1. Urban policy. 2. Cities and towns. 3. City planning. I. LeGates, Richard T. II. Stout, Frederic, 1943– III. Series: Routledge urban reader series.
HT151.C586 2003
307.76–dc21   2003002134

ISBN 978–0–415–77083–5 (hbk)
ISBN 978–0–415–77084–2 (pbk)

To Joanne Fraser
and Lisa Ryan
significant others

COMMENTS ON PREVIOUS EDITIONS

"Comprehensive and deep, this collection embodies the grand tradition, both classical and contemporary, of the urban field. It is a course itself; or a great lode for reference."

Robert J.S. Ross, Ph.D., *Professor of Sociology and Director of International Studies Stream, Clark University*

"This is the most useful reader on the market for students of cities. LeGates and Stout have refined the selections with each edition. My students tell me that the introductory notes and references make the readings more meaningful."

Ben Kohl, *Assistant Professor of Geography and Urban Studies, Temple University, Philadelphia*

"*The City Reader* brings together key works on the urban experience, problems, and policy alternatives in an engaging, accessibly structured and informative way. It draws together classic works and recent scholarship, capturing the dynamism of cities, urban processes and our interpretations of urban life. This is an impressive, comprehensive resource."

Dr Niall Majury, *School of Geography, Queen's University Belfast*

"*The City Reader* is a survival kit for practitioners who seek a better understanding of the towns and cities in which they work."

Steve Crocker, *Senior Skills Advisor, Neighbourhood Renewal Unit, Office of the Deputy Prime Minister, London*

"*The City Reader* weaves urban studies classics and modern writings in a masterful anthology. Editors' introductions to each section and piece make it an effective and accessible classroom tool."

Verrdie A. Craig, *Department of Geography, Rutgers University*

". . . remarkably well-chosen articles and excerpts . . . provides a solid introduction to the main currents of thought about urban form and processes in the twentieth century. . . ."

Environment and Planning

"An excellent, wide-ranging, stimulating reader; attractively presented and easy to read."

Brian Whalley, *Department of Built Environment, De Montfort University*

"A comprehensive mapping of the terrain of Urban Studies, old and new."

Jamie Peck, *Department of Geography, University of Manchester*

"An excellent overview, real breadth of coverage. Particularly valuable as a collection of key contributions which give a real flavour for the temporal development of Urban Studies."

David Valler, *Department of Town and Regional Planning, University of Sheffield*

"An excellent comprehensive overview of urban development and source material."

Allan Bryce, *School of Architecture, University of Dundee*

"This volume is a most welcome collection, without precedent in range and quality."

Alan Simpson, *Urban Design Associates*

"An excellent bringing together of the most important papers and ideas that are relevant to the study of the urban environment."

K.J. Bussey, *Department of Land Economy, Paisley University*

"A real achievement. This book brings together 99 percent of the prominent names in Urban Studies."

Ian Robert Douglas, *Watson Institute for International Studies, Brown University, USA*

"An excellent range of texts. *The City Reader* gathers together some central classics of urban theory, with a few surprises and a number of other pieces, which can be difficult to acquire. Editors' comments are consistently illuminating."

Nick Freeman, *Department of English, University of Bristol*

"This is an essential reader for teaching about the cities and Urban Planning in developing countries."

Horng-Chang Hsieh, *Urban Planning Department, Taiwan University*

"I think this is a splendid selection of writings which illustrate the development of modern thinking on urban problems. This is by far the best book of its type."

Dr Tom Begg, *Queen Margaret College, Edinburgh*

"Provides an international overview of urban design issues and a historical perspective on visionary planners who have shaped thinking about development."

Andrew McCafferty, *Department of Built Environment, Northumbria University*

Contents

Plates

Contributors

Jane Addams (1860–1935) was a pioneering leader of the American settlement house movement whose work at Hull House in Chicago inspired a generation of social workers.

Donald Appleyard (1928–1982) was a professor of urban design in the department of City and Regional Planning at the University of California, Berkeley.

Sherry Arnstein (d. 1997) was the chief adviser on citizen participation in the Model Cities Program at the United States Department of Housing and Urban Development in the late 1960s and early 1970s.

Ibn Battuta (1307–1377) was a Moroccan writer and traveler who recorded what he witnessed on his travels in Africa, the Middle East, and the Far East.

Timothy Beatley is Teresa Heinz Professor of Sustainable Communities in the Department of Urban and Environmental Planning at the University of Virginia and an authority on sustainable urban development and green urbanism.

James Bryce (1838–1922) was a British lord and keen observer of American institutions whose monumental study, *The American Commonwealth*, established the stereotype of American bosses and machines.

Ernest W. Burgess (1886–1966) was a sociology professor at the University of Chicago, and core member of the talented first generation of "Chicago school" sociologists, best known for his concentric zone theory of the internal structure of the city.

Ian Buruma is Luce Professor of Human Rights and Journalism at Bard College.

Peter Calthorpe is an architect practicing in Berkeley, California, and one of the founders of the New Urbanism movement.

Manuel Castells is a research professor at the Open University of Catalonia in Barcelona, Professor Emeritus of Sociology and City and Regional Planning at the University of California, Berkeley and holds the Wallis Annenberg Chair in Communication Technology and Society at the University of Southern California.

V. Gordon Childe (1892–1957) was an Australian professor of archaeology at the University of Edinburgh and director of the Institute of Archaeology at the University of London who studied the urban revolution in Mesopotamia.

Paul Davidoff (1930–1984) was a lawyer, urban planner, professor, and civil rights activist who worked to integrate housing as director of the Suburban Action Institute and proposed the "advocacy planning" model of urban planning.

Kingsley Davis (1908–1996) was Distinguished Professor of Sociology at the University of Southern California, Ford Professor of Sociology and Comparative Studies Emeritus at the University of California, Berkeley, and a senior research fellow at the Hoover Institution at Stanford University.

Mike Davis is a writer and social critic distinguished by his apocalyptic vision of postmodern Los Angeles and take-no-prisoners prose style.

Bernal Diaz del Castillo (1492–1581) was a Spanish conquistador, chronicler of the Cortes expedition, who described Tenochtitan and other Aztec cities, and later became a provincial governor.

Anthony Downs is a senior fellow at the Brookings Institution where he writes about metropolitan policy, real estate economics, housing, transportation, smart growth, urban policies, and other topics.

Andrés Duany is a Miami-based architect and planner who, with his wife Elizabeth Plater-Zyberk, is a leader in the "new urbanist" school of planning, and whose firm does regional and downtown plans, new towns, urban infill, villages and resort villages, transit-oriented development, suburban retrofits, campuses, housing, and civic buildings.

W.E.B. Du Bois (1868–1963) was a professor, editor, author, novelist, playwright, political activist, the first African–American to receive a Ph.D. degree from Harvard, and one of the preeminent intellectuals of his generation.

Albrecht Dürer (1471–1528) was a German-born artist who was court painter to the Holy Roman Emperors Maximilian I and Charles V.

Friedrich Engels (1820–1895) was a friend, partner and supporter of Karl Marx and one of the founders of the international Communist movement who, early in life, chronicled the deplorable conditions of the working class in Manchester in 1844.

Robert Fishman is a historian and professor of architecture and urban planning at the University of Michigan who has studied the rise of technoburbs.

Richard Florida is the Heinz Professor of Economic Development at Carnegie Mellon University, a senior fellow at the Brookings Institution, and the founder of the Creativity Group and Catalytix consulting firms who believes "creative class" individuals are essential to economic development.

John Forester is a professor of city and regional planning at Cornell University who writes about conflict resolution, mediation, and urban planning theory.

William Fulton is a journalist and urban planner and president of the Solimar Research Group of Ventura, California who writes about urban and regional planning issues.

David R. Godschalk is an emeritus professor of urban and regional planning at the University of North Carolina.

Peter Hall is a professor of planning at the Bartlett School of Architecture and Planning, University College London, and former professor of city and regional planning at the University of California, Berkeley, whose sweeping writings on planning history, theory, and practice have influenced the teaching and practice of urban planning worldwide.

David Harvey is a professor of geography and environmental engineering at Johns Hopkins University who has written extensively on Marxist and postmodernist urban theory.

Ebenezer Howard (1850–1928) was a British social reformer and the founder of the Garden City movement.

John Brinckerhoff Jackson (1909–1996) was the founder and editor of the journal *Landscape* and a professor at Harvard and the University of California, Berkeley, who developed a unique approach to understanding the cultural meaning of the American landscape.

Kenneth T. Jackson is the Jacques Barzun Professor of History and the Social Sciences at Columbia University who has studied the history of American suburbs.

Allan Jacobs is an emeritus professor of city and regional planning from the University of California, Berkeley, former San Francisco planning director, and a leading urban design theorist.

Jane Jacobs (1916–2006) was a New York City community activist turned urban theorist whose defense of disorderly and creative urban life, and attacks on rigid urban planning insensitive to human values, had a major impact on humanizing urban planning.

Edward J. Kaiser is an emeritus professor of urban and regional planning at the University of North Carolina where he taught land use planning.

George L. Kelling is a senior fellow at the Manhattan Institute and a professor of criminal justice at Rutgers University who, with James Q. Wilson, developed the "broken windows" theory of crime and community policing.

H.D.F. Kitto (1897–1982) was a professor of classics at the University of Bristol and was widely regarded as the preeminent classicist of his day.

Joel Kotkin is an Irvine Senior Fellow with the New America Foundation in Washington, DC, a senior fellow at the Newman Institute at Baruch College of the City University of New York, and a lecturer at the Southern California Institute of Architecture.

Aprodicio A. Laquian is Professor Emeritus of Community and Regional Planning and former director of the Centre for Human Settlements at the University of British Columbia.

Le Corbusier (Charle-Eduouard Jeanneret, 1887–1965) was an architect, an urban visionary, and an important force in the modernist movement.

Richard T. LeGates is a professor of urban studies at San Francisco State University and a lecturer in the University of California Berkeley Department of City and Regional Planning.

Kevin Lynch (1918–1994) was a professor of urban planning at the Massachusetts Institute of Technology whose influential writings on urban design provide the theoretical basis for modern urban design.

Ali Madanipour is a professor of urban design at the University of Newcastle in England, where he teaches courses on the relationship between urban society and urban design.

Avishai Margalit is Schulman Professor of Philosophy at the Hebrew University of Jerusalem.

William J. Mitchell is a professor of architecture and media arts at the Massachusetts Institute of Technology who has written about how information technology is changing cities.

Lewis Mumford (1895–1990) was a cultural historian, biographer, architectural critic, occasional academic, distinguished urbanist, and, many argue, the last great American public intellectual.

Frederick Law Olmsted (1822–1903) was a social reformer, landscape architect, and founder of the parks movement in America.

Myron Orfield is a law professor at the University of Minnesota Law School, former Minnesota State Congressman and Senator, and GIS expert, whose writings on metropolitics illuminate issues of metropolitan planning and governance.

Elizabeth Plater-Zyberk is an architect and planner, who, with her husband, designed Seaside, Florida, and helped found the "new urbanism" movement in architecture and planning.

George Washington Plunkitt (1842–1924) was a boss of New York City's Tammany Hall political machine whose blunt political advice on how to succeed in machine politics was immortalized by journalist William Riordan.

Marco Polo (1254–1324) was a Venetian merchant and traveler whose descriptions of Kin Sai (today's Hangzhou) and other Chinese cities astonished his European contemporaries.

Michael Porter is the Bishop William Lawrence University Professor of Business Administration at Harvard Business School and director of Harvard's Institute for Strategy and Competitiveness.

Robert D. Putnam is the Peter and Isabel Malkin Professor of Public Policy at Harvard University, most noted for his studies of the decline of civic engagement in America.

William Riordan (1841–1914) was a journalist who immortalized New York political boss George Washington Plunkitt's advice on how to succeed in machine politics in a book paraphrasing Plunkitt's colorful views.

Saskia Sassen is the Ralph Lewis Professor of Sociology at the University of Chicago and Centennial Visiting Professor at the London School of Economics, best known for her studies of global city systems.

Camillo Sitte (1843–1903) was an Austrian architect whose careful on-site studies of European public spaces inspired an urban design movement in the early twentieth century.

Edward Soja is a professor of geography at the University of California, Los Angeles and a member of the "Los Angeles School" of postmodernist scholars who consider Los Angeles the prototype of future cities.

Frederic Stout is a lecturer at Stanford University's Urban Studies Program and director of the co-terminal degree program in Stanford's School of Education.

Wilbur Thompson is an economist whose 1965 text *A Preface to Urban Economics* established the field of urban economics.

Melvin M. Webber is Professor Emeritus of City and Regional Planning at the University of California, Berkeley.

Stephen Wheeler is an assistant professor in the Lanscape Architecture Program at the University of California, Davis, and an authority on sustainable urban development.

William H. Whyte (1918–1999) was a sociologist whose perceptive studies of the way in which people use parks, plazas, and other public space in cities have influenced urban design practice.

James Q. Wilson is the Ronald Reagan Professor of Public Policy at Pepperdine University in California and an authority on criminal justice who, along with George L. Kelling, developed the "broken windows" theory of crime and community policing.

William Julius Wilson is the Lewis P. and Linda L. Geyser University Professor at Harvard University, past president of the American Sociological Association, and an acute observer of the plight of the Black underclass.

Louis Wirth (1897–1952) was a professor of sociology at the University of Chicago who studied the impact of urbanization on the human personality.

The **World Commission on Environment and Development** was a United Nations body chaired by Ms Gro Brundtland, Prime Minister of Norway, whose influential 1987 report brought the concept of sustainable development into the mainstream of public policy discourse.

Frank Lloyd Wright (1867–1959) was, according to many, the greatest American architect of all time and the originator of the "Broadacre City" vision for the form urban settlements might take in the twentieth century.

Acknowledgements

We received constant encouragement and many valuable suggestions, both for selections to include and approaches to critical commentary, from our colleagues. We wish particularly to thank Andrew Mould, our editor at Routledge, for his support, encouragement, and helpful suggestions. Kate O'Hearn at Routledge provided invaluable help in securing permissions. Alan Fidler did a first-rate job of copy-editing the manuscript. Our talented student assistants – June Park of Stanford and Ryan Dodge and Erika Lew of San Francisco State University – were especially helpful, supportive, and resourceful in the provision of expert research assistance.

Editors of the Routledge Urban Reader Series provided inspiration, advice, and assistance in selecting and commenting on selections within their domains of expertise: Timothy Beatley (University of Virgina), Eugenie Birch (University of Pennsylvania), Iain Borden (University College London), Neil Brenner (New York University), Nicholas Fyfe (University of Dundee), Stephen Graham (Newcastle University), Tim Hall (University College London), Roger Keil (York University), Judith Kenny (University of Minnesota), Michael Larice (University of Pennsylvania), Jan Lin (Occidental College), Elizabeth Macdonald (University of California, Berkeley), Chris Mele (University of Hong Kong), Malcolm Miles (University College London), John Mollenkopf (City University of New York), Elizabeth Strom (University of South Florida), Stephen Wheeler (University of New Mexico).

Deborah LeVeen, Raquel Pinderhughes, and Ayse Pamuk at San Francisco State University; Paul Turner, Leonard Ortolano, Linda Darling-Hammond, Ray McDermott, Milbrey McLaughlin, Luis Fraga, and Gerry Gast at Stanford University; Cheyney Ryan at the University of Oregon, Dan Lewis of Northwestern University, and Chester Hartman at the Poverty and Race Research and Action Council – all gave us many valuable suggestions. Alexander Garvin of Yale University, Peter Calthorpe of Calthorpe Associates, and Chris McGee of San Francisco State University were generous in sharing their insights about what visual images to include and their own copies of images they had assembled over the years. We are grateful to Peter Bialobrzeski for permission to use the stunning cover image of Shanghai from his book *Neon Tigers* (Ostfildern (Ruit), Germany: Hatje Cantz Publishers, 2004). Lisa Ryan contributed her artistic talents to creating visual images that appear at the beginning of the prologue, epilogue, and other sections of the reader. Many others, too numerous to mention, made helpful suggestions.

A number of people helped us with technical support throughout the writing and editing process. Particularly helpful were Alex Keller, Jason Stone, and Andrew Roderick at San Francisco State University. All errors and infelicities are, of course, ours.

INTRODUCTION

During the last thirty-five years our students in urban studies and city planning courses at San Francisco State University, Stanford University, and the University of California, Berkeley have often asked us what is the best writing on a given topic or what one single new writing captures current thinking about an important topic in urban studies or urban planning right now. Since there was no one source to which we could refer them, each of us accumulated photocopies of what we consider to be essential writings and bibliographic references to many more. As time passed our colleagues began to come to us for suggested course readings, and we in turn added other selections they have found most useful to our list. We realized that a systematic organization of the best writings we use to meet both requests would make a good anthology to introduce students of urban studies and city planning to the field and to supplement course texts used in these and other courses concerned with cities. Accordingly we set to work to produce *The City Reader*. The contents of the first edition of *The City Reader* were further enriched by members of a distinguished review panel who added their own suggestions for inclusion to our list of selections. Accordingly the first edition of *The City Reader*, published by Routledge in 1996, contained a generous selection of both kinds of essential readings – enduring writings and the exciting new writings that we, our colleagues, and expert reviewers considered would best introduce students to cities.

The first edition was well received and we learned a great deal more about what readings students and faculty find most useful from using the first edition in our own courses and receiving feedback from faculty colleagues. Our only regret was that space limitations made it impossible to include as many of the writings we had accumulated, and which reviewers suggested, as we would have liked.

In 2000 Routledge published an expanded and improved second edition of *The City Reader* that quickly established itself as required reading in courses in urban studies, urban and regional planning, urban geography, and urban sociology worldwide. Based on the success of the second edition Routledge suggested that we act as general editors for a series of urban readers modeled on *The City Reader*. We saw this as a way to draw on the expertise of scholars that went far beyond our own and to make many of the excellent selections we could not fit in *The City Reader* accessible to students worldwide. We enthusiastically agreed to oversee a series of urban readers organized around disciplinary perspectives (such as *The Urban Sociology Reader* and *The Urban Geography Reader*) and around important substantive themes (such as *The Sustainable Urban Development Reader* and *The Global Cities Reader*). Between 2001 and 2003 we assembled a talented team of scholars to edit nine additional readers in what has now become the Routledge Urban Reader Series.

This fourth edition of *The City Reader* continues and expands the tradition established in the earlier editions. Faculty and students familiar with earlier editions will find all of the classic and contemporary selections that have proven most useful, as well as exciting new material on green urbanism, urban economics, sustainable urban development, information technology, spatial equity, globalization, regionalism, responses to urban terrorism, the creative class, social capital, megacities, and much more. The fourth edition has a new section of first-person accounts of nineteenth-century bosses and urban political machines, and an epilogue that provides a comprehensive overview of the literature in urban studies and planning.

The City Reader, fourth edition, serves as the anchor text in the Routledge Urban Reader Series and articulates with the other nine volumes in the series (all of which will be in print when the fourth edition is published). Experts in the academic disciplines most related to the study of cities and in topical areas of greatest interest to urbanists have now published readers under our direction with a format similar to *The City Reader* to provide in-depth coverage of material in their respective disciplines and subject matter areas. Routledge published a third edition of *The City Reader* in 2003. In 2004 three additional titles joined the series: a second edition of *The City Cultures Reader* edited by Malcolm Miles and Tim Hall with Iain Borden, *The Sustainable Urban Development Reader* edited by Stephen Wheeler and Timothy Beatley, and *The Cybercities Reader* edited by Stephen Graham. In 2005 *The Urban Geography Reader* edited by Nick Fyfe and Judith Kenny, *The Urban Sociology Reader* edited by Jan Lin and Christopher Mele, and *The Global Cities Reader* edited by Neil Brenner and Roger Keil were published. As this introduction is written in May, 2006, two of the remaining readers in the series – *The Urban Politics Reader* edited by Elizabeth Strom and John Mollenkopf, and *The Urban Design Reader* edited by Michael Larice and Elizabeth Macdonald – were in production and scheduled to be in print in fall, 2006. The final volume in the Routledge Urban Reader Series, *The City and Regional Planning Reader* edited by Eugenie Birch, is scheduled for publication early in 2007.

It is a great satisfaction that the reader series provides space to include many more selections, covering topics introduced in *The City Reader* in much greater depth, and selections covering many additional topics beyond our subject matter expertise. Our talented team of editors has vastly leveraged our original concept and created a comprehensive compendium for understanding cities. From our first edition with fifty-five selections, the urban reader concept has expanded to include almost four hundred selections in ten volumes.

The City Reader and the other readers in the series focus on *essential* writings. We and the other editors picked enduring issues in urban studies and planning across different cultures and times. In our courses, we have found that H.D.F. Kitto's "The Polis" raises fundamental questions about individuals' relations to their communities that are as relevant today as they were 2,400 years ago; that Louis Wirth's seventy-year-old essay on "Urbanism as a Way of Life" speaks to our students trying to understand contemporary urban violence, economic dislocation, homelessness, and anomie; and that our students are excited by William Julius Wilson's theories on the Black underclass and Manuel Castells' reflections on the rise of the network society. Most writings in this edition of *The City Reader* are from twentieth-century writers, and more than half were written very recently.

This is an *international* anthology. In an increasingly global world students must learn from writers beyond the borders of their country of origin. In addition to writers from the United States, the fourth edition now contains writings by scholars from Austria, Australia, Belgium, England, France, Germany, Iran, Israel, Italy, Morocco, the Netherlands, the Philippines, Scotland, Spain, and Switzerland. Many of the writers included, from the great fourteenth-century Islamic wanderer Ibn Battuta to today's Peter Hall, are truly world citizens. The fourth edition of *The City Reader* includes new material on cities in Asia, and others of the readers now include a wealth of international material. Students studying globalization or comparative urbanism may particularly benefit from paired use of *The City Reader* and *The Global Cities Reader*.

The City Reader is an *interdisciplinary* anthology. The disciplines and professional fields represented in *The City Reader* include anthropology, architecture, archaeology, city planning, classics, culture studies, demography, economics, geography, history, landscape architecture, law, photography, political science, sociology, and urban design. Many of the writers blend insights from more than one discipline. Some of the best writings in *The City Reader* don't fit in conventional disciplinary boxes at all.

Cities are best studied from both interdisciplinary and disciplinary perspectives. The *disciplinary* Routledge Urban Readers contain writings by scholars from the respective academic disciplines that bring to bear their disciplinary expertise and provide depth in the literature of the specific discipline beyond what is possible in *The City Reader*. Pairing *The City Reader* and one of the Routledge Urban Disciplinary Readers will provide students in courses dealing with cities in a specific discipline both the broad interdisciplinary perspective of *The City Reader* and the disciplinary perspective of the disciplinary reader. Thus, for example, using both *The City Reader* and *The Urban Sociology Reader* will give students in urban sociology courses both an interdisciplinary understanding of cities and in-depth coverage of urban sociology topics written primarily by urban sociologists. Using both *The City Reader* and *The Urban Geography Reader* as course texts will give students in urban geography courses both a broad interdisciplinary introduction to scholarship on cities and in-depth coverage of the best scholarship by geographers on topics of particular interest to urban geographers. Using both *The City Reader* and *The Urban Politics Reader* will give students in urban politics courses both an interdisciplinary understanding of cities and in-depth coverage of urban politics topics written by political scientists.

The City Reader emphasizes the connection between the built environment of cities and the natural environment. As the world's population soars and urbanization continues, the imperative to design sustainable cities becomes ever more important. Readings by the World Commission on Environment and Development (The Brundtland Commission), Timothy Beatley, Stephen Wheeler, Peter Calthorpe, Andrés Duany, and Elizabeth Plater-Zyberk introduce students to sustainable urban development, green urbanism, and the New Urbanism. Courses on sustainable urban development, green urbanism, ecological design, and related topics may benefit from pairing *The City Reader* with *The Sustainable Urban Development Reader*.

The writings in this anthology include writings dealing with both urban theory and urban planning practice. Accordingly the fourth edition contains an epilogue that summarizes the literature in urban studies and planning. Courses on urban planning and urban design would benefit from pairing *The City Reader* with the *Urban and Regional Planning Reader* and *The Urban Design Reader*, or both.

An anthology of essential writings on cities should have a *flexible organization*. There is no one best way to organize material on cities. The content of urban studies and city planning courses varies widely and courses are organized in as many different ways as there are courses. This dictates a flexible structure for *The City Reader*. Readings are grouped into eight broad categories: The Evolution of Cities; Urban Culture and Society; Urban Space; Urban Politics, Governance, and Economics; Urban Planning History and Visions; Urban Planning Theory and Practice; Perspectives on Urban Design; and The Future of Cities. Professors may choose to assign some or all of these readings in whatever order best fits the logic of their own course. If *The City Reader* is paired with one or more of the other Routledge Urban Reader Series readers the possibilities for mixing and matching selections greatly increase.

Two other goals in picking the selections were to expose students to great thinking and clear writing. Almost everything written on the emergence of cities acknowledges a debt to the meticulous empirical research, creative theory building, and clear writing of Australian archaeologist V. Gordon Childe, on the internal structure of cities to Ernest W. Burgess's concentric zone model, and on urban design to Kevin Lynch's brilliant and lucid writings on the image of the city. Students can learn a great deal from the way Childe, Burgess, Lynch, and other great urban scholars think and write beyond the substantive content of the selections.

In the fourth edition of *The City Reader* we have said a good deal about the role of visions in urban studies and planning. We close with our own vision of how this anthology will be used. *The City Reader* is aimed primarily at students who will encounter many of the writers and writings for the first time. We hope the writings touch responsive chords and will inspire all the students who use *The City Reader* to think more deeply and read more widely about cities. To that end for each selection we point the way to other related writings by the same authors and other writers on the same subject matter. The epilogue is a more extensive synthesis of the literature on urban studies and planning. *The City Reader* will work well for students in general education courses who do not pursue urban studies further. For students who will take additional coursework in urban studies, city and regional planning, geography, sociology, political science, history, or other academic disciplines and professional fields we have designed *The City Reader* to provide an interdisciplinary overview of cities. In disciplinary courses *The City Reader* will work best if it is paired with another of the Routledge Urban Reader Series readers. *The City Reader* will provide a broad overview, an interdisciplinary perspective, and the best possible selections for each subject matter area that space permits. The disciplinary readers – edited by experts from the varying disciplines and containing many more writings by scholars from the respective discipline – will provide much more depth and a clear focus on the respective disciplinary areas.

We hope *The City Reader* will prove to be a book that students, professors, and practitioners will keep and periodically reread. One test to which we put each of the essential writings included is that it should still be relevant to re-read and enjoy for many years to come.

Richard LeGates
Frederic Stout
San Francisco, May 2006

Prologue

Prologue

Richard T. LeGates

Studying cities is a vast and never-ending enterprise. There is too much material for any one individual to master and always more to learn. Fortunately many fine scholars, past and present, have focused their attention on cities. We now know a great deal about how cities evolved, their social structures, urban culture, how cities are organized spatially, what economic functions they perform, how they are governed, how they are (and might be) planned, and their possible futures. *The City Reader*, fourth edition, contains fifty-seven selections by writers, past and present, who have contributed most to our understanding of cities. The selections introduce readers to the best scholarship on urban history, geography, politics, economics, sociology, planning, urban design, and futurism.

The City Reader is the anchor text in the ten-volume Routledge Urban Reader Series described in the front matter to this book. Nine additional volumes in the series, edited by experts in disciplines and topic areas most important for the study of cities, greatly extend the material introduced in *The City Reader* with entirely new topic areas, many more selections, and much greater depth of coverage. Five of the readers are organized around the core disciplines that contribute to urban studies and urban planning carefully chosen by scholars who are leaders in the disciplines of sociology, geography, and political science, and the professional fields of urban planning and urban design. Four other volumes in the series – chosen by leading scholars in key subject matter areas – bring together essential readings in four topical areas that are fundamental for understanding cities: sustainable urban development, urban culture, cybercities, and global cities. Together the ten-volume series affords students, professors, practitioners, and anyone interested in cities access to nearly four hundred of the best classic and contemporary writings on cities.

This selection provides a general roadmap for readers to the fields of urban studies and planning that can orient them before they read the selections in *The City Reader*. The epilogue at the end of the book is a more detailed roadmap to the literature of urban studies and planning intended to further clarify how *The City Reader* selections fit into the complete set of selections in all the readers and beyond.

While academic teaching about cities occurs in courses as different as English literature and civil engineering, most urban scholarship can be grouped under the heading of "urban studies," "urban planning," or one of the social science disciplines such as "urban sociology." A description of these fields and disciplines, and how they fit into universities, is useful.

DISCIPLINARY AND INTERDISCIPLINARY TEACHING ABOUT CITIES

Almost all modern universities organize teaching and research into academic units called *schools* or *colleges* such as a college of Social Science or a school of Architecture and Urban Planning. Schools and colleges contain academic *departments* and, often, *programs*. Professors educated in different academic *disciplines* are located within the departments and programs: historians in the history department, economists in the economics department, and sociologists in the sociology department.

Most universities also encourage research and teaching that crosses disciplinary boundaries. For example, a university may encourage a historian to teach a course that serves students in an urban studies department or the urban studies department may include the economics department's urban economics course as a required or elective course for the urban studies major. Departments of urban studies and city planning are quite interdisciplinary. They may include faculty trained in city planning, architecture, geography, sociology, economics, political science, and other disciplines.

While professors from many different academic disciplines as well as interdisciplinary scholars study cities, most scholarship about cities – and most of the readings in *The City Reader* and the Routledge Urban Reader Series – has been written by *social scientists*, faculty trained to study different aspects of human society systematically. Some of the contributions in *The City Reader*, and many in *The Urban Design Reader*, come from scholars in fields related to design – urban design, architecture, and landscape architecture.

Most universities have a school or college of social science. Schools of social science contain social science departments where professors trained in the social science disciplines of geography, sociology, economics, political science and anthropology teach. History departments are sometimes located within schools of social science; sometimes within schools of humanities. Within these social science departments professors interested in cities teach urban courses from the point of view of their disciplines: courses on *urban* geography, *urban* sociology, *urban* politics, etc. Professors in these discipline-based courses may include material from other academic disciplines in their courses. For example, a geography professor may use content and methods developed by economists and sociologists in her urban geography course. Urban Studies programs are usually located within colleges of social science.

The borderlines between the objects of study, methods, theory, and writings of different academic disciplines and professional fields are fuzzy. Urban economists tend to study the economics of cities using quantitative methods and urban sociologists tend to study social aspects of the city using qualitative methods. But some urban economists use qualitative methods and some urban sociologists are very quantitative. Political scientists interested in cities focus their attention on urban politics. But lawyer Myron Orfield (p. 287) mapped metropolitan areas like a geographer to develop his theory of metropolitics, which has become influential among political scientists. Sociologist Saskia Sassen's (p. 197) writings on the global system of cities are widely read by planners, economists, and political scientists.

Disciplines have the advantage that they are based on more or less agreed-upon methods for acquiring knowledge and a more or less agreed-upon body of knowledge. All history professors, for example, in order to get their history Ph.D. degree, must study methods of historical research that historians use. All history professors will have taken enough different history courses that they have a good overall knowledge of history in addition to their specialties in one or more specific time periods, issue areas, or methods of historical inquiry.

A disadvantage of disciplines is that they encourage rigid thinking within the four corners of the discipline itself. There is a danger that professors who are rigorously trained in economics, for example, will see only economic factors as important when they teach about urban issues. Because they have been trained in the importance of economics they may neglect political, social, and spatial aspects of the phenomena they are studying.

The strength of interdisciplinary approaches is that, done properly, they provide for a richer, more holistic, more varied understanding of multiple dimensions of the phenomena being studied. Urban issues have multiple dimensions. There are social, spatial, and political aspects to urban poverty as well as economic ones.

The danger of interdisciplinary approaches is that they may become so loose and standardless that they lack intellectual rigor. Well-trained and specialized disciplinary scholars are often justifiably critical of colleagues who do wide but shallow interdisciplinary teaching, research, and writing.

Some professors in academic disciplines value only theory and look down on applied research and writing intended to produce solutions to actual urban problems as derivative and inferior to theory building – a kind of vocational education that is not worthy of true scholars. This is silly. Urban research lends itself well to applied research, and good scholarship directed at problem solving can be just as theoretically subtle and methodologically sophisticated as pure theoretical research. Michael Porter's thoughtful prescriptions for urban poverty (p. 274), James Q. Wilson and George L. Kelling's "broken windows" theory of community policing (p. 256), and John Forester's insights about urban planning in the face of conflict (p. 387) are as intellectually rigorous as any of the less applied selections in *The City Reader*.

METHODS FOR STUDYING CITIES

Scholars who study cities use both *quantitative* and *qualitative* social science *research methods* to study cities. Both approaches can contribute to understanding cities, and the best urban research often combines both quantitative and qualitative research. *Quantitative* methods involve analyzing data using statistical methods. Today most quantitative analysis is assisted by computers. A professor of urban politics doing statistical analysis of city voting data to see if recent immigrants feel differently about immigration than longer-term residents would be doing quantitative urban research. Training in applied statistics is a regular part of most curricula that educate students to study cities. Students learn to use computerized statistical packages to do quantitative analysis. Because many urban phenomena have a spatial dimension, geographical information systems (GIS) software that permits users to map data is also very important in studying cities. GIS is taught in geography, urban planning, and other departments.

Qualitative research need not involve numbers or statistical analysis. William Whyte's study of how people use urban parks and plazas (p. 448) is a notable example of *observation*. Urban sociologist William Julius Wilson conducted exhaustive *field research* in Black ghetto areas of Chicago (p. 110). Urban designer Kevin Lynch and his students *interviewed* residents of Boston and other cities to understand how they perceived the city image (p. 438).

An urban anthropologist interviewing Cambodian immigrants in New York City to see how they feel about immigration would be doing qualitative research. If the political scientist who was doing statistical analysis of opinions about immigration and the anthropologist co-authored an academic paper reporting on what they found out using both quantitative and qualitative methods, they would be doing interdisciplinary research using multiple methods to understand an important dimension of city life.

THEORY AND PRACTICE

Theory in the social sciences is intended to *explain* phenomena and provide a framework for understanding. A professor like Saskia Sassen (p. 197) who is interested in the way in which information technology is affecting the size and relative importance of cities uses quantitative and qualitative methods to develop a body of *theory* explaining what is happening to the global system of cities. Urban studies and urban planning are applied fields and much scholarship is directed towards the practical end of solving problems. Policy makers and urban planners might draw on Sassen's theory to help plan cities and implement policy for them. The relationship between theory and practice is reciprocal. Theory can inform practice and practice can inform theory. John Forester (p. 387) developed his theories about how urban planners manage conflict by talking to practitioners. The theory he developed is in turn helpful to practitioners.

INTRODUCTION TO PART ONE

Cities are civilization. Humankind's rise to urban civilization took tens of thousands of years, but ever since the first true cities arose in Mesopotamia sometime between 4000 and 3000 BCE, the influence of city-based cultures and the steady spread and increase of urban populations around the world have been the central facts of human history.

As Kingsley Davis (p. 17) points out, "urbanization" and "the growth of cities" are not the same thing. "Urbanization," as Davis defines it, is the *increase in the proportion* of a population that is urban as opposed to rural. That such an increase could take place without the growth of cities *per se* (for example, by the death of vast numbers of the rural population) or that city populations could grow without an increase in urbanization (as when the total population, urban and rural, increases at a similar rate) are important concepts that underlie the history of urban life. This definition of urbanization helps to explain, for example, how immigration from the countryside to the city has repeatedly been the key factor in the history of urban development, as it continues to be today.

The first stage of urban history is what the Australian archaeologist V. Gordon Childe (p. 27) called the "Urban Revolution," the momentous shift from simple tribal communities and village-based agricultural production to the complex social, economic, and political systems that characterized the earliest cities of Mesopotamia, Egypt, and the Indus Valley. True, the earliest cities, in the ancient Near East and elsewhere, grew out of accumulated neolithic knowledge, and certain extensive neolithic communities such as Catal Hüyük in Anatolia pre-date the Mesopotamian cities by several millennia and may be regarded as at least proto-urban. For Childe, however, the development of writing was a crucial cultural element of true urbanism, and the emergence of the cities of the ancient Near East, where writing began, constituted the second of a series of massive transformations that gave shape to the whole of human evolutionary development. Although the successive stages overlapped, each of Childe's three "revolutions" (the agricultural, the urban, and the industrial) totally changed the world as it had been before.

In certain important respects, all the ancient cities are remarkably similar. Most are walled – except in Egypt, where the surrounding deserts may have been regarded as sufficient defenses – and all contain a distinct citadel precinct, separately walled, encompassing a temple, a palace, and the central granary. Most of the earliest cities also boasted some sort of pyramid or ziggurat. And, as Karl Wittfogel pointed out in *Oriental Despotism* (1957), almost all were located along major rivers and based their power (and that of their rulers) on the control of massive irrigation systems serving the surrounding countryside.

Thus, both the physical structure and socio-economic complexity of the earliest cities are unlike anything that had come before. Whereas the neolithic village had been ruled by a council of elders, the cities were mostly ruled by totalitarian god-kings and their attendant priests who formed a class totally apart from the rest of the citizenry. And whereas neolithic communities may have built earthen enclosures as ceremonial centers for ritual pageantry and hill forts for refuge and defense, the ancient cities – from Ur and Babylon on the Euphrates to Teotihuacan and Tenochtitlán in the Valley of Mexico – transformed these institutions into elaborate structures so massive that their remains are still visible today.

Many of the ancient cities elsewhere – in Shang and Chou China, sub-Saharan Africa, Southeast Asia, and Mesoamerica – arose quite independently of the cities of the ancient Near East. Still, what is remarkable and in need of massive efforts at renewal and redevelopment. Although the first modern suburbs were built along inter-urban railroad lines, the newer suburbs, especially those developments built after World War II, were automobile-based and created the "sprawl" that characterizes more and more cities worldwide. The new tract-home developments have spawned a vast literature, much of it criticizing suburbia as a cultural wasteland and a segregated sanctuary of class privilege. In *The Levittowners* (1967), Herbert Gans presented a rather sympathetic view of the community of tract-homes built by developer Arthur Levitt on Long Island, New York. He described a family-oriented community of skilled workers and mid-level managers – that is, a true *middle* class, not an upper-middle-class elite. But the more general view of automobile-dominated suburbia, a view that subjected sprawl to cultural as well as design criticism, is ably summarized in Kenneth T. Jackson's "The Drive-in Culture of Contemporary America" from *The Crabgrass Frontier* (p. 59, 1985).

Beginning as sprawl suburbia but quickly transcending suburbia's initial limitations, a new city type arose in California during the early decades of the twentieth century that signaled a new phase in the history of urbanism worldwide. Sometimes dismissed as a mere conurbation of suburbs "in search of a city" and frequently derided as the ultimate in mindless, post-urban chaos, Los Angeles did indeed break all the existing rules and natural boundaries of urban development but emerged finally as a new, radically decentralized urban paradigm: the contemporary multi-nucleated metropolis poised on the edge of post-modernity. The essential characteristics of Los Angeles – a city that grew from less than 600,000 in 1920 to over ten million today – were present almost from the beginning, particularly its sense of "spatial freedom" and its preference for the middle-class single-family dwelling as an "expression of its design for living." For good or ill, these characteristics were further emphasized by a grid pattern of freeways and a reliance – many would say over-reliance – on the automobile that replaced a once-extensive network of streetcars and created a metropolis without a single downtown. Today, Los Angeles is a true world city, and its products – both industrial and cultural – are influential around the globe.

In the nineteenth century, middle-class suburbs developed outside major urban centers, spaced along commuter rail lines. In the twentieth century, the influence of the automobile turned once-attractive small-scale suburbs into an endless, congested sprawl. These first two stages in the development of suburbia depended on the existence of a vital central city, both as a center for production and employment and for cultural amenities. With the emergence of Los Angeles, however, that pattern began to change, and today the new "Edge City" suburban ring is clearly different from the earlier suburban developments in size, complexity, and even function. This is where most of the new houses, most of the new jobs, and even most of the new cultural centers are located. Increasingly, the major commute pattern is not from suburb to central city, but from suburb to suburb. Indeed, as Robert Fishman (p. 69) argues in *Bourgeois Utopias: The Rise and Fall of Suburbia* (1987), the new "Edge City" suburbs are not suburbs at all, but a fundamentally new kind of decentralized city that he calls "technoburbs."

What the future holds for urban civilization is infinitely debatable. Will central cities disappear? Will "Edge Cities" take over the primary urban functions? Will the urbanization process itself reverse direction, as Melvin Webber predicted in 1968 (p. 473), and lead to counter-urbanization and a general dispersal of the human population? Or will certain urban centers become worldwide command and control centers, internally characterized by the uneasy side-by-side coexistence of corporate power and service-sector marginality, as Saskia Sassen (p. 197) argues in "The Impact of the New Technologies and Globalization on Cities"? No one knows for certain, of course, but it increasingly appears that urban history is on the cusp of a major new sharp break with the past. Looming on the horizon is a new paradigm, a new discontinuity in urban history, variously designated as post-modernism or post-urbanism, that will almost certainly be characterized by telecommunications, techno-virtuality, global systems of economic exchange, new ecological constraints demanding an increased concern with issues of sustainability, and, since September 11, 2001, a gnawing sense of terrorized social insecurity. But urban society will continue to be central to the history of humanity. The new urban paradigm that seems to be emerging now – perhaps part-regional sprawl/part-technoburb/part-virtual metropolis – promises to be a major new stage in the history of civilization, and in the on-going evolution of the human community.

"The Urbanization of the Human Population"

Scientific American (1965)

Kingsley Davis

Editors' Introduction

The demographics of the urbanization process are the foundation of all urban history. Demography – from the Greek *demos*: "people" – is the study of human populations. Kingsley Davis (1908–1996) pioneered the study of historical urban demography and was particularly fascinated by the history of world urbanization: that is, the increase over time of the proportion of the total human population that is urban as opposed to rural.

The following selection synthesizes Davis's conclusions about how urbanization has occurred throughout the world during all of human history. He raises fundamental issues and lays out a clear framework for understanding population dynamics and urban growth. Davis's careful distinctions of possible sources of urbanization are fundamental. He concludes that urbanization is caused by rural–urban migration, not because of other possible factors such as differential birth and mortality rates.

Davis's extraordinary data on how tiny European urban settlements were after the fall of Rome, and how slowly they grew throughout the Middle Ages and early modern period, provides the demographic backdrop for the historic growth of European urbanization. During the long period of medieval urbanization the proportion of the population that was urban as opposed to rural changed very slowly. In sharp contrast, Davis concludes that as the Industrial Revolution occurred in England, rapid population growth combined with rural–urban shifts changed both the proportion of the population living in cities and absolute city size very quickly. Friedrich Engels (p. 50) describes in horrifying detail what this revolution in urban demography meant to the impoverished urban proletariat of Manchester and other nineteenth-century industrial cities. His analysis is extremely relevant in assessing prospects for the twenty-first century as the advanced industrial societies and eventually the world reach what some environmental analysts regard as the full "carrying capacity" of the globe.

Davis argues that urbanization follows an attenuated S curve in which pre-industrial cities urbanize very slowly at the long bottom of the S, shoot up at the middle of the S as they industrialize, and then level off at the top of the S. He observes that advanced industrialized countries are now reaching the top part of an S curve, many rapidly urbanizing Third World countries are at the steep middle of the S, and other emerging countries are still moving along the long slowly rising bottom of the S.

The developing countries of Asia, South America, and Africa already have many huge and rapidly growing cities. As the twenty-first century progresses, it appears likely the human population will increasingly live in "megacities" of ten million inhabitants and more, often flowing together in vast urban conurbations sometimes called "mega-urban regions" (p. 489).

Davis concludes that there will be an end to urbanization – but not necessarily to absolute population growth, the physical size of cities or the absolute number of people cities contain. He found that the rural population in Third World countries today continues to grow as these countries urbanize, unlike European cities in the nineteenth century where industrialization led to depopulation of rural areas. His vision of Third World societies unable to sustain their populations helps to explain Saskia Sassen's description of growing poverty and inequality worldwide and the growth of large, poorly-paid immigrant labor forces in the largest cities in the developed world (p. 197).

Research and scholarly debate continues on the nature and causes of world urbanization. Historians continue to shed light on the growth of cities, but because the records from which they work are often fragmentary and incomplete not everyone agrees with Davis or any other standard account. Debate continues on the relative importance of war, plague, medical advances, trade, technology, religion, and ideology on urban growth. And debate is even more intense in the normative area – about what, if anything, governments should do about population growth and urbanization.

Davis stressed the impact of overall population growth (which he saw as a real danger) on world urbanization and implies that family planning is essential if cities are to meet human needs. But many governments reject family planning on religious or policy grounds, and some European countries now face declining populations and are currently debating the desirability of enacting family-friendly policies to reward child-bearing.

Davis's other writings include many articles and studies on demographics and natural resources as well as two anthologies: *Cities: Their Origin, Growth and Human Impact* (San Francisco: W.H. Freeman, 1973) and, with Mikhail S. Bernstram, *Resources, Environment and Population: Present Knowledge, Future Options* (New York: Population Council, Oxford University Press, 1991).

Data on world urbanization are contained in Tertius Chandler and Gerald Fox, *3000 Years of Urban Growth* (New York: Academic Press, 1974) and Tertius Chandler, *Four Thousand Years of Urban Growth: An Historical Census* (Lewiston: Edwin Mellen Press, 1987). An excellent description of urbanization in early modern Europe is Jan de Vries, *European Urbanization, 1500–1800* (Cambridge: Harvard University Press, 1984). Further insight on demography and urbanization can be found in World Bank, *World Development Indicators, 2005* (Washington, DC: World Bank, 2005), Ad van der Woude, Akira Hayami, and Jan de Vries (eds), *Urbanization in History: A Process of Dynamic Interaction* (Oxford: Oxford University Press, 1990), and the frequent revisions of *World Population Prospects* published by the Population Division of the United Nations.

For recent developments in Third World urbanization, see Alan Gilbert (ed.), *The Mega-City in Latin America* (New York: United Nations University Press, 1996), Carole Rakodi (ed.), *The Urban Challenge in Africa* (New York: United Nations University Press, 1997), and Fu-chen Lo and Yue-man Yeung (eds), *Emerging World Cities in Pacific Asia* (New York: United Nations University Press, 1996).

For an environmental view of world urbanization, consult Cedric Pugh (ed.), *Sustainability, the Environment, and Urbanization* (London: Earthscan, 1996). Lester R. Brown and Jodi L. Jacobson provide a somewhat alarming summary of recent world population studies and reflections on the future in *The Future of Urbanization: Facing the Ecological and Economic Constraints* (New York: Worldwatch Paper No. 77, 1987). The future of urbanization is of course an important issue for policy planners. For a fascinating review of population-related policy issues, including the possibility of a "world population implosion," see Nicholas Eberstadt, *Prosperous Paupers and Other Population Problems* (New Brunswick, NJ: Transaction, 2000).

For more on the possibility of declining populations in the future, consult Phillip Longman, *The Empty Cradle: How Falling Birthrates Threaten World Prosperity and What To Do About It* (New York: Basic Books, 2004), Ben J. Wattenburg, *Fewer: How the Demography of Depopulation Will Shape Our Future* (Chicago: Ivan R. Dee, 2004), and, for the special case of the effects of China's "One child" policy, Valerie Hudson, *Bare Essentials: The Security Implications of Asia's Surplus Male Population* (Cambridge: MIT Press, 2004).

■ ■ ■ ■ ■ ■

Urbanized societies, in which a majority of the people live crowded together in towns and cities, represent a new and fundamental step in man's social evolution. Although cities themselves first appeared some 5,500 years ago, they were small and surrounded by an overwhelming majority of rural people; moreover, they relapsed easily to village or small-town status. The urbanized societies of today, in contrast, not only have urban agglomerations of a size never before attained but also have a high proportion of their population concentrated in such agglomerations. In 1960, for example, nearly 52 million Americans lived in only 16 urbanized areas. Together these areas covered less land than one of the smaller counties (Cochise) of Arizona. According to one definition used by the U.S. Bureau of the Census, 96 million people – 53 percent of the nation's population – were concentrated in 213 urbanized areas that together occupied only 0.7 percent of the nation's land. Another definition used by the bureau puts the urban population at about 70 percent. The large and dense agglomerations comprising the urban population involve a degree of human contact and of social complexity never before known. They exceed in size the communities of any other large animal; they suggest the behavior of communal insects rather than of mammals.

Neither the recency nor the speed of this evolutionary development is widely appreciated. Before 1850 no society could be described as predominantly urbanized, and by 1900 only one – Great Britain – could be so regarded. Today, only 65 years later, all industrial nations are highly urbanized, and in the world as a whole the process of urbanization is accelerating rapidly.

Some years ago my associates and I at Columbia University undertook to document the progress of urbanization by compiling data on the world's cities and the proportion of human beings living in them; in recent years the work has been continued in our center – International Population and Urban Research – at the University of California at Berkeley. The data obtained in these investigations . . . show the historical trend in terms of one index of urbanization: the proportion of the population living in cities of 100,000 or larger. Statistics of this kind are only approximations of reality, but they are accurate enough to demonstrate how urbanization has accelerated. Between 1850 and 1950 the index changed at a much higher rate than from 1800 to 1850, but the rate of change from 1950 to 1960 was twice that of the preceding 50 years! If the pace of increase that obtained between 1950 and 1960 were to remain the same, by 1990 the fraction of the world's people living in cities of 100,000 or larger would be more than half. Using another index of urbanization – the proportion of the world's population living in urban places of all sizes – we found that by 1960 the figure had already reached 33 percent.

Clearly the world as a whole is not fully urbanized, but it soon will be. This change in human life is so recent that even the most urbanized countries still exhibit the rural origins of their institutions. Its full implications for man's organic and social evolution can only be surmised.

In discussing the trend – and its implications insofar as they can be perceived – I shall use the term "urbanization" in a particular way. It refers here to the proportion of the total population concentrated in urban settlements, or else to a rise in this proportion. A common mistake is to think of urbanization as simply the growth of cities. Since the total population is composed of both the urban population and the rural, however, the "proportion urban" is a function of both of them. Accordingly, cities can grow without any urbanization, provided that the rural population grows at an equal or a greater rate.

Historically, urbanization and the growth of cities have occurred together, which accounts for the confusion. As the reader will soon see, it is necessary to distinguish the two trends. In the most advanced countries today, for example, urban populations are still growing, but their proportion of the total population is tending to remain stable or to diminish. In other words, the process of urbanization – the switch from a spread-out pattern of human settlement to one of concentration in urban centers – is a change that has a beginning and an end, but the growth of cities has no inherent limit. Such growth could continue even after everyone was living in cities, through sheer excess of births over deaths.

The difference between a rural village and an urban community is of course one of degree; a precise operational distinction is somewhat arbitrary, and it varies from one nation to another. Since data are available for communities of various sizes, a dividing line can be chosen at will. One convenient index of urbanization, for example, is the proportion of people living in places of 100,000 or more. In the following analysis I shall depend on two indexes: the one just mentioned and the proportion of population classed as "urban" in the official statistics of each country. In practice the two indexes are highly correlated; therefore either one can be used as an index of urbanization.

Actually the hardest problem is not that of determining the "floor" of the urban category but of ascertaining the boundary of places that are clearly urban by any definition. How far east is the boundary of Los Angeles? Where along the Hooghly River does

Calcutta leave off and the countryside begin? In the past the population of cities and towns has usually been given as the number of people living within the political boundaries. Thus the population of New York is frequently given as around eight million, this being the population of the city proper. The error in such a figure was not large before World War I, but since then, particularly in the advanced countries, urban populations have been spilling over the narrow political boundaries at a tremendous rate. In 1960 the New York–Northeastern New Jersey urbanized area, as delineated by the Bureau of the Census, had more than 14 million people. That delineation showed it to be the largest city in the world and nearly twice as large as New York City proper.

As a result of the outward spread of urbanites, counts made on the basis of political boundaries alone underestimate the city populations and exaggerate the rural. For this reason our office delineated the metropolitan areas of as many countries as possible for dates around 1950. These areas included the central, or political, cities and the zones around them that are receiving the spillover.

This reassessment raised the estimated proportion of the world's population in cities of 100,000 or larger from 15.1 percent to 16.7 percent. As of 1960 we have used wherever possible the "urban agglomeration" data now furnished to the United Nations by many countries. The U.S., for example, provides data for "urbanized areas," meaning cities of 50,000 or larger and the built-up agglomerations around them.

. . . My concern is with the degree of urbanization in whole societies. It is curious that thousands of years elapsed between the first appearance of small cities and the emergence of urbanized societies in the nineteenth century. It is also curious that the region where urbanized societies arose – northwestern Europe – was not the one that had given rise to the major cities of the past; on the contrary, it was a region where urbanization had been at an extremely low ebb. Indeed, the societies of northwestern Europe in medieval times were so rural that it is hard for modern minds to comprehend them. Perhaps it was the non-urban character of these societies that erased the parasitic nature of towns and eventually provided a new basis for a revolutionary degree of urbanization.

At any rate, two seemingly adverse conditions may have presaged the age to come: one the low productivity of medieval agriculture in both per-acre and per-man terms, the other the feudal social system.

The first meant that towns could not prosper on the basis of local agriculture alone but had to trade and to manufacture something to trade. The second meant that they could not gain political dominance over their hinterlands and thus become warring city-states. Hence they specialized in commerce and manufacture and evolved local institutions suited to this role. Craftsmen were housed in the towns, because there the merchants could regulate quality and cost. Competition among towns stimulated specialization and technological innovation. The need for literacy, accounting skills and geographical knowledge caused the towns to invest in secular education.

Although the medieval towns remained small and never embraced more than a minor fraction of each region's population, the close connection between industry and commerce that they fostered, together with their emphasis on technique, set the stage for the ultimate breakthrough in urbanization. This break-through came only with the enormous growth in productivity caused by the use of inanimate energy and machinery. How difficult it was to achieve the transition is agonizingly apparent from statistics showing that even with the conquest of the New World the growth of urbanization during three postmedieval centuries in Europe was barely perceptible. I have assembled population estimates at two or more dates for 33 towns and cities in the sixteenth century, 46 in the seventeenth and 61 in the eighteenth. The average rate of growth during the three centuries was less than 0.6 percent per year. Estimates of the growth of Europe's population as a whole between 1650 and 1800 work out to slightly more than 0.4 percent. The advantage of the towns was evidently very slight. Taking only the cities of 100,000 or more inhabitants, one finds that in 1600 their combined population was 1.6 percent of the estimated population of Europe; in 1700, 1.9 percent; and in 1800, 2.2 percent. On the eve of the industrial revolution Europe was still an overwhelmingly agrarian region.

With industrialization, however, the transformation was striking. By 1801 nearly a tenth of the people of England and Wales were living in cities of 100,000 or larger. This proportion doubled in 40 years and doubled again in another 60 years. By 1900 Britain was an urbanized society. In general, the later each country became industrialized, the faster was its urbanization. The change from a population with 10 percent of its members in cities of 100,000 or larger to one in which 30 percent lived in such cities took about 79 years in

England and Wales, 66 in the U.S., 48 in Germany, 36 in Japan and 26 in Australia. The close association between economic development and urbanization has persisted: . . . in 199 countries around 1960 the proportion of the population living in cities varied sharply with per capita income.

Clearly, modern urbanization is best understood in terms of its connection with economic growth, and its implications are best perceived in its latest manifestations in advanced countries. What becomes apparent as one examines the trend in these countries is that urbanization is a finite process, a cycle through which nations go in their transition from agrarian to industrial society. The intensive urbanization of most of the advanced countries began within the past hundred years; in the underdeveloped countries it got under way more recently. In some of the advanced countries its end is now in sight. The fact that it will end, however, does not mean that either economic development or the growth of cities will necessarily end.

The typical cycle of urbanization can be represented by a curve in the shape of an attenuated S. Starting from the bottom of the S, the first bend tends to come early and to be followed by a long attenuation. In the United Kingdom, for instance, the swiftest rise in the proportion of people living in cities of 100,000 or larger occurred from 1811 to 1851. In the U.S. it occurred from 1820 to 1890, in Greece from 1879 to 1921. As the proportion climbs above 50 percent the curve begins to flatten out; it falters, or even declines, when the proportion urban has reached about 75 percent. In the United Kingdom, one of the world's most urban countries, the proportion was slightly higher in 1926 (78.7 percent) than in 1961 (78.3 percent).

At the end of the curve some ambiguity appears. As a society becomes advanced enough to be highly urbanized it can also afford considerable suburbanization and fringe development. In a sense the slowing down of urbanization is thus more apparent than real: an increasing proportion of urbanites simply live in the country and are classified as rural. Many countries now try to compensate for this ambiguity by enlarging the boundaries of urban places; they did so in numerous censuses taken around 1960. Whether in these cases the old classification of urban or the new one is erroneous depends on how one looks at it; at a very advanced stage the entire concept of urbanization becomes ambiguous.

The end of urbanization cannot be unraveled without going into the ways in which economic devel-

opment governs urbanization. Here the first question is: where do the urbanites come from? The possible answers are few: the proportion of people in cities can rise because rural settlements grow larger and are reclassified as towns or cities; because the excess of births over deaths is greater in the city than in the country, or because people move from the country to the city.

The first factor has usually had only slight influence. The second has apparently never been the case. Indeed, a chief obstacle to the growth of cities in the past has been their excessive mortality. London's water in the middle of the nineteenth century came mainly from wells and rivers that drained cesspools, graveyards and tidal areas. The city was regularly ravaged by cholera. Tables for 1841 show an expectation of life of about 36 years for London and 26 for Liverpool and Manchester, as compared to 41 for England and Wales as a whole. After 1850, mainly as a result of sanitary measures and some improvement in nutrition and housing, city health improved, but as late as the period 1901–1910 the death rate of the urban counties in England and Wales, as modified to make the age structure comparable, was 33 percent higher than the death rate of the rural counties. As Bernard Benjamin, a chief statistician of the British General Register Office, has remarked: "Living in the town involved not only a higher risk of epidemic and crowd diseases . . . but also a higher risk of degenerative disease – the harder wear and tear of factory employment and urban discomfort." By 1950, however, virtually the entire differential had been wiped out.

As for birth rates, during rapid urbanization in the past they were notably lower in cities than in rural areas. In fact, the gap tended to widen somewhat as urbanization proceeded in the latter half of the nineteenth century and the first quarter of the twentieth. In 1800 urban women in the U.S. had 36 percent fewer children than rural women did; in 1840, 38 percent and in 1930, 41 percent. Thereafter the difference diminished.

With mortality in the cities higher and birth rates lower, and with reclassification a minor factor, the only real source for the growth in the proportion of people in urban areas during the industrial transition was rural–urban migration. This source had to be plentiful enough not only to overcome the substantial disadvantage of the cities in natural increase but also, above that, to furnish a big margin of growth in their populations. If, for example, the cities had a death rate a third higher and a birth rate a third lower than the rural rates (as was typical in the latter half of the nineteenth century), they would require each year perhaps 40 to 45 migrants from elsewhere per 1,000 of their population to maintain a growth rate of 3 percent per year. Such a rate of migration could easily be maintained as long as the rural portion of the population was large, but when this condition ceased to obtain, the maintenance of the same urban rate meant an increasing drain on the countryside.

Why did the rural–urban migration occur? The reason was that the rise in technological enhancement of human productivity, together with certain constant factors, rewarded urban concentration. One of the constant factors was that agriculture uses land as its prime instrument of production and hence spreads out people who are engaged in it, whereas manufacturing, commerce and services use land only as a site. Moreover, the demand for agricultural products is less elastic than the demand for services and manufactures. As productivity grows, services and manufactures can absorb more manpower by paying higher wages. Since nonagricultural activities can use land simply as a site, they can locate near one another (in towns and cities) and thus minimize the fraction of space inevitably involved in the division of labor. At the same time, as agricultural technology is improved, capital costs in farming rise and manpower becomes not only less needed but also economically more burdensome. A substantial portion of the agricultural population is therefore sufficiently disadvantaged, in relative terms, to be attracted by higher wages in other sectors.

In this light one sees why a large *flow* of people from farms to cities was generated in every country that passed through the industrial revolution. One also sees why, with an even higher proportion of people already in cities and with the inability of city people to replace themselves by reproduction, the drain eventually became so heavy that in many nations the rural population began to decline in absolute as well as relative terms. In Sweden it declined after 1920, in England and Wales after 1861, in Belgium after 1910.

Realizing that urbanization is transitional and finite, one comes on another fact – a fact that throws light on the circumstances in which urbanization comes to an end. A basic feature of the transition is the profound switch from agricultural to nonagricultural employment. This change is associated with urbanization but not identical with it. The difference emerges

particularly in the later stages. Then the availability of automobiles, radios, motion pictures and electricity, as well as the reduction of the workweek and the workday, mitigate the disadvantages of living in the country. Concurrently the expanding size of cities makes them more difficult to live in. The population classed as "rural" is accordingly enlarged, both from cities and from true farms. For these reasons the "rural" population in some industrial countries never did fall in absolute size. In all the industrial countries, however, the population dependent on agriculture – which the reader will recognize as a more functional definition of the nonurban population than mere rural residence – decreased in absolute as well as relative terms. In the U.S., for example, the net migration from farms totaled more than 27 million between 1920 and 1959 and thus averaged approximately 700,000 a year. As a result the farm population declined from 32.5 million in 1916 to 20.5 million in 1960, in spite of the large excess of births in farm families. In 1964, by a stricter American definition classifying as "farm families" only those families actually earning their living from agriculture, the farm population was down to 12.9 million. This number represented 6.8 percent of the nation's population; the comparable figure for 1880 was 44 percent. In Great Britain the number of males occupied in agriculture was, at its peak, 1.8 million, in 1851; by 1961 it had fallen to 0.5 million.

In the later stages of the cycle, then, urbanization in the industrial countries tends to cease. Hence the connection between economic development and the growth of cities also ceases. The change is explained by two circumstances. First, there is no longer enough farm population to furnish a significant migration to the cities. (What can 12.9 million American farmers contribute to the growth of the 100 million people already in urbanized areas?) Second, the rural nonfarm population, nourished by refugees from the expanding cities, begins to increase as fast as the city population. The effort of census bureaus to count fringe residents as urban simply pushes the definition of "urban" away from the notion of dense settlement and in the direction of the term "nonfarm." As the urban population becomes more "rural," which is to say less densely settled, the advanced industrial peoples are for a time able to enjoy the amenities of urban life without the excessive crowding of the past.

Here, however, one again encounters the fact that a cessation of urbanization does not necessarily mean a cessation of city growth. An example is provided by

New Zealand. Between 1945 and 1961 the proportion of New Zealand's population classed as urban – that is, the ratio between urban and rural residents – changed hardly at all (from 61.3 percent to 63.6 percent) but the urban population increased by 50 percent. In Japan between 1940 and 1950 urbanization actually decreased slightly, but the urban population increased by 13 percent.

The point to be kept in mind is that once urbanization ceases, city growth becomes a function of general population growth. Enough farm-to-city migration may still occur to redress the difference in natural increase. The reproductive rate of urbanites tends, however, to increase when they live at lower densities, and the reproductive rate of "urbanized" farmers tends to decrease; hence little migration is required to make the urban increase equal the national increase.

I now turn to the currently underdeveloped countries. With the advanced nations having slackened their rate of urbanization, it is the others – representing three-fourths of humanity – that are mainly responsible for the rapid urbanization now characterizing the world as a whole. In fact, between 1950 and 1960 the proportion of the population in cities of 100,000 or more rose about a third faster in the underdeveloped regions than in the developed ones. Among the underdeveloped regions the pace was slow in eastern and southern Europe but in the rest of the underdeveloped world the proportion in cities rose twice as fast as it did in the industrialized countries, even though the latter countries in many cases broadened their definitions of urban places to include more suburban and fringe residents.

Because of the characteristic pattern of urbanization, the current rates of urbanization in underdeveloped countries could be expected to exceed those now existing in countries far advanced in the cycle. On discovering that this is the case one is tempted to say that the underdeveloped regions are now in the typical stage of urbanization associated with early economic development. This notion, however, is erroneous. In their urbanization the underdeveloped countries are definitely not recreating past history. Indeed, the best grasp of their present situation comes from analyzing how their course differs from the previous pattern of development.

The first thing to note is that today's underdeveloped countries are urbanizing not only more rapidly than the industrial nations are now but also more rapidly than the industrial nations did in the

heyday of their urban growth. The difference, however, is not large. In 40 underdeveloped countries for which we have data in recent decades, the average gain in the proportion of the population urban was 20 percent per decade; in 16 industrial countries, during the decades of their most rapid urbanization (mainly in the nineteenth century), the average gain per decade was 15 percent.

This finding that urbanization is proceeding only a little faster in underdeveloped countries than it did historically in the advanced nations may be questioned by the reader. It seemingly belies the widespread impression that cities throughout the nonindustrial parts of the world are bursting with people. There is, however, no contradiction. One must recall the basic distinction between a change in the proportion of the population urban, which is a ratio, and the absolute growth of cities. The popular impression is correct: the cities in underdeveloped areas are growing at a disconcerting rate. They are far outstripping the city boom of the industrializing era in the nineteenth century. If they continue their recent rate of growth, they will double their population every 15 years.

In 34 underdeveloped countries for which we have data relating to the 1940s and 1950s, the average annual gain in the urban population was 4.5 percent. The figure is remarkably similar for the various regions: 4.7 percent in seven countries of Africa, 4.7 percent in 15 countries of Asia and 4.3 percent in 12 countries of Latin America. In contrast, in nine European countries during their period of fastest urban population growth (mostly in the latter half of the nineteenth century) the average gain per year was 2.1 percent. Even the frontier industrial countries – the U.S., Australia–New Zealand, Canada and Argentina – which received huge numbers of immigrants had a smaller population growth in towns and cities: 4.2 percent per year. In Japan and the U.S.S.R. the rate was respectively 5.4 and 4.3 percent per year, but their economic growth began only recently.

How is it possible that the contrast in growth between today's underdeveloped countries and yesterday's industrializing countries is sharper with respect to the absolute urban population than with respect to the urban share of the total population? The answer lies in another profound difference between the two sets of countries – a difference in total population growth, rural as well as urban. Contemporary underdeveloped populations have been growing since 1940 more than twice as fast as industrialized populations, and their increase far exceeds the growth of the latter at the peak of their expansion. The only rivals in an earlier day were the frontier nations, which had the help of great streams of immigrants. Today the underdeveloped nations – already densely settled, tragically impoverished and with gloomy economic prospects – are multiplying their people by sheer biological increase at a rate that is unprecedented. It is this population boom that is overwhelmingly responsible for the rapid inflation of city populations in such countries. Contrary to popular opinion both inside and outside those countries, the main factor is not rural–urban migration.

This point can be demonstrated easily by a calculation that has the effect of eliminating the influence of general population growth on urban growth. The calculation involves assuming that the total population of a given country remained constant over a period of time but that the percentage urban changed as it did historically. In this manner one obtains the growth of the absolute urban population that would have occurred if rural–urban migration were the only factor affecting it. As an example, Costa Rica had in 1927 a total population of 471,500, of which 88,600, or 18.8 percent, was urban. By 1963 the country's total population was 1,325,200 and the urban population was 456,600, or 34.5 percent. If the total population had remained at 471,500 but the percentage urban had still risen from 18.8 to 34.5, the absolute urban population in 1963 would have been only 162,700. That is the growth that would have occurred in the urban population if rural–urban migration had been the only factor. In actuality the urban population rose to 456,600. In other words, only 20 percent of the rapid growth of Costa Rica's towns and cities was attributable to urbanization per se; 44 percent was attributable solely to the country's general population increase, the remainder to the joint operation of both factors. Similarly, in Mexico between 1940 and 1960, 50 percent of the urban population increase was attributable to national multiplication alone and only 22 percent to urbanization alone.

The past performance of the advanced countries presents a sharp contrast. In Switzerland between 1850 and 1888, when the proportion urban resembled that in Costa Rica recently, general population growth alone accounted for only 19 percent of the increase of town and city people, and rural–urban migration alone accounted for 69 percent. In France between 1846 and 1911 only 21 percent of the growth in the absolute urban population was due to general growth alone.

The conclusion to which this contrast points is that one anxiety of governments in the underdeveloped nations is misplaced. Impressed by the mushrooming in their cities of shanty-towns filled with ragged peasants, they attribute the fantastically fast city growth to rural–urban migration. Actually this migration now does little more than make up for the small difference in the birth rate between city and countryside. In the history of the industrial nations, as we have seen, the sizable difference between urban and rural birth rates and death rates required that cities, if they were to grow, had to have an enormous influx of people from farms and villages. Today in the underdeveloped countries the towns and cities have only a slight disadvantage in fertility, and their old disadvantage in mortality not only has been wiped out but also in many cases has been reversed. During the nineteenth century the urbanizing nations were learning how to keep crowded populations in cities from dying like flies. Now the lesson has been learned, and it is being applied to cities even in countries just emerging from tribalism. In fact, a disproportionate share of public health funds goes into cities. As a result, throughout the nonindustrial world people in cities are multiplying as never before, and rural–urban migration is playing a much lesser role.

The trends just described have an important implication for the rural population. Given the explosive overall population growth in underdeveloped countries, it follows that if the rural population is not to pile up on the land and reach an economically absurd density, a high rate of rural–urban migration must be maintained. Indeed, the exodus from rural areas should be higher than in the past. But this high rate of internal movement is not taking place, and there is some doubt that it could conceivably do so.

To elaborate, I shall return to my earlier point that in the evolution of industrialized countries the rural citizenry often declined in absolute as well as relative terms. The rural population of France – 26.8 million in 1846 – was down to 20.8 million by 1926 and 17.2 million by 1962, notwithstanding a gain in the nation's total population during this period. Sweden's rural population dropped from 4.3 million in 1910 to 3.5 million in 1960. Since the category "rural" includes an increasing portion of urbanites living in fringe areas, the historical drop was more drastic and consistent specifically in the farm population. In the U.S., although the "rural" population never quite ceased to grow, the farm contingent began its long descent shortly after the turn of the century; today it is less than two-fifths of what it was in 1910.

This transformation is not occurring in contemporary underdeveloped countries. In spite of the enormous growth of their cities, their rural populations – and their more narrowly defined agricultural populations – are growing at a rate that in many cases exceeds the rise of even the urban population during the evolution of the now advanced countries. The poor countries thus confront a grave dilemma. If they do not substantially step up the exodus from rural areas, these areas will be swamped with underemployed farmers. If they do step up the exodus, the cities will grow at a disastrous rate.

The rapid growth of cities in the advanced countries, painful though it was, had the effect of solving a problem – the problem of the rural population. The growth of cities enabled agricultural holdings to be consolidated, allowed increased capitalization and in general resulted in greater efficiency. Now, however, the underdeveloped countries are experiencing an even more rapid urban growth – and are suffering from urban problems – but urbanization is not solving their rural ills.

A case in point is Venezuela. Its capital, Caracas, jumped from a population of 359,000 in 1941 to 1,507,000 in 1963; other Venezuelan towns and cities equaled or exceeded this growth. Is this rapid rise denuding the countryside of people? No, the Venezuelan farm population increased in the decade 1951–1961 by 11 percent. The only thing that declined was the amount of cultivated land. As a result the agricultural population density became worse. In 1950 there were some 64 males engaged in agriculture per square mile of cultivated land; in 1961 there were 78. (Compare this with 4.8 males occupied in agriculture per square mile of cultivated land in Canada, 6.8 in the U.S. and 15.6 in Argentina.) With each male occupied in agriculture there are of course dependants. Approximately 225 persons in Venezuela are trying to live from each square mile of cultivated land. Most of the growth of cities in Venezuela is attributable to overall population growth. If the general population had not grown at all, and internal migration had been large enough to produce the actual shift in the proportion in cities, the increase in urban population would have been only 28 percent of what it was and the rural population would have been reduced by 57 percent.

The story of Venezuela is being repeated virtually everywhere in the underdeveloped world. It is not only Caracas that has thousands of squatters living in self-constructed junk houses on land that does not belong to them. By whatever name they are called, the squatters are to be found in all major cities in the poorer countries. They live in broad gullies beneath the main plain in San Salvador and on the hillsides of Rio de Janeiro and Bogotá. They tend to occupy with implacable determination parks, school grounds and vacant lots. Amman, the capital of Jordan, grew from 12,000 in 1958 to 247,000 in 1961. A good part of it is slums, and urban amenities are lacking most of the time for most of the people. Greater Baghdad now has an estimated 850,000 people; its slums, like those in many other underdeveloped countries, are in two zones: the central part of the city and the outlying areas. Here are the *sarifa* areas, characterized by self-built reed huts; these areas account for about 45 percent of the housing in the entire city and are devoid of amenities, including even latrines. In addition to such urban problems, all the countries struggling for higher living levels find their rural population growing too and piling up on already crowded land. I have characterized urbanization as a transformation that, unlike economic development, is finally accomplished and comes to an end. At the 1950–1960 rate the term "urbanized world" will be applicable well before the end of the century. One should scarcely expect, however, that mankind will complete its urbanization without major complications. One sign of trouble ahead turns on the distinction I made at the start between urbanization and city growth *per se*. Around the globe today city growth is disproportionate to urbanization. The discrepancy is paradoxical in the industrial nations and worse than paradoxical in the nonindustrial.

It is in this respect that the nonindustrial nations, which still make up the great majority of nations, are far from repeating past history. In the nineteenth and early twentieth centuries the growth of cities arose from and contributed to economic advancement. Cities took surplus manpower from the countryside and put it to work producing goods and services that in turn helped to modernize agriculture. But today in underdeveloped countries, as in present-day advanced nations, city growth has become increasingly unhinged from economic development and hence from rural–urban migration. It derives in greater degree from overall population growth, and this growth in nonindustrial lands has become unprecedented because of modern health techniques combined with high birth rates.

The speed of world population growth is twice what it was before 1940, and the swiftest increase has shifted from the advanced to the backward nations. In the latter countries, consequently, it is virtually impossible to create city services fast enough to take care of the huge, never-ending cohorts of babies and peasants swelling the urban masses. It is even harder to expand agricultural land and capital fast enough to accommodate the enormous natural increase on farms. The problem is not urbanization, not rural–urban migration, but human multiplication. It is a problem that is new in both its scale and its setting, and runaway city growth is only one of its painful expressions.

As long as the human population expands, cities will expand too, regardless of whether urbanization increases or declines. This means that some individual cities will reach a size that will make nineteenth-century metropolises look like small towns. If the New York urbanized area should continue to grow only as fast as the nation's population (according to medium projections of the latter by the Bureau of the Census), it would reach 21 million by 1985 and 30 million by 2010. I have calculated that if India's population should grow as the U.N. projections indicate it will, the largest city in India in the year 2000 will have between 36 and 66 million inhabitants.

What is the implication of such giant agglomerations for human density? In 1950 the New York–Northeastern New Jersey urbanized area had an average density of 9,810 persons per square mile. With 30 million people in the year 2010, the density would be 24,000 per square mile. Although this level is exceeded now in parts of New York City (which averages about 25,000 per square mile) and many other cities, it is a high density to be spread over such a big area; it would cover, remember, the suburban areas to which people moved to escape high density. Actually, however, the density of the New York urbanized region is dropping, not increasing, as the population grows. The reason is that the territory covered by the urban agglomeration is growing faster than the population: it grew by 51 percent from

1950 to 1960, whereas the population rose by 15 percent.

If, then, one projects the rise in population and the rise in territory for the New York urbanized region one finds the density problem solved. It is not solved for long, though, because New York is not the only city in the region that is expanding. So are Philadelphia, Trenton, Hartford, New Haven and so on. By 1960 a huge stretch of territory about 600 miles long and 30 to 100 miles wide along the eastern seaboard contained some 37 million people. (I am speaking of a longer section of the seaboard than the Boston-to-Washington conurbation referred to by some other authors.) Since the whole area is becoming one big polynucleated city, its population cannot long expand without a rise in density. Thus persistent human multiplication promises to frustrate the ceaseless search for space – for ample residential lots, wide-open suburban school grounds, sprawling shopping centers, one-floor factories, broad freeways.

How people feel about giant agglomerations is best indicated by their headlong effort to escape them. The bigger the city, the higher the cost of space; yet the more the level of living rises, the more people are willing to pay for low-density living. Nevertheless, as urbanized areas expand and collide, it seems probable that life in low-density surroundings will become too dear for the great majority.

One can of course imagine that cities may cease to grow and may even shrink in size while the population in general continues to multiply. Even this dream, however, would not permanently solve the problem of space. It would eventually obliterate the distinction between urban and rural, but at the expense of the rural.

It seems plain that the only way to stop urban crowding and to solve most of the urban problems besetting both the developed and the underdeveloped nations is to reduce the overall rate of population growth. Policies designed to do this have as yet little intelligence and power behind them. Urban planners continue to treat population growth as something to be planned for, not something to be itself planned. Any talk about applying brakes to city growth is therefore purely speculative, overshadowed as it is by the reality of uncontrolled population increase.

"The Urban Revolution"

Town Planning Review (1950)

V. Gordon Childe

Editors' Introduction

V. Gordon Childe (1892–1957) is arguably the single most influential archaeologist of the twentieth century. Born in Australia, Childe won a scholarship to Queen's College, Oxford, returned to Australia where he briefly pursued a career in left-wing politics, then returned to the UK as Professor of Archaeology at the University of Edinburgh and, later, Director of the Institute of Archaeology at the University of London.

Childe's most important book, the one that revolutionized the world of archaeological research by laying out an entirely new theoretical framework for understanding the phases of human development throughout history and pre-history, is *Man Makes Himself* (1936). In that pioneering work, Childe threw out the "three age system" (Stone Age, Bronze Age, Iron Age) that had been left over from nineteenth-century conceptions of human historical development. In its place he proposed a series of four stages (paleolithic, neolithic, urban, industrial) punctuated by three "revolutions" (or, as we might term them today, "paradigm shifts").

According to Childe, the first revolution – from old Stone Age hunter–gatherer cultures to settled agriculture – was the Neolithic Revolution. The second – the movement from neolithic agriculture to complex, hierarchical systems of manufacturing and trade that began during the fourth and third millennia BCE – was the Urban Revolution. And the third major shift in the record of human cultural and historical development – the only truly new development since the rise of cities – was the Industrial Revolution of the eighteenth and nineteenth centuries.

Childe is best known for his writings on the first cities, which arose in Mesopotamia (present-day Iraq) beginning about 4000 BCE. These cities sprang up in the area bounded by the Tigris and Euphrates Rivers – often referred to as "the Fertile Crescent." Plate 1, "The Palace of Sargon II of Khorsabad," illustrates the form these first cities took. Monumental gates, massive mud-brick walls, courtyards, residences for priest-kings, and a ziggurat.

Childe's work continues to figure prominently in ongoing debates about when, where, and why the first cities arose and in the antecedent debate about what a city is. Not everyone has accepted Childe's notion that the shift from neolithic to urban was a total break with the past. Evidence of ancient earthworks, wells, irrigation systems, and even continental trade networks have been traced back as far as 10,000 years in a number of areas in both the Old World and the New. Archaeologist James Mellaart has argued that evidence from the great neolithic communities of Catal Hüyük and Hacilar in ancient Turkey, which predate the earliest Mesopotamian cities by some thousands of years, calls the entire Childe theory into question. Still, it is clear that in most locations agriculture generally predated the rise of the first cities by not just thousands but tens of thousands of years, and that the full elaboration of those cultural institutions we associate with urban life only emerged with the rise of the Mesopotamian cities.

Not everyone agrees with Childe's definition of a city. Archaeologists excavating older, smaller, less culturally advanced settlements than the Mesopotamian cities Childe studied often argue that these settlements were urban enough to qualify as true cities. Scholars working in South and Central America point out that many of the cultural features Childe believed essential to the definition of a city (including the wheel, writing, and the plow) did not exist in large and culturally advanced Amerindian settlements that appear truly urban in other respects.

In the selection from *Town Planning Review* reprinted here, Childe details the constituent elements of the Urban Revolution that accompanied the initial rise of complex civilizations in Mesopotamia and elsewhere in the ancient Near East. Childe felt that the major factors motivating the transformation were rooted in the material base of the society: its means of production and its available physical and technological resources. Thus, the economic division of labor, the elaboration of socio-political hierarchies, and even the emergence of basic religious and intellectual patterns of thought characteristic of urban civilizations all rested on the underlying need to increase food production through massive irrigation systems and to protect the communities themselves through the erection of massive walls and fortifications.

Many modern scholars question the deterministic Marxist categories Childe employed. Although he stresses the importance of writing as an element of any truly urban society, Childe has been faulted for his apparent disregard of the primacy of non-material aspects of culture. His system has very little room for what Lewis Mumford (p. 85) called "the urban drama" or what Jane Jacobs (p. 98) called the "street ballet." Still, no one has ever called Childe's vision limited or ideologically cramped. On the contrary, he provided an expansive macro-historical foundation upon which generations of others have built.

A tireless researcher and writer, Childe produced a veritable stream of books, many of which are still classics. Among the most notable are: *The Dawn of European Civilization* (London: Routledge & Kegan Paul, 1925), *The Most Ancient East* (New York: Grove Press, 1928), *What Happened in History* (Harmondsworth: Penguin, 1942), *Social Evolution* (London: Watts, 1951). Other books on Mesopotamian cities include Nicholas Postgate and J.N. Postgate, *Early Mesopotamia: Society and Economy at the Dawn of History* (London and New York: Routledge, 1994), Georges Roux, *Ancient Iraq* (New York: Penguin, 1993), and C. Leonard Woolley's classics: *The Sumerians* (Oxford: Oxford University Press, 1928) and *Ur of the Chaldees* (Oxford: Oxford University Press, 1929).

For surveys of the current state of research into cities in the ancient world, see Gwendolyn Leick, *Mesopotamia* (Penguin, 2003) and Charles Gates, *Ancient Cities: The Archaeology of Urban Life in the Ancient Near East and Egypt, Greece, and Rome* (London and New York: Routledge, 2003).

Earlier studies on the rise of the earliest cities elsewhere in the world that are still of interest include Mortimer Wheeler, *Civilizations of the Indus Valley and Beyond* (London: Thames & Hudson, 1966), Karl Wittfogel, *Oriental Despotism* (New Haven: Yale University Press, 1957), Basil Davidson, *The Lost Cities of Africa* (Boston: Little, Brown, 1959), Richard E.W. Adams, *Prehistoric Mesoamerica* (Norman: University of Oklahoma Press, 1991), Sylvanus G. Morely and George W. Brainerd, *The Ancient Maya* (Stanford: Stanford University Press, 1956), Jacques Soustelle, *The Daily Life of the Aztecs* (New York: Macmillan, 1962), James Mellaart, *Earliest Civilizations of the Near East* (New York: McGraw-Hill, 1965) and *Catal Hüyük* (New York: McGraw-Hill, 1967), and Paul Wheatley, *The Pivot of the Four Quarters: A Preliminary Inquiry into the Origins and Character of the Ancient Chinese City* (Chicago: Aldine, 1971).

The concept of 'city' is notoriously hard to define. The aim of the present essay is to present the city historically – or rather prehistorically – as the resultant and symbol of a 'revolution' that initiated a new economic stage in the evolution of society. The word 'revolution' must not of course be taken as denoting a sudden violent catastrophe; it is here used for the culmination of a progressive change in the economic structure and social organization of communities that caused, or was accompanied by, a dramatic increase in the population affected – an increase that would appear as an obvious bend in the population graph were vital statistics available. Just such a bend is observable at the time of the Industrial Revolution in England. Though not demonstrable statistically, comparable changes of direction must have occurred at two earlier points in the demographic history of Britain and other regions. Though perhaps less sharp and less durable, these too should indicate equally revolutionary changes in economy. They may then be regarded likewise as marking transitions between stages in economic and social development.

Sociologists and ethnographers last century classified existing pre-industrial societies in a hierarchy of three evolutionary stages, denominated respectively 'savagery,' 'barbarism' and 'civilization.' If they be defined by suitably selected criteria, the logical hierarchy of stages can be transformed into a temporal sequence of ages, proved archaeologically to follow one another in the same order wherever they occur. Savagery and barbarism are conveniently recognized and appropriately defined by the methods adopted for procuring food. Savages live exclusively on wild food obtained by collecting, hunting or fishing. Barbarians on the contrary at least supplement these natural resources by cultivating edible plants and – in the Old World north of the Tropics – also by breeding animals for food.

Throughout the Pleistocene Period – the Palaeolithic Age of archaeologists – all known human societies were savage in the foregoing sense, and a few savage tribes have survived in out of the way parts to the present day. In the archaeological record barbarism began less than ten thousand years ago with the Neolithic Age of archaeologists. It thus represents a later, as well as a higher stage, than savagery. Civilization cannot be defined in quite such simple terms. Etymologically the word is connected with 'city,' and sure enough life in cities begins with this stage. But 'city' is itself ambiguous so archaeologists like to use 'writing' as a criterion of civilization; it should be easily recognizable and proves to be a reliable index to more profound characters. Note, however, that, because a people is said to be civilized or literate, it does not follow that all its members can read and write, nor that they all lived in cities. Now there is no recorded instance of a community of savages civilizing themselves, adopting urban life or inventing a script. Wherever cities have been built, villages of preliterate farmers existed previously (save perhaps where an already civilized people have colonized uninhabited tracts). So civilization, wherever and whenever it arose, succeeded barbarism.

We have seen that a revolution as here defined should be reflected in the population statistics. In the case of the Urban Revolution the increase was mainly accounted for by the multiplication of the numbers of persons living together, i.e., in a single built-up area. The first cities represented settlement units of hitherto unprecedented size. Of course it was not just their size that constituted their distinctive character. We shall find that by modern standards they appeared ridicu-

lously small and we might meet agglomerations of population today to which the name city would have to be refused. Yet a certain size of settlement and density of population is an essential feature of civilization.

Now the density of population is determined by the food supply which in turn is limited by natural resources, the techniques for their exploitation and the means of transport and food-preservation available. The last factors have proved to be variables in the course of human history, and the technique of obtaining food has already been used to distinguish the consecutive stages termed savagery and barbarism. Under the gathering economy of savagery population was always exceedingly sparse. In aboriginal America the carrying capacity of normal unimproved land seems to have been from .05 to .10 per square mile. Only under exceptionally favourable conditions did the fishing tribes of the Northwest Pacific coast attain densities of over one human to the square mile. As far as we can guess from the extant remains, population densities in Palaeolithic and pre-neolithic Europe were less than the normal American. Moreover such hunters and collectors usually live in small roving bands. At best several bands may come together for quite brief periods on ceremonial occasions such as the Australian corroborees. Only in exceptionally favoured regions can fishing tribes establish anything like villages. Some settlements on the Pacific coasts comprised thirty or so substantial and durable houses, accommodating groups of several hundred persons. But even these villages were only occupied during the winter; for the rest of the year their inhabitants disposed in smaller groups. Nothing comparable has been found in pre-neolithic times in the Old World.

The Neolithic Revolution certainly allowed an expansion of population and enormously increased the carrying capacity of suitable land. On the Pacific Islands neolithic societies today attain a density of 30 or more persons to the square mile. In pre-Columbian North America, however, where the land is not obviously restricted by surrounding seas, the maximum density recorded is just under 2 to the square mile.

Neolithic farmers could of course, and certainly did, live together in permanent villages, though, owing to the extravagant rural economy generally practised, unless the crops were watered by irrigation, the villages had to be shifted at least every twenty years. But on the whole the growth of population was not reflected so much in the enlargement of the settlement unit as in a multiplication of settlements. In ethnography neolithic villages can boast only a few hundred inhabitants (a couple of 'pueblos' in New Mexico house over a thousand, but perhaps they cannot be regarded as neolithic). In prehistoric Europe the largest neolithic village yet known, Barkaer in Jutland, comprised 52 small, one-roomed dwellings, but 16 to 30 houses was a more normal figure; so the average, local group in neolithic times would average 200 to 400 members.

These low figures are of course the result of technical limitations. In the absence of wheeled vehicles and roads for the transport of bulky crops men had to live within easy walking distance of their cultivations. At the same time the normal rural economy of the Neolithic Age, what is now termed slash-and-burn or *jhumming*, condemns much more than half the arable land to lie fallow so that large areas were required. As soon as the population of a settlement rose above the numbers that could be supported from the accessible land, the excess had to hive off and found a new settlement.

The Neolithic Revolution had other consequences beside increasing the population, and their exploitation might in the end help to provide for the surplus increase. The new economy allowed, and indeed required, the farmer to produce every year more food than was needed to keep him and his family alive. In other words it made possible the regular production of a social surplus. Owing to the low efficiency of neolithic technique, the surplus produced was insignificant at first, but it could be increased till it demanded a reorganization of society.

Now in any Stone Age society, palaeolithic or neolithic, savage or barbarian, everybody can at least in theory make at home the few indispensable tools, the modest cloths and the simple ornaments everyone requires. But every member of the local community, not disqualified by age, must contribute actively to the communal food supply by personally collecting, hunting, fishing, gardening or herding. As long as this holds good, there can be no full-time specialists, no persons nor class of persons who depend for their livelihood on food produced by others and secured in exchange for material or immaterial goods or services.

We find indeed today among Stone Age barbarians and even savages expert craftsmen (for instance flint-knappers among the Ona of Tierra del Fuego), men who claim to be experts in magic, and even chiefs. In Palaeolithic Europe too there is some evidence for magicians and indications of chieftainship in pre-neolithic times. But on closer observation we discover

that today these experts are not full-time specialists. The Ona flintworker must spend most of his time hunting; he only adds to his diet and his prestige by making arrowheads for clients who reward him with presents. Similarly a pre-Columbian chief, though entitled to customary gifts and services from his followers, must still personally lead hunting and fishing expeditions and indeed could only maintain his authority by his industry and prowess in these pursuits. The same holds good of barbarian societies that are still in the neolithic stage, like the Polynesians where industry in gardening takes the place of prowess in hunting. The reason is that there simply will not be enough food to go round unless every member of the group contributes to the supply. The social surplus is not big enough to feed idle mouths.

Social division of labour, save those rudiments imposed by age and sex, is thus impossible. On the contrary community of employment, the common absorption in obtaining food by similar devices guarantees a certain solidarity to the group. For co-operation is essential to secure food and shelter and for defence against foes, human and subhuman. This identity of economic interests and pursuits is echoed and magnified by identity of language, custom and belief; rigid conformity is enforced as effectively as industry in the common quest for food. But conformity and industrious co-operation need no State organization to maintain them. The local group usually consists either of a single clan (persons who believe themselves descended from a common ancestor or who have earned a mystical claim to such descent by ceremonial adoption) or a group of clans related by habitual inter-marriage. And the sentiment of kinship is reinforced or supplemented by common rites focused on some ancestral shrine or sacred place. Archaeology can provide no evidence for kinship organization, but shrines occupied the central place in preliterate villages in Mesopotamia, and the long barrow, a collective tomb that overlooks the presumed site of most neolithic villages in Britain, may well have been also the ancestral shrine on which converged the emotions and ceremonial activities of the villagers below. However, the solidarity thus idealized and concretely symbolized, is really based on the same principles as that of a pack of wolves or a herd of sheep; Durkheim has called it 'mechanical.'

Now among some advanced barbarians (for instance tattooers or woodcarvers among the Maori) still technologically neolithic we find expert craftsmen tending towards the status of full-time professionals, but only at the cost of breaking away from the local community. If no single village can produce a surplus large enough to feed a full-time specialist all the year round, each should produce enough to keep him a week or so. By going round from village to village an expert might thus live entirely from his craft. Such itinerants will lose their membership of the sedentary kinship group. They may in the end form an analogous organization of their own – a craft clan, which, if it remain hereditary, may become a caste, or, if it recruit its members mainly by adoption (apprenticeship throughout Antiquity and the Middle Ages was just temporary adoption), may turn into a guild. But such specialists by emancipation from kinship ties, have also forfeited the protection of the kinship organization which alone under barbarism, guaranteed to its members security of person and property. Society must be reorganized to accommodate and protect them.

In pre-history specialization of labour presumably began with similar itinerant experts. Archaeological proof is hardly to be expected, but in ethnography metal-workers are nearly always full-time specialists. And in Europe at the beginning of the Bronze Age metal seems to have been worked and purveyed by perambulating smiths who seem to have functioned like tinkers and other itinerants of much more recent times. Though there is no such positive evidence, the same probably happened in Asia at the beginning of metallurgy. There must of course have been in addition other specialist craftsmen whom, as the Polynesian example warns us, archaeologists could not recognize because they worked in perishable materials. One result of the Urban Revolution will be to rescue such specialists from nomadism and to guarantee them security in a new social organization.

About 5,000 years ago irrigation cultivation (combined with stockbreeding and fishing) in the valleys of the Nile, the Tigris-Euphrates and the Indus had begun to yield a social surplus, large enough to support a number of resident specialists who were themselves released from food-production. Water-transport, supplemented in Mesopotamia and the Indus valley by wheeled vehicles and even in Egypt by pack animals, made it easy to gather food stuffs at a few centres. At the same time dependence on river water for the irrigation of the crops restricted the cultivable areas while the necessity of canalizing the waters and protecting habitations against annual floods encouraged the aggregation of population. Thus arose the first cities

– units of settlement ten times as great as any known neolithic village. It can be argued that all cities in the old world are offshoots of those of Egypt, Mesopotamia and the Indus basin. So the latter need not be taken into account if a minimum definition of civilization is to be inferred from a comparison of its independent manifestations.

But some three millennia later cities arose in Central America, and it is impossible to prove that the Mayas owed anything directly to the urban civilizations of the Old World. Their achievements must therefore be taken into account in our comparison, and their inclusion seriously complicates the task of defining the essential preconditions for the Urban Revolution. In the Old World the rural economy which yielded the surplus was based on the cultivation of cereals combined with stock-breeding. But this economy had been made more efficient as a result of the adoption of irrigation (allowing cultivation without prolonged fallow periods) and of important inventions and discoveries – metallurgy, the plough, the sailing boat and the wheel. None of these devices was known to the Maya; they bred no animals for milk or meat; though they cultivated the cereal maize, they used the same sort of slash-and-burn method as neolithic farmers in prehistoric Europe or in the Pacific Islands today. Hence the minimum definition of a city, the greatest factor common to the Old World and the New will be substantially reduced and impoverished by the inclusion of the Maya. Nevertheless ten rather abstract criteria, all deducible from archaeological data, serve to distinguish even the earliest cities from any older or contemporary village.

(1) In point of size the first cities must have been more extensive and more densely populated than any previous settlements, although considerably smaller than many villages today. It is indeed only in Mesopotamia and India that the first urban populations can be estimated with any confidence or precision. There excavation has been sufficiently extensive and intensive to reveal both the total area and the density of building in sample quarters and in both respects has disclosed significant agreement with the less industrialized Oriental cities today. The population of Sumerian cities, thus calculated, ranged between 7,000 and 20,000; Harappa and Mohenjo-daro in the Indus valley must have approximated to the higher figure. We can only infer that Egyptian and Maya cities were of comparable magnitude from the scale of public works, presumably executed by urban populations.

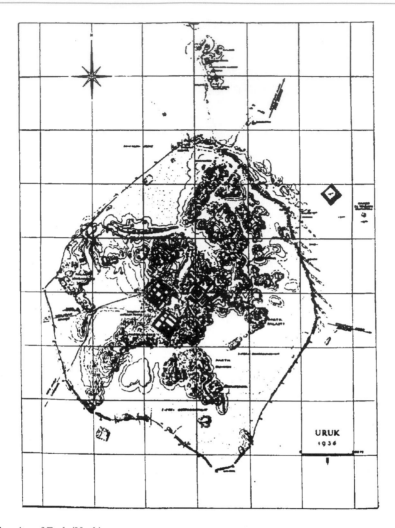

Figure 1 Plan of the city of Erek (Uruk)

(2) In composition and function the urban population already differed from that of any village. Very likely indeed most citizens were still also peasants, harvesting the lands and waters adjacent to the city. But all cities must have accommodated in addition classes who did not themselves procure their own food by agriculture, stock-breeding, fishing or collecting – full-time specialist craftsmen, transport workers, merchants, officials and priests. All these were of course supported by the surplus produced by the peasants living in the city and in dependent villages, but they did not secure their share directly by exchanging their products or services for grains or fish with individual peasants.

(3) Each primary producer paid over the tiny surplus he could wring from the soil with his still very limited technical equipment as tithe or tax to an imaginary deity or a divine king who thus concentrated the surplus. Without this concentration, owing to the low productivity of the rural economy, no effective capital would have been available.

(4) Truly monumental public buildings not only distinguish each known city from any village but also symbolize the concentration of the social surplus. Every Sumerian city was from the first dominated by one or more stately temples, centrally situated on a brick platform raised above the surrounding dwellings and usually connected with an artificial mountain, the staged tower or ziggurat. But attached to the temples were workshops and magazines, and an important appurtenance of each principal temple was a great granary. Harappa, in the Indus basin, was dominated

by an artificial citadel, girt with a massive rampart of kiln-baked bricks, containing presumably a palace and immediately overlooking an enormous granary and the barracks of artisans. No early temples nor palaces have been excavated in Egypt, but the whole Nile valley was dominated by the gigantic tombs of the divine pharaohs while royal granaries are attested from the literary record. Finally the Maya cities are known almost exclusively from the temples and pyramids of sculptured stone round which they grew up.

Hence in Sumer the social surplus was first effectively concentrated in the hands of a god and stored in his granary. That was probably true in Central America while in Egypt the pharaoh (king) was himself a god. But of course the imaginary deities were served by quite real priests who, besides celebrating elaborate and often sanguinary rites in their honour, administered their divine masters' earthly estates. In Sumer indeed the god very soon, if not even before the revolution, shared his wealth and power with a mortal viceregent, the 'City-King,' who acted as civil ruler and leader in war. The divine pharaoh was naturally assisted by a whole hierarchy of officials.

(5) All those not engaged in food-production were of course supported in the first instance by the surplus accumulated in temple or royal granaries and were thus dependent on temple or court. But naturally priests, civil and military leaders and officials absorbed a major share of the concentrated surplus and thus formed a 'ruling class.' Unlike a palaeolithic magician or a neolithic chief, they were, as an Egyptian scribe actually put it, 'exempt from all manual tasks.' On the other hand, the lower classes were not only guaranteed peace and security, but were relieved from intellectual tasks which many find more irksome than any physical labour. Besides reassuring the masses that the sun was going to rise next day and the river would flood again next year (people who have not five thousand years of recorded experience of natural uniformities behind them are really worried about such matters!), the ruling classes did confer substantial benefits upon their subjects in the way of planning and organization.

(6) They were in fact compelled to invent systems of recording and exact, but practically useful, sciences. The mere administration of the vast revenues of a Sumerian temple or an Egyptian pharaoh by a perpetual corporation of priests or officials obliged its members to devise conventional methods of recording that should be intelligible to all their colleagues

and successors, that is, to invent systems of writing and numeral notation. Writing is thus a significant, as well as a convenient, mark of civilization. But while writing is a trait common to Egypt, Mesopotamia, the Indus valley and Central America, the characters themselves were different in each region and so were the normal writing materials – papyrus in Egypt, clay in Mesopotamia. The engraved seals or stelae that provide the sole extant evidence for early Indus and Maya writing no more represent the normal vehicles for the scripts than do the comparable documents from Egypt and Sumer.

(7) The invention of writing – or shall we say the inventions of scripts – enabled the leisured clerks to proceed to the elaboration of exact and predictive sciences – arithmetic, geometry and astronomy. Obviously beneficial and explicitly attested by the Egyptian and Maya documents was the correct determination of the tropic year and the creation of a calendar. For it enabled the rulers to regulate successfully the cycle of agricultural operations. But once more the Egyptian, Maya and Babylonian calendars were as different as any systems based on a single natural unit could be. Calendrical and mathematical sciences are common features of the earliest civilizations and they too are corollaries of the archaeologists' criterion, writing.

(8) Other specialists, supported by the concentrated social surplus, gave a new direction to artistic expression. Savages even in palaeolithic times had tried, sometimes with astonishing success, to depict animals and even men as they saw them – concretely and naturalistically. Neolithic peasants never did that; they hardly ever tried to represent natural objects, but preferred to symbolize them by abstract geometrical patterns which at most may suggest by a few traits a fantastical man or beast or plant. But Egyptian, Sumerian, Indus and Maya artist-craftsmen – full-time sculptors, painters, or seal-engravers – began once more to carve, model or draw likenesses of persons or things, but no longer with the naive naturalism of the hunter, but according to conceptualized and sophisticated styles which differ in each of the four urban centres.

(9) A further part of the concentrated social surplus was used to pay for the importation of raw materials, needed for industry or cult and not available locally. Regular 'foreign' trade over quite long distances was a feature of all early civilizations and, though common enough among barbarians later, is not certainly attested in the Old World before 3000 B.C. nor in the New

before the Maya 'empire.' Thereafter regular trade extended from Egypt at least as far as Byblos on the Syrian coast while Mesopotamia was related by commerce with the Indus valley. While the objects of international trade were at first mainly 'luxuries,' they already included industrial materials, in the Old World notably metal, the place of which in the New was perhaps taken by obsidian. To this extent the first cities were dependent for vital materials on long distance trade as no neolithic village ever was.

(10) So in the city, specialist craftsmen were both provided with raw materials needed for the employment of their skill and also guaranteed security in a State organization based now on residence rather than kinship. Itinerancy was no longer obligatory. The city was a community to which a craftsman could belong politically as well as economically.

Yet in return for security they became dependent on temple or court and were relegated to the lower classes. The peasant masses gained even less material advantages; in Egypt for instance metal did not replace the old stone and wood tools for agricultural work. Yet, however imperfectly, even the earliest urban communities must have been held together by a sort of solidarity missing from any neolithic village. Peasants, craftsmen, priests and rulers form a community, not only by reason of identity of language and belief, but also because each performs mutually complementary functions, needed for the well-being (as redefined under civilization) of the whole. In fact the earliest cities illustrate a first approximation to an organic solidarity based upon a functional complementarity and interdependence between all its members such as subsist between the constituent cells of an organism. Of course this was only a very distant approximation. However necessary the concentration of the surplus really was with the existing forces of production, there seemed a glaring conflict on economic interests between the tiny ruling class, who annexed the bulk of the social surplus, and the vast majority who were left with a bare subsistence and effectively excluded from the spiritual benefits of civilization. So solidarity had still to be maintained by the ideological devices appropriate to the mechanical solidarity of barbarism as expressed in the pre-eminence of the temple or the sepulchral shrine, and now supplemented by the force of the new State organization. There could be no room for sceptics or sectaries in the oldest cities.

These ten traits exhaust the factors common to the oldest cities that archaeology, at best helped out

with fragmentary and often ambiguous written sources, can detect. No specific elements of town planning for example can be proved characteristic of all such cities; for on the one hand the Egyptian and Maya cities have not yet been excavated; on the other neolithic villages were often walled, an elaborate system of sewers drained the Orcadian hamlet of Skara Brae; two-storeyed houses were built in pre-Columbian pueblos, and so on.

The common factors are quite abstract. Concretely Egyptian, Sumerian, Indus and Maya civilizations were as different as the plans of their temples, the signs of their scripts and their artistic conventions. In view of this divergence and because there is so far no evidence for a temporal priority of one Old World centre (for instance, Egypt) over the rest nor yet for contact between Central America and any other urban centre, the four revolutions just considered may be regarded as mutually independent. On the contrary, all later civilizations in the Old World may in a sense be regarded as lineal descendants of those of Egypt, Mesopotamia or the Indus.

But this was not a case of like producing like. The maritime civilizations of Bronze Age Crete or classical Greece for example, to say nothing of our own, differ more from their reputed ancestors than these did among themselves. But the urban revolutions that gave them birth did not start from scratch. They could and probably did draw upon the capital accumulated in the three allegedly primary centres. That is most obvious in the case of cultural capital. Even today we use the Egyptians' calendar and the Sumerians' divisions of the day and the hour. Our European ancestors did not have to invent for themselves these divisions of time nor repeat the observations on which they are based; they took over – and very slightly improved – systems elaborated 5,000 years ago! But the same is in a sense true of material capital as well. The Egyptians, the Sumerians and the Indus people had accumulated vast reserves of surplus food. At the same time they had to import from abroad necessary raw materials like metals and building timber as well as 'luxuries.' Communities controlling these natural resources could in exchange claim a slice of the urban surplus. They could use it as capital to support full-time specialists – craftsmen or rulers – until the latters' achievement in technique and organization had so enriched barbarian economics that they too could produce a substantial surplus in their turn.

"The Polis"
from *The Greeks* (1951)

H.D.F. Kitto

Editors' Introduction

At its peak ancient Athens had only about as many residents as Peoria, Illinois (1990 population 113,504) – not a city that leaps out as a great center of world civilization. But British classicist Humphrey Davy Findley Kitto (1897–1982) reminds us not to commit the vulgar error of confusing size with significance. During its golden age, Athens and the 700 or so other tiny settlements of ancient Greece made a monumental contribution to human culture. What the Greeks achieved in philosophy, literature, drama, poetry, art, logic, mathematics, sculpture, and architecture has exercised a profound influence on Western civilization.

A Greek invention of enduring interest to urbanists is the polis. Since we have not got the thing, which the Greeks called "the polis," Kitto notes, we do not possess an equivalent word. "City-state" or, perhaps, "self-governing community" come closest.

The classical Greek polis came of age in the fifth century BCE, about halfway between the emergence of the great Mesopotamian cities Childe describes and the present time. The physical form of the polis stressed public space. Private houses were low and turned away from the street. In contrast the Greeks emphasized public temples, stadiums, the agora (a combined marketplace and public forum), and theaters like Athens' magnificent Theater of Dionysus illustrated in Plate 2. In the larger poleis, like Athens, these public buildings were spacious and often beautifully constructed of marble. Even in the smaller ones the community devoted many of its resources to them.

If the physical form of the polis was often stunning, it was the social organization of the polis that remains of particular fascination. The polis represents a form of community, which has exerted a powerful fascination for more than two millennia.

In the following selection Kitto describes how the polis made it possible for each citizen to realize his spiritual, moral, and intellectual capacities. The polis was a living community; almost an extended family. While the Greeks were very private in many ways, Kitto notes that their public life was essentially communistic. The polis as a social institution defined the very nature of being human for its citizens.

Not that the polis supported development of every resident: women and slaves were not citizens and did not participate in much of the life of the polis. Foreigners could attend plays in the Greek theater, but were barred from many institutions reserved to the (free, non-foreign, male) citizens. Sir Peter Hall in *Cities in Civilization* further questions the extent to which many citizens actually participated in public affairs. He hypothesizes that only a small percentage of those eligible to participate in public decision-making actually did so. He also notes that while farmers and other of the least educated and least articulate citizens of the Greek polis may have been physically present and possessed the same voting rights as educated upper-class Athenians it is unlikely that they participated very effectively compared to the higher classes. For the most part, Hall believes, they were passive spectators rather than active participants in the public affairs.

While a balanced view of the polis must acknowledge the existence of slavery, exclusion of women from civic life, limitations on the rights of foreigners, and the influence of education and class on social relations, Athens and the other Greek poleis were astonishingly democratic compared to any other urban civilization that preceded them. It is easy to dismiss Kitto as a hopeless romantic and his depiction of the Greek polis as an ivory tower depiction of a Camelot that never was. But that may be too harsh. The Greek polis as a social institution did represent a remarkable advance over social relations in any previous society. And the values it represented for its citizens are of enduring importance in an imperfect world.

In the debate about why the polis arose in Greece when it did, Kitto rejects deterministic answers such as the argument by geographical and economic determinists that the mountainous terrain required little, separate city-states. Rather, Kitto attributes the rise of the polis to the *character* of the Greeks themselves.

Kitto expresses nostalgia for human qualities of life in the polis that appear threatened today. Compare the vision of the polis as a supportive, humanistic, structure for human fulfillment with the vision of large modern cities as centers of alienation and anomie depicted by Louis Wirth (p. 90), or ghettos housing the Black underclass as described by William Julius Wilson (p. 110). Note the connections between humanistic values Kitto felt that the polis nurtured and Robert Putnam's concept of "social capital" growing out of civic engagement (p. 120), idealized life in J.B. Jackson's "almost perfect town" (p. 184), and the return to human-scale community Peter Calthorpe and William Fulton advocate in *The Regional City* (p. 342).

Other books helpful in understanding the polis and its significance are Christian Meier, *Athens: A Portrait of the City in its Golden Age* (New York: Metropolitan Books, 1998), Cecil Maurice Bowra, *The Greek Experience* (London: Weidenfeld & Nicolson, 1957), and a new edition of classic writings by a great classicist – Jacob Burckhardt, *The Greeks and Greek Civilization* (New York: St Martin's, 1998). Also of interest are Lisa Nevett, *House and Society in the Ancient Greek World* (Cambridge: Cambridge University Press, 1999) and Nicholas Cahill, *Household and City Organization at Olynthus* (New Haven: Yale University Press, 2002). Two excellent studies of Greek democracy are James O'Neil, *The Origins and Development of Ancient Greek Democracy* (Lanham: Rowman and Littlefield, 1995) and Josiah Ober, *Political Dissent in Democratic Athens* (Princeton: Princeton University Press, 1998).

For accounts of Greek city planning see Richard Ernest Wycherley, *How the Greeks Built Cities*, 2nd edn (London: Macmillan, 1963) and *The Stones of Athens* (Princeton: Princeton University Press, l978). Dora Crouch's *Water Management in Ancient Greek Cities* (New York: Oxford University Press, 1993) is a gem, and Spiro Kostof, "Polis and Akropolis," Chapter 7 of *A History of Architecture* (New York: Oxford University Press, 1980) provides insight on classical Greek architecture.

Two masterful accounts of the role of cities in civilization give particular emphasis to the contribution of the Greek polis. See Lewis Mumford's chapter on "The Emergence of the Polis" and "Citizen Versus Ideal City" in *The City in History* (New York: Harcourt Brace Jovanovich, 1961), and "The Fountainhead," the second chapter of Sir Peter Hall's *Cities in Civilization* (New York: Pantheon Books, 1998).

"Polis" is the Greek word which we translate as "city-state". It is a bad translation, because the normal polis was not much like a city, and was very much more than a state. But translation, like politics, is the art of the possible; since we have not got the thing which the Greeks called "the polis", we do not possess an equivalent word. From now on, we will avoid the misleading term "city-state", and use the Greek word instead . . . We will first inquire how this political system arose, then we will try to reconstitute the word "polis" and recover its real meaning by watching it in

action. It may be a long task, but all the time we shall be improving our acquaintance with the Greeks. Without a clear conception what the polis was, and what it meant to the Greek, it is quite impossible to understand properly Greek history, the Greek mind, or the Greek achievement.

First then, what was the polis? . . .

. . . In Crete . . . we find over fifty quite independent poleis, fifty small "states" . . . What is true of Crete is true of Greece in general, or at least of those parts which play any considerable part in Greek history . . .

It is important to realize their size. The modern reader picks up a translation of Plato's *Republic* or Aristotle's *Politics*; he finds Plato ordaining that his ideal city shall have 5,000 citizens, and Aristotle that each citizen should be able to know all the others by sight; and he smiles, perhaps, at such philosophic fantasies. But Plato and Aristotle are not fantasts. Plato is imagining a polis on the normal Hellenic scale; indeed he implies that many existing Greek poleis are too small — for many had less than 5,000 citizens. Aristotle says, in his amusing way . . . that a polis of ten citizens would be impossible, because it could not be self-sufficient, and that a polis of a hundred thousand would be absurd, because it could not govern itself properly . . . Aristotle speaks of a hundred thousand citizens; if we allow each to have a wife and four children, and then add a liberal number of slaves and resident aliens, we shall arrive at something like a million — the population of Birmingham; and to Aristotle an independent "state" as populous as Birmingham is a lecture-room joke . . .

In fact, only three poleis had more than 20,000 citizens: Syracuse and Acragas (Girgenti) in Sicily, and Athens. At the outbreak of the Peloponnesian War the population of Attica was probably about 350,000, half Athenian (men, women and children), about a tenth resident aliens, and the rest slaves. Sparta, or Lacedaemon, had a much smaller citizen-body, though it was larger in area. The Spartans had conquered and annexed Messenia, and possessed 3,200 square miles of territory. By Greek standards this was an enormous area: it would take a good walker two days to cross it. The important commercial city of Corinth had a territory of 330 square miles . . . The island of Ceos, which is about as big as Bute, was divided into four poleis. It had therefore four armies, four governments, possibly four different calendars, and, it may be, four different currencies and systems of measures — though this is less likely. Mycenae was in historical times a shrunken relic of Agamemnon's capital, but still independent. She sent an army to help the Greek cause against Persia at the battle of Plataea; the army consisted of eighty men. Even by Greek standards this was small, but we do not hear that any jokes were made about an army sharing a cab.

To think on this scale is difficult for us, who regard a state of ten million as small, and are accustomed to states which, like the U.S.A. and the U.S.S.R., are so big that they have to be referred to by their initials; but when the adjustable reader has become accustomed to the scale, he will not commit the vulgar error of confusing size with significance . . .

But before we deal with the nature of the polis, the reader might like to know how it happened that the relatively spacious pattern of pre-Dorian Greece became such a mosaic of small fragments. The Classical scholar too would like to know; there are no records, so that all we can do is to suggest plausible reasons. There are historical, geographical and economic reasons; and when these have been duly set forth, we may conclude perhaps that the most important reason of all is simply that this is the way in which the Greeks preferred to live.

[Here Kitto describes the evolution of the Greek acropolis from a fortified hilltop strong-point built for protection against Dorian invaders to a place of assembly, religion, and commerce.]

At this point we may invoke the very sociable habits of the Greeks, ancient or modern. The English farmer likes to build his house on his land, and to come into town when he has to. What little leisure he has he likes to spend on the very satisfying occupation of looking over a gate. The Greek prefers to live in the town or village, to walk out to his work, and to spend his rather ampler leisure talking in the town or village square. Therefore the market becomes a market-town, naturally beneath the acropolis. This became the center of the communal life of the people — and we shall see presently how important that was.

But why did not such towns form larger units? This is the important question.

There is an economic point. The physical barriers which Greece has so abundantly made the transport of goods difficult, except by sea, and the sea was not yet used with any confidence. Moreover, the variety of which we spoke earlier enabled quite a small area to be reasonably self-sufficient for a people who made such small material demands on life as the Greek. Both of these facts tend in the same direction; there was in Greece no great economic interdependence, no reciprocal pull between the different parts of the country, strong enough to counteract the desire of the Greek to live in small communities.

There is a geographical point. It is sometimes asserted that this system of independent poleis was imposed on Greece by the physical character of the country. The theory is attractive, especially to those who like to have one majestic explanation of any phenomenon, but it does not seem to be true. It is of course obvious that the physical subdivision of the country helped; the system could not have existed, for example, in Egypt, a country which depends entirely on the proper management of the Nile flood, and therefore must have a central government. But there are countries cut up quite as much as Greece – Scotland, for instance – which have never developed the polis-system; and conversely there were in Greece many neighbouring poleis, such as Corinth and Sicyon, which remained independent of each other although between them there was no physical barrier that would seriously incommode a modern cyclist. Moreover, it was precisely the most mountainous parts of Greece that never developed poleis, or not until later days – Arcadia and Aetolia, for example, which had something like a canton-system. The polis flourished in those parts where communications were relatively easy. So that we are still looking for our explanation.

Economics and geography helped, but the real explanation is the character of the Greeks . . . As it will take some time to deal with this, we may first clear out of the way an important historical point. How did it come about that so preposterous a system was able to last for more than twenty minutes?

The ironies of history are many and bitter, but at least this must be put to the credit of the gods, that they arranged for the Greeks to have the Eastern Mediterranean almost to themselves long enough to work out what was almost a laboratory-experiment to test how far, and in what conditions, human nature is capable of creating and sustaining a civilization . . . this lively and intelligent Greek people was for some centuries allowed to live under the apparently absurd system which suited and developed its genius instead of becoming absorbed in the dull mass of a large empire, which would have smothered its spiritual growth . . . no history of Greece can be intelligible until one has understood what the polis meant to the Greek; and when we have understood that, we shall also understand why the Greeks developed it, and so obstinately tried to maintain it. Let us then examine the word in action.

It meant at first that which was later called the Acropolis, the stronghold of the whole community and the centre of its public life . . . "polis" very soon meant either the citadel or the whole people which, as it were, "used" this citadel. So we read in Thucydides, "Epidamnus is a polis on the right as you sail into the Ionian gulf." This is not like saying "Bristol is a city on the right as you sail up the Bristol Channel", for Bristol is not an independent state which might be at war with Gloucester, but only an urban area with a purely local administration. Thucydides' words imply that there is a town – though possibly a very small one – called Epidamnus, which is the political centre of the Epidamnians, who live in the territory of which the town is the centre – not the "capital" – and are Epidamnians whether they live in the town or in one of the villages in this territory.

Sometimes the territory and the town have different names. Thus, Attica is the territory occupied by the Athenian people; it comprised Athens – the "polis" in the narrower sense – the Piraeus, and many villages; but the people collectively were Athenians, not Attics, and a citizen was an Athenian in whatever part of Attica he might live.

In this sense "polis" is our "state" . . . The actual business of governing might be entrusted to a monarch, acting in the name of all according to traditional usages, or to the heads of certain noble families, or to a council of citizens owning so much property, or to all the citizens. All these and many modifications of them, were natural forms of "polity"; all were sharply distinguished by the Greek from Oriental monarchy, in which the monarch is irresponsible, not holding his powers in trust by the grace of god, but being himself a god. If there were irresponsible government there was no polis . . .

. . . [T]he size of the polis made it possible for a member to appeal to all his fellow citizens in person, and this he naturally did if he thought that another member of the polis had injured him. It was the common assumption of the Greeks that the polis took its origin in the desire for Justice. Individuals are lawless, but the polis will see to it that wrongs are redressed. But not by an elaborate machinery of state-justice, for such a machine could not be operated except by individuals, who may be as unjust as the original wrongdoer. The injured party will be sure of obtaining Justice only if he can declare his wrongs to the whole polis. The word therefore now means "people" in actual distinction from state.

[. . .]

. . . Demosthenes the orator talks of a man who, literally, "avoids the city" – a translation which might lead the unwary to suppose that he lived in something corresponding to the Lake District, or Purley. But the phrase "avoids the polis" tells us nothing about his domicile; it means that he took no part in public life – and was therefore something of an oddity. The affairs of the community did not interest him.

We have now learned enough about the word polis to realize that there is no possible English rendering of such a common phrase as, "It is everyone's duty to help the polis." We cannot say "help the state", for that arouses no enthusiasm; it is "the state" that takes half our incomes from us. Not "the community", for with us "the community" is too big and too various to be grasped except theoretically. One's village, one's trade union, one's class, are entities that mean something to us at once, but "work for the community", though an admirable sentiment, is to most of us vague and flabby. In the years before the war, what did most parts of Great Britain know about the depressed areas? How much do bankers, miners and farmworkers understand each other? But the "polis" every Greek knew; there it was, complete, before his eyes. He could see the fields which gave it its sustenance – or did not, if the crops failed; he could see how agriculture, trade and industry dovetailed into one another; he knew the frontiers, where they were strong and where weak; if any malcontents were planning a *coup*, it was difficult for them to conceal the fact. The entire life of the polis, and the relation between its parts, were much easier to grasp, because of the small scale of things. Therefore to say "It is everyone's duty to help the polis" was not to express a fine sentiment but to speak the plainest and most urgent common sense. Public affairs had an immediacy and a concreteness which they cannot possibly have for us.

[. . .]

Pericles' Funeral Speech, recorded or recreated by Thucydides, will illustrate this immediacy, and will also take our conception of the polis a little further. Each year, Thucydides tells us, if citizens had died in war – and they had, more often than not – a funeral oration was delivered by "a man chosen by the polis". Today, that would be someone nominated by the Prime Minister, or the British Academy, or the BBC [British Broadcasting Corporation]. In Athens it meant that someone was chosen by the Assembly who had often spoken to that Assembly; and on this occasion Pericles spoke from a specially high platform, that his voice might reach as many as possible. Let us consider two phrases that Pericles used in that speech.

He is comparing the Athenian polis with the Spartan, and makes the point that the Spartans admit foreign visitors only grudgingly, and from time to time expel all strangers, "while we make our polis common to all". "Polis" here is not the political unit; there is no question of naturalizing foreigners – which the Greeks

did rarely, simply because the polis was so intimate a union. Pericles means here: "We throw open to all our common cultural life", as is shown by the words that follow, difficult though they are to translate: "nor do we deny them any instruction or spectacle" – words that are almost meaningless until we realize that the drama, tragic and comic, the performance of choral hymns, public recitals of Homer, games, were all necessary and normal parts of "political" life. This is the sort of thing Pericles has in mind when he speaks of "instruction and spectacle", and of "making the polis open to all".

But we must go further than this. A perusal of the speech will show that in praising the Athenian polis Pericles is praising more than a state, a nation, or a people: he is praising a way of life; he means no less when, a little later, he calls Athens the "school of Hellas". – And what of that? Do not we praise "the English way of life"? The difference is this; we expect our State to be quite indifferent to "the English way of life" – indeed, the idea that the State should actively try to promote it would fill most of us with alarm. The Greeks thought of the polis as an active, formative thing, training the minds and characters of the citizens; we think of it as a piece of machinery for the production of safety and convenience. The training in virtue, which the medieval state left to the Church, and the polis made its own concern, the modern state leaves to God knows what.

"Polis", then, originally "citadel", may mean as much as "the whole communal life of the people, political, cultural, moral" – even "economic", for how else are we to understand another phrase in this same speech, "the produce of the whole world comes to us, because of the magnitude of our polis"? This must mean "our national wealth".

Religion too was bound up with the polis – though not every form of religion. The Olympian gods were indeed worshipped by Greeks everywhere, but each polis had, if not its own gods, at least its own particular cults of these gods . . . But beyond these Olympians, each polis had its minor local deities, "heroes" and nymphs, each worshipped with his immemorial rite, and scarcely imagined to exist outside the particular locality where the rite was performed. So . . . there is a sense in which it is true to say that the polis is an independent religious, as well as political, unit . . .

[. . .]

. . . Aristotle made a remark which we most inadequately translate "Man is a political animal." What

Aristotle really said is "Man is a creature who lives in a polis"; and what he goes on to demonstrate, in his *Politics*, is that the polis is the only framework within which man can fully realize his spiritual, moral and intellectual capacities.

Such are some of the implications of this word . . . The polis was a living community, based on kinship, real or assumed – a kind of extended family, turning as much as possible of life into family life, and of course having its family quarrels, which were the more bitter because they were family quarrels.

This it is that explains not only the polis but also much of what the Greek made and thought, that he was essentially social. In the winning of his livelihood he was essentially individualist: in the filling of his life he was essentially "communist". Religion, art, games, the discussion of things – all these were needs of life that could be fully satisfied only through the polis – not, as with us, through voluntary associations of like-minded people, or through entrepreneurs appealing to individuals. (This partly explains the difference between Greek drama and the modern cinema.) Moreover, he wanted to play his own part in running the affairs of the community. When we realize how many of the necessary, interesting and exciting activities of life the Greek enjoyed through the polis, all of them in the open air, within sight of the same acropolis, with the same ring of mountains or of sea visibly enclosing the life of every member of the state – then it becomes possible to understand Greek history, to understand that in spite of the promptings of common sense the Greek could not bring himself to sacrifice the polis, with its vivid and comprehensive life, to a wider but less interesting unity . . .

[. . .]

First-Person Accounts of Great Cities of the Medieval and Early Modern World

Marco Polo, Ibn Battuta, Bernal Diaz, and Albrecht Dürer

Editors' Introduction

The period between the fall of the Roman Empire in Western Europe and the rise of the modern nation-states during the Renaissance was once called the medieval age, or even the "dark ages," but is now more often referred to as the pre-modern or early pre-modern period. Whatever the terminology, these centuries represented a profound transition in the history of technology, economy, religion, and social life . . . and in the emergence of modern urban civilization.

The early modern period was also an extraordinary age of travel and discovery, of cultures intersecting in ways that would determine the course of world history. The Venetian merchant Marco Polo (1254–1324) was not the first European to encounter China, but the account of his travels to the Mongol Empire of Kublai Khan and of his participation in the imperial Chinese government, first published as *The Book of Marco Polo* in 1299, was nothing less than a revelation. Accompanied by his father and his uncle, Polo left for the East in 1275 and did not return until 1295. During the intervening years, he experienced the wonders of a vast and advanced civilization that knew printing, used paper money, had invented gunpowder, and had developed an intricate and efficient form of political and social organization that Europeans could only wonder at.

In the passage here reprinted – "Of the Noble and Magnificent City of Kin-sai" – Polo may be criticized for possible exaggerations and a self-aggrandizing habit of referring to himself in the third person. But his observations on the present-day city of Hangzhou are dazzling, full of wonder, and imbued with a quickening sense of possibility. It is worth comparing Kin-Sai of Marco Polo's day with the "mega-urban regions" of China today (p. 489).

An even more extraordinary traveler of this period was Ibn Battuta (1307–1377). Born in Morocco, Ibn Battuta spent the better part of his life exploring and writing about the full extent of the Muslim world. In 1326, he traveled across North Africa to Cairo, Palestine, Syria, and Mecca. In the years following, up until 1354 when he wrote *The Rihla* (My Travels), he visited Persia and Iraq, Arabia and East Africa, Anatolia, the Central Asian steppes and the Empire of the Golden Horde, India and Ceylon, Malaysia and China, and finally Spain and West Africa. In this passage, Ibn Battuta describes the city of Constantinople, the capital of the Eastern Roman Empire.

A century later, in 1453, Constantinople would fall to the Ottoman Turks and be reconstituted as the Muslim city of Istanbul. But when Ibn Battuta encounters the city, the Byzantine emperor still rules (although Battuta refers to him as a "sultan"), and the court and the city are described as foreign, even exotic.

Bernal Diaz del Castillo (1492–1581?) came to the Aztec city of Tenochtitlán not as a traveler but as a conqueror, a conquistador. Born in Spain in the year of the first Columbus voyage, Diaz pursued gold and glory in Panama and the Yucatán before joining Hernán Cortés on the historic expedition to the Valley of Mexico (1519–1521) that brought the Aztec Empire under Spanish control.

As with Marco Polo's description of Kin-sai, Diaz's account of Mexico, especially the city of Tenochtitlán (now Mexico City), is full of astonishment and wonder. He and the other soldiers, some of whom had seen Constantinople and other European capitals, had never seen anything to compare to the magnificence and splendor of the Aztec city. As Diaz notes in his *Historia Verdadera de la Conquista de la Nueva España* (first published in 1636), the marketplaces, the temples, the court of Montezuma – all leave the Spanish awestruck. The biggest wonder of all, as Diaz freely admits, is that such an empire could fall to a desperate band of less than 400 soldiers!

Our final traveler is not an adventurer or a conqueror but a German artist. Albrecht Dürer (1471–1528) was one of the great painters and engravers of the Renaissance. Born in Nuremberg, Dürer studied art in Venice and became the court painter to two Holy Roman Emperors, Maximilian I and Charles V. He also traveled throughout Western Europe at a time when travel was still difficult and indulged in mostly by itinerant merchants. During his travels to the Low Countries, Dürer, something of an art merchant himself, visited other artists and sought out commissions for portraits from prosperous burghers.

Dürer also observed the cities and social customs of the people he visited, and his description of the Great Procession before the cathedral in Antwerp, Belgium, impressed Lewis Mumford (p. 85) as a notable example of the "urban drama." In *The City in History* (1960), Mumford writes, "Note the vast number of people arrayed in this procession. As in the church itself, the spectators were also communicants and participants: they engaged in the spectacle, watching it from within, not just from without . . . Prayer, mass, pageant, life-ceremony . . . the city itself was stage for these separate scenes of the drama . . ." Each in his own way, all of the travelers represented in this modest selection experienced a heightened sense of "urban drama" as diverse urban civilizations, formerly kept apart by history and geography, begin to interact. In the record of these contacts, we see the beginning of a process that today reaches an unprecedented level of intensity in the instantaneous communications between all parts of a globalized economy dominated by global urban command and control centers.

THE ITALIAN MERCHANT MARCO POLO DESCRIBES THE MAGNIFICENT CHINESE CITY OF KIN-SAI (1299)

Marco Polo, "Of the Noble and Magnificent City of Kin-sai," *The Travels of Marco Polo* (1299), translated by William Marsden and Manuel Komroff

Upon leaving Va-giu you pass, in the course of three days' journey, many towns, castles, and villages, all of them well inhabited and opulent. The people have abundance of provisions. At the end of three days you reach the noble and magnificent city of Kin-sai capital; Hang-chau, a name that signifies 'The Celestial City', and which it merits from its pre-eminence to all others in the world, in point of grandeur and beauty, as well as from its abundant delights, which might lead an inhabitant to imagine himself in paradise.

This city was frequently visited by Marco Polo, who carefully and diligently observed and inquired into every circumstance respecting it, all of which he recorded in his notes, from whence the following particulars are briefly stated. According to common estimation, this city is a hundred miles in circuit. Its streets and canals are extensive, and there are squares, or market-places, which being necessarily proportioned in size to the prodigious concourse of people by whom they are frequented, are exceedingly spacious. It is situated between a lake of fresh and very clear water on the one side, and a river of great magnitude on the other, the waters of which, by a number of canals, large and small, are made to run through every quarter of the city, carrying with them all the filth into the lake, and ultimately to the sea. This furnishes a communication by water, in addition to that by land, to all parts of the town. The canals and the streets being of sufficient width to allow boats on the one, and carriages on the other, to pass easily with articles necessary for the inhabitants.

It is commonly said that the number of bridges, of all sizes, amounts to twelve thousand. Those which are thrown over the principal canals and are connected with the main streets, have arches so high, and built with so much skill, that vessels with their masts can pass under them. At the same time, carts and horses can pass over, so well is the slope from the street graded to the height of the arch. If they were not so numerous, there would be no way of crossing from one place to another.

[. . .]

There are within the city ten principal squares or market-places, besides innumerable shops along the streets. Each side of these squares is half a mile in length, and in front of them is the main street, forty paces in width, and running in a direct line from one extremity of the city to the other. It is crossed by many low and convenient bridges. These market-squares are at the distance of four miles from each other. In a direction parallel to that of the main street, but on the opposite side of the squares runs a very large canal, on the nearer bank of which capacious warehouses are built of stone, for the accommodation of the merchants who arrive from India and other parts with their goods and effects. They are thus conveniently situated with respect to the market-places. In each of these, upon three days in every week, there is an assemblage of from forty to fifty thousand persons, who attend the markets and supply them with every article of provision that can be desired.

[. . .]

At all seasons there is in the markets a great variety of herbs and fruits, and especially pears of an extraordinary size, weighing ten pounds each, that are white in the inside, like paste, and have a very fragrant smell. There are peaches also, in their season, both of the yellow and white kind, and of a delicious flavour. Grapes are not produced there, but are brought in a dried state, and very good, from other parts. This applies also to wine, which the natives do not hold in estimation, being accustomed to their own liquor prepared from rice and spices. From the sea, which is fifteen miles distant, there is daily brought up the river, to the city, a vast quantity of fish; and in the lake also there is abundance, which gives employment at all times to persons whose sole occupation it is to catch them. The sorts are various according to the season of the year. At the sight of such an importation of fish, you would think it impossible that it could be sold; and yet, in the course of a few hours, it is all taken off, so great is the number of inhabitants, even of those classes which can afford to indulge in such luxuries, for fish and flesh are eaten at the same meal.

Each of the ten market-squares is surrounded with high dwelling-houses, in the lower part of which are shops, where every kind of manufacture is carried on, and every article of trade is sold; such, amongst others, as spices, drugs, trinkets, and pearls. In certain shops nothing is vended but the wine of the country, which they are continually brewing, and serve out fresh to their customers at a moderate price. The streets connected with the market-squares are numerous, and in some of them are many cold baths, attended by servants of both sexes. The men and women who frequent them have from their childhood been accustomed at all times to wash in cold water, which they reckon highly conducive to health. At these bathing places, however, they have apartments provided with warm water, for the use of strangers, who cannot bear the shock of the cold. All are in the daily practice of washing their persons, and especially before their meals.

In other streets are the quarters of the courtesans who are here in such numbers as I dare not venture to report. Not only near the squares, which is the situation usually appropriated for their residence, but in every part of the city they are to be found, adorned with much finery, highly perfumed, occupying well-furnished houses, and attended by many female domestics. These women are accomplished, and are perfect in the arts of caressing and fondling which they accompany with expressions adapted to every description of person. Strangers who have once tasted of their charms, remain in a state of fascination, and become so enchanted by their wanton arts, that they can never forget the impression. Thus intoxicated with sensual pleasures, when they return to their homes they report that they have been in Kin-sai, or The Celestial City, and look forward to the time when they may be enabled to revisit this paradise.

In other streets are the dwellings of the physicians and the astrologers, who also give instructions in reading and writing, as well as in many other arts. They have apartments also amongst those which surround the market-squares. On opposite sides of each of these squares there are two large edifices, where officers appointed by the Great Khan are stationed, to take immediate notice of any differences that may happen to arise between the foreign merchants, or amongst the inhabitants of the place. It is their duty likewise to see that the guards upon the several bridges in their respective vicinities are duly placed, and in cases of neglect, to punish the offenders at their discretion.

[. . .]

The inhabitants of the city are idolaters, and they use paper money as currency. The men as well as the women have fair complexions, and are handsome. The greater part of them are always clothed in silk, in consequence of the vast quantity of that material produced in the territory of Kin-sai, exclusively of what the merchants import from other provinces.

[. . .]

The natural disposition of the native inhabitants of Kin-sai is peaceful, and by the example of their former kings, who were themselves unwarlike, they have been accustomed to habits of tranquility. The management of arms is unknown to them, nor do they keep any in their houses. They conduct their mercantile and manufacturing concerns with perfect candour and honesty. They are friendly towards each other, and persons who inhabit the same street, both men and women, from the mere circumstance of neighbourhood, appear like one family.

In their domestic manners they are free from jealousy or suspicion of their wives, to whom great respect is shown, and any man would be accounted infamous who should presume to use indecent expressions to a married woman. To strangers also, who visit their city in the way of commerce, they give proofs of cordiality, inviting them freely to their houses, showing them friendly attentions, and furnishing them with the best advice and assistance in their mercantile transactions. On the other hand, they dislike the sight of soldiery, not excepting the guards of the Great Khan, for they remind them that they were deprived of the government of their native kings and rulers.

[. . .]

By a regulation which his Majesty has established, there is a guard of ten watchmen stationed, under cover, upon all the principal bridges, of whom five do duty by day and five by night. Each of these guards is provided with a sonorous wooden instrument as well as one of metal, together with a water device, by means of which the hours of the day and night are ascertained. As soon as the first hour of the night is expired, one of the watchmen gives a single stroke upon the wooden instrument, and also upon the metal gong, which announces to the people of the neighbouring streets that it is the first hour. At the expiration of the second, two strokes are given; and so on progressively, increasing the number of strokes as the hours advance. The guard is not allowed to sleep, and must be always on the alert. In the morning, as soon as the sun begins to appear, a single stroke is again struck, as in the evening, and so onwards from hour to hour.

Some of these watchmen patrol the streets, to observe whether any person has a light or fire burning after the hour appointed for extinguishing them. Upon making the discovery, they affix a mark to the door, and in the morning the owner of the house is taken before the magistrates, by whom, if he cannot assign a legitimate excuse for his offence, he is punished. Should they find any person abroad at an unseasonable hour, they arrest and confine him, and in the morning he is carried before the same tribunal. If they notice any person who from lameness or other infirmity is unable to work, they place him in one of the hospitals, of which there are several in every part of the city, founded by the ancient kings, and liberally endowed. When cured, he is obliged to work at some trade.

Immediately upon the appearance of fire breaking out in a house, they give the alarm by beating on the wooden machine, when the watchmen from all the bridges within a certain distance assemble to extinguish it, as well as to save the effects of the merchants and others, by removing them to the stone towers that have been mentioned. The goods are also sometimes put into boats, and conveyed to the islands in the lake. Even on such occasions the inhabitants dare not stir out of their houses, when the fire happens in the night, and only those can be present whose goods are actually being removed, together with the guard collected to assist, which seldom amounts to a smaller number than from one to two thousand men.

In cases also of tumult or insurrection amongst the citizens, the services of this police guard are necessary; but, independently of them, his Majesty always keeps on foot a large body of troops, both infantry and cavalry, in the city and its vicinity, the command of which he gives to his ablest officers.

[. . .]

At the distance of twenty-five miles from this city, in a direction to the northward of east, lies the sea, where there is an extremely fine port, frequented by all the ships that bring merchandise from India.

Marco Polo, happening to be in the city of Kin-sai at the time of making the annual report to his Majesty's commissioners of the amount of revenue and the number of inhabitants, had an opportunity of observing that the latter were registered at one hundred and sixty tomans of fire-places, that is to say, of families dwelling under the same roof; and as a toman is ten thousand, it follows that the whole city must have contained one million six hundred thousand families.

THE NORTH AFRICAN MUSLIM TRAVELLER IBN BATTUTA DESCRIBES BYZANTINE CONSTANTINOPLE (1354)

Ibn Battuta, "Constantinople the Great," from *The Travels of Ibn Battuta, AD 1325–1354*, H.A.R. Gibb (ed.), Hakluyt Society, 1962

Our entry into Constantinople was made about noon or a little later, and they beat their church gongs until the very skies shook with the mingling of their sounds. When we reached the first of the gates of the king's palace, we found it guarded by about a hundred men, who had an officer of theirs with them on top of a platform, and I heard them saying 'Sarakinu, Sarakinu' which means 'Muslims'. They would not let us enter, and when the members of the khatun's [the Byzantine Emperor's daughter] party told them that we had come in her suite they answered, 'They cannot enter except by permission', so we stayed by the gate. One of the khatun's party sent a messenger to tell her of this while she was still with her father. She told him about us, whereupon he gave orders to admit us and assigned us a house near the residence of the khatun. He wrote also on our behalf an order that we should not be molested wheresoever we might go in the city, and this order was proclaimed in the bazaars. We remained indoors for three nights, during which hospitality-gifts were sent us of flour, bread, sheep, fowls, ghee, fruit, fish, money and rugs, and on the fourth day we had an audience of the sultan [Emperor Andronicus III].

Account of the sultan of Constantinople

His name is Takfur, son of the sultan Jirjis. [Here Ibn Battuta uses Arabic names and Muslim titles to refer to the Eastern Roman Emperor.] His father . . . was still in the bond of life, but had renounced the world and become a monk, devoting himself to religious exercises in the churches, and had resigned the kingship to his son. . . . On the fourth day from our arrival at Constantinople, the khatun sent her page Sumbul the Indian to me, and he took my hand and led me into the palace. We passed through four gateways, each of which had porticoes in which were footsoldiers with their weapons, their officer being on a carpeted platform. When we reached the fifth gateway, the page Sumbul left me, and going inside returned with four

Greek pages, who searched me to see that I had no knife on my person. The officer said to me, 'This is a custom of theirs. Every person who enters the king's presence, be he noble or commoner, foreigner or native, must be searched'. The same practice is observed in the land of India.

Then, after they had searched me, the man in charge of the gate rose, took me by the hand, and opened the door. Four of the men surrounded me, two holding my sleeves and two behind me, and brought me into a large audience-hall, whose walls were of mosaic work, in which were pictured figures of creatures, both animate and inanimate. In the centre of it was a water-channel with trees on either side of it, and men were standing to right and left, silent, not one of them speaking. In the midst of the hall were three men standing, to whom those four men delivered me. These took hold of my garments as the others had done and so on a signal from another man led me forward. One of them was a Jew and he said to me in Arabic, 'Don't be afraid for this is their custom that they use with every visitor. I am the interpreter and I am originally from Sytia'. So I asked him how I should salute, and he told me to say 'al-salamu alaikum'.

I came then to a great pavilion; the sultan was there on his throne, with his wife, the mother of the khatun. Before him and at the foot of the throne were the khatun and her brothers. To the right of him were six men, to his left four and behind him four, every one of them armed. He signed to me, before I had saluted and reached him, to sit down for a moment, so that my apprehension might be calmed, and I did so. Then I approached him and saluted him, and he signed me to sit down, but I did not do so. He questioned me about Jerusalem, the Sacred Rock, the Church called al Qumama the cradle of Jesus [the Church of the Holy Sepulchre] and Bethlehem, and about the city of al-Khalil (peace be upon him) [Hebron], then about Damascus, Cairo, al-Iraq and the land of al-Rum, and I answered him on all his questions, the Jew interpreting between us. He was pleased with my replies and said to his sons, 'Honour this man and ensure his safety'. He then bestowed on me a robe of honour and ordered for me a horse with saddle and bridle, and a parasol of the kind that the king has carried above his head, that being a sign of protection. I asked him to designate someone to ride about the city with me every day, that I might see its wonders and curious sights and tell of them in my own country, and he designated such a guide for me. It is one of the customs among them that anyone who wears the king's robe of honour and rides on his horse is paraded through the city bazaars with trumpets, fifes and drums, so that the people may see him . . .

Account of the city

It is enormous in magnitude and divided into two parts, between which there is a great river in which there is a flow and ebb of tide [the Golden Horn] just as in the wadi of Sala in the country of the Maghrib [North Africa]. In the former time there was a bridge over it, built of stone, but the bridge has fallen into ruin and nowadays it is crossed in boats. The name of the river is Absumi. One of the two parts of the city is called Astanbiil; it is on the eastern bank of the river and includes the places of residence of the sultan, his officers of state, and the rest of the population. Its bazaars and streets are spacious and paved with flagstones, and the members of each craft have a separate place, no others sharing it with them. Each bazaar has gates which are closed upon it at night, and the majority of the artisans and sellers in them are women. The city is at the foot of a hill that projects about nine miles into the sea, and its breadth is the same or more. On top of the hill is a small citadel and the sultan's palace. This hill is surrounded by the city wall, which is a formidable one and cannot be taken by assault on the side of the sea. Within the wall are about thirteen inhabited villages. The principal church too is in the midst of this section of the city.

As for the other section of it, it is called al-Ghalata, and lies on the western bank of the river, somewhat like Ribat al-Fath in its proximity to the river. This section is reserved for the Christians of the Franks dwelling there. They are of different kinds, including Genoese, Venetians, men of Rome and people of France, and they are under the government of the King of Constantinople, who appoints over them one of their number whom they approve, and him they call the Qums. They are required to pay a tax every year to the King of Constantinople, but they often rebel against his authority and then he makes war on them until the Pope restores peace between them. They are all men of commerce, and their port is one of the greatest of ports; I saw in it about a hundred galleys, such as merchant vessels and other large ships, and as for the small ships they were too numerous to be counted. The bazaars in this section are good, but

overlaid with all kinds of filth, and traversed by a small, dirty and filth-laden stream. Their churches too are dirty and mean.

Account of the great church

I can describe only its exterior; as for its interior I did not see it. It is called in their language Aya Sufiya, and the story goes that it was an erection of Asaf the son of Barakhya, who was the son of the maternal aunt of Solomon (on whom be peace). It is one of the greatest churches of the Greeks; around it is a wall which encircles it so that it looks like a city in itself. Its gates are thirteen in number, and it has a sacred enclosure, which is about a mile long and closed by a great gate. No one is prevented from entering the enclosure, and in fact I went into it with the king's father, who will be mentioned later; it is like an audience-hall, paved with marble and traversed by a water-channel which issues from the church. This flows between two walls about a cubit high, constructed in marble inlaid with pieces of different colours and cut with the most skilful art, and trees are planted in rows on both sides of the channel. From the gate of the church to the gate of this hall there is a lofty pergola made of wood, covered with grape-vines and at the foot with jasmine and scented herbs. Outside the gate of this hall is a large wooden pavilion containing platforms, on which the guardians of this gate sit, and to the right of the 435 pavilions are benches and booths, mostly of wood, in which sit their qadis and the recorders of their bureaux. In the middle of the booths is a wooden pavilion, to which one ascends by a flight of wooden steps; in this pavilion is a great chair swathed in woollen cloth on which their qadi sits. We shall speak of him later. To the left of the pavilion which is at the gate of this hall is the bazaar of the druggists. The canal that we have described divides into two branches, one of which passes through the bazaar of the druggists and the other through the bazaar where the judges and the scribes sit.

At the door of the church there are porticoes where the attendants sit who sweep its paths, light its lamps and close its doors. They allow no person to enter it until he prostrates himself to the huge cross at their place, which they claim to be a relic of the wood on which the double of Jesus (on whom be peace) was crucified. This is over the door of the church, set in a golden frame about ten cubits in height, across which they have placed a similar golden frame so that it forms a cross. This door is covered with plaques of silver and gold, and its two rings are of pure gold. I was told that the number of monks and priests in this church runs into thousands, and that some of them are descendants of the Apostles, also that inside it is another church exclusively for women, containing more than a thousand virgins consecrated to religious devotions, and a still greater number of aged and widowed women. It is the custom of the king, his officers of state, and the rest of the inhabitants to come to visit this church every morning, and the Pope comes to it once in the year. When he is at a distance of four nights' journey from the town the king goes out to meet him and dismounts before him; when he enters the city, the king walks on foot in front of him, and comes to salute him every morning and evening during the whole period of his stay in Constantinople until he departs.

THE SPANISH CONQUISTADOR BERNAL DIAZ DESCRIBES TENOCHTITLÁN, CAPITAL OF AZTEC MEXICO (1521)

Bernal Diaz del Castillo, "About the Great and Solemn Reception which the Great Montezuma Gave Cortés upon Entering the Great City of Mexico," from *The Conquest of Mexico* (1521)

Gazing on such wonderful sights, we did not know what to say, or whether what appeared before us was real, for on one side, on the land, there were great cities, and in the lake ever so many more, and the lake itself was crowded with canoes, and in the Causeway were many bridges at intervals, and in front of us stood the great City of Mexico, and we – we did not even number four hundred soldiers! . . .

Let the curious readers consider whether there is not much to ponder over in this that I am writing. What men have there been in the world who have shown such daring? But let us get on, and march along the Causeway. When we arrived where another small causeway branches off . . . where there were some buildings like towers, which are their oratories, many more chieftains and Caciques approached clad in very rich mantles, the brilliant liveries of one chieftain differing from those of another, and the causeways were crowded with them. The Great Montezuma had sent these great Caciques in advance to receive us, and when they came before Cortés they bade us welcome

in their language, and as a sign of peace, they touched their hands against the ground, and kissed the ground with the hand.

[. . .]

As soon as we arrived and entered into the great court, the Great Montezuma took our Captain by the hand, for he was there awaiting him, and led him to the apartment and saloon where he was to lodge, which was very richly adorned according to their usage . . . A sumptuous dinner was provided for us according to their use and custom, and we ate it at once. So this was our lucky and daring entry into the great city of Tenochtitlán Mexico on the 8th day of November the year of our Saviour Jesus Christ 1519.

[. . .]

As I am almost tired of writing about this subject and my interested readers will be even more so, I will stop talking about it and tell how our Cortés in company with our captains and soldiers went to see Tlaltelolco, which is the great market-place of Mexico, and how we ascended the great Cue [temple pyramid] where stand the Idols Tezcatepuca and Huichilobos. This was the first time that our Captain went out to see the City, and I will relate what else happened.

As we had already been four days in Mexico and neither the Captain nor any of us had left our lodgings except to go to the houses and gardens, Cortés said to us that it would be well to go to the great Plaza and see the great Temple of Huichilobos, and that he wished to consult the Great Montezuma and have his approval . . . When Montezuma knew his wishes he sent to say that we were welcome to go; on the other hand, as he was afraid that we might do some dishonour to his idols, he determined to go with us himself with many of his chieftains . . . So he went on, and ascended the great Cue accompanied by many priests, and he began to burn incense and perform other ceremonies to Huichilobos . . .

When we arrived at the market-place, called Tlaltelolco, we were astounded at the number of people and the quantity of merchandise that it contained, and at the good order and control that was maintained, for we had never seen such a thing before. The chieftains who accompanied us acted as guides. Each kind of merchandise was kept by itself and had its fixed place marked out. Let us begin with the dealers in gold, silver, and precious stones, feathers, mantles, and embroidered goods. Then there were other wares consisting of Indian slaves both men and women; . . . and they brought them along tied to long poles, with collars round their necks so that they could not escape, and others they left free. Next there were other traders who sold great pieces of cloth and cotton, and articles of twisted thread, and there were cacahuateros who sold cacao . . . There were those who sold cloths of henequen and ropes and cotaras with which they are shod, which are made from the same plant, and sweet cooked roots, and other tubers which they get from this plant, all were kept in one part of the market in the place assigned to them. In another part there were skins of tigers and lions, of otters and jackals, deer and other animals and badgers and mountain cats, some tanned and others untanned, and classes of merchandise.

Let us go on and speak of those who sold beans and sage and other vegetables and herbs in another part, and to those who sold fowls, cocks with wattles, rabbits, hares, deer, mallards, young dogs and other things of that sort in their part of the market, and let us also mention the fruiterers, and the women who sold cooked food, dough and tripe in their own part of the market; then every sort of pottery made in a thousand different forms from great water jars to little jugs, these also had a place to themselves; then those who sold honey and honey paste and other dainties like nut paste, and those who sold lumber, boards, cradles, beams, blocks and benches, each article by itself, and the vendors of ocote firewood, and other things of a similar nature . . . But why do I waste so many words in recounting what they sell in that great market, for I shall never finish if I tell it all in detail. Paper, which in this country is called Amal, and reeds scented with liquidambar, and full of tobacco, and yellow ointments and things of that sort are sold by themselves, and much cochineal is sold under the arcades which are in that great market place, and there are many vendors of herbs and other sorts of trades. There are also buildings where three magistrates sit in judgement and there are executive officers like Alguacils who inspect the merchandise. I am forgetting those who sell salt, and those who make the stone knives, and how they split them off the stone itself; and the fisherwomen and others who sell some small cakes made from a sort of ooze which they get out of the great lake, which curdles, and from this they make a bread having a flavour something like cheese. There are for sale axes of brass and copper and tin, and gourds and gaily painted jars made of wood. I could wish that I had finished telling of all the things which are sold there, but they are so numerous and of such different quality and the great market-place with its surrounding arcades was so crowded with people, that one would not have been able to see and inquire about it all in two days.

[. . .]

Now let us leave the great market-place . . . and arrive at the great courts and walls where the great Cue stands. Before reaching the great Cue there is a great enclosure of courts, it seems to me larger than the plaza of Salamanca, with two walls of masonry surrounding it and the court itself all paved with very smooth great white flagstones. And where there were not these stones it was cemented and burnished and all very clean, so that one could not find any dust or a straw in the whole place.

When we arrived near the great Cue and before we had ascended a single step of it, the Great Montezuma sent down from above, where he was making his sacrifices, six priests and two chieftains to accompany our Captain. On ascending the steps, which are one hundred and fourteen in number, they attempted to take him by the arms so as to help him to ascend (thinking that he would get tired), as they were accustomed to assist their lord Montezuma, but Cortés would not allow them to come near him. When we got to the top of the great Cue, on a small plaza which has been made on the top where there was a space like a platform with some large stones placed on it, on which they put the poor Indians for sacrifice, there was a bulky image like a dragon and other evil figures and much blood shed that very day.

When we arrived there Montezuma came out of an oratory where his cursed idols were, at the summit of the great Cue, and two priests came with him, and after paying great reverence to Cortés and to all of us he said: "You must be tired . . . from ascending this our great Cue," and Cortés replied through our interpreters who were with us that he and his companions were never tired by anything. Then Montezuma took him by the hand and told him to look at his great city and all the other cities that were standing in the water, and the many other towns on the land round the lake . . .

So we stood looking about us, for that huge and cursed temple stood so high that from it one could see over everything very well, and we saw the three causeways which led into Mexico . . . and we saw the fresh water that comes from Chapultepec which supplies the city, and we saw the bridges on the three causeways which were built at certain distances apart through which the water of the lake flowed in and out from one side to the other, and we beheld on that great lake a great multitude of canoes, some coming with supplies of food and others returning loaded with cargoes of merchandise; and we saw that from every house of that great city and of all the other cities that were built in the water it was impossible to pass from house to house except by drawbridges which were made of wood or in canoes; and we saw in those cities Cues and oratories like towers and fortresses and all gleaming white, and it was a wonderful thing to behold . . .

After having examined and considered all that we had seen we turned to look at the great market-place and the crowds of people that were in it, some buying and others selling, so that the murmur and hum of their voices and words that they used could be heard more than a league off. Some of the soldiers among us who had been in many parts of the world, in Constantinople, and all over Italy, and in Rome, said that so large a market-place and so full of people, and so well regulated and arranged, they had never beheld before.

THE GERMAN ARTIST ALBRECHT DÜRER DESCRIBES A COMMUNAL PROCESSION IN ANTWERP, BELGIUM (1519)

Albrecht Dürer, "Visit to Antwerp," from *Journal of a Voyage to Belgium* (1527)

The Church of our Lady (the Cathedral) at Antwerp is so very large that many masses are sung in it at one time without interfering with each other. The altars have wealthy endowments, and the best musicians are employed that can be had. The church has many devout services, much stone-work, and in particular a beautiful tower. I have also been into the rich Abbey of St Michael. There are, in the choir there, splendid stalls of sculptured stone-work. But at Antwerp they spare no cost on such things, for there is money enough . . .

On the Sunday after our dear Lady's Assumption I saw the great Procession from the Church of our Lady at Antwerp, when the whole town of every craft and rank was assembled, each dressed in his best according to his rank. And all ranks and guilds had their signs, by which they might be known. In the intervals great costly pole-candles were borne, and their long old Frankish trumpets of silver. There were also in the German fashion many pipers and drummers. All the instruments were loudly and noisily blown and beaten.

I saw the Procession pass along the street, the people being arranged in rows, each man some distance from his neighbour, but the rows close one behind another. There were the Goldsmiths, the Painters, the Masons, the Broderers, the Sculptors, the Joiners, the Carpenters, the Sailors, the Fishermen, the Butchers, the Leatherers, the Clothmakers, the Bakers, the Tailors, the Cordwainers – indeed workmen of all kinds, and many craftsmen and dealers who work for their livelihood. Likewise the shop-keepers and merchants and their assistants of all kinds were there. After these came the shooters with guns, bows, and crossbows and the horsemen and foot-soldiers also. Then followed the watch of the Lords Magistrates. Then came a fine troop all in red, nobly and splendidly clad. Before them however went all the religious Orders and the members of some Foundations very devoutly, all in their different robes. A very large company of widows also took part in this procession. They support themselves with their own hands and observe a special rule. They were all dressed from head to foot in white linen garments, made expressly for the occasion, very sorrowful to see. Among them I saw some very stately persons. Last of all came the Chapter of our Lady's Church with all their clergy, scholars, and treasurers. Twenty persons bore the image of the Virgin Mary with the Lord Jesus, adorned in the costliest manner, to the honour of the Lord God.

In this Procession very many delightful things were shown, most splendidly got up. Wagons were drawn along with masques upon ships and other structures. Behind them came the company of the Prophets in their order and scenes from the New Testament, such as the Annunciation, the Three Holy Kings riding on great camels and on other rare beasts, very well arranged; also how our Lady fled to Egypt – very devout – and many other things, which for shortness I omit. At the end came a great Dragon which St Margaret and her maidens led by a girdle; she was especially beautiful. Behind her came St George with his squire, a very goodly knight in armour. In this host also rode boys and maidens most finely and splendidly dressed in the costumes of many lands, representing various Saints. From beginning to end the Procession lasted more than two hours before it was gone past our house. And so many things were there that I could never write them all in a book, so I let it well alone.

"The Great Towns"

from *The Condition of the Working Class in England in 1844* (1845)

Friedrich Engels

Editors' Introduction

It was the peculiar fate of Friedrich Engels (1820–1895) to live most of his adult life in the shadow of his better-known friend and partner Karl Marx and to be remembered as a fiercely bearded icon of International Communism. It was, however, a more humanly accessible Engels who, full of youthful idealism at the age of only 24, came face to face with the social horrors of the Industrial Revolution. Young Engels was sent by his industrialist father to learn business management in the factories of Manchester in the English North West. The unintended consequences of that particular paternal decision was *The Condition of the Working Class in England in 1844* (1845), a book that ranks as one of the earliest masterpieces of urban socio-politics.

By the 1840s, the Industrial Revolution had transformed conditions in many English cities. Manchester, which Engels observed in detail, was emblematic of what the new industrial cities were like. Plate 4 from Augustus Welby Pugin's *Contrasts* (1841) compares the skyline of a fifteenth-century city dominated by church steeples to the same town in 1840. In the second view mills, factories, and a huge prison dominate the scene.

In the selection "The Great Towns" reprinted here, Engels employs a peripatetic method. Although he summarizes the socialist theory of the origin and historic role of the industrial working class, and although he quotes from many contemporaneous sources to bolster his analysis, Engels constructs the bulk of his argument by merely walking around the city and reporting what he sees. Quickly growing impatient with *telling* his readers about the social misery of working-class life, Engels begins *showing* them the horrors of industrial urbanism by conducting them on a tour of Manchester's working-class districts. As in Dante's *Inferno*, the tour descends deeper and deeper into the filth, misery, and despair that constitute the greater part of the Manchester conurbation.

Engels wrote just as the first daguerreotypes of cities were being produced, but unfortunately he did not illustrate his book with actual pictures of the conditions that he described. Later in the nineteenth century Jacob Riis, Lewis Hine, and other photographers were to document slum conditions. The response of these and other photographers to the new reality of the nineteenth-century city is discussed in the selection by Frederic Stout, "Visions of a New Reality" (p. 141).

No one can read *The Condition of the Working Class* without acknowledging that Engels had come to know the various neighborhoods of proletarian Manchester – Old Town, Irish Town, Long Millgate, and Salford – intimately and that his observations were acute and objective. Of particular interest are his descriptions of the public health consequences (in terms of air and water pollution) of unrestrained overbuilding. In this, Engels anticipated many of the points made by environmental reformers like Frederick Law Olmsted (p. 307) and utopian planners like Ebenezer Howard (p. 314). He may even be said to have laid the groundwork for the arguments of the sustainable planning advocates like Stephen Wheeler (p. 499), the World Commission on Environment and Development (p. 337), and Timothy Beatley (p. 411).

Responding to the spatial arrangements of the class segregation of urban-industrialism, Engels observed that the façades of the main thoroughfares mask the horrors that lie beyond from the eyes of the factory owners and the middle-class managers who commute into the city from outlying suburbs. Plate 5, from *Frank Leslie's Illustrated Newspaper* in 1887, contrasts the belching factories and slums that comprise the bulk of the city with a spacious main thoroughfare in the city.

The entire tradition of twentieth-century urban planning, capitalist and socialist alike, owes an enormous debt to Engels. The connection he draws between the physical decrepitude of the urban infrastructure and the alienation and despair of the urban poor remains valid to the present day. The urban parks movement and the construction of ideal company towns – Saltaire and Port Sunlight in the UK, Lowell and Pullman in the United States – as well as more recent attempts at inner-city redevelopment, all address issues first identified by Engels. What Engels may not have understood is that he was also describing the birth of the modern capitalist society of the West, an institution that 150 years later would excite the hatred of the non-Western militants of fundamentalist Islam described by Ian Buruma and Avishai Margalit (p. 136). To the Islamists, Occidentalist capitalism and socialism would be equally suspect and decadent.

The conditions described by Engels form the basis for the social realist tradition in literature, a tradition that begins with Charles Dickens and Mrs Gaskell in England and is continued in the works of Upton Sinclair and Theodore Dreiser in the United States. One can wonder whether the cultural impact of the industrial working class that Engels and the social realists describe will in any way be paralleled by the cultural impact of the post-industrial "creative class" that Richard Florida describes (p. 129).

See "Manchester 1760–1840," chapter 10 of Peter Hall's *Cities in Civilization* (New York: Pantheon, 1998) for more on early Manchester. Other significant investigations of urban poverty in England include Henry Mayhew, *London Labour and the London Poor* (four volumes, 1851–1862), Charles Booth, *Conditions and Occupation of the People in East London and Hackney* (*Journal of the Royal Historical Society*, 1887), Jack London, *People of the Abyss* (New York: Macmillan, 1903), and George Orwell, *Down and Out in Paris and London* (London: Secker & Warburg, 1933).

In America, key works include Jacob Riis, *How the Other Half Lives* (New York: Scribners, 1903), Upton Sinclair, *The Jungle* (New York: Doubleday, 1906), and a whole series of reports on conditions in the African–American ghettos such as W.E.B. Du Bois, *The Philadelphia Negro* (p. 103), St. Clair Drake and Horace Cayton, *Black Metropolis* (Chicago: University of Chicago Press, 1945), and William Julius Wilson, *When Work Disappears* (p. 110).

For a sampling of Engels's most important writings, as well as those of his lifelong friend Karl Marx, see Robert C. Tucker, *The Marx–Engels Reader* (New York: W.W. Norton, 1978). For an excellent summary of nineteenth-century urban poverty conditions and the broader socio-political context, consult "The City of Dreadful Night" in Peter Hall, *Cities of Tomorrow* (London: Basil Blackwell, 1988), Eric Hobsbawm, *The Age of Revolution, 1789–1848* (New York: Vintage, 1996), and Kenneth Morgan, *The Birth of Industrial Britain, 1750–1850* (London: Longman, 1999). For further information on Engels's Manchester, and its connections to an emerging social realism in fiction, consult literary historian Steven Marcus's magisterial *Engels, Manchester, and the Working Class* (New York: Random House, 1974). Also of interest is Robert Roberts's extraordinary *The Classic Slum* (Manchester: University of Manchester Press, 1971), a first-person account of growing up in Salford during the early years of the twentieth century.

A town, such as London, where a man may wander for hours together without reaching the beginning of the end, without meeting the slightest hint which could lead to the inference that there is open country within reach, is a strange thing. This colossal centralization, this heaping together of two and a half millions of human beings at one point, has multiplied the power of this two and a half millions a hundred-fold; has raised London to the commercial capital of the world, created the giant docks and assembled the thousand vessels that continually cover the Thames. I know nothing more imposing than the view which the Thames offers during the ascent from the sea to London Bridge. The masses of buildings, the wharves on both sides, especially from Woolwich upwards, the countless ships along both shores, crowding ever closer and closer together, until, at last, only a narrow passage remains in the middle of the river, a passage through which hundreds of steamers shoot by one another; all this is so vast, so impressive, that a man cannot collect himself, but is lost in the marvel of England's greatness before he sets foot upon English soil.

But the sacrifices which all this has cost become apparent later. After roaming the streets of the capital a day or two, making headway with difficulty through the human turmoil and the endless lines of vehicles, after visiting the slums of the metropolis, one realizes for the first time that these Londoners have been forced to sacrifice the best qualities of their human nature, to bring to pass all the marvels of civilization which crowd their city; that a hundred powers which slumbered within them have remained inactive, have been suppressed in order that a few might be developed more fully and multiply through union with those of others. The very turmoil of the streets has something repulsive, something against which human nature rebels. The hundreds of thousands of all classes and ranks crowding past each other, are they not all human beings with the same qualities and powers, and with the same interest in being happy? And have they not, in the end, to seek happiness in the same way, by the same means? And still they crowd by one another as though they had nothing in common, nothing to do with one another, and their only agreement is the tacit one, that each keep to his own side of the pavement, so as not to delay the opposing streams of the crowd, while it occurs to no man to honour another with so much as a glance. The brutal indifference, the unfeeling isolation of each in his private interest becomes the more repellent and offensive, the more these individuals are crowded together, within a limited space. And, however much one may be aware that this isolation of the individual, this narrow self-seeking is the fundamental principle of our society everywhere, it is nowhere so shamelessly barefaced, so self-conscious as just here in the crowding of the great city. The dissolution of mankind into monads, of which each one has a separate principle and a separate purpose, the world of atoms, is here carried out to its utmost extreme.

Hence it comes, too, that the social war, the war of each against all, is here openly declared. . . . , people regard each other only as useful objects; each exploits the other, and the end of it all is, that the stronger treads the weaker under foot, and that the powerful few, the capitalists, seize everything for themselves, while to the weak many, the poor, scarcely a bare existence remains.

What is true of London, is true of Manchester, Birmingham, Leeds, is true of all great towns. Everywhere barbarous indifference, hard egotism on one hand, and nameless misery on the other, everywhere social warfare, every man's house in a state of siege, everywhere reciprocal plundering under the protection of the law, and all so shameless, so openly avowed that one shrinks before the consequences of our social state as they manifest themselves here undisguised, and can only wonder that the whole crazy fabric still hangs together.

Since capital, the direct or indirect control of the means of subsistence and production, is the weapon with which this social warfare is carried on, it is clear that all the disadvantages of such a state must fall upon the poor. For him no man has the slightest concern. Cast into the whirlpool, he must struggle through as well as he can. If he is so happy as to find work, i.e. if the bourgeoisie does him the favour to enrich itself by means of him, wages await him which scarcely suffice to keep body and soul together; if he can get no work he may steal, if he is not afraid of the police, or starve, in which case the police will take care that he does so in a quiet and inoffensive manner. During my residence in England, at least twenty or thirty persons have died of simple starvation under the most revolting circumstances, and a jury has rarely been found possessed of the courage to speak the plain truth in the matter. Let the testimony of the witnesses be never so clear and unequivocal, the bourgeoisie, from which the jury is selected, always finds some backdoor through which to escape the frightful verdict, death from starvation. The bourgeoisie dare not speak the truth in these cases, for it would speak its own condemnation. But indirectly, far more than directly, many have died of starvation, where long continued want of proper nourishment has called forth fatal illness, when it has produced such debility that causes which might otherwise have remained inoperative, brought on severe illness and death. The English working-men call this social murder, and accuse our whole society of perpetrating this crime perpetually. Are they wrong?

True, it is only individuals who starve, but what security has the working-man that it may not be his turn tomorrow? Who assures him employment, who vouches for it that, if for any reason or no reason his lord and master discharges him tomorrow, he can struggle along with those dependent upon him, until he may find some one else "to give him bread"? Who guarantees that willingness to work shall suffice to obtain work, that uprightness, industry, thrift, and the rest of the virtues recommended by the bourgeoisie, are really his road to happiness? No one. He knows that he has something today, and that it does not depend upon himself whether he shall have something tomorrow. He knows that every breeze that blows, every whim of his employer, every bad turn of trade may hurl him back into the fierce whirlpool from which he has temporarily saved himself, and in which it is hard and often impossible to keep his head above water. He knows that, though he may have the means of living today, it is very uncertain whether he shall tomorrow.

[. . .]

Manchester lies at the foot of the southern slope of a range of hills, which stretch hither from Oldham, their last peak, Kersallmoor, being at once the race-course and the Mons Sacer of Manchester. Manchester proper lies on the left bank of the Irwell, between that stream and the two smaller ones, the Irk and the Medlock, which here empty into the Irwell. On the right bank of the Irwell, bounded by a sharp curve of the river, lies Salford, and farther westward Pendleton; northward from the Irwell lie Upper and Lower Broughton; northward of the Irk, Cheetham Hill; south of the Medlock lies Hulme; farther east Chorlton on Medlock; still farther, pretty well to the east of Manchester, Ardwick. The whole assemblage of buildings is commonly called Manchester, and contains about four hundred thousand inhabitants, rather more than less. The town itself is peculiarly built, so that a person may live in it for years, and go in and out daily without coming into contact with a working-people's quarter or even with workers; that is, so long as he confines himself to his business or to pleasure walks. This arises chiefly from the fact, that by unconscious tacit agreement, as well as with outspoken conscious determination, the working-people's quarters are sharply separated from the sections of the city reserved for the middle class; or, if this does not succeed, they are concealed with the cloak of charity. Manchester contains, at its heart, a rather extended commercial district, perhaps half a mile long and about as broad, and consisting almost wholly of offices and warehouses. Nearly the whole district is abandoned by dwellers, and is lonely and deserted at night; only watchmen and policemen traverse its narrow lanes with their dark lanterns. This district is cut through by certain main thoroughfares upon which the vast traffic concentrates, and in which the ground level is lined with brilliant shops. In these streets the upper floors are occupied, here and there, and there is a good deal of life upon them until late at night. With the exception of this commercial district, all Manchester proper, all Salford and Hulme, a great part of Pendleton and Chorlton, two-thirds of Ardwick, and single stretches of Cheetham Hill and Broughton are all unmixed working-people's quarters, stretching like a girdle, averaging a mile and a half in breadth, around the commercial district. Outside, beyond this girdle, lives the upper and middle bourgeoisie, the middle bourgeoisie in regularly laid out streets in the vicinity of the working quarters, especially in Chorlton and the lower-lying portions of Cheetham Hill; the upper bourgeoisie in remoter villas with gardens in Chorlton and Ardwick or on the breezy heights of Cheetham Hill, Broughton and Pendleton, in free, wholesome country air, in fine, comfortable homes, passed once every half or quarter hour by omnibuses going into the city. And the finest part of the arrangement is this, that the members of this money aristocracy can take the shortest road through the middle of all the labouring districts to their places of business, without ever seeing that they are in the midst of the grimy misery that lurks to the right and the left. For the thoroughfares leading from the Exchange in all directions out of the city are lined, on both sides, with an almost unbroken series of shops, and are so kept in the hands of the middle and lower bourgeoisie, which, out of self-interest, cares for a decent and cleanly external appearance and *can* care for it. True, these shops bear some relation to the districts which lie behind them, and are more elegant in the commercial and residential quarters than when they hide grimy working-men's dwellings; but they suffice to conceal from the eyes of the wealthy men and women of strong stomachs and weak nerves the misery and grime which form the complement of their wealth. So, for instance, Deansgate, which leads from the Old Church directly southward, is lined first with mills and warehouses, then with second-rate shops and alehouses; farther south, when it leaves the commercial district, with less inviting shops, which grow dirtier and more interrupted by beerhouses and gin palaces the farther one goes, until at the southern end the appearance of the shops leaves no doubt that workers and workers only are their customers. So Market Street running south east from the Exchange; at first brilliant shops of the best sort, with counting-houses or warehouses above; in the continuation, Piccadilly, immense hotels and warehouses; in the farther continuation, London Road, in the neighbourhood of the Medlock, factories, beerhouses, shops for the humbler bourgeoisie and the working population; and from this point onward, large gardens and villas of the wealthier merchants and manufacturers. In this way any one who knows Manchester can infer the adjoining districts, from the appearance of the thoroughfare, but one is seldom in a position to catch from the street a glimpse of the real labouring districts. I know very well that this hypocritical plan is more or less common to all great cities; I know, too, that the retail dealers are forced by the nature of their business to take possession of the great highways; I know that there are more good buildings than bad ones upon such streets everywhere, and that the value of land is greater near them than in remoter districts; but at the same time I have never seen so systematic a shutting out of the working-class from the thoroughfares, so tender a concealment of everything which might affront the eye and the nerves of the bourgeoisie, as in Manchester. And yet, in other respects, Manchester is less built according to a plan, after official regulations, is more an outgrowth of accident, than any other city; and when I consider in this connection the eager assurances of the middle-class, that the working-class is doing famously, I cannot help feeling that the liberal manufacturers, the "Big Wigs" of Manchester, are not so innocent after all, in the matter of this sensitive method of construction.

I may mention just here that the mills almost all adjoin the rivers or the different canals that ramify throughout the city, before I proceed at once to describe the labouring quarters. First of all, there is the Old Town of Manchester [Figure 1], which lies between the northern boundary of the commercial district and the Irk. Here the streets, even the better ones, are narrow and winding, as Todd Street, Long Millgate, Withy Grove, and Shude Hill, the houses dirty, old, and tumble-down, and the construction of the side streets utterly horrible. Going from the Old Church to Long Millgate, the stroller has at once a row of old-fashioned houses at the right, of which not one has kept its original level; these are remnants of the old pre-manufacturing Manchester, whose former inhabitants have removed with their descendants into better-built districts, and have left the houses, which were not good enough for them, to a working-class population strongly mixed with Irish blood. Here one is in an almost undisguised working-men's quarter, for even the shops and beer-

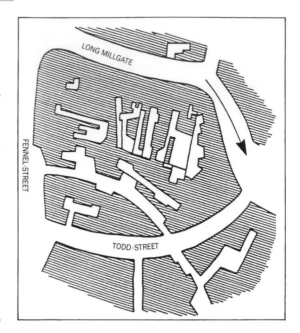

Figure 1

houses hardly take the trouble to exhibit a trifling degree of cleanliness. But all this is nothing in comparison with the courts and lanes which lie behind, to which access can be gained only through covered passages in which no two human beings can pass at the same time. Of the irregular cramming together of dwellings in ways which defy all rational plan, of the tangle in which they are crowded literally one upon the other, it is impossible to convey an idea. And it is not the buildings surviving from the old times of Manchester which are to blame for this; the confusion has only recently reached its height when every scrap of space left by the old way of building has been filled up and patched over until not a foot of land is left to be further occupied.

[. . .]

The south bank of the Irk is here very steep and between fifteen and thirty feet high. On this declivitous hillside there are planted three rows of houses, of which the lowest rise directly out of the river, while the front walls of the highest stand on the crest of the hill in Long Millgate. Among them are mills on the river; in short, the method of construction is as crowded and disorderly here as in the lower part of Long Millgate. Right and left a multitude of covered passages lead from the main street into numerous courts, and he who turns in thither gets into a filth and disgusting grime the equal of which is not to be found – especially in the courts which lead down to the Irk and which contain unqualifiedly the most horrible dwellings which I have yet beheld. In one of these courts there stands directly at the entrance, at the end of the covered passage, a privy without a door, so dirty that the inhabitants can pass into and out of the court only by passing through foul pools of stagnant urine and excrement. This is the first court on the Irk above Ducie Bridge – in case any one should care to look into it. Below it on the river there are several tanneries which fill the whole neighbourhood with the stench of animal putrefaction. Below Ducie Bridge the only entrance to most of the houses is by means of narrow dirty stairs and over heaps of refuse and filth. The first court below Ducie Bridge, known as Allen's Court, was in such a state at the time of the cholera that the sanitary police ordered it evacuated, swept, and disinfected with chloride of lime. Dr. Kay gives a terrible description of the state of this court at that time. Since then it seems to have been partially torn away and rebuilt; at least looking down from Ducie Bridge, the passer-by sees several ruined walls and heaps of debris with some newer houses. The view from this bridge, mercifully concealed from mortals of small stature by a parapet as high as a man, is characteristic for the whole district. At the bottom flows, or rather stagnates, the Irk, a narrow, coal-black, foul-smelling stream full of debris and refuse, which it deposits on the shallower right bank. In dry weather, a long string of the most disgusting, blackish-green, slime pools are left standing on this bank, from the depths of which bubbles of miasmatic gas constantly arise and give forth a stench unendurable even on the bridge forty or fifty feet above the surface of the stream. But besides this, the stream itself is checked every few paces by high weirs, behind which slime and refuse accumulate and rot in thick masses. Above the bridge are tanneries, bonemills, and gasworks, from which all drains and refuse find their way into the Irk, which receives further the contents of all the neighbouring sewers and privies. It may be easily imagined, therefore, what sort of residue the stream deposits. Below the bridge you look upon the piles of debris, the refuse, filth, and offal from the courts on the steep left bank; here each house is packed close behind its neighbour and a piece of each is visible, all black, smoky, crumbling, ancient, with broken panes and window-frames. The background is furnished by old barrack-like factory buildings. On the lower right bank stands a long row of houses and mills; the second house being a ruin without a roof, piled with debris; the third

stands so low that the lowest floor is uninhabitable, and therefore without windows or doors. Here the background embraces the pauper burial-ground, the station of the Liverpool and Leeds railway, and, in the rear of this, the Workhouse, the "Poor-Law-Bastille" of Manchester, which, like a citadel, looks threateningly down from behind its high walls and parapets on the hilltop, upon the working-people's quarter below.

Above Ducie Bridge, the left bank grows more flat and the right bank steeper, but the condition of the dwellings on both banks grows worse rather than better. He who turns to the left here from the main street, Long Millgate, is lost; he wanders from one court to another, turns countless corners, passes nothing but narrow, filthy nooks and alleys, until after a few minutes he has lost all clue, and knows not whither to turn. Everywhere half or wholly ruined buildings, some of them actually uninhabited, which means a great deal here; rarely a wooden or stone floor to be seen in the houses, almost uniformly broken, ill-fitting windows and doors, and a state of filth! Everywhere heaps of debris, refuse, and offal; standing pools for gutters, and a stench which alone would make it impossible for a human being in any degree civilized to live in such a district. The newly-built extension of the Leeds railway, which crosses the Irk here, has swept away some of these courts and lanes, laying others completely open to view. Immediately under the railway bridge there stands a court, the filth and horrors of which surpass all the others by far, just because it was hitherto so shut off, so secluded that the way to it could not be found without a good deal of trouble. I should never have discovered it myself, without the breaks made by the railway, though I thought I knew this whole region thoroughly. Passing along a rough bank, among stakes and washing-lines, one penetrates into this chaos of small one-storeyed, one-roomed huts, in most of which there is no artificial floor; kitchen, living and sleeping room all in one. In such a hole, scarcely five feet long by six broad, I found two beds – and such bedsteads and beds! – which, with a staircase and chimney-place, exactly filled the room. In several others I found absolutely nothing, while the door stood open, and the inhabitants leaned against it. Everywhere before the doors refuse and offal; that any sort of pavement lay underneath could not be seen but only felt, here and there, with the feet. This whole collection of cattle-sheds for human beings was surrounded on two sides by houses and a factory, and on the third by the river, and besides the narrow stair up the bank, a narrow doorway alone led out into another almost equally ill-built, ill-kept labyrinth of dwellings.

Enough! The whole side of the Irk is built in this way, a planless, knotted chaos of houses, more or less on the verge of uninhabitableness, whose unclean interiors fully correspond with their filthy external surroundings. And how could the people be clean with no proper opportunity for satisfying the most natural and ordinary wants? Privies are so rare here that they are either filled up every day, or are too remote for most of the inhabitants to use. How can people wash when they have only the dirty Irk water at hand, while pumps and water pipes can be found in decent parts of the city alone? In truth, it cannot be charged to the account of these helots of modern society if their dwellings are not more clean than the pig-sties which are here and there to be seen among them. The landlords are not ashamed to let dwellings like the six or seven cellars on the quay directly below Scotland Bridge, the floors of which stand at least two feet below the low-water level of the Irk that flows not six feet away from them; or like the upper floor of the corner-house on the opposite shore directly above the bridge, where the ground-floor, utterly uninhabitable, stands deprived of all fittings for doors and windows, a case by no means rare in this region, when this open ground-floor is used as a privy by the whole neighbourhood for want of other facilities!

If we leave the Irk and penetrate once more on the opposite side from Long Millgate into the midst of the working-men's dwellings, we shall come into a somewhat newer quarter, which stretches from St. Michael's Church to Withy Grove and Shude Hill. Here there is somewhat better order. In place of the chaos of buildings, we find at least long straight lanes and alleys or courts, built according to a plan and usually square. But if, in the former case, every house was built according to caprice, here each lane and court is so built, without reference to the situation of the adjoining ones. The lanes run now in this direction, now in that, while every two minutes the wanderer gets into a blind alley, or, on turning a corner, finds himself back where he started from; certainly no one who has not lived a considerable time in this labyrinth can find his way through it.

If I may use the word at all in speaking of this district, the ventilation of these streets and courts is, in consequence of this confusion, quite as imperfect as in the Irk region; and if this quarter may, nevertheless, be said to have some advantage over that of the Irk, the houses being newer and the streets occasionally having gutters, nearly every house has, on the other hand, a cellar dwelling, which is rarely found in the Irk district, by reason of the greater age and more careless construction of the houses. As for the rest the filth, debris, and offal heaps, and the pools in the streets are common to both quarters, and in the district now under discussion, another feature most injurious to the cleanliness of the inhabitants, is the multitude of pigs walking about in all the alleys, rooting into the offal heaps, or kept imprisoned in small pens. Here, as in most of the working-men's quarters of Manchester, the pork-raisers rent the courts and build pigpens in them. In almost every court one or even several such pens may be found into which the inhabitants of the court throw all refuse and offal, whence the swine grow fat; and the atmosphere, confined on all four sides, is utterly corrupted by putrefying animal and vegetable substances. Through this quarter, a broad and measurably decent street has been cut, Millers Street, and the background has been pretty successfully concealed. But if any one should be led by curiosity to pass through one of the numerous passages which lead into the courts, he will find this piggery repeated at every twenty paces.

Such is the Old Town of Manchester, and on re-reading my description, I am forced to admit that instead of being exaggerated, it is far from black enough to convey a true impression of the filth, ruin, and uninhabitableness, the defiance of all considerations of cleanliness, ventilation, and health which characterize the construction of this single district, containing at least twenty to thirty thousand inhabitants. And such a district exists in the heart of the second city of England, the first manufacturing city of the world. If any one wishes to see in how little space a human being can move, how little air – and *such* air! – he can breathe, how little of civilization he may share and yet live, it is only necessary to travel hither. True, this is the *Old* Town, and the people of Manchester emphasize the fact whenever any one mentions to them the frightful condition of this Hell upon Earth; but what does that prove? Everything which here arouses horror and indignation is of recent origin, belongs to the *industrial* epoch. The couple of hundred houses, which belong to old Manchester, have been long since abandoned by their original inhabitants; the industrial epoch alone has crammed into them the swarms of workers whom they now shelter; the industrial epoch alone has built up every spot between these old houses to win a covering for the masses whom it has conjured hither from the agricultural districts and from Ireland; the industrial epoch alone enables the owners of these cattlesheds to rent them for high prices to human beings, to plunder the poverty of the workers, to undermine the health of thousands, in order that they *alone*, the owners, may grow rich. In the industrial epoch alone has it become possible that the worker scarcely freed from feudal servitude could be used as mere material, a mere chattel; that he must let himself be crowded into a dwelling too bad for every other, which he for his hard-earned wages buys the right to let go utterly to ruin. This manufacture has achieved, which, without these workers, this poverty, this slavery could not have lived. True, the original construction of this quarter was bad, little good could have been made out of it; but, have the landowners, has the municipality done anything to improve it when re-building? On the contrary, wherever a nook or corner was free, a house has been run up; where a superfluous passage remained, it has been built up; the value of land rose with the blossoming out of manufacture, and the more it rose, the more madly was the work of building up carried on, without reference to the health or comfort of the inhabitants, with sole reference to the highest possible profit on the principle that *no hole is so bad but that some poor creature must take it who can pay for nothing better.*

[. . .]

It may not be out of place to make some general observations just here as to the customary construction of working-men's quarters in Manchester. We have seen how in the Old Town pure accident determined the grouping of the houses in general. Every house is built without reference to any other, and the scraps of space between them are called courts for want of another name. In the somewhat newer portions of the same quarter, and in other working-men's quarters, dating from the early days of industrial activity, a somewhat more orderly arrangement may be found. The space between two streets is divided into more regular, usually square courts.

These courts were built in this way from the beginning, and communicate with the streets by means of covered passages. If the totally planless construction is injurious to the health of the workers by preventing ventilation, this method of shutting them up in courts surrounded on all sides by buildings is far more so. The air simply cannot escape; the chimneys of the houses are the sole drains for the imprisoned atmosphere of the courts, and they serve the purpose only so long as fire is kept burning. Moreover, the houses surrounding such courts are usually built back to back, having the rear wall in common; and this alone suffices to prevent any sufficient through ventilation. And, as the police charged with care of the streets does not trouble itself about the condition of these courts, as everything quietly lies where it is thrown, there is no cause for wonder at the filth and heaps of ashes and offal to be found here. I have been in courts, in Millers Street, at least half a foot below the level of the thoroughfare, and without the slightest drainage for the water that accumulates in them in rainy weather! More recently another different method of building was adopted, and has now become general. Working-men's cottages are almost never built singly, but always by the dozen or score; a single contractor building up one or two streets at a time. These are then arranged as follows: One front is formed of cottages of the best class, so fortunate as to possess a back door and small court, and these command the highest rent. In the rear of these cottages runs a narrow alley, the back street, built up at both ends, into which either a narrow roadway or a covered passage leads from one side. The cottages which face this back street command least rent, and are most neglected. These have their rear walls in common with the third row of cottages which face a second street, and command less rent than the first row and more than the second. The streets are laid out somewhat as in [Figure 2].

By this method of construction, comparatively good ventilation can be obtained for the first row of cottages, and the third row is no worse off than in the former

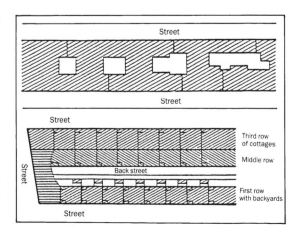

Figure 2

method. The middle row, on the other hand, is at least as badly ventilated as the houses in the courts, and the back street is always in the same filthy, disgusting condition as they. The contractors prefer this method because it saves them space, and furnishes the means of fleecing better-paid workers through the higher rents of the cottages in the first and third rows. These three different forms of cottage building are found all over Manchester and throughout Lancashire and Yorkshire, often mixed up together, but usually separate enough to indicate the relative age of parts of towns. The third system, that of the back alleys, prevails largely in the great working-men's district east of St. George's Road and Ancoats Street, and is the one most often found in the other working-men's quarters of Manchester and its suburbs.

[. . .]

Such are the various working-people's quarters of Manchester as I had occasion to observe them personally during twenty months. If we briefly formulate the result of our wanderings, we must admit that 350,000 working-people of Manchester and its environs live, almost all of them, in wretched, damp, filthy cottages, that the streets which surround them are usually in the most miserable and filthy condition, laid out without the slightest reference to ventilation, with reference solely to the profit secured by the contractor. In a word, we must confess that in the working-men's dwellings of Manchester, no cleanliness, no convenience, and consequently no comfortable family life is possible; that in such dwellings only a physically degenerate race, robbed of all humanity, degraded, reduced morally and physically to bestiality, could feel comfortable and at home.

[. . .]

To sum up briefly the facts thus far cited. The great towns are chiefly inhabited by working-people, since in the best case there is one bourgeois for two workers, often for three, here and there for four; these workers have no property whatsoever of their own, and live wholly upon wages, which usually go from hand to mouth. Society, composed wholly of atoms, does not trouble itself about them; leaves them to care for themselves and their families, yet supplies them no means of doing this in an efficient and permanent manner. Every working-man, even the best, is therefore constantly exposed to loss of work and food, that is, to death by starvation, and many perish in this way. The dwellings of the workers are everywhere badly planned, badly built, and kept in the worst condition, badly ventilated, damp, and unwholesome. The inhabitants are confined to the smallest possible space, and at least one family usually sleeps in each room. The interior arrangement of the dwellings is poverty-stricken in various degrees, down to the utter absence of even the most necessary furniture. The clothing of the workers, too, is generally scanty, and that of great multitudes is in rags. The food is, in general, bad; often almost unfit for use, and in many cases, at least at times, insufficient in quantity, so that, in extreme cases, death by starvation results. Thus the working-class of the great cities offers a graduated scale of conditions in life, in the best cases a temporarily endurable existence for hard work and good wages, good and endurable, that is, from the worker's standpoint; in the worst cases, bitter want, reaching even homelessness and death by starvation. The average is much nearer the worst case than the best. And this series does not fall into fixed classes, so that one can say, this fraction of the working-class is well off, has always been so, and remains so. If that is the case here and there, if single branches of work have in general an advantage over others, yet the condition of the workers in each branch is subject to such great fluctuations that a single working-man may be so placed as to pass through the whole range from comparative comfort to the extremest need, even to death by starvation, while almost every English working-man can tell a tale of marked changes of fortune.

"The Drive-in Culture of Contemporary America"

from *The Crabgrass Frontier: The Suburbanization of the United States* (1985)

Kenneth T. Jackson

Editors' Introduction

As Friedrich Engels (p. 50), Robert Fishman (p. 69), and others have shown, suburbia has a long history, extending back at least as far as the European and American railway suburbs that arose as retreats from the polluted industrial cities for the comfortable middle class. But suburbanization took on new form and historical significance in the 1920s and in the years following World War II. The initial locus was America, and the catalyzing technology was the automobile. In *The Crabgrass Frontier*, Kenneth T. Jackson, the dean of American urban historians, provides a sweeping overview of the "suburban revolution" in the United States. In the chapter entitled "The Drive-in Culture of Contemporary America" he lays out a devastating critique of the mostly negative social and cultural effects that the private automobile has had on urban society.

Kenneth Jackson did not originate the critique of suburbia. Indeed, the suburban developments of the 1940s, 1950s, and 1960s in America and elsewhere gave birth to a massive literature, most of it highly critical. Damned as culturally dead and socially/racially segregated, the post-World War II suburbs were called "sprawl" and stigmatized as "anti-cities" (to use Lewis Mumford's term to describe Los Angeles). Titles such as John Keats's *The Crack in the Picture Window* (1956), Richard Gordon's *The Split-level Trap* (1961), Mark Baldassare's *Trouble in Paradise* (1986), and Robert Fogelson's *Bourgeois Nightmares* (2005) capture the tone of much of the commentary. Indeed, James Howard Kunstler in *The Geography of Nowhere* (1993) calls the automobile suburbs "the evil empire," Joel S. Hirschhorn titles his analysis *Sprawl Kills* (2005), and another radical analysis screams *Bomb the Suburbs* (2001)! Nevertheless, Jackson's measured and well-documented analysis of suburban culture – everything from three-car garages to drive-in churches – stands as the definitive statement on how the automobile transformed both the structure and social life of modern cities.

Kenneth Jackson is the Jacques Barzun Professor of History and the Social Sciences at Columbia University. He earned his Ph.D. at the University of Chicago and is a past president of the Urban History Association and the Organization of American Historians. He is the editor of the *Encyclopedia of New York* (1995) and the author of several influential books, including *The Ku Klux Klan in the City, 1915–1930* (1967) and *Cities in American History* (1972). Jackson has been called an "urban pessimist" because of his dark view of suburbanization – a current work in progress on American transportation policy is entitled *The Road To Hell* – but in a recent interview he noted that he now sees "a ray of hope" for cities "after a long decline."

Classic works on suburbia include Herbert Gans, *The Levittowners* (New York: Pantheon, 1967), remarkable for its overall positive view of middle-class tract-home life, Sam Bass Warner, Jr., *Streetcar Suburbs: The Process of Growth in Boston, 1870–1900* (Cambridge: Harvard University Press, 1962), and Robert Fishman, *Bourgeois Utopias: The Rise and Fall of Suburbia* (New York: Basic Books, 1987).

Other recent studies of suburbia include J. Eric Oliver, *Democracy in Suburbia* (Princeton: Princeton University Press, 2001), Mark Salzman, *Lost in Place: Growing Up Absurd in Suburbia* (New York: Vintage, 1995), Valerie C. Johnson, *Black Power in the Suburbs* (New York: SUNY Press, 2002), Becky Nocolaides, *My Blue Heaven: Life and Politics in the Working-class Suburbs of Los Angeles* (Chicago: University of Chicago Press, 2002), Adam Rome, *The Bulldozer and the Countryside: Suburban Sprawl and the Rise of American Environmentalism* (Cambridge: Cambridge University Press, 2001), Oliver Gillham, *The Limitless City: A Primer on the Urban Sprawl Debate* (Washington, D.C.: Island Press, 2002), and Dolores Hayden, *Building Suburbia: Green Fields and Urban Growth, 1820–2000* (New York: Pantheon, 2003). For a spirited defense of sprawl as an age-old natural process of urban expansion, see Robert Bruegmann, *Sprawl: A Compact History* (Chicago: University of Chicago Press, 2005).

The postwar years brought unprecedented prosperity to the United States, as color televisions, stereo systems, frost-free freezers, electric blenders, and automatic garbage disposals became basic equipment in the middle-class American home. But the best symbol of individual success and identity was a sleek, air-conditioned, high-powered, personal statement on wheels. Between 1950 and 1980, when the American population increased by 50 percent, the number of their automobiles increased by 200 percent. In high school the most important rite of passage came to be the earning of a driver's license and the freedom to press an accelerator to the floor. Educational administrators across the country had to make parking space for hundreds of student vehicles. A car became one's identity, and the important question was: "What does he drive?" Not only teenagers, but also millions of older persons literally defined themselves in terms of the number, cost, style, and horsepower of their vehicles. "Escape," thinks a character in a novel by Joyce Carol Oates. "As long as he had his own car he was an American and could not die."

Unfortunately, Americans did die, often behind the wheel. On September 9, 1899, as he was stepping off a streetcar at 74th Street and Central Park West in New York, Henry H. Bliss was struck and killed by a motor vehicle, thus becoming the first fatality in the long war between flesh and steel. Thereafter, the carnage increased almost annually until Americans were sustaining about 50,000 traffic deaths and about 2 million nonfatal injuries per year. Automobility proved to be far more deadly than war for the United States. It was as if a Pearl Harbor attack took place on the highways every two weeks, with crashes becoming so commonplace that an entire industry sprang up to provide medical, legal, and insurance services for the victims.

The environmental cost was almost as high as the human toll. In 1984 the 159 million cars, trucks, and buses on the nation's roads were guzzling millions of barrels of oil every day, causing traffic jams that shattered nerves and clogged the cities they were supposed to open up and turning much of the countryside to pavement. Not surprisingly, when gasoline shortages created long lines at the pumps in 1974 and 1979, behavioral scientists noted that many people experienced anger, depression, frustration, and insecurity, as well as a formidable sense of loss.

Such reactions were possible because the automobile and the suburb have combined to create a drive-in culture that is part of the daily experience of most Americans . . . Moreover, the American people have proven to be no more prone to motor vehicle purchases than the citizens of other lands. After World War II, the Europeans and the Japanese began to catch up, and by 1980 both had achieved the same level of automobile ownership that the United States had reached in 1950. In automotive technology, American dominance slipped away in the postwar years as German, Swedish, and Japanese engineers pioneered the development of diesel engines, front-wheel drives, disc brakes, fuel-injection, and rotary engines.

Although it is not accurate to speak of a uniquely American love affair with the automobile, and although John B. Rae claimed too much when he wrote in 1971 that "modern suburbia is a creature of the automobile and could not exist without it," the motor vehicle has fundamentally restructured the pattern of everyday life in the United States. As a young man, Lewis Mumford advised his countrymen to "forget the damned motor car and build cities for lovers and friends." As it was, of course, the nation followed a different pattern. Writing in the *American Builder* in 1929, the critic Willard Morgan noted that the building of drive-in structures to serve a motor-driven population had ushered in "a completely new architectural form."

THE INTERSTATE HIGHWAY

The most popular exhibit at the New York World's Fair in 1939 was General Motors' "Futurama." Looking twenty-five years ahead, it offered a "magic Aladdin-like flight through time and space." Fair-goers stood in hour-long lines, waiting to travel on a moving sidewalk above a huge model created by designer Norman Bel Geddes. Miniature superhighways with 50,000 automated cars wove past model farms en route to model cities . . . The message of "Futurama" was as impressive as its millions of model parts: "The job of building the future is one which will demand our best energies, our most fruitful imagination; and that with it will come greater opportunities for all."

The promise of a national system of impressive roadways attracted a diverse group of lobbyists, including the Automobile Manufacturers Association, state-highway administrators, motor-bus operators, the American Trucking Association, and even the American Parking Association – for the more cars on the road, the more cars would be parked at the end of the journey. Truck companies, for example, promoted legislation to spend state gasoline taxes on highways, rather than on schools, hospitals, welfare, or public transit. In 1943 these groups came together as the American Road Builders Association, with General Motors as the largest contributor, to form a lobbying enterprise second only to that of the munitions industry. By the mid-1950s, it had become one of the most broad-based of all pressure groups, consisting of the oil, rubber, asphalt, and construction industries; the car dealers and renters; the trucking and bus concerns; the banks and advertising agencies that depended upon the companies involved; and the labor unions. On the local level, professional real estate groups and home-builders associations joined the movement in the hope that highways would cause a spurt in housing turnover and a jump in prices. They envisaged no mere widening of existing roads, but the creation of an entirely new superhighway system and the initiation of the largest peacetime construction project in history.

[. . .]

Sensitive to mounting political pressure, President Dwight Eisenhower appointed a committee in 1954 to "study" the nation's highway requirements. Its conclusions were foregone, in part because the chairman was Lucius D. Clay, a member of the board of directors of General Motors. The committee considered no alternative to a massive highway system, and it suggested a major redirection of national policy to benefit the car and the truck. The Interstate Highway Act became law in 1956, when the Congress provided for a 41,000-mile (eventually expanded to a 42,500-mile) system, with the federal government paying 90 percent of the cost. President Eisenhower gave four reasons for signing the measure: current highways were unsafe; cars too often became snarled in traffic jams; poor roads saddled business with high costs for transportation; and modern highways were needed because "in case of atomic attack on our key cities, the road net must permit quick evacuation of target areas." Not a single word was said about the impact of highways on cities and suburbs, although the concrete thoroughfares and the thirty-five-ton tractor-trailers which used them encouraged the continued outward movement of industries toward the beltways and interchanges. Moreover, the interstate system helped continue the downward spiral of public transportation and virtually guaranteed that future urban growth would perpetuate a centerless sprawl . . .

[. . .]

The inevitable result of the bias in American transport funding, a bias that existed for a generation before the Interstate Highway program was initiated, is that the United States now has the world's best road system and very nearly its worst public transit offerings. Los Angeles, in particular, provides the nation's most dramatic example of urban sprawl tailored to the mobility of the automobile. Its vast, amorphous conglomeration of housing tracts, shopping centers, industrial parks, freeways, and independent towns blend into each other in a seamless fabric of concrete and asphalt, and nothing over the years has succeeded in gluing this automobile-oriented civilization into any kind of cohesion – save that of individual routine. Los Angeles's basic shape comes from three factors, all of which long preceded the freeway system. The first was cheap land (in the 1920s rather than 1970s) and the desire for single-family houses. In 1950, for example, nearly two-thirds of all the dwelling units in the Los Angeles area were fully detached, a much higher percentage than in Chicago (28 percent), New

York City (20 percent), or Philadelphia (15 percent), and its residential density was the lowest of major cities. The second was the dispersed-location of its oil fields and refineries, which led to the creation of industrial suburbs like Whittier and Fullerton and of residential suburbs like La Habra, which housed oil workers and their families. The third was its once excellent mass transit system, which at its peak included more than 1,100 miles of track and constituted the largest electric interurban railway in the world.

[...]

THE GARAGE

The drive-in structure that is closest to the hearts, bodies, and cars of the American family is the garage. It is the link between the home and the outside world. The word is French, meaning storage space, but its transformation into a multipurpose enclosure internally integrated with the dwelling is distinctively American.

[..]

After World War I, house plans of the expensive variety began to include garages, and by the mid-1920s driveways were commonplace and garages had become important selling points. The popular 1928 *Home Builders* pattern book offered designs for fifty garages in wood, Tudor, and brick varieties. In affluent sections, such large and efficiently planned structures included housing above for the family chauffeur. In less pretentious neighborhoods, the small, single-purpose garages were scarcely larger than the vehicles themselves . . . Although there was a tendency to move garages closer to the house, they typically remained at the rear of the property before 1925, often with access via an alley which ran parallel to the street. The car was still thought of as something similar to a horse – dependable and important, but not something that one needed to be close to in the evening.

By 1935, however, the garage was beginning to merge into the house itself, and in 1937 the *Architectural Record* noted that "the garage has become a very essential part of the residence." The tendency accelerated after World War II, as alleys went the way of the horse-drawn wagon, as property widths more often exceeded fifty feet, and as the car became not only a status symbol, but almost a member of the family, to be cared for and sheltered. The introduction

of a canopied and unenclosed structure called a "car port" represented an inexpensive solution to the problem, particularly in mild climates, but in the 1950s the enclosed garage was back in favor and a necessity even in a tract house. Easy access to the automobile became a key aspect of residential design, and not only for the well-to-do. By the 1960s garages often occupied about 400 square feet (about one-third that of the house itself) and usually contained space for two automobiles and a variety of lawn and woodworking tools. Offering direct access to the house (a conveniently placed door usually led directly into the kitchen), the garage had become an integrated part of the dwelling, and it dominated the front facades of new houses. In California garages and driveways were often so prominent that the house could almost be described as accessory to the garage. Few people, however, went to the extremes common in England, where the automobile was often so precious that living rooms were often converted to garages.

THE MOTEL

As the United States became a rubber-tire civilization, a new kind of roadside architecture was created to convey an instantly recognizable image to the fast-moving traveler. Criticized as tasteless, cheap, forgettable, and flimsy by most commentators, drive-in structures did attract the attention of some talented architects, most notably Los Angeles's Richard Neutra. For him, the automobile symbolized modernity, and its design paralleled his own ideals of precision and efficiency. This correlation between the structure and the car began to be celebrated in the late 1960s and 1970s when architects Robert Venturi, Denise Scott Brown, and Steven Izenour developed such concepts as "architecture as symbol" and the "architecture of communication." Their book, *Learning From Las Vegas*, was instrumental in encouraging a shift in taste from general condemnation to appreciation of the commercial strip and especially of the huge and garish signs which were easily recognized by passing motorists.

A ubiquitous example of the drive-in culture is the motel. In the middle of the nineteenth century, every city, every county seat, every aspiring mining town, every wide place in the road with aspirations to larger size, had to have a hotel. Whether such structures were grand palaces on the order of Boston's Tremont House or New York's Fifth Avenue Hotel, or whether they were jerry-built shacks, they were typically located at the center of the business district, at the focal point of community activities. To a considerable extent, the hotel was the place for informal social interaction and business, and the very heart and soul of the city.

Between 1910 and 1920, however, increasing numbers of traveling motorists created a market for overnight accommodation along the highways. The first tourists simply camped wherever they chose along the road. By 1924, several thousand municipal camp-grounds were opened which offered cold water spigots and outdoor privies. Next came the "cabin camps," which consisted of tiny, white clapboard cottages arranged in a semicircle and often set in a grove of trees. Initially called "tourist courts," these establishments were cheap, convenient, and informal, and by 1926 there were an estimated two thousand of them, mostly in the West and in Florida.

[. . .]

It was not until 1952 that Kemmons Wilson and Wallace E. Johnson opened their first "Holiday Inn" on Summer Avenue in Memphis. But long before that, in 1926, a San Luis Obispo, California, proprietor had coined a new word, "motel," to describe an establishment that allowed a guest to park his car just outside his room . . .

Motels began to thrive after World War II, when the typical establishment was larger and more expensive than the earlier cabins. Major chains set standards for prices, services, and respectability that the traveling public could depend on. As early as 1948, there were 26,000 self-styled motels in the United States. Hard-won respectability attracted more middle-class families, and by 1960 there were 60,000 such places, a figure that doubled again by 1972. By that time an old hotel was closing somewhere in downtown America every thirty hours. And somewhere in sub-urban America, a plastic and glass Shangri-La was rising to take its place.

[. . .]

THE DRIVE-IN THEATER

The downtown movie theaters and old vaudeville houses faced a similar challenge from the automobile. In 1933 Richard M. Hollinshead set up a 16-mm projector in front of his garage in Riverton, New Jersey, and then settled down to watch a movie. Recognizing

a nation addicted to the motorcar when he saw one, Hollinshead and Willis Smith opened the world's first drive-in movie in a forty-car parking lot in Camden on June 6, 1933. Hollinshead profited only slightly from his brainchild, however, because in 1938 the United States Supreme Court refused to hear his appeal against Loew's Theaters, thus accepting the argument that the drive-in movie was not a patentable item. The idea never caught on in Europe, but by 1958 more than four thousand outdoor screens dotted the American landscape. Because drive-ins offered bargain-base-ment prices and double or triple bills, the theaters tended to favor movies that were either second-run or second-rate. Horror films and teenage romance were the order of the night . . . Pundits often commented that there was a better show in the cars than on the screen.

In the 1960s and 1970s the drive-in movie began to slip in popularity. Rising fuel costs and a season that lasted only six months contributed to the problem, but skyrocketing land values were the main factor. When drive-ins were originally opened, they were typically out in the hinterlands. When subdivisions and shop-ping malls came closer, the drive-ins could not match the potential returns from other forms of investments. According to the National Association of Theater Owners, only 2,935 open-air theaters still operated in the United States in 1983, even though the total number of commercial movie screens in the nation, 18,772, was at a thirty-five-year high. The increase picked up not by the downtown and the neighborhood theaters, but by new multiscreen cinemas in shopping centers. Realizing that the large parking lots of indoor malls were relatively empty in the evening, shopping center moguls came to regard theaters as an important part of a successful retailing mix.

THE GASOLINE SERVICE STATION

The purchase of gasoline in the United States has thus far passed through five distinct epochs. The first stage was clearly the worst for the motorist, who had to buy fuel by the bucketful at a livery stable, repair shop, or dry goods store. Occasionally, vendors sold gasoline from small tank cars which they pushed up and down the streets. In any event, the automobile owner had to pour gasoline from a bucket through a funnel into his tank. The entire procedure was inefficient, smelly, wasteful, and occasionally dangerous.

The second stage began about 1905, when C.H. Laessig of St. Louis equipped a hot-water heater with a glass gauge and a garden hose and turned the whole thing on its end. With this simple maneuver, he invented an easy way to transfer gasoline from a storage tank to an automobile without using a bucket. Later in the same year, Sylvanus F. Bowser invented a gasoline pump which automatically measured the outflow. The entire assembly was labeled a "filling station." At this stage, which lasted until about 1920, such an apparatus consisted of a single pump outside a retail store which was primarily engaged in other businesses and which provided precious few services for the motorist . . .

Between 1920 and 1950, service stations entered into a third phase and became, as a group, one of the most widespread kinds of commercial buildings in the United States. Providing under one roof all the functions of gasoline distribution and normal automotive maintenance, these full-service structures were often built in the form of little colonial houses, Greek temples, Chinese pagodas, and Art Deco palaces. Many were local landmarks and a source of community pride . . .

After 1935 the gasoline station evolved again, this time into a more homogeneous entity that was standardized across the entire country and that reflected the mass-marketing techniques of billion-dollar oil companies. Some of the more familiar designs were innovative or memorable, such as the drum-like Mobile station by New York architect Frederick Frost, which featured a dramatically curving facade while conveying the corporate identity. Another popular service station style was the Texaco design of Walter Dorwin Teaguea, smooth white exterior with elegant trim and the familiar red star and bold red lettering. Whatever the product or design, the stations tended to be operated by a single entrepreneur and represented an important part of small business in American life.

The fifth stage of gasoline-station development began in the 1970s, with the slow demise of the traditional service-station businessman. New gasoline outlets were of two types. The first was the super station, often owned and operated by the oil companies themselves. Most featured a combination of self-service and full-service pumping consoles, as well as fully equipped "car care centers." Service areas were separated from the pumping sections so that the two functions would not interfere with each other. Mechanics never broke off work to sell gas.

The more pervasive second type might be termed the "mini-mart station." The operators of such establishments have now gone full circle since the early twentieth century. Typically, they know nothing about automobiles and expect the customers themselves to pump the gasoline. Thus, "the man who wears the star" has given way to the teenager who sells six-packs, bags of ice, and pre-prepared sandwiches.

THE SHOPPING CENTER

Large-scale retailing, long associated with central business districts, began moving away from the urban cores between the world wars. The first experiments to capture the growing suburban retail markets were made by major department stores in New York and Chicago in the 1920s . . .

Another threat to the primacy of the central business district was the "string street" or "shopping strip," which emerged in the 1920s and which was designed to serve vehicular rather than pedestrian traffic. These bypass roads encouraged city dwellers with cars to patronize businesses on the outskirts of town. Short parades of shops could already have been found near the streetcar and rapid transit stops, but, as has been noted, these new retailing thoroughfares generally radiated out from the city business district toward low-density, residential areas, functionally dominating the urban street system. They were the prototypes for the familiar highway strips of the 1980s which stretch far into the countryside.

[. . .]

The concept of the enclosed, climate-controlled mall, first introduced at the Southdale Shopping Center near Minneapolis in 1956, added to the suburban advantage . . .

During the 1970s, a new phenomenon – the super regional mall – added a more elaborate twist to suburban shopping. Prototypical of the new breed was Tyson's Corner, on the Washington Beltway in Fairfax County, Virginia. Anchored by Bloomingdale's, it did over $165 million in business in 1983 and provided employment to more than 14,000 persons. Even larger was Long Island's Roosevelt Field, a 180-store, 2.2 million square foot megamall that attracted 275,000 visitors a week and did $230 million in business in 1980. Most elaborate of all was Houston's Galleria, a world-famed setting for 240 prestigious boutiques, a quartet of cinemas, 26 restaurants, an Olympic-sized ice-skating pavilion, and two luxury hotels. There were few windows in these mausoleums of merchandising, and clocks were rarely seen – just as in gambling casinos.

Boosters of such megamalls argue that they are taking the place of the old central business districts and becoming the identifiable collecting points for the rootless families of the newer areas. As weekend and afternoon attractions, they have a special lure for teenagers, who often go there on shopping dates or to see the opposite sex. As one official noted in 1971: "These malls are now their street corners. The new shopping centers have killed the little merchant, closed most movies, and are now supplanting the older shopping centers in the suburbs." They are also especially attractive to mothers with young children and to the elderly, many of whom visit regularly to get out of the house without having to worry about crime or inclement weather.

[. . .]

THE HOUSE TRAILER AND MOBILE HOME

The phenomenon of a nation on wheels is perhaps best symbolized by the uniquely American development of the mobile home. "Trailers are here to stay," predicted the writer Howard O'Brien in 1936. Although in its infancy at that time, the mobile-home industry has flourished in the United States. The house trailer itself came into existence in the teens of this century as an individually designed variation on a truck or a car, and it began to be produced commercially in the 1920s. Originally, trailers were designed to travel, and they were used primarily for vacation purposes. During the Great Depression of the 1930s, however, many people, especially salesmen, entertainers, construction workers, and farm laborers, were forced into a nomadic way of life as they searched for work, any work. They found that these temporary trailers on rubber tires provided the necessary shelter while also meeting their economic and migratory requirements. Meanwhile, Wally Byam and other designers were streamlining the mobile home into the classic tear-drop form made famous by Airstream.

During World War II, the United States government got into the act by purchasing tens of thousands of trailers for war workers and by forbidding their sale to the general public. By 1943 the National Housing Agency alone owned 35,000 of the aluminum boxes, and more than 60 percent of the nation's 200,000 mobile homes were in defense areas . . .

Not until the mid-1950s did the term "mobile home" begin to refer to a place where respectable people could marry, mature, and die. By then it was less a "mobile" than a "manufactured" home. No longer a trailer, it became a modern industrialized residence with almost all the accoutrements of a normal house. By the late 1950s, widths were increased to ten feet, the Federal Housing Administration (FHA) began to recognize the mobile home as a type of housing suitable for mortgage insurance, and the maturities on sales contracts were increased from three to five years.

In the 1960s, twelve-foot widths were introduced, and then fourteen, and manufacturers began to add fireplaces, skylights, and cathedral ceilings. In 1967 two trailers were attached side by side to form the first "double wide." These new dimensions allowed for a greater variety of room arrangement and became particularly attractive to retired persons with fixed incomes. They also made the homes less mobile. By 1979 even the single-width "trailer" could be seventeen feet wide (by about sixty feet long), and according to the Manufactured Housing Institute, fewer than 2 percent were ever being moved from their original site. Partly as a result of this increasing permanence, individual communities and the courts began to define the structures as real property and thus subject to real estate taxes rather than as motor vehicles subject only to license fees.

Although it continued to be popularly perceived as a shabby substitute for "stick" housing (a derogatory word used to describe the ordinary American balloon-frame dwelling), the residence on wheels reflected American values and industrial practices. Built with easily machined and processed materials, such as sheet metal and plastic, it represented a total consumer package, complete with interior furnishings, carpets, and appliances. More importantly, it provided a suburban-type alternative to the inner-city housing that would otherwise have been available to blue-collar workers, newly married couples, and retired persons . . .

A DRIVE-IN SOCIETY

Drive-in motels, drive-in movies, and drive-in shopping facilities were only a few of the many new institutions that followed in the exhaust of the internal-combustion engine. By 1984 mom-and-pop grocery stores had given way almost everywhere to supermarkets, most banks had drive-in windows, and a few funeral homes were making it possible for mourners to view the deceased, sign the register, and pay their respects without emerging from their cars. Odessa Community College in Texas even opened a drive-through registration window.

Particularly pervasive were fast-food franchises, which not only decimated the family-style restaurants but cut deeply into grocery store sales. In 1915, James G. Huneker, a raconteur whose tales of early twentieth-century American life were compiled as *New Cosmopolis*, complained of the infusion of cheap, quick-fire "food hells," and of the replacement of relaxed dining with "canned music and automatic lunch taverns." With the automobile came the notion of "grabbing" something to eat. The first drive-in restaurant, Royce Hailey's Pig Stand, opened in Dallas in 1921, and later in the decade, the first fast-food franchise, "White Tower," decided that families touring in motorcars needed convenient meals along the way. The places had to look clean, so they were painted white. They had to be familiar, so a minimal menu was standardized at every outlet. To catch the eye, they were built like little castles, replete with fake ramparts and turrets. And to forestall any problem with a land lease, the little white castles were built to be moveable.

The biggest restaurant operation of all began in 1954, when Ray A. Kroc, a Chicago area milkshake-machine salesman, joined forces with Richard and Maurice McDonald, the owners of a fast-food emporium in San Bernardino, California. In 1955 the first of Mr. Kroc's "McDonald's" outlets was opened in Des Plaines, a Chicago suburb long famous as the site of an annual Methodist encampment. The second and third, both in California, opened later in 1955 . . . [T]he McDonald's enterprise is based on free parking and drive-in access, and its methods have been copied by dozens of imitators. Late in 1984, on an interstate highway north of Minneapolis, McDonald's began construction of the most complete drive-in complex in the world. To be called McStop, it will feature a motel, gas station, convenience store, and, of course, a McDonald's restaurant.

[. . .]

THE CENTERLESS CITY

More than anyplace else, California became the symbol of the postwar suburban culture. It pioneered the booms in sports cars, foreign cars, vans, and motor homes, and by 1984 its 26 million citizens owned almost 19 million motor vehicles and had access to the world's most extensive freeway system. The result has been a new type of centerless city, best exemplified by once sleepy and out-of-the-way Orange County, just south and east of Los Angeles. After Walt Disney came down from Hollywood, bought out the ranchers, and opened Disneyland in 1955, Orange County began to evolve from a rural backwater into a suburb and then into a collection of medium and small towns. It had never had a true urban focus, in large part because its oil-producing sections each spawned independent suburban centers, none of which was particularly dominant over the others. The tradition continued when the area became a subdivider's dream in the 1960s and 1970s. By 1980 there were twenty-six Orange County cities, none with more than 225,000 residents. Like the begats of the Book of Genesis, they merged and multiplied into a huge agglomeration of two million people with its own Census Bureau metropolitan area designation – Anaheim, Santa Ana, Garden Grove. Unlike the traditional American metropolitan region, however, Orange County lacked a commutation focus, a place that could obviously be accepted as the center of local life. Instead, the experience of a local resident was typical: "I live in Garden Grove, work in Irvine, shop in Santa Ana, go to the dentist in Anaheim, my husband works in Long Beach, and I used to be the president of the League of Women Voters in Fullerton."

A centerless city also developed in Santa Clara County, which lies forty-five miles south of San Francisco and which is best known as the home of Silicon Valley. Stretching from Palo Alto on the north to the garlic and lettuce fields of Gilroy to the south, Santa Clara County has the world's most extensive concentration of electronics concerns. In 1940, however, it was best known for prunes and apricots, and it was not until after World War II that its largest city, San Jose, also became the nation's largest suburb. With fewer than 70,000 residents in 1940, San Jose exploded to 636,000 by 1980, superseding San Francisco as the region's largest municipality . . .

The numbers were larger in California, but the pattern was the same on the edges of every American city, from Buffalo Grove and Schaumburg near Chicago, to Germantown and Collierville near Memphis, to Creve Coeur and Ladue near St. Louis. And perhaps more important than the growing number of people living outside of city boundaries was the sheer physical sprawl of metropolitan areas. Between 1950 and 1970, the urbanized area of Washington, DC, grew from 181 to 523 square miles, of Miami from 116 to 429, while in the larger megalopolises of New York, Chicago, and Los Angeles, the region of settlement was measured in the thousands of square miles.

THE DECENTRALIZATION OF FACTORIES AND OFFICES

The deconcentration of post-World War II American cities was not simply a matter of split-level homes and neighborhood schools. It involved almost every facet of national life, from manufacturing to shopping to professional services. Most importantly, it involved the location of the workplace, and the erosion of the concept of suburb as a place from which wage-earners commuted daily to jobs in the center. So far had the trend progressed by 1970 that in nine of the fifteen largest metropolitan areas suburbs were the principal sources of employment, and in some cities, like San Francisco, almost three-fourths of all work trips were by people who neither lived nor worked in the core city. In Wilmington, Delaware, 66 percent of area jobs in 1940 were in the core city; by 1970, the figure had fallen below one-quarter. And despite the fact that Manhattan contained the world's highest concentration of office space and business activity, in 1970, about 78 percent of the residents in the New York suburbs also worked in the suburbs. Many outlying communities thus achieved a kind of autonomy from the older downtown areas . . .

Manufacturing is now among the most dispersed of nonresidential activities. As the proportion of industrial jobs in the United States work force fell from 29 percent to 23 percent of the total in the 1970s, those manufacturing enterprises that survived often relocated either to the suburbs or to the lower-cost South and West . . .

Office functions, once thought to be securely anchored to the streets of big cities, have followed the suburban trend. In the nineteenth century, businesses tried to keep all their operations under one centralized roof. It was the most efficient way to run a company when the mails were slow and uncertain and communication among employees was limited to the distance that a human voice could carry. More recently, the economics of real estate and a revolution in communications have changed these circumstances, and many companies are now balkanizing their accounting departments, data-processing divisions, and billing departments. Just as insurance companies, branch banks, regional sales staffs, and doctors' offices have reduced their costs and presumably increased their accessibility by moving to suburban locations, so also have back-office functions been splitting away from front offices and moving away from central business districts.

[. . .]

Since World War II, the American people have experienced a transformation of the manmade environment around them. Commercial, residential, and industrial structures have been redesigned to fit the needs of the motorist rather than the pedestrian. Garish signs, large parking lots, one-way streets, drive-in windows, and throw-away fast-food buildings – all associated with the world of suburbia – have replaced the slower-paced, neighborhood-oriented institutions of an earlier generation. Some observers of the automobile revolution have argued that the car has created a new and better urban environment and that the change in spatial scale, based upon swift transportation, has formed a new kind of organic entity, speeding up personal communication and rendering obsolete the older urban settings. Lewis Mumford, writing from his small-town retreat in Amenia, New York, has emphatically disagreed. His prize-winning book, *The City in History*, was a celebration of the medieval community and an excoriation of "the formless urban exudation" that he saw American cities becoming. He noted that the automobile megalopolis was not a final stage in city development but an anti-city which "annihilates the city whenever it collides with it."

[. . .]

There are some signs that the halcyon days of the drive-in culture and automobile are behind us. More than one hundred thousand gasoline stations, or about one-third of the American total, have been eliminated in the last decade. Empty tourist courts and boarded-up motels are reminders that the fast pace of change can make commercial structures obsolete within a quarter-century of their erection. Even that suburban bellwether, the shopping center, which revolutionized merchandising after World War II, has come to seem small and out-of-date as newer covered malls attract both the trendy and the family trade. Some older centers have been recycled as bowling alleys or industrial buildings, and some have been remodeled to appeal to larger tenants and better-heeled customers. But others stand forlorn and boarded up. Similarly, the characteristic fast-food emporiums of the 1950s, with uniformed "car hops" who took orders at the automobile window, are now relics of the past. One of the survivors, Delores Drive-in, which opened in Beverly Hills in 1946, was recently proposed as an historic landmark, a sure sign that the species is in danger.

"Beyond Suburbia: The Rise of the Technoburb"

from *Bourgeois Utopias: The Rise and Fall of Suburbia* (1987)

Robert Fishman

Editors' Introduction

Robert Fishman is a professor of history at the University of Michigan who established his academic reputation with his first book, the magisterial *Urban Utopias in the Twentieth Century* (1977), a study of the work of Ebenezer Howard, Le Corbusier, and Frank Lloyd Wright. For his second book, Fishman decided to address a totally prosaic, nonvisionary subject – the history of suburbia – only to discover that "the suburban ideal" was, in the final analysis, yet another form of utopia, the utopia of the middle class.

The real focus of *Bourgeois Utopias* is the suburban ideal, more than suburbia itself, and the logic of Fishman's analysis leads him to many surprising insights and conclusions. In the medieval period and up through the eighteenth century, suburbs were clusters of houses inhabited by poor and/or disreputable people on the outskirts of towns. When suburbs were first established for the upper and middle classes – a phenomenon that has thrived more in North America than in Europe where working-class suburbs and *banlieus* often predominate – the ideal was to create a perfect synthesis of urban sophistication and rural virtue. Here was a conception as utopian as that of any visionary social reformer, but with an important difference: "Where other modern utopias have been collectivist," writes Fishman, "suburbia has built its vision of community on the primacy of private property and the individual family."

What suburbia has evolved into today is "technoburbia," a dominant new urban reality that can no longer be considered suburbia in the traditional sense. In Redmond, Washington (Plate 8), Microsoft Corporation's corporate headquarters mix with residential neighborhoods, retail centers, and even bands of open space to make up a new urban form where city and suburb, urbanized and unurbanized areas, high-tech and conventional development flow together.

To describe this new reality Fishman has coined two new terms: "technoburb" and "techno-city." Fishman defines technoburbs as peripheral zones, perhaps as large as a county, that have emerged as viable socio-economic units. The new technoburbs are spread out along highway growth corridors. Along the highways of metropolitan regions shopping malls, industrial parks, campus-like office complexes, hospitals, schools, and a whole range of housing types succeed each other.

By "techno-city" Fishman means the whole metropolitan region, which has been transformed by the coming of the technoburb. In Fishman's view we may still refer to the New York Metropolitan region as "New York City," but increasingly by "New York City" we mean the entire New York City region. And much of the economic and cultural life of the region no longer resides just in the core city. The old central cities have become increasingly marginal, while the technoburb has emerged as the focus of American life. In Fishman's view, the new technoburbs surrounding the old urban cores do not represent "the suburbanization of the United States," as Kenneth Jackson (p. 59) would have it, but "the end of suburbia in its traditional sense and the creation of a new kind of decentralized city." That suburbia has become the city itself is, perhaps, the final irony of modern urbanism.

Fishman lays out a strong indictment of what is wrong with technoburbs. They consist of an unplanned jumble of discordant elements – housing, industry, commerce, even agriculture – with little coherent pattern or structure. They waste land. Technoburbs are dependent on highway systems, yet their highway systems are in a state of chronic chaos. They have no proper boundaries, but consist of a crazy quilt of separate and overlapping political jurisdictions that make meaningful region-wide planning virtually impossible and, as Myron Orfield observes (p. 287), access to revenue to pay for local government services becomes highly inequitable.

Yet Fishman notes that all new urban forms appeared chaotic in their early stages. Even the most "organic" cityscapes of the past evolved slowly after much chaos and trial and error. For example, it took planners of genius like Frederick Law Olmsted (p. 307) and Ebenezer Howard (p. 314) to create orderly parks and garden suburbs out of the chaos of the nineteenth-century city or to imagine and actually build Garden Cities. Fishman acknowledges that there is a functional logic to sprawl. Perhaps, he speculates, if sprawl is better understood and better managed it might prove to be a positive rather than a negative development. Fishman looks to Frank Lloyd Wrights's Broadacre City vision (p. 331) as an example of how inspired planners may yet devise an aesthetic to tame techoburbia.

The technocity, Fishman concludes, is still under construction both physically and culturally. How technocities will evolve is unclear, although both William J. Mitchell (p. 510) and Richard Florida (p. 129) offer persuasive insights into who will live in the new techno-communities and how they will work and socially interact. The jury is still out on whether technoburbia will ultimately be judged as an advance over earlier urban forms.

This selection is from Robert Fishman, *Bourgeois Utopias: The Rise and Fall of Suburbia* (New York: Basic Books, 1987). Fishman's other major books on cities are *Urban Utopias in the Twentieth Century* (New York: Basic Books, 1977) and *The American Planning Tradition: Culture and Policy* (Washington, DC: Woodrow Wilson Center Press, 2000).

For other views of emerging postmodern suburbia, see journalist Joel Garreau's *Edge City* (New York: Anchor, 1992), Manuel Castells, *The Rise of the Network Society* (Cambridge: Blackwell, 1996), Manuel Castells and Peter Hall, *Technopoles of the World* (London and New York: Routledge, 1994), Edward Soja's studies of the Los Angeles region cited in the selection on "Taking Los Angeles Apart" (p. 166), and Peter Calthorpe and William Fulton's description of the emerging "Regional City" (p. 342). Also of interest are two recent books that call for a reconfiguration of the city–suburb relationship: Myron Orfield, *American Metropolitics: The New Suburban Reality* (Washington: Brookings Institution, 2002) and David Rusk, *Cities Without Suburbs: A Census 2000 Update* (Washington, DC: Woodrow Wilson Center Press, 2003).

■ ■ ■ ■ ■ ■

If the nineteenth century could be called the Age of Great Cities, post-1945 America would appear to be the Age of Great Suburbs. As central cities stagnated or declined in both population and industry, growth was channeled almost exclusively to the peripheries. Between 1950 and 1970 American central cities grew by 10 million people, their suburbs by 85 million. Suburbs, moreover, accounted for at least three-quarters of all new manufacturing and retail jobs generated during that period. By 1970 the percentage of Americans living in suburbs was almost exactly double what it had been in 1940, and more Americans lived in suburban areas (37.6 percent) than in central cities (31.4 percent) or in rural areas (31 percent). In the 1970s central cities experienced a net out-migration of 13 million people, combined with an unprecedented deindustrialization, increasing poverty levels, and housing decay.

[. . .]

From its origins in eighteenth-century London, suburbia has served as a specialized portion of the expanding metropolis. Whether it was inside or outside the political borders of the central city, it was always functionally dependent on the urban core. Conversely, the growth of suburbia always meant a strengthening of the specialized services at the core.

In my view, the most important feature of postwar American development has been the almost simultaneous decentralization of housing, industry, specialized services, and office jobs; the consequent breakaway of the urban periphery from a central city it no longer needs; and the creation of a decentralized environment that nevertheless possesses all the economic and technological dynamism we associate with the city. This phenomenon, as remarkable as it is unique, is not suburbanization but a new city.

Unfortunately, we lack a convenient name for this new city, which has taken shape on the outskirts of all our major urban centers. Some have used the terms "exurbia" or "outer city." I suggest (with apologies) two neologisms: the "technoburb" and the "techno-city." By "technoburb" I mean a peripheral zone, perhaps as large as a county, that has emerged as a viable socio-economic unit. Spread out along its highway growth corridors are shopping malls, industrial parks, campus-like office complexes, hospitals, schools, and a full range of housing types. Its residents look to their immediate surroundings rather than to the city for their jobs and other needs; and its industries find not only the employees they need but also the specialized services.

The new city is a technoburb not only because high tech industries have found their most congenial homes in such archetypal technoburbs as Silicon Valley in northern California and Route 128 in Massachusetts. In most technoburbs such industries make up only a small minority of jobs, but the very existence of the decentralized city is made possible only through the advanced communications technology which has so completely superseded the face-to-face contact of the traditional city. The technoburb has generated urban diversity without traditional urban concentration.

By "techno-city" I mean the whole metropolitan region that has been transformed by the coming of the technoburb. The techno-city usually still bears the name of its principal city, for example, "the New York metropolitan area"; its sports teams bear that city's name (even if they no longer play within the boundaries of the central city); and its television stations appear to broadcast from the central city. But the economic and social life of the region increasingly bypasses its supposed core. The techno-city is truly multicentered, along the pattern that Los Angeles first created. The technoburbs, which might stretch over seventy miles from the core in all directions, are often in more direct communication with one another – or with other techno-cities across the country – than they are with the core. The techno-city's real structure is aptly expressed by the circular superhighways or beltways that serve so well to define the perimeters of the new city. The beltways put every part of the urban periphery in contact with every other part without passing through the central city at all.

[...]

The old central cities have become increasingly marginal, while the technoburb has emerged as the focus of American life. The traditional suburbanite – commuting at ever-increasing cost to a center where the available resources barely duplicate those available much closer to home – becomes increasingly rare. In this transformed urban ecology the history of suburbia comes to an end.

PROPHETS OF THE TECHNO-CITY

Like all new urban forms, the techno-city and its technoburbs emerged not only unpredicted but un-observed. We are still seeing this new city through the intellectual categories of the old metropolis. Only two prophets, I believe, perceived the under-lying forces that would lead to the techno-city at the time of their first emergence. Their thoughts are therefore particularly valuable in understanding the new city.

At the turn of the twentieth century, when the power and attraction of the great city was at its peak, H.G. Wells daringly asserted that the technological forces that had created the industrial metropolis were now moving to destroy it. In his 1900 essay "The Probable Diffusion of Great Cities," Wells argued that the seemingly inexorable concentration of people and resources in the largest cities would soon be reversed. In the course of the twentieth century, he prophesied, the metropolis would see its own resources drain away to decentralized "urban regions" so vast that the very concept of "the city" would become, in his phrase, "as obsolete as 'mailcoach.'"

Wells based his prediction on a penetrating ana-lysis of the emerging networks of transportation and communication. Throughout the nineteenth century, rail transportation had been a relatively simple system favoring direct access to large centers. With the spread of branchlines and electric tramways, however, a complex rail network had been created that could serve as the basis for a decentralized region. (As Wells wrote, Henry E. Huntington was proving the truth of his propositions for the Los Angeles region.)

Wells pictured the "urban region" of the year 2000 as a series of villages with small homes and factories set in the open fields, yet connected by high speed rail transportation to any other point in the region. (It was a vision not very different from those who saw Los Angeles developing into just such a network of villages.) The old cities would not completely disappear, but they would lose both their financial and their industrial functions, surviving simply because of an inherent human love of crowds. The "post-urban" city, Wells predicted, will be "essentially a bazaar, a great gallery of shops and places of concourse and rendezvous, a pedestrian place, its pathways reinforced by lifts and moving platforms, and shielded from the weather, and altogether a very spacious, brilliant, and entertaining agglomeration." In short, the great metropolis will dwindle to what we would today call a massive shopping mall, while the productive life of the society would take place in the decentralized urban region.

Wells's prediction was taken up in the late 1920s and early 1930s by Frank Lloyd Wright, who moved from similar assumptions to an even more radical view. Wright had actually seen the beginnings of the automobile and truck era; he was, perhaps not coincidentally, living mostly in Los Angeles in the late 1910s and early 1920s. Wright, like Wells, argued that "the great city was no longer modern" and that it was destined to be replaced by a decentralized society.

He called this new society Broadacre City. It has often been confused with a kind of universal suburbanization, but for Wright "Broadacres" was the exact opposite of the suburbia he despised. He saw correctly that suburbia represented the essential extension of the city into the countryside, whereas Broadacres represented the disappearance of all previously existing cities.

As Wright envisioned it, Broadacres was based on universal automobile ownership combined with a network of superhighways, which removed the need for population to cluster in a particular spot. Indeed, any such clustering was necessarily inefficient, a point of congestion rather than of communication. The city would thus spread out over the countryside at densities low enough to permit each family to have its own homestead and even to engage in part-time agriculture. Yet these homesteads would not be isolated; their access to the superhighway grid would put them within easy reach of as many jobs and specialized services as any nineteenth-century urbanite. Traveling at more than sixty miles an hour, each citizen would create his own city within the hundreds of square miles he could reach in an hour's drive.

Like Wells, Wright saw industrial production inevitably leaving the cities for the space and con-venience of rural sites. But Wright went one step further in his attempt to envision the way that a radically decentralized environment could generate that diversity and excitement which only cities had possessed.

He saw that even in the most scattered environment, the crossing of major highways would possess a certain special status. These intersections would be the natural sites of what he called the roadside market, a remarkable anticipation of the shopping center: "great spacious roadside pleasure places these markets, rising high and handsome like some flexible form of pavilion – designed as places of cooperative exchange, not only of commodities but of cultural facilities." To the roadside markets he added a range of highly civilized yet small scale institutions: schools, a modern cathedral, a center for festivities, and the like. In such an environment, even the entertainment functions of the city would disappear. Soon, Wright devoutly wished, the centralized city itself would disappear.

Taken together, Wells's and Wright's prophecies constitute a remarkable insight into the decentralizing tendencies of modern technology and society. Both were presented in utopian form, an image of the future presented as somehow "inevitable" yet without any sustained attention to how it would actually be achieved. Nevertheless, something like the transformation that Wells and Wright foresaw has taken place in the United States, a transformation all the more remarkable in that it occurred without a clear recognition that it was happening. While diverse groups were engaged in what they believed was "the suburbanization" of America, they were in fact creating a new city.

[. . .]

TECHNOBURB/TECHNO-CITY: THE STRUCTURE OF THE NEW METROPOLIS

To claim that there is a pattern or structure in the new American city is to contradict what appears to be overwhelming evidence. One might sum up the structure of the technoburb by saying that it goes against every rule of planning. It is based on two extravagances that have always aroused the ire of planners: the waste of land inherent in a single family house with its own yard, and the waste of energy inherent in the use of the personal automobile. The new city is absolutely dependent on its road system, yet that system is almost always in a state of chaos and congestion. The landscape of the technoburb is a hopeless jumble of housing, industry, commerce, and even agricultural uses. Finally, the technoburb has no proper boundaries; however defined, it is divided into a crazy quilt of separate and overlapping political jurisdictions, which make any kind of coordinated planning virtually impossible.

Yet the technoburb has become the real locus of growth and innovation in our society. And there is a real structure in what appears to be wasteful sprawl, which provides enough logic and efficiency for the technoburb to fulfill at least some of its promises.

If there is a single basic principle in the structure of the technoburb, it is the renewed linkage of work and residence. The suburb had separated the two into distinct environments; its logic was that of the massive commute, in which workers from the periphery traveled each morning to a single core and then dispersed each evening. The technoburb, however, contains both work and residence within a single decentralized environment.

By the standards of a preindustrial city where people often lived and worked under the same roof, or even of the turn of the century industrial zones where factories were an integral part of working class neighborhoods, the linkage between work and residence in the technoburb is hardly close. A recent study of New Jersey shows that most workers along the state's growth corridors now live in the same county in which they work. But this relative dispersion must be contrasted to the former pattern of commuting into urban cores like Newark or New York. In most cases traveling time to work diminishes, even when the distances traveled are still substantial; as the 1980 census indicates, the average journey to work appears to be diminishing both in distance and, more importantly, in time.

For commuting within the technoburb is multi-directional, following the great grid of highways and secondary roads that, as Frank Lloyd Wright understood, defines the community. This multiplicity of destinations makes public transportation highly inefficient, but it does remove that terrible bottleneck which necessarily occurred when work was concentrated at a single core within the region. Each house in a technoburb is within a reasonable driving time of a truly "urban" array of jobs and services, just as each workplace along the highways can draw upon an "urban" pool of workers.

Those who believed that the energy crisis of the 1970s would cripple the technoburb failed to realize that the new city had evolved its own pattern of transportation in which a multitude of relatively short automobile journeys in a multitude of different directions substitutes for that great tidal wash in and out of a single urban core which had previously defined commuting. With housing, jobs, and services all on the periphery, this sprawl develops its own form of relative efficiency. The truly inefficient form would be any attempted revival of the former pattern of long distance mass transit commuting into a core area. To account for the new linkage of work and residence in the technoburb, we must first confront this paradox: the new city required a massive and coordinated relocation of housing, industry, and other "core" functions to the periphery; yet there were no coordinators directing the process. Indeed, the technoburb emerged in spite of, not because of, the conscious purposes motivating the main actors. The postwar housing boom was an attempt to escape from urban conditions; the new highways sought to channel traffic into the cities; planners attempted to limit peripheral growth; the government programs that did the most to destroy the hegemony of the old industrial metropolis were precisely those designed to save it.

This paradox can be seen clearly in the area of transportation policy. Wright had grasped the basic point in his Broadacre City plan: a fully developed highway grid eliminates the primacy of a central business district. It creates a whole series of highway crossings, which can serve as business centers while promoting the multidirectional travel that prevents any single center from attaining unique importance. Yet, from the time of Robert Moses to the present, highway planners have imagined that the new roads, like the older rail transportation, would enhance the importance of the old centers by funneling cars and trucks into the downtown area and the surrounding industrial belt. At most, the highways were to serve traditional suburbanization; in other words, the movement from the periphery to the core during morning rush hours and the reverse movement in the afternoon. The beltways, those crucial "Main Streets" of the technoburb, were designed simply to allow interstate traffic to avoid going through the central cities.

The history of the technoburb, therefore, is the history of those deeper structural features of modern society first described by Wells and Wright taking precedence over conscious intentions. For purposes of clarity I shall now divide this discussion of the making of the techno-city into two interrelated topics: housing and job location.

Housing

The great American postwar housing boom was perhaps the purest example of the suburban dream in action, yet its ultimate consequence was to render suburbia obsolete. Between 1950 and 1970, on the average, 1.2 million housing units were built each year, the vast majority as suburban single family dwellings; the nation's housing stock increased by 21 million units or over 50 percent. In the 1970s the boom continued even more strongly: twenty million more new units were added, almost as many as in the previous two decades. It was precisely this vast production of new residences that shifted the center of gravity in the United States from the urban core to the periphery and thus ensured that these vital and expanding areas could no longer remain simply bedroom communities.

This great building boom, which seems so characteristic of post-1945 conditions, in fact had its origins early in the twentieth century in the first attempts to universalize suburbia throughout the United States. It can be seen essentially as a continuation of the 1920s building boom, which had been cut off for two decades by the Depression and the war. As George Sternlieb reminds us, the American automobile industry in 1929 was producing as many cars per capita as it did in the 1980s, and real estate developers had already plotted out subdivisions in out-lying areas that were only built up in the 1960s and 1970s.

[. . .]

Even the late 1970s combination of stagnant real income with high interest rates, gasoline prices, and land values did not diminish the desirability of the new single family house. In 1981 a median American family earned only 70 percent of what was needed to make the payments on the median priced house; by 1986, the median family could once again afford the median house. Single family houses still constitute 67 percent of all occupied units, down only 2 percent since 1970 despite the increase in costs; moreover, a survey of potential home buyers in 1986 showed that 85 percent intended to purchase a detached, single family suburban house, while only 15 percent were

looking at condominium apartments or townhouses. The "single," as builders call it, is still alive and well on the urban periphery.

This continuing appeal of the single should not, however, obscure the crucial changes that have transformed the meaning and context of the house. The new suburban house of the 1950s, like its predecessors for more than a century, existed precisely to isolate women and the family from urban economic life; it defined an exclusive zone of residence between city and country. Now a new house might adjoin a landscaped office park with more square feet of new office space than in a downtown building, or might be just down the highway from an enclosed shopping mall with a sales volume that exceeds those of the downtown department stores, or might overlook a high tech research laboratory making products that are exported around the world. No longer a refuge, the single family detached house on the periphery is preferred as a convenient base from which both spouses can rapidly reach their jobs.

Without the simultaneous movement of jobs along with housing, the great "suburban" boom would surely have exhausted itself in ever longer journeys to workplaces in a crowded core on overburdened highways and mass transit facilities. And the new peripheral communities would have been in reality the "isolation wards" for women that critics have called them, instead of becoming the setting for the reintegration of middle-class women into the work force as they have. The unchanging image of the suburban house and the suburban bedroom community has obscured the crucial importance of this transformation in work location, the subject of the next section.

Job location

As those who have tried to plan the process have painfully learned, job location has its own autonomous rules. The movement of factories away from the urban core after 1945 took place independently of the housing boom and probably would have occurred without it. Nevertheless, the simultaneous movement of housing and jobs in the 1950s and 1960s created an unforeseen "critical mass" of entrepreneurship and expertise on the perimeters, which allowed the technoburb to challenge successfully the two century long economic dominance of the central city.

[. . .]

At the same time, the growing importance of trucking meant that factories were no longer as dependent on the confluence of rail lines which existed only in the old factory zones. Workers had their automobiles, so factories could scatter along the periphery without concern about the absence of mass transit. (The scattering of aircraft plants and other factories in Los Angeles in the 1930s prefigured this trend.) The process gained momentum as a result of thousands of uncoordinated decisions in which managers allowed their inner city plants to run down and directed new investment toward the outskirts . . .

These changes in job location during the 1950s and 1960s were, however, only a prelude to the real triumph of the technoburb: the luring of both managerial office employment and advanced technological laboratories and production facilities from the core to the peripheries. This process may be divided into three parts. First came the establishment of "high tech" growth corridors in such diverse locations as Silicon Valley, California; Silicon Prairie, between Dallas and Fort Worth; the Atlanta Beltway; Route 1 between Princeton and New Brunswick, New Jersey; Westchester County, New York; Route 202 near Valley Forge, Pennsylvania; and Route 128 outside Boston. The second step was the movement of office bureaucracies, especially the so-called back office, from center city high-rises to technoburb office parks; and the final phase was the movement of production-service employment – banks, accountants, lawyers, advertising agencies, skilled technicians, and the like – to locations within the technoburb, thus creating that vital base of support personnel for larger firms.

Indeed, this dramatic surge toward the technoburb has been so sweeping that we must now ask whether Wright's ultimate prophecy will be fulfilled: the disappearance of the old urban centers. Is the present-day boom in downtown office construction and inner city gentrification simply a last hurrah for the old city before deeper trends in decentralization lead to its ultimate decay?

In my view, the final diffusion that Wells and Wright predicted is unlikely, if only because both underestimated the forces of economic and political centralization that continue to exist in the late twentieth century. If physical decentralization had indeed meant economic decentralization, then the urban cores would by now be ghost towns. But large and powerful organizations still seek out a central location that validates their importance, and the historic core of great cities still meets that need better than the office complexes on the outskirts. Moreover, the corporate and government headquarters in the core still attract a wide variety of specialized support services – law firms, advertising, publishing, media, restaurants, entertainment centers, museums, and more – that continue to make the center cities viable.

The old factory zones around the core have also survived, but only in the painfully anomalous sense of housing those too poor to earn admission to the new city of prosperity at the periphery. The big city, therefore, will not disappear in the foreseeable future, and residents of the technoburbs will continue to confront uneasily both the economic power and elite culture of the urban core and its poverty. Nevertheless, the technoburb has become the true center of American society.

THE MEANING OF THE NEW CITY

Beyond the structure of the techno-city and its technoburbs, there is the larger question: what is the impact of this decentralized environment on our culture? Can anyone say of the technoburb, as Olmsted said of the suburb a century ago, that it represents "the most attractive, the most refined, and the most soundly wholesome forms of domestic life, and the best application of the arts of civilization to which mankind has yet attained"? Most planners in fact say the exact opposite. Their indictment can be divided into two parts. First, decentralization has been a social and economic disaster for the old city and for the poor, who have been increasingly relegated to its crowded, decayed zones. It has resegregated American society into an affluent outer city and an indigent inner city, while erecting ever higher barriers that prevent the poor from sharing in the jobs and housing of the technoburbs.

Second, decentralization has been seen as a cultural disaster. While the rich and diverse architectural heritage of the cities decays, the technoburb has been built up as a standardized and simplified sprawl, consuming time and space, destroying the natural landscape. The wealth that postindustrial America has generated has been used to create an ugly and wasteful pseudocity, too spread out to be efficient, too superficial to create a true culture. The truth of both indictments is impossible to deny, yet it must be

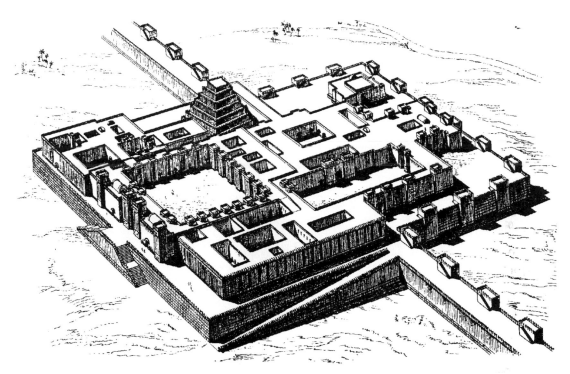

Plate 1 **Palace of Sargon II of Khorsabad.** The first cities arose about 4000 BCE in Mesopotamia between the Tigris and Euphrates Rivers during what V. Gordon Childe termed the "Urban Revolution." Note the mud brick walls, monumental gate, and ziggurat in this very substantial Mesopotamian palace.

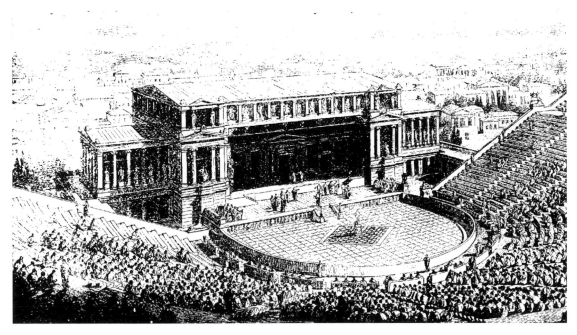

Plate 2 **Theater of Dionysus, Athens, Greece, fifth century BCE.** The Greek polis stressed public over private life. Citizens conversed, shopped, and settled disputes in public agora, exercised and competed in public stadia and gymnasia, and shared in the political and cultural life of the community in large open-air structures like the magnificent Theater of Dionysus.

Plate 5 **Bird's-eye view of La Crosse, Wisconsin, showing the principal business street.** One of the features which Friedrich Engels noted in mid-nineteenth-century Manchester, England, was the appearance of large streets where the upper classes could travel without coming into contact with the squalid living conditions of the residential slums which comprised the greater part of the city. This plate from a popular American illustrated weekly depicted with pride the principal business street in a Midwestern industrial city. The belching factories and packed residential slums, which then comprised much of La Crosse, are not mentioned in the caption.

Plate 6 Levittown, New York, 1947. While suburbs have a long and varied history, it was during the period after World War II that many of the suburbs surrounding US cities arose. Levittown, New York (and its counterpart Levittown, Pennsylvania), provided entirely new communities of affordable, cute, single-family houses on individual lots to returning GIs and other first-time (white) homebuyers.

Plate 7 The auto-centered metropolis. Developers show off the location of a proposed new "50 Foot Boulevard" in the Westwood Village area of Santa Monica in 1922. Capitalizing on the mass production of Henry Ford's Model-T, they and thousands of their counterparts built communities in Southern California that required a car. (Security Pacific Collection. Los Angeles Public Library.)

Plate 8 Technoburbia. Robert Fishman uses the term "technoburbia" to describe the form of urban development that jumbles business, residential, commercial, and other uses together in the area surrounding older core cities. Here Microsoft's Corporate Headquarters in Redmond, Washington, forms part of a pattern that includes residential areas, open space, shopping centers, and other uses.

PART TWO

Urban culture and society

INTRODUCTION TO PART TWO

As Shakespeare wrote, and as urbanists ever since have never tired of quoting, "the people are the city." Urban history expresses the progressive evolution of the city as an institution. Urban form and design describe the physical appearance and infrastructural layout of cities. But it is the people of the city – their individual aspirations and collective struggles, their day-to-day lives and their moments of heightened awareness – that constitute the core subject of urban studies and the final purpose of city planning.

In turning to the people of the cities themselves we move to a consideration of the subtle and ever-shifting interplay between society, community, and culture. This section addresses how urban society affects urban culture and how culture affects the daily lives and the life prospects of city dwellers. It asks what culture is in an urban context and how it expresses itself in different social contexts – either as high culture or popular culture. Finally, it analyzes what community is in an urban context and speculates about what it could be. These and other aspects of urban culture are explored in *The City Cultures Reader*, second edition, edited by Malcolm Miles, Tim Hall, and Iain Borden (London and New York: Routledge, 2004).

In studying the people of the city, the key methodology is sociology, that "science of society" that arose alongside the emergence of the modern industrial city itself. Urban sociology has always been allied with urban anthropology, and in recent years new formulations – social studies, social theory, social relations, culture studies – have joined the discipline either as adjuncts or rivals. But the basic sociological vision remains central to all investigations of people in cities.

There is no better person with whom to begin a discussion of "Urban Culture and Society" than Lewis Mumford (p. 85), one of the great public intellectuals of the twentieth century. Mumford never lost sight of the human dimension of cities. For over sixty years he sparred with those who argued that cities arose and prospered for purely economic reasons or that cities were best defined in terms of size and density. Not so! thundered Mumford: cities are expressions of the human spirit, and cities exist to contribute to the ever-evolving human personality. This perspective comes through loud and clear in "What Is a City?" To Mumford, defining a city in terms of population size, or density, or attributes of the built environment is inadequate. Rather, the human side of cities is their very essence, and city streets are a stage on which life's drama is played out. Like William Whyte (p. 448) and Jane Jacobs (p. 98), Mumford takes real delight in city life. For him, cities reflect and enlarge the human spirit, and he argues that creating better, more human cities will enrich civilization itself.

Mumford, of course, was not alone in focusing on the connections between urban life and the human personality. In "Urbanism as a Way of Life," Louis Wirth (p. 90) asked the fundamental question "What does it mean to be urban?" and concluded that an urban "way of life" resulted in an "urban type" of character and personality. Wirth was one of a gifted group of sociologists at the University of Chicago who, in the 1920s and 1930s, developed a pioneering body of urban sociological theory that still shapes the field of urban sociology today. Studying rural migrants to Chicago from the peasant societies of Southern and Eastern Europe, Wirth perceived that the whole way of life in modern cities was fundamentally different from the way of life in rural cultures. In "Urbanism as a Way of Life" he attempts to abstract the essential characteristics of urban as opposed to rural life and to find the sources of the widely perceived urban characteristics of brusqueness and impersonality. As the face-to-face transactions of static rural village life are replaced by the distanced and mediated transactions of a large city, human personalities are transformed, and the new urbanites respond to each other and to society as a whole in entirely different ways than they did in their rural folk communities.

While Wirth's is a *theoretical* study, it is important to remember that his theories were generated by empirical observations that he and his colleagues conducted in Chicago during the early decades of the twentieth century. Architectural critic and urban community activist Jane Jacobs (p. 98) did her own kind of street-level social observation in writing *The Death and Life of Great American Cities* (1961), a book that shook the complacent world of establishment planning by re-introducing the values of community to the design of urban spaces. Architecture and urban design may not *determine* human behavior, but bad design can numb the human spirit and good design can have powerful, positive influences on human beings. Of the many values designers seek to build into their designs perhaps none is more important than fostering community and human interaction. To Jacobs, traffic engineering should be only one consideration in designing a street. In "The Uses of Sidewalks: Safety," she argues that a street designed so that people can see their children from house windows and will want to congregate on the front door stoops – one very much like her own Hudson Street in Greenwich Village – will be much more user-friendly. It will also be much safer than one that moves traffic efficiently, but is inhospitable to neighborhood life and insensitive to the potential for street life to reduce urban crime. Jacobs stresses the importance of designing streets to promote safety, particularly for women. A safer environment, she argues, is essential to the creation and preservation of community.

The Black ghettoes of the United States have been the subject of an enormous body of sociological research that both celebrates the distinctive culture and analyzes the social pathology of segregated, poverty-ridden inner-city communities. In multi-ethnic societies, racial divisions compounded class distinctions to create an even greater crisis of community in the form of racially segregated neighborhoods that have remained as symbols of inequality and oppression. *The Philadelphia Negro* (1899) by W.E.B. Du Bois (p. 103) specifically describes the African–American district of Philadelphia, Pennsylvania, as it developed in the years following the American Civil War, but the social and cultural dynamic of housing segregation and racial discrimination in the workplace that Du Bois describes can be applied to ghetto and barrio experiences throughout the United States, to the "social exclusion" experienced by residents of Third World immigrant communities worldwide (p. 158). In the years since Du Bois first surveyed the life of the racially segregated ghetto community, conditions have in some ways grown worse: so much so that the persistence of racial segregation, and the emergence of an "underclass" population radically disconnected from the rest of the urban community, threatens the social stability of some of the largest, wealthiest cities in the world.

Beginning with Du Bois's study of *The Philadelphia Negro* and continuing through the work of E. Franklin Frazier (*The Negro Family in the United States*, 1939), St. Clair Drake and Horace Cayton (*Black Metropolis*, 1945), and Kenneth B. Clark (*Dark Ghetto*, 1965), African–American scholars have taken the lead in examining the social and cultural dynamics of ghetto communities in America's northern cities. More recently, an important debate, called the "underclass" debate, arose concerning the plight of the mostly Black residents of American inner-city ghettos. One of the principals in that debate was William Julius Wilson (p. 110), an African–American sociologist.

In "From Institutional to Jobless Ghettos" from *When Work Disappears* (1996), Wilson illustrates the role of ideology in shaping urban theory. As a leading liberal voice, he argues that the situation of the poorest urban Blacks in the United States has grown worse during the last generation and that poor ghetto Blacks today, especially youth, are in deep trouble. But why is this so . . . and what to do about it? Wilson stresses the loss of jobs accessible to unskilled Black youth. It is this loss of jobs, Wilson argues, that has destroyed ghetto family structure and lies at the root of crime, substance abuse, and other ghetto ills. As Wilson sees it, many young Black males are not "marriageable" because they lack minimum job skills, have substance abuse problems, or are in prison. Without "marriageable" males, many young Black women cannot form two-parent nuclear families, and teen pregnancies, out-of-wedlock births, and female-headed, welfare-dependent, single-parent households result.

Perhaps the biggest difference between liberals and conservatives in the underclass debate lies in their views on the role of government in solving social problems. Wilson would like to see government intervene with a universal, not race-specific, full employment program. If the poor of all races are employed, he reasons, family stability will return, and substance abuse and criminal behavior will drop. Conservatives argue that the last thing the poor need is more government assistance. For example, Charles Murray, author of *Losing Ground* (1984) and *The Bell Curve* (1994), argues that patronizing government programs sapped initiative from the poor and created perverse incentives to stay out of the labor market. And Michael Porter (p. 274) argues that programs that do not address the real competitive advantages of inner cities are counter-productive. Murray's remedy: a form of low-level guaranteed national income in place of the existing liberal programs of Social Security and welfare. Porter's remedy: redirect government aid and corporate philanthropy to economic development programs that will employ low-skilled urban residents in jobs that modern economies need.

The "underclass debate" changed direction when, in the mid-1990s, the American Congress and President Clinton changed "welfare as we know it" by radically cutting back on many welfare support programs in favor of decentralized "workfare." On balance, the results have seemed promising, with welfare rolls declining and employment trending upward. At the same time, however, another concern emerged – or rather re-emerged – about the fundamental quality of civic culture in contemporary urban society. The leading voice raising this concern was that of Robert Putnam (p. 120), author of *Bowling Alone* (2000). Drawing on evidence from both Europe and the United States, and without down-playing the importance of the issues of race and poverty, Putnam asked urban leaders and members of the urban middle class to confront the evident social reality that people are no longer as connected to the basic institutions of their communities – the neighborhood groups, the fraternal organizations, even the political parties – as they once were. Putnam attributes the growing lack of community participation – what he calls the decline in "social capital" and loss of civic engagement – to many factors: the movement of women into the workforce, increased social and geographical mobility, and "the technological transformation of leisure." Another explanation might be that contemporary urban society is deeply divided and that every major city has now become culturally contested terrain. Culture used to mean "high culture" – the world of the symphony, opera, and ballet. But culture, of course, is not just the product of an identifiable artistic or intellectual class. Working-class neighborhoods and inner-city ghettos also produce formal poetry and the linguistic inventiveness of street talk, rhapsodies and jazz music, paintings and graffiti. These are surely elements of "the urban drama," but who will dominate a city's culture? Whose "social capital" will be hegemonic in the socially "contested city"? And what exactly constitutes "civic engagement"?

Richard Florida (p. 129) addresses another aspect of the debate on the future of urban economies – the role of the "creative class." In *The Rise of the Creative Class* (2002), Florida argues that a creative environment – or at least one that is compatible with creativity and creative people – is essential to urban life, especially in the post-modern Information Age. Consisting of "information managers" and "symbolic analysts" such as engineers, artists, software programmers, writers, and corporate strategists involved in the new post-industrial economy, these new urban dwellers are knowledge workers who add value to their enterprises, and to society as a whole, by exercising their creative imaginations. More than just an educated class of high-end service workers, members of the creative class bring a new vitality to the city and transform urban culture through their commitment to the values of individuality, meritocracy, and diversity.

Sometimes cultures clash within an urban environment. As socio-economic classes – or racial, ethnic, or new immigrant groups – compete for benefits and status within the urban order, short-term confrontations and long-term accommodations tend to take place as urban society slowly evolves. In the new globalized urban world, however, some culture clashes can take on a more desperate character, and this is nowhere more in evidence than in the conflict between the broadly "liberal" and tolerant values of the modern Western city and the more narrowly prescribed moral values of radical fundamentalist Islam that forms the basis of today's War on Terrorism. When the events of September 11, 2001 took place in New York, many cultural analysts attempted to explain what had happened. Some, like Samuel Huntington, posited a "clash of civilizations." Others, like Ian Buruma and Avishai Margalit, explored how many in the non-dominant, non-Western world conceived of the West, especially the modern Western city, as the locus and source of ungodliness and moral depravity. In *Orientalism* (1978), Edward Said had introduced the concept of "the other" to explain the lens of exotic foreignness through which Westerners had viewed the East ever since the Middle Ages. Buruma and Margalit turned the tables on this analysis by describing the way the West is seen as "the other" in the eyes of the fundamentalist Islamic world. In *Occidentalism: The West in the Eyes of Its Enemies* (2004, p. 136), Buruma and Margalit describe the way in which the Western city of modernism, capitalism, and secularism, symbolized by the libertine and prostitute, embodies all that is anathema to the intensely devout of any religious tradition, but especially post-colonial fundamentalist Islam. With the Islamic world extending from Indonesia to Morocco, and with growing Islamic populations in the cities of Europe and North America, the culture clash defined by the War on Terrorism has already had deep effects on city life almost everywhere and has introduced elements of fear, uncertainty, and heightened security measures into the daily round of urban life.

Finally, in exploring urban society and culture, academic analysts of the urban condition have been trained to investigate a range of measurable conditions and observable social behaviors, and to analyze their findings in terms of class, race, ethnicity, and socio-economic status. But urban culture has other dimensions that do not lend themselves so easily to quantitative analysis or narrative description. Among these are the expressions of the creative arts and the other cultural productions that emerge from urban communities everywhere, indeed from the very conditions of modern urban life. In "Visions of a New Reality: The City and the Emergence of Modern Visual Culture," the image-essay that ends Part Two, Frederic Stout (p. 141) argues that popular illustrated journalism, photography, and cinema are among the most characteristic of urban cultural genres and that visual culture generally is an artifact of modern urban society.

"What is a City?"

Architectural Record (1937)

Lewis Mumford

Editors' Introduction

Lewis Mumford (1895–1990) has been called America's last great public intellectual. Beginning with his first book (on the idea of Utopia) in 1922 and continuing throughout a career that saw the publication of some twenty-five influential volumes, Mumford made signal contributions to social philosophy, American literary and cultural history, the history of technology and, preeminently, the history of cities and urban planning practice.

Mumford saw the urban experience as an integral component in the development of human culture and the human personality. He consistently argued that the physical design of cities and their economic functions are secondary to their relationship to the natural environment and to the spiritual values of human community. Mumford applied these principles to his architectural criticism for *The New Yorker* magazine in the 1920s, his work with the Regional Planning Association of America, his campaign against plans to build a highway through Washington Square in New York's Greenwich Village in the 1950s, and his life-long championing of the Garden City ideals of Ebenezer Howard.

In "What Is a City?" Mumford lays out his fundamental propositions about city planning and the human potential, both individual and social, of urban life. The city, he writes, is "a theater of social action," and everything else – art, politics, education, commerce – only serve to make the "social drama . . . more richly significant, as a stage-set, well-designed, intensifies and underlines the gestures of the actors and the action of the play." It was a theme and an image to which Mumford would return over and over again. In his chapter on "The Nature of the Ancient City" in *The City in History* (1961), he wrote that the city is "above all things a theater" and, as if commenting on the cultural conformity of the 1950s, warned that an urban civilization that has lost its sense of dramatic dialogue "is bound to have a fatal last act."

Mumford's influence on modern urban planning theory can hardly be overstated. His "urban drama" idea clearly resonates with an entire line of urban cultural analysts. Jane Jacobs, for example, talks about "street ballet" (p. 98). William Whyte (p. 448) says that a good urban plaza should function like a stage. Allan Jacobs and Donald Appleyard (p. 456) urge planners to fulfill human needs for "fantasy and exoticism." The city, they write, "has always been a place of excitement; it is a theater, a stage upon which citizens can display themselves and be seen by others." And economist Richard Florida (p. 129) has argued for the importance to urban culture of a "creative class."

As a historian, Mumford's emphasis on community values and the city's role in enlarging the potential of the human personality connects him with a long line of urban theorists that includes Louis Wirth (p. 90) and many others. *The City in History* (1961) is undoubtedly Mumford's masterpiece, but an earlier version of the same material, *The Culture of Cities* (1938), is still of interest. *The Urban Prospect* (1968) is an outstanding collection of his essays on urban planning and culture, and *The Myth of the Machine* (1967) and *The Pentagon of Power* (1970) are excellent analyses of the influence of technology on human culture. The magisterial *The Transformations of Man* (1956) invites comparison with V. Gordon Childe's theory of the urban revolution (p. 27). A sampling of

Mumford's writings are included in Donald L. Miller (ed.), *The Lewis Mumford Reader* (Athens: University of Georgia Press, 1995).

Mumford's illuminating correspondence with Patrick Geddes is contained in Frank G. Novak, *Lewis Mumford and Patrick Geddes: The Correspondence* (London: Routledge, 1995). His correspondence with Frank Lloyd Wright is contained in *Frank Lloyd Wright and Lewis Mumford: Thirty Years of Correspondence* (New York: Princeton Architectural Press, 2001), and his writings for *The New Yorker* are contained in Robert Wojtowicz (ed.), *Sidewalk Critic: Lewis Mumford's Writings on New York* (New York: Princeton Architectural Press, 1998). Also of interest in Bruce Brooks Pfeiffer *et al.*, *Frank Lloyd Wright and Lewis Mumford: Thirty Years of Correspondence* (New York: Princeton Architectural Press, 2001).

Mumford is being rediscovered by the current generation of environmental planners. Examples of recent books applying his perspective to current ecological issues are Mark Luccarelli Lewis, *Mumford and the Ecological Region: The Politics of Planning* (New York: Guilford Press, 1997) and Robert Wojtowicz, *Lewis Mumford and American Modernism: Eutopian Theories for Architecture and Urban Planning* (Cambridge: Cambridge University Press, 1998).

Biographies of Lewis Mumford are Donald L. Miller, *Lewis Mumford: A Life* (New York: Weidenfeld & Nicolson, 1989), Thomas P. Hughes and Agatha C. Hughes (eds), *Lewis Mumford: Public Intellectual* (Oxford: Oxford University Press, 1990), and Frank G. Novak, *Lewis Mumford* (New York: Twayne Publishers, 1998). An excellent bibliography of Mumford's writings is Elmer S. Newman, *Lewis Mumford: A Bibliography, 1914–1970* (New York: Harcourt Brace Jovanovich, 1971).

■ ■ ■ ■ ■ ■

Most of our housing and city planning has been handicapped because those who have undertaken the work have had no clear notion of the social functions of the city. They sought to derive these functions from a cursory survey of the activities and interests of the contemporary urban scene. And they did not, apparently, suspect that there might be gross deficiencies, misdirected efforts, mistaken expenditures here that would not be set straight by merely building sanitary tenements or straightening out and widening irregular streets.

The city as a purely physical fact has been subject to numerous investigations. But what is the city as a social institution? The earlier answers to these questions, in Aristotle, Plato, and the Utopian writers from Sir Thomas More to Robert Owen, have been on the whole more satisfactory than those of the more systematic sociologists: most contemporary treatises on "urban sociology" in America throw no important light upon the problem. One of the soundest definitions of the city was that framed by John Stow, an honest observer of Elizabethan London, who said:

Men are congregated into cities and commonwealths for honesty and utility's sake, these shortly be the commodities that do come by cities, commonalties and corporations. First, men by this

nearness of conversation are withdrawn from barbarous fixity and force, to certain mildness of manners, and to humanity and justice . . . Good behavior is yet called urbanitas because it is rather found in cities than elsewhere. In sum, by often hearing, men be better persuaded in religion, and for that they live in the eyes of others, they be by example the more easily trained to justice, and by shamefastness restrained from injury.

And whereas commonwealths and kingdoms cannot have, next after God, any surer foundation than the love and good will of one man towards another, that also is closely bred and maintained in cities, where men by mutual society and companying together, do grow to alliances, commonalties, and corporations.

It is with no hope of adding much to the essential insight of this description of the urban process that I would sum up the sociological concept of the city in the following terms:

The city is a related collection of primary groups and purposive associations: the first, like family and neighborhood, are common to all communities, while the second are especially characteristic of city life. These varied groups support themselves through economic organizations that are likewise of a more or less corporate, or at least publicly regulated, character; and they are all housed in permanent structures, within a relatively limited area. The essential physical means of a city's existence are the fixed site, the durable shelter, the permanent facilities for assembly, interchange, and storage; the essential social means are the social division of labor, which serves not merely the economic life but the cultural processes. The city in its complete sense, then, is a geographic plexus, an economic organization, an institutional process, a theater of social action, and an aesthetic symbol of collective unity. The city fosters art and is art; the city creates the theater and is the theater. It is in the city, the city as theater, that man's more purposive activities are focused, and work out, through conflicting and cooperating personalities, events, groups, into more significant culminations.

Without the social drama that comes into existence through the focusing and intensification of group activity there is not a single function performed in the city that could not be performed – and has not in fact been performed – in the open country. The physical organization of the city may deflate this drama or make it frustrate; or it may, through the deliberate efforts of art, politics, and education, make the drama more richly significant, as a stage-set, well-designed, intensifies and underlines the gestures of the actors and the action of the play. It is not for nothing that men have dwelt so often on the beauty or the ugliness of cities: these attributes qualify men's social activities. And if there is a deep reluctance on the part of the true city dweller to leave his cramped quarters for the physically more benign environment of a suburb – even a model garden suburb! – his instincts are usually justified: in its various and many-sided life, in its very opportunities for social disharmony and conflict, the city creates drama; the suburb lacks it.

One may describe the city, in its social aspect, as a special framework directed toward the creation of differentiated opportunities for a common life and a significant collective drama. As indirect forms of association, with the aid of signs and symbols and specialized organizations, supplement direct face-to-face intercourse, the personalities of the citizens themselves become many-faceted: they reflect their specialized interests, their more intensively trained aptitudes, their finer discriminations and selections: the personality no longer presents a more or less unbroken traditional face to reality as a whole. Here lies the possibility of personal disintegration; and here lies

the need for reintegration through wider participation in a concrete and visible collective whole. What men cannot imagine as a vague formless society, they can live through and experience as citizens in a city. Their unified plans and buildings become a symbol of their social relatedness; and when the physical environment itself becomes disordered and incoherent, the social functions that it harbors become more difficult to express.

One further conclusion follows from this concept of the city: social facts are primary, and the physical organization of a city, its industries and its markets, its lines of communication and traffic, must be subservient to its social needs. Whereas in the development of the city during the last century we expanded the physical plant recklessly and treated the essential social nucleus, the organs of government and education and social service, as mere afterthought, today we must treat the social nucleus as the essential element in every valid city plan: the spotting and inter-relationship of schools, libraries, theaters, community centers is the first task in defining the urban neighborhood and laying down the outlines of an integrated city.

In giving this sociological answer to the question: What is a City? one has likewise provided the clue to a number of important other questions. Above all, one has the criterion for a clear decision as to what is the desirable size of a city – or may a city perhaps continue to grow until a single continuous urban area might cover half the American continent, with the rest of the world tributary to this mass? From the standpoint of the purely physical organization of urban utilities – which is almost the only matter upon which metropolitan planners in the past have concentrated – this latter process might indeed go on indefinitely. But if the city is a theater of social activity, and if its needs are defined by the opportunities it offers to differentiated social groups, acting through a specific nucleus of civic institutes and associations, definite limitations on size follow from this fact.

In one of Le Corbusier's early schemes for an ideal city, he chose three million as the number to be accommodated: the number was roughly the size of the urban aggregate of Paris, but that hardly explains why it should have been taken as a norm for a more rational type of city development. If the size of an urban unit, however, is a function of its productive organization and its opportunities for active social intercourse and culture, certain definite facts emerge as to adequate ratio of population to the process to be served. Thus, at the present level of culture in America, a million people are needed to support a university. Many factors may enter which will change the size of both the university and the population base; nevertheless one can say provisionally that if a million people are needed to provide a sufficient number of students for a university, then two million people should have two universities. One can also say that, other things being equal, five million people will not provide a more effective university than one million people would. The alternative to recognizing these ratios is to keep on overcrowding and overbuilding a few existing institutions, thereby limiting, rather than expanding, their genuine educational facilities.

What is important is not an absolute figure as to population or area: although in certain aspects of life, such as the size of city that is capable of reproducing itself through natural fertility, one can already lay down such figures. What is more important is to express size *always as a function of the social relationships to be served* . . . There is an optimum numerical size, beyond which each further increment of inhabitants creates difficulties out of all proportion to the benefits. There is also an optimum area of expansion, beyond which further urban growth tends to paralyze rather than to further important social relationships. Rapid means of transportation have given a regional area with a radius of from forty to a hundred miles, the unity that London and Hampstead had before the coming of the underground railroad. But the activities of small children are still bounded by a walking distance of about a quarter of a mile; and for men to congregate freely and frequently in neighborhoods the maximum distance means nothing, although it may properly define the area served for a selective minority by a university, a central reference library, or a completely equipped hospital. The area of potential urban settlement has been vastly increased by the motor car and the airplane; but the necessity for solid contiguous growth, for the purposes of intercourse, has in turn been lessened by the telephone and the radio. In the Middle Ages a distance of less than a half a mile from the city's center usually defined its utmost limits. The block-by-block accretion of the big city, along its corridor avenues, is in all important respects a denial of the vastly improved type of urban grouping that our fresh inventions have brought in. For all occasional types of intercourse, the region is the unit of social life but the region cannot function effectively, as a well-knit unit, if the entire area is densely filled with people – since

their very presence will clog its arteries of traffic and congest its social facilities.

Limitations on size, density, and area are absolutely necessary to effective social intercourse; and they are therefore the most important instruments of rational economic and civic planning. The unwillingness in the past to establish such limits has been due mainly to two facts: the assumption that all upward changes in magnitude were signs of progress and automatically "good for business," and the belief that such limitations were essentially arbitrary, in that they proposed to "decrease economic opportunity" – that is, opportunity for profiting by congestion – and to halt the inevitable course of change. Both these objections are superstitious.

Limitations on height are now common in American cities; drastic limitations on density are the rule in all municipal housing estates in England: that which could not be done has been done. Such limitations do not obviously limit the population itself: they merely give the planner and administrator the opportunity to multiply the number of centers in which the population is housed, instead of permitting a few existing centers to aggrandize themselves on a monopolistic pattern. These limitations are necessary to break up the functionless, hypertrophied urban masses of the past. Under this mode of planning, the planner proposes to replace the "mononucleated city," as Professor Warren Thompson has called it, with a new type of "polynucleated city," in which a cluster of communities, adequately spaced and bounded, shall do duty for the badly organized mass city. Twenty such cities, in a region whose environment and whose resources were adequately planned, would have all the benefits of a metropolis that held a million people, without its ponderous disabilities: its capital frozen into unprofitable utilities, and its land values congealed at levels that stand in the way of effective adaptation to new needs.

Mark the change that is in process today. The emerging sources of power, transport, and communication do not follow the old highway network at all. Giant power strides over the hills, ignoring the limitations of wheeled vehicles; the airplane, even more liberated, flies over swamps and mountains, and terminates its journey, not on an avenue, but in a field. Even the highway for fast motor transportation abandons the pattern of the horse-and-buggy era. The new highways, like those of New Jersey and Westchester, to mention only examples drawn locally, are based more or less on a system definitively formulated by Benton MacKaye in his various papers on the Townless Highway. The most complete plans form an independent highway network, isolated both from the adjacent countryside and the towns that they by-pass: as free from communal encroachments as the railroad system. In such a network no single center will, like the metropolis of old, become the focal point of all regional advantages: on the contrary, the "whole region" becomes open for settlement.

Even without intelligent public control, the likelihood is that within the next generation this dissociation and decentralization of urban facilities will go even farther. The Townless Highway begets the Highwayless Town in which the needs of close and continuous human association on all levels will be uppermost. This is just the opposite of the earlier mechanocentric picture of Roadtown, as pictured by Edgar Chambless and the Spanish projectors of the Linear City. For the highwayless town is based upon the notion of effective zoning of functions through initial public design, rather than by blind legal ordinances. It is a town in which the various functional parts of the structure are isolated topographically as urban islands, appropriately designed for their specific use with no attempt to provide a uniform plan of the same general pattern for the industrial, the commercial, the domestic, and the civic parts.

The first systematic sketch of this type of town was made by Messrs. Wright and Stein in their design for Radburn in 1929; a new type of plan that was repeated on a limited scale – and apparently in complete independence – by planners in Köln and Hamburg at about the same time. Because of restrictions on design that favored a conventional type of suburban house and stale architectural forms, the implications of this new type of planning were not carried very far in Radburn. But in outline the main relationships are clear: the differentiation of foot traffic from wheeled traffic in independent systems, the insulation of residence quarters from through roads; the discontinuous street pattern; the polarization of social life in specially spotted civic nuclei, beginning in the neighborhood with the school and the playground and the swimming pool. This type of planning was carried to a logical conclusion in perhaps the most functional and most socially intelligent of all Le Corbusier's many urban plans: that for Nemours in North Africa, in 1934.

Through these convergent efforts, the principles of the polynucleated city have been well established. Such plans must result in a fuller opportunity for the primary group, with all its habits of frequent direct meeting and face-to-face intercourse: they must also result in a more complicated pattern and a more comprehensive life for the region, for this geographic area can only now, for the first time, be treated as an instantaneous whole for all the functions of social existence. Instead of trusting to the mere massing of population to produce the necessary social concentration and social drama, we must now seek these results through deliberate local nucleation and a finer regional articulation. The words are jargon; but the importance of their meaning should not be missed. To embody these new possibilities in city life, which come to us not merely through better technical organization but through acuter sociological understanding, and to dramatize the activities themselves in appropriate individual and urban structures, forms the task of the coming generation.

"Urbanism as a Way of Life"

American Journal of Sociology (1938)

Louis Wirth

Editors' Introduction

Louis Wirth (1897–1952) was a member of the famed "Chicago School" of urban sociology that included such academic luminaries as Ernest W. Burgess (author of "The Growth of the City," p. 150), Robert E. Park, St Clair Drake, and Horace Cayton. Together, these scholars at the University of Chicago set out to reinvent modern sociology by taking academic research to the streets and by using the city of Chicago itself as a "living laboratory" for the study of urban problems and social processes.

Wirth's major contribution to urban sociology was the formulation of nothing less fundamental than a meaningful and logically coherent "sociological definition" of urban life. As he lays it out in the magnificent synthesis that is his 1938 essay "Urbanism as a Way of Life," a "sociologically significant definition of the city" looks beyond the mere physical structure of the city, or its economic product, or its characteristic cultural institutions – however important all these may be – to discover those underlying "elements of urbanism which mark it as a distinctive mode of human group life."

Wirth argues that three key characteristics of cities – large population size, social heterogeneity, and population density – contribute to the development of a peculiarly "urban way of life" and, indeed, a distinct "urban personality." For centuries, at least as far back as Aesop's fable of the city mouse and the country mouse, casual observers have noted sharp personality differences between urban and rural people and between nature-based and machine-based styles of living. Wirth attempts to explain those differences in terms of the functional responses of urban dwellers to the characteristic environmental conditions of modern urban society. If, for example, city people are regarded as rather more socially tolerant that rural people – and, at the same time, more impersonal and seemingly less friendly – these are merely adaptations to the experience of living in large, dense, socially diverse urban environments. Wirth's analysis invites comparison with Georg Simmel's "The Metropolis and Mental Life," delivered as a lecture in 1903 and reprinted in Jan Lin and Christopher Mele (eds), *The Urban Sociology Reader* (London and New York: Routledge, 2005).

Although some see Wirth's explanation of the sociology of urban life as nothing more than the social scientific verification of the obvious, others have argued that there is actually no such thing as an "urban personality" or an "urban way of life." Sociologist Herbert Gans, for example, argues that both inner-city "urban villagers" and suburbanites tend to maintain their preexisting cultures and personalities, and Oscar Lewis's work on "the culture of poverty" – along with a whole body of Marxist analysis – suggests that culture and personality types differ widely with socio-economic class, not merely being "urban." Wirth's work, however, led to the development of a whole school of urban social ecology, and Wirth's basic ideas about personality and adaptation to urban conditions – many of them quite pessimistic – inform the full range of more recent urban planning theories and the planning practitioners who attempt to create and nurture a sense of community in the urban environment.

Other books by Louis Wirth include *Contemporary Social Problems* (University of Chicago Press, 1940), *The Effect of War on American Minorities* (New York: Social Science Research Council, 1943), *Community*

Life and Social Policy (University of Chicago Press, 1956), and The Ghetto (University of Chicago Press, 1956). Louis Wirth on Cities and Social Life: Selected Papers (University of Chicago Press, 1964) is a useful collection. Also of interest is Roger A. Salerno, Louis Wirth: A Bio-Bibliography (Westport: Greenwood Publishing, 1987).

For other important analyses of the relationship between urban life and the human personality, see Erving Goffman, The Presentation of Self in Everyday Life (Garden City: Doubleday, 1959) and Richard Sennett, The Uses of Disorder: Personal Identity and City Life (New York: W.W. Norton, 1970). And of related interest are Sylvia Fleis Fava, "Suburbanism As a Way of Life" (American Sociological Review, 21(1), 1956), and Fred Dewey, "Cyberurbanism As a Way of Life" from Architecture of Fear (Princeton: Princeton University Press, 1997), reprinted in Stephen Graham (ed.), The Cybercities Reader (London and New York: Routledge, 2004).

THE CITY AND CONTEMPORARY CIVILIZATION

Just as the beginning of Western civilization is marked by the permanent settlement of formerly nomadic peoples in the Mediterranean basin, so the beginning of what is distinctively modern in our civilization is best signalized by the growth of great cities. Nowhere has mankind been farther removed from organic nature than under the conditions of life characteristic of great cities . . . The city and the country may be regarded as two poles in reference to one or the other of which all human settlements tend to arrange themselves. In viewing urban-industrial and rural-folk society as ideal types of communities, we may obtain a perspective for the analysis of the basic models of human association as they appear in contemporary civilization.

A SOCIOLOGICAL DEFINITION OF THE CITY

Despite the preponderant significance of the city in our civilization, however, our knowledge of the nature of urbanism and the process of urbanization is meager. Many attempts have indeed been made to isolate the distinguishing characteristics of urban life. Geographers, historians, economists, and political scientists have incorporated the points of view of their respective disciplines into diverse definitions of the city. While it is in no sense intended to supersede these, the formulation of a sociological approach to the city may incidentally serve to call attention to the interrelations between them by emphasizing the peculiar characteristics of the city as a particular form of human

association. A sociologically significant definition of the city seeks to select those elements of urbanism which mark it as a distinctive mode of human group life.

[. . .]

While urbanism, or that complex of traits which makes up the characteristic mode of life in cities, and urbanization, which denotes the development and extensions of these factors, are thus not exclusively found in settlements which are cities in the physical and demographic sense, they do, nevertheless, find their most pronounced expression in such areas, especially in metropolitan cities. In formulating a definition of the city it is necessary to exercise caution in order to avoid identifying urbanism as a way of life with any specific locally or historically conditioned cultural influences which, while they may significantly affect the specific character of the community, are not the essential determinants of its character as a city.

It is particularly important to call attention to the danger of confusing urbanism with industrialism and modern capitalism. The rise of cities in the modern world is undoubtedly not independent of the emergence of modern power-driven machine technology, mass production, and capitalistic enterprise. But different as the cities of earlier epochs may have been by virtue of their development in a preindustrial and precapitalistic order from the great cities of today, they were, nevertheless, cities.

For sociological purposes a city may be defined as a relatively large, dense, and permanent settlement of socially heterogeneous individuals. On the basis of the postulates which this minimal definition suggests, a theory of urbanism may be formulated in the light of existing knowledge concerning social groups.

A THEORY OF URBANISM

In the rich literature on the city we look in vain for a theory of urbanism presenting in a systematic fashion the available knowledge concerning the city as a social entity. We do indeed have excellent formulations of theories on such special problems as the growth of the city viewed as a historical trend and as a recurrent process, and we have a wealth of literature presenting insights of sociological relevance and empirical studies offering detailed information on a variety of particular aspects of urban life. But despite the multiplication of research and textbooks on the city, we do not as yet have a comprehensive body of competent hypotheses which may be derived from a set of postulates implicitly contained in a sociological definition of the city, and from our general sociological knowledge which may be substantiated through empirical research. The closest approximations to a systematic theory of urbanism that we have are to be found in a penetrating essay, "Die Stadt," by Max Weber, and a memorable paper by Robert E. Park titled "The City: Suggestions for the Investigation of Human Behavior in the Urban Environment." But even these excellent contributions are far from constituting an ordered and coherent framework of theory upon which research might profitably proceed.

In the pages that follow, we shall seek to set forth a limited number of identifying characteristics of the city. Given these characteristics we shall then indicate what consequences or further characteristics follow from them in the light of general sociological theory and empirical research. We hope in this manner to arrive at the essential propositions comprising a theory of urbanism. Some of these propositions can be supported by a considerable body of already available research materials; others may be accepted as hypotheses for which a certain amount of presumptive evidence exists, but for which more ample and exact verification would be required. At least such a procedure will, it is hoped, show what in the way of systematic knowledge of the city we now have and what are the crucial and fruitful hypotheses for future research.

[. . .]

There are a number of sociological propositions concerning the relationship between (a) numbers of population, (b) density of settlement, (c) heterogeneity of inhabitants and group life, which can be formulated on the basis of observation and research.

SIZE OF THE POPULATION AGGREGATE

Ever since Aristotle's *Politics*, it has been recognized that increasing the number of inhabitants in a settlement beyond a certain limit will affect the relationships between them and the character of the city. Large numbers involve, as has been pointed out, a greater range of individual variation. Furthermore, the greater the number of individuals participating in a process of interaction, the greater is the potential differentiation between them. The personal traits, the occupations, the cultural life, and the ideas of the members of an urban community may, therefore, be expected to range between more widely separated poles than those of rural inhabitants.

That such variations should give rise to the spatial segregation of individuals according to color, ethnic heritage, economic and social status, tastes and preferences, may readily be inferred. The bonds of kinship, of neighborliness, and the sentiments arising out of living together for generations under a common folk tradition are likely to be absent or, at best, relatively weak in an aggregate the members of which have such diverse origins and backgrounds. Under such circumstances competition and formal control mechanisms furnish the substitutes for the bonds of solidarity that are relied upon to hold a folk society together.

[. . .]

The multiplication of persons in a state of interaction under conditions which make their contact as full personalities impossible produces that segmentalization of human relationships which has sometimes been seized upon by students of the mental life of the cities as an explanation for the "schizoid" character of urban personality. This is not to say that the urban inhabitants have fewer acquaintances than rural inhabitants, for the reverse may actually be true; it means rather that in relation to the number of people whom they see and with whom they rub elbows in the course of daily life, they know a smaller proportion, and of these they have less intensive knowledge.

Characteristically, urbanites meet one another in highly segmental roles. They are, to be sure, dependent upon more people for the satisfactions of their life-needs than are rural people and thus are associated with a greater number of organized groups, but they are less dependent upon particular persons, and their dependence upon others is confined to a highly fractionalized aspect of the other's round of activity. This is essentially what is meant by saying that the city is characterized by secondary rather than primary contacts. The contacts of the city may indeed be face to face, but they are nevertheless impersonal, superficial, transitory, and segmental. The reserve, the indifference, and the blasé outlook which urbanites manifest in their relationships may thus be regarded as devices for immunizing themselves against the personal claims and expectations of others.

The superficiality, the anonymity, and the transitory character of urban social relations make intelligible, also, the sophistication and the rationality generally ascribed to city-dwellers. Our acquaintances tend to stand in a relationship of utility to us in the sense that the role which each one plays in our life is over-whelmingly regarded as a means for the achievement of our own ends. Whereas, therefore, the individual gains, on the one hand, a certain degree of eman-cipation or freedom from the personal and emotional controls of intimate groups, he loses, on the other hand, the spontaneous self-expression, the morale, and the sense of participation that comes with living in an integrated society. This constitutes essentially the state of anomie or the social void to which Durkheim alludes in attempting to account for the various forms of social disorganization in technological society.

The segmental character and utilitarian accent of interpersonal relations in the city find their institutional expression in the proliferation of specialized tasks which we see in their most developed form in the professions. The operations of the pecuniary nexus lead to predatory relationships, which tend to obstruct the efficient functioning of the social order unless checked by professional codes and occupational eti-quette. The premium put upon utility and efficiency suggests the adaptability of the corporate device for the organization of enterprises in which individuals can engage only in groups. The advantage that the corporation has over the individual entrepreneur and the partnership in the urban-industrial world derives not only from the possibility it affords of centralizing the resources of thousands of individuals or from the legal privilege of limited liability and perpetual succession, but from the fact that the corporation has no soul.

[. . .]

DENSITY

As in the case of numbers, so in the case of concen-tration in limited space certain consequences of relevance in sociological analysis of the city emerge. Of these only a few can be indicated.

As Darwin pointed out for flora and fauna and as Durkheim noted in the case of human societies, an increase in numbers when area is held constant (i.e. an increase in density) tends to produce differentiation and specialization, since only in this way can the area support increased numbers. Density thus reinforces the effect of numbers in diversifying men and their activities and in increasing the complexity of the social structure.

On the subjective side, as Simmel has suggested, the close physical contact of numerous individuals necessarily produces a shift in the mediums through which we orient ourselves to the urban milieu, espe-cially to our fellow-men. Typically, our physical contacts are close but our social contacts are distant. The urban world puts a premium on visual recognition. We see the uniform which denotes the role of the functionaries and are oblivious to the personal eccen-tricities that are hidden behind the uniform. We tend to acquire and develop a sensitivity to a world of artifacts and become progressively farther removed from the world of nature.

We are exposed to glaring contrasts between splendor and squalor, between riches and poverty, intelligence and ignorance, order and chaos. The com-petition for space is great, so that each area generally tends to be put to the use which yields the greatest economic return. Place of work tends to become dissociated from place of residence, for the proximity of industrial and commercial establishments makes an area both economically and socially undesirable for residential purposes.

Density, land values, rentals, accessibility, health-fulness, prestige, aesthetic consideration, absence of nuisances such as noise, smoke, and dirt determine the desirability of various areas of the city as places of settlement for different sections of the population . . . The different parts of the city thus acquire specialized functions. The city consequently tends to resemble a mosaic of social worlds in which the transition from one to the other is abrupt. The juxtaposition of divergent personalities and modes of life tends to produce a relativistic perspective and a sense of toleration of differences which may be regarded as prerequisites for rationality and which lead toward the secularization of life.

The close living together and working together of individuals who have no sentimental and emotional ties foster a spirit of competition, aggrandizement, and mutual exploitation. To counteract irresponsibility and potential disorder, formal controls tend to be resorted to. Without rigid adherence to predictable routines a large, compact society would scarcely be able to maintain itself. The clock and the traffic signal are symbolic of the basis of our social order in the urban world. Frequent close physical contact, coupled with great social distance, accentuates the reserve of unattached individuals toward one another and, unless compensated for by other opportunities for response, gives rise to loneliness. The necessary frequent movement of great numbers of individuals in a congested habitat gives occasion to friction and irritation. Nervous tensions which derive from such personal frustrations are accentuated by the rapid tempo and the complicated technology under which life in dense areas must be lived.

HETEROGENEITY

The social interaction among such a variety of personality types in the urban milieu tends to break down the rigidity of caste lines and to complicate the class structure, and thus induces a more ramified and differentiated framework of social stratification than is found in more integrated societies. The heightened mobility of the individual, which brings him within the range of stimulation by a great number of diverse individuals and subjects him to fluctuating status in the differentiated social groups that compose the social structure of the city, tends toward the acceptance of instability and insecurity in the world at large as a norm. This fact helps to account, too, for the sophistication and cosmopolitanism of the urbanite. No single group has the undivided allegiance of the individual. The groups with which he is affiliated do not lend themselves readily to a simple hierarchical arrangement. By virtue of his different interests arising out of different aspects of social life, the individual acquires membership in widely divergent groups, each of which functions only with reference to a single segment of his personality. Nor do these groups easily permit of a concentric arrangement so that the narrower ones fall within the circumference of the more inclusive ones, as is more likely to be the case in the rural community

or in primitive societies. Rather the groups with which the person typically is affiliated are tangential to each other or intersect in highly variable fashion.

Partly as a result of the physical footlooseness of the population and partly as a result of their social mobility, the turnover in group membership generally is rapid. Place of residence, place and character of employment, income and interests fluctuate, and the task of holding organizations together and maintaining and promoting intimate and lasting acquaintanceship between the members is difficult. This applies strikingly to the local areas within the city into which persons become segregated more by virtue of differences in race, language, income, and social status, than through choice or positive attraction to people like themselves. Overwhelmingly the city-dweller is not a home-owner, and since a transitory habitat does not generate binding traditions and sentiments, only rarely is he truly a neighbor. There is little opportunity for the individual to obtain a conception of the city as a whole or to survey his place in the total scheme. Consequently he finds it difficult to determine what is to his own "best interests" and to decide between the issues and leaders presented to him by the agencies of mass suggestion. Individuals who are thus detached from the organized bodies which integrate society comprise the fluid masses that make collective behavior in the urban community so unpredictable and hence so problematical.

Although the city, through the recruitment of variant types to perform its diverse tasks and the accentuation of their uniqueness through competition and the premium upon eccentricity, novelty, efficient performance, and inventiveness, produces a highly differentiated population, it also exercises a leveling influence. Wherever large numbers of differently constituted individuals congregate, the process of depersonalization also enters . . . Individuality under these circumstances must be replaced by categories. When large numbers have to make common use of facilities and institutions, an arrangement must be made to adjust the facilities and institutions to the needs of the average person rather than to those of particular individuals. The services of the public utilities, of the recreational, educational, and cultural institutions, must be adjusted to mass requirements. Similarly, the cultural institutions, such as the schools, the movies, the radio, and the newspapers, by virtue of their mass clientele, must necessarily operate as leveling influences. The political process as it appears in urban life could not be understood without taking account of the mass appeals made through modern propaganda techniques. If the individual would participate at all in the social, political, and economic life of the city, he must subordinate some of his individuality to the demands of the larger community and in that measure immerse himself in mass movements.

THE RELATION BETWEEN A THEORY OF URBANISM AND SOCIOLOGICAL RESEARCH

By means of a body of theory such as that illustratively sketched above, the complicated and many-sided phenomena of urbanism may be analyzed in terms of a limited number of basic categories. The sociological approach to the city thus acquires an essential unity and coherence enabling the empirical investigator not merely to focus more distinctly upon the problems and processes that properly fall in his province but also to treat his subject matter in a more integrated and systematic fashion. A few typical findings of empirical research in the field of urbanism, with special reference to the United States, may be indicated to substantiate the theoretical propositions set forth in the preceding pages, and some of the crucial problems for further study may be outlined.

On the basis of the three variables, number, density of settlement, and degree of heterogeneity, of the urban population, it appears possible to explain the characteristics of urban life and to account for the differences between cities of various sizes and types.

Urbanism as a characteristic mode of life may be approached empirically from three interrelated perspectives: (1) as a physical structure comprising a population base, a technology, and an ecological order; (2) as a system of social organization involving a characteristic social structure, a series of social institutions, and a typical pattern of social relationships; and (3) as a set of attitudes and ideas, and a constellation of personalities engaging in typical forms of collective behavior and subject to characteristic mechanisms of social control.

URBANISM IN ECOLOGICAL PERSPECTIVE

Since in the case of physical structure and ecological processes we are able to operate with fairly objective indices, it becomes possible to arrive at quite precise and generally quantitative results. The dominance of the city over its hinterland becomes explicable through the functional characteristics of the city which derive in large measure from the effect of numbers and density. Many of the technical facilities and the skills and organizations to which urban life gives rise can grow and prosper only in cities where the demand is sufficiently great. The nature and scope of the services rendered by these organizations and institutions and the advantage which they enjoy over the less developed facilities of smaller towns enhances the dominance of the city and the dependence of ever wider regions upon the central metropolis.

The urban population composition shows the operation of selective and differentiating factors. Cities contain a larger proportion of persons in the prime of life than rural areas which contain more old and very young people. In this, as in so many other respects, the larger the city the more this specific characteristic of urbanism is apparent. With the exception of the largest cities, which have attracted the bulk of the foreign-born males, and a few other special types of cities, women predominate numerically over men. The heterogeneity of the urban population is further indicated along racial and ethnic lines. The foreign born and their children constitute nearly two-thirds of all the inhabitants of cities of one million and over. Their proportion in the urban population declines as the size of the city decreases, until in the rural areas they comprise only about one-sixth of the total population. The larger cities similarly have attracted more Negroes and other racial groups than have the smaller communities. Considering that age, sex, race, and ethnic origin are associated with other factors such as occupation and interest, it becomes clear that one major characteristic of the urban-dweller is his dissimilarity from his fellows. Never before have such large masses of people of diverse traits as we find in our cities been thrown together into such close physical contact as in the great cities of America. Cities generally, and American cities in particular, comprise a motley of peoples and cultures, of highly differentiated modes of life between which there often is only the faintest communication, the greatest indifference and the broadest tolerance, occasionally bitter strife, but always the sharpest contrast.

The failure of the urban population to reproduce itself appears to be a biological consequence of a combination of factors in the complex of urban life, and the decline in the birth-rate generally may be regarded as one of the most significant signs of the urbanization of the Western world. While the proportion of deaths in cities is slightly greater than in the country, the outstanding difference between the failure of present-day cities to maintain their population and that of cities of the past is that in former times it was due to the exceedingly high death-rates in cities, whereas today, since cities have become more livable from a health standpoint, it is due to low birth-rates. These biological characteristics of the urban population are significant sociologically, not merely because they reflect the urban mode of existence but also because they condition the growth and future dominance of cities and their basic social organization. Since cities are the consumers rather than the producers of men, the value of human life and the social estimation of the personality will not be unaffected by the balance between births and deaths. The pattern of land use, of land values, rentals, and ownership, the nature and functioning of the physical structures, of housing, of transportation and communication facilities, of public utilities – these and many other phases of the physical mechanism of the city are not isolated phenomena unrelated to the city as a social entity, but are affected by and affect the urban mode of life.

URBANISM AS A FORM OF SOCIAL ORGANIZATION

The distinctive features of the urban mode of life have often been described sociologically as consisting of the substitution of secondary for primary contacts, the weakening of bonds of kinship, and the declining social significance of the family, the disappearance of the neighborhood, and the undermining of the traditional basis of social solidarity. All these phenomena can be substantially verified through objective indices. Thus, for instance, the low and declining urban reproduction rates suggest that the city is not conducive to the traditional type of family life, including the rearing of children and the maintenance of the home as the locus of a whole round of vital activities. The transfer of industrial, educational, and recreational activities to specialized institutions outside the home has deprived the family of some of its most characteristic historical functions. In cities mothers are more likely to be employed, lodgers are more frequently part of the household, marriage tends to be postponed,

and the proportion of single and unattached people is greater. Families are smaller and more frequently without children than in the country. The family as a unit of social life is emancipated from the larger kinship group characteristic of the country, and the individual members pursue their own diverging interests in their vocational, educational, religious, recreational, and political life.

[. . .]

On the whole, the city discourages an economic life in which the individual in time of crisis has a basis of subsistence to fall back upon, and it discourages self-employment. While incomes of city people are on the average higher than those of country people, the cost of living seems to be higher in the larger cities. Home ownership involves greater burdens and is rarer. Rents are higher and absorb a large proportion of the income. Although the urban-dweller has the benefit of many communal services, he spends a large proportion of his income for such items as recreation and advancement and a smaller proportion for food. What the communal services do not furnish the urbanite must purchase, and there is virtually no human need which has remained unexploited by commercialism. Catering to thrills and furnishing means of escape from drudgery, monotony, and routine thus become one of the major functions of urban recreation, which at its best furnishes means for creative self-expression and spontaneous group association, but which more typically in the urban world results in passive spectatorism on the one hand, or sensational record-smashing feats on the other.

Being reduced to a stage of virtual impotence as an individual, the urbanite is bound to exert himself by joining with others of similar interest into organized groups to obtain his ends. This results in the enormous multiplication of voluntary organizations directed toward as great a variety of objectives as there are human needs and interests. While on the one hand the traditional ties of human association are weakened, urban existence involves a much greater degree of interdependence between man and man and a more complicated, fragile, and volatile form of mutual inter-relations over many phases of which the individual as such can exert scarcely any control. Frequently there is only the most tenuous relationship between the economic position or other basic factors that determine the individual's existence in the urban world and the voluntary groups with which he is affiliated. While in a primitive and in a rural society it is generally possible to predict on the basis of a few known factors who will belong to what and who will associate with whom in almost every relationship of life, in the city we can only project the general pattern of group formation and affiliation, and this pattern will display many incongruities and contradictions.

URBAN PERSONALITY AND COLLECTIVE BEHAVIOR

It is largely through the activities of the voluntary groups, be their objectives economic, political, educational, religious, recreational, or cultural, that the urbanite expresses and develops his personality, acquires status, and is able to carry on the round of activities that constitute his life-career. It may easily be inferred, however, that the organizational framework which these highly differentiated functions call into being does not of itself insure the consistency and integrity of the personalities whose interests it enlists. Personal disorganization, mental breakdown, suicide, delinquency, crime, corruption, and disorder might be expected under these circumstances to be more prevalent in the urban than in the rural community. This has been confirmed insofar as comparable indices are available; but the mechanisms underlying these phenomena require further analysis.

Since for most group purposes it is impossible in the city to appeal individually to the large number of discrete and differentiated individuals, and since it is only through the organizations to which men belong that their interests and resources can be enlisted for a collective cause, it may be inferred that social control in the city should typically proceed through formally organized groups. It follows, too, that the masses of men in the city are subject to manipulation by symbols and stereotypes managed by individuals working from afar or operating invisibly behind the scenes through their control of the instruments of communication. Self-government either in the economic, the political, or the cultural realm is under these circumstances reduced to a mere figure of speech or, at best, is subject to the unstable equilibrium of pressure groups. In view of the ineffectiveness of actual kinship ties we create fictional kinship groups. In the face of the disappearance of the territorial unit as a basis of social solidarity we create interest units. Meanwhile the city as a community resolves itself into a series of tenuous segmental relationships superimposed upon a territorial base with a definite center but without a definite periphery and upon a division of labor which far transcends the immediate locality and is world-wide in scope. The larger the number of persons in a state of interaction with one another the lower is the level of communication and the greater is the tendency for communication to proceed on an elementary level, i.e. on the basis of those things which are assumed to be common or to be of interest to all.

It is obviously, therefore, to the emerging trends in the communication system and to the production and distribution technology that has come into existence with modern civilization that we must look for the symptoms which will indicate the probable future development of urbanism as a mode of social life. The direction of the ongoing changes in urbanism will for good or ill transform not only the city but the world. Some of the more basic of these factors and processes and the possibilities of their direction and control invite further detailed study.

It is only insofar as the sociologist has a clear conception of the city as a social entity and a workable theory of urbanism that he can hope to develop a unified body of reliable knowledge, which passes as "urban sociology" is certainly not at the present time. By taking his point of departure from a theory of urbanism such as that sketched in the foregoing pages to be elaborated, tested, and revised in the light of further analysis and empirical research, it is to be hoped that the criteria of relevance and validity of factual data can be determined. The miscellaneous assortment of disconnected information which has hitherto found its way into sociological treatises on the city may thus be sifted and incorporated into a coherent body of knowledge. Incidentally, only by means of some such theory will the sociologists escape the futile practice of voicing in the name of sociological science a variety of often unsupportable judgments concerning such problems as poverty, housing, city-planning, sanitation, municipal administration, policing, marketing, transportation, and other technical issues. While the sociologist cannot solve any of these practical problems – at least not by himself – he may, if he discovers his proper function, have an important contribution to make to their comprehension and solution. The prospects for doing this are brightest through a general, theoretical, rather than through an *ad hoc* approach.

"The Uses of Sidewalks: Safety"

from *The Death and Life of Great American Cities* (1961)

Jane Jacobs

Editors' Introduction

Jane Jacobs (1916–2006) started writing about city life and urban planning as a neighborhood activist, not as a trained professional. Dismissed as the original "little old lady in tennis shoes" and derided as a political amateur more concerned about personal safety issues than state-of-the-art planning techniques, she nonetheless struck a responsive chord with a 1960s public eager to believe the worst about arrogant city planning technocrats and just as eager to rally behind movements for neighborhood control and community resistance to bulldozer redevelopment.

The Death and Life of Great American Cities hit the world of city planning like an earthquake when it appeared in 1961. The book was a frontal attack on the planning establishment, especially on the massive urban renewal projects that were being carried out by powerful redevelopment bureaucrats like Robert Moses in New York. Jacobs derided urban renewal as a process that only served to create instant slums. She questioned universally accepted articles of faith – for example that parks were good and that crowding was bad. Indeed she suggested that parks were often dangerous and that crowded neighborhood sidewalks were the safest places for children to play. Jacobs ridiculed the planning establishment's most revered historical traditions as "the Radiant Garden City Beautiful" – an artful phrase that not only airily dismissed the contributions of Le Corbusier (p. 322), Ebenezer Howard (p. 314), and Daniel Burnham but lumped them together as well! Lewis Mumford's "Home Remedies for Urban Cancer" (1962), reprinted in Michael Larice and Elizabeth McDonald (eds), *The Urban Design Reader* (London and New York: Routledge, 2006), praises Jacobs' humanity and obvious love of city life but savages her attack on city planners like Ebenezer Howard and Patrick Geddes that Mumford had championed for decades.

The selection from *The Death and Life of Great American Cities* reprinted here presents Jane Jacobs at her very best. In "The Uses of Sidewalks: Safety," she outlines her basic notions of what makes a neighborhood a community and what makes a city livable. Safety – particularly for women and children – comes from "eyes on the street," the kind of involved neighborhood surveillance of public space that modern planning practice in the Corbusian tradition had destroyed with its insistence on superblocks and skyscraper developments. A sense of personal belonging and social cohesiveness comes from well-defined neighborhoods and narrow, crowded, multi-use streets. And, finally, basic urban vitality comes from residents' participation in an intricate "street ballet," a diurnal pattern of observable and comprehensible human activity that is possible only in places like Jacobs' own Hudson Street in her beloved Greenwich Village.

It was this last quality, her unabashed love of cities and urban life, that was Jane Jacobs' most obvious and enduring characteristic. *The Death and Life of Great American Cities* was a scathing attack on the planning establishment – and, in many ways, it was a grassroots political call to arms – but it was also a loving invitation to experience the joys of city living that led many young, college-educated people to seek out neighborhoods like Greenwich Village as places to live, struggle, and raise families. In one sense, the book encouraged and justified middle-class gentrification of formerly working-class neighborhoods. In another, it found itself oddly reflected in the fantasy-nostalgia of "Sesame Street." But in all ways it was committedly urban, never sub-urban, at a time when inner-city communities were being increasingly abandoned to the forces of poverty, decay, and neglect.

Contrast Louis Wirth's theory of how population size, density, and heterogeneity in cities create a distinct urban personality (p. 90) with Jacobs' argument that these very same city characteristics may create neighborhood vitality, social cohesion, and the perception and reality of safety. Jacobs' notion of the "street ballet" invites comparison with Lewis Mumford's idea of the "urban drama" (p. 85), William Whyte's emphasis on the importance of public plazas (p. 448), and Robert Putnam's emphasis on "social capital" (p. 120). Jacobs' community activism in resistance to urban renewal places her within a long tradition that includes Paul Davidoff's "Advocacy and Pluralism in Planning" (p. 400) and Sherry Arnstein's "A Ladder of Citizen Participation" (p. 233).

Other important works by Jane Jacobs include *The Economy of Cities* (New York: Random House, 1969) and *Systems of Survival* (New York: Random House, 1992). In the former book Jacobs again turns conventional explanation on its head by arguing that the rise of cities may have proceeded, and even accounted for, rural agricultural development. The latter is a Platonic dialogue on "the moral foundations of commerce and politics." More recently, she published *Dark Age Ahead* (New York: Random House, 2004), a study of contemporary cultural decay and a call for renewal of the key institutions of civilization: family, community, education, science, and the learned professions. Max Allen, *Ideas that Matter: The Worlds of Jane Jacobs* (New York: Ginger Press, 1997) is of interest, and Robert A. Caro's masterful study of Jacobs's archenemy, *The Power Broker: Robert Moses and the Fall of New York* (New York: Random House, 1975), is essential.

▨ ▨ ▨ ▨ ▨ ▨

Streets in cities serve many purposes besides carrying vehicles, and city sidewalks – the pedestrian parts of the streets – serve many purposes besides carrying pedestrians. These uses are bound up with circulation but are not identical with it and in their own right they are at least as basic as circulation to the proper workings of cities.

A city sidewalk by itself is nothing. It is an abstraction. It means something only in conjunction with the buildings and other uses that border it, or border other sidewalks very near it. The same might be said of streets, in the sense that they serve other purposes besides carrying wheeled traffic in their middles. Streets and their sidewalks, the main public places of a city, are its most vital organs. Think of a city and what comes to mind? Its streets. If a city's streets look interesting, the city looks interesting; if they look dull, the city looks dull.

More than that, and here we get down to the first problem, if a city's streets are safe from barbarism and fear, the city is thereby tolerably safe from barbarism and fear. When people say that a city, or a part of it, is dangerous or is a jungle what they mean primarily is that they do not feel safe on the sidewalks. But sidewalks and those who use them are not passive beneficiaries of safety or helpless victims of danger. Sidewalks, their bordering uses, and their users, are active participants in the drama of civilization versus barbarism in cities. To keep the city safe is a fundamental task of a city's streets and its sidewalks.

This task is totally unlike any service that sidewalks and streets in little towns or true suburbs are called upon to do. Great cities are not like towns, only larger. They are not like suburbs, only denser. They differ from towns and suburbs in basic ways, and one of these is that cities are, by definition, full of strangers. To any one person, strangers are far more common in big cities than acquaintances. More common not just in places of public assembly, but more common at a man's own doorstep. Even residents who live near each other are strangers, and must be, because of the sheer number of people in small geographical compass.

The bedrock attribute of a successful city district is that a person must feel personally safe and secure on the street among all these strangers. He must not feel automatically menaced by them. A city district that fails in this respect also does badly in other ways and lays up for itself, and for its city at large, mountain on mountain of trouble.

Today barbarism has taken over many city streets, or people fear it has, which comes to much the same thing in the end. "I live in a lovely, quiet residential area," says a friend of mine who is hunting another place to live. "The only disturbing sound at night is the occasional scream of someone being mugged." It does not take many incidents of violence on a city street, or in a city district, to make people fear the streets . . . And as they fear them, they use them less, which makes the streets still more unsafe.

To be sure, there are people with hobgoblins in their heads, and such people will never feel safe no matter what the objective circumstances are. But this is a different matter from the fear that besets normally prudent, tolerant and cheerful people who show nothing more than common sense in refusing to venture after dark – or in a few places, by day – into streets where they may well be assaulted, unseen or unrescued until too late. The barbarism and the real, not imagined, insecurity that gives rise to such fears cannot be tagged a problem of the slums. The problem is most serious, in fact, in genteel-looking "quiet residential areas" like that my friend was leaving.

It cannot be tagged as a problem of older parts of cities. The problem reaches its most baffling dimensions in some examples of rebuilt parts of cities, including supposedly the best examples of rebuilding, such as middle-income projects. The police precinct captain of a nationally admired project of this kind (admired by planners and lenders) has recently admonished residents not only about hanging around outdoors after dark but has urged them never to answer their doors without knowing the caller. Life here has much in common with life for the three little pigs or the seven little kids of the nursery thrillers. The problem of sidewalk and doorstep insecurity is as serious in cities which have made conscientious efforts at rebuilding as it is in those cities that have lagged. Nor is it illuminating to tag minority groups, or the poor, or the outcast with responsibility for city danger. There are immense variations in the degree of civilization and safety found among such groups and among the city areas where they live. Some of the safest sidewalks in New York City, for example, at any time of day or night, are those along which poor people or minority groups live. And some of the most dangerous are in streets occupied by the same kinds of people. All this can also be said of other cities.

[. . .]

The first thing to understand is that the public peace – the sidewalk and street peace – of cities is not kept primarily by the police, necessary as police are. It is kept primarily by an intricate, almost unconscious, network of voluntary controls and standards among the people themselves, and enforced by the people

themselves. In some city areas – older public housing projects and streets with very high population turnover are often conspicuous examples – the keeping of public sidewalk law and order is left almost entirely to the police and special guards. Such places are jungles. No amount of police can enforce civilization where the normal, casual enforcement of it has broken down.

The second thing to understand is that the problem of insecurity cannot be solved by spreading people out more thinly, trading the characteristics of cities for the characteristics of suburbs. If this could solve danger on the city streets, then Los Angeles should be a safe city because superficially Los Angeles is almost all suburban. It has virtually no districts compact enough to qualify as dense city areas. Yet Los Angeles cannot, any more than any other great city, evade the truth that, being a city, it is composed of strangers not all of whom are nice. Los Angeles' crime figures are flabbergasting. Among the seventeen standard metropolitan areas with populations over a million, Los Angeles stands so pre-eminent in crime that it is in a category by itself. And this is markedly true of crimes associated with personal attack, the crimes that make people fear the streets.

[. . .]

This is something everyone already knows: A well-used city street is apt to be a safe street. A deserted city street is apt to be unsafe. But how does this work, really? And what makes a city street well used or shunned? . . . What about streets that are busy part of the time and then empty abruptly?

A city street equipped to handle strangers, and to make a safety asset, in itself, out of the presence of strangers, as the streets of successful city neighborhoods always do, must have three main qualities:

First, there must be a clear demarcation between what is public space and what is private space. Public and private spaces cannot ooze into each other as they do typically in suburban settings or in projects.

Second, there must be eyes upon the street, eyes belonging to those we might call the natural proprietors of the street. The buildings on a street equipped to handle strangers and to insure the safety of both residents and strangers must be oriented to the street. They cannot turn their backs or blank sides on it and leave it blind.

And third, the sidewalk must have users on it fairly continuously, both to add to the number of effective eyes on the street and to induce the people in buildings along the street to watch the sidewalks in sufficient numbers. Nobody enjoys sitting on a stoop or looking out a window at an empty street. Almost nobody does such a thing. Large numbers of people entertain themselves, off and on, by watching street activity.

In settlements that are smaller and simpler than big cities, controls on acceptable public behavior, if not on crime, seem to operate with greater or lesser success through a web of reputation, gossip, approval, disapproval and sanctions, all of which are powerful if people know each other and word travels. But a city's streets, which must control the behavior not only of the people of the city but also of visitors from suburbs and towns who want to have a big time away from the gossip and sanctions at home, have to operate by more direct, straightforward methods. It is a wonder cities have solved such an inherently difficult problem at all. And yet in many streets they do it magnificently.

It is futile to try to evade the issue of unsafe city streets by attempting to make some other features of a locality, say interior courtyards, or sheltered play spaces, safe instead. By definition again, the streets of a city must do most of the job of handling strangers, for this is where strangers come and go. The streets must not only defend the city against predatory strangers, they must protect the many, many peaceable and well-meaning strangers who use them, insuring their safety too as they pass through. Moreover, no normal person can spend his life in some artificial haven, and this includes children. Everyone must use the streets.

On the surface, we seem to have here some simple aims: to try to secure streets where the public space is unequivocally public, physically unmixed with private or with nothing-at-all space, so that the area needing surveillance has clear and practicable limits; and to see that these public street spaces have eyes on them as continuously as possible.

But it is not so simple to achieve these objects, especially the latter. You can't make people use streets they have no reason to use. You can't make people watch streets they do not want to watch. Safety on the streets by surveillance and mutual policing of one another sounds grim, but in real life it is not grim. The safety of the street works best, most casually, and with least frequent taint of hostility or suspicion precisely where people are using and most enjoying the city streets voluntarily and are least conscious, normally, that they are policing.

The basic requisite for such surveillance is a substantial quantity of stores and other public places

sprinkled along the sidewalks of a district; enterprises and public places that are used by evening and night must be among them especially. Stores, bars and restaurants, as the chief examples, work in several different and complex ways to abet sidewalk safety.

First, they give people – both residents and strangers – concrete reasons for using the sidewalks on which these enterprises face.

Second, they draw people along the sidewalks past places which have no attractions to public use in themselves but which become traveled and peopled as routes to somewhere else; this influence does not carry very far geographically, so enterprises must be frequent in a city district if they are to populate with walkers those other stretches of street that lack public places along the sidewalk. Moreover, there should be many different kinds of enterprises, to give people reasons for crisscrossing paths.

Third, storekeepers and other small businessmen are typically strong proponents of peace and order themselves; they hate broken windows and hold-ups; they hate having customers made nervous about safety. They are great street watchers and sidewalk guardians if present in sufficient numbers.

Fourth, the activity generated by people on errands, or people aiming for food or drink, is itself an attraction to still other people.

This last point, that the sight of people attracts still other people, is something that city planners and city architectural designers seem to find incomprehensible. They operate on the premise that city people seek the sight of emptiness, obvious order and quiet. Nothing could be less true. People's love of watching activity and other people is constantly evident in cities everywhere.

[. . .]

Under the seeming disorder of the old city, wherever the old city is working successfully, is a marvelous order for maintaining the safety of the streets and the freedom of the city. It is a complex order. Its essence is intricacy of sidewalk use, bringing with it a constant succession of eyes. This order is all composed of movement and change, and although it is life, not art, we may fancifully call it the art form of the city and liken it to the dance – not to a simple-minded precision dance with everyone kicking up at the same time, twirling in unison and bowing off en masse, but to an intricate ballet in which the individual dancers and ensembles all have distinctive parts which miraculously reinforce each other and compose an orderly whole. The ballet of the good city sidewalk never repeats itself from place to place, and in any one place is always replete with new improvisations.

The stretch of Hudson Street where I live is each day the scene of an intricate sidewalk ballet. I make my own first entrance into it a little after eight when I put out the garbage can, surely a prosaic occupation, but I enjoy my part, my little clang, as the droves of junior high school students walk by the center of the stage dropping candy wrappers. (How do they eat so much candy so early in the morning?)

While I sweep up the wrappers I watch the other rituals of morning: Mr. Halpert unlocking the laundry's handcart from its mooring to a cellar door, Joe Cornacchia's son-in-law stacking out the empty crates from the delicatessen, the barber bringing out his sidewalk folding chair, Mr. Goldstein arranging the coils of wire which proclaim the hardware store is open, the wife of the tenement's superintendent depositing her chunky 3-year-old with a toy mandolin on the stoop, the vantage point from which he is learning the English his mother cannot speak. Now the primary children, heading for St. Luke's, dribble through to the south; the children for St. Veronica's cross, heading to the west, and the children for P.S. 41, heading toward the east. Two new entrances are being made from the wings: well-dressed and even elegant women and men with briefcases emerge from doorways and side streets . . . Most of these are heading for the bus and subways, but some hover on the curbs, stopping taxis which have miraculously appeared at the right moment, for the taxis are part of a wider morning ritual: having dropped passengers from midtown in the downtown financial district, they are now bringing downtowners up to midtown. Simultaneously, numbers of women in housedresses have emerged and as they crisscross with one another they pause for quick conversations that sound with either laughter or joint indignation; never, it seems, anything between. It is time for me to hurry to work too, and I exchange my ritual farewell with Mr. Lofaro, the short, thick-bodied, white-aproned fruit man who stands outside his doorway a little up the street, his arms folded, his feet planted, looking solid as earth itself. We nod; we each glance quickly up and down the street then look back to each other and smile.

We have done this many a morning for more than ten years, and we both know what it means: All is well.

[. . .]

I know the deep night ballet and its seasons best from waking; long after midnight to tend a baby and, sitting in the dark, seeing the shadows and hearing the sounds of the sidewalk. Mostly it is a sound like infinitely pattering snatches of party conversation and, about three in the morning, singing, very good singing. Sometimes there is sharpness and anger or sad, sad weeping, or a flurry of search for a string of beads broken. One night, a young man came roaring along, bellowing terrible language at two girls whom he had apparently picked up and who were disappointing him. Doors opened; a wary semicircle formed around him, not too close, until the police came. Out came the heads, too, along Hudson Street, offering opinion, "Drunk . . . Crazy . . . A wild kid from the suburbs." (He turned out to be a wild kid from the suburbs. Sometimes, on Hudson Street, we are tempted to believe the suburbs must be a difficult place to bring up children.)

I have made the daily ballet of Hudson Street sound more frenetic than it is, because writing it telescopes it. In real life, it is not that way. In real life, to be sure, something is always going on, the ballet is never at a halt, but the general effect is peaceful and the general tenor even leisurely. People who know well such animated city streets will know how it is. I am afraid people who do not will always have it a little wrong in their heads like the old prints of rhinoceroses made from travelers' descriptions of rhinoceroses. On Hudson Street, the same as in the North End of Boston or in any other animated neighborhoods of great cities, we are not innately more competent at keeping the sidewalks safe than are the people who try to live off the hostile truce of Turf in a blind-eyed city. We are the lucky possessors of a city order that makes it relatively simple to keep the peace because there are plenty of eyes on the street. But there is nothing simple about that order itself, or the bewildering number of components that go into it. Most of those components are specialized in one way or another. They unite in their joint effect upon the sidewalk, which is not specialized in the least. That is its strength.

"The Negro Problems of Philadelphia," "The Question of Earning a Living," and "Color Prejudice"

from *The Philadelphia Negro* (1899)

W.E.B. Du Bois

Editors' Introduction

William Edward Burghardt Du Bois (1868–1963) was one of the preeminent intellectuals of his generation. As a professor, editor, author, novelist, playwright, and politician he made notable contributions in history, sociology, ethnic studies, literature, politics, and other fields. A brilliant student, Du Bois excelled at Fisk University in Nashville, Tennessee, the University of Berlin where he studied with the great sociologist Max Weber, and at Harvard University, where in 1895 he obtained the first Ph.D. degree Harvard had awarded to an African–American.

Du Bois defies easy classification. He was always an independent and critical thinker. During his long and varied career he was a pan-Africanist who advocated solidarity among Black Africans and Blacks elsewhere in the world; a radical pacifist who was indicted, tried, and acquitted as an unregistered foreign agent during the McCarthy era for circulating the Stockholm peace plan; a humanist who wrote novels and plays and published many of the writers of the "Harlem Renaissance"; a civil rights leader who founded the National Association for the Advancement of Colored People (NAACP) publication *Crisis* in 1910 and served as its influential editor until 1934; a writer of children's books that taught Black pride; and a world political figure who urged United Nations protection for Black Americans as a nation within a nation. Du Bois joined the Communist Party at age 93 and became a Ghanaian citizen just before his death in 1963.

At the time that Du Bois completed his education, Philadelphia had the largest and oldest settlement of African–Americans in the Northern United States. The settlement house movement was under way, and some well-intentioned Philadelphians were concerned to understand "the Negro problem" and to help the many poor Blacks in the city. Two wealthy leaders of Philadelphia society suggested a study of Negroes in the Seventh Ward, the city's Black ghetto.

Du Bois was given a one-year appointment as an assistant instructor in the Sociology Department at the University of Pennsylvania. Living with his bride of three months in one room over a cafeteria in the worst part of Philadelphia's worst Black ghetto, with no contact with students and little with faculty, Du Bois wrote *The Philadelphia Negro* from which the following selection is taken. He was only 31 when his monumental study was published.

While Du Bois found many problems in Philadelphia's African–American community in the 1890s, there was work available for able-bodied laborers, no evidence of drug use, substantial homeownership, middle- and upper-income craftspeople, businessmen, and professionals to serve the community and act as role models, and little Black-on-Black violent crime. This is in marked contrast to William Julius Wilson's description of poor Black ghetto areas of Chicago in the 1980s (p. 110). Wilson describes "underclass" ghettos in Chicago consisting almost entirely of renters (many in public housing), with very few employed residents, extremely high concentrations of single-parent families, welfare dependency, drug use, and violent crime.

Ethnographic studies by sociologists and anthropologists often shed light on variations within communities, which are viewed as homogeneous by outsiders. While white Philadelphians who never visited the Seventh Ward tended to view the area as homogeneous and all African–Americans as similar, Du Bois found a physical and social structure within the neighborhood – alleys peopled by criminals, loafers, and prostitutes separate from streets of the working poor and still other streets where an established group of Black middle-class homeowners lived.

In addition to *The Philadelphia Negro* (Philadelphia: University of Pennsylvania Press, 1899) from which the following selection is taken, Du Bois's writings include *Suppression of the Slave Trade to the United States of America* (New York: Longmans, Green & Co., 1896), *Souls of Black Folk* (Chicago: A.C. McClurg & Co., 1903), *The Negro* (New York: Henry Holt & Co., 1915), *Black Reconstruction* (New York: Harcourt, Brace, 1935), and *The World and Africa* (New York: Viking Press, 1947). There are many anthologies of Du Bois's writings and speeches. Perhaps the best is David Levering Lewis (ed.), *W.E.B. Du Bois: A Reader* (New York: Henry Holt, 1995). Also by Lewis, and of great interest, is *W.E.B. Du Bois: The Fight for Equality and the American Century, 1919–1963* (New York: Henry Holt, 1995).

For more by and about W.E.B. Du Bois see *The Autobiography of W.E.B. Du Bois* (New York: International Publishers, 1968), Francis L. Broderick, *W.E.B. Du Bois* (Palo Alto: Stanford University Press, 1959), Walter Wilson (ed.), *The Selected Writings of W.E.B. Du Bois* (New York: New American Library, 1970), Henry Lee Moon, *The Emerging Thought of W.E.B Du Bois* (New York: Simon & Schuster, 1972), Marable Manning, *W.E.B. DuBois, Black Radical Democrat* (Boston: Twayne Publishers, 1986; new edition published by Paradigm, 2005), and Patricia and Fredrick McKissack, *W.E.B. Du Bois* (New York: Franklin Watts, 1990). For readings on the current state of Black America, see the bibliographical references in the Editors' Introduction to William Julius Wilson's "From Institutional to Jobless Ghettos."

4. THE NEGRO PROBLEMS OF PHILADELPHIA

In Philadelphia, as elsewhere in the United States, the existence of certain peculiar social problems affecting the Negro people are plainly manifest. Here is a large group of people – perhaps forty-five thousand, a city within a city – who do not form an integral part of the larger social group. This in itself is not altogether unusual; there are other unassimilated groups: Jews, Italians, even Americans; and yet in the case of the Negroes the segregation is more conspicuous, more patent to the eye, and so intertwined with a long historic evolution, with peculiarly pressing social problems of poverty, ignorance, crime and labor, that the Negro problem far surpasses in scientific interest and social gravity most of the other race or class questions.

The student of these questions must first ask, What is the real condition of this group of human beings? Of whom is it composed, what sub-groups and classes exist, what sort of individuals are being considered? Further, the student must clearly recognize that a complete study must not confine itself to the group, but must specially notice the environment; the physical environment of city, sections and houses, the far mightier social environment – the surrounding world of custom, wish, whim and thought which envelops this group and powerfully influences its social development.

[. . .]

The Seventh Ward starts from the historic center of Negro settlement in the city, South Seventh street and Lombard, and includes the long narrow strip, beginning at South Seventh and extending west, with South and Spruce streets as boundaries, as far as the Schuylkill River. The colored population of this ward numbered 3,621 in 1860, 4,616 in 1870, and 8,861 in 1890. It is a thickly populated district of varying character; north of it is the residence and business section of the city; south of it a middle class and workingmen's residence section; at the east end it joins Negro, Italian and Jewish slums; at the west end, the wharves of the river and an industrial section separating it from the grounds of the University of Pennsylvania and the residence section of West Philadelphia.

Starting at Seventh street and walking along Lombard, let us glance at the general character of the ward. Pausing a moment at the corner of Seventh and Lombard, we can at a glance view the worst Negro slums of the city. The houses are mostly brick, some wood, not very old, and in general uncared for rather than dilapidated. The blocks between Eighth, Pine, Sixth, and South have for many decades been the center of Negro population. Here the riots of the thirties took place, and here once was a depth of poverty and degradation almost unbelievable. Even today there are many evidences of degradation . . . The alleys near, as Ratcliffe street, Middle alley, Brown's court, Barclay street, etc., are haunts of noted criminals, male and female, of gamblers and prostitutes, and at the same time of many poverty-stricken people, decent but not energetic. There is an abundance of political clubs, and nearly all the houses are practically lodging houses, with a miscellaneous and shifting population. The corners, night and day, are filled with Negro loafers – able-bodied young men and women, all cheerful, some with good natured, open faces, some with traces of crime and excess, a few pinched with poverty. They are mostly gamblers, thieves and prostitutes, and few have fixed and steady occupation of any kind. Some are stevedores, porters, laborers and laundresses. On its face this slum is noisy and dissipated, but not brutal, although now and then highway robberies and murderous assaults in other parts of the city are traced to its denizens. Nevertheless a stranger can usually walk about here day and night with little fear of being molested if he be not too inquisitive.

Passing up Lombard, beyond Eighth, the atmosphere suddenly changes, because these next two blocks have few alleys and the residences are good-sized and pleasant. Here some of the best Negro families of the ward live. Some are wealthy in a small way, nearly all are Philadelphia born, and they represent an early wave of emigration from the old slum section . . .

[. . .]

21. THE QUESTION OF EARNING A LIVING

For a group of freedmen the question of economic survival is the most pressing of all questions; the problem as to how, under the circumstances of modern life, any group of people can earn a decent living, so as to maintain their standard of life, is not always easy to answer. But when the question is complicated by the fact that the group has a low degree of efficiency on account of previous training; is in competition with well-trained, eager and often ruthless competitors; is more or less handicapped by a somewhat wide-reaching discrimination; and finally is seeking not merely to maintain a standard of living but steadily to raise it to a higher plane – such a situation presents baffling problems to the sociologist and philanthropist.

Of the men 21 years of age and over, there were in gainful occupations, the following:

In the learned professions	61	2.0 per cent
Conducting business on their own account	207	6.5
In the skilled trades	236	7.0
Clerks, etc.	159	5.0
Laborers, better class ... 602		
Laborers, common class ... 852	1454	45.0
Servants	1079	34.0
Miscellaneous	11	0.5
	3207	100 per cent
Total male population 21 and over	3850	

Taking the occupations of women 21 years of age and over, we have:

Domestic servants	1262	37.0 per cent
Housewives and day laborers	937	27.0
Housewives	568	17.0
Day laborers, maids, etc.	297	9.0
In skilled trades	221	6.0
Conducting businesses	63	2.0
Clerks, etc.	40	1.0
Learned professions	37	1.0
	3425	100 per cent
Total female population 21 and over	3740	

47. COLOR PREJUDICE

Incidentally throughout this study the prejudice against the Negro has been again and again mentioned. It is time now to reduce this somewhat indefinite term to something tangible. Everybody speaks of the matter, everybody knows that it exists, but in just what form it shows itself or how influential it is few agree. In the Negro's mind, color prejudice in Philadelphia is that widespread feeling of dislike for his blood, which keeps him and his children out of decent employment, from certain public conveniences and amusements, from hiring houses in many sections, and in general, from being recognized as a man. Negroes regard this prejudice as the chief cause of their present unfortunate condition. On the other hand most white people are quite unconscious of any such powerful and vindictive feeling; they regard color prejudice as the easily explicable feeling that intimate social intercourse with a lower race is not only undesirable but impractical if our present standards of culture are to be maintained, and although they are aware that some people feel the aversion more intensely than others, they cannot see how such a feeling has much influence on the real situation or alters the social condition of the mass of Negroes.

As a matter of fact, color prejudice in this city is something between these two extreme views: it is not today responsible for all, or perhaps the greater part of the Negro problems, or of the disabilities under which the race labors; on the other hand it is a far more powerful social force than most Philadelphians realize. The practical results of the attitude of most of the inhabitants of Philadelphia towards persons of Negro descent are as follows:

1. As to getting work:

No matter how well trained a Negro may be, or how fitted for work of any kind, he cannot in the ordinary course of competition hope to be much more than a menial servant.

He cannot get clerical or supervisory work to do save in exceptional cases.

He cannot teach save in a few of the remaining Negro schools.

He cannot become a mechanic except for small transient jobs, and cannot join a trades union.

A Negro woman has but three careers open to her in this city: domestic service, sewing, or married life.

2. As to keeping work:

The Negro suffers in competition more severely than white men.

Change in fashion is causing him to be replaced by whites in the better-paid positions of domestic service.

Whim and accident will cause him to lose a hard-earned place more quickly than the same things would affect a white man.

Being few in number compared with the whites the crime or carelessness of a few of his race is easily imputed to all, and the reputation of the good, industrious, and reliable suffer thereby.

Because Negro workmen may not often work side by side with white workmen, the individual black workman is rated not only by his own efficiency, but by the efficiency of a whole group of black fellow workmen which may often be low.

Because of these difficulties which virtually increase competition in his case, he is forced to take lower wages for the same work than white workmen.

3. As to entering new lines of work:

Men are used to seeing Negroes in inferior positions; when, therefore, by any chance a Negro gets in a better position, most men immediately conclude that he is not fitted for it, even before he has a chance to show his fitness.

If, therefore, he set up a store, men will not patronize him.

If he is put into public position men will complain.

If he gain a position in the commercial world, men will quietly secure his dismissal or see that a white man succeeds him.

4. As to his expenditure:

The comparative smallness of the patronage of the Negro, and the dislike of other customers, makes it usual to increase the charges or difficulties in certain directions in which a Negro must spend money.

He must pay more house-rent for worse houses than most white people pay.

He is sometimes liable to insult or reluctant service in some restaurants, hotels and stores, at public resorts, theaters and places of recreation; and at nearly all barber shops.

5. As to his children:

The Negro finds it extremely difficult to rear children in such an atmosphere and not have them either cringing or impudent: if he impresses upon them patience with their lot, they may grow up satisfied with their condition; if he inspires them with ambition to rise, they may grow to despise their own people, hate the whites, and become embittered with the world.

His children are discriminated against, often in public schools.

They are advised when seeking employment to become waiters and maids.

They are liable to species of insult and temptation peculiarly trying to children.

6. As to social intercourse:

In all walks of life the Negro is liable to meet some objection to his presence or some discourteous treatment; and the ties of friendship or memory seldom are strong enough to hold across the color line.

If an invitation is issued to the public for any occasion, the Negro can never know whether he would be welcomed or not; if he goes he is liable to have his feelings hurt and get into unpleasant altercation; if he stays away, he is blamed for indifference.

If he meet a lifelong white friend on the street, he is in a dilemma; if he does not greet the friend he is put down as boorish and impolite; if he does greet the friend he is liable to be flatly snubbed.

If by chance he is introduced to a white woman or man, he expects to be ignored on the next meeting, and usually is.

White friends may call on him, but he is scarcely expected to call on them, save for strictly business matters.

If he gain the affections of a white woman and marry her he may invariably expect that slurs will be thrown on her reputation and on his, and that both his and her race will shun their company. When he dies he cannot be buried beside white corpses.

7. The result:

Any one of these things happening now and then would not be remarkable or call for especial comment; but when one group of people suffer all these little differences of treatment and discriminations and insults continually, the result is either discouragement, or bitterness, or over-sensitiveness, or recklessness. And a people feeling thus cannot do their best.

Presumably the first impulse of the average Philadelphian would be emphatically to deny any such marked and blighting discrimination as the above against a group of citizens in this metropolis. Every one knows that in the past color prejudice in the city was deep and passionate; living men can remember when a Negro could not sit in a street car or walk many streets in peace. These times have passed, however, and many imagine discrimination against the Negro has passed with them. Careful inquiry will convince any such one of his error. To be sure a colored man to-day can walk the streets of Philadelphia without personal insult; he can go to theaters, parks and some places of amusement without meeting more than stares and discourtesy; he can be accommodated at most hotels and restaurants, although his treatment in some would not be pleasant. All this is a vast advance and augurs much for the future. And yet all that has been said of the remaining discrimination is but too true.

During the investigation of 1896 there was collected a number of actual cases, which may illustrate the discriminations spoken of. So far as possible these have been sifted and only those which seem undoubtedly true have been selected:

I. As to getting work

It is hardly necessary to dwell upon the situation of the Negro in regard to work in the higher walks of life: the white boy may start in the lawyer's office and work himself into a lucrative practice; he may serve a physician as office boy or enter a hospital in a minor position, and have his talent alone between him and affluence and fame; if he is bright in school, he may make his mark in a university, become a tutor with some time and much inspiration for study, and eventually fill a professor's chair. All these careers are at the very outset closed to the Negro on account of his color; what lawyer would give even a minor case to a Negro assistant? What university would appoint a promising young Negro as tutor? Thus the young white man starts in life knowing that within some limits and barring accidents, talent and application will tell. The young Negro starts knowing that on all sides his advance is made difficult if not wholly shut off by his color. Let us come, however, to ordinary occupations which concern more nearly the mass of Negroes. Philadelphia is a great industrial and business center with thousands of foremen, managers and clerks – the lieutenants of industry who direct its progress. They are paid for thinking and for skill to direct, and naturally such positions are coveted because they are well paid, well thought-of and carry some authority. To such positions Negro boys and girls may not aspire no matter what their qualifications. Even as teachers and ordinary clerks and stenographers they find almost no openings. Let us note some actual instances:

A young woman who graduated with credit from the Girls Normal School in 1892 has taught in the kindergarten, acted as substitute, and waited in vain for a permanent position. Once she was allowed to substitute in a school with white teachers; the principal commended her work, but when the permanent appointment was made a white woman got it.

A girl who graduated from a Pennsylvania high school and from a business college sought work in the city as a stenographer and typewriter. A prominent lawyer undertook to find her a position; he went to friends and said, "Here is a girl that does excellent work and is of good character; can you not give her work?" Several immediately answered yes. "But," said the lawyer, "I will be perfectly frank with you and tell you she is colored"; and not in the whole city could he find a man willing to employ her. It happened, however, that the girl was so light in complexion that few not knowing would have suspected her descent. The lawyer therefore gave her temporary work in his own office until she found a position outside the city. "But," said he, "to this day I have not dared to tell my clerks that they worked beside a Negress." Another woman graduated from the high school and the Palmer College of Shorthand, but all over the city has met with nothing but refusal of work.

Several graduates in pharmacy have sought three years' required apprenticeship in the city and in only one case did one succeed, although they offered to work for nothing. One young pharmacist came from Massachusetts and for weeks sought in vain for work here at any price; "I wouldn't have a darky to clean out my store, much less to stand behind the counter," answered one druggist.

A colored man answered an advertisement for a clerk in the suburbs. "What do you suppose we'd want of a nigger?" was the plain answer. A graduate of the University of Pennsylvania in mechanical engineering, well recommended, obtained work in the city, through an advertisement, on account of his excellent record. He worked a few hours and then was discharged because he was found to be colored. He is now a waiter at the University Club, where his white fellow graduates dine. Another young man attended Spring Garden Institute and studied drawing for lithography. He had good references from the institute and elsewhere, but application at the five largest establishments in the city could secure him no work. A telegraph operator has hunted in vain for an opening, and two graduates of the Central High School have sunk to menial labor. "What's the use of an education?" asked one. Mr. A— has elsewhere been employed as a traveling salesman. He applied for a position here by letter and was told he could have one. When they saw him they had no work for him.

Such cases could be multiplied indefinitely. But that is not necessary; one has but to note that, notwithstanding the acknowledged ability of many colored men, the Negro is conspicuously absent from all places of honor, trust, emolument, as well as from those of respectable grade in commerce and industry.

Even in the world of skilled labor the Negro is largely excluded. Many would explain the absence of Negroes from higher vocations by saying that while a few may now and then be found competent, the great mass are not fitted for that sort of work and are destined for some time to form a laboring class. In the matter of the trades, however, there can be raised no serious question of ability; for years the Negroes filled satisfactorily the trades of the city, and to-day in many parts of the South they are still prominent. And yet in Philadelphia a determined prejudice, aided by public opinion, has succeeded nearly in driving them from the field:

A——, who works at a bookbinding establishment on Front street, has learned to bind books and often does so for his friends. He is not allowed to work at the trade in the shop, however, but must remain a porter at a porter's wages.

B——is a brushmaker; he has applied at several establishments, but they would not even examine his testimonials. They simply said: "We do not employ colored people."

C——is a shoemaker; he tried to get work in some of the large department stores. They "had no place" for him.

D——was a bricklayer, but experienced so much trouble in getting work that he is now a messenger.

E——is a painter, but has found it impossible to get work because he is colored.

F——is a telegraph line man, who formerly worked in Richmond, Va. When he applied here he was told that Negroes were not employed.

G——is an iron puddler, who belonged to a Pittsburgh union. Here he was not recognized as a union man and could not get work except as a stevedore.

H——was a cooper, but could get no work trials, and is now a common laborer.

I——is a candy-maker, but has never been able to find employment in the city; he was always told the white help would not work with him.

J——is a carpenter; he can only secure odd jobs or work where only Negroes are employed.

K——was an upholsterer, but could get no work save in the few colored shops which had workmen; he is now a waiter on a dining car.

L——was a first-class baker; he applied for work some time ago near Green street and was told shortly, "We don't work no niggers here."

[. . .]

"From Institutional to Jobless Ghettos"

from *When Work Disappears: The World of the New Urban Poor* (1996)

William Julius Wilson

Editors' Introduction

Harvard sociologist William Julius Wilson spent much of his career at the University of Chicago. Like the earlier Chicago School sociologists Ernest W. Burgess (p. 150) and Louis Wirth (p. 90) writing in the 1920s and 1930s, and St. Clair Drake and Horace Cayton writing in the 1940s, Wilson uses careful empirical studies of Chicago to generate important urban theory. An African–American, Wilson has been particularly concerned about the situation of poor Blacks in America's decaying central city neighborhoods.

Wilson is critical of timid liberals who avoid confronting tough questions about race and poverty because they are afraid anything negative they say about Blacks will appear racist. He argues that there is an *urban underclass* (a term many liberals will not use) and that residents of poor Black ghettos today are socially isolated and caught in a tangle of pathology characterized by unemployment, crime (including violent Black-on-Black crime), teenage pregnancy, out-of-wedlock births, welfare dependency, and drug use. Wilson feels that the situation of Blacks, particularly poor urban Blacks, requires objective research and honest reportage. He is unwilling to let conservatives dominate theoretical discourse about the causes of and cures for Black poverty.

Wilson's research has convinced him that conditions for poor Black ghetto residents are far worse in many ways than a century ago when W.E.B. Du Bois studied *The Philadelphia Negro* (p. 103) or in 1945 when Drake and Cayton published *Black Metropolis*, their study of the African–American community of Chicago at that time. Both Du Bois's and Drake and Cayton's studies found more Blacks employed and role models of upward mobility present, a higher proportion of nuclear families, less Black-on-Black violence, and much less drug use than exist in the poorest Black ghettos today.

Wilson argues that the changed structure of the US economy is more responsible for the plight of poor Blacks today than is racism. A generation ago, Wilson argues, an able-bodied unskilled Black man could readily find work sufficient to support himself and a family – albeit often physically hard, racially segregated, and dirty work. But in the last generation unskilled manual urban jobs have largely disappeared. Without work, Black males cannot support a family. Hence, Wilson argues, Black ghettos today contain few "marriageable" Black males capable of supporting a family. A high incidence of out-of-wedlock births, family dissolution, and welfare-dependent female-headed households follow directly from that fact. With little sense of self-worth, Wilson argues, unemployed Blacks naturally turn to drug dependence and crime.

One of Wilson's most controversial contentions is that race and racism are declining in importance as causes of Black distress. Paradoxically, he argues, less racial discrimination has made matters in Black ghettos worse. According to Wilson, as upwardly mobile Blacks move out of Black ghetto areas, community leadership and positive role models disappear and pathology is concentrated.

Wilson is skeptical that *race-specific* policies like affirmative action will address problems as pervasive and profound as he describes. Rather he favors universal social policies aimed at improving the lot of all poor people regardless of race: new education, training, and particularly full employment policy.

This selection is from *When Work Disappears: The World of the New Urban Poor* (New York: Knopf, 1996). Wilson's two most influential prior books are *The Declining Significance of Race: Blacks and Changing American Institutions*, second edition (Chicago: University of Chicago Press, 1980) and *The Truly Disadvantaged* (Chicago: University of Chicago Press, 1987). Also of interest is *The Bridge Over the Racial Divide: Rising Inequality and Coalition Politics* (Los Angeles: University of California Press, 2001).

The literature on urban social inequality, welfarism, and the "underclass debate" is vast. Classic studies include Richard A. Cloward and Frances Fox Piven, *Regulating the Poor: The Functions of Public Welfare*, second edition (New York: Vintage, 1993), Michael B. Katz, *In the Shadow of the Poorhouse: A Social History of Welfare in America* (New York: Basic Books, 1987) and *The Undeserving Poor: From the War on Poverty to the War on Welfare* (New York: Random House, 1990), and Christopher Jencks, *Rethinking Social Policy: Race, Poverty, and the Underclass* (New York: Harpers, 1993). More recent studies of interest include Benjamin I. Page and James Roy Simmons, *What Government Can Do: Dealing with Poverty and Inequality* (Chicago: University of Chicago Press, 2000), R. Kent Weaver and Michael H. Armacost, *Ending Welfare As We Know It* (Washington, DC: Brookings, 2000), and Theda Skocpol and Richard C. Leone, *The Missing Middle: Working Families and the Future of American Social Policy* (New York: W.W. Norton, 2001) Two recent studies examine the continued poor economic performance of inner-city Blacks through the 1990s: Ronald Mincy (ed.), *Black Males Left Behind* (Washington, DC: Urban Institute Press, 2006) and Peter Edelman and Paul Offner, *Reconnecting Disadvantaged Young Men* (Washington, DC: Urban Institute Press, 2006). Also of interest, taking radically different positions, are Douglas Massey and Nancy Denton, *American Apartheid: Segregation and the Making of the Underclass* (Cambridge: Harvard University Press, 1998) and John McWhorter, *Losing the Race: Self-Sabotage in Black America* (New York: Free Press, 2000). And two books by sociologist Elijah Anderson break new ground: *Streetwise: Race, Class and Change in an Urban Community* (Chicago: University of Chicago Press, 1992) and *Code of the Street: Decency, Violence, and the Moral Life of the Inner City* (New York: Norton, 2000).

For a recent overview of the literature on the underclass debate and writings by Latino scholars exploring the relevance and limitations of underclass theory for America's varied Latino communities, see Joan Moore and Raquel Pinderhughes, *In the Barrios: Latinos and the Underclass Debate* (New York: Russell Sage Foundation, 1993). Conservative explanations of inequality and suggested public policy regarding poverty and race include Edward Banfield, *The Unheavenly City Revisited* (Boston: Little Brown, 1974) and Charles Murray, *Losing Ground* (New York: Basic Books, 1984).

■ ■ ■ ■ ■ ■

An elderly woman who has lived in one inner-city neighborhood on the South Side of Chicago for more than forty years reflected:

I've been here since March 21, 1953. When I moved in, the neighborhood was intact. It was intact with homes, beautiful homes, mini mansions, with stores, laundromats, with cleaners, with Chinese [cleaners]. We had drugstores. We had hotels. We had doctors over on Thirty-ninth Street. We had doctors' offices in the neighborhood. We had the middle class and upper middle class. It has gone from affluent to where it is today. And I would like to see it come back, that we can have some of the things we had. Since I came in young, and I'm a senior citizen now,

I would like to see some of the things come back so I can enjoy them like we did when we first came in.

[. . .]

A 91-year-old woman spoke of safety concerns: "It's not safe anymore because the streets aren't. When all the black businesses and shows closed down, the economy went to the dogs. The stores, the businesses, the shows, everywhere was lighted, the stores and businesses have disappeared."

The negative social forces triggered a decision by a concerned mother to send her son away.

I have a 13-year-old. I sent him away when he was nine because the gangs was at him so tough, because he wouldn't join – he's a basketball player. That's all he ever cared about. They took his gym shoes off his feet. They took his clothes. Made him walk home from school. Jumped on him every day. Took his jacket off his back in subzero weather. You know, and we only live two blocks from the school . . . A boy pulled a gun to his head and told him, "If you don't join, next week you won't be here." I had to send him out of town. His father stayed out of town. He came here last week for a week. He said, "Mom, I want to come home so bad," I said no!

The social deterioration of ghetto neighborhoods is the central concern expressed in the testimony of these residents. As a representative from the media put it, the ghetto has gone "from bad to worse." Few observers of the urban scene in the late 1960s anticipated the extensive breakdown of social institutions and the sharp rise in rates of social dislocation that have since swept the ghettos and spread to other neighborhoods that were once stable. For example, in the neighborhood of Woodlawn, located on the South Side of Chicago, there were over eight hundred commercial and industrial establishments in 1950. Today, it is estimated that only about a hundred are left, many of them represented by "tiny catering places, barber shops, and thrift stores with no more than one or two employees." As Löic Wacquant, a member of the Urban Poverty and Family Life Study research team, put it:

The once-lively streets – residents remember a time, not so long ago, when crowds were so dense at rush hour that one had to elbow one's way to the train station – now have the appearance of an empty, bombed-out war zone. The commercial strip has been reduced to a long tunnel of charred stores, vacant lots littered with broken glass and garbage, and dilapidated buildings left to rot in the shadow of the elevated train line. At the corner of Sixty-third Street and Cottage Grove Avenue, the handful of remaining establishments that struggle to survive are huddled behind wrought iron bars . . . The only enterprises that seem to be thriving are liquor stores and currency exchanges, these "banks of the poor" where one can cash checks, pay bills and buy money orders for a fee.

[. . .]

In 1950, almost two-thirds of Woodlawn's population was white; by 1960 the white population had declined to just 10 percent. Despite the sudden white exodus, the number of residents in the neighborhood increased slightly during this period. After 1960, however, a sizable exodus of black residents followed, including a significant number of working- and middle-class families. The population of the neighborhood declined from over 80,000 in 1960 to 53,814 in 1970; it further slipped to 36,323 in 1980 and finally to 24,473 in 1990. The loss of residents was accompanied by a substantial reduction in the economic, social, and political resources that make a community vibrant. Woodlawn is only one of a growing number of poor black neighborhoods in Chicago plagued by depopulation and social and economic deterioration.

When the black respondents in our large . . . survey were asked to rate their neighborhood as a place to live, only a third said that their area was a good or very good place to live and only 18 percent of those in the ghetto poverty census tracts felt that their neighborhood was a desirable place to live. (The Bureau of the Census defines a census tract as "a relatively homogeneous area with respect to population characteristics, economic status, and living conditions with an average population of 4,000." Poverty tracts are those in which at least 20 percent of the residents are poor, and ghetto poverty tracts are those in which at least 40 percent are poor.)

[. . .]

Many of the respondents described the negative effects of their neighborhood on their own personal outlook. An unmarried, employed clerical worker from a ghetto poverty census tract on the West Side stated:

There is a more positive outlook if you come from an upwardly mobile neighborhood than you would here. In this type of neighborhood, all you hear is negative [things] and that can kind of bring you down when you're trying to make it. So your neighborhood definitely has something to do with it.

This view was shared by a 17-year-old college student and part-time worker from an impoverished West Side neighborhood.

I'd say about 40 percent in my neighborhood . . . I'd say 40 percent are alcoholics . . . And . . . only 5 percent of the alcoholics have homes. Then you got the other 35 percent who are in the street . . . They probably live somewhere, but they in street, on the corner every day, same old thing, because they don't have no chance in life. They live based on today. [They say] "Oh, we gonna get high today." "Oh, whoopee!" "What you gonna do tomorrow, man?" "I don't know, man, I don't know." You can ask any of 'em: "What you gonna do tomorrow?" "I don't know, man. I know when it gets here." And I can really understand, you know, being in that state. If you around totally negative people, people who are not doing anything, that's the way you gonna be regardless.

The state of the inner-city public schools was another major concern expressed by our respondents. The complaints ranged from overcrowded conditions to unqualified and uncaring teachers. Sharply voicing her views on these subjects, a 25-year-old married mother of two children from a South Side census tract that just recently became poor stated: "My daughter ain't going to school here, she was going to a nursery school where I paid and of course they took the time and spent it with her, 'cause they was getting the money. But the public schools, no! They are overcrowded and the teachers don't care."

[. . .]

The respondents were also asked whether their neighborhoods had changed as a place to live over the years. Seventy-one percent of the African–American respondents felt that their neighborhoods had either stayed the same or had gotten worse.

An unemployed black man from a West Side housing project felt that the only thing that had changed in his neighborhood was that it was "going down instead of going back up." He further stated, "It ain't like it used to be. They laid off a lot of people. There used to be a time when you got a broken window, you call up housing and they send someone over to fix it, but it ain't like that no more."

Respondents frequently made statements about the increase in drug trafficking and drug consumption when discussing how their neighborhood had changed. "Well, OK, I realize there was drugs when I was growing up but they weren't as open as they are now," stated a divorced telephone dispatcher and mother of five children from a neighborhood that recently changed from a nonpoverty to a poverty area. "It's nothing to see a 10-year-old kid strung out or a 10-year-old kid selling drugs. I mean, when they were

doing it back then they were sneaking around doing it. It's like an open thing now."

[. . .]

The feelings of many of the respondents in our study were summed up by a 33-year-old married mother of three from a very poor West Side neighborhood:

If you live in an area in your neighborhood where you have people that don't work, don't have no means of support, you know, don't have no jobs, who're gonna break into your house to steal what you have, to sell to get them some money, then you can't live in a neighborhood and try to concentrate on tryin' to get ahead, then you get to work and you have to worry if somebody's breakin' into your house or not. So, you know, it's best to try to move in a decent area, to live in a community with people that works.

In 1959, less than one-third of the poverty population in the United States lived in metropolitan central cities. By 1991, the central cities included close to half of the nation's poor. Many of the most rapid increases in concentrated poverty have occurred in African–American neighborhoods. For example, in the ten community areas that represent the historic core of Chicago's Black Belt (see Figure 1), eight had rates of poverty in 1990 that exceeded 45 percent, including three with rates higher than 50 percent and three that surpassed 60 percent. Twenty-five years earlier, in 1970, only two of these neighborhoods had poverty rates above 40 percent.

In recent years, social scientists have paid particular attention to the increases in urban neighborhood poverty. "Defining an urban neighborhood for analytical purposes is no easy task." The community areas of Chicago referred to in Figure 1 include a number of adjacent census tracts. The seventy-seven community areas within the city of Chicago represent statistical units derived by urban sociologists at the University of Chicago for the 1930 census in their effort to analyze varying conditions within the city. These delineations were originally drawn up on the basis of settlement and history of the area, local identification and trade patterns, local institutions, and natural and artificial barriers. There have been major shifts in population and land use since then. But these units remain useful in tracing changes over time, and they continue to capture much of the contemporary reality of Chicago neighborhoods.

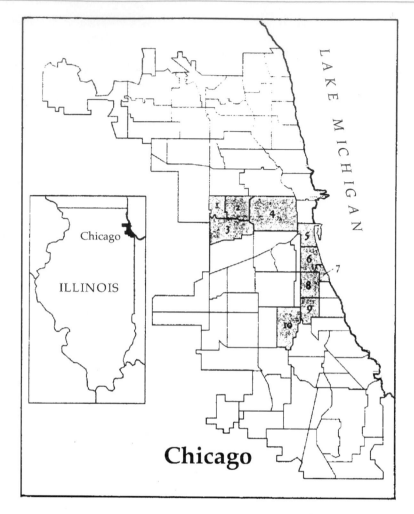

1. West Garfield Park
2. East Garfield Park
3. North Lawndale
4. Near West Side
5. Near South Side
6. Douglas
7. Oakland
8. Grand Boulevard
9. Washington Park
10. Englewood

Figure 1 Community areas in Chicago's Black Belt

Other cities, however, do not have such convenient classifications of neighborhoods, which means that comparison across cities cannot be drawn using community areas. The measurable unit considered most appropriate to represent urban neighborhoods is the census tract. In attempts to examine this problem of ghetto poverty across the nation empirically, social scientists have tended to define ghetto neighborhoods as those located in the *ghetto poverty* census tracts. As indicated earlier, ghetto poverty census tracts are those in which at least 40 percent of the residents are poor. For example, Paul Jargowsky and Mary Jo Bane state: "Visits to various cities confirmed that the 40 percent criterion came very close to identifying areas that looked like ghettos in terms of their housing conditions. Moreover, the areas selected by the 40 percent criterion corresponded closely with the neighborhoods that city officials and local Census Bureau officials considered ghettos." The ghetto poor in Jargowsky and Bane's study are therefore designated as those among the poor who live in these ghetto poverty areas. Three-quarters of all the ghetto poor in metropolitan areas reside in one hundred of the nation's largest central cities; however, it is important to remember that the ghetto areas in these central cities also include a good many families and individuals who are not poor.

In the nation's one hundred largest central cities, nearly one in seven census tracts is at least 40 percent poor. The number of such tracts has more than doubled since 1970 – indeed, it is alarming that 579 tracts fell to ghetto poverty level in these cities between 1970 and 1980, and 624 additional tracts joined these ranks in the following decade.

Paul Jargowsky's research reveals that a vast majority of people (almost seven out of eight) living in metropolitan-area ghettos in 1990 were minority group members. The number of African–Americans in these ghettos grew by more than one-third from 1980 to 1990, reaching nearly 6 million. Most of this growth involved poor people. The proportion of metropolitan blacks who live in ghetto areas climbed from more than a third (37 percent) to almost half (45 percent). Indeed, the metropolitan black poor are becoming increasingly isolated. The poverty rate among metropolitan blacks who reside in ghettos increased while the rate among those who live in nonghettos decreased.

The increase in the *number* of ghetto blacks is related to the *geographical spread* of the ghetto. Jargowsky and Bane found that in the cities they studied (Philadelphia, Cleveland, Milwaukee, and Memphis) areas that had become ghettos by 1980 had been mixed-income tracts in 1970 – but tracts that were contiguous to areas identified as ghettos. The exodus of the nonpoor from mixed-income areas was a major factor in the spread of ghettos in these cities in the 1970s. Since 1980, ghetto census tracts have increased in a substantial majority of the metropolitan areas in the country, including those with fewer people living in them. Nine new ghetto census tracts were added in Philadelphia, even though it experienced one of the largest declines in the proportion of people living in ghetto tracts. In a number of other cities, including Baltimore, Boston, and Washington, D.C., a smaller percentage of poor blacks live in a larger number of ghetto census tracts. Chicago had a 61.5 percent increase in the number of ghetto census tracts from 1980 to 1990, even though the number of poor residing in those areas increased only slightly.

Jargowsky reflects on the significance of the substantial spread of ghetto areas:

The geographic size of a city's ghetto has a large effect on the perception of the magnitude of the problem associated with ghetto poverty. How big an area of the city do you consider off limits? How far out of your way will you drive not to go through a dangerous area? Indeed, the lower density exacerbated the problem. More abandoned buildings mean more places for crack dens and criminal enterprises. Police trying to protect a given number of citizens have to be stretched over a wider number of square miles, making it less likely that criminals will be caught. Lower density also makes it harder for a sense of community to develop, or for people to feel that they can find safety in numbers. From the point of view of local political officials, the increase in the size of the ghetto is a disaster. Many of those leaving the ghetto settle in non-ghetto areas outside the political jurisdiction of the central city. Thus, geographic size of the ghetto is expanding, cutting a wider swath through the hearts of our metropolitan areas.

In sum, the 1970s and 1980s witnessed a sharp growth in the number of census tracts classified as ghetto poverty areas, an increased concentration of the poor in these areas, and sharply divergent patterns of poverty concentration between racial minorities and whites. One of the legacies of historic racial and class subjugation in America is a unique and growing concentration of minority residents in the most impoverished areas of the nation's metropolises.

Some have argued that this concentration of poverty is not new but mirrors conditions prevalent in the 1930s. According to Douglas Massey and Nancy Denton, during the Depression poverty was just as concentrated in the ghettos of the 1930s as in those of the 1970s. The black communities of the 1930s and those of the 1970s shared a common experience: a high degree of racial segregation from the larger society. Massey and Denton argue that "concentrated poverty is created by a pernicious interaction between a group's overall rate of poverty and its degree of segregation in society. When a highly segregated group experiences a high or rising rate of poverty, geographically concentrated poverty is the inevitable result." However convincing the logic of that argument, it does not explain the following: In the ten neighborhoods that make up Chicago's Black Belt, the poverty rate increased almost 20 percent between 1970 and 1980 (from 32.5 to 50.4 percent) despite the fact that the overall black poverty rate for the city of Chicago increased only 7.5 percent during this same period (from 25.1 to 32.6 percent).

Concentrated poverty may be the inevitable result when a highly segregated group experiences an increase in its overall rate of poverty. But segregation does not explain why the concentration of poverty in *certain* neighborhoods of this segregated group should increase to nearly three times the group's *overall* rate of poverty increase. There is no doubt that the disproportionate concentration of poverty among African–Americans is one of the legacies of historic racial segregation. It is also true that segregation often compounds black vulnerability in the face of other changes in the society, including, as we shall soon see, economic changes. Nonetheless, to focus mainly on segregation to account for the growth of concentrated poverty is to overlook some of the dynamic aspects of the social and demographic changes occurring in cities like Chicago. Given the existence of segregation, we must consider the way in which other changes in society have interacted with segregation to produce the dramatic social transformation of inner-city neighborhoods, especially since 1970.

For example, the communities that make up the Black Belt in Chicago have been overwhelmingly black for the last four decades, yet they lost almost half their residents between 1970 and 1980. This rapid depopulation has had profound consequences for the social and economic deterioration of segregated Black Belt neighborhoods, including increases in concentrated poverty and joblessness. If comparisons are made strictly between the Depression years of the 1930s and the 1980s, rates of ghetto poverty and joblessness in these neighborhoods will indeed be similar. But such a comparison obscures significant changes that have occurred in these neighborhoods across the fifty-year span between those two points.

. . . Many of the gains made in inner-city neighborhoods following the Depression were wiped out after 1970. To maintain that concentrated black poverty in the 1970s or in the 1980s is equivalent in severity and pervasiveness to that which occurred during the Depression does not explain its dramatic rise since 1970; nor does it address a far more fundamental problem that is at the heart of the extraordinary increases in and spread of concentrated poverty – namely, the rapid growth of joblessness, which accelerated through these two decades. The problems reported by the residents of poor Chicago neighborhoods are not a consequence of poverty alone. Something far more devastating has happened that can only be attributed to the emergence of concentrated and persistent joblessness and its crippling effects on neighborhoods, families, and individuals. The city of Chicago epitomizes these changes.

[. . .]

The most fundamental difference between today's inner-city neighborhoods and those studied by Drake and Cayton [in the 1940s] is the much higher levels of joblessness. Indeed, there is a new poverty in our nation's metropolises that has consequences for a range of issues relating to the quality of life in urban areas, including race relations.

By "the new urban poverty," I mean poor, segregated neighborhoods in which a substantial majority of individual adults are either unemployed or have dropped out of the labor force altogether. For example, in 1990 only one in three adults aged 16 and over in the twelve Chicago community areas with ghetto poverty rates held a job in a typical week of the year. Each of these community areas, located on the South and West Sides of the city, is overwhelmingly black. We can add to these twelve high-jobless areas three additional predominantly black community areas, with rates approaching ghetto poverty, in which only 42 percent of the adult population were working in a typical week in 1990. Thus, in these fifteen black community areas – comprising a total population of 425,125 – only 37 percent of all the adults were gainfully employed in a typical week in 1990. By contrast, 54 percent of the adults in the seventeen other predominantly black community areas in Chicago – a total population of 545,408 – worked in a typical week in 1990. This was close to the citywide employment figure of 57 percent for all adults. Finally, except for one Asian community area with an employment rate of 46 percent, and one Latino community area with an employment rate of 49 percent, a majority of the adults held a job in a typical week in each of the remaining forty-five community areas of Chicago.

But Chicago is by no means the only city that features new poverty neighborhoods. In the ghetto census tracts of the nation's one hundred largest central cities, there were only 65.5 employed persons for every hundred adults who did not hold a job in a typical week in 1990. In contrast, the nonpoverty areas contained 182.3 employed persons for every hundred of those not working. In other words, the ratio of employed to jobless persons was three times greater in census tracts not marked by poverty.

Looking at Drake and Cayton's Bronzeville, I can illustrate the magnitude of the changes that have occurred in many, inner-city ghetto neighborhoods in recent years. A majority of adults held jobs in the three Bronzeville areas in 1950, but by 1990 only four in ten in Douglas worked in a typical week, one in three in Washington Park, and one in four in Grand Boulevard. In 1950, 69 percent of all males 14 and over who lived in the Bronzeville neighborhoods worked in a typical week, and in 1960, 64 percent of this group were so employed. However, by 1990 only 37 percent of all males 16 and over held jobs in a typical week in these three neighborhoods.

Upon the publication of the first edition of *Black Metropolis* in 1945, there was much greater class integration within the black community. As Drake and Cayton pointed out, Bronzeville residents had limited success in "sorting themselves out into broad community areas designated as 'lower class' and 'middle class' . . . Instead of middle-class areas, Bronzeville tends to have middle-class buildings in all areas, or a few middle-class blocks here and there." Though they may have lived on different streets, blacks of all classes in inner-city areas such as Bronzeville lived in the same community and shopped at the same stores. Their children went to the same schools and played in the same parks. Although there was some class antagonism, their neighborhoods were more stable than the inner-city neighborhoods of today; in short, they featured higher levels of what social scientists call "social organization."

When I speak of social organization I am referring to the extent to which the residents of a neighborhood are able to maintain effective social control and realize their common goals. There are three major dimensions of neighborhood social organization: (1) the prevalence, strength, and interdependence of social networks; (2) the extent of collective supervision that the residents exercise and the degree of personal responsibility they assume in addressing neighborhood problems; and (3) the rate of resident participation in voluntary and formal organizations. Formal institutions (e.g., churches and political party organizations), voluntary associations (e.g., block clubs and parent–teacher organizations), and informal networks (e.g., neighborhood friends and acquaintances, coworkers, marital and parental ties) all reflect social organization. Neighborhood social organization depends on the extent of local friendship ties, the degree of social cohesion, the level of resident participation in formal and informal voluntary associations, the density and stability of formal organizations, and the nature of informal social controls. Neighborhoods in which adults are able to interact in terms of obligations, expectations, and relationships are in a better position to supervise and control the activities and behavior of children. In neighborhoods with high levels of social organization, adults are empowered to act to improve the quality of neighborhood life – for example, by breaking up congregations of youths on street corners and by supervising the leisure activities of youngsters.

Neighborhoods plagued by high levels of joblessness are more likely to experience low levels of social organization: the two go hand in hand. High rates of joblessness trigger other neighborhood problems that undermine social organization, ranging from crime, gang violence, and drug trafficking to family breakups and problems in the organization of family life.

Consider, for example, the problems of drug trafficking and violent crime. As many studies have revealed, the decline in legitimate employment opportunities among inner-city residents has increased incentives to sell drugs. The distribution of crack in a neighborhood attracts individuals involved in violence and lawlessness. Between 1985 and 1992, there was a sharp increase in the murder rate among men under the age of 24; for men 18 years old and younger, murder rates doubled. Black males in particular have been involved in this upsurge in violence. For example, whereas the homicide rate for white males between 14 and 17 increased from 8 per 100,000 in 1984 to 14 in 1991, the rate for black males tripled during that time (from 32 per 100,000 to 112). This sharp rise in violent crime among younger males has accompanied the widespread outbreak of addiction to crack cocaine. The association is especially strong in inner-city ghetto neighborhoods plagued by joblessness and weak social organization.

Violent persons in the crack-cocaine marketplace have a powerful impact on the social organization of a neighborhood. Neighborhoods plagued by high levels of joblessness, insufficient economic opportunities, and high residential mobility are unable to control the volatile drug market and the violent crimes related to it. As informal controls weaken, the social processes that regulate behavior change.

As a result, the behavior and norms in the drug market are more likely to influence the action of others in the neighborhood, even those who are not involved in drug activity. Drug dealers cause the use and spread of guns in the neighborhood to escalate, which in turn raises the likelihood that others, particularly the youngsters, will come to view the possession of weapons as necessary or desirable for self-protection, settling disputes, and gaining respect from peers and other individuals.

Moreover, as Alfred Blumstein pointed out, the drug industry actively recruits teenagers in the neighborhood "partly because they will work more cheaply than adults, partly because they may be less vulnerable to the punishments imposed by the adult criminal justice system, partly because they tend to be daring and willing to take risks that more mature adults would eschew." Inner-city black youths with limited prospects for stable or attractive employment are easily lured into drug trafficking and therefore increasingly find themselves involved in the violent behavior that accompanies it.

A more direct relationship between joblessness and violent crime is revealed in recent research by Delbert Elliott of the University of Colorado, a study based on National Longitudinal Youth Survey data collected from 1976 to 1989, covering ages 11 to 30. As Elliott points out, the transition from adolescence to adulthood usually results in a sharp drop in most crimes as individuals take on new adult roles and responsibilities. "Participation in serious violent offending behavior (aggravated assault, forcible rape, and robbery) increases [for all males] from ages 11 and 12 to ages 15 and 16, then declines dramatically with advancing age." Although black and white males reveal similar age curves, "the negative slope of the age curve for blacks after age 20 is substantially less than that of whites."

The black–white differential in the proportion of males involved in serious violent crime, although almost even at age 11, increases to 3 : 2 over the remaining years of adolescence, and reaches a differential of nearly 4 : 1 during the late twenties. However, when Elliott compared only *employed* black and white males, he found no significant differences in violent behavior patterns among the two groups by age 21. Employed black males, like white males, experienced a precipitous decline in serious violent behavior following their adolescent period. Accordingly, a major reason for the racial gap in violent behavior after adolescence is joblessness; a large proportion of jobless black males do not assume adult roles and responsibilities, and their

serious violent behavior is therefore more likely to extend into adulthood. The new poverty neighborhoods feature a high concentration of jobless males and, as a result, suffer rates of violent criminal behavior that exceed those in other urban neighborhoods.

. . . In 1990, 37 percent of Woodlawn's 27,473 adults were employed and only 23 percent of Oakland's 4,935 adults were working. When asked how much of a problem unemployment was in their neighborhood, 73 percent of the residents in Woodlawn and 76 percent in Oakland identified it as a major problem. The responses to the survey also revealed the residents' concerns about a series of related problems, such as crime and drug abuse, that are symptomatic of severe problems of social organization. Indeed, crime was identified as a major problem by 66 percent of the residents in each neighborhood. Drug abuse was cited as a major problem by as many as 86 percent of the adult residents in Oakland and 79 percent of those in Woodlawn.

Although high-jobless neighborhoods also feature concentrated poverty, high rates of neighborhood poverty are less likely to trigger problems of social organization if the residents are working. This was the case in previous years when the working poor stood out in areas like Bronzeville. Today, the nonworking poor predominate in the highly segregated and impoverished neighborhoods.

The rise of new poverty neighborhoods represents a movement away from what the historian Allan Spear has called an institutional ghetto – whose structure and activities parallel those of the larger society, as portrayed in Drake and Cayton's description of Bronzeville – toward a jobless ghetto, which features a severe lack of basic opportunities and resources, and inadequate social controls.

What can account for the growing proportion of jobless adults and the corresponding increase in problems of social organization in innercity communities such as Bronzeville? An easy answer is racial segregation. However, a race-specific argument is not sufficient to explain recent changes in neighborhoods like Bronzeville. After all, Bronzeville was just as *segregated by skin color in 1950* as it is today, yet the level of employment was much higher then.

Nonetheless, racial segregation does matter. If large segments of the African–American population had not been historically segregated in inner-city ghettos, we would not be talking about the new urban poverty. The segregated ghetto is not the result of voluntary or positive decisions on the part of the residents who live there. As Massey and Denton have carefully documented, the segregated ghetto is the product of systematic racial practices such as restrictive covenants, redlining by banks and insurance companies, zoning, panic peddling by real estate agents, and the creation of massive public housing projects in low-income areas.

Segregated ghettos are less conducive to employ ment and employment preparation than are other areas of the city. Segregation in ghettos exacerbates employment problems because it leads to weak informal employment networks and contributes to the social isolation of individuals and families, thereby reducing their chances of acquiring the human capital skills, including adequate educational training, that facilitate mobility in a society. Since no other group in society experiences the degree of segregation, isolation, and poverty concentration as do African–Americans, they are far more likely to be disadvantaged when they have to compete with other groups in society, including other despised groups, for resources and privileges.

To understand the new urban poverty, one has to account for the ways in which segregation interacts with other changes in society to produce the recent escalating rates of joblessness and problems of social organization in inner-city ghetto neighborhoods.

"Bowling Alone: America's Declining Social Capital"

Journal of Democracy (1995)

Robert D. Putnam

Editors' Introduction

Robert Putnam has been called "the most influential academic in the world today," and his work has been praised by political leaders Bill Clinton, Tony Blair, and George W. Bush. As a social critic, he stands in a tradition that includes Alexis de Tocqueville (*Democracy in America*, 1835), Paul Goodman (*Growing Up Absurd*, 1960), and Philip Slater (*The Pursuit of Loneliness: American Culture at the Breaking Point*, 1970). In this article from the *Journal of Democracy*, and in a subsequent book of the same title, Putnam asserts the "bowling alone" phenomenon as a metaphor for what urban life has become in contemporary middle-class America and for millions of others worldwide living in an increasingly materialistic and solipsistic culture of corporate work, obsessive consumption, and over-determined leisure.

Whereas cities once held out the promise of a wider, higher form of human community, Putnam argues that contemporary urbanites now follow a path of less, not more, civic engagement and that our collective stock of "social capital" – the meaningful human contacts of all kinds that characterize true communities – is so dangerously eroded that it verges on depletion.

In some ways, Putnam's critique is an updated version of Louis Wirth's 1938 essay "Urbanism as a Way of Life" (p. 90). Urban dwellers are not connected to one another through collective action as they once were. Instead, virtually all measures of social engagement – voting, participation in social organizations, active church membership, even friendships and family ties – seem to grow weaker every year. In part, this phenomenon arises from well-known causes: social and geographic mobility, the decreasing importance of families as women join the corporate workforce, and the technological transformation of leisure. However understandable, Putnam argues, these forces have now risen to the level of social crisis and must be addressed by conscious policies to increase civic engagement of all sorts and to strengthen the connection between people in their roles as neighbors, co-workers, and fellow citizens.

Putnam's ideas about urban civic engagement emerge from a deep philosophical tradition that includes Peter Kropotkin's *Mutual Aid* (New York: Knopf, 1922, originally published in 1902) and John Gardner's "Building Community" (*Independent Sector*, 1991), and they raise questions as to whether any modern urban society can hope to regain the intimacy of the ancient Greek polis as described by H.D.F. Kitto (p. 35), or if the planning practices of New Urbanists such as Peter Calthorpe and William Fulton (p. 342) or Andrés Duany and Elizabeth Plater-Zyberk (p. 192) can truly lead to the heightened degrees of "community" that Jane Jacobs describes in *The Death and Life of Great American Cities* (p. 98). These issues take on increased importance when considering the social and technological futures of society discussed by William J. Mitchell in "The Teleserviced City" (p. 510) and Richard Florida in *The Rise of the Creative Class* (p. 129).

Robert Putnam is the Peter and Isabel Malkin Professor of Public Policy at Harvard University and has written widely in the fields of politics, comparative politics, international relations, and public policy. He attended

Swarthmore and Balliol College, Oxford, before taking his Ph.D. at Yale in 1970. He is the author of *The Beliefs of Politicians: Ideology, Conflict, and Democracy in Britain and Italy* (New Haven: Yale University Press, 1973) and, with Robert Leonardi and Rafaella Nanetti, *Making Democracy Work: Civic Traditions in Modern Italy* (Princeton: Princeton University Press, 1993). Following the wide popular success of *Bowling Alone* (New York: Simon & Schuster, 2000), Putnam founded the Saguaro Seminar on Civic Engagement in America where he brings together community leaders and policy makers to develop working plans for increasing civic connectedness in specific urban contexts. He is also the editor of *Democracies in Flux: The Evolution of Social Capital in Contemporary Society* (New York: Oxford University Press, 2002).

Other sources of information on the idea of urban social capital include James S. Coleman, "Social Capital in the Creation of Human Capital" (*American Journal of Sociology*, 1988) and Robert Wuthnow, *Sharing the Journey: Support Groups and America's New Quest for Community* (New York: Free Press, 1994). A number of elaborations and critiques of the social capital idea can be found in Robert M. Silverman (ed.), *Community-Based Organizations: The Intersection of Social Capital and Local Context in Contemporary Urban Society* (Detroit: Wayne State University Press, 2004). Among the more interesting are Ivan Light, "Social Capital for What?" and Randy Stoeker, "The Mystery of the Missing Social Capital and the Ghost of Social Structure: Why Community Development Can't Win." See also James DeFilippis, "The Myth of Social Capital in Community Development," (*Housing Policy Debate*, 12(4), 2001).

■ ■ ■ ■ ■ ■

Many students of the new democracies that have emerged over the past decade and a half have emphasized the importance of a strong and active civil society to the consolidation of democracy. Especially with regard to the postcommunist countries, scholars and democratic activists alike have lamented the absence or obliteration of traditions of independent civic engagement and a widespread tendency toward passive reliance on the state. To those concerned with the weakness of civil societies in the developing or postcommunist world, the advanced Western democracies and above all the United States have typically been taken as models to be emulated. There is striking evidence, however, that the vibrancy of American civil society has notably declined over the past several decades.

Ever since the publication of Alexis de Tocqueville's *Democracy in America* [1835], the United States has played a central role in systematic studies of the links between democracy and civil society. Although this is in part because trends in American life are often regarded as harbingers of social modernization, it is also because America has traditionally been considered unusually "civic" (a reputation that, as we shall later see, has not been entirely unjustified).

When Tocqueville visited the United States in the 1830s, it was the Americans' propensity for civic association that most impressed him as the key to their unprecedented ability to make democracy work.

"Americans of all ages, all stations in life, and all types of disposition," he observed, "are forever forming associations. There are not only commercial and industrial associations in which all take part, but others of a thousand different types – religious, moral, serious, futile, very general and very limited, immensely large and very minute. Nothing, in my view, deserves more attention than the intellectual and moral associations in America."

Recently, American social scientists of a neo-Tocquevillean bent have unearthed a wide range of empirical evidence that the quality of public life and the performance of social institutions (and not only in America) are indeed powerfully influenced by norms and networks of civic engagement. Researchers in such fields as education, urban poverty, unemployment, the control of crime and drug abuse, and even health have discovered that successful outcomes are more likely in civically engaged communities. Similarly, research on the varying economic attainments of different ethnic groups in the United States has demonstrated the importance of social bonds within each group. These results are consistent with research in a wide range of settings that demonstrates the vital importance of social networks for job placement and many other economic outcomes.

Meanwhile, a seemingly unrelated body of research on the sociology of economic development has also focused attention on the role of social networks. Some of this work is situated in the developing countries, and some of it elucidates the peculiarly successful "network capitalism" of East Asia. Even in less exotic Western economies, however, researchers have discovered highly efficient, highly flexible "industrial districts" based on networks of collaboration among workers and small entrepreneurs. Far from being paleo-industrial anachronisms, these dense interpersonal and interorganizational networks undergird ultramodern industries, from the high tech of Silicon Valley to the high fashion of Benetton.

The norms and networks of civic engagement also powerfully affect the performance of representative government. That, at least, was the central conclusion of my own 20-year, quasi-experimental study of sub-national governments in different regions of Italy. Although all these regional governments seemed identical on paper, their levels of effectiveness varied dramatically. Systematic inquiry showed that the quality of governance was determined by longstanding traditions of civic engagement (or its absence). Voter turnout, newspaper readership, membership in choral societies and football clubs – these were the hallmarks of a successful region. In fact, historical analysis suggested that these networks of organized reciprocity and civic solidarity, far from being an epiphenomenon of socioeconomic modernization, were a precondition for it.

No doubt the mechanisms through which civic engagement and social connectedness produce such results – better schools, faster economic development, lower crime, and more effective government – are multiple and complex. While these briefly recounted findings require further confirmation and perhaps qualification, the parallels across hundreds of empirical studies in a dozen disparate disciplines and subfields are striking. Social scientists in several fields have recently suggested a common framework for under-standing these phenomena, a framework that rests on the concept of social capital. By analogy with notions of physical capital and human capital – tools and training that enhance individual productivity – "social capital" refers to features of social organ-ization such as networks, norms, and social trust that facilitate coordination and cooperation for mutual benefit.

For a variety of reasons, life is easier in a community blessed with a substantial stock of social capital. In the first place, networks of civic engagement foster sturdy norms of generalized reciprocity and encourage the emergence of social trust. Such networks facilitate coordination and communication, amplify reputations, and thus allow dilemmas of collective action to be resolved. When economic and political negotiation is embedded in dense networks of social interaction, incentives for opportunism are reduced. At the same time, networks of civic engagement embody past success at collaboration, which can serve as a cultural template for future collaboration. Finally, dense net-works of interaction probably broaden the participants' sense of self, developing the "I" into the "we," or (in the language of rational-choice theorists) enhancing the participants' "taste" for collective benefits.

I do not intend here to survey (much less contribute to) the development of the theory of social capital. Instead, I use the central premise of that rapidly growing body of work – that social connections and civic engagement pervasively influence our public life, as well as our private prospects – as the starting point for an empirical survey of trends in social capital in contemporary America. I concentrate here entirely on the American case, although the developments I portray may in some measure characterize many contemporary societies.

WHATEVER HAPPENED TO CIVIC ENGAGEMENT?

We begin with familiar evidence on changing patterns of political participation, not least because it is imme-diately relevant to issues of democracy in the narrow sense. Consider the well-known decline in turnout in national elections over the last three decades. From a relative high point in the early 1960s, voter turnout had by 1990 declined by nearly a quarter; tens of millions of Americans had forsaken their parents' habitual readiness to engage in the simplest act of citizenship. Broadly similar trends also characterize participation in state and local elections.

It is not just the voting booth that has been increas-ingly deserted by Americans. A series of identical questions posed by the Roper Organization to national samples ten times each year over the last two decades reveals that since 1973 the number of Americans who report that "in the past year" they have "attended a public meeting on town or school affairs" has fallen by more than a third (from 22 percent in 1973 to 13 percent in 1993). Similar (or even greater) relative declines are evident in responses to questions about attending a political rally or speech, serving on a committee of some local organization, and working for a political party. By almost every measure, Americans' direct engagement in politics and government has fallen steadily and sharply over the last generation, despite the fact that average levels of education – the best individual-level predictor of political participation – have risen sharply throughout this period. Every year over the last decade or two, millions more have withdrawn from the affairs of their communities.

Not coincidentally, Americans have also disengaged psychologically from politics and government over this era. The proportion of Americans who reply that they "trust the government in Washington" only "some of the time" or "almost never" has risen steadily from 30 percent in 1966 to 75 percent in 1992.

These trends are well known, of course, and taken by themselves would seem amenable to a strictly political explanation. Perhaps the long litany of political tragedies and scandals since the 1960s (assassinations, Vietnam, Watergate, Irangate, and so on) has triggered an understandable disgust for politics and government among Americans, and that in turn has motivated their withdrawal. I do not doubt that this common interpretation has some merit, but its limitations become plain when we examine trends in civic engagement of a wider sort.

Our survey of organizational membership among Americans can usefully begin with a glance at the aggregate results of the General Social Survey, a scientifically conducted, national-sample survey that has been repeated 14 times over the last two decades. Church-related groups constitute the most common type of organization joined by Americans; they are especially popular with women. Other types of organizations frequently joined by women include school-service groups (mostly parent–teacher associations), sports groups, professional societies, and literary societies. Among men, sports clubs, labor unions, professional societies, fraternal groups, veterans' groups, and service clubs are all relatively popular.

Religious affiliation is by far the most common associational membership among Americans. Indeed, by many measures America continues to be (even more than in Tocqueville's time) an astonishingly "churched" society. For example, the United States has more houses of worship per capita than any other nation on Earth. Yet religious sentiment in America seems to be becoming somewhat less tied to institutions and more self-defined.

How have these complex crosscurrents played out over the last three or four decades in terms of Americans' engagement with organized religion? The general pattern is clear: the 1960s witnessed a significant drop in reported weekly churchgoing – from roughly 48 percent in the late 1950s to roughly 41 percent in the early 1970s. Since then, it has stagnated or (according to some surveys) declined still further. Meanwhile, data from the General Social Survey show a modest decline in membership in all "church related groups" over the last 20 years. It would seem, then, that net participation by Americans, both in religious services and in church-related groups, has declined modestly (by perhaps a sixth) since the 1960s.

For many years, labor unions provided one of the most common organizational affiliations among American workers. Yet union membership has been falling for nearly four decades, with the steepest decline occurring between 1975 and 1985. Since the mid-1950s, when union membership peaked, the unionized portion of the nonagricultural workforce in America has dropped by more than half, falling from 32.5 percent in 1953 to 15.8 percent in 1992. By now, virtually all of the explosive growth in union membership that was associated with the New Deal has been erased. The solidarity of union halls is now mostly a fading memory of aging men.

The parent–teacher association (PTA) has been an especially important form of civic engagement in twentieth-century America because parental involvement in the educational process represents a particularly productive form of social capital. It is, therefore, dismaying to discover that participation in parent–teacher organizations has dropped drastically over the last generation, from more than 12 million in 1964 to barely 5 million in 1982 before recovering to approximately 7 million now.

Next, we turn to evidence on membership in (and volunteering for) civic and fraternal organizations. These data show some striking patterns. First, membership in traditional women's groups has declined more or less steadily since the mid-1960s. For example, membership in the national Federation of Women's Clubs is down by more than half (59 percent) since 1964, while membership in the League of Women Voters (LWV) is off 42 percent since 1969.

Similar reductions are apparent in the numbers of volunteers for mainline civic organizations, such as the Boy Scouts (off by 26 percent since 1970) and the Red Cross (off by 61 percent since 1970). But what about the possibility that volunteers have simply switched their loyalties to other organizations? Evidence on "regular" (as opposed to occasional or "drop-by") volunteering is available from the Labor Department's Current Population Surveys of 1974 and 1989. These estimates suggest that serious volunteering declined by roughly one-sixth over these 15 years, from 24 percent of adults in 1974 to 20 percent in 1989. The multitudes of Red Cross aides and Boy Scout troop leaders now missing in action have apparently not been offset by equal numbers of new recruits elsewhere.

Fraternal organizations have also witnessed a substantial drop in membership during the 1980s and 1990s. Membership is down significantly in such groups as the Lions (off 12 percent since 1983), the Elks (off 18 percent since 1979), the Shriners (off 27 percent since 1979), the Jaycees (off 44 percent since 1979), and the Masons (down 39 percent since 1959). In sum, after expanding steadily throughout most of this century, many major civic organizations have experienced a sudden, substantial, and nearly simultaneous decline in membership over the last decade or two.

The most whimsical yet discomfiting bit of evidence of social disengagement in contemporary America that I have discovered is this: more Americans are bowling today than ever before, but bowling in organized leagues has plummeted in the last decade or so. Between 1980 and 1993 the total number of bowlers in America increased by 10 percent, while league bowling decreased by 40 percent. (Lest this be thought a wholly trivial example, I should note that nearly 80 million Americans went bowling at least once during 1993, nearly a third more than voted in the 1994 congressional elections and roughly the same number as claim to attend church regularly. Even after the 1980s' plunge in league bowling, nearly 3 percent of American adults regularly bowl in leagues.) The rise of solo bowling threatens the livelihood of bowling-lane proprietors because those who bowl as members of leagues consume three times as much beer and pizza as solo bowlers, and the money in bowling is in the beer and pizza, not the balls and shoes. The broader social significance, however, lies in the social interaction and even occasionally civic conversations over beer and pizza that solo bowlers forgo. Whether or not bowling beats balloting in the eyes of most Americans, bowling teams illustrate yet another vanishing form of social capital.

COUNTERTRENDS

At this point, however, we must confront a serious counterargument. Perhaps the traditional forms of civic organization whose decay we have been tracing have been replaced by vibrant new organizations. For example, national environmental organizations (like the Sierra Club) and feminist groups (like the National Organization for Women) grew rapidly during the 1970s and 1980s and now count hundreds of thousands of dues-paying members. An even more dramatic example is the American Association of Retired Persons (AARP), which grew exponentially from 400,000 card-carrying members in 1960 to 33 million in 1993, becoming (after the Catholic Church) the largest private organization in the world. The national administrators of these organizations are among the most feared lobbyists in Washington, in large part because of their massive mailing lists of presumably loyal members.

These new mass-membership organizations are plainly of great political importance. From the point of view of social connectedness, however, they are sufficiently different from classic "secondary associations" that we need to invent a new label – perhaps "tertiary associations." For the vast majority of their members, the only act of membership consists in writing a check for dues or perhaps occasionally reading a newsletter. Few ever attend any meetings of such organizations, and most are unlikely ever (knowingly) to encounter any other member. The bond between any two members of the Sierra Club is less like the bond between any two members of a gardening club and more like the bond between any two Red Sox fans (or perhaps any two devoted Honda owners): they root for the same team and they share some of the same interests, but they are unaware of each other's existence. Their ties, in short, are to common symbols, common leaders, and perhaps common ideals, but not to one another. The theory of social capital argues that associational membership should, for example, increase social trust, but this prediction is much less straightforward with regard to membership in tertiary associations. From the point of view of social connectedness, the Environmental Defense Fund and a bowling league are just not in the same category.

If the growth of tertiary organizations represents one potential (but probably not real) counterexample to my thesis, a second countertrend is represented by the growing prominence of nonprofit organizations, especially nonprofit service agencies. This so-called third sector includes everything from Oxfam and the Metropolitan Museum of Art to the Ford Foundation and the Mayo Clinic. In other words, although most secondary associations are nonprofits, most nonprofit agencies are not secondary associations. To identify trends in the size of the nonprofit sector with trends in social connectedness would be another fundamental conceptual mistake.

A third potential countertrend is much more relevant to an assessment of social capital and civic engagement. Some able researchers have argued that the last few decades have witnessed a rapid expansion in "support groups" of various sorts. Robert Wuthnow reports that fully 40 percent of all Americans claim to be "currently involved in [a] small group that meets regularly and provides support or caring for those who participate in it." Many of these groups are religiously affiliated, but many others are not. For example, nearly 5 percent of Wuthnow's national sample claim to participate regularly in a "self-help" group, such as Alcoholics Anonymous, and nearly as many say they belong to book-discussion groups and hobby clubs.

The groups described by Wuthnow's respondents unquestionably represent an important form of social capital, and they need to be accounted for in any serious reckoning of trends in social connectedness. On the other hand, they do not typically play the same role as traditional civic associations. As Wuthnow emphasizes,

Small groups may not be fostering community as effectively as many of their proponents would like. Some small groups merely provide occasions for individuals to focus on themselves in the presence of others. The social contract binding members together asserts only the weakest of obligations. Come if you have time. Talk if you feel like it. Respect everyone's opinion. Never criticize. Leave quietly if you become dissatisfied . . . We can imagine that [these small groups] really substitute for families, neighborhoods, and broader community attachments that may demand lifelong commitments, when, in fact, they do not.

All three of these potential countertrends – tertiary organizations, nonprofit organizations, and support groups – need somehow to be weighed against the erosion of conventional civic organizations. One way of doing so is to consult the General Social Survey.

Within all educational categories, total associational membership declined significantly between 1967 and 1993. Among the college-educated, the average number of group memberships per person fell from 2.8 to 2.0 (a 26 percent decline); among high-school graduates, the number fell from 1.8 to 1.2 (32 percent); and among those with fewer than 12 years of education, the number fell from 1.4 to 1.1 (25 percent). In other words, at all educational (and hence social) levels of American society, and counting all sorts of group memberships, the average number of associational memberships has fallen by about a fourth over the last quarter-century. Without controls for educational levels, the trend is not nearly so clear, but the central point is this: more Americans than ever before are in social circumstances that foster associational involvement (higher education, middle age, and so on), but nevertheless aggregate associational membership appears to be stagnant or declining.

Broken down by type of group, the downward trend is most marked for church-related groups, for labor unions, for fraternal and veterans' organizations, and for school-service groups. Conversely, membership in professional associations has risen over these years, although less than might have been predicted, given sharply rising educational and occupational levels. Essentially the same trends are evident for both men and women in the sample. In short, the available survey evidence confirms our earlier conclusion: American social capital in the form of civic associations has significantly eroded over the last generation.

GOOD NEIGHBORLINESS AND SOCIAL TRUST

I noted earlier that most readily available quantitative evidence on trends in social connectedness involves formal settings, such as the voting booth, the union hall, or the PTA. One glaring exception is so widely discussed as to require little comment here: the most fundamental form of social capital is the family, and the massive evidence of the loosening of bonds within the family (both extended and nuclear) is well known. This trend, of course, is quite consistent with – and may help to explain – our theme of social decapitalization.

A second aspect of informal social capital on which we happen to have reasonably reliable time-series data involves neighborliness. In each General Social Survey since 1974 respondents have been asked, "How often do you spend a social evening with a neighbor?" The proportion of Americans who socialize with their neighbors more than once a year has slowly but steadily declined over the last two decades, from 72 percent in 1974 to 61 percent in 1993. (On the other hand, socializing with "friends who do not live in your neighborhood" appears to be on the increase, a trend that may reflect the growth of workplace-based social connections.)

Americans are also less trusting. The proportion of Americans saying that most people can be trusted fell by more than a third between 1960, when 58 percent chose that alternative, and 1993, when only 37 percent did. The same trend is apparent in all educational groups; indeed, because social trust is also correlated with education and because educational levels have risen sharply, the overall decrease in social trust is even more apparent if we control for education.

Our discussion of trends in social connectedness and civic engagement has tacitly assumed that all the forms of social capital that we have discussed are themselves coherently correlated across individuals. This is in fact true. Members of associations are much more likely than nonmembers to participate in politics, to spend time with neighbors, to express social trust, and so on.

The close correlation between social trust and associational membership is true not only across time and across individuals, but also across countries. Evidence from the 1991 World Values Survey demonstrates the following: across the 35 countries in this survey, social trust and civic engagement are strongly correlated; the greater the density of associational membership in a society, the more trusting its citizens. Trust and engagement are two facets of the same underlying factor – social capital.

America still ranks relatively high by cross-national standards on both these dimensions of social capital. Even in the 1990s, after several decades' erosion, Americans are more trusting and more engaged than people in most other countries of the world. The trends of the past quarter-century, however, have apparently moved the United States significantly lower in the international rankings of social capital. The recent deterioration in American social capital has been sufficiently great that (if no other country changed its position in the meantime) another quarter-century of change at the same rate would bring the United States, roughly speaking, to the midpoint among all these countries, roughly equivalent to South Korea, Belgium, or Estonia today. Two generations' decline at the same rate would leave the United States at the level of today's Chile, Portugal, and Slovenia.

WHY IS US SOCIAL CAPITAL ERODING?

As we have seen, something has happened in America in the last two or three decades to diminish civic engagement and social connectedness. What could that "something" be? Here are several possible explanations, along with some initial evidence on each.

The movement of women into the labor force. Over these same two or three decades, many millions of American women have moved out of the home into paid employment. This is the primary, though not the sole, reason why the weekly working hours of the average American have increased significantly during these years. It seems highly plausible that this social revolution should have reduced the time and energy available for building social capital. For certain organizations, such as the PTA, the League of Women Voters, the Federation of Women's Clubs, and the Red Cross, this is almost certainly an important part of the story. The sharpest decline in women's civic participation seems to have come in the 1970s; membership in such "women's" organizations as these has been virtually halved since the late 1960s. By contrast, most of the decline in participation in men's organizations occurred about 10 years later; the total decline to date has been approximately 25 percent for the typical organization. On the other hand, the survey data imply that the aggregate declines for men are virtually as great as those for women. It is logically possible, of course, that the male declines might represent the knock-on effect of women's liberation, as dishwashing crowded out the lodge, but time-budget studies suggest that most husbands of working wives have assumed only a minor part of the housework. In short, something besides the women's revolution seems to lie behind the erosion of social capital.

Mobility: The "re-potting" hypothesis

Numerous studies of organizational involvement have shown that residential stability and such related phenomena as homeownership are clearly associated with greater civic engagement. Mobility, like frequent re-potting of plants, tends to disrupt root systems, and it takes time for an uprooted individual to put down new roots. It seems plausible that the automobile, suburbanization, and the movement to the Sun Belt have reduced the social rootedness of the average American, but one fundamental difficulty with this hypothesis is apparent: the best evidence shows that residential stability and homeownership in America have risen modestly since 1965, and are surely higher now than during the 1950s, when civic engagement and social connectedness by our measures was definitely higher.

Other demographic transformations

A range of additional changes have transformed the American family since the 1960s – fewer marriages, more divorces, fewer children, lower real wages, and so on. Each of these changes might account for some of the slackening of civic engagement, since married, middle-class parents are generally more socially involved than other people. Moreover, the changes in scale that have swept over the American economy in these years – illustrated by the replacement of the corner grocery by the supermarket and now perhaps of the supermarket by electronic shopping at home, or the replacement of community-based enterprises by outposts of distant multinational firms – may perhaps have undermined the material and even physical basis for civic engagement.

The technological transformation of leisure

There is reason to believe that deep-seated techno-logical trends are radically "privatizing" or "individual-izing" our use of leisure time and thus disrupting many opportunities for social-capital formation. The most obvious and probably the most powerful instrument of this revolution is television. Time-budget studies in the 1960s showed that the growth in time spent watching television dwarfed all other changes in the way Americans passed their days and nights. Tele-vision has made our communities (or, rather, what we experience as our communities) wider and shallower. In the language of economics, electronic technology enables individual tastes to be satisfied more fully, but at the cost of the positive social externalities associated with more primitive forms of entertainment. The same logic applies to the replacement of vaudeville by the movies and now of movies by the VCR. The new "virtual reality" helmets that we will soon don to be entertained in total isolation are merely the latest extension of this trend. Is technology thus driving a wedge between our individual interests and our collective interests? It is a question that seems worth exploring more systematically.

WHAT IS TO BE DONE?

The last refuge of a social-scientific scoundrel is to call for more research. Nevertheless, I cannot forbear from suggesting some further lines of inquiry.

We must sort out the dimensions of social capital, which clearly is not a unidimensional concept, despite language (even in this essay) that implies the contrary. What types of organizations and networks most effectively embody – or generate – social capital, in the sense of mutual reciprocity, the resolution of dilemmas of collective action, and the broadening of social identities? In this essay I have emphasized the density of associational life. In earlier work I stressed the structure of networks, arguing that "horizontal" ties represented more productive social capital than vertical ties.

Another set of important issues involves macro-sociological crosscurrents that might intersect with the trends described here. What will be the impact, for example, of electronic networks on social capital? My hunch is that meeting in an electronic forum is not the equivalent of meeting in a bowling alley – or even in a saloon – but hard empirical research is needed. What about the development of social capital in the workplace? Is it growing in counterpoint to the decline of civic engagement, reflecting some social analogue of the first law of thermodynamics – social capital is neither created nor destroyed, merely redistributed? Or do the trends described in this essay represent a deadweight loss?

A rounded assessment of changes in American social capital over the last quarter-century needs to count the costs as well as the benefits of community engagement. We must not romanticize small-town, middle-class civic life in the America of the 1950s. In addition to the deleterious trends emphasized in this essay, recent decades have witnessed a substantial decline in intolerance and probably also in overt discrimination, and those beneficent trends may be related in complex ways to the erosion of traditional social capital. Moreover, a balanced accounting of the social-capital books would need to reconcile the insights of this approach with the undoubted insights offered by Mancur Olson and others who stress that closely knit social, economic, and political organizations are prone to inefficient cartelization and to what political economists term "rent seeking" and ordinary men and women call corruption.

Finally, and perhaps most urgently, we need to explore creatively how public policy impinges on (or might impinge on) social-capital formation. In some well-known instances, public policy has destroyed highly effective social networks and norms. American slum-clearance policy of the 1950s and 1960s, for example, renovated physical capital, but at a very high cost to existing social capital. The consolidation of country post offices and small school districts has promised administrative and financial efficiencies, but full-cost accounting for the effects of these policies on social capital might produce a more negative verdict. On the other hand, such past initiatives as the county agricultural-agent system, community colleges, and tax deductions for charitable contributions illustrate that government can encourage social-capital formation. Even a recent proposal in San Luis Obispo, California,

to require that all new houses have front porches illustrates the power of government to influence where and how networks are formed.

The concept of "civil society" has played a central role in the recent global debate about the preconditions for democracy and democratization. In the newer democracies this phrase has properly focused attention on the need to foster a vibrant civic life in soils traditionally inhospitable to self-government. In the established democracies, ironically, growing numbers of citizens are questioning the effectiveness of their public institutions at the very moment when liberal democracy has swept the battlefield, both ideologically and geopolitically. In America, at least, there is reason to suspect that this democratic disarray may be linked to a broad and continuing erosion of civic engagement that began a quarter-century ago. High on our scholarly agenda should be the question of whether a comparable erosion of social capital may be under way in other advanced democracies, perhaps in different institutional and behavioral guises. High on America's agenda should be the question of how to reverse these adverse trends in social connectedness, thus restoring civic engagement and civic trust.

"The Creative Class"

from *The Rise of the Creative Class: And How It's Transforming Work, Leisure, Community and Everday Life* (2002)

Richard Florida

Editors' Introduction

In *The Condition of the Working Class in England in 1844* (p. 50), and in subsequent collaborations with his partner Karl Marx, Friedrich Engels announced the emergence of a new social class – the proletariat or industrial working class – that was destined to have a world-historical impact on the shape and content of human society at the time of the Industrial Revolution and the rise of the industrial city. In *The Rise of the Creative Class*, Richard Florida describes the emergence of a new socio-economic class, one that creates ideas and innovations rather than products and is the driving force of post-industrialism rather than industrialism. Florida asks us to ask ourselves: Will the new "creative class" have as important and revolutionary an impact on the twenty-first-century information-based economy and society as the working class had in the nineteenth and twentieth centuries?

According to Florida, there are two layers to the creative class. First, there is a "Super-Creative Core" consisting of "scientists and engineers, university professors, poets and novelists, artists, entertainers, actors, designers and architects, as well as the thought leadership of modern society: nonfiction writers, editors, cultural figures, think-tank researchers, analysts and other opinion-makers." Second, there are "creative professionals" – those who "work in a wide range of knowledge-intensive industries such as high-tech sectors, financial services, the legal and health care professions, and business management" – as well as many technicians and paraprofessionals who now add "creative value" to an enterprise by having to think for themselves. All these, taken together, constitute a true economic class that "both underpins and informs its members' social, cultural and lifestyle choices."

Florida is quick to note that he is not using the term "class" to denote "the ownership of property, capital or the means of production." On the contrary, he argues that if we use those old Marxist categories we are still talking about old-style, bourgeoisie-and-proletarian capitalism. In the new postmodern, post-industrial economic order, the "members of the Creative Class do not own and control any significant property in the physical sense. Their property – which stems from their creative capacity – is an intangible because it is literally in their heads."

Many cities have embraced Florida's thesis about the creative class. Eager to attract "creative class" residents, some cities have sponsored special arts districts and diversity festivals as a part of their redevelopment policies in an attempt to jump-start lagging economies. In some cases, such as Denver's LoDo neighborhood, arts-friendly policies – along with new light-rail transit and a downtown baseball stadium – succeeded in revivifying what had been a decaying warehouse district. In other cases, such as San Francisco's adoption of planning regulations supporting the building of "live-work" lofts for artists and other creatives, was arguably just another ploy on the part of housing developers with connections at City Hall.

Inevitably, a critical opposition to "creative class" theory has developed. Writing in the *Wall Street Journal*, Steven Malanga, a senior fellow at the conservative Manhattan Institute, has called Florida the peddler of "economic snake oil" and the developer of "trendy, New Age theories" that are just "plain wrong." It is lower taxes and public safety, not arts festivals and lively gay neighborhoods, according to Malanga, that attract the industries that bring high employment and robust tax revenues for municipalities.

Richard Florida is the Hirst Professor of Public Policy at George Mason University He has taught at MIT, Harvard, and Carnegie Mellon and is the founder of the Creative Group and Catalytix, two business communications and strategy consulting firms. *The Rise of the Creative Class* received the *Washington Monthly*'s Political Book Award for 2002 and was praised by the *Harvard Business Review* as one of the most important "breakthrough ideas" of recent socio-economic analysis. Florida has also published *The Flight of the Creative Class: The New Global Competition for Talent* (New York: HarperCollins, 2005) and *Cities and the Creative Class* (London and New York: Routledge, 2005).

For intellectual sources of Florida's thesis about the creative class, see Fritz Machlup, *The Production and Distribution of Knowledge in the United States* (Princeton: Princeton University Press, 1962), Daniel Bell, *The Coming of Post-Industrial Society* (New York: Basic Books, 1973), and Peter Drucker, *Post-Capitalist Society* (New York: Harper Business, 1995).

■ ■ ■ ■ ■ ■

The rise of the Creative Economy has had a profound effect on the sorting of people into social groups or classes. Others have speculated over the years on the rise of new classes in the advanced industrial economies. During the 1960s, Peter Drucker and Fritz Machlup described the growing role and importance of the new group of workers they dubbed "know-ledge workers." Writing in the 1970s, Daniel Bell pointed to a new, more meritocratic class structure of scientists, engineers, managers and administrators brought on by the shift from a manufacturing to a "postindustrial" economy. The sociologist Erik Olin Wright has written for decades about the rise of what he called a new "professional-managerial" class. Robert Reich more recently advanced the term "symbolic analysts" to describe the members of the workforce who manipulate ideas and symbols. All of these observers caught economic aspects of the emerging class structure that I describe here.

Others have examined emerging social norms and value systems. Paul Fussell presciently captured many that I now attribute to the Creative Class in his theory of the "X Class." Near the end of his 1983 book *Class* – after a witty romp through status markers that delineate, say, the upper middle class from "high proles" – Fussell noted the presence of a growing "X" group that seemed to defy existing categories:

[Y]ou are not born an X person . . . you earn X-personhood by a strenuous effort of discovery in which curiosity and originality are indispensable. . . . The young flocking to the cities to devote them-selves to "art," "writing," "creative work" – anything, virtually, that liberates them from the presence of a boss or superior – are aspirant X people. . . . If, as [C. Wright] Mills has said, the middle-class person is "always somebody's man," the X person is nobody's. . . . X people are independent-minded. . . . They adore the work they do, and they do it until they are finally carried out, "retirement" being a concept meaningful only to hired personnel or wage slaves who despise their work.

Writing in 2000, David Brooks outlined the blending of bohemian and bourgeois values in a new social grouping he dubbed the Bobos. My take on Brooks's synthesis . . . is rather different, stressing the very trans-cendence of these two categories in a new creative ethos.

The main point I want to make here is that the basis of the Creative Class is economic. I define it as an economic class and argue that its economic func-tion both underpins and informs its members' social, cultural and lifestyle choices. The Creative Class consists of people who add economic value through their creativity. It thus includes a great many know-ledge workers, symbolic analysts and professional and technical workers, but emphasizes their true role in the economy. My definition of class emphasizes the way people organize themselves into social groupings and common identities based principally on their economic function. Their social and cultural prefer-ences, consumption and buying habits, and their social identities all flow from this.

I am not talking here about economic class in terms of the ownership of property, capital or the means of production. If we use class in this traditional Marxian sense, we are still talking about a basic structure of capitalists who own and control the means of production, and workers under their employ. But little analytical utility remains in these broad categories of bourgeoisie and proletarian, capitalist and worker. Most members of the Creative Class do not own and control any significant property in the physical sense. Their property — which stems from their creative capacity – is an intangible because it is literally in their heads. And it is increasingly clear from my field research and interviews that while the members of the Creative Class do not yet see themselves as a unique social grouping, they actually share many similar tastes, desires and preferences. This new class may not be as distinct in this regard as the industrial Working Class in its heyday, but it has an emerging coherence.

THE NEW CLASS STRUCTURE

The distinguishing characteristic of the Creative Class is that its members engage in work whose function is to "create meaningful new forms." I define the Creative Class as consisting of two components. The Super-Creative Core of this new class includes scientists and engineers, university professors, poets and novelists, artists, entertainers, actors, designers and architects, as well as the thought leadership of modern society: nonfiction writers, editors, cultural figures, think-tank researchers, analysts and other opinion-makers. Whether they are software programmers or engineers, architects or filmmakers, they fully engage in the creative process. I define the highest order of creative work as producing new forms or designs that are readily transferable and widely useful – such as designing a product that can be widely made, sold and used; coming up with a theorem or strategy that can be applied in many cases; or composing music that can be performed again and again. People at the core of the Creative Class engage in this kind of work regularly; it's what they are paid to do. Along with problem solving, their work may entail problem finding: not just building a better mousetrap, but noticing first that a better mousetrap would be a handy thing to have.

Beyond this core group, the Creative Class also includes "creative professionals" who work in a wide range of knowledge-intensive industries such as high-tech sectors, financial services, the legal and health care professions, and business management. These people engage in creative problem solving, drawing on complex bodies of knowledge to solve specific problems. Doing so typically requires a high degree of formal education and thus a high level of human capital. People who do this kind of work may sometimes come up with methods or products that turn out to be widely useful, but it's not part of the basic job description. What they *are* required to do regularly is think on their own. They apply or combine standard approaches in unique ways to fit the situation, exercise a great deal of judgment, perhaps try something radically new from time to time. Creative Class people such as physicians, lawyers and managers do this kind of work in dealing with the many varied cases they encounter. In the course of their work, they may also be involved in testing and designing new techniques, new treatment protocols, or new management methods and even develop such things themselves. As a person continues to do more of this latter work, perhaps through a career shift or promotion, that person moves up to the Super-Creative Core: producing transferable, widely usable new forms is now their primary function.

[. . .]

As the creative content of other lines of work increases – as the relevant body of knowledge becomes more complex, and people are more valued for their ingenuity in applying it – some now in the Working Class or Service Class may move into the Creative Class and even the Super-Creative Core. Alongside the growth in essentially creative occupations, then, we are also seeing growth in creative content across other occupations. A prime example is the secretary in today's pared-down offices. In many cases this person not only takes on a host of tasks once performed by a large secretarial staff, but becomes a true office manager – channeling flows of information, devising and setting up new systems, often making key decisions on the fly. This person contributes more than "intelligence" or computer skills. She or he adds creative value. Everywhere we look, creativity is increasingly valued. Firms and organizations value it for the results that it can produce and individuals value it as a route to self-expression and job satisfaction. Bottom line: As creativity becomes more valued, the Creative Class grows.

Not all workers are on track to join, however. For instance in many lower-end service jobs we find the trend running the opposite way; the jobs continue to be "de-skilled" or "de-creatified." For a counter worker at a fast-food chain, literally every word and move is dictated by a corporate template: "Welcome to Food Fix, sir, may I take your order? Would you like nachos with that?" This job has been thoroughly taylorized – the worker is given far less latitude for exercising creativity than the waitress at the old, independent neighborhood diner enjoyed. Worse yet, there are many people who do not have jobs, and who are being left behind because they do not have the background and training to be part of this new system.

[. . .]

COUNTING THE CREATIVE CLASS

It is one thing to provide a compelling description of the changing class composition of society, as writers like Bell, Fussell or Reich have done. But I believe it is also important to calibrate and quantify the magnitude of the change at hand. . . . Let's take a look at the key trends.

• The *Creative Class* now includes some *38.3* million Americans, roughly 30 percent of the entire U.S. workforce. It has grown from roughly *3* million workers in 1900, an increase of more than tenfold. At the turn of the twentieth century, the Creative Class made up just 10 percent of the workforce, where it hovered until 1950 when it began a slow rise; it held steady around 20 percent in the 1970s and *1980s*. Since that time, this new class has virtually exploded, increasing from less than *20* million to its current total, reaching *25* percent of the working population in 1991 before climbing to *30* percent by *1999*.

• At the heart of the Creative Class is the *Super-Creative Core*, comprising 15 million workers, or 12 percent of the workforce. It is made up of people who work in science and engineering, computers and mathematics, education, and the arts, design and entertainment, people who work in directly creative activity, as we have seen. Over the past century, this segment rose from less than 1 million workers in 1900 to *2.3* million in 1950 before crossing 10 million in 1991. In doing so, it increased its share of the workforce from *2.5* percent in 1900 to *5* percent in 1960, 8 percent in 1980 and 9 percent in 1990, before reaching 12 percent by 1999.

• The traditional *Working Class* has today *33* million workers, or a quarter of the U.S. workforce. It consists of people in production operations, transportation and materials moving, and repair and maintenance and construction work. The percentage of the workforce in working-class occupations peaked at 40 percent in 1920, where it hovered until 1950, before slipping to *36* percent in 1970, and then declining sharply over the past two decades.

• The *Service Class* includes 55.2 million workers or 43 percent of the U.S. workforce, making it the largest group of all. It includes workers in lower-wage, lower-autonomy service occupations such as health care, food preparation, personal care, clerical work and other lower-end office work. Alongside the decline of the Working Class, the past century has seen a tremendous rise in the Service Class, from 5 million workers in 1900 to its current total of more than ten times that amount.

It's also useful to look at the changing composite picture of the U.S. class structure over the twentieth century. In 1900, there were some 10 million people in the Working Class, compared to 2.9 million in the Creative Class and 4.8 million in the Service Class. The Working Class was thus larger than the two other classes combined. Yet the largest class at that time was agricultural workers, who composed nearly 40 percent of the workforce but whose numbers rapidly declined to just a very small percentage today. In 1920, the Working Class accounted for 40 percent of the workforce, compared to slightly more than 12 percent for the Creative Class and 21 percent for the Service Class.

In 1950, the class structure remained remarkably similar. The Working Class was still in the majority, with *25* million workers, some *40* percent of the workforce, compared to 10 million in the Creative Class (16.5 percent) and 18 million in the Service Class (30 percent). In relative terms, the Working Class was as large as it was in 1920 and bigger than it was in 1900. Though the Creative Class had grown slightly in percentage terms, the Service Class had grown considerably, taking up much of the slack coming from the steep decline in agriculture.

The tectonic shift in the U.S. class structure has taken place over the past two decades. In 1970, the Service Class pulled ahead of the Working Class, and by 1980 it was much larger (46 versus 32 percent), marking the first time in the twentieth century that the Working Class was not the dominant class. By 1999, both the Creative Class and the Service Class had pulled ahead of the Working Class. The Service Class, with 55 million workers (43.4 percent), was bigger in relative terms than the Working Class had been at any time in the past century.

These changes in American class structure reflect a deeper, more general process of economic and social change. The decline of the old Working Class is part and parcel of the decline of the industrial economy on which it was based, and of the social and demographic patterns upon which that old society was premised. The Working Class no longer has the hand it once did in setting the tone or establishing the values of American life – for that matter neither does the 1950s managerial class. Why, then, have the social functions of the Working Class not been taken over by the new largest class, the Service Class? As we have seen, the Service Class has little clout and its rise in numbers can be understood only alongside the rise of the Creative Class. The Creative Class – and the modern Creative Economy writ large – depends on this ever-larger Service Class to "outsource" functions that were previously provided within the family. The Service Class exists mainly as a supporting infrastructure for the Creative Class and the Creative Economy. The Creative Class also has considerably more economic power. Members earn substantially more than those in other classes. In 1999, the average salary for a member of the Creative Class was nearly $50,000 ($48,752) compared to roughly $28,000 for a Working Class member and $22,000 for a Service Class worker. . . .

I see these trends vividly played out in my own life. I have a nice house with a nice kitchen but it's often mostly a fantasy kitchen – I eat out a lot, with "servants" preparing my food and waiting on me. My house is clean, but I don't clean it, a housekeeper does. I also have a gardener and a pool service; and (when I take a taxi) a chauffeur. I have, in short, just about all the servants of an English lord except that they're not mine full-time and they don't live below stairs; they are part-time and distributed in the local area. Not all of these "servants" are lowly serfs. The person who cuts my hair is a very creative stylist much in demand, and drives a new BMW. The woman who cleans my house is a gem: I trust her not only to clean but to rearrange and suggest ideas for redecorating; she takes on these things in an entrepreneurial manner. Her husband drives a Porsche. To some degree, these members of the Service Class have adopted many of the functions along with the tastes and values of the Creative Class, with which they see themselves sharing much in common. Both my hairdresser and my housekeeper have taken up their lines of work to get away from the regimentation of large organizations; both of them relish creative pursuits. Service Class people such as these are close to the mainstream of the Creative Economy and prime candidates for reclassification.

CREATIVE CLASS VALUES

The rise of the Creative Class is reflected in powerful and significant shifts in values, norms and attitudes. Although these changes are still in process and certainly not fully played out, a number of key trends have been discerned by researchers who study values, and I have seen them displayed in my field research across the United States. Not all of these attitudes break with the past: Some represent a melding of traditional values and newer ones. They are also values that have long been associated with more highly educated and creative people. On the basis of my own interviews and focus groups, along with a close reading of statistical surveys conducted by others, I cluster these values along three basic lines.

Individuality. The members of the Creative Class exhibit a strong preference for individuality and self-statement. They do not want to conform to organizational or institutional directives and resist traditional group-oriented norms. This has always been the case among creative people from "quirky" artists to "eccentric" scientists. But it has now become far more pervasive. In this sense, the increasing nonconformity to organizational norms may represent a new mainstream value. Members of the Creative Class endeavor to create individualistic identities that reflect their creativity. This can entail a mixing of multiple creative identities.

Meritocracy. Merit is very strongly valued by the Creative Class, a quality shared with Whyte's class of organization men. The Creative Class favors hard work, challenge and stimulation. Its members have a propensity for goal-setting and achievement. They want to get ahead because they are good at what they do.

Creative Class people no longer define themselves mainly by the amount of money they make or their position in a financially delineated status order. While money may be looked upon as a marker of achievement, it is not the whole story. In interviews and focus groups, I consistently come across people valiantly trying to defy an economic class into which they were born. This is particularly true of the young descendants of the truly wealthy – the capitalist class – who frequently describe themselves as just "ordinary" creative people working on music, film or intellectual endeavors of one sort or another. Having absorbed the Creative Class value of merit, they no longer find true status in their wealth and thus try to downplay it.

There are many reasons for the emphasis on merit. Creative Class people are ambitious and want to move up based on their abilities and effort. Creative people have always been motivated by the respect of their peers. The companies that employ them are often under tremendous competitive pressure and thus cannot afford much dead wood on staff: everyone has to contribute. The pressure is more intense than ever to hire the best people regardless of race, creed, sexual preference or other factors.

But meritocracy also has its dark side. Qualities that confer merit, such as technical knowledge and mental discipline, are socially acquired and cultivated. Yet those who have these qualities may easily start thinking they were born with them, or acquired them all on their own, or that others just "don't have it." By papering over the causes of cultural and educational advantage, meritocracy may subtly perpetuate the very prejudices it claims to renounce. On the bright side, of course, meritocracy ties into a host of values and beliefs we'd all agree are positive – from faith that virtue will be rewarded, to valuing self-determination and mistrusting rigid caste systems. Researchers have found such values to be on the rise, not only among the Creative Class in the United States, but throughout our society and other societies.

Diversity and Openness. Diversity has become a politically charged buzzword. To some it is an ideal and rallying cry, to others a Trojan-horse concept that has brought us affirmative action and other liberal abominations. The Creative Class people I study use the word a lot, but not to press any political hot buttons. Diversity is simply something they value in all its manifestations. This is spoken of so often, and so matter-of-factly, that I take it to be a fundamental marker of Creative Class values. As my focus groups and interviews reveal, members of this class strongly favor organizations and environments in which they feel that anyone can fit in and can get ahead.

Diversity of peoples is favored first of all out of self-interest. Diversity can be a signal of meritocratic norms at work. Talented people defy classification based on race, ethnicity, gender, sexual preference or appearance. One indicator of this preference for diversity is reflected in the fact that Creative Class people tell me that at job interviews they like to ask if the company offers same-sex partner benefits, even when they are not themselves gay. What they're seeking is an environment open to differences. Many highly creative people, regardless of ethnic background or sexual orientation, grew up feeling like outsiders, different in some way from most of their schoolmates. They may have odd personal habits or extreme styles of dress. Also, Creative Class people are mobile and tend to move around to different parts of the country; they may not be "natives" of the place they live even if they are American-born. When they are sizing up a new company and community, acceptance of diversity and of gays in particular is a sign that reads "nonstandard people welcome here." It also registers itself in changed behaviors and organizational policies. For example, in some Creative Class centers like Silicon Valley and Austin, the traditional office Christmas party is giving way to more secular, inclusive celebrations. The big event at many firms is now the Halloween party: Just about anyone can relate to a holiday that involves dressing up in costume.

While the Creative Class favors openness and diversity, to some degree it is a diversity of elites, limited to highly educated, creative people. Even though the rise of the Creative Class has opened up new avenues of advancement for women and members of ethnic minorities, its existence has certainly failed to put an end to long-standing divisions of race and gender. Within high-tech industries in particular these divisions still seem to hold. The world of high-tech creativity doesn't include many African-Americans. Several of my interviewees noted that a typical high-tech company "looks like the United Nations minus the black faces." This is unfortunate but not surprising. For several reasons, U.S. blacks are under-represented in many professions, and this may be compounded today by the so-called digital divide – black families in the United States tend to be poorer than average, and thus their children are less likely to have access to computers. My own research shows a negative statistical correlation between concentrations of high-tech firms in a region and nonwhites as a percentage of the population, which is particularly disturbing in light of my other findings on the positive relationship between high-tech and other kinds of diversity – from foreign-born people to gays.

There are intriguing challenges to the kind of diversity that the members of the Creative Class are drawn to. Speaking of a small software company that had the usual assortment of Indian, Chinese, Arabic and other employees, an Indian technology professional said: "That's not diversity! They're all software engineers." Yet despite the holes in the picture, distinctive value changes are indeed afoot. . . .

T
W
O

"The Occidental City"

from *Occidentalism: The West in the Eyes of Its Enemies* (2004)

Ian Buruma and Avishai Margalit

Editors' Introduction

The destruction of the World Trade Center twin towers by al-Qaeda terrorists on September 11, 2001, was a turning point in the history of the world and ushered in a period of struggle that is likely to continue for decades, with the outcome still uncertain. It may also have been a turning point in the history of cities and urban civilization. There will be no mass exodus from the global urban centers. That prediction was made – and turned out to be unfounded – after World War II and the bombings of Dresden, Berlin, Hiroshima, and Nagasaki. But as long as the "war" between the liberal democracies and the forces of Islamist fundamentalism continues, the sense of security of urban dwellers in places like New York, London, Paris, Madrid – cities everywhere, but especially Western and Western-influenced cities – will forever be altered, tipped into a kind of psychological imbalance.

Following 9/11/2001, many theorists have attempted of make sense of militant Islamic fundamentalism and its hostility to the non-Muslim world, especially America and Europe. Some have posited a "clash of civilizations" and a continuation of the medieval crusades. Others have seen the conflict as one taking place mostly within the world of Islam, a struggle between hyper-conservative fundamentalists and modernizing moderates, with the attacks on Western cities being only secondary effects. Ian Buruma and Avishai Margalit contend that much of the hostility fueling Islamist terrorism is a deep-seated hatred of the urban culture of the West.

In 1978, Palestinian scholar and activist Edward Said published *Orientalism*, a ground-breaking cultural study that explored the way Westerners viewed the East as the mysterious and exotic "other." In *Occidentalism*, Buruma and Margalit argue that the Muslim East has just as many misapprehensions about the West, and that it is the Occidental city, conceived as a wicked and cynical harlot, that more than anything else embodies the evil tendencies of liberalism and global capitalism. The image of the harlot, the prostitute, is revealing because it combines notions of amoral economic exchange with sexual license – both of which are anathema to the devout and the puritanical mind of Islamic fundamentalists. Interestingly, this image of the city as the corrupting location of sexual depravity and moral ambiguity is a common trope in Western literature, from the *Epic of Gilgamesh* and the Bible through the poetry of T.S. Eliot and the gangster films of Hollywood. The attack on the World Trade Center, they write, was "at once a real and a symbolic attack, on New York, on America, and on the idea of America and the West it represents." Thus, a "deliberate act of mass murder played into an ancient myth – the myth about the destruction of the sinful city."

Dutch-born cultural critic Ian Buruma – who has been a fellow at the Berlin Wissenschaftscolleg, the Woodrow Wilson Center in Washington, and St Anthony's College, Oxford – is currently the Henry R. Luce Professor of Human Rights and Journalism at Bard College. Formerly the cultural editor of the *Far Eastern Economic Review* and foreign editor of *The Spectator*, he is a frequent contributor to the *New York Review of Books* and other major journals of liberal opinion. He is the author of *The Wages of Guilt: Memories of War in Germany and Japan* (New York: Farrar Straus & Giroux, 1994), *The Missionary and the Libertine* (London and Boston: Faber, 1996),

Bad Elements: Chinese Rebels from Los Angeles to Beijing (New York: Random House, 2001), and *Inventing Japan, 1853–1964* (New York: Modern Library Chronicles, 2003).

Academic and activist Avishai Margalit is the author of *The Ethics of Memory* (Cambridge: Harvard University Press, 2002) and, with Naomi Goldblum, *The Decent Society* (Cambridge: Harvard University Press, 1998). He was one of the founders of "Peace Now," and his essays and reviews regularly appear in the *New York Review of Books*. Born and raised in Israel/Palestine, he graduated from the Hebrew University of Jerusalem where he is now the Schlman Professor of Philosophy. He has taught and conducted research at Oxford, Harvard, the Free University of Berlin, the Max Planck Institute, Princeton, the European University at Florence, and the Central European University at Prague. In 1999, he delivered the Horkheimer Lectures at the University of Frankfurt. In 2001, he was awarded the Spinoza Lens Prize for his "significant contribution to the normative debate on society." And in 2005, he was the Tanner Lecturer at Stanford.

The literature on the War on Terrorism is vast, much of it is contentious and tendentious. For more information specifically about the effects of terrorism on cities, consult Stephen Graham (ed.), *Cities, War and Terrorism: Towards an Urban Geopolitics* (Oxford: Blackwell, 2004), Jon Coaffee, *Terrorism, Risk and the City: The Making of a Contemporary Urban Landscape* (Aldershot: Ashgate, 2003), James Harrigan and Philippe Martin, "Terrorism and the Resilience of Cities," (*Economic Policy Review*, 8(2), 30 November 2002), and Peter Marcuse, "The 'Threat of Terrorism' and the Right to the City," (*Fordham Urban Law Journal*, 32(4), 1 July 2005). For one analysis of how terrorism will affect the future of cities, consult Joel Kotkin, "The Urban Future" (p. 517). One specific change in the way cities will operate under the threat of both terrorism and domestic crime is increased security and the use of electronic surveillance. For more on this, see David Lyon, "Surveillance in the City," and Peter Huber and Mark Mills, "How Technology Will Defeat Terrorism," both in Stephen Graham (ed.), *The Cybercities Reader* (London and New York: Routledge, 2004), also Nicholas Fyfe and Jon Bannister, "City Watching: Closed Circuit Television Surveillance in Public Spaces" in Nicholas Fyfe and Judith Kenny (eds), *The Urban Geography Reader* (London and New York: Routledge, 2005).

The values of this Western civilization under the leadership of America have been destroyed. Those awesome symbolic towers that speak of liberty, human rights, and humanity have been destroyed. They have gone up in smoke.

– Osama Bin Laden

Soon after the two jumbo jets smashed into lower Manhattan, bringing the World Trade Center down in a blaze, videotapes went on sale in China showing the horrific highlights, spliced together with scenes from Hollywood disaster movies. It was as though the real thing – two flaming skyscrapers collapsing on thousands of people – were not dramatic enough, and only fantasy could capture the true flavor of such catastrophes, which most of us know only from the movies.

The deliberate conflation of reality and fantasy left an impression that the victims were not real human beings, but actors. And most were kept invisible anyway by the uncharacteristic modesty of the television networks, which refused to show suffering in close-up.

For at least a few seconds, unreality was the impression many people got when they switched on their television sets. To pretend it wasn't real was a convenient way of distancing oneself from the horror. For a distressingly large number of people, not only in China, the idea that this was a kind of movie, a purely imaginary event, an act of theater, also made it easier to feel something more sinister. The destruction of the towers – symbols of U.S. power and wealth; symbols of imperial, global, capitalist dominance; symbols of New York City, our contemporary Babylon; symbols of everything American that people both hate and long for – the destruction of all that, in less than two hours, gave some people, not only in China, a feeling of deep satisfaction.

In this peculiar sense, the annihilation of the Twin Towers and the people inside them was a huge success. For it was part of Osama bin Laden's war on the West, both physical and metaphysical; it was at once a real and a symbolic attack, on New York, on America, and on an idea of America, and the West it represents. A deliberate act of mass murder played into an ancient myth – the myth about the destruction of the sinful city. That purification was foremost in the minds of the attackers is clear from the last testimony of one of their leaders, a young Egyptian named Mohammed Atta. He expressed a horror of women and sexuality: "The one who will wash my body should wear gloves so that my genital parts should not be touched. . . . I don't want pregnant women or a person who is not clean to come and say goodbye to me because I don't approve of it."

Consumers of the Chinese tapes were not, however, so far as one can gather, poor villagers with a hatred of Americans, or even of city slickers, but young men in Shanghai, Beijing, and other large cities whose skyscrapers reach ever higher to rival those of New York. The West in general, and America in particular, provokes envy and resentment more among those who consume its images, and its goods, than among those who can barely imagine what the West is like. The killers who brought the towers down were well-educated young men who had spent considerable time living in the West, training for their mission. Mohammed Atta received a university degree in architecture in Cairo before writing a thesis on modernism and tradition in city planning at the Technical University in Hamburg. Bin Laden himself was once a civil engineer. If nothing else, the Twin Towers exemplified the technological hubris of modern engineers. Its destruction was plotted by one of their own.

The reaction in many places to the American disaster was, in any case, more than schadenfreude over the misfortunes of a great and sometimes overbearing power, and went deeper than mere dissatisfaction with U.S. foreign policy. There were echoes of more ancient hatreds and anxieties, which recur through history in different guises. Whenever men have built great cities, the fear of vengeance, wreaked by God, or King Kong, or Godzilla, or the barbarians at the city gates, has haunted them. Since ancient times, humans have lived in terror of being punished for their effrontery in challenging the gods, by stealing fire, or gaining too much knowledge, or creating too much wealth, or building towers that reach for the skies. The prob-lem is not with the city per se, but with cities given to commerce and pleasure instead of religious worship. In the case of Osama bin Laden and Mohammed Atta this religious impulse curdled into a dangerous madness.

Hubris, empire building, secularism, individualism, and the power and attraction of money – all these are connected to the idea of the sinful City of Man. Myths of their destruction have existed as long as men built cities in which to trade, accumulate wealth, gain knowledge, and live in comfort.

The fear of punishment for challenging the power of God, for the hubris of thinking we can go it alone, is common to most religions. The story of Babylon and its great tower is one of the oldest known to man. After the great Flood, Nimrod built the city of Babylon. In another account, by the historian Diodorus Siculus, a powerful queen named Semiramis was its builder. She was later associated with a mother goddess cult. Perhaps it was the relative sexual freedom of Babylonian women that prompted pious Jews and Christians to describe their city as "the mother of prostitutes and of the abominations of the earth" (Rev. 17:5). The people of Babylon, rather like the citizens of fourteenth-century Florence, or twenty-first-century New York, lusted after worldly fame. "Come," they said, "let us build ourselves a city, and a tower with its top in the heavens, and let us make a name for ourselves" (Gen. 11:4).

The Tower of Babel was probably a ziggurat, a building with a circular staircase that followed signs of the zodiac. The Babylonians were keen astrologers. Astrology was their way of exploring the workings of nature, of gaining knowledge. God decided to punish these uppity infidels. "Behold," God said, "they are one people, and they all have one language; and this is only the beginning of what they will do; and nothing that they propose to do will be impossible for them" (Gen. 11:6). Then God had his revenge: one language became many languages, the people were scattered all over the world, and the tower was abandoned.

Nebuchadnezzar, a later ruler of Babylon who conquered Jerusalem (very much the City of God), enslaved the Jews, and had visions of a kingdom of gold, was punished for his hubris as well and driven away to "eat grass like cattle." It is not the least of history's ironies that the Jews, who wrote this tale of vengeance against the City of Man, would in later centuries themselves be scattered around the world, speak many languages, and be described by their enemies as rootless cosmopolitans addicted to visions of wealth.

That many people in Muslim countries believe that the destruction of the Twin Towers was in fact the work of Mossad, the Israeli intelligence agency thought to be at the heart of a worldwide Jewish conspiracy, is as unhinged as the religious extremism of Al Qaeda activists. But it is not entirely unexpected. Jews have often been blamed for their own persecution, and anti-Semitism can wrap itself around strange paradoxes. Capitalist conspiracies were associated with the Elders of Zion, but so was communism. There is a possible link. Both capitalism and Bolshevism, opposites in almost every respect, might still be described as attempts to replace the world of God with the world of Man.

The image of the metropolis as a whore is not just a reflection of female sexuality, so feared and loathed by puritans such as Mohammed Atta, but also a comment on a society that revolves around trade. In the city, conceived as a giant marketplace, everything and everyone is for sale. Hotels, brothels, and department stores sell fantasies of the good life. Money allows people to behave in all manners to which they were not born. City people are seen as liars. In Juvenal's satire on ancient Rome, a city of flatterers, robbers, and traders from all over the empire, we find the following sentence: "What can I do in Rome? I never learnt how to lie." Rome, to Juvenal, was a city where "of all gods it's Wealth that compels our deepest reverence," a city where foreigners mixed freely with natives: "Filthy lucre it was that first brought loose foreign morals amongst us, effeminate wealth that with vile self-indulgence destroyed us over the years." Juvenal reserved his greatest bile for Greeks and Jews, and for women, "high-born or not," who would do anything to satisfy "their hot wet groins."

The most symbolic figure of commodified human relations, relations based on flattery, illusion, immorality, and cash, is the prostitute. The trade in sex is perhaps the most basic form of urban commerce. No wonder, then, that hostile visions of the City of Man always come back to this. One of the cliches of erotic trade is that you can buy a person's body, but never her soul. The whore, in her (or his) professional capacity, is soulless, and thus not really human. In their journals, the Goncourt brothers describe a famous courtesan in Paris named Païva. She plied her trade in the 1860s, the dawn of the industrial age: "She was coming forward between the chairs like an automaton, as if she was worked by a spiral spring, without a gesture, without expression . . . [a] rolling puppet from a dance macabre . . . a vampire with the blood of the living on her purple mouth while all the rest was livid, glazed and in dissolution." Here we have it, the Occidentalist view of the city, of capitalism, and of Western "machine civilization": the soulless whore as a greedy automaton.

[. . .]

The question is: when did the idea of the city as a wicked symbol of greed, godlessness, and rootless cosmopolitanism become almost totally associated with the West? When did the Western metropolis become the prime focus of Occidentalist loathing? After all, the great city, containing many races, was hardly an exclusively European or American phenomenon. Muslims, traditionally, were not haters of big cities. On the contrary, in early Islam, urbanism was promoted as a way to break away from nomadic ignorance. For centuries Baghdad and Constantinople had been centers of trade, learning, and pleasure. Farther east, the wealth and opulence of Beijing dazzled a traveler from thirteenth-century Venice. Compared with the refinements of China, seventeenth-century Amsterdam, with all its wealth, had the modest allure of a provincial town. Until the late nineteenth century, Edo, the Japanese capital, was bigger and more densely populated than any European city, including London.

And yet the modern idea of Babylon is now firmly rooted in the West, for the first Occidentalists were Europeans. Richard Wagner once wrote, about his Germanic hero Tannhäuser's sojourn amid the dangerous seductions of the Venusberg: "I agree with Friedrich Dieckmann's argument that the Venusberg stands for 'Paris, Europe, the West': that frivolous, commercialized, and corrupt world in which 'freedom and also alienation' are more advanced than in our 'provincial Germany with its comfortable backwardness.'"

Wagner's sentiments about Paris reflect more than a distaste for French frivolity. People may dislike cities for all kinds of reasons. But the antiurban bias of Occidentalists goes further. It sees the great city as inhuman, a zoo of depraved animals, consumed by lust. The city dweller, from this perspective, has literally lost his soul.

It was the age of empires, spurred by an extraordinary burst of scientific, industrial, and commercial enterprise that made Europe into the metropolitan center, dominating the periphery to which much of the rest of the world had been reduced. Wagner's antipathies against France – and his notion of Germany as the provincial periphery – were a legacy of Napoleon's domination, but the empires that reached the height of their power in the latter half of the nineteenth century were commercial empires, driven by the pursuit of wealth more than a desire for military conquest or spreading God's word. The greatest metropole of all, the commercial imperial capital of the nineteenth-century world, was London. And the greatest industrial city, the capital of dark satanic mills, was Manchester. Paris rivaled London as a cosmopolitan center, and Berlin was always desperately trying to catch up. All these cities inspired fear as well as envy and, like New York two centuries later, came to stand for something particularly hateful in the eyes of those who sought to eradicate the impurities of urban civilization with dreams of spiritual or racial purity.

[. . .]

The story of the lonely outsider, ignored or abused in the big city, is common everywhere. It is often told in the darker tales of Hollywood, where the big city is New York, Chicago, or L.A. It became a cliché in films made in India, Thailand, and Japan in the 1950s. Many of them are gangster pictures. The young man leaves his village, driven by hunger or ambition, his head filled with stories of vast riches and easy women. What he finds instead are the uncaring crowds, and the tricksters and cheats who rob him of his tiny savings. Finally he loses his dignity too, when he learns how to become a robber himself. Sometimes he joins a criminal gang, where some of his traditional village codes of conduct are reenacted in perverse ways, and sometimes he tries to survive alone. But almost always he loses in the end, exploited by a gang boss or some other person he thought he could trust. The climax is an explosion of suicidal violence, when the long-suffering outsider, like Samson among the Philistines, brings down the city pillars in a final act of catastrophic vengeance.

A common feature of the bad, cold, calculating, rich villains in these morality tales is not just their sexual depravity, their greed, or their dishonesty, but their flashy Western ways. In European gangster films, the bad *guys* dress and behave like Americans; in non-Western movies, they behave like phony white men. The wicked gangsters in 1950s Japanese movies use guns, drink whisky, and wear suits, while the kimonoed heroes fight only with traditional samurai swords. In most countries, the typical gangster movie is hostile to the modern world. So of course is the typical American western, where the villains are city slickers from "out east," who come to build cities in the western plains, connected by the new railroads. Old relations of trust between the honest rural folks are replaced by dodgy contracts drawn up by men in suits. It is a universal story, this clash between old and new, authentic culture and metropolitan chicanery and artifice, country and city.

[. . .]

When Sayyid Qutb, one of the most influential Islamist thinkers of the last century, arrived in New York from his native Egypt in 1948, he felt miserable in the city, which appeared to him as a "huge workshop," "noisy" and "clamoring." Even the pigeons looked unhappy in the urban chaos. He longed for a conversation that was not about "money, movie stars or car models." In his letters home, Qutb was particularly distressed by the "seductive atmosphere," the shocking sensuality of daily life, and the immodest behavior of American women. A church dance in remote Greeley, Colorado, hardly a metropolitan place, struck him as wickedly lascivious. Qutb was a defender of the ideal of a pure Islamic community, against what he saw as the empty, idolatrous materialism of the Occident . . . Life in America simply confirmed his prejudices. But like all dreams of purity, his ideal of the spiritual community was a fantasy, which contained the seeds of violence and destruction.

[. . .]

"Visions of a New Reality: The City and the Emergence of Modern Visual Culture"

(1999)

Frederic Stout

A fundamental precept of Marxist cultural analysis is that superstructures of thought and artistic expression rest upon and derive from a material base rooted in social and economic reality. Thus, each historical era creates characteristic forms of expression and explanatory discourse that reflect, indeed construct, the social reality of the period. Lewis Mumford spoke to this process when, in *The City in History* (1961), he wrote that in "the Book of Job, one beholds Jerusalem; in Plato, Sophocles, and Euripides, Athens; in Shakespeare and Marlowe . . . Elizabethan London."

For the cities of the Industrial Revolution, a number of forms of expression and modes of critical analysis arose to make sense of the dramatic and rapidly changing social reality. In literature, Balzac and Dickens pioneered a tradition of social realism particularly focused on the struggles of the urban poor which later attracted such followers as Emile Zola, Theodore Dreiser, Mrs. Gaskell, Upton Sinclair . . . and thousands more. In the realm of social and political analysis, the use of statistical evidence based on survey data compiled by government commissions and philanthropic organizations, the construction of great explanatory theories (such as those of Marx and Weber), and the development of social science methodologies helped observers of the urban milieu make sense of the revolutionary changes that were taking place around them. And in the realm of art, a whole new kind of visual culture emerged rooted in the observation of the new urban reality, both social and physical. It was a culture that began with nineteenth-century popular illustrated journalism and evolved to include mainstream traditions in the twentieth-century history of photography and cinema.

In the history of the fine arts, urban social realism plays a brief but highly visible role. Although much artistic social activism was largely confined to the satirical printmakers (except in those countries where state-sponsored socialist realism became the prescribed orthodoxy), works like William Hogarth's depictions of eighteenth-century London and Honoré Daumier's glimpses of nineteenth-century Paris are as influential as they are familiar. In painting, however, the realism of Courbet and Millet quickly give way to the impressionist celebration of light and the post-impressionist analysis of pure form. In America, the urban imagery of The Eight (dubbed "The Ashcan School" by a hostile critic) was revolutionary and shocking at their group show in 1907 but strictly old hat by the time modern abstraction was introduced at the Armory Show in 1912. One explanation for this is that The Eight stood not at the beginning but at the end of a fifty-year social realist tradition in the visual arts, a tradition that took place not in the salons of the easel painters but in the pages of the illustrated newspapers and magazines that flourished as a ubiquitous element of urban popular culture during the last half of the nineteenth century. Several of the Ashcan School painters – among them John Sloan, William Glackens, Everett Shinn, and Edward Hopper – had previously worked as journalistic illustrators, and it is to the pages of the popular newspapers that one must turn to see the earliest representations of the modern industrial city.

Illustrated journalism began in England as early as the 1820s, and in America in the 1840s, in response to a growing demand for information and entertainment on the part of a marginally literate but fully enfranchised working class and petit bourgeoisie. The first images of cities were almost always long-range or bird's-eye views (similar to the view of La Crosse, Wisconsin, from an 1887 issue of *Frank Leslie's Illustrated Newspaper* that appears as Plate 5). These static images were imitative of a received landscape tradition that had been commonly applied to rural and wild nature scenes, but they conveyed relatively little information about urban subject matter – note the inclusion of a street-scene cut-out in the La Crosse picture – and were soon replaced by more kinetic images and vignettes. "Contrast pictures" – both dual images such as "Music in the Street/Music in the Parlor" (Plate 9) from an 1868 issue of *The Illustrated News* and single images such as "The Hearth-stone of the Poor" (Plate 10) from an 1876 issue of *Harper's Weekly* by the exceptionally talented Sol Eytinge, Jr. – were particularly useful for comparing the lives of the urban rich and poor. Soon, a new kind of composite image emerged – "City Sketches" by the illustrator C.A. Barry from an 1855 issue of *Ballou's Pictorial Drawing-Room Companion* (Plate 11) is an interesting early example – that more accurately reflected the diverse, jumbled-together chaos that was the common experience of urban street life in the nineteenth century. Eventually, such composite pictures became a staple of popular illustrated journalism, and the step-by-step depiction of industrial processes became a favored topic, as in "Bicycles and Tricycles – How They Are Made" (Plate 12) from an 1887 issue of *Frank Leslie's Illustrated Newspaper*.

Handdrawn popular illustration died out around the turn of the century when a new technology of visual representation, photography, took over the journalistic duties of the newspaper illustrators. As a genre, popular illustration had accomplished much. It had created a visual culture embedded in the social reality of urban life and had urged visual art generally away from landscape toward cityscape, from stasis to kinesis. The contribution of photography would be even greater. Here was a technology of visual representation perfectly suited to its age. In the hands of journalists, it created a new and powerfully accurate kind of documentary record. In the hands of social activists, it created images of passion and outrage that could not be ignored. In the hands of the popular masses, it created snapshots and memories, both social and personal. And in the hands of artists – and who with a camera in his hands was not an artist? – it created art.

Photography is arguably the single most characteristic medium of expression and representation of the modern period, and a complete history of photography would include consideration of its use in science, medicine, criminology, education, and commerce as well as art. But as a medium of artistic expression, photography perfectly exemplified the spirit of the modern age, addressing every known genre and inventing new ones of its own. Nature and landscape, still life and the nude, staged heroic set pieces and candid genre scenes – all found new vitality in the hands of the masters of the new technology of visual representation. But it was urban subjects more than any others – the inescapable social and physical reality of the modern city – that captured photography's unique potential as an expressive medium and gave photography its historic and artistic *raison d'être*.

Although photography did not completely supplant handdrawn illustrations in journalism until the 1890s and early 1900s, the very first news photo may well have been "Burning Mills, Oswego, New York" (Plate 13), a daguerreotype by George N. Barnard that was printed in the *Oswego Daily Times* in 1853. While pictures of factory fires had been a staple of popular illustration, photography brought a new realism and intensity to the images presented to the public. Even today, dramatic pictures of big fires, even if they are from cities far away, have a special fascination and are often prominently placed in newspapers and the evening television news broadcasts.

The spectacle of the burning factory was a special case in a larger melodrama: the ongoing, day-to-day reality of modern urban life. Many of the very earliest photographs by Daguerre, Nadar, and Fox Talbot were simple street scenes made by the photographer simply turning his camera out the studio window. Like the "city sketches" type of popular illustration, these photographic street scenes conveyed a special sense of the vitality of urban life. Edward Anthony's "A Rainy Day on Broadway" (Plate 14) shows New York in 1859. The motion of those on the street and on the sidewalks is frozen in time – clearly Anthony's shutter speed was fast enough to eliminate most of the blurring that was so common in early photographs of moving subjects – but the image itself does not seem frozen. Rather, the image speaks directly to the viewer's own experience of street life, and the interaction of the viewed image and the viewer's imagination almost magically captures the hustle and bustle of New York street life.

Not all streets in the modern city were so pleasant. In "Bandits' Roost, 39½ Mulberry Street," of 1889 (Plate 15), Jacob Riis invites the viewer to look into one of New York's most notorious and dangerous back alleys. The author of *How the Other Half Lives* (1890), Riis was a crusading social reformer who had observed the dark side of the city as a police reporter for the local newspapers and he used his skills as a photographer to supplement his activism. Many middle-class viewers were undoubtedly repelled by images such as this one. It was the kind of view that one might catch fleetingly as one passed along a main thoroughfare, something to glance at quickly and then turn away. But Riis the reformer would have us not look away, and photography rivets the gaze on an unpleasant but inescapable social reality. In this, Riis's photographic work is directly linked to Friedrich Engels's descriptions of the Manchester slums in *The Condition of the Working Class in England in 1844.*

Jacob Riis was but one in a long and distinguished line of photographers who turned their art and their talents to the ends of political activism and social protest. One of the greatest of these visual documentors of the modern city was Lewis Hine. Hine's picture of a young girl tending a massive cloth-weaving machine in a North Carolina mill has become a nearly universal icon of industrial child labor, and his photographs of high-steel construction workers casually toiling hundreds of feet above the city are gut-wrenching images of working-class heroism. For a time, Hine worked at Ellis Island, the New York port of entry for immigrants to America. His portraits of the immigrants (Plate 16) are masterpieces of photographic humanism. Portraiture holds an important place in the history of photography. Whereas oil-painting portraiture had served the interests of the aristocracy and the haute bourgeoisie almost exclusively, the new medium allowed members of the broad urban middle class, and the new industrial working class as well, to record their images for posterity. And whereas the images of immigrants in popular illustration had tended toward racial and ethnic caricatures, Hine's portraits reveal the individuality and humanism behind the stereotypes. Another photograph depicting the immigrant condition – all the more powerful for its juxtaposition of social commentary and pure artistic composition – was Alfred Stieglitz's "The Steerage" of 1907 (Plate 17). Here, the interplay between the immigrants, mostly enveloped in shadows, and the physical structures of gangway and ladder combine simultaneously to fascinate the eye and touch the soul.

The Great Depression of the 1930s strengthened the documentary and social activist tendencies in photography that had been pioneered by Riis, Hine, and the photo-journalists, especially in work that had been commissioned by the Works Progress Administration and other New Deal agencies. The photographs of Southern sharecroppers by Walker Evans and the images of Harlem streetlife by Gordon Parkes are notable examples of the intersection of art and social protest. Dorothea Lange was another activist photographer who, like Evans, was drawn to the documentation of rural poverty. Her pictures of dispossessed dust bowl emigres are among the most famous and most powerful images of the Depression era. And in her picture of a roadside sign (Plate 18), published in a 1939 Farm Security Administration report on rural migration, Lange captured both the bleakness of the Depression and the kinds of fears and insecurities that underlay Frank Lloyd Wright's "Broadacre City" proposal – a homestead of at least one acre per person – and other back-to-the-land schemes.

The people of the city and their problems were the dominant, but not the sole, subject of urban documentary photography. The very physical presence of the modern city was often evoked through images of architecture and infrastructure. Stieglitz's moody impression of the Flatiron Building cloaked in mist is one well-known example, and Margaret Bourke-White's formalist, monumental depiction of Hoover Dam that appeared on the cover of the first *Life* magazine is another. Charles Sheeler's "Ford Plant, Detroit" of 1927 (Plate 19) demonstrates the way in which pure architectural form – an assemblage of strong verticals and diagonals revealed through light and shadow – could be read both as pure cubist composition and powerful, if somewhat distanced, social documentation.

If the architecture of the city could present images of foreboding power or lyrical freedom, it was nonetheless the lives of the people of the city themselves that became the constantly recurring theme of photojournalists and artists alike. The work of Weegee (Arthur Fellig) is a case in point. Although most of his pictures were made in the 1940s and 1950s and featured startling images of auto wrecks, transvestite bars, and gangland killings, the individual faces communicate directly to the viewer with a sense of direct, immediate reality that is the transcendent essence of urban life. "The Critic" (Plate 20), for example, was taken in 1943, but the people and the streetlife situation – so like the contrast pictures in popular illustration – are instantly recognizable as big-city icons of the painful urban conflation of luxury and poverty, indifference and despair.

The project of comprehending the modern city visually played, and continues to play, a central role in the history of art and consciousness. The great themes of the city – its kinetic activity, its juxtapositions and ironies, its massive forms and tiny details – provided the artist with a subject matter that could not be ignored and pioneered modes of visual perception and communication that were to fundamentally transform the nature of social life. Whereas popular illustrated journalism liberated visual images from the control of privileged elites, photography completed the democratization of the visual by placing it in the hands of the masses.

Eventually, the new modes of visual perception would be connected to narrative – particularly to the "urban narratives" of young rural innocents encountering the experience of the city for the first time, a plot line as old as Gilgamesh and given renewed meaning as a result of the millions of immigration stories that fueled the urbanization process during the nineteenth and twentieth centuries. And out of the interconnection of urban narrative and visual representation would come cinema, the ultimate realization of the kinetic imagery that urban life characterized and photography mirrored. In King Vidor's *The Crowd* of 1926 (Plate 21), we see the once-optimistic individual lost in the enveloping urban mass. In Fritz Lang's *Metropolis* of 1929 (Plate 22), we see the modern citadels of faceless power looming over the dehumanizing structures of class segregation and oppression. In both cases, the viewer confronts and subconsciously confirms the artist's perception of the reality of modern urban life. The confrontation may be discomfiting, but the ubiquity of modern visual culture makes the confrontations inescapable and memorable, perhaps more memorable even than actual encounters on the streets of the city. Writing about the historic city, Lewis Mumford found that they were characterized by an "urban drama" and that the dramatic dialogue was "one of the ultimate expressions of life in the city." In the modern city, it is the image – sometimes celebratory, sometimes haunting, always definitive in its explanatory value – that is paramount both as spectacle and revelation.

MUSIC IN THE STREET.

MUSIC IN THE PARLOR.—p. 15.

Plate 9 **"Music in the Street/Music in the Parlor"** (Unknown artist, 1868, *The Illustrated News*).

Vol. XX.—No. 998.]   NEW YORK, SATURDAY, FEBRUARY 12, 1876.   [WITH A SUPPLEMENT.
PRICE TEN CENTS.

Entered according to Act of Congress, in the Year 1876, by Harper & Brothers, in the Office of the Librarian of Congress, at Washington.

THE HEARTH-STONE OF THE POOR—WASTE STEAM NOT WANTED.—DRAWN BY SOL EYTINGE, JUN.—[SEE PAGE 151.]

Plate 10   **"The Hearth-stone of the Poor"** (Sol Eytinge, Jr., 1876, *Harper's Weekly*).

Plate 11 **"City Sketches"** (C.A. Barry, 1855, *Ballou's Pictorial Drawing-Room Companion*).

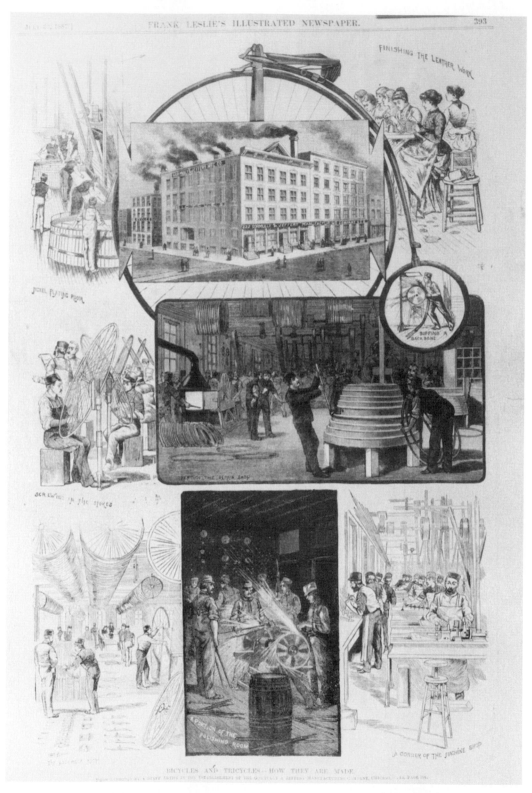

Plate 12 **"Bicycles and Tricycles – How They Are Made"** (Unknown artist, 1887, *Frank Leslie's Illustrated Newspaper*).

Plate 13 **"Burning Mills, Oswego, New York"** (George N. Barnard, 1853, *Oswego Daily Times*). (Courtesy of George Eastman House.)

Plate 14 **"A Rainy Day on Broadway"** (Edward Anthony, 1859). (Courtesy of George Eastman House.)

Plate 15 **"Bandits' Roost, 39½ Mulberry Street"** (Jacob Riis, New York, 1889). (The Jacob A. Riis collection # 101. Copyright © Museum of the City of New York.)

WELSH COAL MINER

SLAVIC STEEL-WORKER

Photos by Hine

ITALIAN LABORER

IRISH IRON WORKER

Plate 16 Ellis Island immigrant portraits (Lewis Hine, *c.* 1910). In Grove S. Dow, *Social Problems of Today* (New York: Thomas Y. Crowell, 1925). Public domain.

Plate 17 **"The Steerage"** (Alfred Stieglitz, 1907). (Courtesy of the Museum of Modern Art, New York.)

Plate 18 **Untitled (roadside sign)** (Dorothea Lange, 1939). Farm Security Administration, US Department of Agriculture. Printed in C.E. Lively and Conrad Tauber, *Rural Migration in the United States* (Washington, DC: Works Progress Administration, 1939).

Plate 19 **"Ford Plant, Detroit"** (Charles Sheeler, 1927). (Courtesy of the Museum of Modern Art, New York.)

Plate 20 **"The Critic"** (Weegee (Arthur Fellig), 1943). (Courtesy of the International Center of Photography, New York. Bequest of Wilma Wilcox.)

Plate 21 Still from ***The Crowd*** (King Vidor, 1926). (Courtesy of the Museum of Modern Art, New York, Film Stills Archive.)

Plate 22 Still from *Metropolis* (Fritz Lang, 1929). (Courtesy of the Museum of Modern Art, New York, Film Stills Archive.)

PART THREE

Urban space

INTRODUCTION TO PART THREE

The physical form of cities described in Part One on "The Evolution of Cities" ranged from Ur's mud brick walls and ziggurat to the marble agora of the Greek polis; from the polluted slums of nineteenth-century Manchester to the high-tech research "campuses," malls, and residential areas of today's technoburbs. The size, density, spatial distribution of functions, and physical and cultural characteristics of cities described in that section display an equally astonishing variety over time.

This section is concerned with urban space. It contains classic and contemporary writings on the spatial aspects of cities. Of all the social science disciplines geography is most centrally concerned with urban space. This section introduces material on urban geography that is much more fully developed in another reader in this series: Nicholas Fyfe and Judith Kenny (eds) *Urban Geography Reader* (London and New York: Routledge, 2005). Parts Five, Six, and Seven of this reader on "Urban Planning History and Visions", "Urban Planning Theory and Practice" and "Perspectives on Urban Design" return to material on urban space, but extend the discussion to interventions to shape the physical form and design of cities and regions.

For millennia geography focused on the mapping of topographical and political features. It was not until the 1950s that urban geography emerged as an important subfield of geography. Today urban geography has been reinvigorated by a new generation of scholars who define the discipline broadly to include a range of topics. Geographical Information Systems (GIS) software permits geographers and other social scientists to do computerized spatial analysis with a power and precision unimaginable in the past. The *Urban Geography Reader* contains selections dealing with topics as varied as globalization; restructuring; politics, governance, and inequality; difference; form and symbolism; and technologies (Fyfe and Kenny, 2005).

Scholars from many disciplines have contributed to our understanding of urban space. In addition to geographers such as Edward Soja and David Harvey, this selection includes writings from the disciplines of sociology (Ernest Burgess and Saskia Sassen), landscape architecture (J.B. Jackson), urban planning (Andrés Duany and Elizabeth Plater-Zyberk), and urban design (Ali Madanipour). Writers from these and other disciplines throughout this book inform our understanding of urban geography.

Many cities have grown organically with little or no explicit overall plan or centrally controlled regulation. They contain vernacular architecture created by ordinary people without specialized training in the design professions. Other cities display a mix of organic growth and development planned at different periods in their history. The old crooked streets and irregular lots that constitute most of central Boston, Massachusetts, contrast with the regularity of Back Bay. There a swampy area was developed during the latter part of the nineteenth century following a plan by Frederick Law Olmsted. Disorderly districts of London that have evolved since the Middle Ages contrast with the area rebuilt after the great fire of 1666. Only a few entire cities, like Australia's capital city Canberra and Brazil's capital city Brasilia, have been built according to a consistent master plan – and even they show vernacular building that does not fit the plan.

It was sociologists, not geographers, who pioneered the modern field of urban geography. In addition to path-breaking work on the social structure of cities by Louis Wirth (p. 90), Robert Park, Ernest W. Burgess and others members of the Chicago School of urban sociology developed the first systematic theory about the physical form of cities. The first selection in this section, Ernest W. Burgess's "The Growth of the City" (p. 150), advances many provocative hypotheses. Burgess was interested in understanding the internal structure of a single city; not the way in which multiple cities were related to each other. Burgess argued that there was an underlying logic to the physical form of cities. In his view they were organized in a series of concentric rings moving out from a central business district. Each ring had a distinct set of residents and functions. Physical form and human life were intimately linked. According to Burgess, cities were not static. Chicago School sociologists were strongly influenced by theory about the way in which plant and animal communities evolved and changed. Drawing on the social ecology perspective, Burgess envisioned cities as dynamic organisms with a constant flow of new residents coming into the inner rings and a flow from these rings outward over time. Each part of an urban system had distinct characteristics and played a unique role in the total system. But each was related to the others and the whole system was in a state of constant flux.

Sociologists inspired by Burgess, land economists, and many geographers have studied city form during the last seventy-five years. Even in the large postmodern Los Angeles region, geographer Edward Soja (p. 166) draws on this early theory and discerns a concentric and sectoral logic. Burgess's "concentric zone" theory is just one of a number of theories that seek to describe the internal structure of cities theoretically. It is often juxtaposed with theories developed by real estate economist Homer Hoyt in the 1930s postulating an organization of cities in sectors radiating out along transportation corridors, and with an essay by geographers Chauncy Harris and Edward Ullman written in 1945 concluding that most cities have multiple nuclei rather than either a concentric zone or sectoral organization.

Burgess wrote about the way in which a *single city* is organized. A related topic in the study of urban form concerns *systems of cities*. Central place theory, developed by German geographer Walter Christaller, focuses on how cities within an entire region relate to each other. In this tradition Saskia Sassen (p. 197) argues that a world system of cities is emerging. As computerized statistical analysis and geographical information systems for analyzing and mapping data rapidly increase our ability to understand urban form, the debates about the internal structure of cities and systems of cities continue.

The selection by landscape architect J.B. Jackson (p. 184) also notes an underlying logic to the physical form, social structure, and function of cities. According to Jackson even small cities that evolved without city planners and architects display an underlying regularity and logic. For Jackson the built environment of every place expresses the belief system and values of the humans who created it. Despite the fact that until recently most cities have grown with little or no formal planning or professional design, little scholarly attention was devoted to studying vernacular urban form and popular design until J.B. Jackson pioneered the study of *vernacular* landscapes, the environments built by common people without trained architects, planners, or other professionals. Jackson was fascinated with the physical form of barns, fences, billboards, and grain silos; pioneer settlements with stumps in the field and muddy roads; dying former railroad towns where the train no longer stops; humble mobile homes in rural New Mexico; and gas stations and frozen custard stands along the principal artery across the American Southwest, Highway 66. Jackson was one of the first to argue that a study of vernacular architecture provides important understanding of the culture and values of the people who have built the built environment. According to Jackson, a close look at Highway 66 in the 1950s reveals a great deal about the worldview of the highway builders and residents along the highway. Today, largely as a result of Jackson's influence, there is a large literature describing and analyzing the function and cultural meaning of Las Vegas casinos, White Tower hamburger stands, suburban tract homes, billboards, and other vernacular architecture.

Urban design professor Ali Madanipour is interested in the relationship between society and the built environment. In his essay in this section (p. 158) Madanipour is particularly concerned with the way in which Europeans may knowingly or unwittingly exclude people from other cultures from the full benefits of their cities. Madanipour distinguishes between *economic discrimination*, in which members of a group are excluded from access to employment, *political discrimination*, in which they are excluded from political power, and *cultural exclusion* in which the group members are marginalized from the symbols, meanings, rituals, and discourses of the dominant culture. Since exclusion often has a spatial dimension he suggests a number of strategies to break down spatial exclusion and increase inclusion. Subsidized housing, for example, may permit low-income foreign immigrants to live in parts of a city they could not otherwise afford, giving them access to job opportunities and better education for their children.

University of California, Los Angeles, geography professor Edward Soja uses the Los Angeles metropolitan region as the basis for a discussion of the postmodern metropolis (p. 166). He is bemused by the bewildering contrasts to be found in the Sixty-Mile Circle surrounding downtown Los Angeles. The pieces of the Los Angeles landscape Soja takes apart display astonishing contrasts: huge military bases and tiny eighteenth-century missions; gated communities for the rich and squalid ghettos for the poor; buildings and whole regions engaged solely in post-industrial research and managerial activities; and a thriving industrial sector employing low-wage immigrant labor. Soja emphasizes the fantastical imagery and theme park quality of the Los Angeles region. But underlying his irony is a serious message. Beneath the glitz and glitter, Los Angeles has a dark side. According to Soja, police stations, courts, and prisons are central downtown features. The region's industry could not exist without poorly paid immigrant labor. Violent and depressing ghettos are threaded through the Los Angeles region. The natural ecology of this congested and smog-blanketed metropolis is in deep trouble.

In "Fortress L.A." (p. 178), social critic Mike Davis addresses the same issue – the dark side of the postmodern metropolis – and he also uses Los Angeles as his exemplar city. Davis describes a built environment complete with surveillance cameras, barrel-shaped park benches designed to keep people from sleeping on them, overhead sprinklers to douse the homeless, windowless concrete hotel walls facing streets, entrances to public buildings reminiscent of the fortifications in front of medieval European castles, and gated communities where the rich can be (or at least feel) safe from outsiders. All these artifacts provide a disturbing glimpse of Los Angeles and, by implication, postmodern cities emerging around the world.

It is not just homeless people and local criminals that inspire fear in cities around the world today. Terrorism poses a threat of deadly attacks not only in Baghdad and the Gaza Strip but, as the attacks on the World Trade Center and the London Underground tragically prove, in cities worldwide.

Andrés Duany and Elizabeth Plater-Zyberk (p. 192) are interested in urban space because as architects and urban planners space is critical to their work. Their selection is an excellent illustration of how professionals who create the built environment can use insights from the academic and theoretical writings of geographers, sociologists, and anthropologists to create humane and interesting cities. Duany and Plater-Zyberk provide an antidote to Mike Davis's cynicism and practical suggestions to build places that do not have the worst features Davis and Soja deplore in twenty-first-century Los Angeles. Duany and Plater-Zyberk anticipate material on urban planning and urban design in the following three sections of *The City Reader*. They are a husband and wife team who run an architectural and urban planning practice in Florida. Duany and Plater-Zyberk are founders of a movement in architecture and planning called the New Urbanism. There are strong neo-traditionalist elements in their design thinking. Like J.B. Jackson they see positive values in small American towns of the past. They have carefully studied vernacular architecture and incorporate into their urban plans insights from the way in which people without architectural training built small towns.

The final selection in this section, "The Impact of the New Technologies and Globalization on Cities," by Saskia Sassen (p. 197), moves from Soja's and Davis's discussions of the contradictions in postmodern cities like Los Angeles to an analysis of the underlying economic characteristics of global cities. Based on her research on the economy of world cities, Sassen concludes that a truly global system of cities has emerged. New York, London, and Tokyo are the main bastions of international finance and control. Decisions made in these three cities affect the residents of cities around the world. Sassen concludes that globalization and information technology are the driving forces behind the new world spatial order, but rejects some facile stereotypes. Some writers argue that because phones, fax machines, e-mail, and other more advanced communication technologies make it possible to communicate easily and instantaneously anywhere in the world, place no longer matters and large cities will wither away. Sassen finds no evidence for this view. Rather she notes further growth of population and power in the commanding global cities. She believes that information technology is already restructuring the physical space of global city regions. According to Sassen, huge new office complexes like La Défense in Paris and nodes of activity following "information highways" are two of the emerging models by which global city metropolitan regions are being restructured. While Sassen agrees that the new world economic order and the rise of global cities are creating enormous inequality she dismisses facile divisions of the world into a developed center and an undeveloped margin. Rather, she argues, there are *de facto* Third World communities in developed global cities and outposts of global cities throughout the Third World.

"The Growth of the City: An Introduction to a Research Project"

from Robert Park *et al.*, *The City* (1925)

Ernest W. Burgess

Editors' Introduction

Ernest W. Burgess (1886–1966) was a member of the famed sociology department at the University of Chicago that included such luminaries as his officemate Rober E. Park and Louis Wirth, author of "Urbanism as a Way of Life" (p. 90). Together, these scholars set out to reinvent modern sociology by taking academic research to the streets and by using the city of Chicago itself as a "living laboratory" for the study of urban problems and social dynamics.

Throughout a long and productive career, Burgess addressed a whole series of issues that connected the social dynamics of the city as a whole with the lives of its citizens. He wrote extensively on issues related to marriage and the family, the relation of personality to social groups, and, in the final decades of his life, problems of the elderly. His most famous contribution to the study of the city was the 1925 essay reprinted here: "The Growth of the City."

Subtitled "An Introduction to a Research Project," Burgess's seminal analysis of the interrelation of the social growth and the physical expansion of modern cities served generations of other urban sociologists, geographers, and planners as a kind of prolegomenon to any future study of the city. Seeking to describe what he called "the pulse of the community," Burgess devised a theory that was thoroughly organic, dynamic, and developmental. "In the expansion of the city," he wrote, "a process of distribution takes place, which sifts and sorts and relocates individuals and groups by residence and occupation." And it was this dynamic process – "process" was one of Burgess's favorite words – that "gives form and character to the city." Burgess focused on patterns within a single city (what is referred to as the internal structure of the city) , rather than relationships among cities (systems of cities).

Central to Burgess's analysis of urban growth was his famous model based on a series of concentric circles that divided the city into five zones. On one level, the concentric zone model was merely a static map of contemporaneous Chicago. On another level, the model was a theoretical diagram of a dynamic process Burgess called "succession," a term he borrowed from the science of plant ecology to describe "urban metabolism and mobility." By borrowing terminology from the natural sciences, and by drawing analogies between the urban and the natural worlds, Burgess established the study of "social ecology" as a distinct approach to understanding the underlying patterns of urban growth and development.

Following the publication of Burgess's essay, a number of urban theorists offered modifications and even refutations of the simple elegance of the concentric zone model. In 1939, real estate economist Homer Hoyt proposed a "sectoral model" for modern capitalist cities based on "wedges of activity" extending outward from the city center along transportation corridors. In 1945, geographers Chauncy Harris and Edward Ullman suggested a "multiple nuclei model," arguing that cities developed around several, not just one, center of economic activity. These authors' descriptions of their alternative models are described in *The Urban Geography Reader* (Fyfe and Kenny, 2005).

Geographer Edward Soja, discussing postmodern Los Angeles (p. 166), notes that even in the polycentric archipelago of modern Los Angeles "population densities do mound up around the center of cities" and that "there is an accompanying concentric residential rhythm" to the region, just as Burgess had found in the city of Chicago eighty years earlier. Soja also notes "the emanation of fortuitous wedges or sectors starting from the center," suggesting that Homer Hoyt's sectoral model still has some explanatory power even in postmodern Los Angeles. Soja notes that the Wilshire corridor, for example, "extends the citadels of the central city almost twenty miles westwards to the Pacific." Finally Soja describes many different centers of activity in the Los Angeles metropolitan region that Harris and Ullman would characterize as distinct nucleii.

Burgess's view may be compared to J.B. Jackson's notion that there is an underlying logic to urban form that occurs even in the absence of formal planning (p. 184). The work of dozens of urban sociologists – for example William Julius Wilson's study of the urban underclass (p. 110) and Ali Madanipour's analysis of "Social Exclusion and Space" (p. 158) – owe a profound debt to Burgess.

Computer technology, including statistical packages and geographic information systems (GIS) software, now makes it possible for present-day geographers and sociologists to summarize vast amounts of data and map the internal structure of cities in ever more sophisticated ways. Thus, understanding of the relationship between social groups and urban form pioneered by Burgess continues to advance by leaps and bounds.

Robert E. Park, Ernest W. Burgess, and Roderick D. McKenzie, *The City* (University of Chicago Press, 1984) – originally published in 1925 – from which this selection is taken, is the best introductory collection of the works of the Chicago School of urban sociology. The "Foundations" section of Nicholas Fyfe and Elizabeth Kenny's *Urban Geography Reader* (London and New York: Routledge, 2005) contains writings by Burgess, Hoyt, and Harris and Ullman on their models of the internal structure of cities.

■ ■ ■ ■ ■ ■

The outstanding fact of modern society is the growth of great cities. Nowhere else have the enormous changes which the machine industry has made in our social life registered themselves with such obviousness as in the cities. In the United States the transition from a rural to an urban civilization, though beginning later than in Europe, has taken place, if not more rapidly and completely, at any rate more logically in its most characteristic forms.

All the manifestations of modern life which are peculiarly urban – the skyscraper, the subway, the department store, the daily newspaper, and social work – are characteristically American. The more subtle changes in our social life, which in their cruder manifestations are termed "social problems," problems that alarm and bewilder us, such as divorce, delinquency, and social unrest, are to be found in their most acute forms in our largest American cities. The profound and "subversive" forces which have wrought these changes are measured in the physical growth and expansion of cities. That is the significance of the comparative statistics of Weber, Bucher, and other students.

These statistical studies, although dealing mainly with the effects of urban growth, brought out into clear relief certain distinctive characteristics of urban as compared with rural populations. The larger proportion of women to men in the cities than in the open country, the greater percentage of youth and middle-aged, the higher ratio of the foreign-born, the increased heterogeneity of occupation increase with the growth of the city and profoundly alter its social structure. These variations in the composition of population are indicative of all the changes going on in the social organization of the community. In fact, these changes are a part of the growth of the city and suggest the nature of the processes of growth.

The only aspect of growth adequately described by Bucher and Weber was the rather obvious process of the aggregation of urban population. Almost as overt a process, that of expansion, has been investigated from a different and very practical point of view by groups interested in city planning, zoning, and regional surveys. Even more significant than the increasing density of urban population is its correlative tendency to overflow, and so to extend over wider areas, and to incorporate these areas into a larger communal life. This paper, therefore, will treat first of the expansion of the city, and then of the less-known processes of urban metabolism and mobility which are closely related to expansion.

EXPANSION AS PHYSICAL GROWTH

The expansion of the city from the standpoint of the city plan, zoning, and regional surveys is thought of almost wholly in terms of its physical growth. Traction studies have dealt with the development of transportation in its relation to the distribution of population throughout the city. The surveys made by the Bell Telephone Company and other public utilities have attempted to forecast the direction and the rate of growth of the city in order to anticipate the future demands for the extension of their services. In the city plan the location of parks and boulevards, the widening of traffic streets, the provision for a civic center, are all in the interest of the future control of the physical development of the city.

This expansion in area of our largest cities is now being brought forcibly to our attention by the Plan for the Study of New York and Its Environs, and by the formation of the Chicago Regional Planning Association, which extends the metropolitan district of the city to a radius of 50 miles, embracing 4,000 square miles of territory. Both are attempting to measure expansion in order to deal with the changes that accompany city growth. In England, where more than one-half of the inhabitants live in cities having a population of 100,000 and over, the lively appreciation of the bearing of urban expansion on social organization is thus expressed by C.B. Fawcett:

> One of the most important and striking developments in the growth of the urban populations of the more advanced peoples of the world during the last few decades has been the appearance of a number of vast urban aggregates, or conurbations, far larger and more numerous than the great cities of any preceding age. These have usually been formed by the simultaneous expansion of a number of neighboring towns, which have grown out toward each other until they have reached a practical coalescence in one continuous urban area. Each such conurbation still has within it many nuclei of denser town growth, most of which represent the central areas of the various towns from which it has grown, and these nuclear patches are connected by the less densely urbanized areas which began as suburbs of these towns. The latter are still usually rather less continuously occupied by buildings, and often have many open spaces.

These great aggregates of town dwellers are a new feature in the distribution of man over the earth. At the present day there are from thirty to forty of them, each containing more than a million people, whereas only a hundred years ago there were, outside the great centers of population on the waterways of China, not more than two or three. Such aggregations of people are phenomena of great geographical and social importance; they give rise to new problems in the organization of the life and well-being of their inhabitants and in their varied activities. Few of them have yet developed a social consciousness at all proportionate to their magnitude, or fully realized themselves as definite groupings of people with many common interests, emotions and thoughts.

In Europe and America the tendency of the great city to expand has been recognized in the term "the metropolitan area of the city," which far overruns its political limits, and, in the case of New York and Chicago, even state lines. The metropolitan area may be taken to include urban territory that is physically contiguous, but it is coming to be defined by that facility of transportation that enables a business man to live in a suburb of Chicago and to work in the loop, and his wife to shop at Marshall Field's and attend grand opera in the Auditorium.

EXPANSION AS A PROCESS

No study of expansion as a process has yet been made, although the materials for such a study and intimations of different aspects of the process are contained in city planning, zoning, and regional surveys. The typical processes of the expansion of the city can best be illustrated, perhaps, by a series of concentric circles, which may be numbered to designate both the successive zones of urban extension and the types of areas differentiated in the process of expansion [Figure 1].

[Figure 1] represents an ideal construction of the tendencies of any town or city to expand radially from its central business district – on the map "the Loop" (I). Encircling the downtown area there is normally an area in transition, which is being invaded by business and light manufacture (II). A third area (III) is inhabited by the workers in industries who have escaped from the area of deterioration (II) but who desire to live within easy access of their work. Beyond this zone is the "residential area" (IV) of high-class apartment buildings

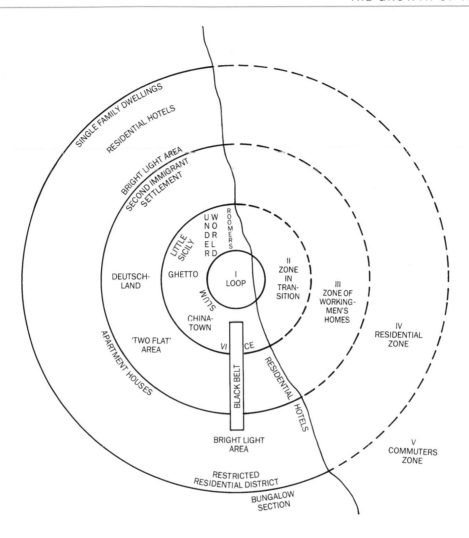

Figure 1

or of exclusive "restricted" districts of single family dwellings. Still farther, out beyond the city limits, is the commuters' zone: suburban areas, or satellite cities, within a thirty- to sixty-minute ride of the central business district.

This [figure] brings out clearly the main fact of expansion, namely, the tendency of each inner zone to extend its area by the invasion of the next outer zone. This aspect of expansion may be called *succession*, a process which has been studied in detail in plant ecology. If this [figure] is applied to Chicago, all four of these zones were in its early history included in the circumference of the inner zone, the present business district. The present boundaries of the area of deterioration were not many years ago those of the zone now inhabited by independent wage-earners, and within the memories of thousands of Chicagoans contained the residences of the "best families." It hardly needs to be added that neither Chicago nor any other city fits perfectly into this ideal scheme. Complications are introduced by the lake front, the Chicago River, railroad lines, historical factors in the location of industry, the relative degree of the resistance of communities to invasion, etc.

Besides extension and succession, the general process of expansion in urban growth involves the antagonistic and yet complementary processes of concentration and decentralization. In all cities there is the natural tendency for local and outside transportation to converge in the central business district.

In the downtown section of every large city we expect to find the department stores, the skyscraper office buildings, the railroad stations, the great hotels, the theaters, the art museum, and the city hall. Quite naturally, almost inevitably, the economic, cultural, and political life centers here. The relation of centralization to the other processes of city life may be roughly gauged by the fact that over half a million people daily enter and leave Chicago's "loop." More recently sub-business centers have grown up in outlying zones. These "satellite loops" do not, it seems, represent the "hoped for" revival of the neighborhood, but rather a telescoping of several local communities into a larger economic unity. The Chicago of yesterday, an agglomeration of country towns and immigrant colonies, is under-going a process of reorganization into a centralized decentralized system of local communities coalescing into sub-business areas visibly or invisibly dominated by the central business district. The actual processes of what may be called centralized decentralization are now being studied in the development of the chain store, which is only one illustration of the change in the basis of the urban organization.

Expansion, as we have seen, deals with the physical growth of the city, and with the extension of the technical services that have made city life not only livable, but comfortable, even luxurious. Certain of these basic necessities of urban life are possible only through tremendous development of communal existence. Three millions of people in Chicago are dependent upon one unified water system, one giant gas company, and one huge electric light plant. Yet, like most of the other aspects of our communal urban life, this economic co-operation is an example of co-operation without a shred of what the "spirit of co-operation" is commonly thought to signify. The great public utilities are a part of the mechanization of life in great cities, and have little or no other meaning for social organization.

Yet the processes of expansion, and especially the rate of expansion, may be studied not only in the physical growth and business development, but also in the consequent changes in the social organization and in personality types. How far is the growth of the city, in its physical and technical aspects, matched by a natural but adequate readjustment in the social organization? What, for a city, is a normal rate of expansion, a rate of expansion with which controlled changes in the social organization might successfully keep pace?

SOCIAL ORGANIZATION AND DISORGANIZATION AS PROCESSES OF METABOLISM

These questions may best be answered, perhaps, by thinking of urban growth as a resultant of organization and disorganization analogous to the anabolic and katabolic processes of metabolism in the body. In what way are individuals incorporated into the life of a city? By what process does a person become an organic part of his society? The natural process of acquiring culture is by birth. A person is born into a family already adjusted to a social environment – in this case the modern city. The natural rate of increase of population most favorable for assimilation may then be taken as the excess of the birth-rate over the death-rate, but is this the normal rate of city growth? Certainly, modern cities have increased and are increasing in population at a far higher rate. However, the natural rate of growth may be used to measure the disturbances of metabolism caused by any excessive increase, as those which followed the great influx of southern Negroes into northern cities since the war. In a similar way all cities show deviations in composition by age and sex from a standard population such as that of Sweden, unaffected in recent years by any great emigration or immigration. Here again, marked variations, as any great excess of males over females, or of females over males, or in the proportion of children, or of grown men or women, are symptomatic of abnormalities in social metabolism.

Normally the processes of disorganization and organization may be thought of as in reciprocal relationship to each other, and as co-operating in a moving equilibrium of social order toward an end vaguely or definitely regarded as progressive. So far as disorganization points to reorganization and makes for more efficient adjustment, disorganization must be conceived not as pathological, but as normal. Disorganization as preliminary to reorganization of attitudes and conduct is almost invariably the lot of the newcomer to the city, and the discarding of the habitual, and often of what has been to him the moral, is not infrequently accompanied by sharp mental conflict and sense of personal loss. Oftener, perhaps, the change gives sooner or later a feeling of emancipation and an urge toward new goals.

In the expansion of the city a process of distribution takes place which sifts and sorts and relocates individuals and groups by residence and occupation. The resulting differentiation of the cosmopolitan American city into areas is typically all from one pattern, with only interesting minor modifications. Within the central business district or on an adjoining street is the "main stem" of "hobohemia," the teeming Rialto of the homeless migratory man of the Middle West. In the zone of deterioration encircling the central business section are always to be found the so-called "slums" and "bad lands," with their submerged regions of poverty, degradation, and disease, and their under-worlds of crime and vice. Within a deteriorating area are rooming-house districts, the purgatory of "lost souls." Nearby is the Latin Quarter, where creative and rebellious spirits resort. The slums are also crowded to over flowing with immigrant colonies – the Ghetto, Little Sicily, Greek town, Chinatown – fascinatingly combining old world heritages and American adaptations. Wedging out from here is the Black Belt with its free and disorderly life. The area of deterioration, while essentially one of decay, of stationary or declining population, is also one of regeneration, as witness the mission, the settlement, the artists' colony, radical centers – all obsessed with the vision of a new and better world.

The next zone is also inhabited predominatingly by factory and shop workers, but skilled and thrifty. This is an area of second immigrant settlement, generally of the second generation. It is the region of escape from the slum, the *Deutschland* of the aspiring Ghetto family. For *Deutschland* (literally "Germany") is the name given, half in envy, half in derision, to that region beyond the Ghetto where successful neighbors appear to be imitating German Jewish standards of living. But the inhabitant of this area in turn looks to the "Promised Land" beyond, to its residential hotels, its apartment-house region, its "satellite loops," and its "bright light" areas.

This differentiation into natural economic and cultural groupings gives form and character to the city. For segregation offers the group, and thereby the individuals who compose the group, a place and a role in the total organization of city life. Segregation limits development in certain directions, but releases it in others. These areas tend to accentuate certain traits, to attract and develop their kind of individuals, and so to become further differentiated.

The division of labor in the city likewise illustrates disorganization, reorganization and increasing differentiation. The immigrant from rural communities in Europe and America seldom brings with him economic skill of any great value in our industrial, commercial, or professional life. Yet interesting occupational selection has taken place by nationality, explainable more by racial temperament or circumstance than by old-world economic background as Irish policemen, Greek ice-cream parlors, Chinese laundries, Negro porters, Belgian janitors, etc.

The facts that in Chicago one million (996,589) individuals gainfully employed reported 509 occupations, and that over 1,000 men and women in *Who's Who* gave 116 different vocations give some notion of how in the city the minute differentiation of occupation "analyzes and sifts the population, separating and classifying the diverse elements." These figures also afford some intimation of the complexity and complication of the modern industrial mechanism and the intricate segregation and isolation of divergent economic groups. Interrelated with this economic division of labor is a corresponding division into social classes and into cultural and recreational groups. From this multiplicity of groups, with their different patterns of life, the person finds his congenial social world and – what is not feasible in the narrow confines of a village – may move and live in widely separated, and perchance conflicting, worlds. Personal disorganization may be but the failure to harmonize the canons of conduct of two divergent groups.

If the phenomena of expansion and metabolism indicate that a moderate degree of disorganization may and does facilitate social organization, they indicate as well that rapid urban expansion is accompanied by excessive increases in disease, crime, disorder, vice, insanity and suicide, rough indexes of social disorganization. But what are the indexes of the causes, rather than of the effects, of the disordered social metabolism of the city? The excess of the actual over the natural increase of population has already been suggested as a criterion. The significance of this increase consists in the immigration into a metropolitan city like New York and Chicago of tens of thousands of persons annually. Their invasion of the city has the effect of a tidal wave inundating first the immigrant colonies, the ports of first entry, dislodging thousands of inhabitants who overflow into the next zone, and so on and on until the momentum of the wave has spent its force on the last urban zone. The whole effect is to speed up expansion, to speed up industry, to speed up the "junking" process in the area of deterioration (II). These internal movements of the population become the more significant for study.

What movement is going on in the city, and how may this movement be measured? It is easier, of course, to classify movement within the city than to measure it. There is the movement from residence to residence, change of occupation, labor turnover, movement to and from work, movement for recreation and adventure. This leads to the question: what is the significant aspect of movement for the study of the changes in city life? The answer to this question leads directly to the important distinction between movement and mobility.

MOBILITY AS THE PULSE OF THE COMMUNITY

Movement, per se, is not an evidence of change or of growth. In fact, movement may be a fixed and unchanging order of motion, designed to control a constant situation, as in routine movement. Movement that is significant for growth implies a change of movement in response to a new stimulus or situation. Change of movement of this type is called *mobility*. Movement of the nature of routine finds its typical expression in work. Change of movement, or mobility, is characteristically expressed in adventure. The great city, with its "bright lights," its emporiums of novelties and bargains, its palaces of amusement, its underworld of vice and crime, its risks of life and property from accident, robbery, and homicide, has become the region of the most intense degree of adventure and danger, excitement and thrill.

Mobility, it is evident, involves change, new experience, stimulation. Stimulation induces a response of the person to those objects in his environment which afford expression for his wishes. For the person, as for the physical organism, stimulation is essential to growth. Response to stimulation is wholesome so long as it is a correlated integral reaction of the entire personality. When the reaction is segmental, that is, detached from, and uncontrolled by, the organization of personality, it tends to become disorganizing or pathological. That is why stimulation for the sake of stimulation, as in the restless pursuit of pleasure, partakes of the nature of vice.

The mobility of city life, with its increase in the number and intensity of stimulations, tends inevitably to confuse and to demoralize the person. For an essential element in the mores and in personal morality is consistency, consistency of the type that is natural in the social control of the primary group. Where mobility is the greatest, and where in consequence primary controls break down completely, as in the zone of deterioration in the modern city, there develop areas of demoralization, of promiscuity, and of vice.

In our studies of the city it is found that areas of mobility are also the regions in which are found juvenile delinquency, boys' gangs, crime, poverty, wife desertion, divorce, abandoned infants, vice.

These concrete situations show why mobility is perhaps the best index of the state of metabolism of the city. Mobility may be thought of, in more than a fanciful sense, as the "pulse of the community." Like the pulse of the human body, it is a process which reflects and is indicative of all the changes that are taking place in the community, and which is susceptible of analysis into elements which may be stated numerically.

The elements entering into mobility may be classified under two main heads: (1) the state of mutability of the person, and (2) the number and kind of contacts or stimulations in his environment. The mutability of city populations varies with sex and age composition, and the degree of detachment of the person from the family and from other groups. All these factors may be expressed numerically. The new stimulations to which a population responds can be measured in terms of change of movement or of increasing contacts. Statistics on the movement of urban population may only measure routine, but an increase at a higher ratio than the increase of population measures mobility. In 1860 the horse-car lines of New York City carried about 50,000,000 passengers; in 1890 the trolley cars (and a few surviving horse-cars) transported about 500,000,000; in 1921, the elevated, subway, surface, and electric and steam suburban lines carried a total of more than 2,500,000,000 passengers. In Chicago the total annual rides per capita on the surface and elevated lines were 164 in 1890; 215 in 1900; 320 in 1910; and 338 in 1921. In addition, the rides per capita on steam and electric suburban lines almost doubled between 1916 (23) and 1921 (41), and the increasing use of the automobile must not be overlooked. For example, the number of automobiles in Illinois increased from 131,140 in 1915 to 833,920 in 1923.

Mobility may be measured not only by these changes of movement, but also by increase of contacts. While the increase of population of Chicago in 1912–22 was less than 25 percent (23.6 percent), the increase of letters delivered to Chicagoans was double that (49.6 percent) – from 693,048,196 to 1,038,007,854. In 1912 New York had 8.8 telephones; in 1922, 16.9 per 100 inhabitants. Boston had, in 1912, 10.1 telephones; ten years later, 19.5 telephones per 100 inhabitants. In the same decade the figures for Chicago increased from 12.3 to 21.6 per 100 population. But increase of the use of the telephone is probably more significant than increase in the number of telephones. The number of telephone calls in Chicago increased from 606,131,928 in 1914 to 944,010,586 in 1922, an increase of 55.7 percent, while the population increased only 13.4 percent.

Land values, since they reflect movement, afford one of the most sensitive indexes of mobility. The highest land values in Chicago are at the point of greatest mobility in the city, at the corner of State and Madison streets, in the Loop. A traffic count showed that at the rush period 31,000 people an hour, or 210,000 men and women in sixteen and one-half hours, passed the southwest corner. For over ten years land values in the Loop have been stationary but in the same time they have doubled, quadrupled and even sextupled in the strategic corners of the "satellite loops," an accurate index of the changes which have occurred. Our investigations so far seem to indicate that variations in land values, especially where correlated with differences in rents, offer perhaps the best single measure of mobility, and so of all the changes taking place in the expansion and growth of the city.

In general outline, I have attempted to present the point of view and methods of investigation which the department of sociology is employing in its studies in the growth of the city, namely, to describe urban expansion in terms of extension, succession, and concentration; to determine how expansion disturbs metabolism when disorganization is in excess of organization; and, finally, to define mobility and to propose it as a measure both of expansion and metabolism, susceptible to precise quantitative formulation, so that it may be regarded almost literally as the pulse of the community. In a way, this statement might serve as an introduction to any one of five or six research projects under way in the department. The project, however, in which I am directly engaged is an attempt to apply these methods of investigation to a cross-section of the city – to put this area, as it were, under the microscope, and so to study in more detail and with greater control and precision the processes which have been described here in the large. For this purpose the West Side Jewish community has been selected. This community includes the so-called "Ghetto," or area of first settlement, and Lawndale, the so-called "Deutschland," or area of second settlement. This area has certain obvious advantages for this study, from the standpoint of expansion, metabolism, and mobility. It exemplifies the tendency to expansion radially from the business center of the city. It is now relatively a homogeneous cultural group. Lawndale is itself an area in flux, with the tide of migrants still flowing in from the Ghetto and a constant egress to more desirable regions of the residential zone. In this area, too, it is also possible to study how the expected outcome of this high rate of mobility in social and personal disorganization is counteracted in large measure by the efficient communal organization of the Jewish community.

"Social Exclusion and Space"

from Ali Madanipour, Goran Cars, and Judith Allen (eds),
Social Exclusion in European Cities: Processes,
Experiences, and Responses (1998)

Ali Madanipour

Editors' Introduction

Exclusion of groups of city residents from access to all that the city has to offer on the basis of race, religion, income, or national origin has been and continues to be a pressing issue in cities throughout the world. While urban design professor Ali Madanipour analyzes spatial aspects of social exclusion in contemporary European cities in this selection, his analysis is very relevant to understanding cities everywhere in the world both today and in the past.

Some of the most dynamic urban societies have welcomed foreigners and included them in the life of the city. H.D.F. Kitto notes that twenty-five centuries ago foreigners (metics) participated in most aspects of the life of the Greek polis (p. 35). They lived throughout the polis rather than in geographically segregated foreigners' neighborhoods, worked as merchants and tradesmen on an equal footing with Athenian citizens, and contributed significantly to the philosophical, scientific, literary, and artistic achievements of Athens' golden age. But they were not Athenian citizens and were excluded from participation in Athens' otherwise extraordinarily inclusive and democratic political institutions.

In other cities law and/or cultural norms have excluded social groups. W.E.B. Du Bois describes in painful detail how Blacks in late nineteenth-century Philadelphia were spatially isolated in just a few wards of the city and systematically barred from white schools, most public facilities, and well-paying jobs (p. 103). Members of the Chicago School of sociology Louis Wirth (p. 90) and Ernest W. Burgess (p. 150) described the spatial separation and social exclusion of early twentieth-century Chicago as waves of immigrants poured into the city.

As globalization continues to bring people from throughout the world into closer contact, and as the pace of immigration increases, the issue of exclusion becomes ever more pressing. All the major cities of the European Union have significant numbers of immigrants from Third World countries – often with skin color, religion, educational backgrounds, and cultures very different from those of the host countries. In what different ways are these people "excluded" from participation in the life of the cities where they live? How is exclusion expressed in urban space?

Madanipour distinguishes between *economic discrimination*, in which members of a group are excluded from employment, *political discrimination*, in which they are excluded from political power by being denied voting rights or full political representation, and *cultural exclusion* in which the group members are marginalized from the symbols, meanings, rituals, and discourses of the dominant culture. He sees exclusion as a continuum from full integration into society at one end of the spectrum to complete lack of integration at the other.

While some societal rules about exclusion are benign – the right of strangers to enter a person's home at will is unacceptable in almost all cultures – exclusion of groups from the opportunities and advantages that cities possess is painful to members of the group and damaging to the society at large, as evidenced by the racial and religious conflict in cities around the world.

Exclusion frequently has a *spatial* dimension. Members of a group are sometimes excluded from areas of a city by law as when medieval Venetian Jews were restricted to the city's ghetto or Chinese (and dogs!) were prohibited from entering parks in nineteenth-century Shanghai. Even when people are legally free to enter areas of the city, as Mike Davis (p. 178) points out, subtle and not-so-subtle signs and cues may signal that members of a particular group are not welcome.

Madanipour suggests two potentially promising approaches to promote greater Inclusion of marginalized groups into urban space – "decommodifying" space so that the private real estate market plays a less decisive role in where different groups are located within the city and deliberate city planning to de-spatialize social exclusion. Building "inclusionary" housing units for low- and moderate-income households in neighborhoods they could otherwise not afford is an example of the first strategy. Mixed-use zoning to promote social diversity is an example of the second.

Madanipour concludes his analysis by advocating inclusionary practices – to assure that "outsiders" are more fully included in urban societies. He wants to break the trap of socio-spatial exclusion and provide more possibilities for inclusion.

Ali Madanipour teaches at the School of Architecture, Planning, and Landscape at the University of Newcastle in England. His special area of interest is the relationship between society and urban space. His books include *Public and Private Spaces of the City* (London: Routledge, 2003), *Terhan: The Making of a Metropolis* (New York: John Wiley & Sons, 1998), *Design of Urban Space: An Inquiry into a Socio-spatial Process* (New York: John Wiley, 1996) and three co-edited anthologies: *The Governance of Place* (Aldershot, Hampshire: Ashgate, 1996), co-edited with Angela Hull and Patsy Healey; *Urban Governance, Institutional Capacity, and Social Milieux* (Aldershot, Hampshire: Ashgate, 2002), co-edited with Goran Cars and Patsy Healey; and *Social Exclusion in European Cities: Processes, Experiences, and Responses* (London: Jessica Kingsley, 1996), co-edited with Goran Cars and Judith Allen, from which this selection is taken.

For historical background on social exclusion in America see Jon Gjerde, *Major Problems in American Immigration and Ethnic History* (New York: Houghton Mifflin, 1998) and Ronald Takaki, *A Different Mirror: A History of Multicultural America*, reissue edition (Boston: Back Bay Books, 1994). Paula S. Rothenberg, *Race, Class, and Gender in the United States: An Integrated Study*, sixth edition (New York: Worth Publishers, 2003) is a systematic treatment of exclusion. Jonathan Kozol analyzes the persistence of segregated schools in *The Shame of the Nation: The Restoration of Apartheid Schooling in America* (New York: Crown Publishing Group, 2005).

■ ■ ■ ■ ■ ■

This chapter concentrates on the relationship between social exclusion and space, exploring some of the frameworks which institute barriers to spatial practices. Its particular emphasis is on the way these barriers to movement are intertwined with social exclusionary processes. This shows that exclusion should be regarded as a socio-spatial phenomenon.

[. . .]

DIMENSIONS OF SOCIAL EXCLUSION

There is little disagreement on some of the major problems facing European cities. Challenges of competition from a global economy marked by a multiplicity of competitors and the European response in the form of moving into an integrative partnership are both aspects of globalization which have reshaped the social and spatial geography of cities. The re-structuring of cities and societies, however, has been a costly exercise, as it has been parallel with a growing social divide, long-term unemployment and jobless-ness, especially for men, and casualization of work, undermining the quality of life for large groups of the population. These symptoms have led to concerns for the fragmentation of the social world, where some members of society are excluded in the 'mainstream' and where this exclusion is painful for the excluded and harmful for society as a whole.

Yet the concept of social exclusion still appears to be in need of clarification due to the variety of the cultural and political contexts in which it has been used. For some it is the question of poverty which should remain the focus of attention, while for others social exclusion makes sense in the broader perspective of citizenship and integration into the social context. Social exclusion, therefore, is not necessarily equated with economic exclusion, although this form of exclusion is often the cause of a wider suffering and deprivation.

As a concept, social exclusion still suffers from a lack of clarity, as it is interpreted and analysed differently. We come across a degree of ambiguity especially between poverty and social exclusion. Some researchers, who have concentrated on the problems of poverty, find social exclusion a vague concept which, for whatever reason, takes attention away from poverty and deprivation. Furthermore, it is argued that the concept of social exclusion is rooted in a certain intellectual and cultural tradition (Catholic, solidarity) and a particular welfare regime (corporatist) and as such is not shared by other (especially liberal) cultures and welfare regimes. On the other hand, those who find social exclusion a useful concept criticize an emphasis on poverty as too narrow. They seek to open the discussion to accommodate the general issues of social integration and citizenship. To confront this ambiguity and contradiction, we need to clarify the concept of social exclusion first.

The overall constitution of the social world is such that different forms of exclusion are fundamental to any social relationship. For example, the division of social life into public and private spheres means drawing boundaries round some spatial and temporal domains and excluding others from these domains. In this way, exclusion becomes an operating mechanism, an institutionalized form of controlling access: to places, to activities, to resources and information. Individual actions as well as legal, political and cultural structures rely heavily upon this operating mechanism and reproduce it constantly. Institutionally organized or individually improvised, it appears that we are all engaged in exclusionary processes that are essential for our social life.

Yet we know that, whatever their importance, these exclusionary processes work in close relationship with inclusionary activities to maintain a social fabric. Maintaining the continuity of the social world is only possible through a combination of and a fine balance between these two processes. At the individual level, seeking privacy without seeking social interaction would lead to isolation. At the social level, exclusion

without inclusion would lead to a collapse of social structures. What is a negative state of affairs, therefore, is not exclusion in all its forms but an absence of inclusionary processes, a lack of a balance between exclusion and inclusion.

But what are the dimensions of the social world in which inclusion and exclusion take place? It is often mentioned that social exclusion is multidimensional. To be able to identify and analyse these dimensions, we should look at the dimensions of the social world in which exclusion and inclusion take place. We can identify economic, political and cultural arenas as the three broad spheres of social life in which social inclusion and exclusion are manifested and, therefore, can be analysed and understood.

In the economic arena, the main form of inclusion is access to resources, which is normally secured through employment. The main form of exclusion, therefore, is a lack of access to employment. Marginal-ization and long-term exclusion from the labour market lead to an absence of opportunity for production and consumption, which can in turn lead to acute forms of social exclusion.

Exclusion from the economic arena is often considered to be a crucial and painful form of exclusion. Poverty and unemployment are therefore frequently at the heart of most discussions of social exclusion, to the extent that poverty and economic exclusion are equated with social exclusion. There is a tendency in the literature to use these terms interchangeably. It is true that long-term economic exclusion can break down the political and cultural ties of the affected individuals and social groups. It is important, however, to note that there are other forms of social exclusion in political and cultural spheres.

In the political arena, the main form of inclusion is to have a stake in power, to participate in decision making. In European liberal democracies, inclusion is often ensured through voting and other processes associated with it. The most obvious form of social exclusion, therefore, is lack of political representa-tion. This may take various forms: from the under-representation of women in parliaments and govern-ments, to the complete exclusion of immigrant groups from political decision making; from the argument by smaller political parties for a new system of repre-sentation which would allow them a fairer share of power, to a withdrawal from political participation by those excluded in the economic and cultural arenas.

In the cultural arena, the main form of inclusion is to share a set of symbols and meanings. The most powerful of these have historically been language, religion and nationality. Some of the new sets of symbolic relationships include the way individual and group identities are formed through association with patterns of consumption, from necessities of daily life to cultural products. For example, in what has been termed a visual culture, aesthetics of social behaviour has become an essential part of social life. The main form of exclusion in the cultural arena, therefore, becomes a marginalization from these symbols, meanings, rituals and discourses. The forms of cultural exclusion vary widely, as experienced by minorities whose language, race, religion and lifestyle are different from those of the larger society.

Different social groups may experience varying degrees of these different but highly interrelated forms of social exclusion. The most acute forms of social exclusion, however, are those that simultaneously include elements of economic, political and cultural exclusion. The other end of the spectrum is occupied by citizens who are fully integrated in the mainstream of society through these three dimensions. Between these two extremes, there is a wide range of variations in which individuals and groups are included in some areas but excluded in others. A major trend is that more and more people suffer from anxiety and uncertainty, as there are ever larger numbers in transition from inclusion to exclusion.

SPATIALITY OF SOCIAL EXCLUSION

Social exclusion, therefore, should be understood in its political, economic and cultural dimensions. Exclusion from the political arena, i.e. the denial of participation in decision making, can alienate individuals and social groups. In the cultural arena, exclusion from common channels of cultural communication and integration can have similar effects. The exclusion from work and its impacts are widely known as undermining the ability of individuals and households to participate actively in social processes. When combined, these forms of exclusion can create an acute form of social exclusion which keeps the excluded at the very margins of the society, a phenomenon all too often marked by a clear spatial manifestation in deprived inner city or peripheral areas . . .

[. . .]

In the past, this spatiality of social exclusion had led to attempts to dismantle such pockets of deprivation without necessarily dismantling the causes of deprivation or the forces bringing them together in particular enclaves. The dismantling of spatial concentrations of deprivation has been a continuous trend: from Baron Haussmann's wide boulevards in the middle of poor neighbourhoods in the nineteenth century, to the slum clearance programmes and more subtle forms of housing management in the twentieth century. These have been attempts to despatialize social exclusion, which is evidence of its inherent and re-emerging spatiality. The latest form of despatialization and re-spatialization of social exclusion is homelessness, a process in which some groups are cut off from their previous socio-spatial contexts and are apparently without a home base. They, however, have clustered in particular parts of cities, spatializing again what was thought to be despatialized.

SPATIALITY AND DIFFERENCE

The absence of homogeneity is most apparent in cities, as they are sites of difference. Large cities have often grown by attracting people from around the country in which they are located or even from around the world. Cities have always been known as the meeting places of different people. As Aristotle noted: 'A city is composed of different kinds of men; similar people cannot bring a city into existence.' The unprecedented growth of cities since the nineteenth century has permanently brought forward the issue of difference in the city as a feature of urban life. Wirth, in his celebrated theory of urbanism, saw heterogeneity as a determining feature of the city, along with population size and density. For him, the city was a 'melting-pot of races, peoples, and cultures, and a most favourable breeding-ground of new biological and cultural hybrids'. In the city, individual differences have 'not only [been] tolerated but rewarded'. Such emphasis on the heterogeneity of cities has led to conceiving it as a world of strangers.

Two sets of reactions to the diversity in the city can be identified: there are those who have tried to impose an order onto it so that it becomes understandable and manageable and those who promote a celebration of diversity. However, both these reactions, which indeed represent modernist and postmodernist thinking, have been unable to deal with the issue of social marginalization and exclusion. Concentrations of disadvantage have remained in cities, despite the large-scale redevelopment schemes of the rationalist tendency and the more sensitive spatial transformations which followed. On the one hand, emphasis on the eradication of difference and seeing the city as a melting pot has led to undermining sensitivities and to disruption of lives. On the other hand, the emphasis on difference has led to social fragmentation and tribalism. Both have failed to cure the wounds of those living on the edge of the society.

BARRIERS TO SPATIAL PRACTICES

But how do we analyse space? There are many gaps and dilemmas associated with understanding space. From the centuries-old philosophical divide between absolute and relational space, to the gap between mental and real space, between physical and social space, between abstract and differential space, to the relationship between space and mass, space and time, and the variety of perspectives from which space can be studied, all bear the possibility of confusion and collision. It is possible to show, however, that to avoid the gaps and dilemmas associated with understanding space, we need to concentrate on the processes which produce the built environment. By analysing the intersection between space production and everyday life practices, we will be able to arrive at a dynamic understanding of space. We will then be able to understand and explain material space and its social and psychological contexts and attributes.

The question of social exclusion and integration, it can be argued, largely revolves around access. It is access to decision making, access to resources, and access to common narratives, which enable social integration. Many of these forms of access have clear spatial manifestations, as space is the site in which these different forms of access are made possible or denied. There is a direct relationship between our general sense of freedom and well-being with the choices open to us in our spatial practices. The more restricted our social options, the more restricted will be our spatial options, and the more excluded we feel or become. On the other hand, if we have a wide range of social options, we would have a wide range of places to go to, places for living, working and entertainment. Two extreme cases of the existence or absence of spatial freedom may be jetsetting executives versus

prisoners. Whereas for one, the world may be shrinking to seem like a global village open for communication and interaction, for the other the world outside is large and out of reach. For most of us, however, our spaces are a continuum from accessible to non-accessible places. The space around us is a collection of open, closed or controlled places.

But how is the urban space organized and how are spatial practices controlled and regulated? We all have an understanding of the places where we can or cannot go, as over the years through our spatial practices, we have accumulated a knowledge about places and their patterns of accessibility. The physical organization of space, using elements from the natural or the built environment, has been socially and symbolically employed to put visible and strict limits on our spatial practices. For example, topography has always been used to institute difference and segregation, from ancient times when the hilltops were the place of gods for Greeks and Mesopotamians, to our own time when they are the living places of the rich and powerful. There is also a mental space, our perceptions of space. This may be regulated through codes and signs, preventing us from entering some spaces through outright warning or more subtle deterrents. Mental space may also be controlled through our fears and perceptions of activities in places. For example, we may be hesitant to enter an expensive-looking shopping centre if we do not have access to the resources needed for the activities there, even though there may not be any physical barriers which would prevent us from going there. A third form of barrier to our spatial behaviour is social control, which can range from legal prohibitions on entering places to constructing formal barriers along publicly recognized borders. National borders and public–private boundaries are examples of this form. A combination of formalized rules and regulations, informal codes and signs, and fears and desires control our spatial behaviour and alert us to the limitations on our access. Through these, we have come to know whether we can enter a place, are welcomed in another and excluded from others. More restrictions on our access to our surroundings would bring about the feeling of being trapped, alienated and excluded from our social space.

Space has, therefore, a major role in the integration or segregation of urban society. It is a manifestation of social relationships while affecting and shaping the geometries of these relationships. This leads us to the argument that social exclusion cannot be studied without also looking at spatial segregation and exclusion. Social cohesion or exclusion, therefore, are indeed socio-spatial phenomena . . . We know that all human societies have their own forms of social and spatial exclusion. So exclusionary processes per se are not the source of social fragmentation and disintegration. It is the absence of social integration which causes social exclusion, as individuals do not find the possibility and channels of participating in the mainstream society.

GLOBAL AND NATIONAL SPACE

National borders are the largest means of socio-spatial exclusion. The modern nation state exerts an exclusionary process along its boundaries, from lines on maps to barbed wires on the landscape. Those who are left outside need to go through special checks and controls to be allowed in. The same applies to those who are in and want to go out. The control of cross-border movement by the nation state, or by blocks of nation states as in the European Union, is a form of exclusion legitimated openly through political processes. A national territory, therefore, is a spatial manifestation of an institutionalized exclusionary process.

Other administrative boundaries, although potentially exclusionary, do not have such a forceful character, nor are they associated with such a degree of public awareness, such historical significance, or guarded by military might. No other form of exclusion has been associated with such high costs in human life, sacrifice and misery. Attempts to change or to protect national borders have inflicted the highest cost in human lives in the twentieth century, as experienced by two world wars and many regional conflicts. The birth of a nation state, when the multi-ethnic empires and states break up, can be a bloody process in which every means is used to exclude others. The surgical subdivision of national space, whether through external forces as in postwar Germany or by exploding internal forces as in the former Yugoslavia, has been equally difficult for those excluded from what they have regarded as their home.

In the national space enclosed within these boundaries, narratives of nationalism have been employed to legitimize the exclusion of others beyond these boundaries. Indeed, exclusionary narratives, which determine how 'we' are different from others, are often essential in binding individuals together as a group. The most dangerous of these narratives has been the rhetoric of hatred against other nations, races and groups. But there are many such exclusionary narratives which do not necessarily promote violence and hatred and still have a binding power. With these narratives, which often rely on a common historical experience, large groups of people have been associated with each other. The focal point of this association has been the nation state, which holds the power of controlling the national borders.

The narratives of nationalism attempt to create homogeneity out of an enormous diversity. As individuals have come together to create a democratic civil society, such narratives have helped the organization of modern democratic states . . .

[. . .]

NEIGHBOURHOODS, MARKETS AND REGULATION

At the local level, by following two processes, land and property development on the one hand and spatial planning on the other hand, we can see how a socio-spatial geometry of difference and segregation, which is the foundation of exclusion, emerges. We come across the term neighbourhood in a variety of distinct but interrelated usages. In one sense, the term is used loosely to address a locality. This daily usage is based on the images and understandings by individuals and groups of their surroundings. This is a view from below and, as such, can lead us to see a city as a collection of overlapping neighbourhoods. Research on people's perception of neighbourhood shows major differences according to age, gender, class and ethnicity. At the other end of the scale, there is a concept of neighbourhood from above, from the viewpoint of such experts as managers, planners and designers. Here neighbourhood refers to a particular part of a town and is used to understand urban structure and change in urban society. It is also used as a tool for management. From this viewpoint, the city is seen as a collection of segregated neighbourhoods.

Neighbourhoods as constituent parts of cities have long been the focus of attention by urban designers and planners. Drawing upon historic precedents and for practical reasons, neighbourhood has provided them with an intimate scale of the urban whole to understand and to deal with. Historically, neighbourhoods have been the sites and physical manifestations of close social relationships and so have been praised by town planners, especially those who have looked nostalgically to the feudal bonds of the medieval towns and the communal bonds of working-class neighbourhoods in the industrial city. A dichotomy emerged as a result of the unprecedented growth of the cities: between *gesellschaft* and *gemeinschaft*, between the alienation of the big city and the romanticized, small communities of towns and villages. To recreate the social cohesion of these small communities, it was thought, cities should be broken into smaller parts, into neighbourhoods. On the other hand, it was thought that the communitarianism of small neighbourhoods could overcome the individualism of the suburbs, those bourgeois utopias.

It is this association of neighbourhood as a physical entity with neighbourhood as a cohesive social unit that led to a series of reformist ideas throughout the twentieth century. From the widely used, and discredited, concept of neighbourhood unit, to Lynch's districts, which are still promoted to make cities legible, and today's urban villages and new urbanist neighbourhoods, there has been a long line of managerial attempts to promote social cohesion by spatial organization.

Along with this promotion of spatial subdivision by town planning, there has been a promotion of socio-spatial segregation by market forces through the ways in which space is produced, exchanged and used. The producers of space, such as volume housebuilders, tend to build in large-scale housing estates, creating an urban fabric which is a collection of different subdivisions. The land and property markets have operated so as to ensure the segregation of income groups and social classes. Commodification of space has led to different patterns of access to space and hence a differential spatial organization and townscape. Wherever there has been a tendency to decommodify space, as in the postwar social housing schemes, town planners and designers have ensured that a degree of spatial subdivision still prevailed.

We can therefore identify two processes: a land and property market which sees space as a commodity and tends to create socio-spatial segregation through differential access to this commodity, and a town planning and design tendency to regulate and rationalize space production by the imposition of some form of order. When we look at these two processes together, the picture which emerges is a collectivization

of difference, of exclusion, which can lead to enclaves for the rich and the creation of new ghettos for the poor.

[. . .]

PUBLIC AND PRIVATE SPACE

Another form of socio-spatial exclusion, which is enforced with a rigour somewhat similar to the protection of national borders, is the separation between public and private territories. We guard our private spheres from intruders by whatever means, in some countries even legitimately by firearms. Privacy, private property and private space are intertwined, demarcated through a variety of objects and signs: from subtle variations of colour and texture to fences and high walls. Those who are in are entitled to be, excluding those who are not. This is an exclusionary process legitimized through public discourse, through custom or law. Violation of this exclusionary process is regarded as, at best, inconvenience and, at worst, crime. Public space, which is one of the manifestations of society's public sphere, is maintained by public agencies in the public interest and is accessible to the public. Access to public space, however, is subject to exclusionary processes. Public space is guarded from intrusion by private interests, a process which is regarded as essential for the health of the society. Some of the main currents in social and political thought that offer concepts of public space appear to stress the need to keep the public and private spheres distinctive and apart, despite the criticism that this idealizes the distinction.

[. . .]

The changing nature of development companies and the entry of the finance industry into built environment production and management has partly led to what is widely known as the privatization of space. Large-scale developers and financiers expect their commodities to be safe for investment and maintenance, hence their inclination to reduce as much as possible all the levels of uncertainty which could threaten their interests. This trend is parallel with the increasing fear of crime, rising competition from similar developments, and the rising expectations of the consumers, all encouraging the development of totally managed environments. What has emerged is an urban space where increasingly large sections are managed by private companies, as distinctive from those controlled by public authorities. Examples of these fragmented and privatized spaces are gated neighbourhoods, shopping malls and city centre walkways, under heavy private surveillance and separated from the public realm by controlled access and clear boundaries. This total management of parts of the city is in part an attempt to control crime. Crime acts as a counter-claim to space and as such is itself an exclusionary force, keeping many groups vulnerable and marginalized.

CONCLUSION: SOCIAL INTEGRATION AND SPATIAL FREEDOM

Social exclusion combines lack of access to resources, to decision making, and to common narratives. The multidimensional phenomenon of social exclusion finds spatial manifestation, in its acute forms, in deprived inner or peripheral urban areas. This spatiality of social exclusion is constructed through the physical organization of space as well as through the social control of space, as ensured by informal codes and signs and formal rules and regulations. These formal channels act at all scales of space. Global space is fragmented by national spaces, which have a tendency to deny difference and homogenize social groups. At the scale of local space, spatialization of social exclusion takes place through land and property markets.

These markets tend to fragment, differentiate and commodify space through town planning mechanisms which tend to fragment, rationalize and manage space, and also through the legal and customary distinctions between the public and private spheres, with a constant tension between the two and a tendency for the privatization of space.

To break the trap of socio-spatial exclusion, one strategy could be to challenge these deep-seated forms of differentiation. We know, however, that wholesale challenges can be problematic themselves, as exemplified by attempts to redefine the public–private relationship in Eastern Europe. Furthermore, we know that any human society is likely to have some form of exclusionary process in its constitution. Nevertheless, it is true that the form of these exclusionary processes changes over time. A reflexive revisiting of the processes of differentiation is therefore a constantly necessary task. At the same time what is necessary and urgent is to institute and promote inclusionary processes, to strike a balance between exclusion and integration, to provide the possibility of integration and to break the trap of socio-spatial exclusion. We have seen that space is a major component part of social exclusion. Revisiting spatial barriers and promoting accessibility and more spatial freedom can therefore be the way spatial planning can contribute to promoting social integration.

"Taking Los Angeles Apart: Towards a Postmodern Geography"

from *Postmodern Geographies: The Reassertion of Space in Critical Social Theory* (1989)

Edward Soja

Editors' Introduction

Because of its size, fragmentation, diversity, and dynamism, and its role as the epicenter of global image and fantasy, Los Angeles is often held out as the quintessential postmodern city. In this selection University of California geographer and city planning professor Edward Soja "deconstructs" the greater metropolitan Los Angeles region. He takes as his area of enquiry the Sixty-Mile Circle around downtown Los Angeles – a five-county area with 12 million people, 132 cities, and a $250 billion economy at the time this selection was written in 1989.

Soja's description of Los Angeles' Sixty-Mile Circle is rich in complexity and paradox. Freeway offramps lead to huge military establishments like Edwards Air Force Base, to an Indian reservation nearly devoid of Indians, and a condor refuge from where the last condor has been transplanted to a local zoo to avoid lead poisoning. Within the Sixty-Mile Circle are ski slopes and orange groves; Spanish missions and Latino barrios; leisure towns and prisons; planned communities and unplanned sprawl; all-white stealth communities and ethnic neighborhoods; multiple center cities and Edge Cities; science parks, technopoles and technoburbs employing well-paid, largely white knowledge workers, and industrial zones relying almost exclusively on Hispanic and Asian workers.

Among the many contradictions and paradoxes of the greater Los Angeles metropolitan area Soja notes is the area's extraordinary *industrial* dynamism. While William Julius Wilson (p. 110), Saskia Sassen (p. 197), Michael Porter (p. 274) and many other writers emphasize the shift of manufacturing jobs away from developed countries to the Third World, Soja characterizes greater Los Angeles as the premier industrial growth pole of the twentieth century. According to Soja, Los Angeles maybe *post-Fordist*, but it is hardly *postindustrial*. Products may no longer be produced in Los Angeles today in large factories like Ford automobiles were once produced, but Los Angeles still has many factories producing a host of tangible products.

Contrary to appearances, in this bastion of private capitalism, government investment is important to the economic health of the Los Angeles region. Soja emphasizes the role of the state, particularly the military, in the creation of this "federated metro-sea of state-rescued capitalism." Defense and aerospace contracts for huge projects like the space shuttle and high-tech weapons systems fuel the growth of the region.

Despite all this growth and change, Soja sees an underlying logic to the region that reflects classic urban geographers' models. There is still a core central business district (CBD) in Los Angeles displaying some of the attributes of a concentric zone internal structure as Ernest Burgess hypothesized (p. 150). According to Soja there is still a radial logic to the region too. For example, Wilshire Boulevard is a twenty-mile "sector" radiating out from downtown Los Angeles in the form that real estate economist Homer Hoyt described seventy years ago as underlying city development.

Downtown Los Angeles exhibits cultural contradictions similar to those David Harvey describes in postmodern New York City (p. 225). The Los Angeles CBD contains a huge police headquarters building, courts, city hall, the largest women's prison in America, and the central organs of press and pulpit such as the Times/Mirror Building and St Vibiana's cathedral. In the newly emerging "left bank" of Los Angeles, yuppies sip cappuccinos just a short distance from ethnic neighborhoods populated by Vietnamese, Latino, and Chinese immigrant families. Conventioneers in this "dual" city enjoy luxury hotels walled off from panhandlers just outside.

Following an approach favored in postmodern critical theory, Soja seeks to "deconstruct" this dazzling array of elements. He seeks to get beneath the symbolism of the area – the semiotic blanket of words and artifacts created by the culture(s) that make up Los Angeles. In many ways the Los Angeles metropolitan region resembles a theme park. Amidst the whimsy and pastiche, Soja notes that it is tempting to become bemused and detached. But underneath the glitz and glitter, Soja points to what he sees as the hard edge of a capitalist, racist, patriarchal landscape.

This selection is from Edward Soja, *Postmodern Geographies: The Reassertion of Space in Critical Social Theory* (London and New York: Verso, 1989). Other books by Soja include *Postmetropolis: Critical Studies of Cities and Regions* (London: Blackwell, 2000), *The City: Los Angeles and Urban Theory at the End of the Twentieth Century* (Berkeley: University of California Press, 1996) co-edited with Allen J. Scott, and *Thirdspace: Journeys to Los Angeles and Other Real-and-imagined Places* (Oxford: Blackwell, 1996).

Books on urban space and culture in Los Angeles include Steven P. Erie, *Globalizing L.A.: Trade, Infrastructure, and Regional Development* (Palo Alto: Stanford University Press, 2004), Mark Vallianatos, Regina M. Freer, Peter Dreier and Robert Gottlieb (eds), *The Next Los Angeles: The Struggle for a Livable City* (Berkeley: University of California Press, 2005), William Fulton, *The Reluctant Metropolis* (Point Arena: Solano Press, 1998), Norman Smith, *The History of Forgetting: Los Angeles and the Erasure of Memory* (London and New York: Verso, 1997), and Dolores Hayden, *The Power of Place: Urban Landscapes and Public History* (Cambridge: MIT Press, 1997). Los Angeles is further described in Kevin Starr's series on California and the American Dream: *Americans and the California Dream, 1850–1915* (New York: Oxford University Press, 1986), *Inventing the Dream: California Through the Progressive Era* (New York: Oxford University Press, 1985), *Material Dreams: California Through the 1920s* (New York: Oxford University Press, 1990), *Endangered Dreams: The Great Depression in California* (New York: Oxford University Press, 1986), *The Dream Endures: California Enters the 1940s* (New York: Oxford University Press, 1996), and *Coast of Dreams: California on the Edge, 1990–2003* (New York: Knopf, 2004).

Other books that explore postmodern landscapes in southern California and elsewhere include Michael Sorkin (ed.), *Variations on a Theme Park: The New American City and the End of Public Space* (New York: Hill & Wang, 1992), Sharon Zukin, *Landscapes of Power: From Detroit to Disney World* (Berkeley: University of California Press, 1993), and Mark Gottdiener, *The Theming of America: Dreams, Visions, and Commercial Spaces* (New York: HarperCollins, 1997).

▪ ▪ ▪ ▪ ▪ ▪

'The Aleph?' I repeated.

'Yes, the only place on earth where all places are – seen from every angle, each standing clear, without any confusion or blending' (10–11)

. . . Then I saw the Aleph . . . And here begins my despair as a writer. All language is a set of symbols whose use among its speakers assumes a shared past. How, then, can I translate into words the limitless Aleph, which my floundering mind can scarcely encompass? (12–13)

(Jorge Luis Borges, 'The Aleph')

Los Angeles, like Borges's Aleph, is exceedingly tough-to-track, peculiarly resistant to conventional description. It is difficult to grasp persuasively in a temporal narrative for it generates too many conflicting images, confounding historicization, always seeming to stretch laterally instead of unfolding sequentially. At the same time, its spatiality challenges orthodox analysis and interpretation, for it too seems limitless and constantly in motion, never still enough to encompass, too filled with 'other spaces' to be informatively described. Looking at Los Angeles from the inside, introspectively, one tends to see only fragments and immediacies, fixed sites of myopic understanding impulsively generalized to represent the whole. To the more far-sighted outsider, the visible aggregate of the whole of Los Angeles churns so confusingly that it induces little more than illusionary stereotypes or self-serving caricatures – if its reality is ever seen at all.

What is this place? Even knowing where to focus, to find a starting point, is not easy, for, perhaps more than any other place, Los Angeles is everywhere. It is global in the fullest sense of the word. Nowhere is this more evident than in its cultural projection and ideological reach, its almost ubiquitous screening of itself as a rectangular dream machine for the world. Los Angeles broadcasts its self-imagery so widely that probably more people have seen this place – or at least fragments of it – than any other on the planet . . .

[. . .]

. . . What follows then is a succession of fragmentary glimpses, a freed association of reflective and interpretive field notes which aim to construct a critical human geography of the Los Angeles urban region. My observations are necessarily and contingently incomplete and ambiguous, but the target I hope will remain clear: to appreciate the specificity and uniqueness of a particularly restless geographical landscape while simultaneously seeking to extract insights at higher levels of abstraction, to explore through Los Angeles glimmers of the fundamental spatiality of social life, the adhesive relations between society and space, history and geography, the splendidly idiographic and the enticingly generalizable features of a postmodern urban geography.

A ROUND AROUND LOS ANGELES

[. . .]

We must have a place to start, to begin reading the context . . . just such a reductionist mapping has popularly presented itself. It is defined by an embracing circle drawn sixty miles (about a hundred kilometres) out from a central point located in the downtown core of the City of Los Angeles.

The Sixty-Mile Circle, so inscribed, covers the thinly sprawling 'built-up' area of five counties, a population of more than 12 million individuals, at least 132 incorporated cities and, it is claimed, the greatest concentration of technocratic expertise and militaristic imagination in the USA. Its workers produce, when last estimated, a gross annual output worth nearly $250 billion, more than the 800 million people of India produce each year. This is certainly Greater Los Angeles, a dizzying world.

CIRCUMSPECTION

Securing the Pacific rim has been the manifest destiny of Los Angeles, a theme which defines its sprawling urbanization perhaps more than any other analytical construct . . . It is not always easy to see the imprint of this imperial history on the material landscape, but an imaginative cruise directly above the contemporary circumference of the Sixty-Mile Circle can be unusually revealing. Figure 1 will help to find the way.

The Circle cuts the south coast at the border between Orange and San Diego Counties, near one of the key checkpoints regularly set up to intercept the northward flow of undocumented migrants, and not far from the San Clemente 'White House' of Richard Nixon and the fitful SONGS of the San Onofre Nuclear Generating Station. The first rampart to watch, however, is Camp Pendleton Marine Corps Base, the largest military base in California in terms of personnel, the freed spouses of whom have helped to build a growing hightechnology complex in northern San Diego County . . .

Another quick hop over Sunnymead, the Box Spring Mountains, and Redlands takes us to Rampart #3, Norton Air Force Base, next to the city of San Bernardino and just south of the almost empty San Manuel Indian Reservation. The guide books tell us that the primary mission of Norton is military airlifts, just in case. To move on we must rise still higher to pass over the skisloped peaks of the San Bernardino Mountains and National Forest, through Cajon Pass and passing the old Santa Fe Trail, into the picturesque Mojave Desert . . .

The next leg is longer and more serene: over the Antelope Valley and the Los Angeles Aqueduct (tapping the Los Angeles-owned segments of the life-giving but rapidly dying Owens River Valley two hundred miles further away); across Interstate 5 (the main freeway corridor to the north), a long stretch of Los Padres National Forest and the Wild Condor Refuge – the last remaining condors were recently removed to zoos after leadpoisoning threatened their extinction in the 'wild' – to the idyllized town of Ojai (site for the filming of 'Lost Horizon'), and then to the Pacific again at the Mission of San Buenaventura, in Ventura County . . .

It is startling how much of the circumference is owned and preserved by the Federal Government in one way or another. Premeditation may be impossible to ascribe, but postmeditation on the circumscriptive federal presence is certainly in order.

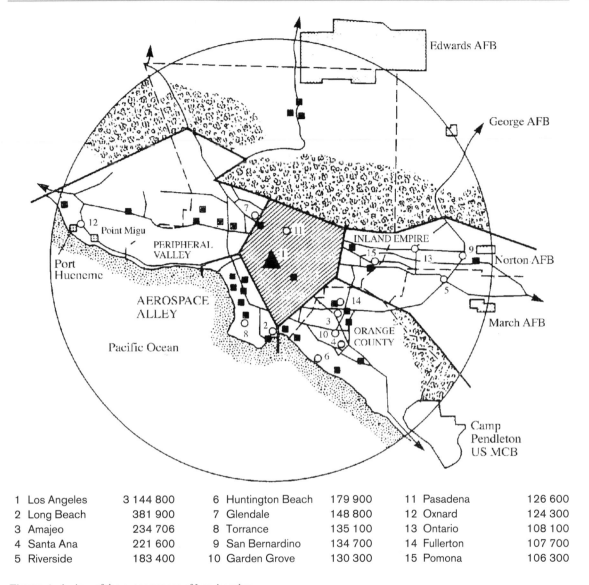

Figure 1 A view of the outer spaces of Los Angeles

The urban core is outlined in the shape of a pentagon, with the Central City denoted by the black triangle. The major military bases on the perimeter of the Sixty-Mile Circle are identified and the black squares are the sites of the largest defence contractors in the region. Also shown are county boundaries, the freeway system outside the central pentagon, and the location of all cities with more than 100,000 inhabitants (small open circles)

1	Los Angeles	3 144 800	6	Huntington Beach	179 900	11	Pasadena	126 600
2	Long Beach	381 900	7	Glendale	148 800	12	Oxnard	124 300
3	Amajeo	234 706	8	Torrance	135 100	13	Ontario	108 100
4	Santa Ana	221 600	9	San Bernardino	134 700	14	Fullerton	107 700
5	Riverside	183 400	10	Garden Grove	130 300	15	Pomona	106 300

ENCLOSURES

[...]

If there has emerged a compelling focus to the recent academic literature on Los Angeles, it is the discovery of extraordinary industrial production ... Yet it is no exaggeration to claim that the Sixty-Mile Circle contains the premier industrial growth pole of the twentieth century, at least within the advanced capitalist countries. Oil, orange groves, films and flying set the scene at the beginning of the century and tend to remain fixed in many contemporary images of industrious, but not industrial, Los Angeles. Since 1930, however, Los Angeles has probably led all other major metropolitan areas in the USA, decade by decade, in the accumulation of new manufacturing employment.

[. . .]

. . . In the past half century, no other area has been so pumped with federal money as Los Angeles, via the Department of Defense to be sure, but also through numerous federal programmes subsidizing suburban consumption (suburbsidizing?) and the development of housing, transportation and water delivery systems. From the last Great Depression to the present, Los Angeles has been the prototypical Keynesian state-city, a federalized metro-sea of state-rescued capitalism enjoying its place in the sunbelt, demonstrating decade by decade its redoubtable ability to go first and multiply the public seed money invested in its promising economic landscape. No wonder it remains so protected. In it are embedded many of the crown jewels of advanced industrial capitalism.

OUTERSPACES

The effulgent Star Wars colony currently blooming around Los Angeles International Airport (LAX) is part of a much larger outer city which has taken shape along the Pacific slope of Los Angeles County. In the context of this landscape, through the story-line of the aerospace industry, can be read the explosive history and geography of the National Security State and what Mike Davis (1984) has called the 'Californianization of Late-Imperial America'.

If there is a single birthplace for this Californianization, it can be found at old Douglas Field in Santa Monica . . . Over half a million people now live in this 'Aerospace Alley', as it has come to be called. During working hours, perhaps 800,000 are present to sustain its global preeminence. Untold millions more lie within its extended orbit.

Attached around the axes of production are the representative locales of the industrialized outer city: the busy international airport; corridors filled with new office buildings, hotels, and global shopping malls; neatly packaged playgrounds and leisure villages; specialized and masterplanned residential communities for the high technocracy; armed and guarded housing estates for top professionals and executives; residual communities of low-pay service workers living in overpriced homes; and the accessible enclaves and ghettoes which provide dependable flows of the cheapest labour power to the bottom bulge of the bi-modal local labour market. The LAX–City compage reproduces the segmentation and segregation of the inner city based on race, class, and ethnicity, but

manages to break it down still further to fragment residential communities according to specific occupational categories, household composition, and a broad range of individual attributes, affinities, desired lifestyles and moods.

This extraordinary differentiation, fragmentation, and social control over specialized pools of labour is expensive. Housing prices and rental costs in the outer city are easily among the highest in the country and the provision of appropriate housing increasingly absorbs the energy not only of the army of real estate agents but of local corporate and community planners as well, often at the expense of longtime residents fighting to maintain their foothold in 'preferred' locations. From the give and take of this competition have emerged peculiarly intensified urban landscapes. Along the shores of the South Bay, for example, part of what Rayner Banham once called 'Surfurbia', there has developed the largest and most homogeneous residential enclave of scientists and engineers in the world. Coincidentally, this beachhead of the high technocracy is also one of the most formidable racial redoubts in the region. Although just a few miles away, across the fortifying boundary of the San Diego freeway, is the edge of the largest and most tightly segregated concentration of Blacks west of Chicago, the sun-belted beach communities stretching south from the airport have remained almost 100 per cent white. The worldview from this highly engineered environment is stereotypically caricatured in Figure 2.

The Sixty-Mile Circle is ringed with a series of these outer cities at varying stages of development, each a laboratory for exploring the contemporaneity of capitalist urbanization . . .

[. . .]

. . . these new territorial complexes seem to be turning the industrial city inside-out, recentring the urban to transform the metropolitan periphery into the core region of advanced industrial production. Decentralization from the inner city has been taking place selectively for at least a century all over the world, but only recently has the peripheral condensation become sufficiently dense to challenge the older urban cores as centres of industrial production, employment nodality, and urbanism. This restructuring process is far from being completed but it is beginning to have some profound repercussions on the way we think about the city, on the words we use to describe urban forms and functions, and on the language of urban theory and analysis.

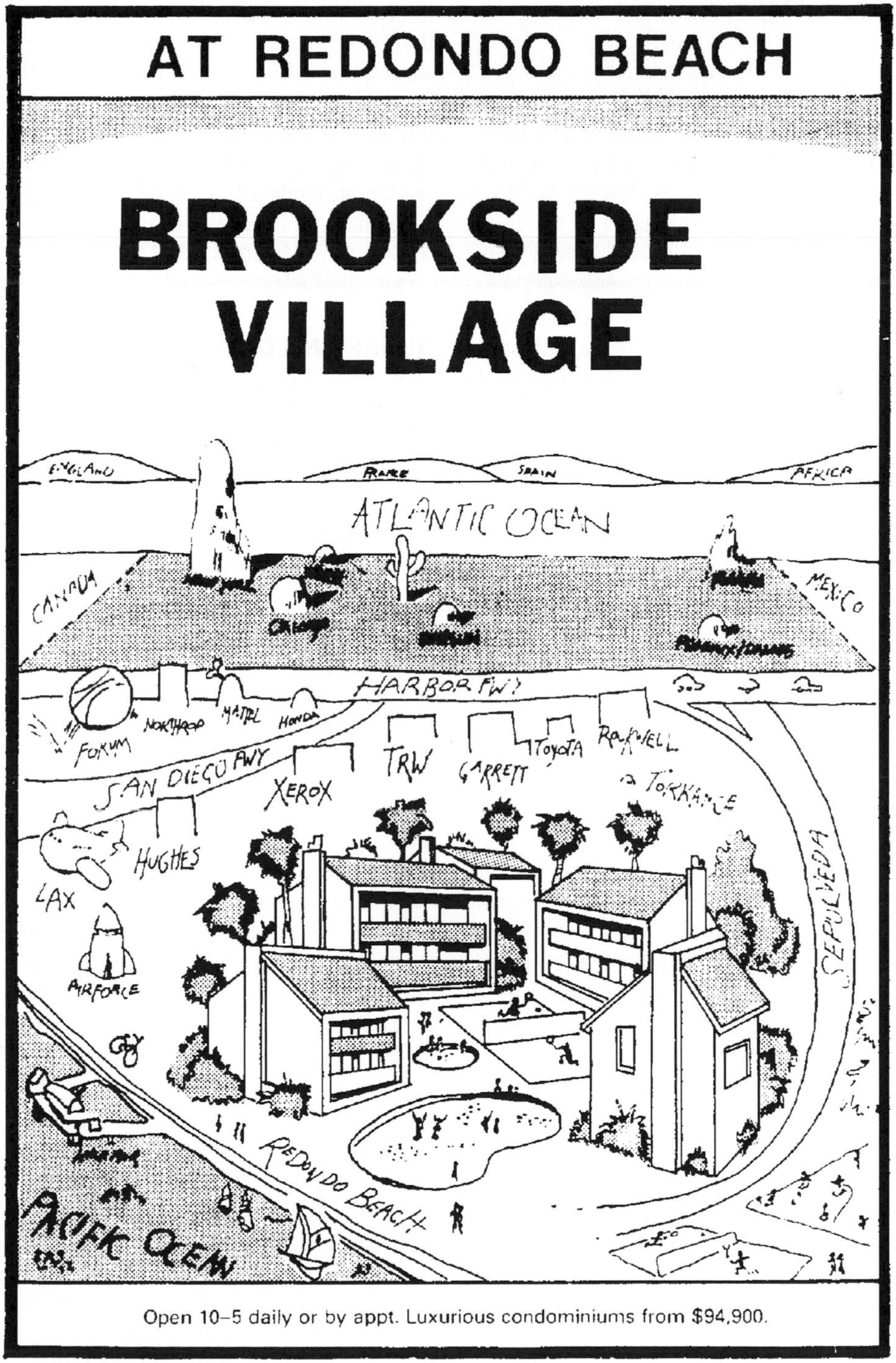

Figure 2 Worldview from Redondo Beach

BACK TO THE CENTRE

[. . .]

To see more of Los Angeles, it is necessary to move away from the riveting periphery and return, literally and figuratively, to the centre of things to the still adhesive core of the urbanized landscape. In Los Angeles as in every city, the nodality of the centre defines and gives substance to the specificity of the urban, its distinctive social and spatial meaning. Urbanization and the spatial divisions of labour associated with it revolve around a socially constructed pattern of nodality and the power of the occupied centres both to cluster and disperse, to centralize and decentralize, to structure spatially all that is social and socially produced. Nodality situates and contextualizes urban society by giving material form to essential social relations. Only with a persistent centrality can there be outer cities and peripheral urbanization. Otherwise, there is no urban at all.

It is easy to overlook the tendential processes of urban structuration that emanate from the centre, especially in the postmodern capitalist landscape. Indeed, in contemporary societies the authoritative and allocative power of the urban centre is purposefully obscured or, alternatively, detached from place, ripped out of context, and given the stretched-out appearance of democratic ubiquity. In addition, as we have seen, the historical development of urbanization over the past century has been marked by a selective dispersal and decentralization, emptying the centre of many of the activities and populations which once aggregated densely around it. For some, this has signalled a negation of nodality, a submergence of the power of central places, perhaps even a Derridean deconstruction of all differences between the 'central' and the 'marginal'.

Yet the centres hold. Even as some things fall apart, dissipate, new nodalities form and old ones are reinforced. The specifying centrifuge is always spinning but the centripetal force of nodality never disappears. And it is the persistent residual of political power which continues to precipitate, specify, and contextualize the urban, making it all stick together.

[. . .]

To maintain adhesiveness, the civic centre has always served as a key surveillant node of the state, supervising locales of production, consumption and exchange. It still continues to do so, even after centuries of urban recomposition and restructuring, after waves of reagglomerative industrialization. It is not production or consumption or exchange in themselves that specifies the urban, but rather their collective surveillance, supervision and anticipated control within the power-filled context of nodality. In Foucauldian terms, cities are the convergent sites of (social) space, knowledge, and power, the headquarters of societal modes of regulation (from *regula* and *regere*, to rule; the root of our keyword: region).

[. . .]

SIGNIFYING DOWNTOWN

The downtown core of the City of Los Angeles, which the signs call 'Central City', is the agglomerative and symbolic nucleus of the Sixty-Mile Circle, certainly the oldest but also the newest major node in the region. Given what is contained within the Circle, the physical size and appearance of downtown Los Angeles seem almost modest, even today after a period of enormous expansion. As usual, however, appearances can be deceptive.

Perhaps more than ever before, downtown serves in ways no other place can as a strategic vantage point, an urban panopticon counterposed to the encirclement of watchful military ramparts and defensive outer cities. Like the central well in Bentham's eminently utilitarian design for a circular prison, the original panopticon, downtown can be seen (when visibility permits) by each separate individual, from each territorial cell, within its orbit. Only from the advantageous outlook of the centre, however, can the surveillant eye see everyone collectively, disembedded but interconnected. Not surprisingly, from its origin, the central city has been an aggregation of overseers, a primary locale for social control, political administration, cultural codification, ideological surveillance, and the incumbent regionalization of its adherent hinterland.

Looking down and out from City Hall, the site is especially impressive to the observer. Immediately below and around is the largest concentration of government offices and bureaucracy in the country outside the federal capital district. To the east, over a pedestrian skyway, are City Hall East and City Hall South, relatively new civic additions enclosing a shopping mall, some murals, a children's museum, and the Triforium, a splashy sixty-foot fountain of water, light, and music entertaining the lunchtime masses. Just beyond is the imposing police administration building, Parker Center, hallowing the name of a former police chief of note. Looking further outside the central well of downtown but within its eastern salient, one can see an area which houses 25 per cent of California's prison population, at least 12,000 inmates held in four jails designed to hold half that number. Included within this carceral wedge are the largest women's prison in the country (Sybil Brand) and the seventh largest men's prison (Men's Central). More enclosures are being insistently planned by the state to meet the rising demand.

On the south, along First Street, are the State Department of Transportation (CALTRANS) with its electronic wall maps monitoring the arterial freeways of the region, the California State Office Building, and the headquarters of the fourth estate, the monumental Times-Mirror building complex, which many have claimed houses the unofficial governing power of Los Angeles, the source of many stories that mirror the times and spaces of the city. Near the spatial sanctum of the *Los Angeles Times* is also St Vibiana's Cathedral, mother church to one of the largest Catholic archdioceses in the world (nearly four million strong) and controller of another estate of significant proportions. The Pope slept here, across the street from Skid Row missions temporarily closed so that he could not see all his adherents.

Looking westward now, toward the Pacific and the smog-hued sunsets which brilliantly paint the nightfalls of Los Angeles, is first the Criminal Courts Building, then the Hall of Records and Law Library, and next the huge Los Angeles County Courthouse and Hall of Administration, major seats of power for what is by far the country's largest county in total population (now over eight million). Standing across Grand Avenue is the most prominent cultural centre of Los Angeles, described by Unique Media Incorporated in their pictorial booster maps of downtown as 'the cultural crown of Southern California, reigning over orchestral music, vocal performance, opera, theatre and dance'. They add that the Music Center 'tops Bunker Hill like a contemporary Acropolis', one which has dominated civil cultural life since it was inaugurated in 1964. Just beyond this cultural crown is the Department of Water and Power (surrounded by usually waterless fountains) and a multi-level extravaganza of freeway interchanges connecting with every corner of the Sixty-Mile Circle, a peak point of accessibility within the regional transportation network. On its edge, one of Japan's greatest architects has designed a Gateway Building to punctuate the teeming sea.

Along the northern flank is the Hall of Justice, the US Federal Courthouse, and the Federal Building, completing the ring of local, city, state and federal government authority which comprises the potent civic centre. Sitting more tranquilly just beyond, cut off by a swathe of freeway, is the preserved remains of the old civic centre, now part of El Pueblo de Los Angeles State Historical Park, additional testimony to the lasting power of the central place. Since the origins of Los Angeles the sites described have served as the political citadel, designed with other citadels to command, protect, socialize and dominate the surrounding urban population.

There is still another segment of the citadel–panopticon which cannot be overlooked. Its form and function may be more specific to the contemporary capitalist city but its mercantile roots entwine historically with the citadels of all urbanized societies. Today, it has become the acknowledged symbol of the urbanity of Los Angeles, the visual evidence of the successful 'search for a city' by the surrounding sea of suburbs. This skylined sight contains the bunched castles and cathedrals of corporate power, the gleaming new 'central business district' of the 'central city', pinned next to its ageing predecessor just to the east. Here too the LA-leph's unending eyes are kept open and reflective, reaching out to and mirroring global spheres of influence, localizing the world that is within its reach.

Nearly all the landmarks of the new LA CBD have been built over the past fifteen years and flashily signify the consolidation of Los Angeles as a world city. Now more than half the major properties are in part or wholly foreign owned, although much of this landed presence is shielded from view. The most visible wardens are the banks which light up their logos atop the highest towers: Security Pacific (there again), First Interstate, Bank of America (co-owner of the sleek-black Arco Towers before their recent purchase by the Japanese), Crocker, Union, Wells Fargo, Citicorp (billing itself as 'the newest city in town'). Reading the skyline one sees the usual corporate panorama: large insurance companies (Manulife, Transamerica, Prudential), IBM and major oil companies, the real estate giant Coldwell Banker, the new offices of the Pacific Stock Exchange, all serving as attachment points for silvery webs of financial and commercial transactions extending practically everywhere on earth.

. . . Contrary to popular opinion, Los Angeles is a tightly planned and plotted urban environment, especially with regard to the social and spatial divisions of labour necessary to sustain its pre-eminent industrialization and consumerism. Planning choreographs Los Angeles through the fungible movements of the zoning game and the flexible staging of supportive community participation (when there are communities to be found), a dance filled with honourable intent, dedicated expertise, and selective beneficence. It has excelled, however, as an ambivalent but nonetheless enriching pipeline and place-maker to the domestic and foreign developers of Los Angeles, using its influential reach to prepare the groundwork and facilitate the selling of specialized locations and populations to suit the needs of the most powerful organizers of the urban space economy.

. . . Through a historic act of preservation and renewal, there now exists around downtown a deceptively harmonized showcase of ethni-cities and specialized economic enclaves which play key roles, albeit somewhat noisily at times, in the contemporary redevelopment and internationalization of Los Angeles. Primarily responsible for this packaged and planned production of the inner city is the Community Redevelopment Agency, probably the leading public entrepreneur of the Sixty-Mile Circle.

There is a dazzling array of sites in this compartmentalized corona of the inner city: the Vietnamese shops and Hong Kong housing of a redeveloping Chinatown; the Big Tokyo financed modernization of old Little Tokyo's still resisting remains; the induced pseudo-SoHo of artists' lofts and galleries hovering near the exhibitions of the 'Temporary Contemporary' art warehouse; the protected remains of El Pueblo along Calmexified Olvera Street and in the renewed Old Plaza; the strangely anachronistic wholesale markets for produce and flowers and jewellery growing bigger while other downtowns displace their equivalents; the foetid sweatshops and bustling merchandise marts of the booming garment district; the Latino retail festival along pedestrian-packed Broadway (another preserved zone and inch-for-inch probably the most profitable shopping street in the region); the capital site of urban homelessness in the CRA-gilded skid row district; the enormous muralled *barrio* stretching eastward to the still unincorporated East Los Angeles; the de-industrializing and virtually resident-less wholesaling City of Vernon to the south filled with chickens and pigs awaiting their slaughter; the Central American and Mexican communities of Pico-Union and Alvarado abutting the high-rises on the west; the obtrusive oil wells and aggressive graffiti in the backyards of predominantly immigrant Temple-Beaudry progressively being eaten away by the spread of Central City West (now being called 'The Left Bank' of downtown); the intentionally yuppifying South Park redevelopment zone hard by the slightly seedy Convention Center; the revenue-milked towers and fortresses of Bunker Hill; the resplendently gentrified pocket of 'Victorian' homes in old Angelino Heights overlooking the citadel; the massive new Koreatown pushing out west and south against the edge of Black Los Angeles; the Filipino pockets to the north-west still uncoalesced into a 'town' of their own; and so much more: a constellation of Foucauldian heterotopias 'capable of juxtaposing in a single real place several spaces, several sites that are in themselves incompatible' but 'function in relation to all the space that remains'.

What stands out from a hard look at the inner city seems almost like an obverse (and perverse) reflection of the outer city, an agglomerative complex of dilapidated and overcrowded housing, low technology workshops, relics and residuals of an older urbanization, a sprinkling of niches for recentred professionals and supervisors, and, above all, the largest concentration of cheap, culturally splintered/occupationally manipulable Third World immigrant labour to be found so tangibly available in any First World urban region. Here in this colonial corona is another of the crown jewels of Los Angeles, carefully watched over, artfully maintained and reproduced to service the continued development of the manufactured region.

The extent and persistence of agglomerated power and ever-watchful eyes in downtown Los Angeles cannot be ignored by either captive participants or outside observers. The industrialization of the urban periphery may be turning the space economy of the region inside-out, but the old centre is more than holding its own as the preeminent political and economic citadel . . .

[. . .]

LATERAL EXTENSIONS

Radiating from the specifying nodality of the central city are the hypothesized pathways of traditional urban theory . . . Formal models of urban morphology have conventionally begun with the assumption of a structuring central place organizing an adherent landscape into discoverable patterns of hinterland development and regionalization . . .

Population densities do mound up around the centres of cities, even in the polycentric archipelago of Los Angeles (where there may be several dozen such mounds, although the most pronounced still falls off from the central city). There is also an accompanying concentric residential rhythm associated with the family life cycle and the relative premiums placed on access to the dense peaks versus the availability of living space in the sparseness of the valleys (at least for those who can afford such freedoms of choice). Land values (when they can be accurately calculated) and some job densities also tend to follow in diminishing peaks outwards from the centre, bringing back to mind those tented webs of the urban geography textbooks.

Adding direction to the decadence of distance reduces the Euclidian elegance of concentric gradations, and many of the most mathematical of urban geometricians have accordingly refused to follow this slightly unsettling path. But direction does indeed induce another fit by pointing out the emanation of fortuitous wedges or sectors from the centre. The sectoral wedges of Los Angeles are especially pronounced once you leave the inner circle around downtown.

The Wilshire Corridor, for example, extends the citadels of the central city almost twenty miles westwards to the Pacific, picking up several other prominent but smaller downtowns en route (the Miracle Mile that initiated this extension, Beverly Hills, Century City, Westwood, Brentwood, Santa Monica). Watching above it is an even lengthier wedge of the wealthiest residences, running with almost staggering homogeneities to the Pacific Palisades and the privatized beaches of Malibu, sprinkled with announcements of armed responsiveness and signs which say that 'trespassers will be shot'. Here are the hearths of the most vocal homeowners' movements, arms raised to slow growth and preserve their putative neighbourhoods in the face of the encroaching, view-blocking, street-clogging, and declassé downtowns.

As if in counterbalance, on the other side of the tracks east of downtown is the salient containing the largest Latino *barrio* in Anglo-America, where many of those who might be shot are carefully barricaded in poverty and there is at least one more prominent wedge, stretching southward from downtown to the twin ports of Los Angeles–Long Beach, still reputed to be one of the largest consistently industrial urban

sectors in the world. This is the primary axis of Ruhral Los Angeles.

A third ecological order perturbs the geometrical neatness still further, punching holes into the monocentric gradients and wedges as a result of the territorial segregation of races and ethnicities. Segregation is so noisy that it overloads the conventional statistical methods of urban factorial ecology with scores of tiny but 'significant' eco-components. In Los Angeles, arguably the most segregated city in the country, these components are so numerous that they operate statistically to obscure the spatiality of social class relations deeply embedded in the zones and wedges of the urban landscape, as if they needed to be obscured any further.

These broad social geometries provide an attractive model of the urban geography of Los Angeles, but like most of the inherited overviews of formal urban theory they are seriously diverting and illusory. They mislead not because there is disagreement over their degree of fit – such regular empiricist arguments merely induce a temporary insensibility by forcing debate on to the usually sterile grounds of technical discourse. Instead, they deceive by involuting explanation, by the legerdemain of making the nodality of the urban explain itself through its mere existence, one outcome explaining another. Geographical covariance in the form of empirico-statistical regularity is elevated to causation and frozen into place without a history – and without a human geography which recognizes that the organization of space is a social product filled with politics and ideology, contradiction and struggle, comparable to the making of history. Empirical regularities are there to be found in the surface geometry of any city, including Los Angeles, but they are not explained in the discovery, as is so often assumed.

Different routes and different roots must be explored to achieve a practical understanding and critical reading of urban landscapes. The illusions of empirical opaqueness must be shattered, along with the other disciplining effects of Modern Geography.

DECONSTRUCTION

Back in the centre, shining from its circular turrets of bronzed glass, stands the Bonaventure Hotel, an amazingly storeyed architectural symbol of the splintered labyrinth that stretches sixty miles around it. Like many other Portmanteaus which dot the eyes of urban citadels in New York and San Francisco, Atlanta and Detroit, the Bonaventure has become a concentrated representation of the restructured spatiality of the late capitalist city: fragmented and fragmenting, homogeneous and homogenizing, divertingly packaged yet curiously incomprehensible, seemingly open in presenting itself to view but constantly pressing to enclose, to compartmentalize, to circumscribe, to incarcerate. Everything imaginable appears to be available in this micro-urb, but real places are difficult to find, its spaces confuse an effective cognitive mapping, its pastiche of superficial reflections bewilder co-ordination and encourage submission instead. Entry by land is forbidding to those who carelessly walk but entrance is nevertheless encouraged at many different levels, from the truly pedestrian skyways above to the bunkered inlets below. Once inside, however, it becomes daunting to get out again without bureaucratic assistance. In so many ways, its architecture recapitulates and reflects the sprawling manufactured spaces of Los Angeles.

[. . .]

From the centre to the periphery, in both inner and outer cities, the Sixty-Mile Circle today encloses a shattered metro-sea of fragmented yet homogenized communities, cultures, and economies confusingly arranged to contingently ordered spatial division of labour and power . . .

Over 130 other municipalities and scores of county-administered areas adhere loosely around the irregular City of Los Angeles in a dazzling, sprawling patchwork mosaic. Some have names which are startlingly self-explanatory. Where else can there be a City of Industry and a City of Commerce, so flagrantly commemorating the fractions of capital which guaranteed their incorporation. In other places, names casually try to recapture a romanticized history (as in the many new communities called Rancho something-or-other) or to ensconce the memory of alternative geographies (as in Venice, Naples, Hawaiian Gardens, Ontario, Manhattan Beach, Westminster). In naming, as in so many other contemporary urban processes, time and space, the 'once' and the 'there', are being increasingly played with and packaged to serve the needs of the here and the now, making the lived experience of the urban increasingly vicarious, screened through *simulacra*, those exact copies for which the real originals have been lost.

A recent clipping from the *Los Angeles Times* tells of the 433 signs which bestow identity within the hyper-space of the City of Los Angeles, described as 'A City Divided and Proud of It.' . . . One of the founders of the programme pondered its development: 'At first, in the early 1960s, the Traffic Department took the position that all the communities were part of Los Angeles and we didn't want cities within cities . . . but we finally gave in. Philosophically it made sense. Los Angeles is huge. The city had to recognize that there were communities that needed identification . . .'

For at least fifty years, Los Angeles has been defying conventional categorical description of the urban, of what is city and what is suburb, of what can be identified as community or neighbourhood, of what co-presence means in the elastic urban context. It has in effect been deconstructing the urban into a confusing collage of signs which advertise what are often little more than imaginary communities and outlandish representations of urban locality. I do not mean to say that there are no genuine neighbourhoods to be found in Los Angeles. Indeed, finding them through car-voyages of exploration has become a popular local pastime, especially for those who have become so isolated from propinquitous community in the repetitive sprawl of truly ordinary-looking landscapes that make up most of the region. But again the urban experience becomes increasingly vicarious, adding more layers of opaqueness to *l'espace vécu*.

Underneath this semiotic blanket there remains an economic order, an instrumental nodal structure, an essentially exploitative spatial division of labour, and this spatially organized urban system has for the past half century been more continuously productive than almost any other in the world. But it has also been increasingly obscured from view, imaginatively mystified in an environment more specialized in the production of encompassing mystifications than practically any other you can name. As has so often been the case in the United States, this conservative deconstruction is accompanied by a numbing depoliticization of fundamental class and gender relations and conflicts. When all that is seen is so fragmented and filled with whimsy and pastiche, the hard edges of the capitalist, racist and patriarchal landscape seem to disappear, melt into air.

With exquisite irony, contemporary Los Angeles has come to resemble more than ever before a gigantic agglomeration of theme parks, a lifespace comprised of Disneyworlds. It is a realm divided into showcases of global village cultures and mimetic American landscapes, all-embracing shopping malls and crafty

Main Streets, corporation-sponsored magic kingdoms, high-technology-based experimental prototype communities of tomorrow, attractively packaged places for rest and recreation all cleverly hiding the buzzing workstations and labour processes which help to keep it together: like the original 'Happiest Place on Earth', the enclosed spaces are subtly but tightly controlled by invisible overseers despite the open appearance of fantastic freedoms of choice . . .

AFTERWORDS

[. . .]

I have been looking at Los Angeles from many different points of view and each way of seeing assists in sorting out the interjacent medley of the subject landscape. The perspectives explored are purposeful, eclectic, fragmentary, incomplete, and frequently contradictory, but so too is Los Angeles and, indeed, the experienced historical geography of every urban landscape . . . The task of comprehensive, holistic regional description may therefore be impossible, as may be the construction of a complete historico-geographical materialism.

There is hope nonetheless. The critical and theoretical interpretation of geographical landscapes has recently expanded into realms that functionally had been spatially illiterate for most of the twentieth century. New and avid readers abound as never before, many are directly attuned to the specificity of the urban, and several have significantly turned their eyes to Los Angeles. Moreover, many practised readers of surface geographies have begun to see through the alternatively myopic and hypermetropic distortions of past perspectives to bring new insight to spatial analysis and social theory. Here too Los Angeles has attracted observant readers after a history of neglect and misapprehension, for it insistently presents itself as one of the most informative palimpsests and paradigms of twentieth-century urban-industrial development and popular consciousness.

As I have seen and said in various ways, everything seems to come together in Los Angeles, the totalizing Aleph. Its representations of spatiality and historicity are archetypes of vividness, simultaneity, and interconnection. They beckon inquiry at once into their telling uniqueness and, at the same time, into their assertive but cautionary generalizability. Not all can be understood, appearances as well as essences persistently deceive, and what is real cannot always be captured even in extraordinary language. But this makes the challenge more compelling, especially if once in a while one has the opportunity to take it all apart and reconstruct the context. The reassertion of space in critical social theory – and in critical political praxis – will depend upon a continued deconstruction of a still occlusive historicism and many additional voyages of exploration into the heterotopias of contemporary postmodern geographies.

"Fortress L.A."

from *City of Quartz: Excavating the Future in Los Angeles* (1990)

Mike Davis

Editors' Introduction

Mike Davis is to contemporary Los Angeles what Friedrich Engels was to mid-nineteenth-century Manchester, England. Engels was a kind of explorer, reporting to an educated, middle-class audience about the horrors of the industrial city and the miserable lives of the new industrial proletarian class of modern capitalism. Similarly, Davis, 150 years later, explores the dark side of the postmodern metropolis and reports on the hopelessness and despair of the *post-industrial* "underclass" (largely defined by race, ethnicity, and gender) to an audience largely comprised of young, disaffected intellectuals and academics. The parallels are striking. Davis may be the heir to Engels simply because the contemporary metropolis – characterized by wealth and homelessness and divided against itself along the fault lines that separate suburban enclaves from inner-city slums – is the heir to the geographical and social class divisions of the cities of the Industrial Revolution.

Mike Davis is an acerbic Southern California social critic who makes his social-class sympathies and anti-establishment bias perfectly clear on every page of *City of Quartz*, which one admirer praised as a "visionary rant."

In *The Condition of the Working Class in England in 1844* (p. 50), Engels noted the boulevards that intersected the city of Manchester and how the façades of those broad thoroughfares served to mask and disguise the hovels of the poor that lay beyond the view of the middle-class commuter. Davis, employing an eclectic "culture studies" methodology, makes a similar but even subtler point about contemporary Los Angeles. The freeways allow middle-class suburbanites to navigate the city as a whole without encountering the lives of the residents of the inner-city neighborhoods. But beyond this, as Madanipour describes in less vitriolic prose (p. 158), the city itself has become a vast and continuous system of signs that we read and obey mostly on a subconscious level. "Today's upscale, pseudo-public spaces," Davis writes, ". . . are full of invisible signs warning off the underclass 'Other.' Although architectural critics are usually oblivious to how the built environment contributes to segregation, pariah groups – whether poor Latino families, young Black men, or elderly homeless white females – read the meaning immediately."

Another similarity between Engels and Davis is that both men, as they proceeded to "read the streets" of their respective paradigm cities, reach the limits of language's ability to describe the physical and psychological conditions being reported. Engels compiled a mountain of personal observations, journalistic reports, and official survey data to create a catalog of social horror. Davis relies on a more emotional strategy. His over-heated rhetorical excesses often overwhelm rational discourse. Davis turned out to be a prophet: the ghettos and barrios of South Central Los Angeles erupted into open rebellion in the Rodney King riots two years after the publication of *City of Quartz*.

Whereas Engels saw massive social dislocations and systematically set about to fashion a theory of revolutionary socialism in response to the observed reality, Davis offers no similarly optimistic solution. Davis *personalizes* the horrors of the contemporary city, and his voice is one of desperation and despair offering no obvious way out of the current impasse. He leaves it to others, like William Julius Wilson (p. 110), Ali Madanipour (p. 158), and Paul

Davidoff (p. 400) to propose how government officials, urban designers, and city planners might advocate on behalf of the poor and oppressed to make cities better.

Davis's insights regarding the underclass "other" suggest comparisons with such classic analysts of ghettoization as W.E.B. Du Bois (p. 103) and William Julius Wilson (p. 110). But it is difficult to compare Davis with social scientists and social policy and planning professionals whose focus is on objectively analyzing reality and influencing the existing city systems by working within them.

This selection is from Mike Davis, *City of Quartz: Excavating the Future in Los Angeles* (London and New York: Verso, 1990). Davis has written interesting, controversial, and despairing books on Las Vegas: *The Grit Beneath the Glitter: Tales from the Real Las Vegas* (Berkeley: University of California Press, 2002), co-edited with Hal Rothman; the abuse of the natural environment by humans: *Dead Cities: A Natural History* (New York: New Press, 2002); natural disasters: *The Ecology of Fear: Los Angeles and the Imagination of Disaster* (New York: Metropolitan Books, 1998); the avian flu pandemic threat: *The Monster at our Door: The Global Threat of Avian Flu* (New York: New Press, 2005); the oppression of immigrants: *Prisoners of the American Dream* (London: Verso, 1986), *Operation Gatekeeper: The Rise of the 'Illegal Alien' and the Reinvention of the US–Mexico Border* (London: Routledge, 2001), *Late Victorian Holocausts: El Niño Families and the Making of the Third World* (London: Verso, 2001); and Third World slums: *Planet of Slums* (London: Verso, 2006). One book by Davis, *Magical Urbanism: Latinos Reinvent the U.S. Big City* (London: Verso, 2001), in addition to characteristic "slash and burn" prose excoriating the oppression of Hispanic immigrants, contains some uncharacteristically positive comments about how large US cities are being improved by immigration.

For more on Los Angeles, consult William Fulton, *The Reluctant Metropolis: The Politics of Urban Growth in Los Angeles* (Baltimore: Johns Hopkins University Press, 2001), R.A. Nadeau, *Los Angeles: From Mission to Modern City* (New York: Longman, 1960) and R. Marchand, *The Emergence of Los Angeles: Population and Housing in the City of Dreams* (London: Pion, 1986). Architectural and cultural analyses of Los Angeles can be found in Rayner Banham, *Los Angeles: The Architecture of Four Ecologies* (London: Penguin, 1971) and Dolores Hayden, *The Power of Place: Urban Landscapes as Public History* (Cambridge: MIT Press, 1995).

For analyses of Los Angeles' troubled race relations, see Mauricio Mazon, *The Zoot-suit Riots* (Austin: University of Texas Press, 1984), Raphael J. Sonenshein, *Politics in Black and White* (Princeton: Princeton University Press, 1993), and Mark Baldassare (ed.), *The Los Angeles Riots: Lessons for the Urban Future* (Boulder: Westview Press, 1994).

▪ ▪ ▪ ▪ ▪ ▪

The carefully manicured lawns of Los Angeles' Westside sprout forests of ominous little signs warning: "Armed Response!" Even richer neighborhoods in the canyons and hillsides isolate themselves behind walls guarded by gun-toting private police and state-of-the-art electronic surveillance. Downtown, a publicly subsidized "urban renaissance" has raised the nation's largest corporate citadel, segregated from the poor neighborhoods around it by a monumental architectural glacis. In Hollywood, celebrity architect Frank Gehry, renowned for his "humanism," apotheosizes the siege look in a library designed to resemble a foreign-legion fort. In the Westlake district and the San Fernando Valley the Los Angeles Police barricade streets and seal off poor neighborhoods as part of their "war on drugs." In Watts, developer Alexander Haagen demonstrates his strategy for recolonizing inner-city retail markets: a panopticon shopping mall surrounded by staked metal fences and a substation of the LAPD in a central surveillance tower. Finally, on the horizon of the next millennium, an ex-chief of police crusades for an anti-crime "giant eye" – a geo-synchronous law enforcement satellite – while other cops discreetly tend versions of "Garden Plot," a hoary but still viable 1960s plan for a law-and-order armageddon.

Welcome to post-liberal Los Angeles, where the defense of luxury lifestyles is translated into a proliferation of new repressions in space and movement, undergirded by the ubiquitous "armed response." This obsession with physical security systems, and, collaterally, with the architectural policing of social boundaries, has become a zeitgeist of urban restructuring, a master narrative in the emerging built environment of the 1990s. Yet contemporary urban theory, whether debating the role of electronic technologies in precipitating "postmodern space," or discussing the dispersion of urban functions across poly-centered metropolitan "galaxies," has been strangely silent about the militarization of city life so grimly visible at the street level. Hollywood's pop apocalypses and pulp science fiction have been more realistic, and politically perceptive, in representing the programmed hardening of the urban surface in the wake of the social polarizations of the Reagan era. Images of carceral inner cities (*Escape from New York, Running Man*), high-tech police death squads (*Blade Runner*), sentient buildings (*Die Hard*), urban bantustans (*They Live!*), Vietnam-like street wars (*Colors*), and so on, only extrapolate from actually existing trends.

Such dystopian visions grasp the extent to which today's pharaonic scales of residential and commercial security supplant residual hopes for urban reform and social integration. The dire predictions of Richard Nixon's 1969 National Commission on the Causes and Prevention of Violence have been tragically fulfilled: we live in "fortress cities" brutally divided between "fortified cells" of affluent society and "places of terror" where the police battle the criminalized poor. The "Second Civil War" that began in the long hot summers of the 1960s has been institutionalized into the very structure of urban space. The old liberal paradigm of social control, attempting to balance repression with reform, has long been superseded by a rhetoric of social warfare that calculates the interests of the urban poor and the middle classes as a zero-sum game. In cities like Los Angeles, on the bad edge of postmodernity, one observes an unprecedented tendency to merge urban design, architecture and the police apparatus into a single, comprehensive security effort.

This epochal coalescence has far-reaching consequences for the social relations of the built environment. In the first place, the market provision of "security" generates its own paranoid demand. "Security" becomes a positional good defined by income access to private "protective services" and membership in some hardened residential enclave or restricted suburb. As a prestige symbol – and sometimes as the decisive borderline between the merely well-off and the "truly rich" – "security" has less to do with personal safety than with the degree of personal insulation, in residential, work, consumption and travel environments, from "unsavory" groups and individuals, even crowds in general.

Secondly, as William Whyte has observed of social intercourse in New York, "fear proves itself." The social perception of threat becomes a function of the security mobilization itself, not crime rates. Where there is an actual rising arc of street violence, as in Southcentral Los Angeles or Downtown Washington D.C., most of the carnage is self-contained within ethnic or class boundaries. Yet white middle-class imagination, absent from any firsthand knowledge of inner-city conditions, magnifies the perceived threat through a demonological lens. Surveys show that Milwaukee suburbanites are just as worried about violent crime as inner-city Washingtonians, despite a twentyfold difference in relative levels of mayhem. The media, whose function in this arena is to bury and obscure the daily economic violence of the city, ceaselessly throw up spectres of criminal underclasses and psychotic stalkers. Sensationalized accounts of killer youth gangs high on crack and shrilly racist evocations of marauding Willie Hortons foment the moral panics that reinforce and justify urban apartheid.

Moreover, the neo-military syntax of contemporary architecture insinuates violence and conjures imaginary dangers. In many instances the semiotics of so-called "defensible space" are just about as subtle as a swaggering white cop. Today's upscale, pseudo-public spaces – sumptuary malls, office centers, culture acropolises, and so on – are full of invisible signs warning off the underclass "Other." Although architectural critics are usually oblivious to how the built environment contributes to segregation, pariah groups – whether poor Latino families, young Black men, or elderly homeless white females – read the meaning immediately.

THE DESTRUCTION OF PUBLIC SPACE

The universal and ineluctable consequence of this crusade to secure the city is the destruction of accessible public space. The contemporary opprobrium attached to the term "street person" is in itself a harrowing index of the devaluation of public spaces. To reduce contact with untouchables, urban redevelopment has converted once vital pedestrian streets into traffic sewers and transformed public parks into temporary receptacles for the homeless and wretched. The American city, as many critics have recognized, is being systematically turned inside out – or, rather, outside in. The valorized spaces of the new megastructures and super-malls are concentrated in the center, street frontage is denuded, public activity is sorted into strictly functional compartments, and circulation is internalized in corridors under the gaze of private police.

The privatization of the architectural public realm, moreover, is shadowed by parallel restructurings of electronic space, as heavily policed, pay-access "information orders," elite databases and subscription cable services appropriate parts of the invisible agora. Both processes, of course, mirror the deregulation of the economy and the recession of non-market entitlements. The decline of urban liberalism has been accompanied by the death of what might be called the "Olmstedian vision" of public space. Frederick Law Olmsted, it will be recalled, was North America's Haussmann, as well as the Father of Central Park. In the wake of Manhattan's "Commune" of 1863, the great Draft Riot, he conceived public landscapes and parks as social safety-valves, mixing classes and ethnicities in common (bourgeois) recreations and enjoyments. As Manfredo Tafuri has shown in his well-known study of Rockefeller Center, the same principle animated the construction of the canonical urban spaces of the La Guardia–Roosevelt era.

This reformist vision of public space – as the emollient of class struggle, if not the bedrock of the American *polis* – is now as obsolete as Keynesian nostrums of full employment. In regard to the "mixing" of classes, contemporary urban America is more like Victorian England than Walt Whitman's or La Guardia's New York. In Los Angeles, once-upon-a-time a demi-paradise of free beaches, luxurious parks, and "cruising strips," genuinely democratic space is all but extinct. The Oz-like archipelago of Westside pleasure domes – a continuum of tony malls, arts centers and gourmet strips – is reciprocally dependent upon the social imprisonment of the third-world service proletariat who live in increasingly repressive ghettoes and barrios. In a city of several million yearning immigrants, public amenities are radically shrinking, parks are becoming derelict and beaches more segregated, libraries and playgrounds are closing, youth congregations of ordinary kinds are banned, and the streets are becoming more desolate and dangerous.

Unsurprisingly, as in other American cities, municipal policy has taken its lead from the security offensive and the middle-class demand for increased spatial and social insulation. De facto disinvestment in traditional public space and recreation has supported the shift of fiscal resources to corporate-defined redevelopment priorities. A pliant city government – in this case ironically professing to represent a bi-racial coalition of liberal whites and Blacks – has collaborated in the massive privatization of public space and the subsidization of new, racist enclaves (benignly described as "urban villages"). Yet most current, giddy discussions of the "postmodern" scene in Los Angeles neglect entirely these overbearing aspects of counter-urbanization and counter-insurgency. A triumphal gloss – "urban renaissance," "city of the future," and so on – is laid over the brutalization of inner-city neighborhoods and the increasing South Africanization of its spatial relations. Even as the walls have come down in Eastern Europe, they are being erected all over Los Angeles.

The observations that follow take as their thesis the existence of this new class war (sometimes a continuation of the race war of the 1960s) at the level of the built environment. Although this is not a comprehensive account, which would require a thorough analysis of economic and political dynamics, these images and instances are meant to convince the reader that urban form is indeed following a repressive function in the political furrows of the Reagan–Bush era. Los Angeles, in its usual prefigurative mode, offers an especially disquieting catalogue of the emergent liaisons between architecture and the American police state.

THE FORBIDDEN CITY

The first militarist of space in Los Angeles was General Otis of the *Times*. Declaring himself at war with labor, he infused his surroundings with an unrelentingly bellicose air:

> He called his home in Los Angeles the Bivouac. Another house was known as the Outpost. The *Times* was known as the Fortress. The staff of the paper was the Phalanx. The *Times* building itself was more fortress than newspaper plant, there were turrets, battlements, sentry boxes. Inside he stored fifty rifles.

A great, menacing bronze eagle was the *Times*'s crown; a small, functional cannon was installed on the hood of Otis's touring car to intimidate onlookers. Not surprisingly, this overwrought display of aggression produced a response in kind. On 1 October 1910 the heavily fortified *Times* headquarters – citadel of the open shop on the West Coast – was destroyed in a catastrophic explosion blamed on union saboteurs.

Eighty years later, the spirit of General Otis has returned to subtly pervade Los Angeles' new "post-modern" Downtown: the emerging Pacific Rim financial complex which cascades, in rows of skyscrapers, from Bunker Hill southward along the Figueroa corridor. Redeveloped with public tax increments under the aegis of the powerful and largely unaccountable Community Redevelopment Agency (CRA), the Downtown project is one of the largest postwar urban designs in North America. Site assemblage and clearing on a vast scale, with little mobilized opposition, have resurrected land values, upon which big developers and off-shore capital (increasingly Japanese) have planted a series of billion-dollar, block-square megastructures: Crocker Center, the Bonaventure Hotel and Shopping Mall, the World Trade Center, the Broadway Plaza, Arco Center, CitiCorp Plaza, California Plaza, and so on. With historical landscapes erased, with megastructures and superblocks as primary components, and with an increasingly dense and self-contained circulation system, the new financial district is best conceived as a single, demonically self-referential hyperstructure, a Miesian skyscape raised to dementia.

Like similar megalomaniac complexes, tethered to fragmented and desolated Downtowns (for instance, the Renaissance Center in Detroit, the Peachtree and Omni Centers in Atlanta, and so on), Bunker Hill and the Figueroa corridor have provoked a storm of liberal objections against their abuse of scale and composition, their denigration of street landscape, and their confiscation of so much of the vital life activity of the center, now sequestered within subterranean concourses or privatized malls. Sam Hall Kaplan, the crusty urban critic of the *Times*, has been indefatigable in denouncing the anti-pedestrian bias of the new corporate citadel, with its fascist obliteration of street frontage. In his view the superimposition of "hermetically sealed fortresses" and air-dropped "pieces of suburbia" has "dammed the rivers of life" Downtown.

Yet Kaplan's vigorous defense of pedestrian democracy remains grounded in hackneyed liberal complaints about "bland design" and "elitist planning practices." Like most architectural critics, he rails against the oversights of urban design without recognizing the dimension of foresight, of explicit repressive intention, which has its roots in Los Angeles' ancient history of class and race warfare. Indeed, when Downtown's new "Gold Coast" is viewed en bloc from the standpoint of its interactions with other social areas and landscapes in the central city, the "fortress effect" emerges, not as an inadvertent failure of design, but as deliberate socio-spatial strategy.

The goals of this strategy may be summarized as a double repression: to raze all association with Downtown's past and to prevent any articulation with the non-Anglo urbanity of its future. Everywhere on the perimeter of redevelopment this strategy takes the form of a brutal architectural edge or glacis that defines the new Downtown as a citadel vis-à-vis the rest of the central city. Los Angeles is unusual amongst major urban renewal centers in preserving, however negligently, most of its circa 1900–30 Beaux Arts commercial core. At immense public cost, the corporate headquarters and financial district was shifted from the old Broadway–Spring corridor six blocks west to the greenfield site created by destroying the Bunker Hill residential neighborhood. To emphasize the "security" of the new Downtown, virtually all the traditional pedestrian links to the old center, including the famous Angels' Flight funicular railroad, were removed.

The logic of this entire operation is revealing. In other cities developers might have attempted to articulate the new skyscape and the old, exploiting the latter's extraordinary inventory of theaters and historic buildings to create a gentrified history – a gaslight district, Faneuil Market or Ghirardelli Square – as a support to middle-class residential colonization. But Los Angeles' redevelopers viewed property values in the old Broadway core as irreversibly eroded by the area's very centrality to public transport, and especially by its heavy use by Black and Mexican poor. In the wake of the Watts rebellion, and the perceived Black threat to crucial nodes of white power (spelled out in lurid detail in the McCone Commission Report), resegregated spatial security became the paramount concern. The Los Angeles Police Department abetted the flight of business from Broadway to the fortified redoubts of Bunker Hill by spreading scare literature typifying Black teenagers as dangerous gang members.

As a result, redevelopment massively reproduced spatial apartheid. The moat of the Harbor Freeway and the regraded palisades of Bunker Hill cut off the new financial core from the poor immigrant neighborhoods that surround it on every side. Along the base of California Plaza, Hill Street became a local Berlin Wall separating the publicly subsidized luxury of Bunker Hill from the lifeworld of Broadway, now reclaimed by Latino immigrants as their primary shopping and entertainment street. Because politically connected speculators are now redeveloping the northern end of the Broadway corridor (sometimes known as "Bunker Hill East"), the CRA is promising to restore pedestrian linkages to the Hill in the 1990s, including the Angels' Flight incline railroad. This, of course, only dramatizes the current bias against accessibility – that is to say, against any spatial interaction between old and new, poor and rich, except in the framework of gentrification or recolonization. Although a few white-collars venture into the Grand Central Market – a popular emporium of tropical produce and fresh foods – Latino shoppers or Saturday strollers never circulate in the Gucci precincts above Hill Street. The occasional appearance of a destitute street nomad in Broadway Plaza or in front of the Museum of Contemporary Art sets off a quiet panic; video cameras turn on their mounts and security guards adjust their belts.

Photographs of the old Downtown in its prime show mixed crowds of Anglo, Black and Latino pedestrians of different ages and classes. The contemporary Downtown "renaissance" is designed to make such heterogeneity virtually impossible. It is intended not just to "kill the street" as Kaplan fears, but to "kill the crowd," to eliminate that democratic admixture on the pavements and in the parks that Olmsted believed was America's antidote to European class polarizations. The Downtown hyperstructure – like some Buckminster Fuller post-Holocaust fantasy – is programmed to ensure a seamless continuum of middle-class work, consumption and recreation, without unwonted exposure to Downtown's working-class street environments. Indeed the totalitarian semiotics of ramparts and battlements, reflective glass and elevated pedways, rebukes any affinity or sympathy between different architectural or human orders. As in Otis's fortress *Times* building, this is the archisemiotics of class war.

"The Almost Perfect Town"

Landscape (1952)

John Brinckerhoff Jackson

Editors' Introduction

John Brinckerhoff Jackson (1909–1997) can best be described as a historian of landscapes as physical, cultural, and conceptual artifacts. To Jackson, the landscape was neither wilderness nor rural nature "out there," but the totality of natural and man-made environments that simultaneously surround and infuse all forms of human activity.

Jackson taught for many years in schools of landscape architecture and geography at Harvard and the University of California, Berkeley. He was the founder of the influential journal *Landscape* and its editor from 1951 to 1968. He may be regarded as an heir to the nineteenth-century French and English landscape gardening traditions and to the pioneering work of Frederick Law Olmsted (p. 307) and the American parks movement. What sets him clearly apart from those traditional roots, however, is that, for Jackson, the urban landscape is not just a city's parks, public gardens, official buildings, and tree-lined boulevards, but its highways, its shopping malls, diners, run-down warehouse districts, two-bedroom tract houses, and its slums. Jackson sees these many elements of the man-made landscape not just as physical objects, but also as social constructs full of meaning and moral implication. It is hardly surprising, then, that J.B. Jackson's approximate version of urban utopia – his *almost* perfect place – is the commonplace, vernacular, Main Street environment of a typical American small town.

In "The Almost Perfect Town," written in the early 1950s, Jackson describes a mythical Optimo City that is located in the American Southwest, but could just as easily have been found almost anywhere in North America, Europe, Australia, and even parts of Asia and Africa beyond the margins of the great metropolitan regions. It is important that Optimo is a small place surrounded by rural land. Jackson observes that the "world of Optimo City is still complete" precisely because "the ties between country and town have not yet been broken." It is also important that Optimo has a history, however slender, that can be read in its architecture, in the layout of its streets, and in its traditional rivalry with Apache Center twenty miles away.

Jackson uses his loving, elegiac description of Optimo City as a way of criticizing developments in city planning after World War II that threatened to destroy local communities in the name of economic progress. He compares Optimo's unexceptional Courthouse Square to the Spanish plazas and the great public squares of the Baroque era that serve as socially unifying communal centers. And he notes with dismay that some of Optimo's business leaders want to tear the courthouse down to build a parking lot and to replace it with the typical "bureaucrat modernism" of so many contemporary civic centers. Jackson would like to preserve positive features of small towns that Peter Calthorpe (p. 342), Andrés Duany and Elizabeth Plater-Zyberg (p. 192), and other architects associated with the "New Urbanism" seek to recapture.

Jackson's Optimo City provides a revealing snapshot of small-town America in the 1950s before freeway construction, mass auto-ownership, and the proliferation of highway-based shopping centers. It also represents a way of thinking about urban space that is timeless.

Among the best of J.B. Jackson's book-length studies and collections of essays are *Landscapes* (Amherst: University of Massachusetts Press, 1970), *American Space* (New York: Norton, 1972), *Discovering the*

Vernacular Landscape (New Haven: Yale University Press, 1984), *The Necessity for Ruins, and Other Topics* (Amherst: University of Massachusetts Press, 1980), and *A Sense of Place, A Sense of Time* (New Haven: Yale University Press, 1994). The journal *Landscape* that Jackson founded and edited for many years contains additional writings by J.B. Jackson himself and articles exploring the meaning of landscape by the many academic and applied landscape architects, urban planners, landscape designers, historians, and others Jackson inspired.

Books by Jackson's student and his successor at Harvard are John R. Stilgoe, *Common Landscape of America, 1580 to 1845* (New Haven: Yale University Press, 1986) and *Outside Lies Magic: Regaining History and Awareness in Everyday Places* (New York: Walker & Co., 1998). Paul F. Groth and Todd Bressi, *Understanding Ordinary Landscapes* (New Haven: Yale University Press, 1997) contains essays in Jackson's tradition.

Other books on vernacular architecture include Robert Venturi, Denise Scott Brown, and Steven Izenour, *Learning from Las Vegas: The Forgotten Symbolism of Architectural Form* (Cambridge: MIT Press, 1972), Dell Upton and John M. Vlach, *Common Places: Readings in American Vernacular* (Atlanta: University of Georgia Press, 1986), Ronald W. Haase, *Classic Cracker: Florida's Wood-frame Vernacular Architecture* (Sarasota: Pineapple Press, 1992), Thomas Carter (ed.), *Images of an American Land: Vernacular Architecture in the Western United States* (Albuquerque: University of New Mexico Press, 1997), Jim Heimann, *California Crazy: Roadside Vernacular Architecture* (San Francisco: Chronicle Books, 1981), Colleen Josephine Sheehy, *The Flamingo in the Garden: American Yard Art and the Vernacular Landscape* (London: Garland, 1998), Fred E.H. Schroeder, *Front Yard America: The Evolution and Meanings of a Vernacular Domestic Landscape* (Bowling Green: Bowling Green State University, 1993), John Chase, *Glitter Stucco and Dumpster Diving: Reflections on Building Production in the Vernacular City* (London: Verso, 1998), Bernard Rudofsky, *Architecture Without Architects: A Short Introduction to Non-pedigreed Architecture* (Albuquerque: University of New Mexico Press, 1987), and Paul Oliver (ed.), *Encyclopedia of Vernacular Architecture of the World* (Cambridge: Cambridge University Press, 1998).

▪ ▪ ▪ ▪ ▪ ▪

Optimo City (pop. 10,783, alt. 2,100 ft.), situated on a small rise overlooking the N. branch of the Apache River, is a farm and ranch center served by a spur of the S.P. County seat of Sheridan Co. Optimo City (originally established in 1843 as Ft. Gaffney). It was the scene of a bloody encounter with a party of marauding Indians in 1857. (See marker on courthouse lawn.) It is the location of a state Insane Asylum, of a sorghum processing plant and an overall factory. Annual County Fair and Cowboy Roundup Sept. 4. The highway now passes through a rolling countryside devoted to grain crops and cattle raising.

Thus would the state guide dispose of Optimo City and hasten on to a more spirited topic if Optimo City as such existed. Optimo City, however, is not one town, it is a hundred or more towns, all very much alike, scattered across the United States from the Alleghenies to the Pacific, most numerous west of the Mississippi and south of the Platte. When, for instance, you travel through Texas and Oklahoma and New Mexico and even parts of Kansas and Missouri, Optimo City is the blur of filling stations and motels you occasionally pass;

the solitary traffic light, the glimpse up a side street of an elephantine courthouse surrounded by elms and sycamores, the brief congestion of mud-spattered pickup trucks that slows you down before you hit the open road once more. And fifty miles farther on Optimo City's identical twin appears on the horizon, and a half dozen more Optimos beyond that until at last, with some relief, you reach the metropolis with its new housing developments and factories and the cluster of downtown skyscrapers.

Optimo City, then, is actually a very familiar feature of the American landscape. But since you have never stopped there except to buy gas, it might be well to know it a little better. What is there to see? Not a great deal, yet more than you would at first suspect.

Optimo, being after all an imaginary average small town, has to have had an average small-town history, or at least a Western version of that average. The original Fort Gaffney (named after some inconspicuous worthy in the U.S. Army) was really little more than a stockade on a bluff overlooking a ford in the river; a few roads or trails in the old days straggled out into the plain (or desert as they called it then), lost heart and disappeared two or three miles from town. Occasionally even today someone digs up a fragment of the palisade or a bit of rust-eaten hardware in the backyards of the houses near the center of town, and the historical society possesses what it claims is the key to the principal gate. But on the whole, Optimo City is not much interested in its martial past. The fort as a military installation ceased to exist during the Civil War, and the last of the pioneers died a half century ago before anyone had the historical sense to take down his story. And when the county seat was located in the town the name was changed from Fort Gaffney with its frontier connotation to Optimo, which means (so the townspeople will tell you) "I hope for the best" in Latin.

What Optimo is really proud of even now is its identity as county seat. Sheridan County (and you will do well to remember that it was NOT named after the notorious Union general but after Horace Sheridan, an early member of the territorial legislature; Optimo still feels strongly about what it calls the War between the States) was organized in the 1870s and there ensued a brief but lively competition for the possession of the courthouse between Optimo and the next largest settlement, Apache Center, twenty miles away. Optimo City won, and Apache Center, a cowtown with one paved street, is not allowed to forget the fact. The football and basketball games between the Optimo Cougars and the Apache Braves are still characterized by a very special sort of rivalry. No matter how badly beaten Optimo City often is, it consoles itself by remembering that it is still the county seat, and that Apache Center, in spite of the brute cunning of its team, has still only one street paved. We shall presently come back to the meaning of that boast.

To get on with the history of Optimo.

THE INFLEXIBLE GRIDIRON

Aided by the state and Army engineers, the city fathers, back in the 1870s, surveyed and laid out the new metropolis. As a matter of course they located a square or public place in the center of the town and eventually they built their courthouse in the middle of the square; such having been the layout of every county seat these Western Americans had ever seen. Streets led from the center of each side of the square, being named Main Street North and South, and Sheridan Street East

and West. Eventually these four streets and the square were surrounded by a gridiron pattern of streets and avenues – all numbered or lettered, and all of them totally oblivious of the topography of the town. Some streets charge up impossibly steep slopes, straight as an arrow; others lead off into the tangle of alders and cottonwoods near the river and get lost.

Strangely enough, this inflexibility in the plan has had some very pleasant results. South Main Street, which leads from the square down to the river, was too steep in the old days for heavily laden wagons to climb in wet weather, so at the foot of it on the flats near the river those merchants who dealt in farm produce and farm equipment built their stores and warehouses. The blacksmith and welder, the hay and grain supply, and finally the auction ring and the farmers' market found South Main the best location in town for their purpose – which purpose being primarily dealing with out-of-town farmers and ranchers. And when, after considerable pressure on the legislature and much resistance from Apache Center (which already had a railroad), the Southern Pacific built a spur to Optimo, the depot was naturally built at the foot of South Main. And of course the grain elevator and the stockyards were built near the railroad. The railroad spur was intended to make Optimo into a manufacturing city, and never did; all that ever came was a small overall factory and a plant for processing sorghum with a combined payroll of about 150. Most of the workers in the two establishments are Mexicans from south of the border – locally referred to in polite circles as "Latinos" or "Hispanos." They have built for themselves flimsy little houses under the cottonwoods and next to the river. "If ever we have an epidemic in Optimo," the men at the courthouse remark, "it will break out first of all in those Latino shacks." But they have done nothing as yet about providing them with better houses, and probably never will.

DOWNTOWN AND UPTOWN

Depot, market, factories, warehouses, slum – these features, combined with the fascination of the river bank and stockyards and the assorted public of railroaders and Latinos and occasional ranch hands – have all given South Main a very definite character: easy-going, loud, colorful, and perhaps during fair week or at shipping time a little disreputable. Boys on the Cougar football squad have specific orders to stay away from South Main, but they don't. Actually the whole of Optimo looks on the section with indulgence and pride; it makes the townspeople feel that they understand metropolitan problems when they can compare South Main with the New York waterfront.

North Main, up on the heights beyond the Courthouse Square and past the two or three blocks of retail stores, is (on the other hand) the very finest part of Optimo. The northwestern section of town, with its tree-shaded streets, its view over the river and the prairie, its summer breezes, has always been identified with wealth and fashion as Optimo understands them. Colonel Ephraim Powell (Confederate Army, Ret., owner of some of the best ranch country in the region) built his bride a handsome limestone house with a slate roof and a tower, and Walter Slymaker, proprietor of Slymaker's Mercantile and of the grain elevator, not to be outdone, built an even larger house farther up Main; so did Hooperson, first president of the bank. There are a dozen such houses in all, stone or Milwaukee brick with piazzas (or galleries, as the old timers still call them) and large, untidy gardens around them. It is worth noting, by the way, that the brightest claim to aristocratic heritage is this: grandfather came out West for his health. New England may have its "Mayflower" and "Arabella," east Texas its Three Hundred Founding Families, New Mexico its Conquistadores; but Optimo is loyal to the image of the delicate young college graduate who arrived by train with his law books, his set of Dickens, his taste for wine, and the custom of dressing for dinner. This legendary figure has about seen his day in the small talk of Optimo society, and the younger generation frankly doubts his having ever existed; but he (or his ghost) had a definite effect on local manners and ways of living. At all events, because of this memory Optimo looks down on those Western mining towns where Sarah Bernhardt and de Reszke and Oscar Wilde seem to have played so many one-night stands in now-vanished opera houses.

A WORLD IN ITSELF

Wickedness – or the suggestion of wickedness – at one end of Main, affluence and respectability at the other. How about Sheridan Street running East and West? That is where you'll find most of the stores; in the first four or five blocks on either side of the Courthouse Square. They form a rampart: narrow brick houses, most of them two stories high with elaborate cornices and long narrow windows; all of them devoid of modern commercial graces of chromium and black glass and artful window display, all of them ugly but all of them pretty uniform; and so you have on Sheridan Street something rarely seen in urban America: a harmonious and restful and dignified business section. Only eight or ten blocks of it in all, to be sure; turn any corner and you are at once in a residential area.

Here there is block after block of one-story frame houses with trees in front and picket fences or hedges; no sidewalk after the first block or so; a hideous church (without a cemetery of course); a small-time auto repair shop in someone's back yard; dirt roadway; and if you follow the road a few blocks more – say to 10th Street (after that there are no more signs) – you are likely to see a tractor turn into someone's drive with wisps of freshly cut alfalfa clinging to the vertical sickle bar. The countryside is that close to the heart of Optimo City, farmers are that much part of the town. And the glimpse of the tractor (like the glimpse of a deer or a fox driven out of the hills by a heavy winter) restores for a moment a feeling for an old kinship that seemed to have been destroyed forever. But this is what makes Optimo, the hundreds of Optimos throughout America, so valuable; the ties between country and town have not yet been broken. Limited though it may well be in many ways, the world of Optimo City is still complete.

The center of this world is Courthouse Square, with the courthouse, ponderous, barbaric, and imposing, in the center of that. The building and its woebegone little park not only interrupts the vistas of Main and Sheridan – it was intended to do this – it also interrupts the flow of traffic in all four directions. A sluggish eddy of vehicles and pedestrians is the result, Optimo's animate existence slowed and intensified. The houses on the four sides of the square are of the same vintage (and the same general architecture) as the monument in their midst: mid-nineteenth-century brick or stone; cornices like the brims of hats, fancy dripstones over the arched windows like eyebrows; painted blood-red or mustard-yellow or white; identical except for the six-story Gaffney Hotel and the classicism of the First National Bank.

Every house has a tin roof porch extending over the sidewalk, a sort of permanent awning which protects passersby and incidentally conceals the motley of store windows and signs. To walk around the square and down Sheridan Street under a succession of these galleries or metal awnings, crossing the strips of bright sunlight between the roofs of different height, is one of the delights of Optimo – one of its amenities in the English use of that word. You begin to understand why the Courthouse Square is such a popular part of town.

SATURDAY NIGHTS – BRIGHT LIGHTS

Saturday, of course, is the best day for seeing the full tide of human existence in Sheridan County. The rows of parked pickups are like cattle in a feed lot; the sidewalks in front of Slymaker's Mercantile, the Ranch Cafe, Sears, the drugstore, resound to the mincing steps of cowboy boots; farmers and ranchers, thumbs in their pants pockets, gather in groups to lament the drought (there is always a drought) and those men in Washington, while their wives go from store to movie house to store. Radios, jukeboxes, the bell in the courthouse tower; the teenagers doing "shave-and-a-haircut; bay rum" on the horns of their parents' cars as they drive round and round the square. The smell of hot coffee, beer, popcorn, exhaust, alfalfa, cow manure. A man is trying to sell a truckload of grapefruit so that he can buy a truckload of cinderblocks to sell somewhere else. Dogs; 10-year-old cowboys firing cap pistols at each other. The air is full of pigeons, floating candy wrappers, the flat strong accent erroneously called Texan.

All these people are here in the center of Optimo for many reasons – for sociability first of all, for news, for the spending and making of money; for relaxation. "Jim Guthrie and wife were in town last week, visiting friends and transacting business," is the way the *Sheridan Sentinel* describes it; and almost all of Jim Guthrie's business takes place in the square. That is one of the peculiarities of Optimo and one of the reasons why the square as an institution is so important. For it is around the square that the oldest and most essential urban (or county) services are established. Here are the firms under local control and ownership, those devoted almost exclusively to the interest of the surrounding countryside. Upstairs are the lawyers, doctors, dentists, insurance firms, the public stenographer, the Farm Bureau. Downstairs are the bank, the prescription drugstore, the newspaper office, and of course Slymaker's Mercantile and the Ranch Cafe.

INFLUENCE OF THE COURTHOUSE

Why have the chain stores not invaded this part of town in greater force? Some have already got a foothold, but most of them are at the far end of Sheridan or even out on the Federal Highway. The presence of the courthouse is partly responsible. The traditional services want to be as near the courthouse as they can, and real-estate values are high. The courthouse itself attracts so many out-of-town visitors that the problem of parking is acute. The only solution that occurs to the enlightened minds of the Chamber of Commerce is to tear the courthouse down, use the place for parking, and build a new one somewhere else. They have already had an architect draw a sketch of a new courthouse to go at the far end of Main Street: a chaste concrete cube with vertical motifs between the windows – a fine specimen of bureaucrat modernism. But the trouble is, where to get the money for a new courthouse when the old one is still quite evidently adequate and in constant use?

If you enter the courthouse you will be amazed by two things: the horrifying architecture of the place, and the variety of functions it fills. Courthouse means of course courtrooms, and there are two of those. Then there is the office of the County Treasurer, the Road Commissioner, the School Board, the Agricultural Agent, the Extension Agent, Sanitary Inspector, and usually a group of Federal agencies as well – PMA, Soil Conservation, FHA and so on. Finally the Red Cross, the Boy Scouts, and the District Nurse. No doubt many of these offices are tiresome examples of government interference in private matters; just the same, they are for better or worse part of almost every farmer's and rancher's business, and the courthouse, in spite of all the talk about county consolidation, is a more important place than ever.

As it is, the ugly old building has conferred upon Optimo a blessing which many larger and richer American towns can envy: a center for civic activity and a symbol for civic pride – something as different from the modern "civic center" as day is from night. Contrast the array of classic edifices, lost in the midst of waste space, the meaningless pomp of flagpoles and war memorials and dribbling fountains of any American city from San Francisco to Washington with the animation and harmony and the almost domestic intimacy of Optimo Courthouse Square, and you have a pretty good measure of what is wrong with much American city planning: civic consciousness has been divorced from everyday life, put in a special zone all by itself. Optimo City has its zones; but they are organically related to one another.

Doubtless the time will never come that the square is studied as a work of art. Why should it be? The craftsmanship in the details, the architecture of the building, the notions of urbanism in the layout of the square itself are all on a very countrified level. Still, such a square as this has dignity and even charm. The charm is perhaps antiquarian – a bit of rural America of seventy-five years ago; the dignity is something else again. It derives from the function of the courthouse and the square, and from its peculiarly national character.

COMMUNAL CENTER

The practice of erecting a public building in the center of an open place is in fact pretty well confined to America – more specifically to nineteenth-century America. The vast open areas favored by eighteenth-century European planners were usually kept free of construction, and public buildings – churches and palaces and law-courts – were located to face these squares; to command them, as it were. But they were not allowed to interfere with the original open effect. Even the plans of eighteenth-century American cities, such as Philadelphia and Reading and Savannah and Washington, always left the square or public place intact. Spanish America, of course, provides the best illustrations of all; the plaza, nine times out of ten, is surrounded by public buildings, but it is left free. Yet almost every American town laid out after (say) 1820 deliberately planted a public building in the center of its square. Sometimes it was a school, sometimes a city hall, more often a courthouse, and it was always approachable from all four sides and always as conspicuous as possible.

Why? Why did these pioneer city fathers go counter to the taste of the past in this matter? One guess is as good as another. Perhaps they were so proud of their representative institutions that they wanted to give their public buildings the best location available. Perhaps frontier America was following an aesthetic movement, already at that date strong in Europe, that held that an open space was improved when it contained some prominent free-standing object – an obelisk or a statue or a triumphal arch. However that may have been, the pioneer Americans went Europe

one better, and put the largest building in town right in the center of the square.

Thus the square ceased to be thought of in nineteenth-century America as a vacant space; it became a container or (if you prefer) a frame. A frame, so it happened, not merely for the courthouse, but for all activity of a communal sort. Few aesthetic experiments have ever produced such brilliantly practical results. A society which had long since ceased to rally around the individual leader and his resi dence and which was rapidly tiring of rallying around the meetinghouse or church all at once found a new symbol: local representative government, or the courthouse. A good deal of flagwaving resulted – as European travelers have always told us – and a good deal of very poor "representational" architecture; but Optimo acquired something to be proud of, something to moderate that American tendency to think of every town as existing entirely for money-making purposes.

SYMBOL OF INDEPENDENCE

At this juncture the protesting voice of the Chamber of Commerce is heard. "One moment. Before you finish with our courthouse you had better hear the other side of the question. If the courthouse were torn down we would not only have more parking space – sorely needed in Optimo – we would also get funds for widening Main Street into a four-lane highway. If Main Street were widened Optimo could attract many new businesses catering to tourists and other transients – restaurants and motels and garages and all sorts of drive-in establishments. In the last ten years" (continues the Chamber of Commerce) "Optimo has grown by twelve hundred. Twelve hundred! At that rate we'll still be a small town of less than twenty thousand in 1999. But if we had new businesses we'd grow fast and have better schools and a new hospital, and the young people wouldn't move to the cities. Or do you expect Optimo to go on depending on a few hundred tight-fisted farmers and ranchers for its livelihood?" The voice, now shaking with emotion, adds something about "eliminating" South Main by means of an embankment and a clover leaf and picnic grounds for tourists under the cottonwoods where the Latinos still reside.

These suggestions are very sensible ones on the whole. Translate them into more general terms and what they amount to is this: if we want to get ahead, the best thing to do is break with our own past, become as independent as possible of our immediate environment and at the same time become almost completely dependent for our well-being on some remote outside resource. Whatever you may think of such a program, you cannot very well deny that it has been successful for a large number of American towns. Think of the hayseed communities which have suddenly found themselves next to an oil field or a large factory or an Army installation, and which have cashed in on their good fortune by transforming themselves overnight, turning their backs on their former sources of income, and tripling their population in a few years! It is true that these towns put all their eggs in one basket, that they are totally at the mercy of some large enterprise quite beyond their control. But think of the freedom from local environment; think of the excitement and the money! Given the same circumstances – and the Southwest is full of surprises still – why should Optimo not do the same?

A COMMON DESTINY

Because there are many different kinds of towns just as there are many different kinds of men, a development which is good for one kind can be death on another. Apache Center (to use that abject community as an example), with its stockyards and its one paved street and its very limited responsibility to the county, as a community might well become a boom-town and no one would be worse off. Optimo seems to have a different destiny. For almost a hundred years – a long time in this part of the world – it has been identified with the surrounding landscape and been an essential part of it. Whatever wealth it possesses has come from the farms and ranches, not from the overall factory or from tourists. The bankers and merchants will tell you, of course, that without their ceaseless efforts and their vision the countryside could never have existed; the farmers and ranchers consider Optimo's prosperity and importance entirely their own creation. Both parties are right to the extent that the town is part of the landscape – one might even say part of every farm, since much farm business takes place in the town itself.

Now if Optimo suddenly became a year-round tourist resort, or the overall capital of the Southwest; what would happen to that relationship, do you sup-

pose? It would vanish. The farmers and ranchers would soon find themselves crowded out, and would go elsewhere for those services and benefits which they now enjoy in Optimo. And as for Optimo itself, it would soon achieve the flow of traffic, the new store fronts, the housing developments, the payrolls and bank accounts it cannot help dreaming about; and in the same process achieve a total social and physical dislocation, and a loss of a sense of its own identity. County Seat of Sheridan County? Yes; but much more important: Southwestern branch of the "American Cloak and Garment Corporation"; or the LITTLE TOWN WITH THE BIG WELCOME – 300 tourist beds which, when empty for one night out of three, threaten bankruptcy to half the town.

As of the present, Optimo remains pretty much as it has been for the last generation. The Federal Highway still bypasses the center (what a roadblock, symbolical as well as actual, that courthouse is!); so if you want to see Optimo, you had better turn off at the top of the hill near the watertower of the lunatic asylum – now called Fairview State Rest Home, and with the hideous high fence around it torn down. The dirt road eventually becomes North Main. The old Slymaker place is still intact. The Powell mansion, galleries and all, belongs to the American Legion, and a funeral home has taken over the Hooperson house. Then comes downtown Optimo; and then the court-house, huge and graceless, in detail and proportion more like a monstrous birdhouse than a monument. Stop here. You'll find nothing of interest in the stores, and no architectural gems down a side street. Even if there were, no one would be able to point them out. The historical society, largely in the hands of ladies, thinks of antiquity in terms of antiques, and art as anything that looks pretty on the mantelpiece.

The weather is likely to be scorching hot and dry, with a wild ineffectual breeze in the elms and sycamores. You'll find no restaurant in town with atmosphere – no chandeliers made out of wagon wheels, no wall decorations of famous brands, no bar disguised as the Hitching Rail or the Old Corral. Under a high ceiling with a two-bladed fan in the middle, you'll eat ham hock and beans, hot bread, iced tea without lemon, and like it or go without. But as compensation of sorts at the next table there will be two ranchers eating with their hats on, and discussing the affairs, public and private, of Optimo City. To hear them talk, you'd think they owned the town.

That's about all. There's the market at the foot

of South Main, the Latino shacks around the overall factory, a grove of cottonwoods, and the Apache River (North Branch) trickling down a bed ten times too big; and then the open country. You may be glad to have left Optimo behind.

Or you may have liked it, and found it pleasantly old-fashioned. Perhaps it is; but it is in no danger of dying out quite yet. As we said to begin with, there is another Optimo City fifty miles farther on. The country is covered with them. Indeed they are so numerous that it sometimes seems as if Optimo and rural America were one and indivisible.

THREE

"The Neighborhood, the District, and the Corridor"

from Peter Katz and Vincent Scully, Jr. (eds), *The New Urbanism: Toward an Architecture of Community* (1993)

Andrés Duany and Elizabeth Plater-Zyberk

Editors' Introduction

Andrés Duany and Elizabeth Plater-Zyberk (p. 192), the husband and wife team, along with Peter Calthorpe and William Fulton (p. 342), are leaders in a current intellectual movement in architecture and urban planning called *the New Urbanism*. They are interested in urban space because they want to reshape urban and suburban patterns to create places where people can interact with their neighbors, enjoy public space, and get from home to work, school, and stores by foot and bicycle as well as by car. Duany and Plater-Zyberk are notable not only for their theorizing about better urban spaces, but also because they have designed and built successful new communities such as Seaside, Florida that actually implement their ideas.

Duany and Plater-Zyberk are neo-traditionalists. They find much of value in the traditional vernacular architecture and people-oriented layout of the kind of small American towns J.B. Jackson lovingly describes (p. 184). Like Jane Jacobs (p. 98) they feel that promoting human interaction and community are as important values as designing for efficiency and are skeptical of the way in which urban planners have segregated land uses and imposed artificial and de-humanizing order on urban spaces. Duany and Plater-Zyberk are often described as postmodernists. They reject the emphasis on machine-age efficiency, large scale, and speed that dominated the thinking of modernist architects and planners like Le Corbusier (p. 322) and other modernist architects and planners in the middle of the twentieth century.

In this essay Duany and Plater-Zyberk describe how architects and planners can design urban elements to create urban spaces much better than the isolating, auto-dependent suburbs common today. Their approach of breaking down urban space into distinct elements – neighborhoods, districts, edges, and corridors – is influenced by Kevin Lynch (p. 438), who identified key "elements" that make up the image of the city and urged design professionals to follow principles of good design for each element.

Duany and Plater-Zyberk are particularly interested in *neighborhoods* – urbanized areas with a balanced mix of uses. Their ideas about the ideal neighborhood are very concrete and elements of their ideal model fit together into a unified vision for an alternative to the pattern of development that evolved after World War II. The ideal neighborhood they envisage is *small* – a five-minute walk from center to edge. It is *diverse*, containing a balanced mix of dwellings, workplaces, shops, parks, and civic institutions such as schools and churches. Their ideal neighborhood has a *center* dominated by civic buildings and public space, and an *edge* of some kind (perhaps defined by fields and forest; perhaps by a highway or rail line). Land use in their ideal neighborhood articulates with a system of pedestrian-friendly and transit-oriented transportation corridors that offer residents opportunities to walk, bicycle, drive, or take public transit to work, school, shopping, and entertainment. This vision stands in sharp contrast to the predominant suburban pattern today: sprawling, auto-dependent, homogeneous areas lacking in public space and without clear centers or edges.

New Urbanists like Duany and Plater-Zyberk believe the modernist values that produced modern suburbia are no longer relevant. While the segregation of housing from the smoke and stench of nineteenth-century factories may have been an improvement over the dreadful living conditions Friedrich Engels described in nineteenth-century Manchester (p. 50), Duany and Plater-Zyberk argue that such a segregation of land uses is no longer necessary in the post-industrial information age. While the street pattern built to accommodate mass auto use allowed people to move efficiently from suburban homes to city-centered work locations, this is no longer the pattern in the technoburbia, as Fishman (p. 69) describes. Duany, Plater-Zyberk, and other New Urbanists feel the auto-oriented corridor system has had a devastating impact on community and the quality of life in suburbia that needs to be changed.

Duany and Plater-Zyberk have a private architectural/urban planning practice in Florida specializing in developments consistent with the design principles in this selection. They are founders of the Congress for New Urbanism – a loose association of like-minded architects, planners, and other professionals. Their most notable accomplishment is the design of Seaside, Florida, an entire new town, built on New Urbanist principles, that has attracted worldwide attention.

This reading is from Peter Katz and Vincent Scully, Jr. (eds), *The New Urbanism: Toward an Architecture of Community* (New York: McGraw-Hill, 1993), a collection of essays and illustrations of New Urbanist developments. Duany and Plater-Zyberk's most recent statement of their critique of suburbia and alternative vision is *Suburban Nation: The Rise of Sprawl and the Decline of the American Dream* (New York: North Point Press, 2001) co-authored with Jeff Speck. An earlier work describing their theories is *Towns and Townmaking Principles* (London: Rizzoli, 1991). Duany co-authored *The Smart Growth Manual* (New York: McGraw-Hill, 2002) with Jeff Speck and edited *New Civic Art: Elements of Town Planning* (London: Rizzoli, 2002).

For a description of how Seaside, Florida was designed and built see David Mohoney and Keasterling Keller (eds), *Seaside: Making a Town in America* (Princeton: Princeton Architectural Press, 1991). Critiques of Seaside – both positive and negative – are contained in Tod Bressi (ed.), *The Seaside Debates: A Critique of the New Urbanism* (London: Rizzoli, 2002).

Other writings on the New Urbanism include Emily Talen, *New Urbanism and American Planning: The Conflict of Cultures* (London and New York: Routledge, 2005), Michael Lecesse and Kathleen McCormick (eds), *Charter of the New Urbanism* (New York: McGraw-Hill, 1999), John A. Dutton, *New American Urbanism: Re-forming the Suburban Metropolis* (New York: Skira, 2001), Peter Calthorpe, *The Next American Metropolis: Ecology, Community, and the American Dream* (New York: Princeton Architectural Press, 1993), and Philip Langdon, *A Better Place to Live: Reshaping the American Suburb* (Amherst: University of Massachusetts Press, 2001). James Howard Kunstler has popularized critiques of suburbia and New Urbanist ideas in *The Geography of Nowhere: The Rise and Decline of America's Man-made Landscape* (New York: Touchstone Books, 1994) and *Home from Nowhere: Remaking Our Everyday World for the Twenty-first Century* (New York: Touchstone Books, 1998).

▪ ▪ ▪ ▪ ▪ ▪

The fundamental organizing elements of the New Urbanism are the neighborhood, the district and the corridor. Neighborhoods are urbanized areas with a balanced mix of human activity; districts are areas dominated by a single activity; corridors are connectors and separators of neighborhoods and districts. Neighborhoods, districts and corridors are urban elements. By contrast, suburbia, which is the result of zoning laws that separate uses, is composed of pods, highways and interstitial spaces.

THE NEIGHBORHOOD

The nomenclature may vary, but there is general agreement regarding the physical composition of the neighborhood. The "neighborhood unit" of the 1929 New York Regional Plan, the "quartier" identified by Leon Krier, the "traditional neighborhood development" (TND) and "transit-oriented development"

A single neighborhood standing free in the landscape is a village. Cities and towns are made up of multiple neighborhoods and districts, organized by corridors of transportation or open space.

(TOD) share similar attributes. They all propose a model of urbanism that is limited in area and structured around a defined center. While the population density may vary, depending on its context, each model offers a balanced mix of dwellings, workplaces, shops, civic buildings and parks.

Like the habitat of any species, the neighborhood possesses a natural logic that can be described in physical terms. The following are the principles of an ideal neighborhood design: (1) the neighborhood has a center and an edge; (2) the optimal size of a neighborhood is a quarter mile from center to edge; (3) the neighborhood has a balanced mix of activities – dwelling, shopping, working, schooling, worshipping and recreating; (4) the neighborhood structures building sites and traffic on a fine network of interconnecting streets; (5) the neighborhood gives priority to public space and to the appropriate location of civic buildings.

The neighborhood has a center and an edge

The combination of a focus and a limit contribute to the social identity of the community. The center is a necessity, the edge not always so. The center is always a public space, which may be a square, a green or an important street intersection. It is near the center of the urban area unless compelled by some geographic circumstance to be elsewhere. Eccentric locations are justified if there is a shoreline, a transportation corridor or a place with an engaging view.

The center is the locus of the neighborhood's public buildings, ideally a post office, a meeting hall, a daycare center, and sometimes religious and cultural institutions. Shops and workplaces are usually associated with the center, especially in a village. In the aggregations of multiple neighborhoods which occur in a town or city, retail buildings and workplaces may be at the edge of the neighborhood, where they can combine with others and intensify commercial and community activity.

Neighborhood edges may vary in character: they can be natural, such as a forest, or man-made, such as infrastructure. In villages, the edge is usually defined by land designated for cultivation such as farms, orchards and nurseries or for conservation in a natural state as woodland, desert, wetland or escarpment. The edge may also be assigned to very low-density residential use with lots of at least 10 acres. When

a community cannot afford to sustain large tracts of public open land, such large private ownerships are a way to maintain a green edge.

In cities and towns, edges can be formed by the systematic accretion between the neighborhoods of recreational open spaces, such as parks, schoolyards and golf courses. It is important that golf courses be confined to the edge of neighborhoods, because fairways obstruct direct pedestrian ways to the neighborhood center. These continuous green edges can be part of a larger network of corridors, connecting urban open space with rural surroundings, as described in the 1920s by Benton McKaye.

In high-density urban areas, the neighborhood edge is often defined by infrastructure, such as rail lines and high traffic thoroughfares that best remain outside the neighborhood. The latter, if generously lined with trees, become parkways that reinforce the legibility of the edge and, over a long distance, form the corridors connecting urban neighborhoods.

The optimal size of a neighborhood is a quarter mile from center to edge

This distance is the equivalent of a five-minute walk at an easy pace. The area thus circumscribed is the neighborhood proper, to differentiate it from the green edge, which extends beyond the discipline of the quarter mile. The limited area gathers the population of a neighborhood within walking distance of many of their daily needs, such as a convenience store, post office, community police post, automatic bank teller, school, daycare center and transit stop.

The stop's location among other neighborhood services and within walking distance of home or work makes the transit system convenient. When an automobile trip is necessary to arrive at a transit stop, most potential users will simply continue driving to their destinations. But the neighborhood, which focuses the required user population within walking distance of the stop, makes transit viable at densities that a suburban pattern cannot sustain.

Pedestrian-friendly and transit-oriented neighborhoods permit a region of cities, towns and villages to be accessible without singular reliance on cars. Such a system gives access to the major cultural and social institutions, the variety of shopping and the broad job base that can only be supported by the larger population of an aggregation of neighborhoods.

The neighborhood has a balanced mix of activities – dwelling, shopping, working, schooling, worshipping and recreating

This is particularly important for those who are unable to drive and thus depend on others for mobility. For instance, the young are able to walk or bicycle to school and other activities, freeing their parents from the responsibility and tedium of chauffeuring. The size of a school should be determined by the number of children who can walk or bicycle to it from adjacent neighborhoods.

And the elderly, who relinquish their willingness to drive before they lose their ability to walk, can age in place with dignity rather than being forced into specialized retirement communities, which are the attendant creations of the suburban pattern.

Even those for whom driving may not be a burden enjoy secondary advantages. The proximity of daily destinations and the convenience of transit reduces the number and length of trips, decreases the private stress of time in traffic and minimizes the public-borne expenses of road construction and atmospheric pollution.

The neighborhood's fine-grained mix of activities includes a range of housing types for a variety of incomes, from the wealthy business owner to the school teacher and the gardener. Suburban areas, which are most commonly segregated by income, do not provide for the full range of society. The true neighborhood, however, offers a variety of affordable housing choices: garage apartments in conjunction with single-family houses, apartments above shops and apartment buildings adjacent to shopping and workplaces. The latter's transitional sites are not provided within the suburban pattern whose rigorous, sanitized segregation of uses precludes them.

But the greatest contribution to affordable housing may be realized by the neighborhood's ability to reduce multiple automobile ownership and many of its associated costs. By enabling households to own one less vehicle, the average annual operating cost of $5,000 can be applied toward an additional $50,000 increment of mortgage financing at 10 percent. No other action of the designer can achieve an improvement in the availability of housing for the middle class comparable to the sensible organization of a good neighborhood plan.

The neighborhood structures building sites and traffic on a fine network of interconnecting streets

Neighborhood streets are configured to create blocks of appropriate building sites and to shorten pedestrian routes. They are designed to keep local traffic off regional roads and to keep through traffic off local streets. An interconnecting pattern of streets provides multiple routes that diffuse traffic congestion.

This contrasts to the easily congested single trajectories standard to the suburban pattern: culs-de-sac spill onto collector streets, which connect at single points to arterials, which in turn supply the highways. The suburban traffic model is more concerned with speeding traffic through a place than with the quality of the place itself; the pedestrian is assumed to be elsewhere on separate "walkways" or nonexistent.

Neighborhood streets of varying types are detailed to provide equitably for pedestrian comfort and for automobile movement. Slowing the automobile and increasing pedestrian activity encourages the casual meetings that form the bonds of community.

The neighborhood gives priority to public space and to the appropriate location of civic buildings

Public spaces and buildings represent community identity and foster civic pride. The neighborhood plan structures its streets and blocks to create a hierarchy of public spaces and locations for public buildings. Squares and streets have their size and geometry defined by the intention to create special places. Public buildings occupy important sites, overlooking a square or terminating a street vista.

The suburban practice of locating government buildings, places of worship, schools and even public art according to the expediencies of land cost is ineffective. The importance of these civic and community structures is enhanced by their suitable siting, without incurring additional costs to the infrastructure.

THE DISTRICT

The district is an urbanized area that is functionally specialized. Although districts preclude the full range of activities necessary for a complete neighborhood, they are not the rigorously single activity zones of suburbia: the office parks, housing subdivisions or shopping centers. The specialization of a district still allows multiple activities to support its primary identity. Typical are theater districts, which have restaurants and bars to support and intensify their nightlife; tourist districts, which concentrate hotels, retail activity and entertainment; and the capitol area and the college campus, which are dominated by a large institution. Others accommodate large-scale transportation or manufacturing, such as airports, container terminals and refineries.

Although a degree of specialization for certain urban areas enhances their character and efficiency, in reality, few pure districts are really justified. Thanks to industrial evolution and environmental regulation, the reasons for segregating uses recede with time. The modern North American workplace is no longer a bad neighbor to dwellings and shops.

The organizational structure of the district parallels that of the neighborhood and similarly, for a good fit within the greater region, relies on its relationship to transit. An identifiable focus encourages the formation of special communities: a park for workers at lunch, a square for theater-goers to meet, a mall for civic gatherings. Clear boundaries and dimensions facilitate the formation of special taxing or management organizations. Interconnected circulation supports the pedestrian, enhances transit viability and ensures security. And like the neighborhood, attention to the character of the public spaces creates a sense of place for its users, even if their home is elsewhere.

THE CORRIDOR

The corridor is at once the connector and the separator of neighborhoods and districts. Corridors include natural and man-made elements, ranging from wildlife trails to rail lines. The corridor is not the haphazardly residual space that remains outside subdivisions and shopping centers in suburbia. Rather, it is an urban element characterized by its visible continuity. It is defined by its adjacent districts and neighborhoods and provides entry to them.

The corridor's location and type is determined by its technological intensity and nearby densities. Heavy rail corridors are tangent to towns and traverse the industrial districts of cities. Light rail and trolleys may occur within a boulevard at the neighborhood edge. As such, they are detailed for pedestrian use and to accommodate the frontages of buildings. Bus corridors can pass through neighborhood centers on conventional streets. All of these should be landscaped to reinforce their continuity.

In low-density areas, the corridor may be the continuous green edge between neighborhoods, providing long-distance walking and bicycle trails, other recreational amenities and a continuous natural habitat.

The corridor is a significant element of the New Urbanism because of its inherently civic nature. In the age of the metropolis, with villages, towns, neighborhoods and districts aggregated in unprecedented quantity, the most universally used public spaces are the corridors that serve connection and mobility. Of the three elements – the neighborhood, the district and the corridor – the latter, in its optimum form, is the most difficult to implement because it requires regional coordination.

CONCLUSION

The conventional suburban practice of segregating uses by zones is the legacy of the "dark satanic mills," which were once genuine hazards to public welfare. The separation of dwelling from workplace in the course of the last century was the great achievement of the nascent planning profession and remains institutionalized in zoning ordinances. The suburbs and cities of today continue to separate the naturally integrated human activities of dwelling, working, shopping, schooling, worshipping and recreating.

The hardship caused by this separation has been mitigated by widespread automobile ownership and use, which in turn has increased the demand for vehicular mobility. The priority given to road building at the expense of other civic programs during the last four decades has brought our country to the multiple crises of environmental degradation, economic bankruptcy and social disintegration.

The New Urbanism offers an alternative future for the building and re-building of regions. Neighborhoods that are compact, mixed-use and pedestrian friendly; districts of appropriate location and character; and corridors that are functional and beautiful can integrate natural environments and man-made communities into a sustainable whole.

"The Impact of the New Technologies and Globalization on Cities"

from Arie Graafland and Deborah Hauptmann (eds),
Cities in Transition (2001)

Saskia Sassen

Editors' Introduction

Saskia Sassen, a professor of sociology at the University of Chicago, has analyzed data on information technology, the economies, and the organization of physical space in the most advanced cities and metropolitan regions in the world. In the following selection she describes how globalization and information technology appear to be creating a new global system of cities, changing relationships among cities, and reconfiguring the physical arrangement of activities within metropolitan space.

Sassen is particularly interested in global *cities* – where international financial functions are disproportionately concentrated and whose economies are most closely integrated with the world economy. Sassen argues that globalization is both concentrating and simultaneously dispersing activity at the global, national, and metropolitan levels. At the global scale, economic power is increasingly *concentrated* in New York, London, and Tokyo. But cities as dispersed as Mexico City, Taipei, Bangkok, Buenos Aires, São Paulo, Frankfurt, Zurich, and Sydney may also be characterized as global cities.

As economic activity is increasingly globalized and as industries need more specialized services, Sassen believes a new world "system of cities" is emerging unlike anything that has previously existed. What Sassen terms "corporate service complexes" – sophisticated networks of high-level financiers, lawyers, accountants, advertising professionals, and other skilled professionals serving international corporations – are clustered in global cities. Decisions made by joint headquarters/corporate services complexes in London, New York, Tokyo, and other global cities affect not only the residents of these cities, but jobs, wages, and the economic health of cities as dispersed around the globe as Kuala Lumpur, Malaysia, Ho Chi Minh City, Vietnam, and Santiago, Chile. If financial analysts advise their corporate clients that Argentina's economy is weakening and lawyers inform them that legal reforms in China present new opportunities for super profits, the corporations may pull billions of dollars out of Argentina and redirect the funds to Shanghai – assisted by a small army of advertising executives touting their products in China's vast emerging markets.

A traditional reason why businesses cluster in large cities has been to be in touch with other businesses and with lawyers, accountants, bankers, advertising firms and other specialized service providers that help them do business. It is efficient to walk next door to a lawyer's office and down the block to do business with a major business partner. Cities had what economists call "agglomeration economies." Now in the information age, as Manuel Castells (p. 478) and William Mitchell (p. 510) discuss, information can travel almost instantly to anywhere in the world. Businessmen no longer need to walk next door to communicate with their lawyers or down the street to meet with a business partner: they can just phone, fax, e-mail, or videoconference next door or to a remote location anywhere in the world, so long as the telecommunications infrastructure permits. Highly developed tele-communications infrastructure in global cities facilitate transmission of information in staggering quantities at lightning speed. International banks in Rio de Janeiro can bounce a year's worth of financial records to a New York bank via satellite in seconds.

Anticipating these trends, Melvin Webber (p. 473) more than three decades ago hypothesized that information technology would make space increasingly irrelevant and as a consequence cities would diminish in importance. But according to Sassen that has not happened. Sassen has found that global cities such as New York, London, and Tokyo have grown in population, wealth, and power since the information revolution. Their status and importance continues to grow, not decline. On the other hand, many cities that historically served as secondary corporate command and control centers are in economic decline as corporate power continues to concentrate in the most advanced global cities. Paris is growing in economic power and wealth; Marseilles is declining.

Sassen points out that production and retailing are becoming more dispersed. Many corporations design products at a headquarters location – perhaps (but not necessarily) in a global city like London. They contract with firms in a developing country like Malaysia to produce the products they have designed. And then they market the finished products worldwide. This kind of production process requires sophisticated support to manage dispersed and rapidly changing operations all over the world – lawyers familiar with British law, accountants who understand Malaysian accounting practices, and advertising executives sensitive to the cultural preferences of German consumers. Corporations rarely have the internal capacity to do all that. Instead they turn to networks of specialized firms located in global cities to provide the services they need.

Sassen questions the whole notion of "rich" countries and "rich" cities; places central to the world economy, and those that exist at the margin. She argues that economic inequality is sharply increasing – particularly in global cities at the center of the world economy. The opportunity for hyper-profits in international finance is creating extraordinary wealth for Wall Street bankers. But many of the low-paid janitors and file clerks who work on Wall Street were born in Third World countries and live in ethnic New York neighborhoods just a short distance away. In São Paulo, wealthy Brazilian nationals and expatriates earn salaries and participate in lifestyles more like those of wealthy New Yorkers than those of the people in São Paulo neighborhoods a few blocks from where they work. An important public policy question facing countries all over the world is how to promote economic equality and help more of their citizens benefit from the new world economic order.

Regions – such as the Asian mega-city regions Aprodicio aquian describes (p. 489) – are also changing as a result of globalization and information technology. Sassen argues that economic activity within the metropolitan areas of global cities manifests a dynamic of both concentration and dispersion just as is occurring in the world system of cities. At the time that Ernest W. Burgess developed his concentric zone theory (p. 150), Chicago and many other cities had a distinct central business district (CBD) where intense economic activity was concentrated. In advanced metropolitan areas today there is often no longer a single, clearly demarcated CBD. Sassen believes four different models are emerging. In some cities something close to a classic CBD still exists. New York City's Wall Street area is an example. In others there is a new pseudo-CBD just outside the historic city center, such as the massive planned office complex named La Défense just outside the center of Paris. In other regions Sassen sees nodes of intense business activity emerging along "cyber routes" or "digital highways." Sassen points out that these spaces along which information flows often follow historic infrastructure for highways, high-speed rail lines, and airports. Twentieth-century infrastructure appears to be shaping the spatial organization of twenty-first-century regions. Sassen also discerns agglomeration and centralization across physical space and within cyberspace – transterritorial centers of intense economic activity and centrality in electronically generated space.

The opportunity for what Sassen calls "super-profits" in global cities may change the entire pattern of land values in these areas most connected to the global economy. Boutiques and expensive restaurants may outbid community-serving local businesses.

Saskia Sassen is the Ralph Lewis Professor of Sociology at the University of Chicago, and Centennial Visiting Professor at the London School of Economics. Her books include *Digital Formations: IT and New Architectures in the Global Realm* (Princeton: Princeton University Press, 2006), *Territory, Authority, Rights: From Medieval to Global Assemblages* (Princeton: Princeton University Press, 2005), *Denationalization: Economy and Polity in a Global Digital Age* (Princeton: Princeton University Press, 2003), *Global Networks/Linked Cities* (London and

New York: Routledge, 2002), *The Global City*, second edition (Princeton: Princeton University Press, 2001), *Guests and Aliens* (New York: New Press, 1999), *Globalization and its Discontents* (New York: New Press, 1998) co-authored with Anthony Appiah, *Losing Control?: Sovereignty in an Age of Globalization* (New York: Columbia University Press, 1996), *Cities in a Global Economy*, third edition (Thousand Oaks: Pine Forge Press, 2006), and *The Mobility of Labor and Capital: A Study in International Investment and Labor Flows* (New York: Cambridge University Press, 1988).

For an excellent set of readings on the impact of globalization on cities see Neil Brenner and Roger Keil (eds), *The Global Cities Reader* (London and New York: Routledge, 2005). Part III of Nicholas Fyfe and Judith Kenny's *The Urban Geography Reader* (London and New York: Routledge, 2005) discusses the impact of global economic and cultural restructuring on cities. Stephen Graham (ed.), *The Cybercities Reader* (London and New York: Routledge, 2003), contains additional writings on how information technology is impacting cities.

Telecommunications and globalization have emerged as major forces shaping the organization of urban space. This reorganization ranges from the spatial virtualization of a growing number of social and economic activities to the reconfiguration of the geography of the built environment for these activities. Whether in electronic space or in the geography of the built environment, this reorganization involves a repositioning of the urban and of urban centrality in particular.

The growth of global markets for finance and specialized services, the need for transnational servicing networks due to sharp increases in international investment, the reduced role of the government in the regulation of international economic activity and the corresponding ascendance of other institutional arenas, notably global markets and corporate headquarters – all these point to the existence of a series of transnational networks of cities. We can see here the formation, at least incipient, of transnational urban systems. To a large extent it seems to me that the major business centers in the world today draw their importance from these transnational networks. There is no such thing as a single global city – and in this sense there is a sharp contrast with the erstwhile capitals of empires.

[. . .]

WORLDWIDE NETWORKS AND CENTRAL COMMAND FUNCTIONS

The geography of globalization contains both a dynamic of dispersal and of centralization, the latter a condition that began receiving recognition only recently. The massive trends towards the spatial dispersal of economic activities at the metropolitan, national and global level which we associate with globalization have contributed to a demand for new forms of territorial centralization of top-level management and control operations. The rapid growth of affiliates illustrates this dynamic. By 1998 firms had about half a million affiliates outside their home countries. The sheer number of dispersed factories and service outlets that are part of a firm's integrated operation creates massive new needs for central coordination and servicing. Thus the spatial dispersal of economic activity made possible by telecommunications and the new legal frameworks for globalization contributes to an expansion of central functions if this dispersal is to take place under the continuing concentration in control, ownership and profit appropriation that characterizes the current economic system.

Another instance today of this negotiation between a global cross-border dynamic and territorially specific sites is that of the global financial markets. The orders of magnitude in these transactions have risen sharply, as illustrated by the $75 U.S. trillion in turnover in the global capital market, a major component of the global economy. These transactions are partly embedded in telecommunications systems that make possible the instantaneous transmission of money/information around the globe. Much attention has gone to the capacity for instantaneous transmission of the new technologies. But the other half of the story is the extent to which the global financial markets are located in particular cities in the highly developed countries; indeed, the degrees of concentration are unexpectedly high, a subject I discuss empirically in a later section.

Stock markets worldwide have become globally integrated. Besides deregulation in the 1980s in all the major European and North American markets, the late 1980s and early 1990s saw the addition of such markets as Buenos Aires, São Paulo, Mexico City, Bangkok, Taipei, etc. The integration of a growing number of stock markets has contributed to raise the capital that can be mobilized through stock markets. Worldwide market value reached over 20 trillion dollars in 1998 . . .

The specific forms assumed by globalization over the last decade have created particular organizational requirements. The emergence of global markets for finance and specialized services, the growth of investment as a major type of international transaction, all have contributed to the expansion in command functions and in the demand for specialized services for firms.

By central functions I do not only mean top-level headquarters; I am referring to all the top-level financial, legal, accounting, managerial, executive, planning functions necessary to run a corporate organization operating in more than one country, and increasingly in several countries. These central functions are partly embedded in headquarters, but also in good part in what has been called the corporate services complex, that is, the network of financial, legal, accounting, advertising firms that handle the complexities of operating in more than one national legal system, national accounting system, advertising culture, etc. and do so under conditions of rapid innovations in all these fields. Such services have become so specialized and complex, that headquarters increasingly buy them from specialized firms rather than producing them in-house. These agglomerations of firms producing central functions for the management and coordination of global economic systems, are disproportionately concentrated in the highly developed countries – particularly, though not exclusively, in the kinds of cities I call global cities . . . Such concentrations of functions represent a strategic factor in the organization of the global economy, and they are situated right here, in New York, in Paris, in Amsterdam.

[. . .]

NEW FORMS OF CENTRALITY

Today there is no longer a simple straightforward relation between centrality and such geographic entities as the downtown, or the central business district. In the past, and up to quite recently in fact, the center

was synonymous with the downtown or the CBD. Today, the spatial correlate of the center can assume several geographic forms. It can be the CBD, as it still is largely in New York City, or it can extend into a metropolitan area in the form of a grid of nodes of intense business activity, as we see in Frankfurt and Zurich. The center has been profoundly altered by telecommunications and the growth of a global economy, both inextricably linked; they have contributed to a new geography of centrality (and marginality). Simplifying, one could identify four forms assumed by centrality today.

First, while there is no longer a simple straightforward relation between centrality and such geographic entities as the downtown, or the central business district as was the case in the past, the CBD remains a key form of centrality. But the CBD in major international business centers is one profoundly reconfigured by technological and economic change.

We may be seeing a difference in the pattern of global city formation in parts of the United States and in parts of Western Europe. In the United States, major cities such as New York and Chicago have large centers that have been rebuilt many times, given the brutal neglect suffered by much urban infrastructure and the imposed obsolescence so characteristic of US cities. This neglect and accelerated obsolescence produce vast spaces for rebuilding the center according to the requirements of whatever regime of urban accumulation or pattern of spatial organization of the urban economy prevails at a given time. In Europe, urban centers are far more protected and they rarely contain significant stretches of abandoned space; the expansion of workplaces and the need for intelligent buildings necessarily will have to take place partly outside the old centers. One of the most extreme cases is the complex of La Défense, the massive, state-of-the-art office complex developed right outside Paris to avoid harming the built environment inside the city. This is an explicit instance of government policy and planning aimed at addressing the growing demand for central office space of prime quality. Yet another variant of this expansion of the "center" onto hitherto peripheral land can be seen in London's Docklands. Similar projects for recentralizing peripheral areas were launched in several major cities in Europe, North America, and Japan during the 1980s.

Second, the center can extend into a metropolitan area in the form of a grid of nodes of intense business activity. One might ask whether a spatial organization characterized by dense strategic nodes spread over a broader region does or does not constitute a new form of organizing the territory of the "center," rather than, as in the more conventional view, an instance of suburbanization or geographic dispersal. Insofar as these various nodes are articulated through cyber-routes or digital highways, they represent a new geographic correlate of the most advanced type of "center." The places that fall outside this new grid of digital highways, however, are peripheralized. This regional grid of nodes represents, in my analysis, a reconstitution of the concept of region. Far from neutralizing geography the regional grid is likely to be embedded in conventional forms of communications infrastructure, notably rapid rail and highways connecting to airports. Ironically perhaps, conventional infrastructure is likely to maximize the economic benefits derived from telematics. I think this is an important issue that has been lost somewhat in discussions about the neutralization of geography through telematics.

Third, we are seeing the formation of a transterritorial "center" constituted via telematics and intense economic transactions . . . The most powerful of these new geographies of centrality at the inter-urban level binds the major international financial and business centers: New York, London, Tokyo, Paris, Frankfurt, Zurich, Amsterdam, Los Angeles, Sydney, Hong Kong, among others. But this geography now also includes cities such as São Paulo and Mexico City. The intensity of transactions among these cities, particularly through the financial markets, trade in services, and investment, has increased sharply, and so have the orders of magnitude involved. At the same time, there has been a sharpening inequality in the concentration of strategic resources and activities between each of these cities and others in the same country. For instance, Paris now concentrates a larger share of leading economic sectors and wealth in France than it did fifteen years ago, while Marseilles, once a major economic hub, has lost its share and is suffering severe decline.

[. . .]

Fourth, new forms of centrality are being constituted in electronically generated spaces. Electronic space is often read as a purely technological event and in that sense a space of innocence. But if we consider for instance that strategic components of the financial industry operate in such space we can see that these are spaces where profits are produced and power is thereby constituted. Insofar as these technologies strengthen the profit-making capability of finance and

make possible the hyper-mobility of finance capital, they also contribute to the often devastating impacts of the ascendance of finance on other industries, on particular sectors of the population, and on whole economies. Cyberspace, like any other space, can be inscribed in a multiplicity of ways, some benevolent or enlightening; others, not. My argument is that structures for economic power are being built in electronic space and that their highly complex configurations contain points of coordination and centralization.

[. . .]

A CONCENTRATION AND THE REDEFINITION OF THE CENTER: SOME EMPIRICAL REFERENTS

The trend towards concentration of top-level management, coordination and servicing functions is evident at the national and international scales in all highly developed countries. For instance, the Paris region accounts for over 40 percent of all producer services in France, and over 80 percent of the most advanced ones. New York City is estimated to account for between a fourth and a fifth of all US producer services exports though it has only 3 percent of the US population. London accounts for 40 percent of all exports of producer services in the UK. Similar trends are also evident in Zurich, Frankfurt, and Tokyo, all located in much smaller countries.

[. . .]

In the financial district in Manhattan, the use of advanced information and telecommunication technologies has had a strong impact on the spatial organization of the district because of the added spatial requirements of "intelligent" buildings. A ring of new office buildings meeting these requirements was built over the last decade immediately around the old Wall Street core, where the narrow streets and lots made this difficult; furthermore, renovating old buildings in the Wall Street core is extremely expensive and often not possible. The new buildings in the district were mostly corporate headquarters and financial services industry facilities. These firms tend to be extremely intensive users of telematics, and the availability of the most advanced forms typically is a major factor in their real estate and locational decisions. They need complete redundancy of telecommunications systems, high carrying capacity, often their own private branch exchange, etc. With this often goes a need for large spaces. For instance, the technical installations backing a firm's trading floor are likely to require additional space equivalent to the size of the trading floor itself.

The case of Sydney illuminates the interaction of a vast, continental economic scale and pressures towards spatial concentration. Rather than strengthening the multipolarity of the Australian urban system, the developments of the 1980s – increased internationalization of the Australian economy, sharp increases in foreign investment, a strong shift towards finance, real estate and producer services – contributed to a greater concentration of major economic activities and actors in Sydney. This included a loss of share of such activities and actors by Melbourne, long the center of commercial activity and wealth in Australia.

[. . .]

THE INTERSECTION OF SERVICE INTENSITY AND GLOBALIZATION

To understand the new or sharply expanded role of a particular kind of city in the world economy since the early 1980s, we need to focus on the intersection of two major processes. The first is the sharp growth in the globalization of economic activity; this has raised the scale and the complexity of transactions, thereby feeding the growth of top-level multinational headquarter functions and the growth of advanced corporate services. It is important to note that even though globalization raises the scale and complexity of these operations, they are also evident at smaller geographic scales and lower orders of complexity, as is the case with firms that operate regionally. Thus while regionally oriented firms need not negotiate the complexities of international borders and the regulations of different countries, they are still faced with a regionally dispersed network of operations that requires centralized control and servicing.

The second process we need to consider is the growing service intensity in the organization of all industries. This has contributed to a massive growth in the demand for services by firms in all industries, from mining and manufacturing to finance and consumer services. Cities are key sites for the production of services for firms. Hence the increase in service intensity in the organization of all industries has had a significant growth effect on cities in the 1980s. It is important to recognize that this growth in services for firms is

evident in cities at different levels of a nation's urban system. Some of these cities cater to regional or subnational markets; others cater to national markets and yet others cater to global markets. In this context, globalization becomes a question of scale and added complexity.

The key process from the perspective of the urban economy is the growing demand for services by firms in all industries and the fact that cities are preferred production sites for such services, whether at the global, national, or regional level. As a result we see in cities the formation of a new urban economic core of banking and service activities that comes to replace the older typically manufacturing-oriented core.

In the case of cities that are major international business centers, the scale, power, and profit levels of this new core suggest that we are seeing the formation of a new urban economy. This is so in at least two regards. First, even though these cities have long been centers for business and finance, since the late 1970s there have been dramatic changes in the structure of the business and financial sectors, as well as sharp increases in the overall magnitude of these sectors and their weight in the urban economy. Second, the ascendance of the new finance and services complex, particularly international finance, engenders what may be regarded as a new economic regime, that is, although this sector may account for only a fraction of the economy of a city, it imposes itself on that larger economy. Most notably, the possibility for super-profits in finance has the effect of devalorizing manufacturing insofar as the latter cannot generate the super-profits typical in much financial activity. This is not to say that everything in the economy of these cities has changed. On the contrary, they still show a great deal of continuity and many similarities with cities that are not global nodes. Rather, the implantation of global processes and markets has meant that the internationalized sector of the economy has expanded sharply and has imposed a new valorization dynamic – that is, a new set of criteria for valuing or pricing various economic activities and outcomes. This has had devastating effects on large sectors of the urban economy. High prices and profit levels in the internationalized sector and its ancillary activities, such as top-of-the-line restaurants and hotels, have made it increasingly difficult for other sectors to compete for space and investments. Many of these other sectors have experienced considerable downgrading and/or displacement, as, for example, neighborhood shops tailored to local needs are replaced by upscale boutiques and restaurants catering to new high-income urban elites.

Though at a different order of magnitude, these trends also became evident during the late 1980s in a number of major cities in the developing world that have become integrated into various world markets: São Paulo, Buenos Aires, Bangkok, Taipei, and Mexico City are only a few examples. Also here the new urban core was fed by the deregulation of financial markets, ascendance of finance and specialized services, and integration into the world markets. The opening of stock markets to foreign investors and the privatization of what were once public sector firms have been crucial institutional arenas for this articulation. Given the vast size of some of these cities, the impact of this new core on the broader city is not always as evident as in central London or Frankfurt, but the transformation is still very real.

[. . .]

The formation of a new production complex

According to standard conceptions about information industries, the rapid growth and disproportionate concentration of producer services in central cities should not have happened. Because they are thoroughly embedded in the most advanced information technologies, producer services could be expected to have locational options that bypass the high costs and congestion typical of major cities. But cities offer agglomeration economies and highly innovative environments. The growing complexity, diversity, and specialization of the services required have contributed to the economic viability of a freestanding specialized service sector.

The production process in these services benefits from proximity to other specialized services. This is especially the case in the leading and most innovative sectors of these industries. Complexity and innovation often require multiple highly specialized inputs from several industries. The production of a financial instrument, for example, requires inputs from accounting, advertising, legal expertise, economic consulting, public relations, designers, and printers. The particular characteristics of production of these services, especially those involved in complex and innovative operations, explain their pronounced concentration in major cities. The commonly heard explanation that high-level professionals require face-to-face interactions needs to be refined in several ways. Producer services, unlike other types of services, are not necessarily dependent on spatial proximity to the consumers, i.e. firms, served. Rather, economies occur in such specialized firms when they locate close to others that produce key inputs or whose proximity makes possible joint production of certain service offerings. The accounting firm can service its clients at a distance, but the nature of its service depends on proximity to specialists, lawyers, programmers. Moreover, concentration arises out of the needs and expectations of the people likely to be employed in these new high-skill jobs, who tend to be attracted to the amenities and lifestyles that large urban centers can offer. Frequently, what is thought of as face-to-face communication is actually a production process that requires multiple simultaneous inputs and feedbacks. At the current stage of technical development, immediate and simultaneous access to the pertinent experts is still the most effective way, especially when dealing with a highly complex product. The concentration of the most advanced telecommunications and computer network facilities in major cities is a key factor in what I refer to as the production process of these industries.

[. . .]

This combination of constraints suggests that the agglomeration of producer services in major cities actually constitutes a production complex. This producer services complex is intimately connected to the world of corporate headquarters; they are often thought of as forming a joint headquarters–corporate services complex. But in my reading, we need to distinguish the two. Although it is true that headquarters still tend to be disproportionately concentrated in cities, over the last two decades many have moved out. Headquarters can indeed locate outside cities, but they need a producer services complex somewhere in order to buy or contract for the needed specialized services and financing. Further, headquarters of firms with very high overseas activity or in highly innovative and complex lines of business tend to locate in major cities. In brief, firms in more routinized lines of activity, with predominantly regional or national markets, appear to be increasingly free to move or install their headquarters outside cities. Firms in highly competitive and innovative lines of activity and/or with a strong world market orientation appear to benefit from being located at the center of major international business centers, no matter how high the costs.

[. . .]

The region in the global information age

The massive use of telematics in the economy and the corresponding possibility for geographic dispersal and mobility of firms suggest that the whole notion of regional specialization and of the region may become obsolete. But there are indications that, as is the case for large cities, so also for regions the hypermobility of information industries and the heightened capacity for geographic dispersal may be only part of the story. The evidence on regional specialization in the US and in other highly developed countries along with new insights into the actual work involved in producing these services point to a different set of outcomes.

What is important from the perspective of the region is that the existence of, for instance, a producer services complex in the major city or cities in a region creates a vast concentration of communications infrastructure which can be of great use to other economic nodes in that region. Such nodes can (and do) connect with the major city or cities in a region and thereby to a worldwide network of firms and markets. The issue from the regional perspective is, then, that somewhere in its territory the region connects with state-of-the-art communication facilities which connect it with the world and which brings foreign firms from all over the world to the region. Given a regional grid of economic nodes, the benefits of this concentration in the major city or cities are no longer confined only to firms located in those cities.

Secondly, given the nature of the production process in advanced information industries, as described in the preceding section, the geographic dispersal of activities has limits. The importance of actual face-to-face transactions means that a metropolitan or regional network of firms will need conventional communications infrastructure, e.g. highways or rapid rail, and locations not farther than something like two hours. One of the ironies of the new information technologies is that to maximize their use we need access to conventional infrastructure. In the case of international networks it takes airports and planes; and in the case of metropolitan or regional networks, trains and cars.

The importance of conventional infrastructure in the operation of economic sectors that are heavy users of telematics has not received sufficient attention. The dominant notion seems to be that telematics obliterates the need for conventional infrastructure. But it is precisely the nature of the production process in advanced industries, whether they operate globally or nationally, which contributes to explain the immense rise in business travel we have seen in all advanced economies over the last decade, the new electronic era. The virtual office is a far more limited option than a purely technological analysis would suggest. Certain types of economic activities can be run from a virtual office located anywhere. But for work processes requiring multiple specialized inputs, considerable innovation and risk taking, the need for direct interaction with other firms and specialists remains a key locational factor. Hence the metropolitanization and regionalization of an economic sector has boundaries that are set by the time it takes for a reasonable commute to the major city or cities in the region. The irony of today's electronic era is that the older notion of the region and older forms of infrastructure re-emerge as critical for key economic sectors. This type of region in many ways diverges from older forms of region. It corresponds rather to the second form of centrality posited above in this paper – a metropolitan grid of nodes connected via telematics. But for this digital grid to work, conventional infrastructure – ideally of the most advanced kind – is also a necessity.

[. . .]

THE INTERSECTION BETWEEN ACTUAL AND DIGITAL SPACE

There is a new topography of economic activity sharply evident in this subeconomy. This topography weaves in and out between actual and digital space. There is today no fully virtualized firm or economic sector. Even finance, the most digitalized, dematerialized and globalized of all activities, has a topography that weaves back and forth between actual and digital space. To different extents in different types of sectors and different types of firms, a firm's tasks now are distributed across these two kinds of spaces; further the actual configurations are subject to considerable transformation as tasks are computerized or standardized, markets are further globalized, etc. More generally, telematics and globalization have emerged as fundamental forces reshaping the organization of economic space.

The question I have for architects here is whether the point of intersection between these two kinds of spaces in a firm's or a dynamic's topography of activity, is one worth thinking about, theorizing, exploring. This intersection is unwittingly, perhaps, thought of as a line that divides two mutually exclusive zones. I would propose, again, to open up this line into an "analytic borderland" which demands its own empirical specification and theorization, and contains its own possibilities for architecture. The space of the computer screen, which one might posit as one version of the intersection, will not do, or is at most a partial enactment of this intersection.

What does contextuality mean in this setting? A networked subeconomy that operates partly in actual space and partly in globe-spanning digital space cannot easily be contextualized in terms of its surroundings. Nor can the individual firms. The orientation is simultaneously towards itself and towards the global. The intensity of its internal transactions is such that it overrides all considerations of the broader locality or region within which it exists. On another, larger scale, in my research on global cities I found rather clearly that these cities develop a stronger orientation towards the global markets than to their hinterlands. Thereby they override a key proposition in the urban systems literature, to wit, that cities and urban systems integrate, articulate national territory. This may have been the case during the period when mass manufacturing and mass consumption were the dominant growth machines in developed economies and thrived on the possibility of a national scale. But it is not today with the ascendance of digitalized, globalized, dematerialized sectors such as finance. The connections with other zones and sectors in its "context" are of a special sort – one that connects worlds that we think of as radically distinct. For instance, the informal economy in several immigrant communities in New York provides some of the low-wage workers for the "other" jobs on Wall Street, the capital of global finance. The same is happening in Paris, London, Frankfurt, Zurich . . .

[. . .]

CONCLUSION

Economic globalization and telecommunications have contributed to produce a spatiality for the urban that pivots on cross-border networks and territorial locations with massive concentrations of resources. This is not a completely new feature. Over the centuries cities have been at the crossroads of major, often worldwide, processes. What is different today is the intensity, complexity and global span of these networks, the extent to which significant portions of economies are now dematerialized and digitalized and hence the extent to which they can travel at great speeds through some of these networks, and, thirdly, the numbers of cities that are part of cross-border networks operating at vast geographic scales.

The new urban spatiality thus produced is partial in a double sense: it accounts for only part of what happens in cities and what cities are about, and it inhabits only part of what we might think of as the space administrative boundaries or in the sense of a city's public imaginary. What stands out, however, is the extent to which the city remains an integral part of these new configurations.

PART FOUR

Urban politics, governance, and economics

"WHO STOLE THE PEOPLE'S MONEY?" — DO TELL . N.Y.TIMES.

'TWAS HIM.

INTRODUCTION TO PART FOUR

The material on urban society and culture in Part Two and urban space in Part Three raises important issues about how cities should be governed and the economy of cities. The conflicts related to gated communities and sprinklers that douse the homeless that Mike Davis discusses (p. 178) pose political questions: what *should* government, particularly local government, do when different groups want to use urban space in different ways? The selection by William Julius Wilson (p. 110) in Part Two illustrates how sociological issues of race and class are intimately related to both economic questions and questions of urban *politics and governance*. Part Four focuses on these questions of urban politics, governance, and economics, beginning with four first-person accounts of urban bosses and machines.

Harvey's selection nicely bridges the material in Part Three on urban space and the material in this section on urban politics, governance, and economics. While Harvey is a geographer, the issues he discusses are essentially political ones. Harvey sees cities as centers of conflict based on ideology, race, gender, and interests. Civil rights groups, feminists, Republicans, environmentalists, businessmen, and punks have different interests and values. The land that businessmen want for a new office building may be the exact same land that environmentalists feel should be a park. This inevitably leads to conflict among different groups in cities. The processes used to resolve these kinds of conflicts are political processes. Harvey is sympathetic to the "militant particularism" of some groups who focus on their specific interest or locality, but feels narrow values are more effective if they are generalized.

Many of the best writings about urban politics are by political scientists. Urban politics is a distinct subfield within the social science discipline of political science. But scholars from other academic disciplines are also interested in urban politics. For example, political sociology is a subfield within sociology. In the 1950s and 1960s, seminal work by sociologist Floyd Hunter and political scientist Robert Dahl began a major debate between two schools of thought. In his book *Community Power Structure* (1953), based on research in Atlanta, Georgia, Hunter concluded that in Atlanta – and by implication other cities – a small, interlocking *elite* consisting of key businessmen and members of established and socially prominent families made all the really important decisions about Atlanta, including the governmental decisions. They sat on the boards of each other's corporations, intermarried, chatted at the same social clubs, and ran things in their interest. Because Hunter concluded that a small elite ruled Atlanta, his model of urban community power is referred to as the *elitist* model of community power. In *Who Governs?* (1961) Dahl reached almost diametrically opposite conclusions. He concluded that local political power in New Haven, Connecticut – and by implication other cities – was fragmented. Dahl reported that many different people from a variety of walks of life were involved in decision-making by the New Haven city government, and many people influenced the outcome of different political decisions. As a result, Dahl advanced a *pluralist* model of urban community power. The competing models of Dahl and Hunter stimulated debate between elitists and pluralists and the further elaboration of theories of urban politics. Some theorists critical of pluralist interpretations of community power focused on structural features of global capitalism. Marxists in particular feel that the global capitalist system really determines what city governments can and cannot do. Structuralists argue that if bankers in New York and London decide to pull investment out of Buenos Aires, Argentina, the city government in Buenos Aires cannot stop them. And if billions of dollars flow out of Buenos Aires, the city government cannot build roads or schools or carry out other parts of their political agenda.

During the 1960s Sherry Arnstein was the chief advisor on citizen participation in the US Department of Housing and Urban Development's Model Cities Program – a federal government program that provided billions of dollars to help lower-income city neighborhoods develop and implement physical and social programs. Arnstein begins her selection on the ladder of citizen participation by noting that "the idea of citizen participation is a little like eating spinach: no one is against it in principle because it is good for you." But whether or not city dwellers participate *effectively* in government programs that affect them and their neighborhoods has varied greatly.

A little background will help clarify Arnstein's selection. Many urban renewal programs in the United States during the 1950s and 1960s were intended to, and did, remove low-income and minority residents and replace their homes and community institutions with office buildings, luxury housing, garages, and other developments totally unrelated to the former residents. The funky, but communal, Italian neighborhood in Boston's West End that Jane Jacobs lovingly describes (p. 98) was leveled by Boston's West End urban renewal project. The "street ballet" she describes has been replaced by yuppies going to Starbucks coffee shops and smart boutiques. In the urban renewal program, invitations to neighborhood residents to help decide what the urban renewal project should be like were usually a sham. In sharp contrast the US "War on Poverty" in the late 1960s emphasized "maximum feasible participation of the poor." Many decisions about how to use federal anti-poverty funds were made by residents themselves. Many local poverty programs lacked the capacity to manage programs or dissolved into internal feuding. While political machines and bosses' extraordinary organization and struggle to dispense patronage that Bryce (p. 215), Plunkitt (p. 219), and Addams (p. 221) describe is a thing of the past, local elected officials' interest in exercising power and controlling jobs and contracts was alive and well in the 1960s as it is today. Cutting local elected officials and established agencies out of the loop to design programs and manage anti-poverty funds created a major political backlash. The Model Cities program, where Arnstein worked, sought to strike a balance between the top-down urban renewal approach in which there was no real citizen participation in decisions and the War on Poverty model in which neighborhood groups often fought among themselves, wasted money, and accomplished little.

Arnstein asks readers to imagine a "ladder" with different "rungs" of citizen participation from lowest to highest. The rungs of the ladder range from non-participation (*manipulation* and *therapy*) at the bottom of the ladder to *citizen control* at the top. While Arnstein herself favors citizen control as the ultimate objective of urban programs, she considers all but the bottom two rungs of her ladder useful to varying degrees. In the decades since Arnstein's classic article was written there has been a succession of urban development programs in the United States, Western European countries, and elsewhere in the world. Many urban revitalization programs call for some degree of citizen participation. Arnstein's ladder is useful in understanding how to create meaningful citizen participation in these urban programs.

Anthony Downs, a policy analyst at the Brookings Institution – a Washington, DC think-tank – focuses on a major problem with the way in which local government within metropolitan regions is organized. While Downs concentrates on local government in the United States, his critique and proposed solutions are relevant for metropolitan regions everywhere in the world. The problem is that government structures that were invented to govern small, separate places before the massive population increase of the recent past are not adequate spatial and political units to manage huge, interconnected metropolitan regions. Most of the 101 local governments in the San Francisco Bay Area, for example, were incorporated before World War II when the population of the region was less than half what it is today. Many of the local governments were no more than villages separated from each other. Little local governments like these are adequate to decide how to patch potholes in the local street or how much of the city budget should go to the local beauty queen pageant. They are poorly equipped to handle complex regional decisions about air and water quality, traffic congestion, and the preservation of endangered species. Dividing up the region into 101 little governments – each protective of its own interests – makes it hard to get low-income housing built or assure that immigrants and low-income people have access to housing, schools, and jobs throughout the region. The same kind of fragmentation of local government authority and obsolete spatial organization of local government is common in many metropolitan areas of Europe, Asia, Africa, and Latin America.

Downs' selection argues strongly for *regional* solutions to metropolitan problems. It proposes a new vision that eliminates the legal/governmental distinctions between impacted inner-city ghettos and comfortable suburban communities. It argues for integrated metropolitan solutions to traffic congestion and other problems that cannot be solved on a city-by-city basis.

One of the most important – and certainly among the most sensitive – local government functions everywhere in the world is police work. In "Broken Windows" (p. 256), James Q. Wilson and George L. Kelling propose a theory about how crime comes to dominate declining neighborhoods and what to do about it. Wilson and Kelling emphasize the importance of citizen perceptions of crime. They argue that whether or not crime has actually increased in a neighborhood, if residents think crime has increased, they will become more reclusive and less involved in the community. And that in turn will open the door to real crime. A single broken window may be a trivial problem, Wilson and Kelling argue, but if it is not fixed the signal that no one cares enough to fix it will lead to more broken windows and then to drug dealing and violent crime. The remedy Wilson and Kelling suggest – which has been widely adopted since this article was written – is community policing where highly visible police regularly patrol marginal neighborhoods on foot. In community policing, officers are assigned to a neighborhood for a long enough time to learn about the neighborhood and its residents. They will come to understand community norms and values. Wilson and Kelling go on to advance the controversial view that these community police should informally enforce community norms of appropriate civil behavior as the neighborhood itself defines them, even if that calls for extralegal or perhaps illegal controls. Wilson and Kelling note that a transitional neighborhood may be inhabited both by "neighborhood regulars" and by strangers. The regulars in turn consist of what Wilson and Kelling term "decent folk," and derelicts and drunks who are not so decent but "know their place." So long as questionable street behavior stays within neighborhood-defined norms, community police will look the other way. Police will leave the well-known and harmless local drunk alone even if he is technically breaking the law. But if rowdy teenagers, prostitutes, dope dealers, or other strangers violate community norms – regardless of whether they violate the law – community police will intervene, perhaps by ordering the strangers to leave even if there is no legal basis for such an order.

Politics, economics, and public finance are intimately connected. The final three selections in this section introduce important ideas about the way in which urban economies work and how urban economics concepts can produce better local government decision-making, greater fiscal equality in metropolitan regions, and help central cities capitalize on the advantages they possess to compete effectively in a global economy.

Wilbur Thompson is largely responsible for creating the field of urban economics. His 1965 *Preface to Urban Economics* was the standard text dozens of economics departments used as they launched urban economics courses in the 1960s and 1970s. In the selection included in this section Thompson introduces concepts that were new at the time, but have since been thoroughly incorporated into urban economics thinking. He argues that the failure to consider the real costs of publicly provided goods and services often leads to inefficient – sometimes irrational – policy. Thompson provides examples of how using pricing can produce better public policy. He classifies the kinds of goods local governments provide into several key types: "public goods," like highways, that need to be provided on a collective, rather than an individual, basis; "merit goods," like polio vaccinations, that society deems so important that they should be free and universally available at the public expense; and payments, like food stamps, made to redistribute income from one group such as well-off taxpayers to another group such as the indigent poor. These are all justifiable bases for public expenditures and Thompson supports them if they are carefully thought out and consciously applied. But where local governments don't think through the rationale for a public expenditure clearly – based on these and related economic concepts – Thompson argues that public programs may be distorted and public funds wasted. Thompson also discusses pricing public goods. Where the goods are scarce – like space on a congested highway – he advocates using price to ration the scarce goods. Thompson pioneered the idea of "congesting pricing" – e.g. charging higher highway and bridge tolls during commuting hours when they are congested. A new idea at the time, congestion pricing is now widely used.

Minnesota Law Professor Myron Orfield (p. 287) turns from the basic economic issues Thompson discusses to the question of regional equity. Metropolitan regions are fragmented into many small local government jurisdictions. These jurisdictions compete with each other in a zero-sum game to attract the most economically desirable land uses – the ones that will produce the greatest net revenue. Since the playing field is far from level, some jurisdictions succeed in attracting most of the revenue-generating land uses. They can maintain low taxes and still provide superior local services. The jurisdictions that lose out in the competition have limited fiscal capacity. They must levy higher taxes, provide fewer services, or both. Orfield believes that stable, cooperative regions are essential for the well-being of society and advances specific policies to increase regional fiscal equity.

The selection in this section by Harvard business professor Michael Porter (p. 274) approaches the issue of urban inequality and what to do about it from a private-sector perspective. Porter believes that the economic and social health of inner cities depends upon economic development by the private sector. To succeed, he argues, economic development must be based on the economic self-interest of private firms rather than phony businesses propped up by government subsidies and preference programs. The key to success, according to Porter, lies in capitalizing on the competitive advantage that inner-city neighborhoods possess: strategic location, local market demand, their capacity to integrate with regional business clusters, and human resources. Porter urges an economic, not a social, model for development. He counsels corporations to shift their philanthropic priorities away from providing social services, such as daycare or food for the homeless, to providing managerial expertise to develop neighborhood economies.

The Urban Politics Reader (London and New York: Routledge, 2006) edited by Elizabeth Strom and John Mollenkopf, contains additional readings on urban politics.

First-Person Accounts of Nineteenth-Century Political Bosses and Machines

James Bryce, William Riordan, and Jane Addams

Editors' Introduction

Tip O'Neil (1912–1994), a quotable Massachusetts politician, is noted for observing that "All politics is local." Local political organizations in all societies take care of the day-to-day needs of local residents and translate national policies and programs into reality.

Urban politics – the political activity that takes place at the city level – is an essential building block in higher-level politics, particularly in democracies. Politicians court voters, by delivering services to a specific neighborhood, meet the needs of distinct voting groups (African–Americans, gays, environmentalists), and try to demonstrate by their words and actions that they listen to and speak for people at the grassroots.

In the United States, during the second half of the nineteenth century and the early part of the twentieth century, a distinct brand of urban politics emerged: political *bosses* and *machines*. Political scientists and historians have been fascinated with city political machines, both for the intrinsic interest of the colorful characters involved, and from the challenge of interpreting how the machines operated and whether they were good or bad.

The following three selections – written in 1893, 1898, and 1905 – are all by astute contemporary observers of US city bosses and machines.

In the nineteenth century a combination of migration, exploitation of natural resources, and industrialization produced rapid urbanization. The majority of the residents in the new, large nineteenth-century US industrial cities were poor immigrants. Local leaders ("bosses") organized voting blocs ("machines") along ethnic and national origin lines. The three writers agreed that machines were true grassroots organizations and that the bosses were very efficient in organizing voters and gaining control of local governments. They see the machines as performing two main functions: obtaining votes and dispensing patronage (jobs, contracts, political favors). They disagree about how corrupt or efficient the machines were and how well they served the interests of city dwellers, including their own constituents.

Viscount James Bryce (1838–1922) was a British aristocrat who published a magisterial treatise on the *American Commonwealth* in 1888. Bryce was a historian, Oxford law professor, member of the British parliament, and, late in life, Britain's ambassador to the United States.

Bryce found much to admire in brash, economically successful, late-nineteenth-century American cities. But he considered local – particularly urban – politics to be the great failure of the American system. Bryce noted that city bosses were often of foreign birth and humble origin. They lacked formal education and their appearance and speech were unpolished. But bosses were hard working and astute. They maintained strict discipline over political machines. Bryce felt the bosses were not wicked men, but he concluded that their moral values and political goals lead to inefficiency and corruption. The bosses valued loyalty and personal help to friends and constituents – even if this crossed legal lines into fixing arrests, accepting bribes, ignoring liquor law violations, distributing jobs to unqualified supporters, or benefiting financially from insider information. As an educated aristocrat Bryce saw the immigrant masses as ignorant and totally unequipped to participate in civic life. He saw American bosses as apolitical "boodlers" in a system with little in common with the intellectual debates and ostensibly rational actions of the British parliament. Bryce's negative stereotype became the accepted view until revisionist historians began to re-examine bosses and political machines in the 1960s.

William Riordan, a journalist, recorded for posterity the colorful sayings of George Washington Plunkitt – a New York City politician who had risen from butcher's boy to head of New York City's powerful Tammany Hall machine. Plunkitt advised aspiring city politicians unfortunate enough to have attended college to unlearn all they had learned there and instead to "study human nature and act accordin'." He advised flattery and attentiveness to the employment, welfare, and cultural needs of the people to line up supporters whose votes he could exchange for influence with the machine. Plunkett's candid views, told with insight and wit, give an excellent feel of all that was good and bad about bosses and machines.

Jane Addams (1860–1935) was a pioneering social worker who led the American settlement house movement. Addams opened Hull House in 1889 in a poor Chicago neighborhood to "provide a center for a higher civic and social life; to institute and maintain educational and philanthropic enterprises and to investigate and improve the conditions in the industrial districts of Chicago." Eventually, a small army of upper-class philanthropic young women at Hull House served more than two thousand people – mostly poor Italian and other immigrants – with kindergarten classes, club meetings for children, and adult education classes. Hull House had two public kitchens, a coffee house, gym, swimming pool, bookbindery, art studio, art gallery, music school, drama group, library, labor museum, and an employment bureau.

Addams considered the South Italian peasants at Hull House "primitive people." While she empathized with their kindness and generosity, she considered them at a low level of moral development who judged politicians by their local good deeds rather than their service to the larger community. Addams shows a similar ambivalence about bosses. She acknowledges that they were often effective in procuring jobs and benefits for their constituents and sincere in their acts of kindness. But she considers the typical boss a boodler "incapable of unselfish action the results of which will not benefit some one of his acquaintances."

While local politics has grown more subtle today, many of the behaviors described in the above selections remain. Voters still respond to a politician's public persona as much as to his or her actions. Politicians still seek votes by distributing political favors to constituents in their voting district or of their national original, ethnic, religious, or sexual identity group.

James Bryce's *American Commonwealth* was first published in 1888 in New York by Macmillan. A 1996 reprint is available from the Liberty Fund in Indianapolis, Indiana.

William Riordan's *Plunkitt of Tammany Hall: A Series of Very Plain Talks on Very Practical Politics* was published in New York by Dutton, 1905. A reprint edition published in 1996 is available from Signet Classics.

Jane Addams's "Why the Ward Boss Rules" was published in *The Outlook* in 1898. It is reprinted in Howard P. Chudakoff and Peter C. Baldwin, *Major Problems in American Urban History*, second edition (New York: Houghton Mifflin, 2004).

The large literature on urban political machines and bosses is summarized in Howard P. Chudakoff and Judith E. Smith, *The Evolution of American Urban Society*, sixth edition (Upper Saddle River: Pearson/Prentice-Hall, 2005), chapter 6, "City Politics in the Era of Transformation." Other books about machines and bosses include: Bruce Stave and Sondra Stave (eds), *Urban Bosses, Machines, and Progressive Reformers*, second edition (Malabar: Krieger Publishing, 1984), John M. Allswang, *Bosses, Machines and Urban Voters: An American Symbiosis*, revised edition (Baltimore: Johns Hopkins University Press, 1986), and Blaine A. Brownell and Warren E. Stickle (eds), *The City Boss in America: An Interpretive Reader* (Oxford: Oxford University Press, 1976).

"RINGS AND BOSSES"

from *The American Commonwealth* (1888)
James Bryce

HOW THE MACHINE WORKS

. . . To understand how [a machine] actually works one must distinguish between two kinds of constituencies or voting areas. One kind is to be found in the great cities – places whose population exceeds, speaking roughly, 100,000 souls, of which there are more than thirty in the Union. The other kind includes constituencies in small cities and rural districts.

. . . The tests by which one may try the results of the system of selecting candidates are two. Is the choice of candidates for office really free – i.e. does it represent the unbiased wish and mind of the voters generally? Are the offices filled by men of probity and capacity sufficient for their duties?

In the country generally, i.e. in the rural districts and small cities, both these tests are tolerably well satisfied.

[. . .]

In large cities the results are different because the circumstances are different. We find there

A vast population of ignorant immigrants

The leading men all intensely occupied with business

Communities so large that people know little of one another, and that the interest of each individual in good government is comparatively small.

Anyone can see how these conditions affect the problem. The immigrants vote, that is, they obtain votes after three or four years' residence at most (often less), but they are not fit for the suffrage. They know nothing of the institutions of the country, of its statesmen, of its political issues. Neither from Central Europe nor from Ireland do they bring much knowledge of the methods of free government, and from Ireland they bring a suspicion of all government. Incompetent to give an intelligent vote, but soon finding that their vote has a value, they fall into the hands of the party organizations, whose officers enroll them in their lists, and undertake to fetch them to the polls. I was taken to watch the process of citizen-making in New York. Droves of squalid men, who looked as if they had just emerged from an emigrant ship, and had perhaps done so only

a few weeks before, for the law prescribing a certain term of residence is frequently violated, were brought up to a magistrate by the ward agent of the party which had captured them, declaring their allegiance to the United States and were forthwith placed on the roll. Such a sacrifice of common sense to abstract principles has seldom been made by any country. Nobody pretends that such persons are fit for civic duty, or will be dangerous if kept for a time in pupilage, but neither party will incur the odium of proposing to exclude them. The real reason for admitting them, besides democratic theory, was that the party which ruled New York expected to gain their votes. It is an afterthought to argue that they will sooner become good citizens by being immediately made full citizens . . .

As party machinery is in great cities most easily perverted, so the temptation to pervert it is there strongest, because the prizes are great. The offices are well paid, the patronage is large, the opportunities for jobs, commissions on contracts, pickings, and even stealings, are enormous. Hence it is well worth the while of unscrupulous men to gain control of the machinery by which these prizes may be won.

. . . Every feature of the machine is the result of patent causes. The elective offices are so numerous that ordinary citizens cannot watch them, and cease to care who gets them. The conventions come so often that busy men cannot serve in them. The minor offices are so unattractive that able men do not stand for them. The primary lists are so contrived that only a fraction of the party get on them; and of this fraction many are too lazy or too busy or too careless to attend. The mass of the voters are ignorant; knowing nothing about the personal merits of the candidates, they are ready to follow their leaders like sheep. Even the better class, however they may grumble, are swayed by the inveterate habit of party loyalty, and prefer a bad candidate of their own party to a (probably no better) candidate of the other party. It is less trouble to put up with impure officials, costly city government, a jobbing State legislature, an inferior sort of congressman, than to sacrifice ones own business in the effort to set things right. Thus the Machine grinds on, and grinds out places, power, and opportunities for illicit gain to those who manage it.

[. . .]

RINGS AND BOSSES

Those who in great cities form the committees and work the Machine are persons whose chief aim in life is to make their living by office. Such a man generally begins by acquiring influence among a knot of voters who live in his neighborhood, or work under the same employer, or frequent the same grog-shop or beer saloon, which perhaps he keeps himself . . . Soon he becomes conspicuous in the primary, being recognized as controlling the votes of others – "owning them" is the technical term . . ." He is appointed to some petty office in one of the city departments and presently is himself nominated for elective office . . . He is one of the small knot of persons who pull the wire for the whole city, controlling the primaries, selecting candidates, "running" conventions, organizing elections, treating on behalf of the party in the city with the leaders of the party in the State . . .

Such a combination is called a Ring.

. . . In a Ring there is usually some one person who holds more strings in his hand than do the others. Like them he has worked himself up to power from small beginnings, gradually extending the range of his influence over the mass of workers, and knitting close bonds with influential men outside as well as inside politics, perhaps with great financiers or railway magnates, whom he can oblige, and who can furnish him with funds. At length his superior skill, courage, and force of will make him, as such gifts always do make their possessor, dominant among his fellows. An army led by a council seldom conquers: it must have a commander-in-chief, who settles disputes, decides in emergencies, inspires fear or attachment. The head of the Ring is such a commander. He dispenses places, rewards the loyal, punishes the mutinous, concocts schemes, negotiates treaties. He generally avoids publicity, preferring the substance to the pomp of power, and is all the more dangerous because he sits, like a spider, hidden in the midst of his web. He is a Boss.

Although the career I have sketched is that whereby most Bosses have risen to greatness, some attain it by a shorter path. There have been brilliant instances of persons stepping at once on to the higher rungs of the ladder in virtue of their audacity and energy, especially if coupled with oratorical power. The first theatre of such a man's successes may have been the stump rather than the primary: he will then become potent in conventions, and either by hectoring or by plausible address, for both have their value, spring into popular favour, and make himself necessary to the party managers. It is of course a gain to a Ring to have among them a man of popular gifts, because he helps to conceal the odious features of their rule, gilding it by his rhetoric, and winning the applause of the masses who stand outside the circle of workers. However, the position of the rhetorical boss is less firmly rooted than that of the intriguing boss, and there have been instances of his suddenly falling to rise no more.

A great city is the best soil for the growth of a Boss, because it contains the largest masses of manageable voters as well as numerous offices and plentiful opportunities for jobbing. But a whole State sometimes falls under the dominion of one intriguer. To govern so large a territory needs high abilities; and the State boss is always an able man, somewhat more of a politician, in the European sense, than a city boss need be. He dictates State nominations, and through his lieutenants controls State and sometimes Congressional conventions, being in diplomatic relations with the chief city bosses and local rings in different parts of the State. His power over them mainly springs from his influence with the Federal executive and in Congress. He is usually, almost necessarily, a member of Congress, probably a senator, and can procure, or at any rate can hinder, such legislation as the local leaders desire or dislike. The President cannot ignore him, and the President's ministers, however little they may like him, find it worth while to gratify him with Federal appointments for persons he recommends, because the local votes he controls may make all the difference to their own prospects of getting some day a nomination for the presidency. Thus he uses his Congressional position to secure State influence, and his State influence to strengthen his Federal position. Sometimes however he is rebuffed by the powers at Washington and then his State thanes fly from him. Sometimes he quarrels with a powerful city boss, and then honest men come by their own.

It must not be supposed that the members of Rings, or the great Boss himself, are wicked men. They are the offspring of a system. Their morality is that of their surroundings. They see a door open to wealth and power, and they walk in. The obligations of patriotism or duty to the public are not disregarded by them, for these obligations have never been present to their minds. A State boss is usually a native American and a person of some education, who avoids the grosser forms of corruption, though he has to wink at them when practiced by his friends. He may be a man of personal integrity. A city boss is often of foreign birth and humble origin; he has grown up in an atmosphere of oaths and cocktails: ideas of honour and purity are as strange to him as ideas about the nature of the currency and the incidence of taxation: politics is merely a means for getting and distributing places. "What," said an ingenuous delegate at one of the National Conventions at Chicago in 1880, "what are we here for except the offices?" It is no wonder if he helps himself from the city treasury and allows his minions to do so . . . And even the city Boss improves as he rises in the world. Like a tree growing out of a dust heap, the higher he gets, the cleaner do his boughs and leaves become. America is a country where vulgarity is sealed off more easily than in England, and where the general air of good nature softens the asperities of power. Some city bosses are men from whose decorous exterior and unobtrusive manners no one would divine either their sordid beginnings or their noxious trade. As for the State boss, whose talents are probably greater to begin with, he must be of very coarse metal if he does not take a certain polish from the society of Washington.

A city Ring works somewhat as follows. When the annual or biennial city or State elections come round, its members meet to discuss the apportionment of offices. Each may desire something for himself, unless indeed he is already fully provided for, and anyhow desires something for his friends. The common sort are provided for with small places in the gift of some official, down to the place of a policeman or door-keeper or messenger, which is thought good enough for a common "ward worker." Better men receive clerkships or the promise of a place in the custom-house or post-office to be obtained from the Federal authorities. Men still more important aspire to the elective posts, seats in the State legislature, a city alder-manship or commissionership, perhaps even a seat in Congress. All the posts that will have to be filled at the coming elections are considered with the object of bringing out a party ticket, i.e. a list of candidates to be supported by the party at the polls when its various nominations have been successfully run through the proper conventions. Some leading man, or probably the Boss himself, sketches out an allotment of places; and when this allotment has been worked out fully, it results in a Slate, i.e. a complete draft list of candidates to be proposed for the various offices. It may happen that the slate does not meet everybody's

wishes. Some member of the ring or some local boss – most members of a ring are bosses each in his own district, as the members of a cabinet are heads of the departments of state, or as the cardinals are bish-ops of dioceses near Rome and priests and deacons of her parish churches – may complain that he and his friends have not been adequately provided for, and may demand more. In that case the slate will probably be modified a little to ensure good feeling and content; and will then be presented to the Convention.

. . . The party must at all hazards be kept together, for the power of a united party is enormous. It has not only a large, but a thoroughly trained and disciplined army in its office-holders and office-seekers; and it can concentrate its force upon any point where opposition is threatened to the regular party nominations. All these office-holders and office-seekers have not only the spirit of self-interest to rouse them, but the bridle of fear to check any stirrings of independence.

Discipline is very strict in this army. Even city politicians must have a moral code and moral standard. It is not the code of an ordinary unprofessional citizen. It does not forbid falsehood, or malversation, or ballot stuffing, or "repeating." But it denounces apathy or cowardice, disobedience, and above all, treason to the party. Its typical virtue is "solidity," unity of heart, mind, and effort among the workers, unquestioning loyalty to the party leaders, and devotion to the party ticket. He who takes his own course is a Kicker or Bolter; and is punished not only sternly but vindictively. The path of promotion is closed to him; he is turned out of the primary, and forbidden to hope for a delegacy to a convention; he is dismissed from any office he holds which the Ring can command . . . What the client was to his patron at Rome, what the vassal was to his lord in the Middle Ages, that the heelers and workers are to their boss in these great transatlantic cities. They render a personal feudal service, which their suzerain repays with the gift of a livelihood; and the relation is all the more cordial because the lord bestows what costs him nothing, while the vassal feels that he can keep his post only by the favour of the lord. European readers must again be cautioned against drawing for themselves too dark a picture of the Boss. He is not a demon. He is not regarded with horror even by those "good citizens" who strive to shake off his yoke. He is not necessarily either corrupt or mendacious, though he grasps at place, power, and wealth. He is a leader to whom certain peculiar social and political conditions have given a character dissimilar from the party leaders whom Europe knows. It is worthwhile to point out in what the dissimilarity consists.

A Boss needs fewer showy gifts than a European demagogue. His special theatre is neither the halls of the legislature nor the platform, but the committee room. A power of rough and ready repartee, or a turn for florid declamation, will help him; but he can dispense with both. What he needs are the arts of intrigue and that knowledge of men which teaches him when to bully, when to cajole, whom to attract by the hope of gain, whom by appeals to party loyalty. Nor are so-called "social gifts" unimportant. The lower sort of city politicians congregate in clubs and bar-rooms; and as much of the cohesive strength of the smaller party organizations arises from their being also social bodies, so also much of the power which liquor dealers exercise is due to the fact that "heelers" and "workers" spend their evenings in drinking places, and that meetings for political purposes are held there. Of the 1007 primaries and conventions of all parties held in New York City preparatory to the elections of 1884, 633 took place in liquor saloons. A Boss ought therefore to be hail fellow well met with those who frequent these places, not fastidious in his tastes, fond of a drink and willing to stand one, jovial in manners, and ready to oblige even a humble friend.

The aim of a Boss is not so much fame as power, and power not so much over the conduct of affairs as over persons. Patronage is what he chiefly seeks, patronage understood in the largest sense in which it covers the disposal of lucrative contracts and other modes of enrichment as well as salaried places. The dependants who surround him desire wealth, or at least a livelihood; his business is to find this for them, and in doing so he strengthens his own position. It is as the bestower of riches that he holds his position, like the leader of a band of condottieri in the fifteenth century.

The interest of a Boss in political questions is usually quite secondary. Here and there one may be found who is a politician in the European sense, who, whether sincerely or not, professes to be interested in some measure affecting the welfare of the country. But the attachment of the ringster is usually given wholly to the concrete party, that is to the men who compose it, regarded as office-holders or office-seekers; and there is often not even a profession of zeal for any party doctrine. As a noted politician once happily observed, "There are no politics in politics." Among

bosses, therefore, there is little warmth of party spirit. The typical boss regards the boss of the other party much as counsel for the plaintiff regards counsel for the defendant. They are professionally opposed, but not necessarily personally hostile. Between bosses there need be no more enmity than results from the fact that the one has got what the other wishes to have. Accordingly it sometimes happens that there is a good understanding between the chiefs of opposite parties in cities; they will even go the length of making a joint "deal," i.e. of arranging for a distribution of offices whereby some of the friends of one shall get places, the residue being left for the friends of the other. A well organized city party has usually a disposable vote which can be so cast under the directions of the managers as to effect this, or any other desired result. The appearance of hostility must, of course, be maintained for the benefit of the public; but as it is for the interest of both parties to make and keep these private bargains, they are usually kept when made, though it is seldom possible to prove the fact.

The real hostility of the Boss is not to the opposite party, but to other factions within his own party.

[. . .]

It has been pointed out that rings and bosses are the product not of democracy, but of a particular form of democratic government, acting under certain peculiar conditions . . . We have seen that these conditions are

The existence of a Spoils System (= paid offices given and taken away for party reasons).

Opportunities for illicit gains arising out of the possession of office.

The presence of a mass of ignorant and pliable voters.

The insufficient participation in politics of the "good citizens."

If these be the true causes or conditions producing the phenomenon, we may expect to find it most fully developed in the places where the conditions exist in fullest measure, less so where they are more limited, absent where they do not exist. A short examination of the facts will show that such is the case.

It may be thought that the Spoils System is a constant, existing everywhere, and therefore not admitting of the application of this method of concomitant variations. That system does no doubt prevail over every State of the Union, but it is not everywhere an equally potent factor, for in some cities the offices are much better paid than in others, and the revenues which their occupants control are larger. In some small communities the offices, or most of them, are not paid at all. Hence this factor varies scarcely less than the others.

We may therefore say with truth that all of the four conditions above named are most fully present in great cities. Some of the offices are highly paid; many give facilities for lucrative jobbing, and the unpaid officers are sometimes the most apt to abuse these facilities. The voters are so numerous that a strong and active organization is needed to drill them; the majority so ignorant as to be easily led. The best citizens are engrossed in business and cannot give to political work the continuous attention it demands. Such are the phenomena of New York, Philadelphia, Chicago, Brooklyn, St. Louis, Cincinnati, San Francisco, Baltimore, and New Orleans. In these cities Ring-and-bossdom has attained its amplest growth, overshadowing the whole field of politics.

"HOW TO BECOME A STATESMAN" AND "TO HOLD YOUR DISTRICT – STUDY HUMAN NATURE AND ACT ACCORDIN'"

from *Plunkitt of Tammany Hall* (1905) William Riordan

HOW TO BECOME A STATESMAN

"There's thousands of young men in this city who will go to the polls for the first time next November. Among them will be many who have watched the careers of successful men in politics, and who are longin' to make names and fortunes for themselves at the same game. It is to these youths that I want to give advice. First, let me say that I am in a position to give what the courts call expert testimony on the subject. I don't think you can easily find a better example than I am of success in politics. After forty years' experience at the game I am – well, I'm George Washington Plunkitt. Everybody knows what figure I cut in the greatest organization on earth, and if you hear people say that I've laid away a million or so since I was a butcher's boy in Washington Market, don't come to me for an indignant denial. I'm pretty comfortable, thank you.

"Now, havin' qualified as an expert, as the lawyers say, I am goin' to give advice free to the young men who are goin' to cast their first votes, and who are lookin' forward to political glory and lots of cash. Some young men think they can learn how to be successful in politics from books, and they cram their heads with all sorts of college rot. They couldn't make a bigger mistake. Now, understand me, I ain't sayin' nothin' against colleges. I guess they'll have to exist as long as there's bookworms, and I suppose they do some good in a certain way, but they don't count in politics. In fact, a young man who has gone through the college course is handicapped at the outset. He may succeed in politics, but the chances are 100 to 1 against him.

"Another mistake; some young men think that the best way to prepare for the political game is to practise speakin' and becomin' orators. That's all wrong. We've got some orators in Tammany Hall, but they're chiefly ornamental. You never heard of Charlie Murphy delivering a speech, did you? Or Richard Croker, or John Kelly, or any other man who has been a real power in the organization? Look at the thirty-six district leaders of Tammany Hall today. How many of them travel on their tongues? Maybe one or two, and they don't count when business is doin' at Tammany Hall. The men who rule have practised keepin' their tongues still, not exercisin' them. So you want to drop the orator idea unless you mean to go into politics just to perform the sky-rocket act.

"Now, I've told you what not to do; I guess I can explain best what to do to succeed in politics by tellin' you what I did. After goin' through the apprenticeship of the business while I was a boy by workin' around the district headquarters and hustlin' about the polls on election day, I set out when I cast my first vote to win fame and money in New York city politics. Did I offer my services to the district leader as a stump-speaker? Not much. The woods are always full of speakers. Did I get up a book on municipal government and show it to the leader? I wasn't such a fool. What I did was to get some marketable goods before goin' to the leaders. What do I mean by marketable goods? Let me tell you: I had a cousin, a young man who didn't take any particular interest in politics. I went to him and said: 'Tommy, I'm goin' to be a politician, and I want to get a followin'; can I count on you?' He said: 'Sure, George.' That's how I started in business. I got a marketable commodity – one vote. Then I went to the district leader and told him I could command two votes on election day, Tommy's and my own. He smiled on me and told me to go ahead. If I had offered him a speech or a bookful of learnin', he would have said, 'Oh, forget it!' That was beginnin' business in a small way, wasn't it? But that is the only way to become a real lastin' statesman. I soon branched out. Two young men in the flat next to mine were school friends. I went to them, just as I went to Tommy, and they agreed to stand by me. Then I had a followin' of three voters and I began to get a bit chesty. Whenever I dropped into district headquarters, everybody shook hands with me, and the leader one day honored me by lightin' a match for my cigar. And so it went on like a snowball rollin' down a hill. I worked the flat-house that I lived in from the basement to the top floor, and I got about a dozen young men to follow me. Then I tackled the next house and so on down the block and around the corner. Before long I had sixty men back of me, and formed the George Washington Plunkitt Association.

"What did the district leader say then when I called at headquarters? I didn't have to call at headquarters. He came after me and said: 'George, what do you want? If you don't see what you want, ask for it. Wouldn't you like to have a job or two in the departments for your friends?' I said: 'I'll think it over; I haven't yet decided what the George Washington Plunkitt Association will do in the next campaign.' You ought to have seen how I was courted and petted then by the leaders of the rival organizations. I had marketable goods and there was bids for them from all sides, and I was a risin' man in politics. As time went on, and my association grew, I thought I would like to go to the Assembly. I just had to hint at what I wanted, and three different organizations offered me the nomination. Afterwards, I went to the Board of Aldermen, then to the State Senate, then became leader of the district, and so on up and up till I became a statesman.

"That is the way and the only way to make a lastin' success in politics. If you are goin' to cast your first vote next November and want to go into politics, do as I did. Get a followin', if it's only one man, and then go to the district leader and say: 'I want to join the organization. I've got one man who'll follow me through thick and thin.' The leader won't laugh at your one-man followin'. He'll shake your hand warmly, offer to propose you for membership in his club, take you down to the corner for a drink, and ask you to call again. But go to him and say: 'I took first prize at college in Aristotle; I can recite all Shakspere forwards and backwards; there ain't nothin' in science that ain't as familiar to me as blockades on the elevated roads and I'm the

real thing in the way of silver-tongued orators.' What will he answer? He'll probably say: 'I guess you are not to blame for your misfortunes, but we have no use for you here.'"

[. . .]

TO HOLD YOUR DISTRICT – STUDY HUMAN NATURE AND ACT ACCORDIN'

"There's only one way to hold a district; you must study human nature and act accordin'. You can't study human nature in books. Books is a hindrance more than anything else. If you have been to college, so much the worse for you. You'll have to unlearn all you learned before you can get right down to human nature, and unlearnin' takes a lot of time. Some men can never forget what they learned at college. Such men may get to be district leaders by a fluke, but they never last.

"To learn real human nature you have to go among the people, see them and be seen. I know every man, woman, and child in the Fifteenth District, except them that's been born this summer and I know some of them, too. I know what they like and what they don't like, what they are strong at and what they are weak in, and I reach them by approachin' at the right side.

"For instance, here's how I gather in the young men. I hear of a young feller that's proud of his voice, thinks that he can sing fine. I ask him to come around to Washington Hall and join our Glee Club. He comes and sings, and he's a follower of Plunkitt for life. Another young feller gains a reputation as a baseball player in a vacant lot. I bring him into our baseball club. That fixes him. You'll find him workin' for my ticket at the polls next election day. Then there's the feller that likes rowin' on the river, the young feller that makes a name as a waltzer on his block, the young feller that's handy with his dukes. I rope them all in by givin' them opportunities to show themselves off. I don't trouble them with political arguments. I just study human nature and act accordin'.

"But you may say this game won't work with the high-toned fellers, the fellers that go through college and then join the Citizens' Union. Of course it wouldn't work. I have a special treatment for them. I ain't like the patent medicine man that gives the same medicine for all diseases. The Citizens' Union kind of a young man! I love him! He's the daintiest morsel of the lot, and he don't often escape me.

"Before telling you how I catch him, let me mention that before the election last year, the Citizens' Union said they had four hundred or five hundred enrolled voters in my district. They had a lovely headquarters, too, beautiful roll-top desks and the cutest rugs in the world. If I was accused of havin' contributed to fix up the nest for them, I wouldn't deny it under oath. What do I mean by that? Never mind. You can guess from the sequel, if you're sharp.

"Well, election day came. The Citizens' Union's candidate for Senator, who ran against me, just polled five votes in the district, while I polled something more than 14,000 votes. What became of the 400 or 500 Citizens' Union enrolled voters in my district? Some people guessed that many of them were good Plunkitt men all along and worked with the Cits just to bring them into the Plunkitt camp by election day. You can guess that way, too, if you want to. I never contradict stories about me, especially in hot weather. I just call your attention to the fact that on last election day 395 Citizens' Union enrolled voters in my district were missin' and unaccounted for.

"I tell you frankly, though, how I have captured some of the Citizens' Union's young men. I have a plan that never fails. I watch the City Record to see when there's civil service examinations for good things. Then I take my young Cit in hand, tell him all about the good thing and get him worked up till he goes and takes an examination. I don't bother about him any more. It's a cinch that he comes back to me in a few days and asks to join Tammany Hall. Come over to Washington Hall some night and I'll show you a list of names on our rolls marked 'C. S.' which means, 'bucked up against civil service.'

"As to the older voters, I reach them, too. No, I don't send them campaign literature. That's rot. People can get all the political stuff they want to read and a good deal more, too in the papers. Who reads speeches, nowadays, anyhow? It's bad enough to listen to them. You ain't goin' to gain any votes by stuffin' the letter boxes with campaign documents. Like as not you'll lose votes, for there's nothin' a man hates more than to hear the letter-carrier ring his bell and go to the letter-box expectin' to find a letter he was lookin' for, and find only a lot of printed politics. I met a man this very mornin' who told me he voted the Democratic State ticket last year just because the Republicans kept crammin' his letter-box with campaign documents.

"What tells in holdin' your grip on your district is to go right down among the poor families and help them in the different ways they need help. I've got a regular system for this. If there's a fire in Ninth, Tenth, or Eleventh Avenue, for example, any hour of the day or night, I'm usually there with some of my election district captains as soon as the fire-engines. If a family is burned out I don't ask whether they are Republicans or Democrats, and I don't refer them to the Charity Organization Society, which would investigate their case in a month or two and decide they were worthy of help about the time they are dead from starvation I just get quarters for them, buy clothes for them if their clothes were burned up, and fix them up till they get things runnin' again. It's philanthropy, but it's politics, too – mighty good politics. Who can tell how many votes one of these fires bring me? The poor are the most grateful people in the world, and, let me tell you, they have more friends in their neighborhoods than the rich have in theirs.

"If there's a family in my district in want I know it before the charitable societies do, and me and my men are first on the ground. I have a special corps to look up such cases. The consequence is that the poor look up to George W. Plunkitt as a father, come to him in trouble – and don't forget him on election day.

"Another thing, I can always get a job for a deservin' man. I make it a point to keep on the track of jobs, and it seldom happens that I don't have a few up my sleeve ready for use. I know every big employer in the district and in the whole city, for that matter, and they ain't in the habit of sayin' no to me when I ask them for a job.

"And the children – the little roses of the district! Do I forget them? Oh, no! They know me, every one of them, and they know that a sight of Uncle George and candy means the same thing. Some of them are the best kind of vote-getters. I'll tell you a case. Last year a little Eleventh Avenue rosebud whose father is a Republican, caught hold of his whiskers on election day and she said she wouldn't let go until he'd promise to vote for me. And she didn't."

"WHY THE WARD BOSS RULES"

from *The Outlook* (1898) Jane Addams

Primitive people, such as the South Italian peasants who live in the Nineteenth Ward, deep down in their hearts admire nothing so much as a good man. The successful candidate must be a good man according to the standards of his constituents. He must not attempt to hold up a morality beyond them, nor must he attempt to reform or change the standard. If he believes what they believe, and does what they are all cherishing a secret ambition to do, he will dazzle them by his success and win their confidence. Any one who has lived among poorer people cannot fail to be impressed with their constant kindness to each other; that unfailing response to the needs and distresses of their neighbors, even when in danger of bankruptcy themselves. This is their reward for living in the midst of poverty. They have constant opportunities for self-sacrifice and generosity, to which, as a rule, they respond. A man stands by his friend when he gets too drunk to take care of himself, when he loses his wife or child, when he is evicted for non-payment of rent, when he is arrested for a petty crime. It seems to such a man entirely fitting that his Alderman should do the same thing on a larger scale – that he should help a constituent out of trouble just because he is in trouble, irrespective of the justice involved.

The Alderman, therefore, bails out his constituents when they are arrested, or says a good word to the police justice when they appear before him for trial; uses his "pull" with the magistrate when they are likely to be fined for a civil misdemeanor, or sees what he can do to "fix up matters" with the State's attorney when the charge is really a serious one.

Because of simple friendliness, the Alderman is expected to pay rent for the hard-pressed tenant when no rent is forthcoming, to find jobs when work is hard to get, to procure and divide among his constituents all the places which he can seize from the City Hall. The Alderman of the Nineteenth Ward at one time made the proud boast that he had two thousand six hundred people in his ward upon the public pay-roll. This, of course, included day-laborers, but each one felt under distinct obligations to him for getting the job.

If we recollect, further, that the franchise-seeking companies pay respectful heed to the applicants backed by the Alderman, the question of voting for the successful man becomes as much an industrial as a political one. An Italian laborer wants a job more than anything else, and quite simply votes for the man who promises him one.

The Alderman may himself be quite sincere in his acts of kindness. In certain stages of moral evolution, a man is incapable of unselfish action the results of which will not benefit some one of his acquaintances;

still more, of conduct that does not aim to assist any individual whatsoever; and it is a long step in moral progress to appreciate the work done by the individual for the community.

The Alderman gives presents at weddings and christenings. He seizes these days of family festivities for making friends. It is easiest to reach people in the holiday mood of expansive good will, but on their side it seems natural and kindly that he should do it. The Alderman procures passes from the railroads when his constituents wish to visit friends or to attend the funerals of distant relatives; he buys tickets galore for benefit entertainments given for a widow or a consumptive in peculiar distress; he contributes to prizes which are awarded to the handsomest lady or the most popular man. At a church bazaar, for instance, the Alderman finds the stage all set for his dramatic performance. When others are spending pennies he is spending dollars. Where anxious relatives are canvassing to secure votes for the two most beautiful children who are being voted upon, he recklessly buys votes from both sides, and laughingly declines to say which one he likes the best, buying off the young lady who is persistently determined to find out, with five dollars for the flower bazaar, the posies, of course, to be sent to the sick of the parish. The moral atmosphere of a bazaar suits him exactly. He murmurs many times, "Never mind; the money all goes to the poor," or, "It is all straight enough if the church gets it."

There is something archaic in a community of simple people in their attitude towards death and burial. Nothing so easy to collect money for as a funeral. If the Alderman seizes upon festivities for expressions of his good will, much more does he seize upon periods of sorrow. At a funeral he has the double advantage of ministering to a genuine craving for comfort and solace, and at the same time of assisting at an important social function.

In addition to this, there is among the poor, who have few social occasions, a great desire for a well-arranged funeral, the grade of which almost determines their social standing in the neighborhood. The Alderman saves the very poorest of his constituents from that awful horror of burial by the county; he provides carriages for the poor, who otherwise could not have them; for the more prosperous he sends extra carriages, so that they may invite more friends and have a longer procession; for the most prosperous of all there will be probably only a large "flower-piece." It may be too much to say that all the relatives and friends who ride in the carriages provided by the Alderman's bounty vote for him, but they are certainly influenced by his kindness, and talk of his virtues during the long hours of the ride back and forth from the suburban cemetery. A man who would ask at such a time where all this money comes from would be considered sinister. Many a man at such a time has formulated a lenient judgment of political corruption and has heard kindly speeches which he has remembered on election day. "Ah, well, he has a big Irish heart. He is good to the widow and the fatherless." "He knows the poor better than the big guns who are always about talking civil service and reform."

Indeed, what headway can the notion of civic purity, of honesty of administration, make against this big manifestation of human friendliness, this stalking survival of village kindness? The notions of the civic reformer are negative and impotent before it. The reformers give themselves over largely to criticisms of the present state of affairs, to writing and talking of what the future must be; but their goodness is not dramatic; it is not even concrete and human.

Such an Alderman will keep a standing account with an undertaker, and telephone every week, and sometimes more than once, the kind of outfit he wishes provided for a bereaved constituent, until the sum may roll up into hundreds a year. Such a man understands what the people want, and ministers just as truly to a great human need as the musician or the artist does. I recall an attempt to substitute what we might call a later standard.

A delicate little child was deserted in the Hull House nursery. An investigation showed that it had been born ten days previously in the Cook County Hospital, but no trace could be found of the unfortunate mother. The little thing lived for several weeks, and then, in spite of every care, died. We decided to have it buried by the county, and the wagon was to arrive by eleven o'clock. About nine o'clock in the morning the rumor of this awful deed reached the neighbors. A half-dozen of them came, in a very excited state of mind, to protest. They took up a collection out of their poverty with which to defray a funeral. We were then comparatively new in the neighborhood. We did not realize that we were really shocking a genuine moral sentiment of the community. In our crudeness, we instanced the care and tenderness which had been expended upon the little creature while it was alive; that it had had every attention from a skilled physician and trained nurse; we even intimated that the excited members of the group

had not taken part in this, and that it now lay with us to decide that the child should be buried, as it had been born, at the county's expense. It is doubtful whether Hull House has ever done anything which injured it so deeply in the minds of some of its neighbors. We were only forgiven by the most indulgent on the ground that we were spinsters and could not know a mother's heart. No one born and reared in the community could possibly have made a mistake like that. No one who had studied the ethical standards with any care could have bungled so completely.

Last Christmas our Alderman distributed six tons of turkeys, and four or more tons of ducks and geese; but each luckless biped was handed out either by himself or one of his friends with a "Merry Christmas." Inevitably, some families got three or four apiece, but what of that? He had none of the nagging rules of the charitable societies, nor was he ready to declare that, because a man wanted two turkeys for Christmas, he was a scoundrel, who should never be allowed to eat turkey again.

The Alderman's wisdom was again displayed in procuring from downtown friends the sum of three thousand dollars wherewith to uniform and equip a boys' temperance brigade which had been formed in the ward a few months before his campaign. Is it strange that the good leader, whose heart was filled with innocent pride as he looked upon these promising young scions of virtue, should decline to enter into a reform in campaign?

The question does, of course, occur to many minds, Where does the money come from with which to dramatize so successfully? The more primitive people accept the truthful statement of its sources without any shock to their moral sense. To their simple minds he gets it "from the rich," and so long as he again gives it out to the poor, as a true Robin Hood, with open hand, they have no objections to offer. Their ethics are quite honestly those of the merry-making foresters. The next less primitive people of the vicinage are quite willing to admit that he leads "the gang" in the City Council, and sells out the city franchises; that he makes deals with the franchise-seeking companies; that he guarantees to steer dubious measures through the Council, for which he demands liberal pay; that he is, in short, a successful boodler. But when there is intellect enough to get this point of view, there is also enough to make the contention that this is universally done; that all the Aldermen do it more or less successfully, but that the Alderman of the Nineteenth Ward is unique in being so generous; that such a state of affairs is to be deplored, of course, but that that is the way business is run, and we are fortunate when a kind-hearted man who is close to the people gets a large share of the boodle; that he serves these franchised companies who employ men in the building and construction of their enterprises, and that they are bound in return to give jobs to his constituency. Even when they are intelligent enough to complete the circle, and to see that the money comes, not from the pockets of the companies' agents, but from the streetcar fares of people like themselves, it almost seems as if they would rather pay two cents more each time they ride than give up the consciousness that they have a big, warm-hearted friend at court who will stand by them in an emergency. The sense of just dealing comes apparently much later than the desire for protection and kindness. The Alderman is really elected because he is a good friend and neighbor.

During a campaign a year and a half ago, when a reform league put up a candidate against our corrupt Alderman, and when Hull House worked hard to rally the moral sentiment of the ward in favor of the new man, we encountered another and unexpected difficulty. Finding that it was hard to secure enough local speakers of the moral tone which we desired, we imported orators from other parts of the town, from the "better element," so to speak. Suddenly we heard it rumored on all sides that, while the money and speakers for the reform candidate were coming from the swells, the money which was backing our corrupt Alderman also came from a swell source; it was rumored that the president of a street-car combination, for whom he performed constant offices in the City Council, was ready to back him to the extent of fifty thousand dollars; that he, too, was a good man, and sat in high places; that he had recently given a large sum of money to an educational institution, and was, therefore, as philanthropic, not to say good and upright, as any man in town; that our Alderman had the sanction of the highest authorities, and that the lecturers who were talking against corruption, and

the selling and buying of franchises, were only the cranks, and not the solid business men who had developed and built up Chicago.

All parts of the community are bound together in ethical development. If the so-called more enlightened members of the community accept public gifts from the man who buys up the Council, and the so-called less enlightened members accept individual gifts from the man who sells out the Council, we surely must take our punishment together.

Another curious experience during that campaign was the difference of standards between the imported speakers and the audience. One man, high in the council of the "better element," one evening used as an example of the philanthropic politician an Alderman of the vicinity, recently dead, who was devotedly loved and mourned by his constituents. When the audience caught the familiar name in the midst of the platitudes, they brightened up wonderfully. But, as the speaker went on, they first looked puzzled, then astounded, and gradually their astonishment turned to indignation. The speaker, all unconscious of the situation, went on, imagining, perhaps, that he was addressing his usual audience, and totally unaware that he was perpetrating an outrage upon the finest feelings of the people who were sitting before him. He certainly succeeded in irrevocably injuring the chances of the candidate for whom he was speaking. The speaker's standard of ethics was upright dealing in positions of public trust. The standard of ethics held by his audience was, being good to the poor and speaking gently of the dead. If he considered them corrupt and illiterate voters, they quite honestly held him a blackguard.

If we would hold to our political democracy, some pains must be taken to keep on common ground in our human experiences, and to some solidarity in our ethical conceptions. And if we discover that men of low ideals and corrupt practice are forming popular political standards simply because such men stand by and for and with the people, then nothing remains but to obtain a like sense of identification before we can hope to modify ethical standards.

"Contested Cities: Social Process and Spatial Form"

from Nick Jewson and Susanne MacGregor (eds),
Transforming Cities (1997)

David Harvey

Editors' Introduction

David Harvey was trained as a geographer and teaches in the Geography Department of Johns Hopkins University, but his prolific and influential writings over the last three decades bridge the concerns of all of the preceding sections of *The City Reader* on urban history, society, culture, and space as well as the selections in this section on urban politics. Harvey writes theoretically at a high level of abstraction. He skillfully blends Marxist and postmodernist approaches to understanding cities. Harvey challenges fundamental assumptions about history, epistemology, social justice, community, ecology, and the physical organization of cities.

Urban politics is a distinct subfield within the discipline of political science and much good writing on urban politics and governance comes from political scientists. But questions of politics and governance also interest geographers like David Harvey, policy analysts like Anthony Downs (p. 245), lawyers like Myron Orfield (p. 287), social workers like Sherry Arnstein (p. 233), and people from other academic disciplines and professional fields.

Harvey starts this selection by reminding us of the massive urbanization of the human population that Kingsley Davis describes (p. 17). Cities are critical to understanding the current human condition. Yet, Harvey notes, cities as a category are strikingly absent from many discussions of modernization, modernity, postmodernity, and capitalist and industrial society. He would like to see cities included in discussions of these topics.

Harvey emphasizes the importance of thinking about cities in terms of *processes* rather than just *things*. For him cities are sites of conflict – based on race, ideology, gender, and other values. Harvey thinks dialectically. He argues that processes are both *shaped by* time and place and *shape* time and place. For example, when communist Chinese leader Mao Tse Tung decided to send 20 million urban Chinese intellectuals out of Shanghai, Beijing, Guangzhou, and other large Chinese cities into the countryside during the Great Proletarian Cultural Revolution, this was a social process shaped by a specific place (China) at a specific time (during the great proletarian cultural revolution). For a while this social process produced a distinct kind of chaotic and austere revolutionary city. Historical sites and buildings with European architecture were destroyed; massive public works projects, bomb shelters, and monuments to revolutionary heroes were built. The cultural revolution effectively ended with Mao's death in 1976. Some of the urban things built (and destroyed) during the great proletarian cultural revolution period are still evident in Chinese cities, but Chinese society and politics are remarkably fluid and much has changed, as witnessed by the modern skyline of Shanghai on the cover to this book.

Some processes lead to very durable outcomes. The decision to build the Three Mile Island nuclear power plant, dam the Yangtze River, or rebuild the World Trade Center site produce physical forms and social structures that last for a very long time. What looks like a good solution to urban development to one generation, such as a massive highway construction program to solve transportation problems, building identical suburban tract housing to deal with housing needs, or creating a nuclear power plant to meet urban energy needs, may not look so good to the next generation. This is why Harvey urges planners and policy makers to design flexible, adjustable cities and to encourage fluid social processes that can change over time.

Two additional themes in Harvey's selection deal with community and the relationship between the natural and built environment. Louis Wirth (p. 90), Robert Putnam (p. 120), Ali Madanipour (p. 158), Jane Jacobs (p. 98), Peter Calthorpe and William Fulton (p. 342), Andrés Duany and Elizabeth Plater-Zyberk (p. 192), and many others are concerned that cities alienate people from each other and break down community. They all favor agendas to build greater community in cities. Harvey raises some important questions about community building. He argues that many community-building projects are recipes for isolation. The positive identification of some groups is often achieved by first defining other groups as the other, the devalued, semi-human. For example, people within a gated community in Los Angeles of the kind that Mike Davis describes (p. 178) may become a community that play golf together and have drinks at their country club, safe within the gates. They will not be alienated and isolated from each other. But this kind of community is based on walling themselves off from lower-income people and people of color outside the gates of their artificial community.

Harvey criticizes the belief that the proper design of things will solve social process problems. Rather, Harvey argues, the social processes underlying even the best-designed, community-enhancing place need to be cultivated and sustained. Just building a remarkable physical community like Ebenezer Howard's garden city of Letchworth, England, or Andrés Duany and Elizabeth Plater-Zyberk's neo-traditionalist new town of Seaside, Florida, will not, Harvey argues, create community.

Harvey notes that community activism is often built around "militant particularisms" in which a group coheres around a value, like social justice or environmental conservation, related to a very narrow and very local concern. For example, groups may focus on getting higher wages for low-paid garment workers at the Ajax garment factory or saving an endangered salamander in the Bolinas lagoon. These militant particularisms may contribute to society, but Harvey argues that they will be much more useful if they spill over into wider, more universal concerns; for example, if militant environmentalists generalize their concern for one breed of salamander into a more universal (and perhaps more balanced) environmental movement or the militant garment workers generalize their concerns into a broad movement for all workers to be paid a living wage.

Harvey critiques the artificial distinction some environmentalists draw between the natural and the built environments. He agrees with Stephen Wheeler (p. 499) and Timothy Beatley (p. 411) that the natural and built environments need to evolve together. The sustainable urban development, green urbanism, design with nature, and ecological design movements all follow this notion of harmony between the natural and built environments.

David Harvey is Professor of Geography and Environmental Engineering at Johns Hopkins University; Senior Research Fellow, St Peter's College, University of Oxford; and Miliband Visiting Fellow, London School of Economics and Political Science. From 1987 to 1993 he was Halford McKinder Professor of Geography at Oxford. Harvey is currently working on questions of environmental justice, globalization, alternative modes of urbanization, and uneven geographical development within a globalizing world.

This selection is from Nick Jewson and Susanne MacGregor (eds), *Transforming Cities* (London and New York: Routledge, 1997). Books by Harvey include *Spaces of Capital* (London and New York: Routledge, 2002), *Spaces of Hope* (Berkeley: University of California Press, 2000), *Limits to Capital* (London: Verso, 1999), *Justice, Nature, and the Geography of Difference* (Oxford: Blackwell, 1996), *The Urban Experience* (Malden: Blackwell, 1990), *The Condition of Postmodernity: An Enquiry into the Origins of Cultural Change* (Oxford: Blackwell, 1989), *The Urbanization of Capital: Studies in the History and Theory of Capitalist Urbanization* (Baltimore: Johns Hopkins University Press, 1985), *Consciousness and the Urban Experience: Studies in the History and Theory of Capitalist Urbanization* (Baltimore: Johns Hopkins University Press, 1985), *The Limits to Capital* (Oxford: Blackwell, 1982), *Social Justice and the City* (Oxford: Basil Blackwell, 1973, new edition 1988), and *Explanation in Geography* (London: Edward Arnold, 1969).

Classical and contemporary writings on urban culture that include postmodernist and Marxist writings are contained in Malcolm Miles and Tim Hall (with Ian Borden), *The City Cultures Reader*, second edition (London and New York: Routledge, 2004).

Other postmodernist writings on social processes and spatial form include Henri Lefebvre, *Production of Space* (Oxford: Blackwell, 1991) and *Writings on Cities* (Oxford: Blackwell, 1995), Edward Soja, *Postmodern*

Geographies (London: Verso, 1997) and *Postmetropolis* (London: Blackwell, 2000), Mike Craig and Nigel Thrift (eds), *Thinking Space* (London and New York: Routledge, 2000), and Fredric Jameson, *Postmodernism or the Cultural Logic of Capitalism* (Durham: Duke University Press, 1992).

At the beginning of this century, there were little more than a dozen or so cities in the world with more than a million people. They were all in the advanced capitalist countries and London, by far the largest of them, had just under 7 million. At the beginning of this century too, no more than 7 per cent of the world's population could reasonably be classified as 'urban'. By the year 2000 there may be as many as 500 cities with more than a million inhabitants. The largest of them (like Tokyo, São Paulo, Bombay and possibly Shanghai) will boast populations of more than 20 million, trailed by a score of cities, mostly in the so-called developing countries, with upwards of 10 million. Sometime early next century, if present trends continue, more than half of the world's population will be classified as urban rather than rural.

The twentieth century has been the century of urbanization. There has been a massive reorganization of the world's population, of its political and its institutional structures and of the very ecology of the earth.

These observations immediately suggest some fundamental questions. First, given these transformations, why is it that the urban so frequently disappears from our discussions of broader political-economic processes and social trends? Most of the writing about our recent history has failed to take into account this massive reorganization and its consequences. The urban rarely appears as a salient category in our analyses. The crucial categories seem to be those of modernization, modernity, post-modernity, capitalist and industrial society. So what has happened to the category 'urban'? This question is important because the qualities of urban living in the next century will define the qualities of life for the mass of humanity. And all political-economic processes we observe are mediated through the filter of urban organization. Discussions of contemporary politics, for example, often proceed as if a concept like that of 'democracy' can remain unaffected by urban transformations when, plainly, there is a huge difference between democracy in ancient Athens and democracy in contemporary São Paulo.

If we think about the likely qualities of life in the next century by projecting forward current trends in our cities, most commentators would end up with a somewhat dystopian view. We are producing marginalization, disempowerment, alienation, pollution and degradation. It might be said that this is nothing new and that, in the nineteenth century, conditions were even worse. In the past, however, urbanization and the consequences of urbanization were taken rather more seriously than they are today. In the late nineteenth century, the bourgeoisie at least had some notion that cities were important places and, therefore, that urban reform was necessary. This generated a bourgeois reform movement – from Birmingham to Chicago – which included figures such as Jane Addams, Octavia Hill, Charles Booth, Patrick Geddes, Ebenezer Howard and many others. All of these had some vision for the future and a clear grasp of the need for reform. The nineteenth century faced the difficulties of the urban in a very positive and powerful way. It blended socialist sentiments, anarchist ideas, notions of bourgeois reformism and social responsibility into a programmatic attempt to clean up the cities. The 'gas and water socialism' of the late nineteenth and early twentieth centuries did a great deal to improve the conditions of urban life for the mass of the population. There are many contemporary analysts who, armed with the insights of Foucault, will assert that these innovations were merely about social control, which indeed in part they were. But having acknowledged this point, I think we have also to recognize that a significant proportion of the population found itself living in better circumstances as a result. Moreover, inherent in these interventions was a visionary notion of an alternative city – a city beautiful with facilities and services that would, indeed, pacify alienated populations.

Some of that concern would be helpful to have back in our cities right now. In the past, capital regarded cities as important places which had to be efficiently organized and where social controls needed to operate in some sort of meaningful way. We now find that capital is no longer concerned about cities. Capital needs fewer workers and much of it can move all over the world, deserting problematic places and populations at will. As a result, the coalition between big capital and bourgeois reformism has disappeared. Moreover, the bourgeoisie itself seems to have lost much of its guilty conscience about cities. It has, I think, concluded there is little to fear from socialist revolution, and so has attenuated its engagement with reformism. Increasingly the wealthy seal themselves off in those fanciful, gated communities – which are being built all over the United States – that enable the bourgeoisie to cut themselves off from what their representatives call by the hateful term 'the underclass'. 'The underclass' is left inside the ghetto, along with drugs, AIDS, epidemics of tuberculosis and much else. In this new politics, the poor no longer matter. The marginalization of the poor is accompanied by a blasé indifference on the part of the rich and powerful.

This blasé indifference is a matter of great concern. Accordingly I would like to highlight some fundamental questions and beliefs about the role of the city in political, economic, social and ecological life. In defining that role, we are also formulating a notion of the kind of cities we would like to construct into the next century.

I would like to begin with a fundamental methodological question: what is the relationship between process and form? . . . In my own work – from the standpoint of historical, geographical materialism and very strongly in the dialectical tradition – one of the rules of engagement which I have always tried to follow is to say that process takes precedence over things. We should focus on processes rather than things and we should think of things as products of processes. From this standpoint, we have to ask some fundamental questions about the nature of the categories we use to describe the world. Most of the categories we use tend to be 'thing' categories. If instead we examine dynamics and processes, we may try to do so by conceiving them as relationships between pre-existing things. But if things too are not pre-existing, but are actually constituted in some way by a process, then you have to have a rather different vision. This transformation in our way of thought seems to me absolutely essential if we are going to get to the heart of what the city is about.

Tony Leeds, an urban anthropologist, towards the end of his life wrote this:

In earlier years I thought of society . . . as a structure of positions, roles, statuses, groups, institutions and so on, all given shape . . . by the cultures on which they draw. Process I saw as 'forces', movement, connection, pressures, taking place in and among these loci or nodes of organization, peopled by individuals. Although this still seems largely true to me, it has also come to seem a static view – more societal order than societal becoming . . . Since it does not seem inherent in nature . . . that these loci exist, it seems unacceptable simply to take them as axiomatic; rather we must search for ways to account for their appearances and forms. More and more, the problems of becoming . . . have led me to look at society as continuous process out of which structure or order precipitates in the forms of the loci listed above.

This, then, is a conceptualization in which process takes priority over things and which focuses on the way in which things get precipitated out of process.

Two terms or words deserve closer examination in our discussions. One is 'urbanization' – which we can convert into the 'urban process' or the 'urbanizing process' or the 'urbanization process'. The other is a 'thing-type' word – 'the city'. It is important to consider the relationship between the urbanizing process and this thing called the city. Now, from a dialectical standpoint, the relationship between process and thing becomes complicated because things, once constituted, have the habit of affecting the very processes which constituted them. The ways that particular 'thing-like structures' (such as political-administrative territories, built environments, fixed networks of social relations) precipitate out of fluid social processes and the fixed forms these things then assume have a powerful influence upon the way that social processes can operate. Moreover, different fixed forms have been precipitated out at different historical moments and assume qualities reflective of social processes at work in particular times and places. The result is an urban environment constituted as a palimpsest, a series of layers constituted and constructed at different historical moments all superimposed upon each other. The question then becomes how does the life process work in and around all of those things which have been constituted at different historical periods? How are new meanings given to them? How are new possibilities constructed? I suggest that attention to this relationship between process and form will help us understand why the urban has been neglected and, furthermore, will enable us to change completely the terms of the debate.

In this vein, I want to suggest that the reduction of the urban – or the portrayal of the city as a minor feature of social organization – can only occur when particular assumptions are made about the nature of space and time or space/time. There are three different ways of understanding spatiality or space/time that are worth noting here. The first way is the absolute notion of space/time – attributable to Newton, Descartes and Kant – in which space and time are mere containers of social action. They are passive, neutral containers. These passive, neutral containers simply allow us to locate the action which is occurring. I would like to suggest that there is a parallel here with thinking that conceives of cities as passive, neutral containers of processes and contests. These ways of thinking focus on contestations occurring within the city – the city happens to be the mere site of a process of contestation (over gender, race, class or whatever). A radically different approach is one which sees the city not so much as a site of contestation but as something to be constructed and in which the contestation is over the construction, or framing, of the city itself. What would that imply about notions of space and time?

There is a well-known alternative to the absolute view of space/time: that is, the relative view attributed mainly to Einstein and worked on by others since. In this view, space and time, although they are still containers, are not neutral with respect to the processes they contain. Metrics of space and time can and do vary depending upon the nature of the processes under consideration. In geography, this idea has been adapted to think of different ways of measuring and mapping distances. Physical distance is different from distance measured in terms of the cost or time taken to move between points, and in the last two cases the space described is not necessarily Euclidean. Different metrics yield different maps of the space/time co-ordinates within which social interaction occurs.

A third perspective on space/time that I have employed – indeed it was incorporated in *Social Justice in the City* more than twenty years ago – is a relational view . . . This view is that space and time do not exist outside of process: process defines space/time. Each particular kind of process will define its own distinctive spatio-temporality. Our studies should, therefore, aim to explain the way in which different processes define spatio-temporality, and then, having defined that spatio-temporality, find themselves bound by its rules in certain kinds of ways. Moreover, our cities are constituted not by one but by multiple spatio-temporalities, producing multiple frameworks within which conflictual social processes are worked out.

From this standpoint, we have to take very seriously the notion . . . that space and time are not simply constituted by but are also constitutive of social processes. This is also true for the urban. The urban and the city are not simply constituted by social processes, they are constitutive of them. We have to understand that dialectic in order to appreciate how urbanization is constructed and produces all of these thing-like configurations which we call cities – with political organization, social organization and physical structures. We have to appreciate better the centrality of that moment of urban construction, which is fundamental to how the social process operates. In exactly the same way, we have to take seriously the idea of that moment of construction of spatio-temporality, which then defines how the system itself will operate. From this standpoint, it is possible to reposition the urban as fundamental in contemporary debates. At the same time we transform our notion of urbanization. We would abandon the view of the urban as simply a site or a container of social action in favour of the idea that it is, in itself, a set of conflictual heterogeneous processes which are producing spatio-temporalities as well as producing things, structures and permanencies in ways which constrain the nature of the social process. Social processes, in giving rise to things, create the things which then enhance the nature of those particular social processes.

One outcome may be that we find ourselves stuck for a very long time with a particular kind of social process. An example would be nuclear power. Once nuclear power stations exist all sorts of things follow. If a nuclear power station goes on the blink, can you imagine calling a town meeting to discuss democratically what to do about it? The answer is no, you can't. In these circumstances, we are immediately driven back to the realms of expert knowledge and expert decision-making. So a thing has been created which for as long as it lasts – which is going to be a very long time – is by its very nature going to be basically undemocratic in terms of the sort of social process that supports it. Here is a social process that has defined a certain spatio-temporality for the next 10,000 years, which in turn implies perpetuation of a certain kind of social order if it is not to unravel in highly destructive ways.

We have to be thinking in these kinds of terms about the nature of cities. What kinds of cities we create, how we create them, how flexible they are, how adjustable they can be: these are the questions we need to ask in order to understand better the relationship between process and thing. Our aim and objective should be to liberate emancipatory processes of social change. In so doing, however, we must understand that liberatory impulses and politics are always going to be contained and constrained by the nature of things which have been produced in the past.

This, then, is my first major point. We have to reconceptualize the urban as the production of space and the production of spatio-temporality, understood as a dialectical relationship between process and thing.

The second major point I would like to make concerns the currently widespread invocation of the word 'community'. It too entails an exploration of process/thing relationships. One of the aspects of much contemporary debate about the urban which I find particularly striking is the tendency when faced with all sorts of difficulties again and again to reach into this bag called 'community', on the assumption that 'community' is going to save us all. Community, endowed with salving powers, is perceived as capable of redeeming the mess which we are creating in our cities. This mode of thinking goes all the way from Prince Charles and the construction of urban villages through to communitarian philosophies that, it is believed, will save us from crass individualistic materialism.

There is here too an issue about the relationship between the thing called community and the processes which constitute it. What kinds of processes constitute community? Is a community, once constituted, going to liberate or imprison further social processes? A lot of community construction projects are, in the end, a recipe for isolation. They isolate groups from the city as a whole. They move them towards a fragmented notion of what the urban process is about. Here I find myself in agreement with Iris Marion Young when she says:

Racism, ethnic chauvinism and class devaluation, I suggest, grow partly from the desire for community . . . Practically speaking, such mutual understanding can be approximated only within a homogeneous group that defines itself by common attributes. Such common identification, however, entails reference also to those excluded. In the dynamics of racism and ethnic chauvinism in the United States today, the positive identification of some groups is often achieved by first defining other groups as the other, the devalued, semihuman.

What, then, are the implications of current notions of community? In answer, I would like to propose a dialectical view of relationships between process and community.

I think it is important to acknowledge that a lot of community activism is absolutely fundamental to many forms of social struggle. As a form of mobilization of power of people in place it can sometimes be extremely important and extremely useful. Community activism can simply be a way of containing discontent but it can also be a very important moment in more general mobilization. In this context, we have to think about the construction of community not as an end in itself but as a moment in a process. Here I refer to critiques of the nineteenth-century thinking which I described earlier. There were two flaws in that thinking. The first was the belief that, somehow or other, the proper design of things would solve all of the problems in the social process. It was assumed that if you could just build your urban village, like Ebenezer Howard, or your Radiant City, like Le Corbusier, then the thing would have the power to keep the process forever in harmonious state. The problem of these thinkers was not that they had a totalizing vision or subscribed to master narratives or indulged in master planning. Their problem was not that they had a conception of the city or the social process as a whole. Their problem was that they took this notion of thing and gave it power over the process. Their second flaw was that they did much the same with community. Much of the ideology that came out of Geddes and Ebenezer Howard was precisely about the construction of community. In particular, the construction of communities which were fixed and had certain qualities with respect to class and gender relations. Once again, the domination of things seems to me to be the fundamental flaw.

What then is the significance of community mobilization? The concept I wish to use here is the one that Raymond Williams tentatively suggested, and which he then shrank away from, but which I want to resurrect. It is what Williams calls 'militant particularism'. This idea suggests that almost all radical movements have their origin in some place, with a particular set of issues which people are pursuing and following. The key issue is whether that militant particularism simply remains localized or whether, at some point or other, it spills over into some more universal construction. Williams suggested that the whole history of socialism had to be read as a series of militant particularisms which generated what he described as the extraordinary claim that there is an alternative kind of society, called socialist, which would be a universal kind of condition to which we could all reasonably aspire. In other words, in this view foundational values and beliefs were discovered in particular struggles and then translated onto a broader terrain of conflict. It seems to me that the notion of community, viewed in this way, can be a positive moment within a political process. However, it is only a positive moment if it ceases to be an end in itself, ceases to be a thing which is going to solve all of our problems, and starts to be a moment in this process of broader construction of a more universal set of values which are going to be about how the city is going to be as a whole.

The third major point I am going to make is this: until very recently there was almost no mention of cities in the ecological literature. Cities were always regarded as the high point of the pollution and plundering of planet Earth. The environment was equated with nature; it was certainly not the built environment of cities. There is something curious about ecological rhetoric here (although I am probably misrepresenting some of the current thinking because it is getting a bit more sophisticated). Ecological rhetoric is committed to a totalizing perspective in the sense that, quite rightly, it perceives that everything relates to everything else. However, it has also failed to address the environment of cities and the 50 per cent of the world's population that are living in urban circumstances.

Why is it that we tend to think of the built environment of cities as somehow or other not being the environment? Where did that separation come from? Again it comes back to the notion that there is a thing called a city, which has various qualities and attributes, that is not part of a process. It seems to me that we have to think of environment and environmental modification as a fundamental process which we have always been engaged in and will always continue to be engaged in. The environmental modification process then has to be understood as producing certain kinds of structures and things, such as fields, forests and cities. That environmental modification process cannot be separated from the whole question of urban living. There is, it seems to me, nothing particularly anti-ecological about cities. Why should we think of them

that way? When does the built, constructed environment end and 'the natural environment' begin? Where does society begin and nature end? Go and look in a field of wheat and say where nature begins and society ends. You can't do it.

Here too, then, there is a dichotomy which works its way through our thinking, which we have to challenge, in which the relationship between processes and things is fundamental. We have to pay serious attention to the nature of the ecological modification process, and understand it not as something which is simply resident in nature. For example, one of the major ecological variables at work in the world right now is money flow. Just think of what would happen to the ecosystems of the world if the money flow stopped. How many ecosystems of this world are actually supported by money flow? Vast areas of the world would undergo radical ecological change if the money supply or commodity exchange was suddenly cut off or stopped. Some radical ecologists appear to relish such an outcome, as a transformation back to some ecologically sustainable condition in which the alienation of self from nature can be overcome by human beings treading far more lightly on the surface of the earth. But I believe we must pursue much more positive ecological politics. Ecological transformations are an inevitable facet of how human beings live their lives and construct their historical geographies. Urbanization is an ecological process and we desperately need creative ways to think and act on that relation. Conversely, it is impossible to talk of ecological politics without concomitantly examining urban processes in all their complexity and fullness.

We have to move the urban, and the urbanizing process, into a more central position in our debates and discussions about ecological, social, political and economic change. From this standpoint, there are a number of myths that we have to confront and contest.

The first myth is the simple idea that when we have got the economy right then we can spend money to get our cities right. This sort of thinking takes the view that cities are relatively unimportant: when we have got enough money and we have organized ourselves right then we can spend a little time fixing them up. From my perspective that is entirely the wrong way round. Getting things economically right in our cities is the path towards economic change and economic development, even to economic growth. To treat the cities as the secondary feature of this whole dynamic is essentially wrong.

The same is true with respect to social relations. We should not wait upon some great political revolution to tell us how to reorganize our cities in a socialist or eco-feminist or some other way. No, what we have to do is to work on the nature of the social relations in the cities. If there is going to be a revolution, it is going to be a long revolution, located within the urban process. That long revolution of social relations is going to have to comprise a steady working out, over a long period of time, of transformations. Here, I think again, community mobilization and the transformation of militant particularism have a vital role to play, enabling us to find the universal concerns that exist within a realm of difference. There is a certain dialectic here of unity and difference, universal and particular, which has to be worked out. We should not retain the notion of community as particularity or difference. We have to transcend those particularities and look for a negotiation of universalities through which to talk about how the cities of the future should be.

The point is not to see cities as anti-ecological. Cities are fundamental ecological features in themselves and the processes that build cities are ecological processes. The world of ecology and that of cities are part and parcel of each other; what we have to do is to link them together much more strongly, in a more programmatic way. It is only in those terms that we can really push towards a full understanding of . . . 'contested cities'. This issue is not simply about contestation inside cities but more importantly concerns contests over the construction and framing of cities – especially what they are going to be in the future.

"A Ladder of Citizen Participation"

Journal of the American Institute of Planners (1969)

Sherry Arnstein

Editors' Introduction

Local government is important and plural actors can influence the outcome of policies and programs that affect their lives. Local government is an important part of a new global order where public, private, and nonprofit sectors often work together in complex regimes. This new order raises local citizens' stakes in having their interests taken into account by local decision makers. But how, exactly, should citizens participate in local government decision making? Guidance as to how this might best be done comes from a classic article by Sherry Arnstein titled "A Ladder of Citizen Participation." Arnstein was the chief adviser on citizen participation in the Model Cities Program at the United States Department of Housing and Urban Development in the late 1960s and early 1970s.

Arnstein uses the metaphor of a ladder to describe gradations of "citizen participation" in urban programs that affect their lives. She makes clear her own personal commitment to a redistribution of power from haves to have-nots by empowering the poor and powerless. "A Ladder of Citizen Participation" has been reprinted more than eighty times and translated into five foreign languages.

At the lowest level of Arnstein's ladder are two forms of nonparticipation, which she terms *manipulation* and *therapy*. According to Arnstein, some governmental organizations have contrived phony forms of participation, which are really aimed at getting citizens to accept a predetermined course of action. While gullible citizens may think they are participating in decision making at these lowest levels of the ladder, Arnstein says they really are not. They are simply being used by decision makers. Almost at the bottom of the ladder is another form of nonparticipation, which Arnstein identifies as *therapy*. Arnstein brands this form of nonparticipation both dishonest and arrogant. Here the intent is to "cure" participants of attitudes and behaviors that local government officials do not like under the guise of seeking their advice.

Legitimate, but low, rungs of the ladder are *informing* and *consultation*. Informing citizens of the facts about a government program and their rights, responsibilities and options is a good first step, particularly if it is designed to go beyond a one-way flow of information. Consultation – getting citizens' opinions – is even better if the process is honest and citizens' opinions are really considered. Surveys, for example, may provide real input from citizens to decision makers, but if that is the only form of participation they would not go far in assuring that citizen views really carry weight. *Placation* – in which government gives in to some citizen demands – goes a step further. But a model in which government throws complaining citizens some crumbs to placate them is not really a satisfactory relationship.

The highest rungs on Arnstein's ladder are *partnership*, three rungs from the top, *delegated power*, one rung below the top, and *citizen control* at the very top of the ladder. During the "War on Poverty" in the 1960s, local government delegated power to run programs to some citizen groups or gave them full control over programs. Delegated power and citizens' control have been rare since that time. Opponents of citizen control advance many of the arguments that Arnstein identifies – that citizen control arguably balkanizes public services, may be costly and inefficient, can reward opportunistic citizen hustlers, and may be symbolic politics.

Today partnerships between public, private, and nonprofit organizations are popular. Arnstein places true partnerships relatively high on her eight-rung ladder. Partnerships represent a redistribution of power arrived at through negotiation. Where the odd bedfellows of local government, private corporations, and neighborhood nonprofit community-based organizations form joint planning and decision-making structures, citizen views can have real weight.

Both Sherry Arnstein and Paul Davidoff (p. 400) were engaged liberals who wrote their classic statements about citizen participation and advocacy planning in the late 1960s. Compare the approach of Davidoff, the lawyer who argues in favor of skilled professionals advocating on behalf of powerless clients, with the approach of Arnstein, the social work professional who favors empowering individuals and communities by involving them directly in planning and decision making.

Other books on citizen participation in urban planning and programs include James L. Breighton, *The Public Participation Handbook: Making Better Decisions Through Citizen Involvement* (San Francisco: Jossey-Bass, 2005), Thomas Ehrlich, *Public Policymaking in a Democratic Society: A Guide to Civic Engagement* (Armonk: M.E. Sharpe, 2002), Henry Sanoff, *Community Participation Methods in Design and Planning* (New York: Wiley, 1999), and John F. Forester, *The Deliberative Practitioner: Encouraging Participatory Planning Processes* (Cambridge: MIT Press, 1999).

Peter Marris and Martin Rein's classic *Dilemmas of Social Reform*, second edition (Chicago: University of Chicago Press, 1982) describes community-based urban programs and articulates a philosophy of social change that influenced US urban policy in the 1960s. Two very different views on the US "War on Poverty" are Sar Levitan, *The Great Society's Poor Law* (Baltimore: Johns Hopkins University Press, 1969), and Daniel Patrick Moynihan, *Maximum Feasible Misunderstanding* (New York: Free Press, 1969). The US Model Cities program, its antecedents, and the initial phase of the successor Community Development Block Grant program are discussed in Bernard J. Frieden and Marshal Kaplan, *The Politics of Neglect: Urban Aid from Model Cities to Revenue Sharing* (Cambridge: MIT Press, 1975).

Books on public participation in urban planning and programs in Europe include James Barlow, *Public Participation in Urban Development: The European Experience* (Washington, DC: Brookings, 1995), Her Majesty's Stationery Office, *Community Involvement in Planning and Development Processes* (London: HMSO, 1995), and Albert Mabileau, *Local Politics and Participation in Britain and France* (Cambridge: Cambridge University Press, 1990).

■ ■ ■ ■ ■ ■

The idea of citizen participation is a little like eating spinach: no one is against it in principle because it is good for you. Participation of the governed in their government is, in theory, the cornerstone of democracy – a revered idea that is vigorously applauded by virtually everyone. The applause is reduced to polite handclaps, however, when this principle is advocated by the have-not blacks, Mexican Americans, Puerto Ricans, Indians, Eskimos, and whites. And when the have-nots define participation as redistribution of power, the American consensus on the fundamental principle explodes into many shades of outright racial, ethnic, ideological, and political opposition.

There have been many recent speeches, articles, and books which explore in detail *who* are the have-nots of our time. There has been much recent documentation of *why* the have-nots have become so offended and embittered by their powerlessness to deal with the profound inequities and injustices pervading their daily lives. But there has been very little analysis of the content of the current controversial slogan: "citizen participation" or "maximum feasible participation." In short: *What* is citizen participation and what is its relationship to the social imperatives of our time?

Citizen participation is citizen power

Because the question has been a bone of political contention, most of the answers have been purposely buried in innocuous euphemisms like "self-help" or "citizen involvement." Still others have been embellished with misleading rhetoric like "absolute control" which is something no one – including the President of the United States – has or can have. Between understated euphemisms and exacerbated rhetoric, even scholars have found it difficult to follow the controversy. To the headline reading public, it is simply bewildering.

My answer to the critical *what* question is simply that citizen participation is a categorical term for citizen power. It is the redistribution of power that enables the have-not citizens, presently excluded from the political and economic processes, to be deliberately included in the future. It is the strategy by which the have-nots join in determining how information is shared, goals and policies are set, tax resources are allocated, programs are operated, and benefits like contracts and patronage are parceled out. In short, it is the means by which they can induce significant social reform which enables them to share in the benefits of the affluent society.

EMPTY REFUSAL VERSUS BENEFIT

There is a critical difference between going through the empty ritual of participation and having the real power needed to affect the outcome of the process. This difference is brilliantly capsulized in a poster painted last spring [1968] by the French students to explain the student-worker rebellion. (See Figure 1.) The poster highlights the fundamental point that participation without redistribution of power is an empty and frustrating process for the powerless. It allows the powerholders to claim that all sides were considered, but makes it possible for only some of those sides to benefit. It maintains the status quo. Essentially, it is what has been happening in most of the 1,000 Community Action Programs, and what promises to be repeated in the vast majority of the 150 Model Cities programs.

Types of participation and "nonparticipation"

A typology of eight *levels* of participation may help in analysis of this confused issue. For illustrative purposes the eight types are arranged in a ladder pattern with each rung corresponding to the extent of citizens' power in determining the end product. (See Figure 2.)

Figure 1 French student poster. In English, "I participate, you participate, he participates, we participate, you participate . . . they profit"

The bottom rungs of the ladder are (1) *Manipulation* and (2) *Therapy*. These two rungs describe levels of "nonparticipation" that have been contrived by some to substitute for genuine participation. Their real objective is not to enable people to participate in planning or conducting programs, but to enable powerholders to "educate" or "cure" the participants. Rungs 3 and 4 progress to levels of "tokenism" that allow the have-nots to hear and to have a voice: (3) *Informing* and (4) *Consultation*. When they are proffered by power-holders as the total extent of participation, citizens may indeed hear and be heard. But under these conditions they lack the power to insure that their views will be *heeded* by the powerful. When participation is restricted to these levels, there is no follow-through, no "muscle," hence no assurance of changing the status quo. Rung (5) *Placation* is simply a higher level tokenism because the groundrules allow have-nots to advise, but retain for the powerholders the continued right to decide.

Further up the ladder are levels of citizen power with increasing degrees of decision-making clout. Citizens can enter into a (6) *Partnership* that enables

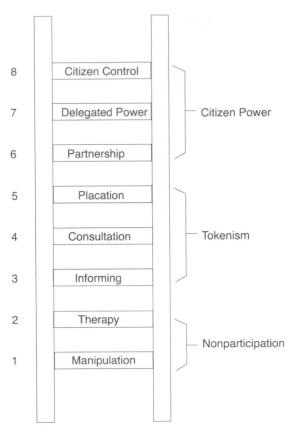

8	Citizen Control
7	Delegated Power
6	Partnership
5	Placation
4	Consultation
3	Informing
2	Therapy
1	Manipulation

Citizen Power (8, 7, 6)

Tokenism (5, 4, 3)

Nonparticipation (2, 1)

Figure 2 Eight rungs on the ladder of citizen participation

them to negotiate and engage in trade-offs with traditional power holders. At the topmost rungs, (7) *Delegated Power* and (8) *Citizen Control*, have-not citizens obtain the majority of decision-making seats, or full managerial power.

Obviously, the eight-rung ladder is a simplification, but it helps to illustrate the point that so many have missed – that there are significant gradations of citizen participation. Knowing these gradations makes it possible to cut through the hyperbole to understand the increasingly strident demands for participation from the have-nots as well as the gamut of confusing responses from the powerholders.

Though the typology uses examples from federal programs such as urban renewal, anti-poverty, and Model Cities, it could just as easily be illustrated in the church, currently facing demands for power from priests and laymen who seek to change its mission; colleges and universities which in some cases have become literal battlegrounds over the issue of student power; or public schools, city halls, and police depart-

ments (or big business which is likely to be next on the expanding list of targets). The underlying issues are essentially the same – "nobodies" in several arenas are trying to become "somebodies" with enough power to make the target institutions responsive to their views, aspirations, and needs.

LIMITATIONS OF THE TYPOLOGY

The ladder juxtaposes powerless citizens with the powerful in order to highlight the fundamental divisions between them. In actuality, neither the have-nots nor the powerholders are homogeneous blocs. Each group encompasses a host of divergent points of view, significant cleavages, competing vested interests, and splintered subgroups. The justification for using such simplistic abstractions is that in most cases the have-nots really do perceive the powerful as a monolithic "system," and powerholders actually do view the have-nots as a sea of "those people," with little comprehension of the class and caste differences among them.

It should be noted that the typology does not include an analysis of the most significant roadblocks to achieving genuine levels of participation. These roadblocks lie on both sides of the simplistic fence. On the powerholders' side, they include racism, paternalism, and resistance to power redistribution. On the have-nots' side, they include inadequacies of the poor community's political socioeconomic infrastructure and knowledge-base, plus difficulties of organizing a representative and accountable citizens' group in the face of futility, alienation, and distrust.

Another caution about the eight separate rungs on the ladder: In the real world of people and programs, there might be 150 rungs with less sharp and "pure" distinctions among them. Furthermore, some of the characteristics used to illustrate each of the eight types might be applicable to other rungs. For example, employment of the have-nots in a program or on a planning staff could occur at any of the eight rungs and could represent either a legitimate or illegitimate characteristic of citizen participation. Depending on their motives, powerholders can hire poor people to coopt them, to placate them, or to utilize the have-nots' special skills and insights. Some mayors, in private, actually boast of their strategy in hiring militant black leaders to muzzle them while destroying their credibility in the black community.

Characteristics and illustrations

It is in this context of power and powerlessness that the characteristics of the eight rungs are illustrated by examples from current federal social programs.

1. MANIPULATION

In the name of citizen participation, people are placed on rubberstamp advisory committees or advisory boards for the express purpose of "educating" them or engineering their support. Instead of genuine citizen participation, the bottom rung of the ladder signifies the distortion of participation into a public relations vehicle by powerholders.

This illusory form of "participation" initially came into vogue with urban renewal when the socially elite were invited by city housing officials to serve on Citizen Advisory Committees (CACs). Another target of manipulation were the CAC subcommittees on minority groups, which in theory were to protect the rights of Negroes in the renewal program. In practice, these subcommittees, like their parent CACs, functioned mostly as letterheads, trotted forward at appropriate times to promote urban renewal plans (in recent years known as Negro removal plans).

At meetings of the Citizen *Advisory* Committees, it was the officials who educated, persuaded, and advised the citizens, not the reverse. Federal guidelines for the renewal programs legitimized the manipulative agenda by emphasizing the terms "information-gathering," public relations," and "support" as the explicit functions of the committees.

This style of nonparticipation has since been applied to other programs encompassing the poor. Examples of this are seen in Community Action Agencies (CAAs) which have created structures called "neighborhood councils" or "neighborhood advisory groups." These bodies frequently have no legitimate function or power. The CAAs use them to "prove" that "grass-roots people" are involved in the program. But the program may not have been discussed with "the people." Or it may have been described at a meeting in the most general terms; "We need your signatures on this proposal for a multiservice center which will house, under one roof, doctors from the health department, workers from the welfare department, and specialists from the employment service."

The signatories are not informed that the $2 million-per-year center will only refer residents to the same old waiting lines at the same old agencies across town. No one is asked if such a referral center is really needed in his neighborhood. No one realizes that the contractor for the building is the mayor's brother-in-law, or that the new director of the center will be the same old community organization specialist from the urban renewal agency.

After signing their names, the proud grassrooters dutifully spread the word that they have "participated" in bringing a new and wonderful center to the neighborhood to provide people with drastically needed jobs and health and welfare services. Only after the ribbon-cutting ceremony do the members of the neighborhood council realize that they didn't ask the important questions, and that they had no technical advisors of their own to help them grasp the fine legal print. The new center, which is open 9 to 5 on weekdays only, actually adds to their problems. Now the old agencies across town won't talk with them unless they have a pink paper slip to prove that they have been referred by "their" shiny new neighborhood center.

Unfortunately, this chicanery is not a unique example. Instead it is almost typical of what has been perpetrated in the name of high-sounding rhetoric like "grassroots participation." This sham lies at the heart of the deep-seated exasperation and hostility of the have-nots toward the powerholders.

One hopeful note is that, having been so grossly affronted, some citizens have learned the Mickey Mouse game, and now they too know how to play. As a result of this knowledge, they are demanding genuine levels of participation to assure them that public programs are relevant to their needs and responsive to their priorities.

2. THERAPY

In some respects group therapy, masked as citizen participation, should be on the lowest rung of the ladder because it is both dishonest and arrogant. Its administrators – mental health experts from social workers to psychiatrists – assume that powerlessness is synonymous with mental illness. On this assumption, under a masquerade of involving citizens in planning, the experts subject the citizens to clinical group therapy. What makes this form of "participation" so invidious is that citizens are engaged in extensive activity, but the focus of it is on curing them of their "pathology" rather than changing the racism and victimization that create their "pathologies."

Consider an incident that occurred in Pennsylvania less than one year ago. When a father took his seriously ill baby to the emergency clinic of a local hospital, a young resident physician on duty instructed him to take the baby home and feed it sugar water. The baby died that afternoon of pneumonia and dehydration. The overwrought father complained to the board of the local Community Action Agency. Instead of launching an investigation of the hospital to determine what changes would prevent similar deaths or other forms of malpractice, the board invited the father to attend the CAA's (therapy) child-care sessions for parents, and promised him that someone would "telephone the hospital director to see that it never happens again."

Less dramatic, but more common examples of therapy, masquerading as citizen participation, may be seen in public housing programs where tenant groups are used as vehicles for promoting control-your-child or cleanup campaigns. The tenants are brought together to help them "adjust their values and attitudes to those of the larger society." Under these ground rules, they are diverted from dealing with such important matters as: arbitrary evictions; segregation of the housing project; or why there is a three-month time lapse to get a broken window replaced in winter.

The complexity of the concept of mental illness in our time can be seen in the experiences of student/civil rights workers facing guns, whips, and other forms of terror in the South. They needed the help of socially attuned psychiatrists to deal with their fears and to avoid paranoia.

3. INFORMING

Informing citizens of their rights, responsibilities, and options can be the most important first step toward legitimate citizen participation. However, too frequently the emphasis is placed on a one-way flow of information – from officials to citizens – with no channel provided for feedback and no power for negotiation. Under these conditions, particularly when information is provided at a late stage in planning, people have little opportunity to influence the program designed "for their benefit." The most frequent tools used for such one-way communication are the news media, pamphlets, posters, and responses to inquiries.

Meetings can also be turned into vehicles for one-way communication by the simple device of providing superficial information, discouraging questions, or giving irrelevant answers. At a recent Model Cities citizen planning meeting in Providence, Rhode Island, the topic was "tot-lots." A group of elected citizen representatives, almost all of whom were attending three to five meetings a week, devoted an hour to a discussion of the placement of six tot-lots. The neighborhood is half black, half white. Several of the black representatives noted that four tot-lots were proposed for the white district and only two for the black. The city official responded with a lengthy, highly technical explanation about costs per square foot and available property. It was clear that most of the residents did not understand his explanation. And it was clear to observers from the Office of Economic Opportunity that other options did exist which, considering available funds, would have brought about a more equitable distribution of facilities. Intimidated by futility, legalistic jargon, and prestige of the official, the citizens accepted the "information" and endorsed the agency's proposal to place four lots in the white neighborhood.

4. CONSULTATION

Inviting citizens' opinions, like informing them, can be a legitimate step toward their full participation. But if consulting them is not combined with other modes of participation, this rung of the ladder is still a sham since it offers no assurance that citizen concerns and ideas will be taken into account. The most frequent methods used for consulting people are attitude surveys, neighborhood meetings, and public hearings.

When powerholders restrict the input of citizens' ideas solely to this level, participation remains just a window-dressing ritual. People are primarily perceived as statistical abstractions, and participation is measured by how many come to meetings, take brochures home, or answer a questionnaire. What citizens achieve in all this activity is that they have "participated in participation." And what powerholders achieve is the evidence that they have gone through the required motions of involving "those people."

Attitude surveys have become a particular bone of contention in ghetto neighborhoods. Residents are increasingly unhappy about the number of times per week they are surveyed about their problems and hopes. As one woman put it: "Nothing ever happens with those damned questions, except the surveyor gets

$3 an hour, and my washing doesn't get done that day." In some communities, residents are so annoyed that they are demanding a fee for research interviews.

Attitude surveys are not very valid indicators of community opinion when used without other input from citizens. Survey after survey (paid for out of anti-poverty funds) has "documented" that poor housewives most want tot-lots in their neighborhood where young children can play safely. But most of the women answered these questionnaires without knowing what their options were. They assumed that if they asked for something small, they might just get something useful in the neighborhood. Had the mothers known that a free prepaid health insurance plan was a possible option, they might not have put tot-lots so high on their wish lists.

A classic misuse of the consultation rung occurred at a New Haven, Connecticut, community meeting held to consult citizens on a proposed Model Cities grant. James V. Cunningham, in an unpublished report to the Ford Foundation, described the crowd as large and mostly hostile:

> Members of The Hill Parents Association demanded to know why residents had not participated in drawing up the proposal. CAA director Spitz explained that it was merely a proposal for seeking Federal planning funds – that once funds were obtained, residents would be deeply involved in the planning. An outside observer who sat in the audience described the meeting this way:
>
> "Spitz and Mel Adams ran the meeting on their own. No representatives of a Hill group moderated or even sat on the stage. Spitz told the 300 residents that this huge meeting was an example of 'participation in planning.' To prove this, since there was a lot of dissatisfaction in the audience, he called for a 'vote' on each component of the proposal. The vote took this form: 'Can I see the hands of all those in favor of a health clinic? All those opposed?' It was a little like asking who favors motherhood."

It was a combination of the deep suspicion aroused at this meeting and a long history of similar forms of "window-dressing participation" that led New Haven residents to demand control of the program.

By way of contrast, it is useful to look at Denver where technicians learned that even the best intentioned among them are often unfamiliar with, and even insensitive to, the problems and aspirations of the poor. The technical director of the Model Cities program has described the way professional planners assumed that the residents, victimized by high-priced local storekeepers, "badly needed consumer education." The residents, on the other hand, pointed out that the local storekeepers performed a valuable function. Although they overcharged, they also gave credit, offered advice, and frequently were the only neighborhood place to cash welfare or salary checks. As a result of this consultation, technicians and residents agreed to substitute the creation of needed credit institutions in the neighborhood for a consumer education program.

5. PLACATION

It is at this level that citizens begin to have some degree of influence though tokenism is still apparent. An example of placation strategy is to place a few hand-picked "worthy" poor on boards of Community Action Agencies or on public bodies like the board of education, police commission, or housing authority. If they are not accountable to a constituency in the community and if the traditional power elite hold the majority of seats, the have-nots can be easily outvoted and outfoxed. Another example is the Model Cities advisory and planning committees. They allow citizens to advise or plan ad infinitum but retain for powerholders the right to judge the legitimacy or feasibility of the advice. The degree to which citizens are actually placated, of course, depends largely on two factors: the quality of technical assistance they have in articulating their priorities; and the extent to which the community has been organized to press for those priorities.

It is not surprising that the level of citizen participation in the vast majority of Model Cities programs is at the placation rung of the ladder or below. Policymakers at the Department of Housing and Urban Development (HUD) were determined to return the genie of citizen power to the bottle from which it had escaped (in a few cities) as a result of the provision stipulating "maximum feasible participation" in poverty programs. Therefore, HUD channeled its physical-social-economic rejuvenation approach for blighted neighborhoods through city hall. It drafted legislation requiring that all Model Cities' money flow to a local City Demonstration Agency (CDA) through the elected city council. As enacted by Congress, this gave local city councils final veto power over planning and programming and ruled out any direct funding relationship between community groups and HUD.

HUD required the CDAs to create coalition, policy-making boards that would include necessary local powerholders to create a comprehensive physical-social plan during the first year. The plan was to be carried out in a subsequent five-year action phase. HUD, unlike OEO, did not require that have-not citizens be included on the CDA decision-making boards. HUD's Performance Standards for Citizen Participation only demanded that "citizens have clear and direct access to the decision-making process."

Accordingly, the CDAs structured their policy-making boards to include some combination of elected officials; school representatives; housing, health, and welfare officials; employment and police department representatives; and various civic, labor, and business leaders. Some CDAs included citizens from the neighborhood. Many mayors correctly interpreted the HUD provision for "access to the decision-making process" as the escape hatch they sought to relegate citizens to the traditional advisory role.

Most CDAs created residents' advisory committees. An alarmingly significant number created citizens' policy boards and citizens' policy committees which are totally misnamed as they have either no policy-making function or only a very limited authority. Almost every CDA created about a dozen planning committees or task forces on functional lines: health, welfare, education, housing, and unemployment. In most cases, have-not citizens were invited to serve on these committees along with technicians from relevant public agencies. Some CDAs, on the other hand, structured planning committees of technicians and parallel committees of citizens.

In most Model Cities programs, endless time has been spent fashioning complicated board, committee, and task force structures for the planning year. But the rights and responsibilities of the various elements of those structures are not defined and are ambiguous. Such ambiguity is likely to cause considerable conflict at the end of the one-year planning process. For at this point, citizens may realize that they have once again extensively "participated" but have not profited beyond the extent the powerholders decide to placate them.

Results of a staff study (conducted in the summer of 1968 before the second round of seventy-five planning grants were awarded) were released in a December 1968 HUD bulletin. Though this public document uses much more delicate and diplomatic language, it attests to the already cited criticisms of non-policy-making policy boards and ambiguous complicated structures, in addition to the following findings:

1. Most CDAs did not negotiate citizen participation requirements with residents.
2. Citizens, drawing on past negative experiences with local powerholders, were extremely suspicious of this new panacea program. They were legitimately distrustful of city hall's motives.
3. Most CDAs were not working with citizens' groups that were genuinely representative of model neighborhoods and accountable to neighborhood constituencies. As in so many of the poverty programs, those who were involved were more representative of the upwardly mobile working-class. Thus their acquiescence to plans prepared by city agencies was not likely to reflect the views of the unemployed, the young, the more militant residents, and the hard-core poor.
4. Residents who were participating in as many as three to five meetings per week were unaware of their minimum rights, responsibilities, and the options available to them under the program. For example, they did not realize that they were not required to accept technical help from city technicians they distrusted.
5. Most of the technical assistance provided by CDAs and city agencies was of third-rate quality, paternalistic, and condescending. Agency technicians did not suggest innovative options. They reacted bureaucratically when the residents pressed for innovative approaches. The vested interests of the old-line city agencies were a major – albeit hidden – agenda.
6. Most CDAs were not engaged in planning that was comprehensive enough to expose and deal with the roots of urban decay. They engaged in "meetingitis" and were supporting strategies that resulted in "projectitis," the outcome of which was a "laundry list" of traditional programs to be conducted by traditional agencies in the traditional manner under which slums emerged in the first place.
7. Residents were not getting enough information from CDAs to enable them to review CDA developed plans or to initiate plans of their own as required by HUD. At best, they were getting superficial information. At worst, they were not even getting copies of official HUD materials.

8. Most residents were unaware of their rights to be reimbursed for expenses incurred because of participation – babysitting, transportation costs, and so on. The training of residents, which would enable them to understand the labyrinth of the federal–state–city systems and networks of sub-systems, was an item that most CDAs did not even consider.

These findings led to a new public interpretation of HUD's approach to citizen participation. Though the requirements for the seventy-five "second-round" Model City grantees were not changed, HUD's twenty-seven-page technical bulletin on citizen participa-tion repeatedly advocated that cities share power with residents. It also urged CDAs to experiment with subcontracts under which the residents' groups could hire their own trusted technicians.

A more recent evaluation was circulated in February 1969 by OSTI, a private firm that entered into a contract with OEO to provide technical assis-tance and training to citizens involved in Model Cities programs in the north-east region of the country. OSTI's report to OEO corroborates the earlier study. In addition it states:

> In practically no Model Cities structure does citizen participation mean truly shared decision-making, such that citizens might view themselves as "the partners in this program . . ."
>
> In general, citizens are finding it impossible to have a significant impact on the comprehensive planning which is going on. In most cases the staff planners of the CDA and the planners of existing agencies are carrying out the actual planning with citizens having a peripheral role of watchdog and, ultimately, the "rubber stamp" of the plan generated. In cases where citizens have the direct responsibility for generating program plans, the time period allowed and the independent technical resources being made available to them are not adequate to allow them to do anything more than generate very traditional approaches to the problems they are attempting to solve.
>
> In general, little or no thought has been given to the means of insuring continued citizen partici-pation during the stage of implementation. In most cases, traditional agencies are envisaged as the implementors of Model Cities programs and few mechanisms have been developed for encouraging

organizational change or change in the method of program delivery within these agencies or for insuring that citizens will have some influence over these agencies as they implement Model Cities programs . . . By and large, people are once again being planned *for*. In most situations the major planning decisions are being made by CDA staff and approved in a formalistic way by policy boards.

6. PARTNERSHIP

At this rung of the ladder, power is in fact redistributed through negotiation between citizens and power-holders. They agree to share planning and decision-making responsibilities through such structures as joint policy boards, planning committees, and mechanisms for resolving impasses. After the groundrules have been established through some form of give-and-take, they are not subject to unilateral change.

Partnership can work most effectively when there is an organized power-base in the community to which the citizen leaders are accountable; when the citizens' group has the financial resources to pay its leaders reasonable honoraria for their time-consuming efforts; and when the group has the resources to hire (and fire) its own technicians, lawyers, and community organizers. With these ingredients, citizens have some genuine bargaining influence over the outcome of the plan (as long as both parties find it useful to maintain the partnership). One community leader described it "like coming to city hall with hat on head instead of in hand."

In the Model Cities program only about fifteen of the so-called first generation of seventy-five cities have reached some significant degree of power-sharing with residents. In all but one of those cities, it was angry citizen demands, rather than city initiative, that led to the negotiated sharing of power. The negotiations were triggered by citizens who had been enraged by previous forms of alleged participation. They were both angry and sophisticated enough to refuse to be "conned" again. They threatened to oppose the awarding of a planning grant to the city. They sent delegations to HUD in Washington. They used abrasive language. Negotiation took place under a cloud of suspicion and rancor.

In most cases where power has come to be shared it was *taken by the citizens*, not given by the city. There is nothing new about that process. Since those who have power normally want to hang onto it, historically it has had to be wrested by the powerless rather than proffered by the powerful.

Such a working partnership was negotiated by the residents in the Philadelphia model neighborhood. Like most applicants for a Model Cities grant, Philadelphia wrote its more than 400-page application and waved it at a hastily called meeting of community leaders. When those present were asked for an endorsement, they angrily protested the city's failure to consult them on preparation of the extensive application. A community spokesman threatened to mobilize a neighborhood protest *against* the application unless the city agreed to give the citizens a couple of weeks to review the application and recommend changes. The officials agreed.

At their next meeting, citizens handed the city officials a substitute citizen participation section that changed the groundrules from a weak citizens' advisory role to a strong shared power agreement. Philadelphia's application to HUD included the citizens' substitution word for word. (It also included a new citizen prepared introductory chapter that changed the city's description of the model neighborhood from a paternalistic description of problems to a realistic analysis of its strengths, weaknesses, and potentials.) Consequently, the proposed policy-making committee of the Philadelphia CDA was revamped to give five out of eleven seats to the residents' organization, which is called the Area Wide Council (AWC). The AWC obtained a subcontract from the CDA for more than $20,000 per month, which it used to maintain the neighborhood organization, to pay citizen leaders $7 per meeting for their planning services, and to pay the salaries of a staff of community organizers, planners, and other technicians. AWC has the power to initiate plans of its own, to engage in joint planning with CDA committees, and to review plans initiated by city agencies. It has a veto power in that no plans may be submitted by the CDA to the city council until they have been reviewed, and any differences of opinion have been successfully negotiated with the AWC. Representatives of the AWC (which is a federation of neighborhood organizations grouped into sixteen neighborhood "hubs") may attend all meetings of CDA task forces, planning committees, or sub-committees.

Though the city council has final veto power over the plan (by federal law), the AWC believes it has a neighborhood constituency that is strong enough to negotiate any eleventh-hour objections the city council might raise when it considers such AWC proposed innovations as an AWC Land Bank, an AWC Economic Development Corporation, and an experimental income maintenance program for 900 poor families.

7. DELEGATED POWER

Negotiations between citizens and public officials can also result in citizens achieving dominant decision-making authority over a particular plan or program. Model City policy boards or CAA delegate agencies on which citizens have a clear majority of seats and genuine specified powers are typical examples. At this level, the ladder has been scaled to the point where citizens hold the significant cards to assure accountability of the program to them. To resolve differences, powerholders need to start the bargaining process rather than respond to pressure from the other end.

Such a dominant decision-making role has been attained by residents in a handful of Model Cities including Cambridge, Massachusetts; Dayton and Columbus, Ohio; Minneapolis, Minnesota; St. Louis, Missouri; Hartford and New Haven, Connecticut; and Oakland, California.

In New Haven, residents of the Hill neighborhood have created a corporation that has been delegated the power to prepare the entire Model Cities plan. The city, which received a $117,000 planning grant from HUD, has subcontracted $110,000 of it to the neighborhood corporation to hire its own planning staff and consultants. The Hill Neighborhood Corporation has eleven representatives on the twenty-one-member CDA board which assures it a majority voice when its proposed plan is reviewed by the CDA.

Another model of delegated power is separate and parallel groups of citizens and powerholders, with provision for citizen veto if differences of opinion cannot be resolved through negotiation. This is a particularly interesting coexistence model for hostile citizen groups too embittered toward city hall – as a result of past "collaborative efforts" – to engage in joint planning.

Since all Model Cities programs require approval by the city council before HUD will fund them, city councils have final veto powers even when citizens have the majority of seats on the CDA Board. In

Richmond, California, the city council agreed to a citizens' counter-veto, but the details of that agreement are ambiguous and have not been tested.

Various delegated power arrangements are also emerging in the Community Action Program as a result of demands from the neighborhoods and OEO's most recent instruction guidelines which urged CAAs "to exceed (the) basic requirements" for resident participation. In some cities, CAAs have issued subcontracts to resident dominated groups to plan and/or operate one or more decentralized neighborhood program components like a multipurpose service center or a Headstart program. These contracts usually include an agreed upon line-by-line budget and program specifications. They also usually include a specific statement of the significant powers that have been delegated, for example: policy-making; hiring and firing; issuing subcontracts for building, buying, or leasing. (Some of the subcontracts are so broad that they verge on models for citizen control.)

8. CITIZEN CONTROL

Demands for community controlled schools, black control, and neighborhood control are on the increase. Though no one in the nation has absolute control, it is very important that the rhetoric not be confused with intent. People are simply demanding that degree of power (or control) which guarantees that participants or residents can govern a program or an institution, be in full charge of policy and managerial aspects, and be able to negotiate the conditions under which "outsiders" may change them.

A neighborhood corporation with no intermediaries between it and the source of funds is the model most frequently advocated. A small number of such experimental corporations are already producing goods and/or social services. Several others are reportedly in the development stage, and new models for control will undoubtedly emerge as the have-nots continue to press for greater degrees of power over their lives.

Though the bitter struggle for community control of the Ocean Hill-Brownsville schools in New York City has aroused great fears in the headline reading public, less publicized experiments are demonstrating that the have-nots can indeed improve their lot by handling the entire job of planning, policy-making, and managing a program. Some are even demonstrating that they can do all this with just one arm because

they are forced to use their other one to deal with a continuing barrage of local opposition triggered by the announcement that a federal grant has been given to a community group or an all black group.

Most of these experimental programs have been capitalized with research and demonstration funds from the Office of Economic Opportunity in cooperation with other federal agencies. Examples include:

1. A $1.8 million grant was awarded to the Hough Area Development Corporation in Cleveland to plan economic development programs in the ghetto and to develop a series of economic enterprises ranging from a novel combination shopping-center-public-housing project to a loan guarantee program for local building contractors. The membership and board of the nonprofit corporation is composed of leaders of major community organizations in the black neighborhood.

2. Approximately $1 million ($595,751 for the second year) was awarded to the Southwest Alabama Farmers' Cooperative Association (SWAFCA) in Selma, Alabama, for a ten-county marketing cooperative for food and livestock. Despite local attempts to intimidate the coop (which included the use of force to stop trucks on the way to market) first year membership grew to 1,150 farmers who earned $52,000 on the sale of their new crops. The elected coop board is composed of two poor black farmers from each of the ten economically depressed counties.

3. Approximately $600,000 ($300,000 in a supplemental grant) was granted to the Albina Corporation and the Albina Investment Trust to create a black-operated, black-owned manufacturing concern using inexperienced management and unskilled minority group personnel from the Albina district. The profitmaking wool and metal fabrication plant will be owned by its employees through a deferred compensation trust plan.

4. Approximately $800,000 ($400,000 for the second year) was awarded to the Harlem Commonwealth Council to demonstrate that a community-based development corporation can catalyze and implement an economic development program with broad community support and participation. After only eighteen months of program development and negotiation, the council will soon launch several large-scale ventures including operation of two

supermarkets, an auto service and repair center (with built-in manpower training program), a finance company for families earning less than $4,000 per year, and a data processing company. The all black Harlem-based board is already managing a metal castings foundry.

Though several citizen groups (and their mayors) use the rhetoric of citizen control, no Model City can meet the criteria of citizen control since final approval power and accountability rest with the city council.

Daniel P. Moynihan argues that city councils are representative of the community, but Adam Walinsky illustrates the nonrepresentativeness of this kind of representation:

> Who . . . exercises "control" through the representative process? In the Bedford-Stuyvesant ghetto of New York there are 450,000 people – as many as in the entire city of Cincinnati, more than in the entire state of Vermont. Yet the area has only one high school, and 80 per cent of its teenagers are dropouts; the infant mortality rate is twice the national average; there are over 8000 buildings abandoned by everyone but the rats, yet the area received not one dollar of urban renewal funds during the entire first 15 years of that program's operation; the unemployment rate is known only to God.
>
> Clearly, Bedford-Stuyvesant has some special needs; yet it has always been lost in the midst of the city's eight million. In fact, it took a lawsuit to win for this vast area, in the year 1968, its first Congressman. In what sense can the representative system be said to have "spoken for" this community, during the long years of neglect and decay?

Walinsky's point on Bedford-Stuyvesant has general applicability to the ghettos from coast to coast. It is therefore likely that in those ghettos where residents have achieved a significant degree of power in the Model Cities planning process, the first-year action plans will call for the creation of some new community institutions entirely governed by residents with a specified sum of money contracted to them. If the groundrules for these programs are clear and if citizens understand that achieving a genuine place in the pluralistic scene subjects them to its legitimate forms of give-and-take, then these kinds of programs might begin to demonstrate how to counteract the various corrosive political and socioeconomic forces that plague the poor.

In cities likely to become predominantly black through population growth, it is unlikely that strident citizens' groups like AWC of Philadelphia will eventually demand legal power for neighborhood self-government. Their grand design is more likely to call for a black city achieved by the elective process. In cities destined to remain predominantly white for the foreseeable future, it is quite likely that counterpart groups to AWC will press for separatist forms of neighborhood government that can create and control decentralized public services such as police protection, education systems, and health facilities. Much may depend on the willingness of city governments to entertain demands for resource allocation weighted in favor of the poor, reversing gross imbalances of the past.

Among the arguments against community control are: it supports separatism; it creates balkanization of public services; it is more costly and less efficient; it enables minority group "hustlers" to be just as opportunistic and disdainful of the have-nots as their white predecessors; it is incompatible with merit systems and professionalism; and ironically enough, it can turn out to be a new Mickey Mouse game for the have-nots by allowing them to gain control but not allowing them sufficient dollar resources to succeed. These arguments are not to be taken lightly. But neither can we take lightly the arguments of embittered advocates of community control – that every other means of trying to end their victimization has failed!

"The Need for a New Vision for the Development of Large U.S. Metropolitan Areas"

Saloman Brothers (1989)

Anthony Downs

Editors' Introduction

In the United States and throughout the world large, interconnected metropolitan regions have arisen. Often dozens of independent local governments exist in a single metropolitan region. Urban development in large metropolitan regions is often driven by the limited and self-serving plans of individual developers or the narrow self-interest of individual local governments without much concern for their neighbors or the public welfare. In a few rare instances regional plans with a clear vision have been prepared and implemented. Portland, Oregon, for example, has an elected metropolitan government that is pursuing a regional vision to develop multiple city centers surrounded by open space. But in most cities visions of a desirable metropolitan future are sadly lacking. In addition to Anthony Downs, new regionalist urban planners and architects such as Peter Calthorpe and William Fulton (p. 342) and Andrés Duany and Elizabeth Plater-Zyberk (p. 192) have worked out principles for regional plans that would address the problems Downs identifies.

In "The Need for a New Vision for the Development of Large U.S. Metropolitan Areas," Downs proposes a radical overhaul of metropolitan planning that would abandon the prevailing vision of single-family homes, an over-reliance on the automobile, and a governance structure that separates the needs of inner cities and suburban communities. In place of this outmoded vision, Downs argues for a more integrated system that provides sub-urban homes and apartments for low-wage workers and creates "governance structures that preserve substantial local authority – but within a framework that compels local governments to act responsibly to meet area-wide needs."

Downs is a senior fellow at the Brookings Institution, a Washington, DC, think-tank. He has written extensively over the past three decades on poverty, race, housing, urban sprawl, traffic congestion, metropolitan planning, and other urban issues. He often takes the logic of strategies that have been suggested or tried on a modest scale to their conceptual limit: What would it *really* take to completely disperse the population of every poor Black ghetto in America? How much would it really cost to completely eliminate traffic congestion? How much housing at what cost would really be required to provide every American household a decent, safe, and sanitary affordable housing unit?

Often, as in this selection, Downs concludes that strategies to achieve broad social goals would require massive government action and significantly increased levels of funding. This emphasis on government programs and spending differentiates Downs from conservatives like Michael Porter (p. 274) who oppose government spending to solve urban problems. But Downs is no bleeding-heart liberal. He advocates triage strategies for urban neighborhoods – leaving the worst neighborhoods to deteriorate while directing scarce resources to poor but salvageable communities. This puts him at odds with many liberals.

The selection reprinted here argues strongly for a sweeping planning strategy that eliminates the legal/governmental distinctions between impacted inner-city ghettos and comfortable suburban communities. It argues for integrated metropolitan solutions to traffic congestion and other problems that cannot be solved on a city-by-city basis. Originally published by the Saloman Brothers brokerage house, "The Need for a New Vision for the Development of Large U.S. Metropolitan Areas" is a visionary, yet hard-headed, approach to solving what has long been an intractable problem and one that threatens the stability of any urban future in a diverse society separated by race and class divisions.

A more extended discussion of the ideas advanced in this selection is: *New Visions for Metropolitan America* (Washington, DC and Cambridge: Brookings Institution and Lincoln Institute of Land Policy, 1992).

Other recent books by Anthony Downs include, *Sprawl Costs: Economic Impacts of Unchecked Development* (Washington, DC: Island Press, 2005), with Robert Burchell, Sahan Mukherji, and Barbara McCann; *Still Stuck in Traffic* (Washington, DC: Brookings Institution, 2004), and an edited volume: *Growth Management and Affordable Housing: Do They Conflict?* (Washington, DC: Brookings Institution, 2004). Particularly notable among Downs's extensive earlier writings are: *Urban Problems and Prospects* (Chicago: Markham, 1970), *Opening Up the Suburbs: An Urban Strategy for America* (New Haven: Yale University Press, 1973), *Inside Bureaucracy* (Boston: Little Brown, 1967), and *An Economic Theory of Democracy* (New York: Harper, 1957).

Many of Downs's essays on urban issues are reprinted in *The Selected Essays of Anthony Downs, vol. 1: Political Theory and Public Choice*, and *vol. 2: Urban Affairs and Urban Policy* (Cheltenham and Northampton: Edward Elgar, 1998).

For more on regional planning solutions to metropolitan problems see Eugenie Birch (ed.), *The City and Regional Planning Reader* (London and New York: Routledge, 2006).

INTRODUCTION

The very success of the long-dominant American ideal vision of how our metropolitan areas should develop now threatens its continuance, as that vision contains several major inner inconsistencies. This paradox is manifest in the rising protest across the nation against worsening suburban traffic congestion and declines in urban environmental quality. Up to now, these protests have been concentrated in our largest and most prosperous metropolitan areas, where the inherent inconsistencies of our dominant ideal vision have become most apparent. But some of that vision's failings will soon be visible in many more metropolitan areas.

Therefore, American society needs a new ideal vision to guide the future development of its large metropolitan areas. For us to formulate and accept such a new vision would be almost equivalent to a paradigm shift concerning urban development. Formulating such a new vision is vitally important to all participants in real estate in large metropolitan areas, including commercial developers, institutional investors, homebuilders, realtors, local officials, and homeowners. This article examines the nature and failings of the still-dominant ideal vision, and proposes an approach to a new alternative.

THE CURRENTLY DOMINANT IDEAL VISION OF DESIRABLE METROPOLITAN DEVELOPMENT

For the past few decades, one major vision about how US metropolitan areas ought to be developed has become totally dominant. This ideal vision reflects the views of the vast majority of American households – especially the more than 45% who live in the suburban portions of US metropolitan areas. The same vision is held by almost all suburban government officials. Hence, it has immensely influenced nearly all their policies concerning land use, transportation, etc.

This dominant ideal vision is built upon four pillars. Each is a key desire or aspiration shared by nearly all American households:

The first pillar is ownership of detached, single-family homes on spacious lots. Repeated polls show that over 90% of all American households would like to own their own homes, and the vast majority want single-family detached units. Hence, ownership of such a unit has become the heart of "the American dream" – the prevailing image of how a household "makes it" in contemporary society. Realization of this aspiration implies the dominance of very low-density settlement patterns.

The second pillar is ownership and use of a personal, private automotive vehicle. Every American wants to be able to leap into his or her own car and zoom off on an uncongested road, to wherever he or she wants to go, in total privacy and great comfort – and to arrive there in not more than 20 minutes. This factor has rapidly escalated total vehicle ownership and use. In 1983, over 87% of all U.S. households owned at least one car or truck, and 53% owned two or more such vehicles. Even among households with 1983 incomes under $10,000, over 60% owned at least one car or truck; among those with incomes of $40,000 and over, 98.9% owned at least one vehicle and 86.7% owned two or more. In 1987, there were 162 million licensed drivers in the United States, but 167 million cars and trucks in use.

During the five years 1983–1987, the United States added 9.2 million persons to its total human population, but more than twice as many cars and trucks – 20.1 million – to its total number of vehicles in use.

The third pillar of the dominant ideal vision involves the structure of suburban workplaces. They are visualized as consisting predominantly of low-rise office or industrial buildings or shopping centers, in attractively landscaped, park-like settings. Each such structure ought to be surrounded by a large supply of its own free parking so that those who work in it or patronize it can conveniently drive and park there without cost.

The fourth pillar for this ideal vision concerns governance. Most Americans want to live in small communities with strong local self-governments. They want those governments to control land use, public schools, and other key elements affecting what they perceive as the quality of neighborhood life. This institutional structure permits existing residents to have a strong voice in controlling their local environments.

The above four elements define the prevailing ideal vision of "the American dream" as it is conceived of by the vast majority of suburbanites and by many city-dwellers. All four elements express what might be termed unconstrained individualism. They represent the pursuit of an environment that maximizes one's own well being, without regard to the collective results of such behavior. This ideal vision has been reinforced over the past few decades by promotional efforts made by parts of the real estate industry and by suburban communities. Such promoters include homebuilders selling new dwellings, realtors reselling them, advertisers highlighting suburban lifestyles, politicians running for election, and planning officials trying to carry out the policies of local politicians and voters. The dominant ideal vision has become so strongly entrenched that it has become almost political suicide to challenge openly any of these four pillars of "the American dream."

THE FIRST FLAW WITHIN THE IDEAL VISION: EXCESSIVE TRAVEL

Unfortunately, the prevailing ideal vision of how metropolitan areas ought to be developed contains four major flaws. They involve basic inconsistencies between its key elements and the real preferences of the citizenry concerning how suburban life should be lived.

The first flaw is that the low-density settlement patterns required by single-family housing and low-density workplaces generate immense travel requirements. Low density inherently spreads both homes and jobs widely across the landscape. That forces people to travel long distances between where they live and where they work, shop, or play. Moreover, low-density settlement patterns cannot efficiently support mass transit; hence they can only be served by private automotive vehicles. Mass transit, whether buses or fixed-rail, can be effective only if at least one end of most journeys is relatively concentrated in a few points.

The resulting requirement for massive daily use of private vehicles is consistent with the second element of the dominant vision but has several negative consequences. The most obvious is traffic congestion. The more fully Americans achieve their ideal vision of widespread low-density homeownership and private vehicle ownership, the more congested their commuting journeys become. In addition, these heavy travel requirements cause severe air pollution, costly demands for road and other infrastructure construction, and high-level consumption of energy.

Most people who believe in this ideal vision have not yet realized that their success in achieving it is causing these adverse outcomes. Moreover, they are reluctant to come to this conclusion. Doing so would force them to admit that their own ideals and behavior are the main causes of their principal problems.

Consequently, most suburbanites suffering from excessive traffic congestion, air pollution, and other results of low-density settlement patterns blame both real estate developers and the latest-arriving residents in their suburban communities and adopt antigrowth or growth-limiting policies. But these policies do not attack the fundamental causes of the problems concerned; hence, they cannot solve such problems.

THE SECOND FLAW: NO HOUSING PROVISION FOR LOW-WAGE WORKERS

The second flaw in the dominant ideal vision is that it contains only relatively high-cost housing, providing no residences where low- and moderate-income households can afford to live. Yet such households are an integral part of American life, because they provide workers for relatively low-wage jobs. Those jobs are vital to the efficient operation of every community, including the wealthiest exurban enclaves. Low-wage workers are essential to many service firms, including but not limited to fast-food establishments, gas stations, laundries, dry cleaners, hospitals, retail stores, shopping centers, construction firms and subcontractors, and gardening and lawn-care firms. And moderate-wage workers are often the backbone of local government services, including police, firefighters, teachers, and administrative personnel.

These households cannot afford to live in the spacious, single-family homes that populate the ideal vision. In fact, many cannot afford to live in any type of single-family home. Among the 88.4 million U.S. households in 1985, 62.2% lived in detached single-family units, and 4.6% lived in attached single-family units. But 5.4% lived in mobile homes, 11.6% lived in

structures containing two and four units, and 16.1% lived in structures containing five or more units. Thus, about one-third of all American households did not live in single-family homes, and over one-third were not homeowners, but renters. The ideal vision of metropolitan development does not contain any dwelling units appropriate to this "forgotten fraction."

True, U.S. suburbs – including the newest ones – contain a lot of housing other than single-family homes. Such housing is in fact being built in our metropolitan areas. Nevertheless, it does not form part of the ideal vision of how those areas should be developed. Therefore, its construction is often strongly opposed by local residents whose views are dominated by that ideal vision.

Moreover, many low- and moderate-income households cannot afford to live in brand new housing units since such units have their highest relative prices when they are first built. At that moment, they contain the most modern and up-to-date amenities and design available, and they have not yet been subjected to wear and tear. Moreover, they are most often located at the edge of the built-up territory within their metropolitan area, which is then usually among the most fashionable neighborhoods in which to live.

As time passes, these units become relatively less desirable. They no longer contain the most up-to-date amenities; they have experienced some normal wear and tear and deterioration; and their neighborhoods are no longer at the edge of the built-up territory, which has now expanded beyond them.

Hence, their relative prices decline, even though their absolute prices may rise along with inflation. If this life cycle continues, these units eventually fall in both relative prices and absolute prices enough so that low- and moderate-income households can afford to live in them. This life-cycle movement of relative prices is part of the "filtering" or "trickle-down" process through which older housing is made available to those households that cannot afford brand new units.

Most American suburbs have been built since World War II. The nation's total suburban population rose from 40.9 million in 1950 to 108.6 million in 1986 – an increase of 165.5%, compared to a 58.7% rise in total population. By 1986, the fraction of all U.S. residents living in suburbs had risen to 44.9% from 27% in 1950. In very large metropolitan areas, the newest suburbs are now located quite far from the older portions of both the central city and older suburbs.

This is particularly true in those metropolitan areas that have experienced the fastest growth in total population, such as many in California and Florida.

Historically, employers in newer suburbs have relied upon the more affordable housing inventory in older neighborhoods to provide dwellings for their low- and moderate-income workers. This arrangement permitted the newest communities to avoid providing any housing directly affordable to all the persons working within their boundaries. Hence, these communities could develop themselves in full accord with the ideal vision of universal homeownership, involving mainly detached single-family units.

However, such reliance upon the "trickle-down" process to house low-wage workers has recently become much less effective, for several reasons. First, jobs have spread out into the suburbs much more than in the past. In fact, many more new jobs are being created in the suburbs than in central cities. They include jobs located in relatively far-out suburbs, as well as close in ones. Second, the sheer size of the suburban portions of large, fast-growing metropolitan areas means that much of the older, less costly housing is many miles from new-growth regions where most new jobs are located. Low- and moderate-income workers who live in older, more central neighborhoods are far from those new jobs. Hence, they have a hard time finding job openings, or commuting to jobs they find (often because of lack of public transportation). Third, soaring land prices in many such metropolitan areas have helped raise housing prices there to extremely high levels. Few middle-income households can afford most new single-family units there. Even the few multifamily units being built there have rents far above the payment ability of most low- or moderate-income households. Fourth, the long economic expansion that began in 1982 has soaked up most available workers. This has forced the unemployment rate well below 6% nationally, and below 2% in some booming suburban areas. In addition, the changes in U.S. age distribution are creating a nationwide shortage of entry-level workers, tightening the market further. The higher wages resulting from this prosperity have become embedded in housing costs, increasing prices beyond the reach of many households.

As a result, far-out suburban portions of many large, fast-growth metropolitan areas are experiencing acute shortages of low-wage and even moderate-wage labor. Workers willing to take such jobs cannot afford to live anywhere within reasonable commuting distance of those jobs. Thus, the ideal vision's exclusive focus upon relatively new single-family housing, which is quite expensive, is inconsistent with the actual need for low- and moderate-wage workers that arises in every community.

THE THIRD FLAW: NO CONSENSUS ON HOW TO FINANCE INFRASTRUCTURES FAIRLY

A third major flaw springs from the absence of any consensus in the dominant vision about how best to finance new infrastructures, more roads, sewage systems, water systems, and school systems. They also include costs of increasing the capacity of existing arterial facilities in established areas through which new residents will pass on their way to and from work or shopping. The absence of a widespread consensus about how to pay for such added facilities "fairly" has two negative consequences: It creates severe political conflicts in many suburban areas, and it often results in gross underfunding – and therefore under-provision – of needed facilities and services. Residents already living in fast-growth areas want all the added facilities but expect services required by newcomers to be paid for by the latter. Existing residents consider it justifiable to load those marginal costs entirely onto new developments through various impact fees, exactions, proffers, and permit fees. They regard this as merely requiring the newcomers to pay "their fair share" of such costs, since newcomers are the main beneficiaries of those facilities. Hence, they vote against any increases in general taxes or general bonding powers to pay for such facilities.

In contrast, developers and potential newcomers to growth areas believe the entire community should share in paying for these added infrastructures. They cite three reasons for such sharing: First, because many existing residents benefited from past general financing of similar infrastructures when their subdivisions were being built, it is unfair of them to change the rules of the game once they have received such benefits. Second, growth creates many gains for society in general – including existing residents. Its main benefit is greater economic prosperity, which aids everyone. Therefore, society should pay for some of the marginal costs of growth. Third, loading all the marginal costs of growth onto new developments raises housing costs. That unfairly reduces the homeownership and rental opportunities of households with low and moderate incomes. From this perspective, existing residents also ought to pay "a fair share" of the costs of added growth.

Political resolution of this controversy is inherently biased in favor of existing residents because potential newcomers do not yet reside in the areas concerned. Consequently, they cannot vote on local government policies that directly affect their welfare, whereas existing residents can. The deck is thus stacked in favor of loading most marginal costs of development onto newcomers, rather than sharing those costs throughout the community.

However, realistically it is quite difficult to force newcomers to pay all the marginal costs of growth through impact fees or other exactions. Some costs of growth spring from more intensive use of existing facilities, such as expressways, arterial streets, sewage-treatment plants, and school systems. Those facilities and services are shared between newcomers and previous residents. Moreover, many infrastructures can only be built in certain minimum sizes that have substantial capacity, thus presenting communities with sizable costs. For example, the addition of 200 households to an area where the sewage-treatment plant has reached its absolute capacity volume of 20,000 households may require construction of a whole new treatment plant with a minimum capacity of 10,000 households. It is not possible to load the entire cost of this new plant onto those 200 newcomers. Instead, existing residents must bear some of that initial cost.

Faced by this situation, existing residents often choose to provide inadequate facilities for newcomers to hold down taxes. They make such choices without fear of being outvoted by potential newcomers, as the latter are not yet present to vote on their own behalf.

As a result, new housing and commercial subdivisions often are built without adequate infrastructures and other supporting facilities. Existing roads, streets, schools, parks, and utility systems become grossly overloaded and congested. This imposes heavy costs not only upon newcomers, but also upon the existing residents themselves. Precisely this sequence of events has been happening in Florida for some time. The Florida State government has been unable to agree on some means of financing the infrastructure to accommodate Florida's rapid population growth.

Such developments aggravate inherent political tensions between residents trying to minimize their taxes and newcomers and developers trying to accom-modate new growth. Many existing residents would like to prohibit all additional growth completely. But doing so is impossible in practice because it violates basic freedoms of movement, private property rights, and contract guaranteed by the U.S. Constitution. The result in many fast growth suburban areas is an undesirable combination of high political tensions and facility overload because of underfunding and congestion. The dominant vision of how metropolitan areas ought to develop contains no means or even guidance concerning how to resolve these issues.

THE FOURTH FLAW: INABILITY TO ACCOMMODATE LULUS

The dominant ideal vision's fourth major flaw is that it contains no effective political mechanisms for resolving inevitable conflicts between the welfare of society as a whole and the welfare of geographically small parts of society. This results from extreme spatial fragmentation of government decision-making powers. Yet such fragmentation is required by the governance arrangements that most Americans prefer. They want to live in geographically small communities, each of which fully controls the land use, public schools, and other environmental aspects within its own boundaries. Such "local sovereignty" is one of the pillars of the ideal vision. But it means that each government has a highly parochial viewpoint. Its officials are concerned only with the welfare of the local residents who elect them. As a result, no public officials are primarily concerned with the welfare of society as a whole, or even of the metropolitan area as a whole.

However, every modern society must contain those facilities that benefit society as a whole, but have negative "spillover effects" upon their immediate surroundings (e.g. airports, expressways, jails, garbage incinerators, and landfills). These essential facilities are known as Locally Undesirable Land Uses (LULUs). Because of their undesirable impacts upon their immediate neighbors, no residents want to permit a LULU to be located near them. This attitude has become known as the Not In My Back Yard (NIMBY) syndrome.

Whenever a LULU has to be built, there is no effective way to find a location for it in a society designed in accordance with the predominant ideal vision. Residents near every potential site pressure their local governments to oppose locating the LULU there.

Those governments have the power to reject the LULU because controls over land uses have been divided up among myriad local entities. And officials within all those entities are motivated to reject the LULU because they are politically responsible only to their own residents. Hence, it becomes almost impossible to find politically acceptable locations for facilities that every metropolitan area needs to operate efficiently. The resulting paralysis has virtually halted both major airport construction and prison expansion in most of the United States, and blocked creation of thousands of other badly needed LULUs.

Furthermore, there is a strong single-family-home bias built into the dominant ideal vision. As a result, nearly all new high-density real estate developments – especially those involving any multifamily housing – are usually considered LULUs, and are therefore extremely difficult to build. But low- and moderate-income households cannot afford to occupy new single-family homes. So the NIMBY syndrome re-inforces the second flaw in the dominant ideal vision: Its failure to provide housing that low- and moderate-income households can afford.

In fact, low- and moderate-income households themselves are regarded as LULUs by many middle- and upper-income households. The latter consequently adopt a NIMBY attitude toward all housing affordable to low- and moderate-income households. They believe that such "undesirable" people near them will reduce the market values of their costly homes. About three-fourths of the financial net worth of the average American household consists of its equity in the home it owns and lives in. Hence, its members are quite sensitive about any changes in their neighborhoods that they believe would reduce the value of those equities, and very little such housing gets built in most U.S. suburbs.

SOCIAL INCONSISTENCIES RESULTING FROM THE IDEAL VISION'S FLAWS

The four major flaws in the ideal vision described above have caused the emergence and persistence of certain negative conditions in suburban America, the very heartland of that ideal vision. Hence, we must conclude that the dominant vision itself contains inherent characteristics that render it less ideal than we previously believed it to be – or than most Americans still believe it to be. These characteristics can be

considered "social inconsistencies," as they are inconsistent with the high quality of life promised by the ideal vision. (Note: This situation illustrates a fundamental problem in democratic societies. They have great difficulty solving problems that arise as the long-run result of a majority's pursuit of policies that produce short-run benefits. Once a majority of citizens has begun receiving those short-run benefits, its members do not want to give them up. But they often fail to agree about how to distribute the long-run costs necessary to sustain those short-run benefits. Each subgroup among the beneficiaries tries to shift as much of its share of the costs as possible onto other subgroups. No consensus arises about how to pay for these costs. In some cases, the costs are not paid at all. That is precisely why the United States has been unable to reduce its huge recurrent Federal deficits.)

THE ALTERNATIVE VISION OFFERED BY THE URBAN PLANNING PROFESSION

The American urban planning profession has long been aware of some of the internal inconsistencies in the dominant ideal vision described above. In particular, urban planners have repeatedly attacked the heavy reliance upon automotive vehicle travel required by very low-density settlement patterns. In fact, many urban planners have advocated an alternative vision that rejects such reliance. But their alternative has never gained substantial public acceptance in America.

This alternative has been based upon the European pattern of urban development. It featured high-density residential settlements, high-density workplaces, tightly circumscribed land use patterns that prevent peripheral sprawl, and massive use of heavily subsidized public transit systems for movement. These traits are only possible under a governance system that centralizes power over the land use patterns in each metropolitan area in a single governing body with authority over the entire area. Accompanying such centralization in Europe is use of large amounts of publicly subsidized rental housing. Such housing accommodates a sizable fraction of the entire population in subsidized multi-family units scattered throughout each metropolitan area.

For decades, U.S. urban planners have been advo-cating an alternative vision of future metropolitan development based largely upon this pattern. But they have been unable to persuade either the public or local government officials to accept it because this alternative directly opposes the dominant vision described earlier that most Americans cherish. It rejects too many elements of that dominant vision, including primary reliance upon owner-occupied single-family housing, widespread use of private automotive vehicles, and decentralized local government.

Furthermore, some of the key elements in the European pattern do not appeal to Americans. For example, placing large clusters of high-rise apartments right around fixed-rail transit stops repels most suburban residents, who do not like high-rise housing. In addition, the immense cost of fixed-rail transit systems, and their lack of flexibility, are not acceptable to many Americans. The only large-scale fixed-rail transit system built in the U.S. since 1945 without major Federal subsidies is the Bay Area Rapid Transit System. All the others probably never would have been created if local residents had been compelled to pay their full costs without massive Federal subsidies.

Moreover, the urban planning profession has been guilty of largely ignoring the second major flaw in the dominant vision (its inability to provide housing for low- and moderate-income households). Few urban planners have ever developed comprehensive plans that specifically indicated where to accommodate poor households. Most comprehensive plans ignore the need for low-rent housing, because explicitly locating it within the plan would arouse immense opposition from nonpoor residents.

Yet the urban planning profession has been correct in claiming that the dominant vision would not work well in the long run. Most suburbanites in fast-growth metropolitan areas are now discovering this fact, to their own dismay.

CRITERIA FOR A NEW IDEAL VISION OF FUTURE METROPOLITAN AREA DEVELOPMENT

The preceding analysis shows that American society needs a new ideal vision of how future development ought to occur in our large metropolitan areas. The current dominant vision contains too many serious inconsistencies and undesirable outcomes. The new ideal vision should conform to several specific criteria. Such conformance would enable this vision both to (1) meet the future needs of American society; and

(2) avoid the problems generated by the currently dominant vision.

As noted earlier, development and widespread acceptance of a new vision would be similar to a paradigm shift in science – that is, the replacement of one fundamentally governing perception by another one. Such replacements may appear to occur suddenly, but they actually build up gradually over time.

The new ideal vision of future metropolitan growth will not require the complete rejection of the pillars on which the existing vision has been based. Rather, it will modify those pillars. This new perspective will be much more conscious of the collective impacts of individual decisions than the currently dominant perspective. Hence, it will embody individualism sensitive to collective behavior patterns, rather than unconstrained individualism. For example, we need not deny the desirability of homeownership or of single-family homes in order to recognize that a balanced society also needs rental housing and much higher average densities than we have considered ideal in the past. Similarly, we will certainly not abandon widespread use of individual vehicles. But those vehicles may be powered differently, and more people will be traveling together in shared rides or transit than at present.

Specific criteria of this new ideal vision are as follows:

First, the new vision must contain sizable areas of at least moderately high-density development, especially of housing, but also of workplaces. This is necessary to achieve three key goals: the creating of housing that is less costly than single-family homes; the reduction of the area required to accommodate both jobs and workers, thereby cutting travel requirements and reducing both congestion and air pollution; the development of more efficient spatial interrelations among buildings within suburban commercial and industrial developments. This would reduce travel requirements within such developments and promote more face-to-face contacts there. The third of these goals is far less important than the first two.

In order to make such higher-density areas politically acceptable to suburban residents, owners of single-family homes must be convinced that they can tolerate multifamily housing nearby. Such housing need not consist of big high-rise towers, but can be a mixture of low-rise and mid-rise structures, with an occasional high-rise. Owners of existing single-family homes must become confident that such housing will neither jeopardize the investments they have made in their own homes, nor expose their families to crime and violence. That will probably require careful studies of the impact of well-designed existing multifamily projects upon the market values of surrounding single-family homes. Owners of single-family homes must also be convinced that moderate-density housing can be designed to look good and therefore not reduce the aesthetic quality of their areas.

Second, the new ideal vision must encourage people to live nearer to where they work. This means that each subarea within a metropolitan area must contain a "balanced blend" of different types and prices of housing, all close to the jobs within the same subarea. Then it will be possible for workers holding jobs at all wage levels to live close to those jobs and still occupy housing they can afford.

True, people cannot be forced to live nearby their workplaces. Today, many people could live closer to where they work, but they prefer traveling longer distances in order to live in what they deem an ideal environment. But this preference is changing as congestion increases the time required to move any specific distance during peak commuting periods. Eventually, more people will decide that suffering the time losses and frustration of such worsened congestion is not worthwhile and will move either their jobs or their homes to reduce the distance between them. But this is possible only if they are compelled to live near their jobs by the appropriate cost of the housing close to those jobs. That, in turn, is possible only if each sizable subregion of the entire metropolitan area contains a mixture of different types of housing available at widely varying occupancy costs.

Achieving such "balanced blends" of housing in all parts of a large metropolitan area will not be easy. The logical spatial units within which such "balancing" should occur will rarely coincide with existing governmental boundaries. Moreover, at least one job cluster in each (i.e. downtown) contains so many jobs that not all of its workers could possibly live nearby. In order to reduce commuting travel, it may be necessary in the long run to "decant" some existing down-towns by shifting some of their existing jobs elsewhere. That is not a happy thought for owners and managers of downtown office buildings and other properties. But the goal of this article is not to determine how these criteria might be met, but rather to formulate them in the hope that others might determine how they might be implemented.

Third, the new ideal vision must contain governance structures that preserve substantial local authority – but within a framework that compels local governments to act responsibly to meet area-wide needs. That probably requires a comprehensive planning framework imposed by state governments. In such a framework, every local government would be responsible for drawing up comprehensive plans for its own future development. But all such plans would have to meet certain general criteria laid down by the state government. Such criteria might include provision of land zoned for low- and moderate-income housing, and arrangements that pushed the choosing of sites for certain region-affecting facilities up to regional or state levels to avoid complete paralysis. Moreover, the state government would have to create mechanisms to review and modify local government plans to insure that they met statewide criteria. Oregon has already put such arrangements into practice, and several other states are doing so. Thus, in considering possible future changes in existing governance within metropolitan areas, one should expect neither true metropolitan government, nor any total abandonment of local sovereignty. Simply, such sovereignty must be placed within a broader framework that enforces wider social responsibility. While this probably can occur only at the state level, where the constitutional power over local government resides, the Federal government might provide financial incentives to get more states to engage in this type of restructuring.

Fourth, the new vision should contain incentive arrangements that encourage individuals and households to take a more realistic account of the collective costs of their behavioral choices. For example, individual auto drivers are not required to pay more for taking trips alone during rush hours than during off-peak hours, because auto drivers are never charged directly for using roads. For decades, economists have pointed out that each driver who enters a well-traveled highway during peak hours creates a substantial social cost. By adding to the prevailing congestion, that driver slows down all other drivers to some degree, thereby imposing a time loss upon them. In contrast, a driver who takes the same trip during nonpeak hours does not generate any such social cost. Therefore, it would be economically efficient for society if drivers were charged a direct fee for driving during peak hours, but not during off-peak hours. Such charges would encourage more drivers to shift to off-peak periods, thereby reducing peak-hour congestion. And those charges could raise money useful in improving the nation's road system.

This type of incentive arrangement is especially important in view of the fact that most peak-hour auto trips made in the nation's largest metropolitan areas are nonwork trips, many of which could be shifted to other times relatively easily.

Unfortunately, politicians in most democracies have consistently refused to adopt differential road-use pricing. They argue that such pricing would penalize low-income drivers and favor higher-income ones, who could more easily afford to pay peak-hour charges. In a democracy where 87% of all households have cars, but most do not have high incomes, differential road pricing has always seemed to be a politically losing proposition.

This situation will change only when major freeways become so overcrowded that traffic literally slows to a crawl for several hours per day. Many southern California freeways are close to such prolonged gridlock today, and are likely to reach it soon. When they do, even low-income auto-driving voters will become frustrated enough to accept almost any proposed solution likely to be effective. Peak-hour road pricing will then be perceived in a much more favorable light because of the realistic incentives it presents to drivers.

So-called "exclusionary zoning" arrangements or "linkage fees" concerning housing could be another form of such incentives. Such exclusionary arrangements are created and sustained by political pressures on local governments from middle- and upper-income homeowners. They believe socioeconomic exclusion helps maximize the market values of their homes and improve the quality of their local environments.

In many suburban areas, exclusionary zoning has made the cost of housing units too great for most low- and moderate-income households. The typical application of this type of zoning has been the "down zoning" of lots where minimum sizes are often two acres and up. In the end, this imposes a cost upon society as a whole by requiring many such households to live far from their suburban jobs and commute long distances each day. The result is more overall traffic congestion and air pollution, plus the imposition of very high travel costs upon low- and moderate-income commuters.

Therefore, those who impose exclusionary arrangements on their suburban areas should be required to offset the social costs they are generating through various arrangements that charge them for engaging in exclusionary practices. Such charges would also discourage exclusion by raising the price of housing to those who engage in it.

Several forms of such incentive arrangements concerning housing have been tried in different areas. Local or state governments could add a fee onto all market-priced housing units built or sold in exclusionary regions, and use the proceeds to subsidize units available at costs affordable to low- and moderate-income households. Another tactic would be to add a transfer tax onto the sale of single-family homes and then put the proceeds into a housing trust fund used to help finance construction of low-rent units in otherwise exclusionary communities. Another approach would be to require each developer to allocate a certain percentage of the units he or she builds for low- and moderate-income occupancy. The resulting higher costs for market-priced units could be offset to some extent by permitting the developer to build at higher-than-usual average densities in the projects concerned. Yet another tactic would be to charge developers of commercial and industrial properties in exclusionary areas "linkage fees" that are used to subsidize housing for low- and moderate-income households.

Not all of these arrangements would have the same negative incentive impact upon exclusionary zoning practices that peak-hour-driving charges would have upon rush-hour trips. But they all would compel persons engaging in exclusionary zoning practices to bear some of the collective costs their individual behavioral choices are creating for society.

Fifth, the new vision should incorporate stable and predictable strategies that would adequately finance infrastructures built to accommodate growth. Such strategies need to be stable and predictable so that all concerned can efficiently plan how to carry out new developments, both public and private. These strategies also must be fair and politically acceptable both to existing residents on the one hand, and to developers and potential newcomers on the other. Financing strategies appropriate to slow-growth areas may be different from those appropriate to fast-growth areas. In all areas, such strategies will probably have to combine some impact fees and exactions upon newcomers, with some increases in general taxes upon all residents.

Another key criterion is that the new ideal vision should not depend upon construction of costly fixed-rail mass transit systems that are politically acceptable in each area only when heavily subsidized with funds from other parts of the nation. Many urban planners will reject this criterion. But this author believes the American people will accept it – as they have for many decades. Perhaps some forms of light rail transit can be created inexpensively enough to be incorporated within viable metropolitan area transportation systems. But most such systems must be based upon freely moving automotive vehicles, whether they are cars, trucks, vans, big buses, or mini-buses. Thus, designing urban transportation systems that will reduce congestion and pollution poses a tremendous challenge to transportation planners.

Further, the ideal vision should incorporate workplace designs that combine efficient interchange among individual workplaces with big enough critical masses of jobs to create productive interchange possibilities. These characteristics are now found within traditional downtowns and within large regional shopping malls. In both places, office buildings or retail shops are located right next to each other along attractive pedestrian walkways, thereby encouraging maximum efficiency of movement among different facilities. Parking is not placed around each structure, but in external zones that are nearby but do not interfere with movements among individual units. Yet enough facilities are massed together to permit efficient comparison shopping or interchanges of services and meetings among persons working in different enterprises.

These traits are now almost totally absent from all suburban office and industrial parks. Each such structure is separated from others nearby by a sea of parking or landscaping. Therefore, meeting this criterion might require forcing owners to build new structures between existing ones, as near Tyson's Corner, in Fairfax County, Virginia, so as to recreate the interchange efficiency of downtown streets. It cannot be determined how this could be done with present fragmented private site ownership. But considering the ingenuity with which Wall Street investment bankers have reorganized existing companies into multiple entities, or created new securities, we should be equally ingenious about land-use controls and forms of ownership and management.

Finally, the new ideal vision should be internally consistent, in the dual sense that (1) the amount of travel it requires does not lead to the levels of traffic congestion or air pollution we are now encountering; and (2) that some homes in the vision are affordable to the low- and moderate-income households needed to do much of the work. We need complex computerized transportation and land-use models to test such relationships. Designing both the models and the transportation elements of the new ideal vision will not be easy. But few aspects of urban life are more important.

CONCLUSION

Is it really possible to create an alternative ideal vision of future large metropolitan area development – a new paradigm – that meets all these criteria? Our society desperately needs to make an attempt to find out. We have overwhelming current evidence that the currently dominant ideal vision does not work effectively. But we cannot displace that vision unless we have some plausible, attractive, and persuasive alternative to offer. The option traditionally offered by urban planners is never going to win widespread acceptance among the American people.

"Broken Windows"

Atlantic Monthly (1982)

James Q. Wilson and George L. Kelling

Editors' Introduction

Why is urban crime a problem in inner-city neighborhoods? What can and should government do about it? That is the subject of the following selection by James Q. Wilson and George L. Kelling.

Wilson and Kelling studied police behavior in troubled neighborhoods – particularly in Newark, New Jersey. One of the authors (Kelling) spent many months walking Newark neighborhoods with local police officers, observing what was going on in the neighborhoods and how the police actually handled neighborhood problems. Wilson, then a Harvard professor, worked with Kelling's empirical observations to jointly develop this classic article. Wilson and Kelling were particularly interested in police discretion and how the police handled troublesome behavior at the borderline between acceptable and unacceptable, legal and illegal. Wilson and Kelling are interested as much in neighborhood residents' perceptions of crime as in crime itself. A particular focus of their research and theory building is to develop a model of policing that neighborhood residents (at least those residents the authors term "decent folk") will support. Based on this research, they advanced a controversial theory of neighborhood transition and an influential model for community policing that has been implemented in many communities.

The central metaphor of this selection is of a single broken window. Imagine a city neighborhood like Boston's West End or New York City's Greenwich Village as described by Jane Jacobs (p. 98) – a diverse, viable, exciting urban neighborhood, but with some crime. Most of the people in the neighborhood Jane Jacobs sees each morning as the "street ballet" begins are regulars – people who live or work in the neighborhood or are frequently in the area. Some are strangers. Most of the people in the neighborhood – the grocer putting out his vegetables, the children on their way to school – are not criminals and do not engage in behavior that harms other people. The behavior of some people in the neighborhood might startle or even scare conventional Bostonians or New Yorkers, but it is understood and tolerated by neighborhood regulars. Some of the people in the area, however, are dangerous and do engage in criminal behavior: drug dealing, prostitution, theft, assault, and other serious crimes. Civil order exists, but it is fragile. Imagine that someone breaks a single window in the neighborhood. According to Wilson and Kelling, how the police respond to that trivial but unacceptable act is fraught with consequences. One response is to do nothing. After all police are busy working on serious crimes and municipal budgets are tight. But doing nothing in response to a broken window, Wilson and Kelling argue, will signal that no one cares about unacceptable conduct. It will be an invitation for people to break more windows. Neighborhood residents will start avoiding one another, stop participating in neighborhood block parties, and eventually cower in their own homes.

Wilson and Kelling argue that historically neighborhood residents judged the success of police activity by whether it succeeded in maintaining order. Governments often tolerated (or encouraged) police to keep order through *informal means* without much regard to legal niceties. A drunk might have a *legal* right to sit on a neighborhood park bench. But if his presence sufficiently disturbed "decent" neighborhood residents, the local cop was encouraged to make him move elsewhere. Gang members in Chicago's crime-ridden Robert Taylor Homes might have a legal right to loiter by a playground, but the authors argue that "decent" project residents would want the local police to get rid of them ("kick ass" is the term Wilson and Kelling use). Wilson and Kelling take the controversial position that broad police discretion to enforce neighborhood standards is acceptable, even desirable.

James Q. Wilson is the Ronald Reagan Professor of Public Policy at Pepperdine University in California. He was awarded the Presidential medal of freedom by President George W. Bush in 2003. From 1961 to 1986 he was a professor of government at Harvard and from 1986 until 1997 the James Collins Professor of Management at UCLA. He has written extensively on politics, economics, and criminology.

George L. Kelling is a Senior Fellow at the Manhattan Institute and a professor in the School of Criminal Justice at Rutgers University.

Compare Wilson and Kelling's approach to Mike Davis's description of how he feels the Los Angeles police oppress low-income minorities (p. 178). How would you draw the line between "decent" and not-so-decent folk in New York's culturally heterogeneous neighborhoods? Whose culture should define neighborhood norms?

The argument in this selection is further developed in George L. Kelling and Catherine M. Coles, *Fixing Broken Windows* (New York: Martin Kessler, 1996). Other books by James Q. Wilson related to crime prevention include *Crime* (San Francisco: ICS Press, 1995), co-edited with Joan Petersilia; *Crime and Public Policy* (San Francisco: ICS Press, 1983); *Drugs and Crime*, co-edited with Michael Tonry (Chicago: University of Chicago Press, 1990); *Families, Schools, and Delinquency Prevention*, co-edited with Glenn C. Loury (New York: Springer-Verlag, 1987); *Understanding and Controlling Crime*, co-edited with David P. Farrington and Lloyd E. Ohlin (New York: Springer-Verlag, 1986); *Thinking About Crime* (New York: Basic Books, 1983), *Varieties of Police Behavior* (Cambridge: Harvard University Press, 1968), and *Crime and Human Nature* (New York: Free Press, 1998).

Other notable books by Wilson include *American Government*, ninth edition (Boston: Houghton Mifflin, 2004) with John J. Dilulio, Jr; *Bureaucracy: What Government Agencies Do and Why They Do It* (New York: Basic Books, reprint edition 2000), and *The Moral Sense* (New York: Free Press, 1997).

Books on community policing include Elizabeth M. Watson, Alfred R. Stone, and Stuart M. DeLuca, *Strategies for Community Policing* (Upper Saddle River: Prentice-Hall, 1998), Wesley G. Skogan and Susan M. Hartnett, *Community Policing, Chicago Style* (New York: Oxford University Press, 1997), Kenneth J. Peak and Ronald W. Glensor, *Community Policing and Problem Solving: Strategies and Practices* (Upper Saddle River: Prentice-Hall, 1996), and Nigel Fielding, *Community Policing* (Oxford and New York: Oxford University Press, 1995).

In the mid-1970s, the state of New Jersey announced a "Safe and Clean Neighborhoods Program," designed to improve the quality of community life in twenty-eight cities. As part of that program, the state provided money to help cities take police officers out of their patrol cars and assign them to walking beats. The governor and other state officials were enthusiastic about using foot patrol as a way of cutting crime, but many police chiefs were skeptical. Foot patrol, in their eyes, had been pretty much discredited. It reduced the mobility of the police, who thus had difficulty responding to citizen calls for service, and it weakened headquarters control over patrol officers.

Many police officers also disliked foot patrol, but for different reasons: it was hard work, it kept them outside on cold, rainy nights, and it reduced their chances for making a "good pinch." In some depart-

ments, assigning officers to foot patrol had been used as a form of punishment. And academic experts on policing doubted that foot patrol would have any impact on crime rates; it was, in the opinion of most, little more than a sop to public opinion. But since the state was paying for it, the local authorities were willing to go along.

Five years after the program started, the Police Foundation, in Washington, DC, published an evaluation of the foot-patrol project. Based on its analysis of a carefully controlled experiment carried out chiefly in Newark, the foundation concluded, to the surprise of hardly anyone, that foot patrol had not reduced crime rates. But residents of the foot-patrolled neighborhoods seemed to feel more secure than persons in other areas, tended to believe that crime had been reduced, and seemed to take fewer steps to protect themselves from crime (staying at home with the doors locked, for example). Morcover, citizens in the foot-patrol areas had a more favorable opinion of the police than did those living elsewhere. And officers walking beats had higher morale, greater job satisfaction, and a more favorable attitude toward citizens in their neighborhoods than did officers assigned to patrol cars.

These findings may be taken as evidence that the skeptics were right – foot patrol has no effect on crime; it merely fools the citizens into thinking that they are safer. But in our view, and in the view of the authors of the Police Foundation study (of whom Kelling was one), the citizens of Newark were not fooled at all. They knew what the foot-patrol officers were doing, they knew it was different from what motorized officers do, and they knew that having officers walk beats did in fact make their neighborhoods safer.

But how can a neighborhood be "safer" when the crime rate has not gone down – in fact, may have gone up? Finding the answer requires first that we understand what most often frightens people in public places. Many citizens, of course, are primarily frightened by crime, especially crime involving a sudden, violent attack by a stranger. This risk is very real, in Newark as in many large cities. But we tend to overlook or forget another source of fear – the fear of being bothered by disorderly people. Not violent people, nor, necessarily, criminals, but disreputable or obstreperous or unpredictable people: panhandlers, drunks, addicts, rowdy teenagers, prostitutes, loiterers, the mentally disturbed.

What foot-patrol officers did was to elevate, to the extent they could, the level of public order in these neighborhoods. Though the neighborhoods were predominantly black and the foot patrolmen were mostly white, this "order-maintenance" function of the police was performed to the general satisfaction of both parties.

One of us (Kelling) spent many hours walking with Newark foot-patrol officers to see how they defined "order" and what they did to maintain it. One beat was typical: a busy but dilapidated area in the heart of Newark, with many abandoned buildings, marginal shops (several of which prominently displayed knives and straight-edged razors in their windows), one large department store, and, most important, a train station and several major bus-stops. Though the area was run-down, its streets were filled with people, because it was a major transportation center. The good order

of this area was important not only to those who lived and worked there but also to many others, who had to move through it on their way home, to supermarkets, or to factories.

The people on the street were primarily black; the officer who walked the street was white. The people were made up of "regulars" and "strangers." Regulars included both "decent folk" and some drunks and derelicts who were always there but who "knew their place." Strangers were, well, strangers, and viewed suspiciously, sometimes apprehensively. The officer – call him Kelly – knew who the regulars were, and they knew him. As he saw his job, he was to keep an eye on strangers, and make certain that the disreputable regulars observed some informal but widely understood rules. Drunks and addicts could sit on the stoops, but could not lie down. People could drink on side streets, but not at the main intersection. Bottles had to be in paper bags. Talking to, bothering, or begging from people waiting at the bus stop was strictly forbidden. If a dispute erupted between a businessman and a customer, the businessman was assumed to be right, especially if the customer was a stranger. If a stranger loitered, Kelly would ask him if he had any means of support and what his business was; if he gave unsatisfactory answers, he was sent on his way. Persons who broke the informal rules, especially those who bothered people waiting at bus stops, were arrested for vagrancy. Noisy teenagers were told to keep quiet.

These rules were defined and enforced in collaboration with the "regulars" on the street. Another neighborhood might have different rules, but these, everybody understood, were the rules for this neighborhood. If someone violated them, the regulars not only turned to Kelly for help but also ridiculed the violator. Sometimes what Kelly did could be described as "enforcing the law," but just as often it involved taking informal or extralegal steps to help protect what the neighborhood had decided was the appropriate level of public order. Some of the things he did probably would not withstand a legal challenge.

A determined skeptic might acknowledge that a skilled foot-patrol officer can maintain order but still insist that this sort of "order" has little to do with the real sources of community fear – that is, with violent crime. To a degree, that is true. But two things must be borne in mind. First, outside observers should not assume that they know how much of the anxiety now endemic in many big-city neighborhoods stems from a fear of "real" crime and how much from a sense that the street is disorderly, a source of distasteful, worrisome encounters. The people of Newark, to judge from their behavior and their remarks to interviewers, apparently assign a high value to public order, and feel relieved and reassured when the police help them maintain that order.

Second, at the community level, disorder and crime are usually inextricably linked, in a kind of developmental sequence. Social psychologists and police officers tend to agree that if a window in a building is broken *and is left unrepaired*, all the rest of the windows will soon be broken. This is as true in nice neighborhoods as in run-down ones. Window-breaking does not necessarily occur on a large scale because some areas are inhabited by determined window-breakers whereas others are populated by window-lovers; rather, one unrepaired broken window is a signal that no one cares, and so breaking more windows costs nothing. (It has always been fun.)

Philip Zimbardo, a Stanford psychologist, reported in 1969 on some experiments testing the broken-window theory. He arranged to have an automobile without license plates parked with its hood up on a street in the Bronx and a comparable automobile on a street in Palo Alto, California. The car in the Bronx was attacked by "vandals" within ten minutes of its "abandonment." The first to arrive were a family – father, mother, and young son – who removed the radiator and battery. Within twenty-four hours, virtually everything of value had been removed. Then random destruction began – windows were smashed, parts torn off, upholstery ripped. Children began to use the car as a playground. Most of the adult "vandals" were well-dressed, apparently clean-cut whites. The car in Palo Alto sat untouched for more than a week. Then Zimbardo smashed part of it with a sledge-hammer. Soon, passersby were joining in. Within a few hours, the car had been turned upside down and utterly destroyed. Again, the "vandals" appeared to be primarily respectable whites.

Untended property becomes fair game for people out for fun or plunder, and even for people who ordinarily would not dream of doing such things and who probably consider themselves law-abiding. Because of the nature of community life in the Bronx – its anonymity, the frequency with which cars are abandoned and things are stolen or broken, the past experience of "no one caring" – vandalism begins much more quickly than it does in staid Palo Alto, where people have come to believe that private possessions are cared for, and that mischievous behavior is costly. But vandalism can occur anywhere once communal barriers – the sense of mutual regard and the obligations of civility – are lowered by actions that seem to signal that "no one cares."

We suggest that "untended" behavior also leads to the breakdown of community controls. A stable neighborhood of families who care for their homes, mind each other's children, and confidently frown on unwanted intruders can change, in a few years or even a few months, to an inhospitable and frightening jungle. A piece of property is abandoned, weeds grow up, a window is smashed. Adults stop scolding rowdy children; the children, emboldened, become more rowdy. Families move out, unattached adults move in. Teenagers gather in front of the corner store. The merchant asks them to move; they refuse. Fights occur. Litter accumulates. People start drinking in front of the grocery; in time, an inebriate slumps to the sidewalk and is allowed to sleep it off. Pedestrians are approached by panhandlers.

At this point it is not inevitable that serious crime will flourish or violent attacks on strangers will occur. But many residents will think that crime, especially violent crime, is on the rise, and they will modify their behavior accordingly. They will use the streets less often, and when on the streets will stay apart from their fellows, moving with averted eyes, silent lips, and hurried steps. "Don't get involved." For some residents, this growing atomization will matter little, because the neighborhood is not their "home" but "the place where they live." Their interests are elsewhere; they are cosmopolitans. But it will matter greatly to other people, whose lives derive meaning and satisfaction from local attachments rather than worldly involvement; for them, the neighborhood will cease to exist except for a few reliable friends whom they arrange to meet.

Such an area is vulnerable to criminal invasion. Though it is not inevitable, it is more likely that here, rather than in places where people are confident they can regulate public behavior by informal controls, drugs will change hands, prostitutes will solicit, and cars will be stripped. That the drunks will be robbed by boys who do it as a lark, and the prostitutes' customers will be robbed by men who do it purposefully and perhaps violently. That muggings will occur.

Among those who often find it difficult to move away from this are the elderly. Surveys of citizens suggest that the elderly are much less likely to be the victims of crime than younger persons, and some have inferred from this that the well-known fear of crime voiced by the elderly is an exaggeration: perhaps we ought not to design special programs to protect older persons; perhaps we should even try to talk them out of their mistaken fears. This argument misses the point. The prospect of a confrontation with an obstrep- erous teenager or a drunken panhandler can be as fear-inducing for defenseless persons as the prospect of meeting an actual robber; indeed, to a defenseless person, the two kinds of confrontation are often in- distinguishable. Moreover, the lower rate at which the elderly are victimized is a measure of the steps they have already taken – chiefly, staying behind locked doors – to minimize the risks they face. Young men are more frequently attacked than older women, not because they are easier or more lucrative targets but because they are on the streets more.

Nor is the connection between disorderliness and fear made only by the elderly. Susan Estrich, of the Harvard Law School, has recently gathered together a number of surveys on the sources of public fear. One, done in Portland, Oregon, indicated that three-fourths of the adults interviewed cross to the other side of a street when they see a gang of teenagers; another survey, in Baltimore, discovered that nearly half would cross the street to avoid even a single strange youth. When an interviewer asked people in a housing project where the most dangerous spot was, they mentioned a place where young persons gathered to drink and play music, despite the fact that not a single crime had occurred there. In Boston public housing projects, the greatest fear was expressed by persons living in the buildings where disorderliness and incivility, not crime, were the greatest. Knowing this helps one understand the significance of such otherwise harmless displays as subway graffiti. As Nathan Glazer has written, the pro- liferation of graffiti, even when not obscene, confronts the subway rider with the "inescapable knowledge that the environment he must endure for an hour or more a day is uncontrolled and uncontrollable, and that anyone can invade it to do whatever damage and mischief the mind suggests."

In response to fear, people avoid one another, weak- ening controls. Sometimes they call the police. Patrol cars arrive, an occasional arrest occurs, but crime continues and disorder is not abated. Citizens complain to the police chief, but he explains that his department is low on personnel and that the courts do not punish petty or first-time offenders. To the residents, the police who arrive in squad cars are either ineffective or uncaring; to the police, the residents are animals who deserve each other. The citizens may soon stop calling the police, because "they can't do anything."

The process we call urban decay has occurred for centuries in every city. But what is happening today is different in at least two important respects. First, in the period before, say, World War II, city dwellers – because of money costs, transportation difficulties, familial and church connections – could rarely move away from neighborhood problems. When movement did occur, it tended to be along public-transit routes. Now mobility has become exceptionally easy for all but the poorest or those who are blocked by racial prejudice. Earlier crime waves had a kind of built-in self-correcting mechanism: the determination of a neighborhood or community to reassert control over its turf. Areas in Chicago, New York, and Boston would experience crime and gang wars, and then normalcy would return, as the families for whom no alternative residences were possible reclaimed their authority over the streets.

Second, the police in this earlier period assisted in that reassertion of authority by acting, sometimes violently, on behalf of the community. Young toughs were roughed up, people were arrested "on suspicion" or for vagrancy, and prostitutes and petty thieves were routed. "Rights" were something enjoyed by decent folk, and perhaps also by the serious professional criminal, who avoided violence and could afford a lawyer.

This pattern of policing was not an aberration or the result of occasional excess. From the earliest days of the nation, the police function was seen primarily as that of a night watchman: to maintain order against the chief threats to order – fire, wild animals, and dis- reputable behavior. Solving crimes was viewed not as a police responsibility but as a private one. In the March, 1969, *Atlantic*, one of us (Wilson) wrote a brief account of how the police role had slowly changed from maintaining order to fighting crimes. The change began with the creation of private detectives (often ex-criminals), who worked on a contingency-fee basis for individuals who had suffered losses. In time, the detectives were absorbed into municipal police agen- cies and paid a regular salary; simultaneously, the responsibility for prosecuting thieves was shifted from the aggrieved private citizen to the professional prose- cutor. This process was not complete in most places until the twentieth century.

In the 1960s, when urban riots were a major problem, social scientists began to explore carefully the order-maintenance function of the police, and to suggest ways of improving it – not to make streets safer (its original function) but to reduce the incidence of mass violence. Order-maintenance became, to a degree, coterminous with "community relations." But, as the crime wave that began in the early 1960s continued without abatement throughout the decade and into the 1970s, attention shifted to the role of the police as crime-fighters. Studies of police behavior ceased, by and large, to be accounts of the order-maintenance function and became, instead, efforts to propose and test ways whereby the police could solve more crimes, make more arrests, and gather better evidence. If these things could be done, social scientists assumed, citizens would be less fearful.

A great deal was accomplished during this transition, as both police chiefs and outside experts emphasized the crime-fighting function in their plans, in the allocation of resources, and in deployment of personnel. The police may well have become better crime-fighters as a result. And doubtless they remained aware of their responsibility for order. But the link between order-maintenance and crime-prevention, so obvious to earlier generations, was forgotten.

That link is similar to the process whereby one broken window becomes many. The citizen who fears the ill-smelling drunk, the rowdy teenager, or the importuning beggar is not merely expressing his distaste for unseemly behavior; he is also giving voice to a bit of folk wisdom that happens to be a correct generalization – namely, that serious street crime flourishes in areas in which disorderly behavior goes unchecked. The unchecked panhandler is, in effect, the first broken window. Muggers and robbers, whether opportunistic or professional, believe they reduce their chances of being caught or even identified if they operate on streets where potential victims are already intimidated by prevailing conditions. If the neighborhood cannot keep a bothersome panhandler from annoying passersby, the thief may reason, it is even less likely to call the police to identify a potential mugger or to interfere if the mugging actually takes place.

Some police administrators concede that this process occurs, but argue that motorized-patrol officers can deal with it as effectively as foot-patrol officers. We are not so sure. In theory, an officer in a squad car can observe as much as an officer on foot; in theory, the former can talk to as many people as the latter. But the reality of police–citizen encounters is powerfully altered by the automobile. An officer on foot cannot separate himself from the street people; if he is approached, only his uniform and his personality can help him manage whatever is about to happen. And he can never be certain what that will be – a request for directions, a plea for help, an angry denunciation, a teasing remark, a confused babble, a threatening gesture.

In a car, an officer is more likely to deal with street people by rolling down the window and looking at them. The door and the window exclude the approaching citizen; they are a barrier. Some officers take advantage of this barrier, perhaps unconsciously, by acting differently if in the car than they would on foot. We have seen this countless times. The police car pulls up to a corner where teenagers are gathered. The window is rolled down. The officer stares at the youths. They stare back. The officer says to one, "C'mere." He saunters over, conveying to his friends by his elaborately casual style the idea that he is not intimidated by authority "What's your name?" "Chuck." "Chuck who?" "Chuck Jones." "What'ya doing, Chuck?" "Nothin'." "Got a P.O. [parole officer]?" "Nah." "Sure?" "Yeah." "Stay out of trouble, Chuckie." Meanwhile, the other boys laugh and exchange comments among themselves, probably at the officer's expense. The officer stares harder. He cannot be certain what is being said, nor can he join in and, by displaying his own skill at street banter, prove that he cannot be "put down." In the process, the officer has learned almost nothing, and the boys have decided the officer is an alien force who can safely be disregarded, even mocked.

Our experience is that most citizens like to talk to a police officer. Such exchanges give them a sense of importance, provide them with the basis for gossip, and allow them to explain to the authorities what is worrying them (whereby they gain a modest but significant sense of having "done something" about the problem). You approach a person on foot more easily, and talk to him more readily, than you do a person in a car. Moreover, you can more easily retain some anonymity if you draw an officer aside for a private chat. Suppose you want to pass on a tip about who is stealing handbags, or who offered to sell you a stolen TV. In the inner city, the culprit, in all likelihood, lives nearby. To walk up to a marked patrol car and lean in the window is to convey a visible signal that you are a "fink."

The essence of the police role in maintaining order is to reinforce the informal control mechanisms of the community itself. The police cannot, without committing extraordinary resources, provide a substitute for that informal control. On the other hand, to reinforce those natural forces the police must accommodate them. And therein lies the problem.

Should police activity on the street be shaped, in important ways, by the standards of the neighborhood rather than by the rules of the state? Over the past two decades, the shift of police from order-maintenance to law-enforcement has brought them increasingly under the influence of legal restrictions, provoked by media complaints and enforced by court decisions and departmental orders. As a consequence, the order-maintenance functions of the police are now governed by rules developed to control police relations with suspected criminals. This is, we think, an entirely new development. For centuries, the role of the police as watchmen was judged primarily not in terms of its compliance with appropriate procedures but rather in terms of its attaining a desired objective. The objective was order, an inherently ambiguous term but a condition that people in a given community recognized when they saw it. The means were the same as those the community itself would employ, if its members were sufficiently determined, courageous, and authoritative. Detecting and apprehending criminals, by contrast, was a means to an end, not an end in itself; a judicial determination of guilt or innocence was the hoped-for result of the law-enforcement mode. From the first, the police were expected to follow rules defining that process, though states differed in how stringent the rules should be. The criminal-apprehension process was always understood to involve individual rights, the violation of which was unacceptable because it meant that the violating officer would be acting as a judge and jury – and that was not his job. Guilt or innocence was to be determined by universal standards under special procedures.

Ordinarily, no judge or jury ever sees the persons caught up in a dispute over the appropriate level of neighborhood order. That is true not only because most cases are handled informally on the street but also because no universal standards are available to settle arguments over disorder, and thus a judge may not be any wiser or more effective than a police officer. Until quite recently in many states, and even today in some places, the police make arrests on such charges as "suspicious person" or "vagrancy" or "public drunken-ness" – charges with scarcely any legal meaning. These charges exist not because society wants judges to punish vagrants or drunks but because it wants an officer to have the legal tools to remove undesirable persons from a neighborhood when informal efforts to preserve order in the streets have failed.

Once we begin to think of all aspects of police work as involving the application of universal rules under special procedures, we inevitably ask what constitutes an "undesirable person" and why we should "criminalize" vagrancy or drunkenness. A strong and commendable desire to see that people are treated fairly makes us worry about allowing the police to rout persons who are undesirable by some vague or parochial standard. A growing and not-so-com-mendable utilitarianism leads us to doubt that any behavior that does not "hurt" another person should be made illegal. And thus many of us who watch over the police are reluctant to allow them to perform, in the only way they can, a function that every neighborhood desperately wants them to perform.

This wish to "decriminalize" disreputable behavior that "harms no one" – and thus remove the ultimate sanction the police can employ to maintain neigh-borhood order – is, we think, a mistake. Arresting a single drunk or a single vagrant who has harmed no identifiable person seems unjust, and in a sense it is. But failing to do anything about a score of drunks or a hundred vagrants may destroy an entire community. A particular rule that seems to make sense in the individual case makes no sense when it is made a universal rule and applied to all cases. It makes no sense because it fails to take into account the con-nection between one broken window left untended and a thousand broken windows. Of course, agencies other than the police could attend to the problems posed by drunks or the mentally ill, but in most communities – especially where the "deinstitutionalization" movement has been strong – they do not.

The concern about equity is more serious. We might agree that certain behavior makes one person more undesirable than another, but how do we ensure that age or skin color or national origin or harmless mannerisms will not also become the basis for dis-tinguishing the undesirable from the desirable? How do we ensure, in short, that the police do not become the agents of neighborhood bigotry?

We can offer no wholly satisfactory answer to this important question. We are not confident that there is a satisfactory answer, except to hope that by their selection, training, and supervision, the police will be inculcated with a clear sense of the outer limit of their discretionary authority. That limit, roughly, is this – the police exist to help regulate behavior, not to maintain the racial or ethnic purity of a neighborhood.

Consider the case of the Robert Taylor Homes in Chicago, one of the largest public-housing projects in the country. It is home for nearly 20,000 people, all black, and extends over ninety-two acres along South State Street. It was named after a distinguished black who had been, during the 1940s, chairman of the Chicago Housing Authority. Not long after it opened, in 1962, relations between project residents and the police deteriorated badly. The citizens felt that the police were insensitive or brutal; the police, in turn, complained of unprovoked attacks on them. Some Chicago officers tell of times when they were afraid to enter the Homes. Crime rates soared.

Today, the atmosphere has changed. Police–citizen relations have improved – apparently, both sides learned something from the earlier experience. Recently, a boy stole a purse and ran off. Several young persons who saw the theft voluntarily passed along to the police information on the identity and residence of the thief, and they did this publicly, with friends and neighbors looking on. But problems persist, chief among them the presence of youth gangs that terrorize residents and recruit members in the project. The people expect the police to "do something" about this, and the police are determined to do just that.

But do what? Though the police can obviously make arrests whenever a gang member breaks the law, a gang can form, recruit, and congregate without breaking the law. And only a tiny fraction of gang-related crimes can be solved by an arrest; thus, if an arrest is the only recourse for the police, the residents' fears will go unassuaged. The police will soon feel helpless, and the residents will again believe that the police "do nothing." What the police in fact do is to chase known gang members out of the project. In the words of one officer, "We kick ass."

Project residents both know and approve of this. The tacit police–citizen alliance in the project is reinforced by the police view that the cops and the gangs are the two rival sources of power in the area, and that the gangs are not going to win.

None of this is easily reconciled with any conception of due process or fair treatment. Since both residents and gang members are black, race is not a factor. But it could be. Suppose a white project confronted a black gang, or vice versa. We would be apprehensive about the police taking sides. But the substantive problem remains the same: how can the police strengthen the informal social-control mechanisms of natural communities in order to minimize fear in public places? Law enforcement, per se, is no answer. A gang can weaken or destroy a community by standing about in a menacing fashion and speaking rudely to passersby without breaking the law.

We have difficulty thinking about such matters, not simply because the ethical and legal issues are so complex but because we have become accustomed to thinking of the law in essentially individualistic terms. The law defines *my* rights, punishes *his* behavior, and is applied by *that* officer because of *this* harm. We assume, in thinking this way, that what is good for the individual will be good for the community, and what doesn't matter when it happens to one person won't matter if it happens to many. Ordinarily, those are plausible assumptions. But in cases where behavior that is tolerable to one person is intolerable to many others, the reactions of the others – fear, withdrawal, flight – may ultimately make matters worse for everyone, including the individual who first professed his indifference.

It may be their greater sensitivity to communal as opposed to individual needs that helps explain why the residents of small communities are more satisfied with their police than are the residents of similar neighborhoods in big cities. Elinor Ostrom and her co-workers at Indiana University compared the perception of police services in two poor, all-black Illinois towns – Phoenix and East Chicago Heights – with those of three comparable all-black neighborhoods in Chicago. The level of criminal victimization and the quality of police–community relations appeared to be about the same in the towns and the Chicago neighborhoods, but the citizens living in their own villages were much more likely than those living in the Chicago neighborhoods to say that they do not stay at home for fear of crime, to agree that the local police have "the right to take any action necessary" to deal with problems, and to agree that the police "look out for the needs of the average citizen." It is possible that the residents and the police of the small towns saw themselves as engaged in a collaborative effort to maintain a certain standard of communal life, whereas those of the big city felt themselves to be simply requesting and supplying particular services on an individual basis.

If this is true, how should a wise police chief deploy his meager forces? The first answer is that nobody knows for certain, and the most prudent course of action would be to try further variations on the Newark experiment, to see more precisely what works in what kinds of neighborhoods. The second answer is also a hedge – many aspects of order-maintenance in neighborhoods can probably best be handled in ways that involve the police minimally, if at all. A busy, bustling shopping center and a quiet, well-tended suburb may need almost no visible police presence. In both cases, the ratio of respectable to disreputable people is ordinarily so high as to make informal social control effective.

Even in areas that are in jeopardy from disorderly elements, citizen action without substantial police involvement may be sufficient. Meetings between teenagers who like to hang out on a particular corner and adults who want to use that corner might well lead to an amicable agreement on a set of rules about how many people can be allowed to congregate, where, and when.

Where no understanding is possible – or if possible, not observed – citizen patrols may be a sufficient response. There are two traditions of communal involvement in maintaining order. One, that of the "community watchmen," is as old as the first settlement of the New World. Until well into the nineteenth century, volunteer watchmen, not policemen, patrolled their communities to keep order. They did so, by and large, without taking the law into their own hands – without, that is, punishing persons or using force. Their presence deterred disorder or alerted the community to disorder that could not be deterred. There are hundreds of such efforts today in communities all across the nation. Perhaps the best known is that of the Guardian Angels, a group of unarmed young persons in distinctive berets and T-shirts, who first came to public attention when they began patrolling the New York City subways but who claim now to have chapters in more than thirty American cities. Unfortunately, we have little information about the effect of these groups on crime. It is possible, however, that whatever their effect on crime, citizens find their presence reassuring, and that they thus contribute to maintaining a sense of order and civility.

The second tradition is that of the "vigilante." Rarely a feature of the settled communities of the East, it was primarily to be found in those frontier towns that grew up in advance of the reach of government. More than 350 vigilante groups are known to have existed; their distinctive feature was that their members did take the law into their own hands, by acting as judge, jury, and often executioner as well as policeman. Today, the vigilante movement is conspicuous by its rarity, despite the great fear expressed by citizens that the older cities are becoming "urban frontiers." But some community-watchmen groups have skirted the line, and others may cross it in the future. An ambiguous case, reported in *The Wall Street Journal*, involved a citizens' patrol in the Silver Lake area of Belleville, New Jersey. A leader told the reporter, "We look for outsiders." If a few teenagers from outside the neighborhood enter it, "we ask them their business," he said. "If they say they're going down the street to see Mrs. Jones, fine, we let them pass. But then we follow them down the block to make sure they're really going to see Mrs. Jones."

Though citizens can do a great deal, the police are plainly the key to order-maintenance. For one thing, many communities, such as the Robert Taylor Homes, cannot do the job by themselves. For another, no citizen in a neighborhood, even an organized one, is likely to feel the sense of responsibility that wearing a badge confers. Psychologists have done many studies on why people fail to go to the aid of persons being attacked or seeking help, and they have learned that the cause is not "apathy" or "selfishness" but the absence of some plausible grounds for feeling that one must personally accept responsibility. Ironically, avoiding responsibility is easier when a lot of people are standing about. On streets and in public places, where order is so important, many people are likely to be "around," a fact that reduces the chance of any one person acting as the agent of the community. The police officer's uniform singles him out as a person who must accept responsibility if asked. In addition, officers, more easily than their fellow citizens, can be expected to distinguish between what is necessary to protect the safety of the street and what merely protects its ethnic purity.

But the police forces of America are losing, not gaining, members. Some cities have suffered substantial cuts in the number of officers available for duty. These cuts are not likely to be reversed in the near future. Therefore, each department must assign its existing officers with great care. Some neighborhoods are so demoralized and crime-ridden as to make foot patrol useless; the best the police can do with limited resources is respond to the enormous number of calls for service. Other neighborhoods are so stable and serene as to make foot patrol unnecessary. The key is to identify neighborhoods at the tipping point where the public order is deteriorating but not unreclaimable, where the streets are used frequently but by apprehensive people, where a window is likely to be broken at any time, and must quickly be fixed if all are not to be shattered.

Most police departments do not have ways of systematically identifying such areas and assigning officers to them. Officers are assigned on the basis of crime rates (meaning that marginally threatened areas are often stripped so that police can investigate crimes in areas where the situation is hopeless) or on the basis of calls for service (despite the fact that most citizens do not call the police when they are merely frightened or annoyed). To allocate patrols wisely, the department must look at the neighborhoods and decide, from first-hand evidence, where an additional officer will make the greatest difference in promoting a sense of safety.

One way to stretch limited police resources is being tried in some public-housing projects. Tenant organizations hire off-duty police officers for patrol work in their buildings. The costs are not high (at least not per resident), the officer likes the additional income, and the residents feel safer. Such arrangements are probably more successful than hiring private watchmen, and the Newark experiment helps us understand why. A private security guard may deter crime or misconduct by his presence, and he may go to the aid of persons needing help, but he may well not intervene – that is, control or drive away someone challenging community standards. Being a sworn officer – a "real cop" – seems to give one the confidence, the sense of duty, and the aura of authority necessary to perform this difficult task.

Patrol officers might be encouraged to go to and from duty stations on public transportation and, while on the bus or subway car, enforce rules about smoking, drinking, disorderly conduct, and the like. The enforcement need involve nothing more than ejecting the offender (the offense, after all, is not one with which a booking officer or a judge wishes to be bothered). Perhaps the random but relentless maintenance of standards on buses would lead to conditions on buses that approximate the level of civility we now take for granted on airplanes.

But the most important requirement is to think that to maintain order in precarious situations is a vital job. The police know this is one of their functions, and they also believe, correctly, that it cannot be done to the exclusion of criminal investigation and responding to calls. We may have encouraged them to suppose, however, on the basis of our oft-repeated concerns about serious, violent crime, that they will be judged exclusively on their capacity as crime-fighters. To the extent that this is the case, police administrators will continue to concentrate police personnel in the highest-crime areas (though not necessarily in the areas most vulnerable to criminal invasion), emphasize their training in the law and criminal apprehension (and not their training in managing street life), and join too quickly in campaigns to decriminalize "harmless" behavior (though public drunkenness, street prostitution, and pornographic displays can destroy a community more quickly than any team of professional burglars).

Above all, we must return to our long-abandoned view that the police ought to protect communities as well as individuals. Our crime statistics and victimization surveys measure individual losses, but they do not measure communal losses. Just as physicians now recognize the importance of fostering health rather than simply treating illness, so the police – and the rest of us – ought to recognize the importance of maintaining, intact, communities without broken windows.

"The City as a Distorted Price System"

Psychology Today (1968)

Wilbur Thompson

Editors' Introduction

It was not until Wilbur Thompson's 1965 book, *A Preface To Urban Economics*, that economics departments began to teach courses in urban economics. Unlike urban sociology, which originated in late nineteenth-century Germany and flowered in the 1920s and 1930s with the Chicago School sociologists like Ernest W. Burgess (p. 150) and Louis Wirth (p. 90), the field of urban economics did not take hold until Thompson defined it.

Scholars from a specific discipline often feel people trained in the discipline will be the best decision makers. Architect Frank Lloyd Wright had the fanciful vision that a county architect should be the key policy maker at the local level (p. 331). Similarly, Thompson proposes the idea of a city economist to shape city policy. While cities don't have a position like that, economists within government and acting as consultants now bring the theory and practice Thompson and others developed to bear on decision making.

Three concepts Thompson introduces are fundamental in urban economics: the idea of collectively consumed public goods, merit goods designed to encourage desired behaviors, and payments to redistribute income.

Public goods are things provided free by government for the use of everyone. Air pollution control, police, and highways are examples. Thompson sees a place for this kind of good. Public goods all cost money. There is a price associated with providing them. But unlike the cost of goods in the private market the price of these public goods is often implicit, rather than explicit and, Thompson argues, failure to think through intended policy and price public goods accordingly leads to muddled and sometimes irrational policy. An example Thompson gives of how inattention to price can lead to bad public policy involves fireproofing. A rational public policy might be to encourage homeowners to invest in fireproofing their homes in order to reduce the risk of fire. This costs money and a fireproofed home will be worth more than it was before the fireproofing. Since homeowners' property taxes are based on the value of their homes, property taxes will go up. Having his property tax go up penalizes the prudent homeowner, whose investment reduces the cost of fireman responding to a fire or the risk of a fire on his property spreading to adjacent properties. If another homeowner in the same neighborhood does nothing to maintain his building and lets it deteriorate into a firetrap, he will be rewarded by lower property taxes. An alternative approach would be to reward the homeowner who fireproofs with some form of tax abatement or tax credit and to penalize the non-performing homeowner by increasing his taxes.

The government provides merit goods free because there has been a collective decision that they are important. Free polio shots are an example. With rare exceptions governments have felt that children should have polio shots. They want to encourage all parents to have their children immunized regardless of ability to pay. Even though Thompson calls merit goods "a case of the majority playing God, and 'coercing' the minority by the use of bribes to change their behavior," he acknowledges this may be a legitimate activity that should not be subject to market pricing in rare cases where the public interest requires the majority to force everyone to cooperate. For example, Thompson would feel that requiring all children to get polio shots is acceptable, because a small minority of holdouts who chose not to have their children immunized could put others at risk.

Payments to redistribute income are a third kind of payment that is not governed by price. The classic example is welfare payments to the indigent, such as a monthly payment to a blind, elderly person with no assets or income. One group (taxpayers) pays; another group (indigent, elderly, blind people) receive the goods.

Thompson feels that if clearly identified and intended these three types of goods – public goods, merit goods, and redistributive payments – can serve a useful role. Often, however, he feels that there is too little thought about the purpose of such payments and conceptual sloppiness about what is intended. If there were a city economist in a city, she might, for example, force a city council to think through just how much of a subsidy they care to give to a museum. While the city council may regard museum visits as a legitimate public good that should be the recipient of public tax dollars, a careful examination, informed by concepts such as public and merit goods, might lead them to decide that museum fees should be set high enough to cover half of the operating expenses of the museum or that there should be a differential fee structure so that most students and senior citizens pay a lower admission fee than the general public. Perhaps the city economist would convince the city council that a low museum entrance fee was necessary when the museum opened in order to lure patrons, but once patronage was established, higher fees would be in order. More extreme examples of public goods that should be, but often are not, carefully scrutinized are municipal golf courses and marinas. To what extent does it make sense for taxpayers to subsidize golfers and yacht owners? Are these considered merit goods? Public goods?

Thompson was an early advocate of using price to ration scarce goods – particularly use of highways and parking. Since this seminal article was written congestion pricing has become common in many parts of the world. Just as Thompson suggested might be done, tolls are set high during peak commute hours. People who value their mobility highly (affluent commuters rushing to get to work on time) will be willing to pay the high toll cost and will appreciate the lack of congestion. People for whom mobility at that time is less important (friends planning to meet get together to work out at the gym) may choose to exercise later and save the toll.

Wilbur Thompson's seminal urban economics book is *A Preface To Urban Economics* (Baltimore: Johns Hopkins Press, 1965). Many urban economics courses in the 1960s and 1970s were built around this book.

The leading contemporary urban economics text is Arthur O'Sullivan, *Urban Economics*, fifth edition (Chicago: McGraw-Hill, 2003). Other urban economics texts include: John McDonald, *Urban Economics: Theory and Policy* (Oxford: Blackwell, 2006), Brendan O'Flaherty, *City Economics* (Cambridge: Harvard University Press, 2005), Robert W. Wassmer (ed.), *Readings in Urban Economics: Issues and Public Policy* (Oxford: Blackwell, 2000).

Jane Jacobs, *The Economy of Cities* (New York: Vintage, 1970) is a provocative, contrarian book that sees cities as incubators of new economic ideas and argues in favor of experimentation.

▪ ▪ ▪ ▪ ▪ ▪

The failure to use price – as an *explicit* system – in the public sector of the metropolis is at the root of many, if not most, of our urban problems. Price, serving its historic functions, might be used to ration the use of existing facilities, to signal the desired directions of new public investment, to guide the distribution of income, to enlarge the range of public choice and to change tastes and behavior. Price performs such functions in the private market place, but it has been virtually eliminated from the public sector. We say "virtually eliminated" because it does exist but in an implicit, subtle, distorted sense that is rarely seen or acknowledged by even close students of the city, much less by public managers. Not surprisingly, this implicit price system results in bad economics.

We think of the property tax as a source of public revenue, but it can be reinterpreted as a price. Most often, the property tax is rationalized on "ability-to-pay" grounds with real property serving as a proxy for income. When the correlation between income and real property is challenged, the apologist for the property tax shifts ground and rationalizes it as a "benefit" tax. The tax then becomes a "price" which the property owner pays for the benefits received – fire protection, for example. But this implicit "price" for fire services is hardly a model of either efficiency or equity. Put in a new furnace and fireproof your building (reduce the likelihood of having a fire) and your property tax (fire service premium) goes up, let your property deteriorate and become a firetrap and your fire protection premium goes down! One bright note is New York City's one-year tax abatement on new pollution-control equipment; a timid step but in the right direction.

Often "urban sprawl" is little more than a color word which reflects (betrays?) the speaker's bias in favor of high population density and heavy interpersonal interaction – his "urbanity." Still, typically, the price of using urban fringe space has been set too low – well below the full costs of running pipes, wires, police cars and fire engines farther than would be necessary if building lots were smaller. Residential developers are, moreover, seldom discouraged (penalized by price) from "leap frogging" over the contiguous, expensive vacant land to build on the remote, cheaper parcels. Ordinarily, a flat price is charged for extending water or sewers to a new household regardless of whether the house is placed near to or far from existing pumping stations.

Again, the motorist is subject to the same license fees and tolls, if any, for the extremely expensive system of streets, bridges, tunnels, and traffic controls he enjoys, regardless of whether he chooses to drive downtown at the rush hour and thereby pushes against peak capacity or at off-peak times when it costs little or nothing to serve him. To compound this distortion of prices, we usually set the toll at zero. And when we do charge tolls, we quite perversely cut the commuter (rush-hour) rate below the off-peak rate.

It is not enough to point out that the motorist supports roadbuilding through the gasoline tax. The social costs of noise, air pollution, traffic control and general loss of urban amenities are borne by the general taxpayer. In addition, drivers during off-peak hours overpay and subsidize rush-hour drivers. Four lanes of expressway or bridge capacity are needed in the morning and evening rush hours where two lanes would have served if movements had been random in time and direction: that is, near constant in average volume. The peak-hour motorists probably should share the cost of the first two lanes and bear the full cost of the other two that they alone require. It is best to begin by carefully distinguishing where market tests are possible and where they are not. Otherwise, the case for applying the principles of price is misunderstood; either the too-ardent advocate overstates his case or the potential convert projects too much. In either case, a "disenchantment" sets in that is hard to reverse.

Much of the economics of the city is "public economies," and the pricing of urban public services poses some very difficult and even insurmountable problems. Economists have, in fact, erected a very elegant rationalization of the public economy almost wholly on the nonmarketability of public goods and services. While economists have perhaps oversold the inapplicability of price in the public sector, let us begin with what we are not talking about.

The public economy supplies "collectively consumed" goods, those produced and consumed in one big indivisible lump. Everyone has to be counted in the system, there is no choice of *in* or *out*. We cannot identify individual benefits, therefore we cannot exact a *quid pro quo*. We cannot exclude those who would not pay voluntarily; therefore we must turn to compulsory payments: taxes. Justice and air-pollution control are good examples of collectively consumed public services.

A second function of the public economy is to supply "merit goods." Sometimes the majority of us become a little paternalistic and decide that we know what is best for all of us. We believe some goods are especially meritorious, like education, and we fear that others might not fully appreciate this truth. Therefore, we produce these merit goods, at considerable cost, but offer them at a zero price. Unlike the first case of collectively consumed goods, we could sell these merit goods. A schoolroom's doors can be closed to those who do not pay, *quite unlike justice*. But we choose to open the doors wide to ensure that no one will turn away from the service because of its cost, and then we finance the service with compulsory payments. Merit goods are a case of the majority playing God, and "coercing" the minority by the use of bribes to change their behavior.

A third classic function of government is the redistribution of income. Here we wish to perform a service for one group and charge another group the cost of that service. Welfare payments are a clear case. Again, any kind of a private market or pricing mechanism is totally inappropriate; we obviously do not expect welfare recipients to return their payments. Again, we turn to compulsory payments: taxes. In sum, the private market may not be able to process certain goods and services (pure "public goods"), or it may give the "wrong" prices ("merit goods"), or we simply do not want the consumer to pay (income-redistributive services).

But the virtual elimination of price from the public sector is an extreme and highly simplistic response to the special requirements of the public sector. Merit goods may be subsidized without going all the way to zero prices. Few would argue for full-cost admission prices to museums, but a good case can be made for moderate prices that cover, say, their daily operating costs, (e.g., salaries of guards and janitors, heat and light).

Unfortunately, as we have given local government more to do, we have almost unthinkingly extended the tradition of "free" public services to every new undertaking, despite the clear trend in local government toward the assumption of more and more functions that do not fit the neat schema above. The provision of free public facilities for automobile movement in the crowded cores of our urban areas can hardly be defended on the grounds that: (a) motorists could not be excluded from the expressways if they refused to pay the toll, or (b) the privately operated motor vehicle is an especially meritorious way to move through densely populated areas, or (c) the motorists cannot afford to pay their own way and that the general (property) taxpayers should subsidize them. And all this applies with a vengeance to municipal marinas and golf courses.

PRICES TO RATION THE USE OF EXISTING FACILITIES

We need to understand better the rationing function of price as it manifests itself in the urban public sector: how the demand for a temporarily (or permanently) fixed stock of a public good or service can be adjusted to the supply. At any given time the supply of street, bridge, and parking space is fixed; "congestion" on the streets and a "shortage" of parking space express demand greater than supply at a zero price, a not too surprising phenomenon. Applying the market solution, the shortage of street space at peak hours ("congestion") could have been temporarily relieved (rationalized) by introducing a short-run rationing price to divert some motorists to other hours of movement, some to other modes of transportation, and some to other activities.

Public goods last a long time and therefore current additions to the stock are too small to relieve shortages quickly and easily. *The rationing function of price is probably more important in the public sector where it is customarily ignored than in the private sector where it is faithfully expressed.*

Rationing need not always be achieved with money, as when a motorist circles the block over and over

looking for a place to park. The motorist who is not willing to "spend time" waiting and drives away forfeits the scarce space to one who will spend time (luck averaging out). The parking "problem" may be reinterpreted as an implicit decision to keep the money price artificially low (zero or a nickel an hour in a meter) and supplement it with a waiting cost or time price. The problem is that we did not clearly understand and explicitly agree to do just that.

The central role of price is to allocate – across the board – scarce resources among competing ends to the point where the value of another unit of any good or service is equal to the incremental cost of producing that unit. Expressed loosely, in the long run we turn from using prices to dampen demand to fit a fixed supply to adjusting the supply to fit the quantity demanded, at a price which reflect the production costs.

Prices which ration also serve to signal desired new directions in which to reallocate resources. If the rationing price exceeds those costs of production which the user is expected to bear directly, more resources should ordinarily be allocated to that activity. And symmetrically a rationing price below the relevant costs indicates an uneconomic provision of that service in the current amounts. Rationing prices reveal the intensity of the users' demands. How much is it really worth to drive into the heart of town at rush hour or launch a boat? In the long run, motorists and boaters should be free to choose, in rough measure, the amount of street and dock space they want and for which they are willing to pay. But, as in the private sector of our economy, free choice would carry with it full (financial) responsibility for that choice.

We need also to extend our price strategy to "factor prices"; we need a sophisticated wage policy for local public employees. Perhaps the key decision in urban development pertains to the recruiting and assignment of elementary- and secondary-school teachers. The more able and experienced teachers have the greater range of choice in post and quite naturally they choose the newer schools in the better neighborhoods, after serving the required apprenticeship in the older schools in the poorer neighborhoods. Such a pattern of migration certainly cannot implement a policy of equality of opportunity.

This author argued six years ago that egalitarianism in the public school system has been overdone; even the army recognizes the role of price when it awards extra "jump pay" to paratroopers, only a slightly more hazardous occupation than teaching behind the lines. Besides, it is male teachers whom we need to attract to slum schools, both to serve as father figures where there are few males at home and to serve quite literally as disciplinarians. It is bad economics to insist on equal pay for teachers everywhere throughout the urban area when males have a higher productivity in some areas and when males have better employment opportunities outside teaching – higher "opportunity costs" that raise their supply price. It is downright silly to argue that "equal pay for equal work" is achieved by paying the same money wage in the slums as in the suburbs.

About a year ago, on being offered premium salaries for service in ghetto schools, the teachers rejected, by name and with obvious distaste, any form of "jump pay." One facile argument offered was that they must protect the slum child from the stigma of being harder to teach, a nicety surely lost on the parents and outside observers. One suspects that the real reason for avoiding salary differentials between the "slums and suburbs" is that the teachers seek to escape the hard choice between the higher pay and the better working conditions. *But that is precisely what the price system is supposed to do: equalize sacrifice.*

PRICES TO GUIDE THE DISTRIBUTION OF INCOME

A much wider application of tolls, fees, fines, and other "prices" would also confer greater control over the distribution of income for two distinct reasons. First, the taxes currently used to finance a given public service create *implicit* and *unplanned* redistribution of income. Second, this drain on our limited supply of tax money prevents local government from undertaking other programs with more *explicit* and *planned* redistributional effects.

More specifically, if upper-middle- and upper-income motorists, golfers, and boaters use subsidized public streets, golf links, and marinas more than in proportion to their share of local tax payments from which the subsidy is paid, then these public activities redistribute income toward greater inequality. Even if these activities were found to be neutral with respect to the distribution of income, public provision of these discretionary services comes at the expense of a roughly equivalent expenditure on the more classic public services: protection, education, public health, and welfare.

Self-supporting public golf courses are so common and marinas are such an easy extension of the same principle that it is much more instructive to test the faith by considering the much harder case of the public museum: "culture." Again, we must recall that it is the middle- and upper-income classes who typically visit museums, so that free admission becomes, in effect, redistribution toward greater inequality, to the extent that the lower-income nonusers pay local taxes (e.g., property taxes directly or indirectly through rent, local sales taxes). The low prices contemplated are not, moreover, likely to discourage attendance significantly and the resolution of special cases (e.g., student passes) seems well within our competence.

Unfortunately, it is not obvious that "free" public marinas as tennis courts pose foregone alternatives – "opportunity costs." If we had to discharge a teacher or policeman every time we built another boat dock or tennis court, we would see the real cost of these public services. But in a growing economy, we need only not hire another teacher or policeman and that is not so obvious. In general, then, given a binding local budget constraint – scarce tax money – to undertake a local public service that is unequalizing or even neutral in income redistribution is to deny funds to programs that have the desired distributional effect, and is to lose control over equity.

Typically, in oral presentations at question time, it is necessary to reinforce this point by rejoining, "No, I would not put turnstiles in the playgrounds in poor neighborhoods, rather it is only because we do put turnstiles at the entrance to the playgrounds for the middle- and upper-income groups that we will be able to 'afford' playgrounds for the poor."

PRICES TO ENLARGE THE RANGE OF CHOICE

But there is more at stake in the contemporary chaos of hidden and unplanned prices than "merely" efficiency and equity. *There is no urban goal on which consensus is more easily gained than the pursuit of great variety and choice – "pluralism."* The great rural to urban migration was prompted as much by the search for variety as by the decline of agriculture and rise of manufacturing. Wide choice is seen as the saving grace of bigness by even the sharpest critics of the metropolis. Why, then, do we tolerate far less variety in our big cities than we could have? We have lapsed into a state of tyranny by the majority, in matters of both taste and choice.

In urban transportation the issue is not, in the final analysis, whether users of core-area street space at peak hours should or should not be required to pay their own way in full. The problem is, rather, that by not forcing a direct *quid pro quo* in money, we implicitly substitute a new means of payment – time in the transportation services "market." The peak-hour motorist does pay in full, through congestion and time delay. But *implicit choices* blur issues and confuse decision making.

Say we were carefully to establish how many more dollars would have to be paid in for the additional capacity needed to save a given number of hours spent commuting. The *majority* of urban motorists perhaps would still choose the present combination of "under-investment" in highway, bridge and parking facilities, with a compensatory heavy investment of time in slow movement over these crowded facilities. Even so, a substantial minority of motorists do prefer a different combination of money and time cost. A more affluent, long-distance commuter could well see the current level of traffic congestion as a real problem and much prefer to spend more money to save time. If economies of scale are so substantial that only one motorway to town can be supported, or if some naturally scarce factor (e.g., bridge or tunnel sites) prevents parallel transportation facilities of different quality and price, then the preferences of the minority must be sacrificed to the majority interest and we do have a real "problem." But, ordinarily, in large urban areas there are a number of near-parallel routes to town, and an unsatisfied minority group large enough to justify significant differentiation of one or more of these streets and its diversion to their use. Greater choice through greater scale is, in fact, what bigness is all about.

The simple act of imposing a toll, at peak hours, on one of these routes would reduce its use, assuming that nearby routes are still available without user charges, thereby speeding movement of the motorists who remain and pay. The toll could be raised only to the point where some combination of moderately rapid movement and high physical output were jointly optimized. Otherwise the outcry might be raised that the public transportation authority was so elitist as to gratify the desire of a few very wealthy motorists for very rapid movement, heavily overloading the "free" routes. It is, moreover, quite possible, even probable, that the newly converted, rapid-flow, toll-route would

handle as many vehicles as it did previously as a congested street and not therefore spin off any extra load on the free routes.

Our cities cater, at best, to the taste patterns of the middle-income class, as well they should, *but not so exclusively*. This group has chosen, indirectly through clumsy and insensitive tax-and-expenditure decisions and ambiguous political processes, to move about town flexibly and cheaply, but slowly, in private vehicles. Often, and almost invariably in the larger urban areas, we would not have to encroach much on this choice to accommodate also those who would prefer to spend more money and less time, in urban movement. In general, we should permit urban residents to pay in their most readily available "currency" – time or money.

Majority rule by the middle class in urban transportation has not only disenfranchised the affluent commuter, but more seriously it has debilitated the low-fare, mass transit system on which the poor depend. The effect of widespread automobile ownership and use on the mass transportation system is an oft-told tale: falling bus and rail patronage leads to less frequent service and higher overhead costs per trip and often higher fares which further reduce demand and service schedules. Perhaps two-thirds or more of the urban residents will tolerate and may even prefer slow, cheap automobile movement. But the poor are left without access to many places of work – the suburbanizing factories in particular – and they face much reduced opportunities for comparative shopping, and highly constrained participation in the community life in general. A truly wide range of choice in urban transportation would allow the rich to pay for fast movement with money, the middle-income class to pay for the privacy and convenience of the automobile with time, and the poor to economize by giving up (paying with) privacy.

A more sophisticated price policy would expand choice in other directions. Opinions differ as to the gravity of the water-pollution problem near large urban areas. The minimum level of dissolved oxygen in the water that is needed to meet the standards of different users differs greatly, as does the incremental cost that must be incurred to bring the dissolved oxygen levels up to successively higher standards. The boater accepts a relatively low level of "cleanliness" acquired at relatively little cost. Swimmers have higher standards attained only at much higher cost. Fish and fisherman can thrive only with very high levels of dissolved oxygen acquired only at the highest cost. Finally, one can imagine an elderly convalescent or an impoverished slum dweller or a confirmed landlubber who is not at all interested in the nearby river. What, then, constitutes "clean"?

A majority rule decision, whether borne by the citizen directly in higher taxes or levied on the industrial polluters and then shifted on to the consumer in higher produce prices, is sure to create a "problem." If the pollution program is a compromise – a halfway measure – the fisherman will be disappointed because the river is still not clean enough for his purposes and the landlubbers will be disgruntled because the program is for "special interests" and he can think of better uses for his limited income. Surely, we can assemble the managerial skills in the local public sector needed to devise and administer a structure of user charges that would extend choice in outdoor recreation, consistent with financial responsibility, with lower charges for boat licenses and higher charges for fishing licenses.

Perhaps the most fundamental error we have committed in the development of our large cities is that we have too often subjected the more affluent residents to petty irritations which serve no great social purpose, then turned right around and permitted this same group to avoid responsibilities which have the most critical and pervasive social ramifications. It is a travesty and a social tragedy that we have prevented the rich from buying their way out of annoying traffic congestion – or at least not helped those who are long on money and short on time arrange such an accommodation. Rather, we have permitted them, through political fragmentation and flight to tax havens, to evade their financial and leadership responsibilities for the poor of the central cities. That easily struck goal, "pluralism and choice," will require much more managerial sophistication in the local public sector than we have shown to date.

PRICING TO CHANGE TASTES AND BEHAVIOR

Urban managerial economies will probably also come to deal especially with "developmental pricing" analogous to "promotional pricing" in business. Prices below cost may be used for a limited period to create a market for a presumed "merit good." The hope would be that the artificially low price would stimulate consumption and that an altered *expenditure pattern* (practice) would lead in time to an altered *taste pattern* (preference), as experience with the new service led to a fuller appreciation of it. Ultimately, the subsidy would be withdrawn, whether or not tastes changed sufficiently to make the new service self-supporting – provided, of course, that no permanent redistribution of income was intended.

For example, our national parks had to be subsidized in the beginning and this subsidy could be continued indefinitely on the grounds that these are "merit goods" that serve a broad social interest. But long experience with outdoor recreation has so shifted tastes that a large part of the costs of these parks could now be paid for by a much higher set of park fees.

It is difficult, moreover, to argue that poor people show up at the gates of Yellowstone Park, or even the much nearer metropolitan area regional parks, in significant number, so that a subsidy is needed to continue provision of this service for the poor. A careful study of the users and the incidence of the taxes raised to finance our parks may even show a slight redistribution of income toward greater inequality.

Clearly, this is not the place for an economist to pontificate on the psychology of prices but a number of very interesting phenomena that seem to fall under this general heading deserve brief mention. A few simple examples of how charging a price changes behavior are offered, but left for others to classify.

In a recent study of depressed areas, the case was cited of a community-industrial-development commission that extended its fund-raising efforts from large business contributors to the general public in a supplementary "nickel and dime" campaign. They hoped to enlist the active support of the community at large, more for reasons of public policy than for finance. But even a trivial financial stake was seen as a means to create broad and strong public identification with the local industrial development programs and to gain their political support.

Again, social-work agencies have found that even a nominal charge for what was previously a free service enhances both the self-respect of the recipient and his respect for the usefulness of the service. Paradoxically, we might experiment with higher public assistance payments coupled to *nominal* prices for selected public health and family services, personal counseling, and surplus foods.

To bring a lot of this together now in a programmatic way, we can imagine a very sophisticated urban public management beginning with below-cost prices on, say, the new rapid mass transit facility during the promotional period of luring motorists from their automobiles and of "educating" them on the advantages of a carefree journey to work. Later, if and when the new facility becomes crowded during rush hours and after a taste of this new transportation mode has become well established, the "city economist" might devise a-three-price structure of fares: the lowest fare for regular off-peak use, the middle fare for regular peak use (tickets for commuters), and the highest fare for the occasional peak-time user. Such a schedule would reflect each class's contribution to the cost of having to carry standby capacity.

If the venture more than covered its costs of operation, the construction of additional facilities would begin. Added social benefits in the form of a cleaner, quieter city or reduced social costs of traffic control and accidents could be included in the cost accounting ("cost-benefit analysis") underlying the fare structure. But below-cost fares, taking care to count social as well as private costs, would not be continued indefinitely except for merit goods or when a clear income-redistribution end is in mind. And, even then, not without careful comparison of the relative efficiency of using the subsidy money in alternative redistributive programs. We need, it would seem, not only a knowledge of the economy of the city, but some very knowledgeable city economists as well.

"The Competitive Advantage of the Inner City"

Harvard Business Review (1995)

Michael Porter

Editors' Introduction

Harvard Business School professor Michael Porter declares that the economic distress of inner-city neighborhoods may be the most pressing issue facing the United States. Like William Julius Wilson (p. 110), Porter sees lack of jobs as the root cause of drug abuse, crime, and other social problems. And like Wilson, Porter feels that government response to the problem of inner-city decline and lack of jobs has been ineffective.

Porter characterizes the current approach as based on a *social model* aimed at the individual. Governments have historically provided "relief" from the effects of economic problems on individuals in the form of welfare payments, subsidized housing, food stamps, and free medical assistance. Corporate philanthropy has often done the same.

Porter feels that government programs to create jobs in inner-city neighborhoods and train local residents for them have been fragmented and inefficient. He notes that they have consisted of subsidies, preference programs, and expensive efforts to stimulate economic activity. Governments have dumped money on small, marginal inner-city businesses, which could not turn a profit without government help. They have hired incompetent workers or required private sector contractors to hire them. Or they have spent large sums to get hopelessly blighted redevelopment areas and badly contaminated brownfield areas back into usable condition. Often the subsidized firms fail, the workers hired through preference programs are fired, and the brownfields and redevelopment project areas sit vacant or are "showcase" projects that burn up money better spent elsewhere.

Porter argues for a new economic model of inner-city revitalization. He favors private, for-profit initiatives based on economic self-interest, not artificial inducements, charity, and government mandates. Such an approach will only work, he says, if it takes advantage of the true competitive advantages of the inner city. Like Wilbur Thompson (p. 266), Porter's approach is grounded in the realities of free market economics and responsive to price.

Porter punctures the myth that inner-city real estate or labor costs are sufficiently lower to make a compelling reason for firms to locate in inner-city neighborhoods rather than in suburban or exurban locations. Rather, he argues, there are four true advantages to inner-city locations: (1) their strategic location, (2) local market demand the areas themselves possess, (3) possibilities of integration with regional job clusters, and (4) an industrious labor force that is eager to work. Firms that can best exploit these inner-city advantages may find inner-city locations the best place to do business. Unlike Wilson, who emphasizes the loss of jobs that people with limited education can perform, Porter argues that there are still firms that can use unskilled labor for warehouse and production-line workers, truck drivers, and other unskilled jobs.

Porter places a large part of the blame for high costs and the difficulties of doing business in inner cities on government regulation and anti-business attitudes. He argues that local governments can improve the economic climate and make inner-city neighborhoods more attractive for private investment by reducing regulation. This has been a theme in British Enterprise Zone and US Empowerment Zone/Enterprise Community Programs that have reduced environmental, occupational health, and other government regulations in distressed neighborhoods in order to lure in business.

What would his Harvard colleague William Julius Wilson (p. 110) think of Porter's characterization of the social model and preference programs? Given historic racism, bad schools, and the lack of skills many inner-city Black ghetto residents possess, aren't preference programs necesssary?

Michael Porter is the Bishop William Lawrence University Professor of Business Administration at the Harvard Business School and the Director of Harvard's Institute for Strategy and Competitiveness. He is a leading authority on competitive strategy and the competitiveness and economic development of nations, states, and regions. Porter is the author of over 125 articles and 17 books including: *Michael E. Porter on Competition* (Boston: Harvard Business School Press, 1998), *Competitive Advantage: Creating and Sustaining Superior Performance* (New York: Free Press, 1998), *Competitive Strategy: Techniques for Analyzing Industries and Competitors* (New York: Free Press, 1998), *On Competition* (Boston: Harvard Business School Press, 1998), and *The Competitive Advantage of Nations* (New York: Free Press, 1990).

Professor Porter extended the ideas advanced in this selection in "New Strategies for Inner-City Economic Development" (*Economic Development Quarterly* 11(1), February 1997).

Books on community-based economic development – generally less oriented to private sector approaches than Porter – include Sammis White, Richard D. Bingham, and Edward W. Hill (eds), *Financing Economic Development in the 21st Century* (Birmingham: M.E. Sharpe, 2003), Edward J. Blakeley and Ted Bradshaw, *Planning Local Economic Development: Theory and Practice* (Thousand Oaks: Sage 2002), Joan Fitzgerald, and Nancey Green Leigh, *Economic Revitalization: Cases and Strategies for City and Suburb* (Thousand Oaks: Sage, 2002), Michael H. Shuman, *Going Local: Creating Self-reliant Communities in a Global Age* (New York: Free Press, 1998), Douglas Henton, John Melville, and Kimberly Walsh, *Grassroots Leaders for a New Economy* (San Francisco: Jossey-Bass, 1997), W. Dennis Keating, Norman Krumholz, and Philip Star (eds), *Revitalizing Urban Neighborhoods* (Kansas City: University Press of Kansas, 1996), and Richard P. Taub, *Community Capitalism* (Boston: Harvard Business School Press, 1994).

■ ■ ■ ■ ■ ■

The economic distress of America's inner cities may be the most pressing issue facing the nation. The lack of businesses and jobs in disadvantaged urban areas fuels not only a crushing cycle of poverty but also crippling social problems, such as drug abuse and crime. And, as the inner cities continue to deteriorate, the debate on how to aid them grows increasingly divisive.

The sad reality is that the efforts of the past few decades to revitalize the inner cities have failed. The establishment of a sustainable economic base – and with it employment opportunities, wealth creation, role models, and improved local infrastructure – still eludes us despite the investment of substantial resources.

Past efforts have been guided by a social model built around meeting the needs of individuals. Aid to inner cities, then, has largely taken the form of relief programs such as income assistance, housing subsidies, and food stamps, all of which address highly visible – and real – social needs.

Programs aimed more directly at economic development have been fragmented and ineffective. These piecemeal approaches have usually taken the form of subsidies, preference programs, or expensive efforts to stimulate economic activity in tangential fields such as housing, real estate, and neighborhood development. Lacking an overall strategy, such programs have treated the inner city as an island isolated from the surrounding economy and subject to its own unique laws of competition. They have encouraged and supported small, subscale businesses designed to serve the local community but ill equipped to attract the community's own spending power, much less export outside it. In short, the social model has inadvertently undermined the creation of economically viable companies. Without such companies and the jobs they create, the social problems will only worsen.

The time has come to recognize that revitalizing the inner city will require a radically different approach. While social programs will continue to play a critical role in meeting human needs and improving education, they must support – and not undermine – a coherent economic strategy. The question we should be asking is how inner-city-based businesses and nearby employment opportunities for inner-city residents can proliferate and grow. A sustainable economic base can be created in the inner city, but only as it has been created elsewhere: through private, for-profit initiatives and investment based on economic self-interest and genuine competitive advantage – not through artificial inducements, charity, or government mandates.

We must stop trying to cure the inner city's problems by perpetually increasing social investment and hoping for economic activity to follow. Instead, an economic model must begin with the premise that inner-city businesses should be profitable and should be positioned to compete on a regional, national, and even international scale. These businesses should be capable not only of serving the local community but also of exporting goods and services to the surrounding economy. The cornerstone of such a model is to identify and exploit the competitive advantages of inner cities that will translate into truly profitable businesses.

Our policies and programs have fallen into the trap of redistributing wealth. The real need – and the real opportunity – is to create wealth.

TOWARDS A NEW MODEL: LOCATION AND BUSINESS DEVELOPMENT

Economic activity in and around inner cities will take root if it enjoys a competitive advantage and occupies a niche that is hard to replicate elsewhere. If companies are to prosper, they must find a compelling competitive reason for locating in the inner city. A coherent strategy for development starts with that fundamental eco-

nomic principle, as the contrasting experiences of the following companies illustrate.

Alpha Electronics (the company's name has been disguised), a 28-person company that designed and manufactured multimedia computer peripherals, was initially based in lower Manhattan. In 1987, the New York City Office of Economic Development set out to orchestrate an economic "renaissance" in the South Bronx by inducing companies to relocate there. Alpha, a small but growing company, was sincerely interested in contributing to the community and eager to take advantage of the city's willingness to subsidize its operations. The city, in turn, was happy that a high-tech company would begin to stabilize a distressed neighborhood and create jobs. In exchange for relocating, the city provided Alpha with numerous incentives that would lower costs and boost profits. It appeared to be an ideal strategy.

By 1994, however, the relocation effort had proved a failure for all concerned. Despite the rapid growth of its industry, Alpha was left with only 8 of its original 28 employees. Unable to attract high-quality employees to the South Bronx or to train local residents, the company was forced to outsource its manufacturing and some of its design work. Potential suppliers and customers refused to visit Alpha's offices. Without the city's attention to security, the company was plagued by theft.

What went wrong? Good intentions notwithstanding, the arrangement failed the test of business logic. Before undertaking the move, Alpha and the city would have been wise to ask themselves why none of the South Bronx's thriving businesses was in electronics. The South Bronx as a location offered no specific advantages to support Alpha's business, and it had several disadvantages that would prove fatal. Isolated from the lower Manhattan hub of computer-design and software companies, Alpha was cut off from vital connections with customers, suppliers, and electronic designers.

In contrast, Matrix Exhibits, a $2.2 million supplier of trade-show exhibits that has 30 employees, is thriving in Atlanta's inner city. When Tennessee-based Matrix decided to enter the Atlanta market in 1985, it could have chosen a variety of locations. All the other companies that create and rent trade show exhibits are based in Atlanta's suburbs. But the Atlanta World Congress Center, the city's major exhibition space, is just a six-minute drive from the inner city, and Matrix chose the location because it provided a real competitive advantage. Today Matrix offers customers superior response time, delivering trade-show exhibits faster than its suburban competitors. Matrix benefits from low rental rates for warehouse space – about half the rate its competitors pay for similar space in the suburbs – and draws half its employees from the local community. The commitment of local police has helped the company avoid any serious security problems. Today Matrix is one of the top five exhibition houses in Georgia.

Alpha and Matrix demonstrate how location can be critical to the success or failure of a business. Every location – whether it be a nation, a region, or a city – has a set of unique local conditions that underpin the ability of companies based there to compete in a particular field. The competitive advantage of a location does not usually arise in isolated companies but in clusters of companies – in other words, in companies that are in the same industry or otherwise linked together through customer, supplier, or similar relationships. Clusters represent critical masses of skill, information, relationships, and infrastructure in a given field. Unusual or sophisticated local demand gives companies insight into customers' needs. Take Massachusetts's highly competitive cluster of information-technology industries: it includes companies specializing in semiconductors, workstations, supercomputers, software, networking equipment, databases, market research, and computer magazines.

Clusters arise in a particular location for specific historical or geographic reasons – reasons that may cease to matter over time as the cluster itself becomes powerful and competitively self-sustaining. In successful clusters such as Hollywood, Silicon Valley, Wall Street, and Detroit, several competitors often push one another to improve products and processes. The presence of a group of competing companies contributes to the formation of new suppliers, the growth of companies in related fields, the formation of specialized training programs, and the emergence of technological centers of excellence in colleges and universities. The clusters also provide newcomers with access to expertise, connections, and infrastructure that they in turn can learn and exploit to their own economic advantage.

If locations (and the events of history) give rise to clusters, it is clusters that drive economic development. They create new capabilities, new companies, and new industries. I initially described this theory of location

in *The Competitive Advantage of Nations* (Free Press, 1990), applying it to the relatively large geographic areas of nations and states. But it is just as relevant to smaller areas such as the inner city. To bring the theory to bear on the inner city, we must first identify the inner city's competitive advantages and the ways inner-city businesses can forge connections with the surrounding urban and regional economies.

THE TRUE ADVANTAGES OF THE INNER CITY

The first step toward developing an economic model is identifying the inner city's true competitive advantages. There is a common misperception that the inner city enjoys two main advantages: low-cost real estate and labor. These so-called advantages are more illusory than real. Real estate and labor costs are often higher in the inner city than in suburban and rural areas. And even if inner cities were able to offer lower-cost labor and real estate compared with other locations in the United States, basic input costs can no longer give companies from relatively prosperous nations a competitive edge in the global economy. Inner cities would inevitably lose jobs to countries like Mexico or China, where labor and real estate are far cheaper.

Only attributes that are unique to inner cities will support viable businesses. My ongoing research of urban areas across the United States identifies four main advantages of the inner city: strategic location, local market demand, integration with regional clusters, and human resources. Various companies and programs have identified and exploited each of those advantages from time to time. To date, however, no systematic effort has been mounted to harness them.

Strategic location

Inner cities are located in what *should* be economically valuable areas. They sit near congested high-rent areas, major business centers, and transportation and communications nodes. As a result, inner cities can offer a competitive edge to companies that benefit from proximity to downtown business districts, logistical infrastructure, entertainment or tourist centers, and concentrations of companies.

Local market demand

The inner-city market itself represents the most immediate opportunity for inner-city-based entre-preneurs and businesses. At a time when most other markets are saturated, inner-city markets remain poorly served – especially in retailing, financial services, and personal services. In Los Angeles, for example, retail penetration per resident in the inner city compared with the rest of the city is 35% in supermarkets, 40% in department stores, and 50% in hobby, toy, and game stores.

The first notable quality of the inner-city market is its size. Even though average inner-city incomes are relatively low, high population density translates into an immense market with substantial purchasing power. Boston's inner city, for example, has an estimated total family income of $3.4 billion.

Spending power per acre is comparable with the rest of the city despite a 21% lower average household income level than in the rest of Boston, and, more significantly, higher than in the surrounding suburbs. In addition, the market is young and growing rapidly, owing in part to immigration and relatively high birth rates.

Integration with regional clusters

The most exciting prospects for the future of inner-city economic development lie in capitalizing on nearby regional clusters: those unique-to-a-region collections of related companies that are competitive nationally and even globally. For example, Boston's inner city is next door to world-class financial-services and health-care clusters. South Central Los Angeles is close to an enormous entertainment cluster and a large logistical-services and wholesaling complex.

The ability to access competitive clusters is a very different attribute – and one much more far reaching in economic implication – than the more generic advantage of proximity to a large downtown area with concentrated activity. Competitive clusters create two types of potential advantages. The first is for business formation. Companies providing supplies, compo-nents, and support services could be created to take advantage of the inner city's proximity to multiple nearby customers in the cluster . . .

The second advantage of these clusters is the potential they offer inner-city companies to compete in downstream products and services. For example, an inner-city company could draw on Boston's strength in financial services to provide services tailored to inner-city needs – such as secured credit cards, factor-ing, and mutual funds – both within and outside the inner city in Boston and elsewhere in the country.

[. . .]

Human resources

The inner city's fourth advantage takes on a number of deeply entrenched myths about the nature of its residents. The first myth is that inner-city residents do not want to work and opt for welfare over gainful employment. Although there is a pressing need to deal with inner-city residents who are unprepared for work, most inner-city residents are industrious and eager to work. For moderate-wage jobs ($6 to $10 per hour) that require little formal education (for instance, warehouse workers, production-line workers, and truck drivers), employers report that they find hard-working, dedicated employees in the inner city. For example, a company in Boston's inner-city neighbor-hood of Dorchester bakes and decorates cakes sold to supermarkets throughout the region. It attracts and retains area residents at $7 to $8 per hour (plus contributions to pensions and health insurance) and has almost 100 local employees. The loyalty of its labor pool is one of the factors that has allowed the bakery to thrive.

Admittedly, many of the jobs currently available to inner-city residents provide limited opportunities for advancement. But the fact is that they are jobs; and the inner city and its residents need many more of them close to home. Proposals that workers com-mute to jobs in distant suburbs – or move to be near those jobs – underestimate the barriers that travel time and relative skill level represent for inner-city residents. Moreover, in deciding what types of busi-nesses are appropriate to locate in the inner city, it is critical to be realistic about the pool of potential employees. Attracting high-tech companies might make for better press, but it is of little benefit to inner-city residents. Recall the contrasting experiences of Alpha Electronics and Matrix Exhibits. In the case of Alpha, there was a complete mismatch between the company's need for highly skilled professionals and the available labor pool in the local community. In contrast, Matrix carefully considered the available workforce when it established its Atlanta office. Unlike the Tennessee headquarters, which custom-designs and creates exhibits for each client, the Atlanta office specializes in rentals made from prefabricated components – work requiring less-skilled labor, which can be drawn from the inner- city. Given the workforce, low-skill jobs are realistic and economically viable: they represent the first rung on the economic ladder for many individuals who otherwise would be unemployed. Over time, successful job creation will trigger a self-reinforcing process that raises skill and wage levels.

The second myth is that the inner city's only entrepreneurs are drug dealers. In fact, there is a real capacity for legitimate entrepreneurship among inner-city residents, most of which has been channeled into the provision of social services. For instance, Boston's inner city has numerous social service providers as well as social, fraternal, and religious organizations. Behind the creation and building of those organizations is a whole cadre of local entrepreneurs who have responded to intense local demand for social services and to funding opportunities provided by government, foundations, and private sector sponsors. The challenge is to redirect some of that talent and energy toward building for-profit businesses and creating wealth.

The third myth is that skilled minorities, many of whom grew up in or near inner cities, have abandoned their roots. Today's large and growing pool of talented minority managers represents a new generation of potential inner-city entrepreneurs. Many have been trained at the nation's leading business schools and have gained experience in the nation's leading companies. Approximately 2,800 African Americans and 1,400 Hispanics graduate from M.B.A. programs every year compared with only a handful 20 years ago. Thousands of highly trained minorities are working at leading companies such as Morgan Stanley, Citibank, Ford, HewlettPackard, and McKinsey & Company. Many of these managers have developed the skills, net-work, capital base, and confidence to begin thinking about joining or starting entrepreneurial companies in the inner city . . .

New Model	Old Model
Economic: create wealth	*Social: redistribute wealth*
Private sector	Government and social service organizations
Profitable businesses	Subsidized businesses
Integration with the regional economy	Isolation from the larger economy
Companies that are export oriented	Companies that serve the local community
Skilled and experienced minorities engaged in building businesses	Skilled and experienced minorities engaged in the social service sector
Mainstream, private sector institutions enlisted	Special institutions created
Inner-city disadvantages addressed directly	Inner-city disadvantages counterbalanced
Government focuses on improving the environment for business	Government directly involved with providing services or funding

Figure 1 Inner-city economic development

THE REAL DISADVANTAGES OF THE INNER CITY

The second step toward creating a coherent economic strategy is addressing the very real disadvantages of locating businesses in the inner city. The inescapable fact is that businesses operating in the inner city face greater obstacles than those based elsewhere. Many of those obstacles are needlessly inflicted by government. Unless the disadvantages are addressed directly, instead of indirectly through subsidies or mandates, the inner city's competitive advantages will continue to erode.

Land

Although vacant property is abundant in inner cities, much of it is not economically usable. Assembling small parcels into meaningful sites can be prohibitively expensive and is further complicated by the fact that a number of city, state, and federal agencies each control land and fight over turf . . .

Building costs

The cost of building in the inner city is significantly higher than in the suburbs because of the costs and delays associated with logistics, negotiations with *community* groups, and strict urban regulations: restrictive zoning, architectural codes, permits, inspections, and government-required union contracts and minority setasides. Ironically, despite the desperate need for new projects, construction in inner cities is far more regulated than it is in the suburbs – a legacy of big city politics and entrenched bureaucracies.

[. . .]

Other costs

Compared with the suburbs, inner cities have high costs for water, other utilities, workers' compensation, health care, insurance, permitting and other fees, real estate and other taxes, OSHA compliance, and neighborhood hiring requirements. For example, Russer Foods, a manufacturing company located in Boston's inner city, operates a comparable plant in upstate New York. The Boston plant's expenses are 55% higher for workers' compensation, 50% higher for family medical insurance, 166% higher for unemployment insurance, 340% higher for water, and 67% higher for electricity. High costs like these drive away companies and hold down wages. Some costs, such as those for workers' compensation, apply to the state or region as a whole. Others, such as real estate taxes, apply citywide. Still others, such as property insurance, are specific to the inner city. All are devastating to maintaining fragile inner-city companies and to attracting new businesses.

It is an unfortunate reality that many cities – because they have a greater proportion of residents dependent on welfare, Medicaid, and other social programs – require higher government spending and, as a result, higher corporate taxes. The resulting tax burden feeds a vicious cycle – driving out more companies while requiring even higher taxes from those that remain. Cities have been reluctant to challenge entrenched bureaucracies and unions, as well as inefficient and outdated government departments, all of which unduly raise city costs.

Finally, excessive regulation not only drives up building and other costs but also hampers almost all facets of business life in the inner city, from putting up an awning over a shop window to operating a pushcart to making site improvements. Regulation also stunts inner-city entrepreneurship, serving as a formidable barrier to small and start-up companies. Restrictive licensing and permitting, high licensing fees, and archaic safety and health regulations create barriers to entry into the very types of businesses that are logical and appropriate for creating jobs and wealth in the inner city.

Security

Both the reality and the perception of crime represent profound impediments to urban economic development. First, crime against property raises costs. For example, the Shops at Church Square, an inner-city strip shopping center in Cleveland, Ohio, spends $2 per square foot more than a comparable suburban center for a full-time security guard, increased lighting, and continuous cleaning – raising overall costs by more than 20%. Second, crime against employees and customers creates an unwillingness to work in and patronize inner-city establishments and restricts companies' hours of operation. Fear of crime ranks among the most important reasons why companies opening new facilities failed to consider inner-city locations and why companies already located in the inner city left. Currently, police devote most of their resources to the security of residential areas, largely overlooking commercial and industrial sites.

Infrastructure

Transportation infrastructure planning, which today focuses primarily on the mobility of residents for shopping and commuting, should consider equally the mobility of goods and the ease of commercial transactions. The most critical aspects of the new economic model – the importance of the location of the inner city, the connections between inner-city businesses and regional clusters, and the development of export-oriented businesses – require the presence of strong logistical links between inner-city business sites and the surrounding economy. Unfortunately, the business infrastructure of the inner city has fallen into disrepair. The capacity of roads, the frequency and location of highway on-ramps and off-ramps, the links to downtown, and the access to railways, airports, and regional logistical networks are inadequate.

Employee skills

Because their average education levels are low, many inner-city residents lack the skills to work in any but the most unskilled occupations. To make matters worse, employment opportunities for less educated workers have fallen markedly. In Boston between 1970 and 1990, for example, the percentage of jobs held by people without high school diplomas dropped from 29% to 7%, while those held by college graduates climbed from 18% to 44%. And the unemployment rate for African–American men aged 16 to 64 with less than a high school education in major northeastern cities rose from 19% in 1970 to 57% in 1990.

Management skills

The managers of most inner-city companies lack formal business training. That problem, however, is not unique to the inner city; it is a characteristic of small businesses in general. Many individuals with extensive work histories but little or no formal mana-

gerial training start businesses. Inner-city companies without well trained managers experience a series of predictable problems that are similar to those that affect many small businesses: weaknesses in strategy development, market segmentation, customer-needs evaluation, introduction of information technology, process design, cost control, securing or restructuring financing, interaction with lenders and government regulatory agencies, crafting business plans, and employee training. Local community colleges often offer management courses, but their quality is uneven, and entrepreneurs are hard-pressed for time to attend them.

Capital

Access to debt and equity capital represents a formidable barrier to entrepreneurship and company growth in inner-city areas.

First, most inner-city businesses still suffer from poor access to debt funding because of the limited attention that mainstream banks paid them historically. Even in the best of circumstances, small business lending is only marginally profitable to banks because transaction costs are high relative to loan amounts. Many banks remain in small-business lending only to attract deposits and to help sell other more profitable products.

The federal government has made several efforts to address the inner city's problem of debt capital. As a result of legislation like the Community Reinvestment Act, passed in order to overcome bias in lending, banks have begun to pay much more attention to inner-city areas. In Boston, for example, leading banks are competing fiercely to lend in the inner city – and some claim to be doing so profitably. Direct financing efforts by government, however, have proved ineffective. The proliferation of government loan pools and quasi-public lending organizations has produced fragmentation, market confusion, and duplication of overhead. Business loans that would provide scale to private sector lenders are siphoned off by these organizations, many of which are high-cost, bureaucratic, and risk-averse. In the end, the development of high-quality private sector expertise in inner-city business financing has been undermined.

Second, equity capital has been all but absent. Inner-city entrepreneurs often lack personal or family savings and networks of individuals to draw on for capital. Institutional sources of equity capital are scarce for minority-owned companies and have virtually ignored inner-city business opportunities.

Attitudes

A final obstacle to companies in the inner city is antibusiness attitudes. Some workers perceive businesses as exploitative, a view that guarantees poor relations between labor and management. Equally debilitating are the antibusiness attitudes held by community leaders and social activists. These attitudes are the legacy of a regrettable history of poor treatment of workers, departures of companies, and damage to the environment. But holding on to these views today is counterproductive. Too often, community leaders mistakenly view businesses as a means of directly meeting social needs; as a result, they have unrealistic expectations for corporate involvement in the community . . .

Demanding linkage payments and contributions and stirring up antibusiness sentiment are political tools that brought questionable results in the past when owners had less discretion about where they chose to locate their companies. In today's increasingly competitive business environment, such tactics will serve only to stunt economic growth.

Overcoming the business disadvantages of the inner city as well as building on its inherent advantages will require the commitment and involvement of business, government, and the nonprofit sector. Each will have to abandon deeply held beliefs and past approaches. Each must be willing to accept a new model for the inner city based on an economic rather than social perspective. The private sector, nongovernment or social service organizations, must be the focus of the new model.

The new role of the private sector

The economic model challenges the private sector to assume the leading role. First, however, it must adopt new attitudes toward the inner city. Most private sector initiatives today are driven by preference programs or charity. Such activities would never stand on their own merits in the marketplace. It is inevitable, then, that they contribute to growing cynicism. The private sector will be most effective if it focuses on what it does best: creating and supporting economically viable businesses built on true competitive advantage. It should pursue four immediate opportunities as it assumes its new role.

1. *Create and expand business activity in the inner city.* The most important contribution companies can make to inner cities is simply to do business there. Inner cities hold untapped potential for profitable businesses. Companies and entrepreneurs must seek out and seize those opportunities that build on the true advantages of the inner city. In particular, retailers, franchisers, and financial services companies have immediate opportunities. Franchises represent an especially attractive model for inner-city entrepreneurship because they provide not only a business concept but also training and support.

Businesses can learn from the mistakes that many outside companies have made in the inner city. One error is the failure of retail and service businesses to tailor their goods and services to the local market . . .

Another common mistake is the failure to build relationships within the community and to hire locally. Hiring local residents builds loyalty from neighborhood customers, and local employees of retail and service businesses can help stores customize their products. Evidence suggests that companies that were perceived to be in touch with the community had far fewer security problems, whether or not the owners lived in the community.

[. . .]

2. *Establish business relationships with inner city companies.* By entering into joint ventures or customer–supplier relationships, outside companies will help inner-city companies by encouraging them to export and by forcing them to be competitive. In the long run, both sides will benefit . . . Such relationships, based not on charity but on mutual self-interest, are sustainable ones; every major company should develop them.

3. *Redirect corporate philanthropy from social services to business-to-business efforts.* Countless companies give many millions of dollars each year to worthy inner-city social-service agencies. But philanthropic efforts will be more effective if they also focus on building business-to-business relationships that, in the long run, will reduce the need for social services.

First, corporations could have a tremendous impact on training. The existing system for job training in the United States is ineffective. Training programs are fragmented, overhead intensive, and disconnected from the needs of industry. Many programs train people for nonexistent jobs in industries with no projected growth. Although reforming training will require the help of government, the private sector must determine how and where resources should be allocated to ensure that the specific employment needs of local and regional businesses are met. Ultimately, employers, not government, should certify all training programs based on relevant criteria and likely job availability.

Training programs led by the private sector could be built around industry clusters located in both the inner city (for example, restaurants, food service, and food processing in Boston) and the nearby regional economy (for example, financial services and health care in Boston). Industry associations and trade groups, supported by government incentives, could sponsor their own training programs in collaboration with local training institutions.

[. . .]

Second, the private sector could make an equally substantial impact by providing management assistance to inner-city companies. As with training, current programs financed or operated by the government are inadequate. Outside companies have much to offer companies in the inner city: talent, know-how, and contacts. One approach to upgrading management skills is to emphasize networking with companies in the regional economy that either are part of the same cluster (customers, suppliers, and related businesses) or have expertise in needed areas. An inner-city company could team up with a partner in the region who provides management assistance; or a consortium of companies with a required expertise, such as information technology, could provide assistance to inner-city businesses in need of upgrading their systems.

[. . .]

4. *Adopt the right model for equity capital investments*. The investment community – especially venture capitalists – must be convinced of the viability of investing in the inner city. There is a small but growing number of minority-oriented equity providers (although none specifically focus on inner cities). A successful model for inner-city investing will probably not look like the familiar venture-capital model created primarily for technology companies. Instead, it may resemble the equity funds operating in the emerging economies of Russia or Hungary – investing in such mundane but potentially profitable projects as super-

markets and laundries. Ultimately, inner-city-based businesses that follow the principles of competitive advantage will generate appropriate returns to investors – particularly if aided by appropriate incentives, such as tax exclusions for capital gains and dividends for qualifying inner-city businesses.

The new role of government

To date, government has assumed primary responsibility for bringing about the economic revitalization of the inner city. Existing programs at the federal, state, and local levels designed to create jobs and attract businesses have been piecemeal and fragmented at best. Still worse, these programs have been based on subsidies and mandates rather than on marketplace realities. Unless we find new approaches, the inner city will continue to drain our rapidly shrinking public coffers.

Undeniably, inner cities suffer from a long history of discrimination. However, the way for government to move forward is not by looking behind. Government can assume a more effective role by supporting the private sector in new economic initiatives. It must shift its focus from direct involvement and intervention to creating a favorable environment for business. This is not to say that public funds will not be necessary. But subsidies must be spent in ways that do not distort business incentives, focusing instead on providing the infrastructure to support genuinely profitable businesses. Government at all levels should focus on four goals as it takes on its new role.

1. *Direct resources to the areas of greatest economic need*. The crisis in our inner cities demands that they be first in line for government assistance. This may seem an obvious assertion. But the fact is that many programs in areas such as infrastructure, crime prevention, environmental cleanup, land development, and purchasing preference spread funds across constituencies for political reasons. For example, most transportation infrastructure spending goes to creating still more attractive suburban areas. In addition, a majority of preference-program assistance does not go to companies located in low-income neighborhoods.

[. . .]

Unfortunately, the qualifying criteria for current government assistance programs are not properly designed to channel resources where they are most needed. Preference programs support business based on the race, ethnicity, or gender of their owners rather than on economic need. In addition to directing resources away from the inner city, such race-based or gender-based distinctions reinforce inappropriate stereotypes and attitudes, breed resentment, and increase the risk that programs will be manipulated to serve unintended populations. Location in an economically distressed area and employment of a significant percentage of its residents should be the qualification for government assistance and preference programs. Shifting the focus to economic distress in this way will help enlist all segments of the private sector in the solutions to the inner city's problems.

2. *Increase the economic value of the inner city as a business location.* In order to stimulate economic development, government must recognize that it is a part of the problem. Today its priorities often run counter to business needs. Artificial and outdated government-induced costs must be stripped away in the effort to make the inner city a profitable location for business. Doing so will require rethinking policies and programs in a wide range of areas . . .

Indeed, there are numerous possibilities for reform. Imagine, for example, policy aimed at eliminating the substantial land and building cost penalties that businesses face in the inner city. Ongoing rent subsidies run the risk of attracting companies for which an inner-city location offers no other economic value. Instead, the goal should be to provide building-ready sites at market prices. A single government entity could be charged with assembling parcels of land and with subsidizing demolition, environmental cleanup, and other costs. The same entity could also streamline all aspects of building – including zoning, permitting, inspections, and other approvals.

That kind of policy would require further progress on the environmental front. A growing number of cities – including Detroit, Chicago, Indianapolis, Minneapolis, and Wichita, Kansas – have successfully developed so-called brownfield urban areas by making environmental cleanup standards more flexible depending on land use, indemnifying land owners against additional costs if contamination is found on a site after a cleanup, and using tax-increment financing to help fund cleanup and redevelopment costs.

Government entities could also develop a more strategic approach to developing transportation and communications infrastructures, which would facilitate the fluid movement of goods, employees, customers, and suppliers within and beyond the inner city. Two

projects in Boston are prime examples: first, a new exit ramp connecting the inner city to the nearby Massachusetts Turnpike, which in turn connects to the surrounding region and beyond; and a direct access road to the harbor tunnel, which connects to Logan International Airport. Though inexpensive, both projects are stalled because the city does not have a clear vision of their economic importance.

3. *Deliver economic development programs and services through mainstream, private sector institutions.* There has been a tendency to rely on small community-based nonprofits, quasi-governmental organizations, and special-purpose entities, such as community development banks and specialized small-business investment corporations, to provide capital and business-related services. Social service institutions have a role, but it is not this. With few exceptions nonprofit and government organizations cannot provide the quality of training, advice, and support to substantial companies that mainstream, private sector organizations can. Compared with private sector entities such as commercial banks and venture capital companies, special-purpose institutions and non-profits are plagued by high overhead costs; they have difficulty attracting and retaining high-quality personnel, providing competitive compensation, or offering a breadth of experience in dealing with companies of scale.

[. . .]

The most important way to bring debt and equity investment to the inner city is by engaging the private sector. Resources currently going to government or quasi-public financing would be better channeled through other private financial institutions or directed at recapitalizing minority-owned banks focusing on the inner city, provided that there were matching private sector investors. Minority-owned banks that have superior knowledge of the inner city market could gain a competitive advantage by developing business-lending expertise in inner-city areas.

As in lending, the best approach to increase the supply of equity capital to the inner cities is to provide private sector incentives consistent with building economically sustainable businesses. One approach would be for both federal and state governments to eliminate the tax on capital gains and dividends from long-term equity investments in inner-city-based businesses or subsidiaries that employ a minimum percentage of inner-city residents. Such tax incentives, which are based on the premise of profit, can play a vital role in speeding up private sector investment. Private sector sources of equity will be attracted to inner-city investment only when the creation of genuinely profitable businesses is encouraged.

4. *Align incentives built into government programs with true economic performance.* Aligning incentives with business principles should be the goal of every government program. Most programs today would fail such a test. For example, preference programs in effect guarantee companies a market. Like other forms of protectionism, they dull motivation and retard cost and quality improvement. A 1988 General Accounting Office report found that within six months of graduating from the Small Business Association's purchasing preference program, 30% of the companies had gone out of business. An additional 58% of the remaining companies claimed that the withdrawal of the SBA's support had had a devastating impact on business. To align incentives with economic performance, preference programs should be rewritten to require an increasing amount of non-set-aside business over time.

Direct subsidies to businesses do not work. Instead, government funds should be used for site assembly, extra security, environmental clean-up, and other investments designed to improve the business environment. Companies then will be left to make decisions based on true profit.

The new role of community-based organizations

Recently, there has been renewed activity among community-based organizations (CBOs) to become directly involved in business development. CBOs can, and must, play an important supporting role in the process. But choosing the proper strategy is critical, and many CBOs will have to change fundamentally the way they operate. While it is difficult to make a general set of recommendations to such a diverse group of organizations, four principles should guide community-based organizations in developing their new role.

1. *Identify and build on strengths.* Like every other player, CBOs must identify their unique competitive advantages and participate in economic development based on a realistic assessment of their capabilities, resources, and limitations. Community-based organizations have played a much-needed role in developing low-income housing, social programs, and civic infrastructure. However, while there have been a few notable successes, the vast majority of businesses owned or managed by CBOs have been failures. Most CBOs lack the skills, attitudes, and incentives to advise, lend to, or operate substantial businesses. They were able to master low-income housing development, in which there were major public subsidies and a vacuum of institutional capabilities. But, when it comes to financing and assisting for-profit business development, CBOs simply can't compete with existing private sector institutions.

Moreover, CBOs naturally tend to focus on community entrepreneurship: small retail and service businesses that are often owned by neighborhood residents. The relatively limited resources of CBOs, as well as their focus on relatively small neighborhoods, is not well-suited to developing the more substantial companies that are necessary for economic vitality.

Finally, the competitive imperatives of for-profit business activity will raise inevitable conflicts for CBOs whose mission rests with the community. Turning down local residents in favor of better qualified outside entrepreneurs, supporting necessary layoffs or the dismissal of poorly performing workers, assigning prime sites for business instead of social uses, and approving large salaries to successful entrepreneurs and managers are only a handful of the necessary choices. Given these organizations' roots in meeting the social needs of neighborhoods, it will be difficult for them to put profit ahead of their traditional mission.

2. *Work to change workforce and community attitudes.* Community-based organizations have a unique advantage in their intimate knowledge of and influence within inner-city communities, and they can use that advantage to help promote business development. CBOs can help create a hospitable environment for business by working to change community and workforce attitudes and acting as a liaison with residents to quell unfounded opposition to new businesses . . .

3. *Create work-readiness and job-referral systems.* Community-based organizations can play an active role in preparing, screening, and referring employees to local businesses. A pressing need among many inner city residents is work-readiness training, which includes communication, self-development, and workplace practices. CBOs, with their intimate knowledge of the local community, are well equipped to provide this service in close collaboration with industry . . .

CBOs can also help inner-city residents by actively developing screening and referral systems. Admittedly, some inner-city-based businesses do not hire many local residents. The reasons are varied and complex but seem to revolve around a few bad experiences that owners have had with individual employees and their work attitudes, absenteeism, false injury claims, or drug use . . .

4. *Facilitate commercial site improvement and development.* Community-based organizations (especially community development corporations) can also leverage their expertise in real estate and act as a catalyst to facilitate environmental cleanup and the development of commercial and industrial property . . .

OVERCOMING IMPEDIMENTS TO PROGRESS

This economic model provides a new and comprehensive approach to reviving our nation's distressed urban communities. However, agreeing on and implementing it will not be without its challenges. The private sector, government, inner-city residents, and the public at large all hold entrenched attitudes and prejudices about the inner city and its problems. These will be slow to change. Rethinking the inner city in economic rather than social terms will be uncomfortable for many who have devoted years to social causes and who view profit and business in general with suspicion. Activists accustomed to lobbying for more government resources will find it difficult to embrace a strategy for fostering wealth creation. Elected officials used to framing urban problems in social terms will be resistant to changing legislation, redirecting resources, and taking on recalcitrant bureaucracies. Government entities may find it hard to cede power and control accumulated through past programs. Local leaders who have built social service organizations and merchants who have run mom-and-pop stores could feel threatened by the creation of new initiatives and centers of power. Local politicians schooled in old-style community organizing and confrontational politics will have to tread unfamiliar ground in facilitating cooperation between business and residents.

These changes will be difficult ones for both individuals and institutions. Nonetheless, they must be made. The private sector, government, and community-based organizations all have vital new parts to play in revitalizing the economy of the inner city. Businesspeople, entrepreneurs, and investors must assume a lead role; and community activists, social service providers, and government bureaucrats must support them. The time has come to embrace a rational economic strategy and to stem the intolerable costs of outdated approaches.

"Fiscal Equity"
American Metropolitics (2002)

Myron Orfield

Editors' Introduction

In this selection University of Minnesota law professor Myron Orfield tackles the very important issue of equity in local government finance and makes the case for equalizing revenue and services across jurisdictions.

Metropolitan areas in the United States are fragmented into dozens or hundreds of separate cities and counties. Fragmentation is often extreme. For example the Pittsburgh, Pennsylvania, metropolitan region is governed by 418 separate local governments.

While he argues forcefully in favor of more regionally based local government finance, Orfield acknowledges some worthy arguments in favor of localism. Small government, close to voters, encourages citizen participation at high rungs of Sherry Arnstein's "ladder of citizen participation" (p. 233). Fragmentation may keep local governments responsible because voters can hold local decision makers accountable for the results of their actions. There is value in diversity and in permitting people to choose from a wide variety of different kinds of places to live.

Despite the tradition of local autonomy and some advantages it may offer, Orfield, Downs (p. 245), Calthorpe and Fulton (p. 342), and many other political scientists, economists, and urban planners deplore the extreme fragmentation of metropolitan government. They point out that this fragmentation encourages local jurisdictions to strive to maximize their own well-being at the expense of that of their neighbors and to put very local interests above wider regional concerns. Accordingly, these critics call for fundamental reform of metropolitan governance and finance.

An important dimension of the debate about how local government should be organized involves finance. Each layer of government in the American federal system is assigned governmental functions and given access to some revenue sources. For example, the federal government receives revenue from the federal income tax and has responsibility for national defense and the US postal service. Some states raise revenue from a state sales tax and spend it on state highways. At the local level cities and counties raise most of their revenue from the local property tax and use it to provide police and fire services, streets, sanitation, parks, libraries, and a host of other local services. School districts are typically separate from city and county government. Most school districts are highly dependent on property tax revenue to pay the cost of K-12 education. Local governments have the legal authority to regulate land use within their boundaries and, as Orfield points out, often use fiscal zoning to encourage land uses they consider desirable and discourage others. For example, a jurisdiction that relies primarily on the local property tax for its revenue may encourage only high-value local land uses such as commercial, industrial, and high-end residential uses. They want new development to decrease current residents' financial burden. They hope neighboring jurisdictions will build apartment houses, mobile homes, and public housing to satisfy regional needs. In states were cities derive substantial revenue from sales tax based on sales within the city, cities will compete to attract retail stores with large sales. Local governments compete with each other for these desirable land uses: a zero sum game that Orfield considers wasteful and shortsighted.

Access to revenue at the local level is not equal. Nor do all citizens in a region pay proportionally or receive comparable local services. Some jurisdictions have much greater tax capacity than others: particularly high value property that can yield a large amount of property tax revenue. This means that they can tax citizens less, provide more and better services, or both. Moreover, it is often the wealthiest jurisdictions with the least need that are best off financially. Thus a technoburb such as Fishman describes (p. 69) consisting almost exclusively of expensive single family homes, high-tech research and development firms, and high-end shopping malls can raise a large amount of property tax per capita even with a low property rate. A neighboring suburb with only slum housing, a scattering of small firms, and marginal retail establishments can't raise nearly as much property tax per capita even if they charge high tax rates. Moreover, there is a vicious cycle in which communities that lose out in the competition for desirable revenue sources early on are unable to compete for desirable development thereafter. Rich jurisdictions grow richer; poor ones are locked into a cycle of decline. Fiscal zoning and tax-base competition encourage concentrations of poor families in communities that are the least able to generate the revenues to provide for social services.

Orfield's concern is how to reduce metropolitan fiscal disparities in order to reduce the burden on the hardest-pressed jurisdictions and increase services for the most underserved populations. He argues that a healthy society needs stable, cooperative regions and that fiscal equity will benefit society as a whole.

Orfield suggests three main policies to promote regional equity: institutional aid to jurisdictions with especially high needs, state revenue sharing programs that distribute state revenues to local governments, and metropolitan tax base-sharing programs that share tax resources within a single region.

Institutional aid to schools is well established. States generally view good education as what Thompson (p. 266) calls a merit good – something that should be free and universally available because it is good for individuals and society. Accordingly, they provide equalization aid to communities with the poorest schools so that the huge disparity in per capita funding for students in rich communities versus poor ones that would otherwise exist becomes much more equal. Accordingly, they also have revenue sharing programs. For example, Michigan distributes a portion of state sales tax revenue to local jurisdictions based on the type of jurisdiction, population, and fiscal capacity.

A final approach – existing only the Minneapolis–St. Paul region in Orfield's home state of Minnesota – is tax sharing. Since 1971 each jurisdiction within the seven-county Minneapolis–St. Paul region must assign a portion of its tax revenue to a regional pool. Currently over $300 million a year in the regional pool is redistributed, significantly reducing local tax-based disparity in the region. Orfield favors regional tax-base sharing, but no other states have replicated the tax-sharing system that has existed in Minneapolis–St. Paul for more than thirty-five years.

Myron Orfield is a law professor at the university of Minnesota Law School, Associate Director of the University of Minnesota's Institute on Race and Poverty, and a non-resident senior fellow at the Brookings Institution in Washington, DC. In 2005/2006 he held the Fesler-Lampert Chair in Urban and Regional Affairs at the University of Minnesota. He is an authority on civil rights, state and local government, state and local finance, land use, regional governance, and the legislative process. Orfield served five terms in the Minnesota House of Representatives and one term in the Minnesota Senate. During his legislative career, Orfield was the architect of a series of important changes in land use, fair housing, and school and local government aid programs.

This selection is from Myron Orfield's *American Metropolitics: The New Suburban Reality* (Washington, DC: Brookings Institution Press, 2002). An earlier related book by Orfield is *Metropolitics: A Regional Agenda for Community and Stability* (Washington, DC: Brookings Institution Press, 1999).

Other books exploring metropolitan issues are Peter Dreier, John Mollenkopf, and Todd Swanstrom, *Place Matters: Metropolitics For The Twenty-First Century*, second revised edition (Kansas City: University Press of Kansas Press, 2005), Robert Lang, *Edgeless Cities: Exploring the Elusive Metropolis* (Washington, DC: Brookings Institution Press, 2002), and Peter Calthorpe, *The Next American Metropolis: Ecology, Community, and the American Dream*, third edition (Princeton: Princeton Architectural Press, 1997).

An essential part of creating a stable, cooperative region is to gradually equalize the resources of local governments with land-use planning powers. In addition to improving equity, which will allow cities, at-risk suburbs, and many bedroom-developing suburbs to lower taxes and improve services, it will reduce the competition between places, give communities real fiscal incentives to cooperate, and make regional land-use planning easier to achieve. The Rubicon of local fiscal equity was crossed decades ago with the nearly universal adoption of state school equalization programs. Furthermore, state aid is already a significant source of municipal revenue in most of the regions studied here. The challenge is to take these ideas and make them gradually more extensive and efficient – a challenge of degree rather than of kind.

All state governments distribute some of their resources to local governments to help finance local public services. How much varies from state to state and across services. The states liberally support local school systems for reasons both economic and idealistic. Educating the labor force increases the country's (and states') competitiveness in the global economy. By enhancing individuals' economic mobility, education provides a means of improving income distribution. Education is also seen as a vehicle for strengthening democracy. State constitutions and the courts have long recognized the important public functions of schools. The courts have played an important role in the lengthy and still incomplete effort to equalize access for all children to at least some minimum standard of education, regardless of their parents' or their communities' economic status.

Traditionally, states have been much more willing to support local public schools than other municipal services. In 1996–97, states financed roughly half the costs of local public schools but only a fifth of the costs of municipal services. Also there was much less variation among states in the extent of their support for local schools than for municipal services. . . . State aid programs for schools also are much more likely to be explicitly designed to reduce inequities resulting from differences among local tax bases.

Equalizing access to municipal services has not received the same attention as education. Local governments have been left much more to their own devices in financing these services, and that is still true. There are several reasons for the difference. To some extent, the legal system is responsible – municipal

services do not receive the same attention as schools in state constitutions and the courts. Those institutions have elevated access to education to a legal right, a status rarely achieved by any of the traditional functions of municipal governments. Similarly, many social scientists (economists in particular) view municipal services much as they do the goods and services individuals purchase privately in a market economy. Although there may be reasons to provide those services through public agencies rather than through private markets, the desirability of accommodating diverse tastes or preferences is often used to argue that such services should be financed locally to maximize consumer choice and to allow voters the greatest voice in determining how and how much of a service is provided.

Nevertheless, a close look at municipal governments' functions in the U.S. economy and society makes it difficult to justify those notions. In 1996–97 roughly half of all municipal government expenditures in the United States went for core services that most people would consider essential: streets, public safety (police and fire protection), and sanitation services (sewerage and water treatment). An additional 14 percent of budgets went for income maintenance, health services, and courts (core functions financed at the county or state level in most states but at the local level in some). Thus about two-thirds of municipal budgets are dedicated to a set of functions that, although they may not attain the stature of public schools in many people's minds, are viewed as the minimum acceptable menu of public services. They are such important contributors to the quality of life in the United States that most Americans view access to a reasonable standard of service as a birthright. Nor is the remainder of the average local budget devoted to luxuries. Parks, recreation, housing, community development, and libraries claim 12 percent; interest on debt, 7 percent; various unallocated expenses, 7 percent; and general administration, 7 percent.

There is virtue in diversity of choice – that is the central strength of a market economy and of a market-like set of local governments competing for residents. However, most of the goods and services provided by local governments are crucial building blocks for what most Americans would regard as a reasonable quality of life, and the arguments are strong for providing them through public agencies. Furthermore, the literature on demand for local government services implies that differences in the extent and quality of those services are determined largely by differences in those governments' economic resources and their costs. In light of the extreme variations in tax capacities and costs . . . the case for reducing those disparities is clear. The question is which policies do so most efficiently and in ways that support other priorities, such as land-use planning, while minimizing intrusions on legitimate local prerogatives.

GOVERNMENT FINANCE AND FISCAL DISPARITIES

Local governments in the United States provide important public services demanded by businesses and residents and financed largely through local taxes (for example, streets, public safety, sanitation, and other basic infrastructure). At the same time, local governments have the authority to regulate land uses within their political boundaries, to determine the location and the extent of residential, commercial, industrial, parkland, and agricultural development.

Though designed for seemingly separate purposes, local tax policy and land-use regulations are closely related. The amount of revenue a local government can generate on its own depends largely on the value and types of land uses within its jurisdiction. Of the three primary sources of local tax revenues – property, sales, and income tax – the property tax is the most common and accounts for the largest share of local tax revenues, an average of 74 percent in 1996. General and selective sales taxes account for another 16 percent, and individual income taxes another 5 percent. Local governments with land-use planning powers (such as municipalities or counties) therefore have direct incentives to develop a land-use plan that maximizes the value of property (if the property tax is the primary revenue source), attracts high-value retail businesses (if sales taxes are important), or tailors the housing stock to high-income residents or employers (if income taxes are important).

Incentives also can be found in local governments' expenditures. Different land uses imply different public service needs. Residential development requires water and gas lines, roads, sewerage and waste management systems, fire and police protection, schools, libraries, social services, and other public infrastructure and services. However, different types of residential development imply different degrees of need for public services. An apartment building with low-income residents is likely to entail greater social service costs than a middle-class family living in a four-bedroom single-family home. Commercial and industrial development requires a different set of local services: utilities, infrastructure, police and fire protection. Again, different types of commercial and industrial development imply different service needs – a busy commercial area with many retail businesses and an industrial park do not entail the same expenses.

Ultimately, what localities worry about is the net effect of specific types of development on revenues and expenditures. That balance (the fiscal dividend) determines whether a new development will increase or decrease current residents' financial burden. The overall incentive is therefore to allow or pursue only the kind of development that generates a positive fiscal dividend – a surplus of revenues over expenditures. Individual local governments in a fragmented local government system compete with each other for these fiscal dividends chiefly through land-use regulations to limit new activity in particular ways (fiscal zoning) and development incentives to attract particular types of activities or firms.

Fiscal zoning

Fiscal zoning is a deliberate attempt by a local government to reap the best fiscal dividend by limiting the types of land uses within its jurisdiction. The types of development that will do this vary from place to place, depending on local revenue sources and the way responsibilities for expenditures are divided among state, county, local, and other public entities. Research findings on this question are not in complete agreement, but some general patterns emerge.

Most notably, some general rules appear to apply when a local government seeks the most "profitable" land uses, including a hierarchy of land uses and fiscal impacts that demonstrates which land uses break even in terms of the fiscal dividend. For example, municipalities generate more revenues than costs with office parks, industrial development, and small (one- or two-bedroom) apartments and condominiums. School districts break even on those types of properties as well as on retail development, two- to three-bedroom townhouses, and expensive single-family homes. The types of development that generate negative fiscal dividends include larger townhouses (three- to four-bedroom units), inexpensive single-family homes, larger apartments (three- to four-bedroom units), and mobile homes.

[. . .]

The overall implication is that, from the standpoint of the fiscal dividend, commercial, industrial, and high-end residential developments are the way to go. Since the ability to generate revenues depends largely on the value of property, high-value commercial, industrial, and residential properties that generate relatively little in public costs are preferred, and lower value properties that create the need for public expenditures are avoided. When determining how to allow their land to be developed, local governments have a direct incentive to favor profitable uses over others and, whenever possible, to exclude unprofitable uses such as large apartments and moderate- to low-value homes.

Local governments have tools for encouraging development of the most profitable uses. Zoning can be deployed to allow particular types of commercial or industrial development and limit others. To attract particular companies, incentives can then be proffered. Tools for regulating residential development (for example, minimum lot size or set-back requirements and construction standards designed to increase the cost of building in the community) are usually less direct.

This fiscal zoning process, when played out over an entire metropolitan area, can significantly influence where people are able to live, the types and quality of public services they receive from their local government, and the presence or absence of employment opportunities near their homes. How binding those limitations are for an individual or family will for the most part be directly related to the value of the home they can afford – in other words, their income.

Competition for tax base

Another aspect of local governments' pursuit of positive fiscal dividends is the competition among them for desirable commercial and industrial properties. That competition, which pervades metropolitan areas throughout the United States, is likely to be both wasteful and biased. It is wasteful because, from a regional point of view, interlocal competition is largely a zero-sum game. When a business is lured to one jurisdiction in a metropolitan area by financial or other types of incentives, its arrival is unlikely to add net new activity to the regional economy. Instead, one community's gain is likely to be another's loss – a direct loss in the case of a business that relocates within the same metropolitan area and an indirect loss in the case of a new business that would have located somewhere in the metropolitan area regardless of the incentives. Communities may also find they have to spend resources to ensure that their existing businesses do not leave for other places in the metropolitan area that offer them incentives. The resources expended in such competition are largely wasted: they do not enhance the over-all regional economy; they only shuffle activity from one place to another.

The competition is biased because it creates the potential for a vicious, self-reinforcing cycle of decline in places that "lose" early in the game. As a locality loses activities that generate positive fiscal dividends, it must either raise taxes on its remaining tax base to maintain services at existing levels or reduce services at existing tax rates. Either choice reduces the locality's ability to compete for additions to its tax base or to keep its existing base, leading to further losses and further declines in the place's ability to compete.

Similarly, as communities become segmented by housing and income type, they also become segmented by the type and amount of commercial and industrial development they can attract. Again, the commercial and industrial developments that generate the highest revenues tend to locate in the relatively few communities that have large, expensive houses. Part of the attraction lies in the low tax rates such places can maintain, but it is also related to their residents' socio-economic status. With sufficient high-end housing available for company executives, better public services, and little social stress, those communities often offer the most attractive location for new businesses. Furthermore, higher-end retail stores also find them desirable because of the purchasing power of their middle- to upper-income residents. Just as they win the competition for revenue-generating housing, those communities win the competition for commercial properties that also generate the largest revenues.

By contrast, the newer, moderate-income communities growing at the edges of the region have little success in attracting the most desirable commercial and industrial properties. If they happen to be located along a major highway or commuter route, some of them may be able to attract retail developments that increase their tax base. Even in those cases, however, the increased costs for the roads, police and fire services, and public utilities that retail developments require tend to outstrip the revenues that they generate. Most edge communities serve primarily as bedroom communities for the rest of the region and develop at lower densities than older suburban communities, generating fewer revenues and higher costs per square mile.

Finally, the older core communities that have become or are becoming home to more and more of the region's low- and moderate-income families experience significant disadvantages in attracting new business development. With higher taxes, aging infrastructure, a lack of available land, and a relatively high degree of social stress, those communities do not usually attract new investment and often lose the businesses that they have. Disinvestment by the business community can hasten community decline, as jobs evaporate and the local tax base continues to shrink. These communities thus fall farther behind other metropolitan communities as they struggle to maintain some degree of stability and competitiveness.

In sum, strong incentives for engaging in fiscal zoning and tax-base competition are built into local fiscal systems. Those processes shape metropolitan areas in important ways: they generate a highly stratified mosaic of localities in a metropolitan region, they encourage short-sighted land-use planning, and they contribute to sprawl.

Stratification of metropolitan areas

Fiscal zoning and tax-base competition encourage concentrations of families and individuals with the greatest need for public services in communities that are the least able to generate the revenue to provide those services. A primary reason can be seen by looking again at the hierarchy of profitable land uses and the types of people that live in different kinds of residential properties. Low- or moderate-income families with children most frequently purchase cost-generating residences, relatively inexpensive housing with three or more bedrooms. Revenue-generating homes – with fewer bedrooms and costing more – are likely to be purchased by high-income families, empty-nesters, and two-income households with no children.

As municipalities in a metropolitan region compete for the limited number of "good" land uses, only a few of them succeed. Localities that attract large, expensive

homes and other revenue-generating land uses see their revenues rise while their costs remain relatively stable. The resulting surplus funds allow them either to provide additional, higher-quality services or to reduce their property tax rates, making them even more attractive to people looking for housing. That increased demand allows them to continue building high-end housing, creating a continuous cycle of increasing revenue-raising capacity and lower taxes . . . [Such] clear winners in the competition are rare and the benefits accrue to a disproportionately small segment of metropolitan populations . . .

Conversely, places that do not attract or keep sufficient high-end housing and commercial-industrial development have a much harder time generating revenues and keeping taxes low. In older communities at the core of the region, expenditures are often related to social stresses: poverty, crime, and aging infrastructure in need of repair. Those communities, already fully developed, have older houses that do not offer the size or amenities that more affluent home-buyers want. They thus become home to the region's lower-income families, with greater need for social services, public housing, and public safety services.

In more rapidly developing edge communities, young families with children purchase relatively affordable, newer three- and four-bedroom homes that allow them to avoid the social stresses of older core communities. School enrollments and the need for new schools expand quickly in those communities, as does the need for new roads, parks, libraries, police and fire protection, and other public services. Unable to keep up, edge communities are perpetually short of cash.

Both types of fiscally strained communities become home to families and individuals with the greatest need for additional spending and include a disproportionate share of a region's cost-generating land uses. Meanwhile, they fall farther and farther behind in their ability to generate revenues to meet their needs. [T]hese communities housed 40 percent of the population in the 25 largest metropolitan areas, nearly six times the share of the high-capacity, low-cost communities. As a group, they commanded only a fraction of the tax capacity per household of the high-capacity, low-cost localities . . .

The middle ground, held by communities thought of as the prototypical suburb, is surprisingly small. They house just a quarter of the metropolitan population in the 25 largest areas, a bit more than a third of the suburban population. These communities, which have relatively low social costs but high growth rates, maintain a fragile balance between costs and revenues. They command roughly average tax capacities but are just managing to maintain that position over time.

Thus the distribution of public resources in metropolitan areas is highly skewed. A small part of metropolitan populations live in communities that are free of fiscal stress. Even if middle-ground suburbs that are just holding their own are included, just a third of the population in the 25 largest areas resides in fairly stress-free communities. Forty percent live in suburban communities with a much lower than average tax capacity that had been growing well below the average rate. The remainder (about 30 percent) of the population lives in central cities that, although they are holding their own in the competition for tax base, have a grossly disproportionate share of regional problems.

Misguided land-use planning

Fiscal zoning and competition for limited "good" land uses also discourage long-term planning that might allow communities to develop in an orderly and efficient way. Because competition for certain land uses can be so intense – and the impact of losing so severe – communities often feel they have to grab all the development they can before it leaves for another community. That is especially true in newly developing communities, trying to build an adequate tax base to pay for their growing needs and to pay off debts on new infrastructure. As property tax increases threaten, tremendous pressures build to spread the costs through growth.

However, they are rarely in a good position to win the competition for the most "profitable" land uses, ending up instead with single-family, moderately priced housing that generates more costs than revenues. Development of this sort at the fringe is also likely to generate long-term environmental costs – it is unlikely to be dense enough to support sewer and water treatment systems in the short term, but at the same time, it is too dense for effective use of septic systems over the longer term.

The imperative to grow in order to spread the costs of past development thus can lead instead to more new costs than new revenues, beginning a recurring cycle of budget shortfalls. In the long run, these communities would often be better off forestalling growth until

metropolitan expansion generates demand for denser, more balanced use of the land – an outcome more consistent with long-term regional and environmental needs. However, they need the fiscal capacity to take the long view, capacity that current development patterns often do not provide. Communities in the economic middle ground face strong incentives to push fiscal zoning and tax-base competition to the limit. If they do not, they become vulnerable to the downward cycle that they see in nearby at-risk communities. Since these places usually occupy the geographic middle ground of their metropolitan area, the resulting planning practices effectively push new, middle-income development even farther out into the metropolitan fringe.

Sprawling development patterns

The cumulative effect of fiscal zoning and tax-base competition worsens two aspects of urban sprawl: low-density development and a shift in population growth to the edges of a metropolitan region. Those characteristics are reinforced over time, fueled partly by the push of community decline in the region's older, developed areas and partly by the pull of rapidly growing communities on the metropolitan fringe. Steadily the gap widens between people who can avoid sharing the costs of public services by moving to another community and those who cannot. As businesses and homeowners migrate outward from a region's core, federal and state investment in highways, sewers, economic development, and schools follow, further diverting resources from the places that arguably need them most. There is a certain irony in the fact that taxes paid by the people left behind in the declining core often subsidize those investments.

Shifting investment in this way to previously undeveloped areas is a waste of taxpayers' limited resources, considering the significant investments in infrastructure and housing that many core areas have already made. Instead of maintaining those investments, metropolitan areas throughout the country are courting new investments – building entirely new cities to replace the ones they are throwing away.

THE PROS AND CONS OF PROMOTING REGIONAL EQUITY

Policies to reduce fiscal inequities in metropolitan areas can bring great benefits. They can narrow the disparity in local governments' capacities to provide public services; they also can reduce incentives for fiscal zoning and inefficient tax-base competition and the negative consequences of those activities. However, those gains do not always come without costs. Poorly designed policies to reduce inequality can also compromise local autonomy, derail efficient provision of local services, and create incentives for inefficient patterns of migration. Finding policy designs that strike the best possible balance between trade-offs is not a trivial exercise.

Potential benefits of reducing inequality

Policies that promote equity in the distribution of local taxes can reduce incentives for fiscal zoning and tax-base competition and their negative outcomes in several ways. By ensuring that all local governments can provide the infrastructure and services communities need to function, equity-enhancing policies can guarantee that all residents of a metropolitan area enjoy at least a minimum standard of service for important local public goods like public safety. By reducing the need for local governments to "steal" revenue-generating land uses from each other, such policies allow them to engage in more thoughtful and beneficial land-use planning. By reducing tax-rate disparities – leveling the playing field in the tax-base competition – these policies encourage reinvestment in the central city and other fiscally stressed communities and reduce the growth of those disparities.

Fair provision of basic services. In a nation committed to equal opportunity for all citizens, all local governments must be able to provide their communities with basic infrastructure and services. However, an insufficient local tax base makes it difficult for local governments to do so at a reasonable tax rate. Equal opportunity is undercut when people living in low-tax-base communities must be taxed at burdensome levels to provide even minimum levels of public infrastructure and services while high-tax-base communities can keep local taxes low.

In every one of the 25 largest metropolitan areas in the United States, more than half the population lives in [an at-risk community]. At-risk places are not limited to central cities and core areas – they include older satellite cities, growing exurbs, and some second-ring suburbs. Nearly everywhere in a metropolitan region where social needs are high or increasing, the local tax base is low and declining. By contrast, places with few social needs generally have a much greater ability to generate revenues from their local tax base, and that base has been growing at higher-than-average rates. Even to approach the goal of guaranteed access to adequate municipal services would involve a significant increase in current equity policies.

Reduction of wasteful competition. Intrametropolitan competition for a limited tax base harms a region. It is a waste of resources for local governments to engage in bidding wars for businesses that already have chosen to locate in the region. In such situations, public monies are used to improve the fiscal position and services of one community at the expense of another, while businesses take unfair advantage of the competition to reduce their social responsibilities. The mere threat of leaving can induce troubled communities to offer a business large public subsidies to stay.

Greater equity frees local governments from the pressure to base land-use decisions primarily on the need for additional revenues. Instead, they can focus on developing land-use plans that accommodate growth efficiently, reduce the effects of concentrated poverty, and respond to the desires of local citizens without fearing that the resources available to them will be diminished. Furthermore, reducing the incentives for competition allows local governments to focus on cooperative efforts to help build a strong, dynamic region that is attractive to employers and residents.

Reinvestment in disadvantaged communities. In most metropolitan areas, older communities at the core live with aging infrastructure, industrial pollution, high concentrations of poverty, and other factors that strain their limited resources. Without sufficient funds, they cannot reinvest to rebuild sewer systems and roads, rehabilitate housing, maintain parks, and clean up polluted land. The expense entailed makes it difficult for such communities to remain competitive with newer communities elsewhere in the region that offer cheaper land, new homes, and more open space.

Greater fiscal equity diminishes the difficulty in two ways. First, it reduces tax-rate disparities, which are an important factor in intrametropolitan tax-base competition. Second, it helps to ensure that central cities and other stressed communities have the resources to maintain or enhance their competitive position by repairing crumbling infrastructure, cleaning up polluted land, and addressing other issues related to their age and social needs. Instead of being forced to leave their community because of inadequate infrastructure or crushing taxes, residents can be assured that the community will have the resources necessary to reinvest in its infrastructure and ensure its stability.

Potential costs of reducing inequality

Policies designed to reduce inequality come with costs. Some may compromise local autonomy and the efficient provision of local services; they may also generate incentives for inefficient patterns of migration.

Reduced local autonomy and efficiency. The U.S. political system places a high value on local autonomy. It is often argued that policies designed to create greater equity would undermine local autonomy and the advantages derived from providing individuals with a wide range of choices. Local control is desirable for several reasons. One argument is that, because local governments are smaller and closer to voters, local service provision encourages residents to participate in the democratic process. Another is that because the actions of local governments have a direct impact on the economic well-being of voters, primarily through their effect on home values, local control creates a powerful incentive for voters to monitor those actions: the services the government produces, for example, and its efficiency in providing them. Local autonomy also rewards localities for accepting land uses with some undesirable effects – for example, commercial activity that generates congestion – by allowing them to reap the full benefit of local taxation. Finally, a system of many local governments provides consumers and voters with a variety of combinations of local services to choose from, limiting the inevitable welfare losses that result from uniform provision of services within a jurisdiction.

Those are all worthy arguments. Designing policies to promote equity in a way that compromises local autonomy the least is one of the primary challenges associated with their development. However, local autonomy does not come without limitations, hence the need for higher levels of government. Some public services (such as wastewater treatment) cannot be provided efficiently at the small scale implied by a highly fragmented system of local governments. In

such cases, the efficiency gained by providing that service on a larger scale – at a higher level of government – clearly outweighs the costs associated with the loss of local autonomy.

Other public services generate costs and benefits over a much larger geographic area than one, two, or several jurisdictions. Many activities currently carried out by local governments have consequences beyond local borders. Natural systems spread the costs and benefits associated with water, sewer, and sewage treatment systems; regional housing markets spread the costs and benefits of local affordable housing programs, land-use restrictions, and income redistribution policies; regional labor markets spread the costs and benefits of economic development and education programs; transportation systems spread the costs and benefits of local street and bridge maintenance and enable nonresidents to enjoy locally maintained amenities such as parks.

When local actions have regional consequences, local and regional interests may diverge. An activity that makes perfect sense on the basis of an assessment of the potential costs and benefits to a locality may be undesirable from a regional perspective because many of the costs may be felt regionally, not locally. Another may not seem worthwhile to a locality but may nevertheless be highly beneficial from a regional standpoint because many of the benefits accrue to residents of surrounding communities. In those cases, some form of regional or state participation in decisionmaking is preferable to complete local autonomy. That participation can take many forms, including assumption of full responsibility for the activity, provision of full or partial financing of the activity through intergovernmental aid, or regulation of local government activities.

Other public services provided at least in part by local governments in most states are an integral part of what most Americans would regard as an acceptable quality of life. Good examples are public safety, a healthy environment, access to the courts or legal representation, access to housing in a variety of locations (facilitating access to jobs and good schools), and various health services. To many people, even "second-tier" local services such as parks and recreation fit into that category. The implication is that all people have the right to a reasonable standard of service in those areas – a standard that a highly fragmented and fiscally stratified system with full local autonomy may not always be able to meet.

Thus the choice is not simply between having or not having local autonomy. Instead, it comes down to balancing the benefits of local autonomy against the costs, policy area by policy area, and finding equity-enhancing policies that compromise the benefits of local autonomy the least.

Negative migration incentives. Conventional wisdom holds that income redistribution is the job of higher levels of government. Local governments pursue income redistribution at their peril because, in most metropolitan areas, it is easy for the losers in a locality that redistributes income to escape to a nearby locality that does not. Policies designed to reduce fiscal inequities in metropolitan areas must recognize that limitation. The clear implication is that the geographic scope of equity-enhancing policies must, at a minimum, include entire metropolitan areas – entire housing and labor markets. Programs that do not include the entire area run the risk of being undermined by migration. A program that affects only the inner portions of a metropolitan region, for instance, may inadvertently encourage sprawl by pushing higher-income residents to outer areas beyond the scope of the program.

Areas that do implement properly scaled programs to promote fiscal equity reduce the incentives for residents to move in pursuit of "personal" fiscal dividends or to flee small changes in the fiscal position of their neighborhood or community. Many people value the result – greater neighborhood and community stability.

POLICIES TO PROMOTE FISCAL EQUITY

State governments have long recognized the need to promote fiscal equity among local governments. They have followed three primary strategies to reduce interlocal fiscal disparities: "institutional" aid to juris-dictions with especially high needs (often central cities); state revenue-sharing programs that distribute state revenues to local governments; and metropolitan tax base–sharing programs that share tax resources within a single region.

Institutional Aid

States can help specific localities balance their capacity to raise revenues and their service needs or costs in ways that do not involve giving them direct financial aid. They can ease the burden on low-capacity or high-need places either by relieving them of financial responsibility for a service that is a local responsibility in their state (reducing their expenditures relative to those of other localities in the state) or by allowing them to levy taxes that other localities in the state are not permitted to levy (increasing their capacity). The capacity-increasing type of aid is most commonly provided to central cities or other large localities where the need for such measures is most evident.

[. . .]

In a regional context, a much more meaningful approach to easing demands on local budgets would be to relieve all localities of all or some of the respon-sibility for activities involving costs and benefits beyond their borders. A variety of activities for which local governments are responsible in many states are candidates for institutional reform of that sort, including wastewater treatment, many transportation-related activities, a variety of income-redistribution programs, and assistance for housing that does not generate positive fiscal dividends in a specific locality but is needed within a region. All such activities can entail a wide divergence of local and regional interests, implying the need for regional action . . .

The second form of institutional aid – special taxing power – is easier to document. . . . Washington, D.C., Minneapolis, St. Paul, and New York have special access to the sales tax. Washington, D.C., Detroit, Kansas City (Missouri), New York, and St. Louis assess local income taxes that other cities in their region generally cannot assess, and Philadelphia is the only city in its region with the power to tax the income of nonresidents who work in the city. These cities receive a real benefit in terms of their ability to raise revenues. Their tax capacities are 121 percent of their regional average, compared with an average of just 95 percent for other central cities. However . . . they do not benefit as much from state aid as the other central cities . . . In effect, some of the increase in capacity from institutional aid is offset by financial aid forgone . . . This form of aid presents an additional drawback: special taxes mark these cities as higher tax places than their suburban counterparts. This distinc-tion puts them at a disadvantage in the competition for economic activity within their metropolitan area.

Overall, the central cities . . . do not appear to benefit greatly, if at all, from institutional aid. Indeed, any advantages gained appear to be more than offset by disadvantages in other areas. More fundamental reforms are called for in the way local services are organized – reforms that explicitly recognize the regional costs and benefits of local activities.

State aid programs

Many states attempt to reduce fiscal inequity among jurisdictions through revenue-sharing programs that distribute a portion of the revenues from one or more state taxes to local governments through a variety of formulas. Massachusetts, Michigan, and Wisconsin are conspicuous examples of states with large revenue-sharing programs. Whether fiscal inequities are reduced in this way depends largely on the formula used to distribute revenues to local governments. Most programs began with a simple per-capita (return-to-origin) approach to distributing funds. As those programs evolved, however, it became clear that such approaches did little to reduce the fiscal inequities among localities.

Most distribution formulas now place greater emphasis on the communities' needs, typically determined by characteristics such as tax base, revenues, spending, or some combination of the three. For instance, in Wisconsin, revenues are distributed on the basis of five criteria, including population, the value of the local tax base, and compensation for the presence of a power company property not taxable locally. In Michigan, the legislature recently revised its revenue-sharing program so that a portion of the revenues collected from the state sales tax is distributed to local governments according to three separate formulas – one accounting for population and the type of jurisdiction, and two based on taxable value per capita. Detroit, which is excluded from those formulas, receives a fixed payment, protecting it from potential decreases due to future population declines. . . . Most aid programs do equalize revenue-raising capacity to some extent.

[. . .]

Tax-base sharing

Tax-base sharing, an alternative way to reduce tax base inequities, has several advantages over the patchwork quilt of aid programs common to most states. First, tax-base sharing provides resources to multiple taxing jurisdictions at the same time. Unlike separate programs that distribute state revenues to counties, cities, townships, and special districts, tax-base sharing simply redistributes the common base from which each local jurisdiction derives its revenues. Second, it helps to equalize the resources available to local governments without removing local control over tax rates. Third, by requiring local governments to relinquish some of their fiscal dividend from new commercial-industrial development, tax-base sharing weakens their incentive to waste taxpayer dollars by stealing it away from other communities. Similarly, including residential property in tax-base sharing dilutes local governments' incentives to use fiscal zoning or its substitutes to restrict residential development to "profitable" types of housing, making cooperative, efficient land-use planning easier. Minnesota's experience with tax-base sharing is evidence of those effects. In the Twin Cities region in the early 1970s, reformers attempting to pass legislation for metropolitan land-use planning used tax-base sharing as a quid pro quo to gain political support in the low-fiscal capacity, developing suburbs. These suburbs initially cried foul when told that an urban service line would be drawn through the middle of their towns and that land outside that boundary would be zoned at agricultural densities. They argued that they desperately needed the land to develop their tax base so that they could keep their tax rates down and still relieve overcrowding in their schools. Compromise and acceptance was reached when they were shown the potential benefits of a tax-base-sharing system: that is, that they would share in the full region's new tax base and would gain fiscal capacity per capita faster than they would by developing lower-value residential property. In the end, low-tax-base communities in the region accepted land-use planning in exchange for tax-base sharing.

Flexibility is a fourth advantage of tax-base sharing. This tool can be designed to offset intraregional variations in the need for or cost of public services as well as variations in revenue-raising capacity. It can also be implemented at the regional level to avoid the problems associated with designing aid formulas to simultaneously accommodate the vastly different needs and costs of urban and rural areas. Furthermore, if properly structured, it avoids migration issues.

With tax-base sharing, a portion of each locality's tax base is contributed to a regional pool and redistributed according to some criteria other than the locality's original contribution to the pool. A community's contribution can be set as a percentage of growth in tax base or as a percentage of current tax base. The tax-base pool can be limited to particular types of tax base (for example, commercial-industrial property), or it can include all types (sales tax, income tax, and property tax). Distributions from the pool can be determined by tax capacity, service cost or need indicators, land-use decisions, or other criteria. The essential features of tax-base sharing are that it distributes tax base or revenues by criteria other than the origin or collection point (unlike piggyback taxes, for instance); it provides resources for the full range of local services (unlike special district assessments); and it provides additional resources for the provision of local services (unlike county or state taxes).

Tax-base sharing across a metropolitan area has been attempted in only one place in the United States: the Minneapolis–St. Paul (Twin Cities) region of Minnesota. Created by the Minnesota Fiscal Disparities Act of 1971 as an alternative to annexation and consolidation of local governments, the Twin Cities' tax-base-sharing program was an attempt to respond to a number of concerns, including increasing property tax rates, tax-base and tax-rate disparities, and interjurisdictional competition for development.

Under the Twin Cities program, each taxing jurisdiction in a seven-county area must contribute to a regional pool 40 percent of the growth in the value of its commercial-industrial tax capacity since 1971. Municipalities are assigned a portion of that pool, based on population and the ratio of the total market value of property per capita in the jurisdiction and the average market value of property per capita in the region. The formula assigns a share of the pool that is greater than their share of population to municipalities with lower-than-average market value per capita; high-market-value localities receive a lower portion than their population share.

The program's overall design balances regional goals with local autonomy. It is designed to narrow interlocal business tax-rate disparities by taxing part of commercial-industry property at a uniform regional rate, but it also allows each locality to set the rate at which the locality taxes its distribution from the pool.

In 2000 the Twin Cities program shared about 28 percent of the region's commercial-industrial tax base, an amount that would have generated roughly $300 million in revenue at the regional average tax rate on commercial-industrial property. That amount represented about 12 percent of total tax base. The program reduced local tax-base disparities by roughly 20 percent . . .

[. . .]

The results [of simulations] show that tax-base sharing is a much more cost-effective means of reducing tax-base inequity than existing aid programs. . . . Although current aid is not designed solely to equalize the tax base, that is still an impressive difference in "bang for the buck."

AN AGENDA FOR REFORM

Policies designed to reduce inequality can be controversial and divisive. On the surface, they create winners and losers: The potential losers often argue that, in addition to their direct losses, equity-enhancing activities generate economic losses in the form of a less efficient regional economy and public sector. However, . . . resources are so skewed in most metropolitan areas that winners are likely to far outnumber losers, even for relatively modest proposals to share the benefits of metropolitan growth. Nor need one appeal to unenlightened self-interest for support of such programs. Properly designed institutional reform, aid, and tax-base-sharing programs offer real efficiency gains in addition to the equity benefits.

The arguments for institutional reform are primarily efficiency arguments. Public services should be designed and financed on a scale commensurate with the scope of their costs and benefits. If they are not, then local and regional interests diverge. Strict local control can result in waste – wasteful tax-base competition, excessive depletion of water and other resources, overconsumption of land, and segregation and concentration of impoverished populations in specific areas, all of which increase the social and public costs of income inequality. A growing body of research implies that suburban growth is tied to the economic health of central cities. Strict local control of land-use planning also increases the physical separation between rich and poor, to the detriment of everyone.

Similarly, aid and tax-base-sharing programs can be designed to enhance the efficiency of both the local and regional economies and the public sector as well as to improve equity. Attenuating the link between growth in particular types of local land uses and the tax base available to produce local services reduces wasteful competition. Providing financial incentives for particular types of development that provide regional benefits but do not generate local fiscal dividends can improve the functioning of regional housing and labor markets.

Reforms in these policy areas need not be radical. All states provide at least some financial support to local governments. A reform agenda can begin with incremental improvements in the way current aid is allocated. Tax-base-sharing programs can be designed to capture a portion of tax-base growth, as occurred in the Twin Cities, rather than part of existing tax bases, allowing regions to reap the efficiency benefits immediately while the redistributive impacts grow more slowly.

Nor does institutional reform necessarily imply designing a regional government from whole cloth. New or existing cooperative arrangements in areas in which the argument for regional control is strongest (water management and transportation, for instance) can be the building blocks for future reform.

F
O
U
R

Plate 23 New Urbanist neo-traditionalism. In Seaside, Florida, designed by the firm of Duany–Plater-Zyberk, a postmodernist streetscape mimics the sense and scale of the traditional American small town.

Plate 24 The mall has it all. The suburban shopping center, sometimes called the "new downtown," is both an extravaganza of commercialism and a new kind of social and entertainment center. The Mall of America in Minneapolis, Minnesota, has shops, restaurants, and even a roller coaster!

Plate 25 Architecture as symbolic power. The Petronas Twin Towers in Kuala Lumpur, designed by Cesar Pelli and Associates, are the tallest buildings in the world and proclaim Malaysia to be one of the new power centers of the global economy. (Photograph by Roger Mellor.)

Plate 26 Shanghai, China: an Asian megacity. As the world's population increases and higher percentages of the population live in cities there are increasing numbers of megacities of ten million residents and more. Photographer Peter Bialobrzeski's images of Asian megacities appear in his book *Neon Tigers*. (Photograph by Peter Bialobrzeski from *Neon Tigers*.)

Plate 27 **The persistence of poverty.** While postmodernism and the global economy advance, streets in working-class neighborhoods continue to serve as market-places, playgrounds, and political venues. Here, a street in central Rome exhibits a lively blending of urban uses and purposes. Note the Communist Party wall poster. (Photograph by Patti Walters.)

Plate 28 **The persistence of decay.** Amid the splendor of new global power centers and the comfort of New Urbanist neighborhoods, inner-city decay persists, even dominates. Desperate graffiti, barred windows, and litter of all sorts mark this sidewalk in San Francisco, California. (Photograph by Derek Metzger.)

Plate 29 The persistence of tradition. Today, many Islamic cities are gleamingly modern, but others retain a traditional, almost medieval flavor. In Fez, Morocco, a crowd fills a walled plaza to watch a street entertainer. (Photograph by Paul V. Turner.)

Plate 30 The streets belong to the kids. Skateboarders may seem like a relatively recent phenomenon, but youth has always found a way to expropriate (adult) public space. (Photograph by Frederic Stout.)

Plate 31 **The streets belong to the people!** Street protests, like this demonstration in Seattle, Washington, have greeted the emergence of the new global economy. The political cause is new, but the use of public space for popular and insurgent politics is an ancient and honorable tradition. (Photograph by Jason Morrison.)

Plate 32 Urban terror. The World Trade Center, New York, September 11, 2001. A new kind of global politics renders the city suddenly vulnerable. (Photograph by Wanda McCormick, Readio.com.)

PART FIVE

Urban planning history and visions

THE CRISIS,

OR THE CHANGE FROM ERROR AND MISERY, TO TRUTH AND HAPPINESS

1832.

IF WE CANNOT YET   RECONCILE ALL OPINIONS,

LET US ENDEAVOUR   TO UNITE ALL HEARTS

IT IS OF ALL TRUTHS THE MOST IMPORTANT, THAT THE CHARACTER OF MAN IS FORMED FOR—NOT BY HIMSELF.

Design of a Community of 2,000 Persons, founded upon a principle, commended by Plato, Lord Bacon, Sir T. More, & R. Owen.

EDITED BY
ROBERT OWEN AND ROBERT DALE OWEN.

London:

PRINTED AND PUBLISHED BY J. EAMONSON, 15, CHICHESTER PLACE
GRAY'S INN ROAD.

STRANGE, PATERNOSTER ROW, PURKISS, OLD COMPTON STREET,
AND MAY BE HAD OF ALL BOOKSELLERS.

INTRODUCTION TO PART FIVE

The effects of urban planning are perhaps the greatest – and, at the same time, the most invisible – influences on human life and culture. In the words of Paul and Percival Goodman, the co-authors of *Communitas: Means of Livelihood and Ways of Life* (1947), we hardly realize as we go about the daily round of our lives "that somebody once drew some lines on a piece of paper who might have drawn otherwise" and that "now, as engineer and architect once drew, people have to walk and live."

When the Sumerian kings built the walls of Eridu and Uruk, they engaged in acts of urban planning and thus determined how their people would "walk and live." The walls provided safety and protection for the people of the city and also defined the new political unity of the city-state. The associated roads, bridges, irrigation systems, and centers for market and ceremonial functions, all served a dual function in that they met the practical social needs of the urban population in general and fulfilled the power aspirations of the urban elites in particular. The ancient citadels were centers of religious meaning, as well as economic and political power, and thus a third component of urban planning – an idealized, often spiritual vision of what constitutes the best possible state of human existence – was present at the very beginning of city building.

The origins of modern urban planning are complex. On one level, modern planning is a direct extension of the ancient and pre-modern models: imposing order on nature for the health, safety, and amenity of the urban masses, for the political benefit of the urban elites, and for a way of expressing each culture's highest spiritual ideals. On another level, however, modern planning is far more complex than anything that had ever gone on before. Modern planning operates, by and large, in a politically and economically pluralistic environment, making every alteration of the physical arrangements of the city a complex negotiation between competing interests. And the practice of modern urban planning also takes place at a stage of human development when the planner's defining goal is no longer merely to impose human order on nature but continuously to impose order on the city itself.

All the goals and functions of planning – both the ancient hold-overs and the modern elaborations – are present in the first planning responses to the urban conditions associated with the Industrial Revolution. As Friedrich Engels (p. 50) and other contemporary observers described, the cities of the new industrialism were characterized by horrendous overcrowding, ubiquitous misery, and despair. There were daily threats to the public health and safety, not just for the impoverished working class but for the capitalist middle class as well. These conditions gave rise to movements for housing reform, to great advances in the technologies of water supply and sewage disposal, and to the emergence of middle-class suburbs. They also led to the construction of model "company towns" by various industrial firms in both Europe and America, and to the development of a modern urban planning profession.

Reviewing the history of urban planning in the nineteenth century, Richard LeGates and Frederic Stout (the co-editors of this volume) have written that "the classic texts of early urban planning history often seem surprisingly modern." An example of the surprising modernity of early urban planning is the nineteenth-century parks movement, especially the work of Frederick Law Olmsted (p. 307), which gave rise to something very like comprehensive urban planning practice. Projects like Central Park in New York (Plate 33) represented a transplantation and democratization of European landscape gardening traditions, to be sure, but Olmsted's goal was not merely to bring nature into the city. Rather, Olmsted repeatedly appealed to the political and economic leadership of American cities to create parks that would achieve a whole range of public benefits: they would contribute to the public health by serving as the "lungs" of the city; they would be practical and necessary additions to the physical infrastructure of the metropolis, providing a general recreation ground; their ponds and reservoirs would serve as adjuncts to municipal water-supply systems; and they would soften and tame human nature by providing wholesome alternatives to the vulgar street amusements that daily tempted poor and working-class youth.

Olmsted was a visionary and a reformer, but he was also a successful businessman and a canny political operative capable of offering his clients useful strategic advice on how to fund and build constituencies in favor of large municipal projects. Somewhat less practical, but even more visionary, were a group of architects, planners, and activists who may be termed, collectively, the utopian modernists. Three of these – Ebenezer Howard, Le Corbusier, and Frank Lloyd Wright – define the mainstream of that utopian tradition.

Not one of them had his utopian vision realized in its entirety, but each had an enormous influence on the way contemporary cities, and city life, developed in the twentieth century. A fourth, the Spanish engineer/planner Arturo Soria y Mata, is influential for his vision of the relationship between transportation systems and land use. Plate 34 illustrates Soria y Mata's vision of a "linear city" developed along a central spine containing an electric streetcar line and other utilities.

Ebenezer Howard (p. 314) prided himself on being "the inventor of the Garden City idea," and his tireless devotion to the project of decongesting the modern metropolis by building small, self-contained, green-belted cities in the rural countryside is one of the marvels of modern urban planning history. Plate 35, from *Garden Cities of To-morrow*, illustrates Howard's vision of "a group of slumless, smokeless cities." Howard originally wanted his Garden Cities to be cooperatively owned. He wanted the surrounding greenbelt to be much larger than the built-up part of the city itself. And he wanted his cities to be economically independent, not commuter suburbs. In the process of actually building Letchworth and Welwyn – the two Garden Cities constructed before his death in 1924 – Howard had to compromise many of his original goals. Building lots and businesses were privately owned; the greenbelt became more of a park than an extensive rural buffer zone; and neither of the original Garden Cities ever became a fully independent economic entity. Nonetheless, as the plan for Welwyn illustrates (Plate 36), these were fully planned communities that embodied many of Howard's ideals. Still, the Garden City experiment gave rise to a larger movement of town planning, and disciples of Howard spread his ideas and his example worldwide.

Charles-Édouard Jeanneret, better known as Le Corbusier (p. 322), was another utopian visionary who never saw his ideal plans fully developed, but who was enormously influential nonetheless. Le Corbusier wanted his "Contemporary City of Three Million," illustrated in Plate 37, to be a series of exquisite towers, geometrically arranged in a surrounding park, and he spent years looking for governmental and industrial sponsors for his plan. Many "Corbusian" high-rise urban developments have been built throughout the world. Indeed, the "International Style" of modern architecture and the principles of the International Congress of Modern Architecture (CIAM), which he pioneered, have become global standards of urban development. But in almost every case the surrounding park has been compromised away in the process of realization. In case after case, the tower in the park has become the tower without the park or, even worse, the tower in the parking lot!

While Le Corbusier was issuing his manifestos and shocking the architectural and planning establishment with his modernist plans, American planners Clarence Stein and Henry Wright were also wrestling with the problem of how to adapt urban form to the automobile. In their influential plan for Radburn, New Jersey, illustrated in Plate 38, Stein and Wright invented and implemented a series of planning concepts, including superblocks and the separation of pedestrian and vehicular traffic. Frank Lloyd Wright (p. 331), the originator of the visionary "Broadacre City" plan, was also responding to the automobile. Wright called for a city composed of family homesteads – one full acre per person – and the withering away of dense and crowded traditional cities. Wright's 1935 plan for Broadacre City is illustrated in Plate 39. The private automobile, Wright thought, would virtually abolish distance and allow for a new kind of community based on individualism and self-reliance. What actually became of Wright's Broadacre was sprawl suburbia. One acre per person became one-eighth acre per family or less; the core cities refused to wither away; the transportation monoculture of the automobile became a new form of dependency, rather than a technology of liberation; and the family-oriented suburban community became problematic at best, an object of ridicule at worst, producing the kind of "drive-in culture" described by Kenneth T. Jackson (p. 59).

The utopian visionaries were more than just planners, if they can be said to be planners at all. Even Ebenezer Howard, the most moderate of the group, was a dreamer. Together, the utopian modernists concerned themselves with great philosophical issues such as the connection between Mankind and Nature, the relationship of city plan to moral reform, and the role of urban design and technologies to the evolutionary transformation of society. It would be left to more practical men and women – the actual members of the urban planning profession as it developed in the twentieth century – to address the real-world problems of ever-changing cities and metropolitan regions. If the utopian modernists established the lofty goals, the professional planners attended to the details.

Still, the role of visionary projections of better lives through better urban planning persists as an important motivating force in contemporary urban planning. Establishing a good planning vision and sticking to it can have profound positive impacts. Plate 40 illustrates the famous "Paseo del Rio" of San Antonio, Texas. Like hundreds of other cities, San Antonio had a blighted area – in this case a river-turned-drainage-ditch – disfiguring the downtown. But unlike other cities, San Antonio developed a vision of turning the problem area into a magnificent location of riverfront amenities and recreational activities. Today, the Paseo provides a pleasant place to sit, stroll, paddle, and shop. Boston, Massachusetts worked the same kind of urban planning magic by collaborating with developer James Rouse to turn a seedy and obsolete market place around Quincy Market into a magnificent center for strolling, shopping, dining, and cultural events. Quincy Market today is illustrated in Plate 41.

In the 1970s and 1980s, a new idea took hold in planning circles – that of sustainable development. Closely allied with environmentalism and "green" politics, sustainability was defined and raised to a level of worldwide prominence by the publication, in 1987, of the Report of the World Commission on Environment and Development, commonly known as the Brundtland Report (p. 337) because the WCED chairperson was Gro Brundtland of Norway.

According to the Brundtland Report, sustainable development is "development that meets the needs of the present without compromising the ability of future generations to meet their own needs." Concerned about "the biosphere's ability to absorb the effect of human activities," the WCED defined humanity's basic needs as "food, clothing, shelter, and jobs" and called on the nations of the world to pay special attention to "the largely unmet needs of the world's poor, which should be given overriding priority." In the wake of the Brundtland Report, the world's nations have been asked to cut back on industrial production and the emission of greenhouse gases that affect the climate, and cities around the world have been asked to adopt "green policies" relating to transportation, energy use, resource management, and sprawl. As Timothy Beatley describes (p. 411), many European cities are now pursuing policies suggested by the Brundtland Report.

Today, the tradition of visionary urban planning – especially the idea of sustainability – is very much alive and well. In addition to visions of sustainable urban development discussed throughout this anthology – most particularly the work of Timothy Beatley and Stephen Wheeler (p. 499) – the contemporary planning movement called the "New Urbanism" is both a profit-making process of real-estate development that emphasizes small-town scale as the basis of new communities from Florida to California and a practical application of visionary ideas. Architect/planner Peter Calthorpe and urban planner William Fulton (p. 342) advocate a vision of a new "Regional City" in which land use and transportation systems are designed together in harmony with the natural environment to eliminate the blight of suburban sprawl and produce livable communities. One of Calthorpe's earliest visions was for "Pedestrian Pockets" such as those illustrated in Plate 42. Over time, these evolved into the kind of transit-oriented developments (TODs) that are currently being built, not as further extensions of traditional suburbia but in and for what he has variously called "The Next American Metropolis" and "The Post-Suburban Metropolis." Calthorpe's Pedestrian Pockets, transit-oriented developments, and Regional City conceptions look backward to the Garden Cities of Ebenezer

Howard – especially in their use of greenbelting and light-rail mass-transit options – but they look forward to an entirely new relationship between city and region, between individual and community. In short, the New Urbanist vision is a response to a future already in the process of becoming, characterized by "a dramatic shift in the nature and location of our workplace and a fundamental change in the character of our increasingly diverse households."

"Public Parks and the Enlargement of Towns"

American Social Science Association (1870)

Frederick Law Olmsted

Editors' Introduction

Frederick Law Olmsted (1822–1903) has been called "America's great pioneer landscape architect," and, during his lifetime, he was widely recognized as one of the most influential public figures in the nation. Along with his business partner, the English-born architect Calvert Vaux, he originated and dominated the urban parks movement, pioneered the development of planned suburbs, and laid out scores of public and private institutions. Central Park in New York, illustrated as it looked in 1863 in Plate 33, remains his best-known masterpiece. The designs for Riverside, Illinois (outside Chicago), the Boston park system, the Capitol grounds in Washington, DC, the 1893 World's Fair, and the campus of Stanford University in California are equally impressive contributions to the built environment.

Olmsted began his career practicing and writing about farming, then turned his talents to journalism and, in the 1850s, published a series of books describing the society and economy of the slave states of the American South (collected into one volume as *The Cotton Kingdom* in 1861). With this background, it is hardly surprising that Olmsted thoroughly imbued his art of landscape architecture with a wide variety of social and political, as well as cultural, concerns.

"Public Parks and the Enlargement of Towns" was originally presented as an address to the American Social Science Association, meeting at the Lowell Institute, Boston, in 1870. In it, Olmsted provides a number of specific guidelines for parks and parkways and suggests ways to overcome political resistance to public funding for parks and planned urban growth. Most importantly, however, he lays out the political and philosophical case for public parks in terms of three great moral imperatives: first, the need to improve public health by sanitation measures and the use of trees to combat air and water pollution; second, the need to combat urban vice and social degeneration, particularly among the children of the urban poor; and third, the need to advance the cause of civilization by the provision of urban amenities that would be democratically available to all.

Both as a practitioner and as a theorist, Olmsted anticipated many of the principal concerns of urban planning, both infrastructural and social, down to the present day. Indeed, behind the somewhat convoluted Victorianisms of his prose lies a strikingly modern mind. In the design of the Garden City, Ebenezer Howard (p. 314) borrowed directly from Olmsted, and even plans so fundamentally different as those of Frank Lloyd Wright (p. 331) and Le Corbusier (p. 322) owe a debt to Olmsted insofar as they recognize and address the central problem of the relationship between nature and the built urban environment. As the father of modern landscape architecture, Olmsted's work and thought invite comparison with all those – including J.B. Jackson (p. 184) and William H. White (p. 448) – who came after him in the profession, either as practitioners or critics. Although the contemporary New Urbanists like Peter Calthorpe and Andrés Duany (pp. 342, 192) and the advocates of sustainable planning like the World Commission on Environment and Development, Timothy Beatley, and Stephen Wheeler (pp. 337, 411, 499) go well beyond the parks movement in their comprehensive vision of the city–nature relationship, Olmsted and the nineteenth-century park builders can still be regarded as the pioneers of a new way at looking at the urban built environment.

A selection of Olmsted's most important writings may be found in S.B. Sutton (ed.), *Civilizing American Cities: Writings on City Landscapes by Frederick Law Olmsted* (Cambridge: MIT Press, 1971). Johns Hopkins University has published most of Olmsted's work in the multi-volume *Collected Papers of Frederick Law Olmsted* (Baltimore: Johns Hopkins University Press, 1977–1992). Biographies of Olmsted and commentary on his work include Laura Wood Roper, *FLO: A Biography of Frederick Law Olmsted* (Baltimore: Johns Hopkins University Press, 1973), Elizabeth Stevenson, *Park Maker: A Life of Frederick Law Olmsted* (New York: Macmillan, 1977), Charles E. Beveridge, Paul Rocheleau, and David Larkin, *Frederick Law Olmsted: Designing the American Landscape* (New York: Rizzoli, 1995), and Witold Rybczynski, *A Clearing in the Distance: Frederick Law Olmsted and America in the Nineteenth Century* (New York: Scribner's, 1999).

Galen Cranz's *The Politics of Park Design: A History of Urban Parks in America* (Cambridge: MIT Press, 1982) is a superb overview that places Olmsted's planning and landscape design achievements in the context of a larger movement for urban social reform. See also Cynthia Zaitzevsky, *Frederick Law Olmsted and the Boston Park System* (Cambridge: Belknap Press, 1992), and Susan L. Klaus, *Modern Arcadia: Frederick Law Olmsted, Jr. and the Plan for Forest Hill Gardens* (Cambridge: MIT Press, 2002). For information on La Villette in Paris and other great European parks, see Topos, *Parks: Green Spaces in European Cities* (New York: Princeton Architectural Press, 2002). Also of interest are Peter Harnik, *Inside City Parks* (Washington, DC: Urban Land Institute, 2000) and Terence Young, *Building San Francisco's Parks, 1850–1930* (Baltimore: Johns Hopkins University Press, 2004).

We have reason to believe, then, that towns which of late have been increasing rapidly on account of their commercial advantages, are likely to be still more attractive to population in the future; that there will in consequence soon be larger towns than any the world has yet known, and that the further progress of civilization is to depend mainly upon the influences by which men's minds and characters will be affected while living in large towns.

Now, knowing that the average length of the life of mankind in towns has been much less than in the country, and that the average amount of disease and misery and of vice and crime has been much greater in towns, this would be a very dark prospect for civilization, if it were not that modern Science has beyond all question determined many of the causes of the special evils by which men are afflicted in towns, and placed means in our hands for guarding against them. It has shown, for example, that under ordinary circumstances, in the interior parts of large and closely built towns, a given quantity of air contains considerably less of the elements which we require to receive through the lungs than the air of the country or even of the outer and more open parts of a town, and that instead of them it carries into the lungs highly corrupt and irritating matters, the action of which tends strongly to vitiate all our sources of vigor – how strongly may perhaps be indicated in the shortest way by the statement that even metallic plates and statues corrode and wear away under the atmosphere influences which prevail in the midst of large towns, more rapidly than in the country.

The irritation and waste of the physical powers which result from the same cause, doubtless indirectly affect and very seriously affect the mind and the moral strength; but there is a general impression that a class of men are bred in towns whose peculiarities are not perhaps adequately accounted for in this way. We may understand these better if we consider that whenever we walk through the denser part of a town, to merely avoid collision with those we meet and pass upon the sidewalks, we have constantly to watch, to foresee, and to guard against their movements. This involves a consideration of their intentions, a calculation of their strength and weakness, which is not so much for their benefit as our own. Our minds are thus brought into close dealings with other minds without any friendly flowing toward them, but rather a drawing from them. Much of the intercourse between men when engaged in the pursuits of commerce has the same tendency – a tendency to regard others in a hard if not always hardening way. Each detail of observation and of the process of thought required in this kind of intercourse or contact of minds is so slight and so common in the experience of towns-people that they are seldom conscious of it. It certainly involves some expenditure nevertheless. People from the country are even conscious of the effect on their nerves and minds of the street contact – often complaining that they feel confused by it; and if we had no relief from it at all during our waking hours, we should all be conscious of suffering from it. It is upon our opportunities of relief from it, therefore, that not only our comfort in town life, but our ability to maintain a temperate, good-natured, and healthy state of mind, depends. This is one of many ways in which it happens that men who have been brought up, as the saying is, in the streets, who have been most directly and completely affected by town influences, so generally show, along with a remarkable quickness of apprehension, a peculiarly hard sort of selfishness. Every day of their lives they have seen thousands of their fellow-men, have met them face to face, have brushed against them, and yet have had no experience of anything in common with them.

[. . .]

It is practically certain that the Boston of today is the mere nucleus of the Boston that is to be. It is practically certain that it is to extend over many miles of country now thoroughly rural in character, in parts of which farmers are now laying out roads with a view to shortening the teaming distance between their wood-lots and a railway station, being governed in their courses by old property lines, which were first run simply with reference to the equitable division of heritages, and in other parts of which, perhaps, some wild speculators are having streets staked off from plans which they have formed with a rule and pencil in a broker's office, with a view, chiefly, to the impressions they would make when seen by other speculators on a lithographed map. And by this manner of planning, unless views of duty or of interest prevail that are not yet common, if Boston continues to grow at its present rate even for but a few generations longer, and then simply holds its own until it shall be as old as the Boston in Lincolnshire now is, more men, women, and children are to be seriously affected in health and morals than are now living on this Continent.

Is this a small matter – a mere matter of taste; a sentimental speculation?

It must be within the observation of most of us that where, in the city, wheel-ways originally twenty-feet wide were with great difficulty and cost enlarged to thirty, the present width is already less nearly adequate to the present business than the former was to the former business; obstructions are more frequent, movements are slower and oftener arrested, and the liability to collision is greater. The same is true of sidewalks. Trees thus have been cut down, porches, bow-windows, and other encroachments removed, but every year the walk is less sufficient for the comfortable passing of those who wish to use it.

It is certain that as the distance from the interior to the circumference of towns shall increase with the enlargement of their population, the less sufficient relatively to the service to be performed will be any given space between buildings.

In like manner every evil to which men are specially liable when living in towns, is likely to be aggravated in the future, unless means are devised and adapted in advance to prevent it.

Let us proceed, then, to the question of means, and with a seriousness in some degree befitting a question, upon our dealing with which we know the misery or happiness of many millions of our fellow-beings will depend.

We will for the present set before our minds the two sources of wear and corruption which we have seen to be remediable and therefore preventible. We may admit that commerce requires that in some parts of a town there shall be an arrangement of buildings, and a character of streets and of traffic in them which will establish conditions of corruption and of irritation, physical and mental. But commerce does not require the same conditions to be maintained in all parts of a town.

Air is disinfected by sunlight and foliage. Foliage also acts mechanically to purify the air by screening it. Opportunity and inducement to escape at frequent intervals from the confined and vitiated air of the commercial quarter, and to supply the lungs with air screened and purified by trees, and recently acted upon by sunlight, together with opportunity and induce-ment to escape from conditions requiring vigilance, wariness, and activity toward other men, – if these could be supplied economically, our problem would be solved.

In the old days of walled towns all tradesmen lived under the roof of their shops, and their children and apprentices and servants sat together with them in the evening about the kitchen fire. But now that the dwelling is built by itself and there is greater room, the inmates have a parlor to spend their evening in; they spread carpets on the floor to gain in quiet, and hang drapery in their windows and papers on their walls to gain in seclusion and beauty. Now that our towns are built without walls, and we can have all the room that we like, is there any good reason why we should not make some similar difference between parts which are likely to be dwelt in, and those which will be required exclusively for commerce?

Would trees, for seclusion and shade and beauty, be out of place, for instance, by the side of certain of our streets? It will, perhaps, appear to you that it is hardly necessary to ask such a question, as throughout the United States trees are commonly planted at the sides of streets. Unfortunately they are seldom so planted as to have fairly settled the question of the desirableness of systematically maintaining trees under these circumstances. In the first place, the streets are planned, wherever they are, essentially alike. Trees are planted in the space assigned for sidewalks, where at first, while they are saplings and the vicinity is rural or suburban, they are not much in the way, but where, as they grow larger, and the vicinity becomes urban, they take up more and more space, while space is more and more required for passage. That is not all. Thousands and tens of thousands are planted every year in a manner and under conditions as nearly certain as possible either to kill them outright, or to so lessen their vitality as to prevent their natural and beautiful development, and to cause premature decrepitude. Often, too, as their lower limbs are found inconvenient, no space having been provided for trees in laying out the street, they are deformed by butcherly amputations. If by rare good fortune they are suffered to become beautiful, they still stand subject to be condemned to death at any time, as obstructions in the highway.

What I would ask is, whether we might not with economy make special provision in some of our streets – in a twentieth or a fiftieth part, if you please, of all – for trees to remain as a permanent furniture of the city? I mean, to make a place for them in which they would have room to grow naturally and gracefully. Even if the distance between the houses should have to be made half as much again as it is required to be in our commercial streets, could not the space be afforded? Out of town space is not costly when measures to secure it are taken early. The assessments for benefit where such streets were provided for, would, in nearly all cases, defray the cost of the land required. The strips of ground required for the trees, six, twelve, twenty feet wide, would cost nothing for paving or flagging.

The change both of scene and of air which would be obtained by people engaged for the most part in the necessarily confined interior commercial parts of the town, on passing into a street of this character after the trees have become stately and graceful, would be worth a good deal. If such streets were made still broader in some parts, with spacious malls, the advantage would be increased. If each of them were given the proper capacity, and laid out with laterals and connections in suitable directions to serve as a convenient trunk line of communication between two large districts of the town or the business centre and the suburbs, a very great number of people might thus be placed every day under influences counteracting those with which we desire to contend.

These, however, would be merely very simple improvements upon arrangements which are in common use in every considerable town. Their advantages would be incidental to the general uses of streets as they are. But people are willing very often to seek recreations as well as receive it by the way. Provisions may indeed be made expressly for public recreations, with certainty that if convenient they will be resorted to.

We come then to the question: what accommodations for recreation can we provide which shall be so agreeable and so accessible as to be efficiently attractive to the great body of citizens, and which, while giving decided gratification, shall also cause those who resort to them for pleasure to subject themselves, for the time being, to conditions strongly counteractive to the special, enervating conditions of the town?

In the study of this question all forms of recreation may, in the first place, be conveniently arranged under two general heads. One will include all of which the predominating influence is to stimulate exertion of any part or parts needing it; the other, all which cause us to receive pleasure without conscious exertion. Games chiefly of mental skill, as chess, or athletic sports, as baseball, are examples of means of recreation of the first class, which may be termed that of *exertive* recreation; music and the fine arts generally of the second or *receptive* division.

Considering the first by itself, much consideration will be needed in determining what classes of exercises may be advantageously provided for. In the Bois de Boulogne there is a race-course; in the Bois de Vincennes a ground for artillery target-practice. Military parades are held in Hyde Park. A few cricket clubs are accommodated in most of the London parks, and swimming is permitted in the lakes at certain hours. In the New York Park, on the other hand, none of these exercises are provided for or permitted, except that the boys of the public schools are given the use on holidays of certain large spaces for ball playing. It is considered that the advantage to individuals which would be gained in providing for them would not compensate for the general inconvenience and expense they would cause.

I do not propose to discuss this part of the subject at present, as it is only necessary to my immediate purpose to point out that if recreations requiring large spaces to be given up to the use of a comparatively small number, are not considered essential, numerous small grounds so distributed through a large town that some one of them could be easily reached by a short walk from every house, would be more desirable than a single area of great extent, however rich in landscape attractions it might be. Especially would this be the case if the numerous local grounds were connected and supplemented by a series of trunk-roads or boulevards such as has already been suggested.

Proceeding to the consideration of receptive recreations, it is necessary to ask you to adopt and bear in mind a further subdivision, under two heads, according to the degree in which the average enjoyment is greater when a large congregation assembles for a purpose of receptive recreation, or when the number coming together is small and the circumstances are favorable to the exercise of personal friendliness.

The first I shall term *gregarious*; the second, *neighborly*. Remembering that the immediate matter in hand is a study of fitting accommodations, you will, I trust, see the practical necessity of this classification.

Purely gregarious recreation seems to be generally looked upon in New England society as childish and savage, because, I suppose, there is so little of what we call intellectual gratification in it. We are inclined to engage in it indirectly, furtively, and with complication. Yet there are certain forms of recreation, a large share of the attraction of which must, I think, lie in the gratification of the gregarious inclination, and which, with those who can afford to indulge in them, are so popular as to establish the importance of the requirement.

If I ask myself where I have experienced the most complete gratification of this instinct in public and out of doors, among trees, I find that it has been in the promenade of the Champs-Élysées. As closely following it I should name other promenades of Europe, and our own upon the New York parks. I have

studiously watched the latter for several years. I have several times seen fifty thousand people participating in them; and the more I have seen of them, the more highly have I been led to estimate their value as means of counteracting the evils of town life.

Consider that the New York Park and the Brooklyn Park are the only places in those associated cities where, in this eighteen hundred and seventieth year after Christ, you will find a body of Christians coming together, and with an evident glee in the prospect of coming together, all classes largely represented, with a common purpose, not at all intellectual, competitive with none, disposing to jealousy and spiritual or intellectual pride toward none, each individual adding by his mere presence to the pleasure of all others, all helping to the greater happiness of each. You may thus often see vast numbers of persons brought closely together, poor and rich, young and old, Jew and Gentile. I have seen a hundred thousand thus congregated, and I assure you that though there have been not a few that seemed a little dazed, as if they did not quite understand it, and were, perhaps, a little ashamed of it, I have looked studiously but vainly among them for a single face completely unsympathetic with the prevailing expression of good nature and light-heartedness.

Is it doubtful that it does men good to come together in this way in pure air and under the light of heaven, or that it must have an influence directly counteractive to that of the ordinary hard, hustling working hours of town life?

You will agree with me, I am sure, that it is not, and that opportunity, convenient, attractive opportunity, for such congregation, is a very good thing to provide for, in planning the extension of a town.

[. . .]

Think that the ordinary state of things to many is at this beginning of the town. The public is reading just now a little book in which some of your streets of which you are not proud are described. Go into one of those red cross streets any fine evening next summer, and ask how it is with their residents. Oftentimes you will see half a dozen sitting together on the door-steps or, all in a row, on the curb-stones, with their feet in the gutter; driven out of doors by the closeness within; mothers among them anxiously regarding their children who are dodging about at their play, among the noisy wheels on the pavement.

Again, consider how often you see young men in knots of perhaps half a dozen in lounging attitudes rudely obstructing the sidewalks, chiefly led in their little conversation by the suggestions given to their minds by what or whom they may see passing in the street, men, women, or children, whom they do not know and for whom they have no respect or sympathy. There is nothing among them or about them which is adapted to bring into play a spark of admiration, of delicacy, manliness, or tenderness. You see them presently descend in search of physical comfort to a brilliantly lighted basement, where they find others of their sort, see, hear, smell, drink, and eat all manner of vile things.

Whether on the curb-stones or in the dram-shops, these young men are all under the influence of the same impulse which some satisfy about the tea-table with neighbors and wives and mothers and children, and all things clean and wholesome, softening, and refining.

If the great city to arise here is to be laid out little by little, and chiefly to suit the views of land-owners, acting only individually, and thinking only of how what they do is to affect the value in the next week or the next year of the few lots that each may hold at the time, the opportunities of so obeying this inclination as at the same time to give the lungs a bath of pure sunny air, to give the mind a suggestion of rest from the devouring eagerness and intellectual strife of town life, will always be few to any, to many will amount to nothing.

But is it possible to make public provision for recreation of this class, essentially domestic and secluded as it is?

It is a question which can, of course, be conclusively answered only from experience. And from experience in some slight degree I shall answer it. There is one large American town, in which it may happen that a man of any class shall say to his wife, when he is going out in the morning: "My dear, when the children come home from school, put some bread and butter and salad in a basket, and go to the spring under the chestnut-tree where we found the Johnsons last week. I will join you there as soon as I can get away from the office. We will walk to the dairy-man's cottage and get some tea, and some fresh milk for the children, and take our supper by the brook-side"; and this shall be no joke, but the most refreshing earnest.

There will be room enough in the Brooklyn Park, when it is finished, for several thousand little family and neighborly parties to bivouac at frequent intervals through the summer, without discommoding one another, or interfering with any other purpose, to say nothing of those who can be drawn out to make a day

of it, as many thousand were last year. And although the arrangements for the purpose were yet very incomplete, and but little ground was at all prepared for such use, besides these small parties, consisting of one or two families, there came also, in companies of from thirty to a hundred and fifty, somewhere near twenty thousand children with their parents, Sunday-school teachers, or other guides and friends, who spent the best part of a day under the trees and on the turf, in recreations of which the predominating element was of this neighborly receptive class. Often they would bring a fiddle, flute, and harp, or other music. Tables, seats, shade, turf, swings, cool spring-water, and a pleasing rural prospect, stretching off half a mile or more each way, unbroken by a carriage road or the slightest evidence of the vicinity of the town, were supplied them without charge and bread and milk and ice-cream at moderate fixed charges. In all my life I have never seen such joyous collections of people. I have, in fact, more than once observed tears of gratitude in the eyes of poor women, as they watched their children thus enjoying themselves.

The whole cost of such neighborly festivals, even when they include excursions by rail from the distant parts of the town, does not exceed for each person, on an average, a quarter of a dollar; and when the arrangements are complete, I see no reason why thousands should not come every day where hundreds come now to use them; and if so, who can measure the value, generation after generation, of such provisions for recreation to the over-wrought, much-confined people of the great town that is to be?

For this purpose neither of the forms of ground we have heretofore considered are at all suitable. We want a ground to which people may easily go after their day's work is done, and where they may stroll for an hour, seeing, hearing, and feeling nothing of the bustle and jar of the streets, where they shall, in effect, find the city put far away from them. We want the greatest possible contrast with the streets and the shops and the rooms of the town which will be consistent with convenience and the preservation of good order and neatness. We want, especially, the greatest possible contrast with the restraining and confining conditions of the town, those conditions which compel us to walk circumspectly, watchfully, jealously, which compel us to look closely upon others without sympathy. Practically, what we most want is a simple, broad, open space of clean greensward, with sufficient play of surface and a sufficient number of trees about it to supply a variety of light and shade. This we want as a central feature. We want depth of wood enough about it not only for comfort in hot weather, but to completely shut out the city from our landscapes.

The word *park*, in town nomenclature, should, I think, be reserved for grounds of the character and purpose thus described.

[. . .]

A park fairly well managed near a large town, will surely become a new center of that town. With the determination of location, size, and boundaries should therefore be associated the duty of arranging new trunk routes of communication between it and the distant parts of the town existing and forecasted.

These may be either narrow informal elongations of the park, varying say from two to five hundred feet in width, and radiating irregularly from it, or if, unfortunately, the town is already laid out in the unhappy way that New York and Brooklyn, San Francisco and Chicago, are, and, I am glad to say, Boston is not, on a plan made long years ago by a man who never saw a spring-carriage, and who had a conscientious dread of the Graces, then we must probably adopt formal Park-ways. They should be so planned and constructed as never to be noisy and seldom crowded, and so also that the straightforward movement of pleasure-car carriages need never be obstructed, unless at abso-

lutely necessary crossings, by slow-going heavy vehicles used for commercial purposes. If possible, also, they should be branched or reticulated with other ways of a similar class, so that no part of the town should finally be many minutes' walk from some one of them; and they should be made interesting by a process of planting and decoration, so that in necessarily passing through them, whether in going to or from the park, or to and from business, some substantial recreative advantage may be incidentally gained. It is a common error to regard a park as something to be produced complete in itself, as a picture to be painted on canvas. It should rather be planned as one to be done in fresco, with constant consideration of exterior objects, some of them quite at a distance and even existing as yet only in the imagination of the painter.

I have thus barely indicated a few of the points from which we may perceive our duty to apply the means in our hands to ends far distant, with reference to this problem of public recreations. Large operations of construction may not soon be desirable, but I hope you will agree with me that there is little room for question, that reserves of ground for the purposes I have referred to should be fixed upon as soon as possible, before the difficulty of arranging them, which arises from private building, shall be greatly more formidable than now.

FIVE

"Author's Introduction" and "The Town–Country Magnet"

from *Garden Cities of To-morrow* (1898)

Ebenezer Howard

Editors' Introduction

A stenographer by trade, Ebenezer Howard (1850–1928) was a quiet, modest, self-effacing man – "a man without credentials or connections," as one biographer put it – who nevertheless managed to change the world. Born in London, Howard early experienced the pollution, congestion, and social dislocations of the modern industrial metropolis. After a year in America (as a homesteader in Nebraska!), he returned to England in 1876 and became involved in political movements and discussion groups addressing what was then termed "the Social Question." Howard was influenced by a number of radical theorists and visionaries, including the social reformer Robert Owen, the utopian novelist Edward Bellamy, and the single-tax advocate Henry George. He published *To-morrow: a Peaceful Path to Real Reform* in 1898 (now better known under its 1902 title, *Garden Cities of To-morrow*) and methodically set about convincing people of the beauty and utility of "the Garden City idea."

Although Howard's plan may seem quaintly Victorian to the modern reader, the ideas he put forward were revolutionary at the time. Indeed, Howard's ideas of urban decentralization, zoning for different uses, the integration of nature into cities, greenbelting, and the development of self-contained "New Town" communities outside crowded central cities, illustrated in Plate 35, laid the groundwork for the entire tradition of modern city planning.

Unlike many other utopian dreamers, Howard lived to see his plans (if in a somewhat compromised form) actually put into action. In his own lifetime, the Garden Cities of Letchworth and Welwyn were built in England. Later, the Garden City idea spread to continental Europe, to America by way of the New Deal, and to much of the rest of the world.

Howard's argument begins with a protest against urban overcrowding; the one issue upon which, he writes, "men of all parties" are "well-nigh universally agreed." He then explains why "the people continue to stream into the already overcrowded cities" by reference to "the town magnet," that combination of jobs and amenities that characterizes the modern metropolis. Arrayed against this urban magnetic force is "the country magnet," the appealing features of the more natural, but increasingly desolate, rural districts. Finally, Howard describes his own plan, a new kind of human community based on "the town–country magnet," which is the best of both worlds.

As detailed in his famous concentric-ring diagram (which, he is careful to warn, is "a diagram only," not an actual site plan), the center of Garden City is to be a central park containing important public buildings and surrounded by a "Crystal Palace" ring of retail stores. The entire city of approximately 1,000 acres is to be encircled by a permanent agricultural greenbelt of some 5,000 acres, and the new cities are to be connected with central "Social Cities" and each other by a system of railroad lines.

Howard's ideas about the evils of overcrowding are similar to those of Friedrich Engels (p. 50), and his solution to the problem invites comparison with the very different solutions proposed by Le Corbusier (p. 322) and Frank Lloyd Wright (p. 331). Direct followers of Howard include Patrick Geddes and Lewis Mumford (p. 85), who helped to spread the Garden City idea throughout Europe and America. More recently, Peter Calthorpe (p. 342) has effectively reinvented the Garden City idea in California as the Regional City in the form of greenbelted, suburban "Pedestrian Pockets" and TODs (transit-oriented developments) linked to central cities (and each other) by a network of light-rail transportation systems.

Garden Cities of To-morrow remains a readable and relevant book. It is available as the second volume of Richard T. LeGates and Frederic Stout (eds), *Early Urban Planning* (nine volumes, London: Routledge/Thoemmes, 1998) and in earlier editions by Attic Books (1985), Eastbourne (1985), MIT Press (1965), and Faber and Faber (1960, 1951, and 1946). The original edition appeared under the title *To-morrow: A Peaceful Path to Real Reform* (Sonnenschein, 1898), and an elegant new edition is now available, under the original title, edited by Peter Hall and Colin Ward (Routledge, 2003).

Biographies of Ebenezer Howard include Robert Beevers, *The Garden City Utopia: A Critical Biography of Ebenezer Howard* (New York: St. Martin's, 1988), and Dugald Macfadyen, *Sir Ebenezer Howard and the Town Planning Movement* (Manchester: Manchester University Press, 1933; reprinted Cambridge: MIT Press, 1970). Excellent accounts of Howard and the Garden City movement may be found in Robert Fishman, *Urban Utopias in the Twentieth Century* (New York: Basic Books, 1977) and Peter Hall, *Cities of Tomorrow* (Oxford: Basil Blackwell, 1988).

Additional books about Ebenezer Howard and the Garden City movement include Standish Meacham, *Regaining Paradise: Englishness and the Early Garden City Movement* (New Haven: Yale University Press, 1999), Peter Geoffrey Hall and Colin Ward, *Sociable Cities: The Legacy of Ebenezer Howard* (New York: John Wiley and Sons, 1998), Stephen V. Ward (ed.), *The Garden City: Past, Present and Future* (London and New York: E & FN Spon, 1992), Stanley Buder, *Visionaries and Planners: The Garden City Movement and the Modern Community* (Oxford University Press, 1990), and Kermit Parsons and David Schuyler (eds), *From the Garden City to Green Cities: The Legacy of Ebenezer Howard* (Baltimore: Johns Hopkins University Press, 2002).

AUTHOR'S INTRODUCTION

In these days of strong party feeling and of keenly contested social and religious issues, it might perhaps be thought difficult to find a single question having a vital bearing upon national life and well-being on which all persons, no matter of what political party, or of what shade of sociological opinion, would be found to be fully and entirely agreed . . .

[. . .]

There is, however, a question in regard to which one can scarcely find any difference of opinion . . . It is wellnigh universally agreed by men of all parties, not only in England, but all over Europe and America and our colonies, that it is deeply to be deplored that the people should continue to stream into the already over-crowded cities, and should thus further deplete the country districts.

All . . . are agreed on the pressing nature of this problem, all are bent on its solution, and though it would doubtless be quite Utopian to expect a similar agreement as to the value of any remedy that may be proposed, it is at least of immense importance that, on a subject thus universally regarded as of supreme importance, we have such a consensus of opinion at the outset. This will be the more remarkable and the more hopeful sign when it is shown, as I believe will be conclusively shown in this work, that the answer to this, one of the most pressing questions of the day, makes of comparatively easy solution many other problems which have hitherto taxed the ingenuity of the greatest thinkers and reformers of our time. Yes, the key to the problem how to restore the people to the land – that beautiful land of ours, with its canopy of sky, the air that blows upon it, the sun that warms it, the rain and dew that moisten it – the very embodiment of Divine love for man – is indeed a *Master Key*, for it is the key to a portal through which, even when scarce ajar, will be seen to pour a flood of light on the problems of intemperance, of excessive toil, of restless anxiety, of grinding poverty – the true limits of Governmental interference, ay, and even the relations of man to the Supreme Power.

It may perhaps be thought that the first step to be taken towards the solution of this question – how to restore the people to the land – would involve a careful consideration of the very numerous causes which have hitherto led to their aggregation in large cities.

Were this the case, a very prolonged enquiry would be necessary at the outset. Fortunately, alike for writer and for reader, such an analysis is not, however, here requisite, and for a very simple reason, which may be stated thus: Whatever may have been the causes which have operated in the past, and are operating now, to draw the people into the cities, those causes may all be summed up as "attractions"; and it is obvious, therefore, that no remedy can possibly be effective which will not present to the people, or at least to considerable portions of them, greater "attractions" than our cities now possess, so that the force of the old "attractions" shall be overcome by the force of new "attractions" which are to be created. Each city may be regarded as a magnet, each person as a needle; and, so viewed, it is at once seen that nothing short of the discovery of a method for constructing magnets of yet greater power than our cities possess can be effective for redistributing the population in a spontaneous and healthy manner.

So presented, the problem may appear at first sight to be difficult, if not impossible, of solution. "What", some may be disposed to ask, "can possibly be done to make the country more attractive to a workaday people than the town – to make wages, or at least the standard of physical comfort, higher in the country than in the town; to secure in the country equal possibilities of social intercourse, and to make the prospects of advancement for the average man or woman equal, not to say superior, to those enjoyed in our large cities?" The issue one constantly finds presented in a form very similar to that. The subject is treated continually in the public press, and in all forms of discussion, as though men, or at least working men, had not now, and never could have, any choice or alternative, but either, on the one hand, to stifle their love for human society – at least in wider relations than can be found in a straggling village – or, on the other hand, to forgo almost entirely all the keen and pure delights of the country. The question is universally considered as though it were now, and for ever must remain, quite impossible for working people to live in the country and yet be engaged in pursuits other than agricultural; as though crowded, unhealthy cities were the last word of economic science; and as if our present form of industry, in which sharp lines divide agricultural from industrial pursuits, were necessarily an enduring one. This fallacy is the very common one of ignoring altogether the possibility of alternatives other than those presented to the mind. There are in reality not only, as is so constantly assumed, two alternatives – town life and country life – but a third alternative, in which all the advantages of the most energetic and active town life, with all the beauty and delight of the country, may be secured in perfect combination; and the certainty of being able to live this life will be the magnet which will produce the effect for which we are all striving – the spontaneous movement of the people from our crowded cities to the bosom of our kindly mother earth, at once the source of life, of happiness, of wealth, and of power. The town and the country may, therefore, be regarded as two magnets, each striving to draw the people to itself – a rivalry which a new form of life, partaking of the nature of both, comes to take part in. This may be illustrated by a diagram (Figure 1) of "The Three Magnets", in which the chief advantages of the Town and of the Country are set forth with their corresponding drawbacks, while the advantages of the Town–Country are seen to be free from the disadvantages of either.

The Town magnet, it will be seen, offers, as compared with the Country magnet, the advantages of high wages, opportunities for employment, tempting prospects of advancement, but these are largely counterbalanced by high rents and prices. Its social opportunities and its places of amusement are very alluring, but excessive hours of toil, distance from work, and the "isolation of crowds" tend greatly to reduce the value of these good things. The well-lit streets are a great attraction, especially in winter, but the sunlight is being more and more shut out, while the air is so vitiated that the fine public buildings, like the sparrows, rapidly become covered with soot, and the very statues are in despair. Palatial edifices and fearful slums are the strange, complementary features of modern cities.

The Country magnet declares herself to be the source of all beauty and wealth; but the Town magnet mockingly reminds her that she is very dull for lack of society, and very sparing of her gifts for lack of capital. There are in the country beautiful vistas, lordly parks, violet-scented woods, fresh air, sounds of rippling water; but too often one sees those threatening words, "Trespassers will be prosecuted". Rents, if estimated by the acre, are certainly low, but such low rents are the natural fruit of low wages rather than a cause of substantial comfort; while long hours and lack of amusements forbid the bright sunshine and the pure air to gladden the hearts of the people. The one industry, agriculture, suffers frequently from excessive rainfalls;

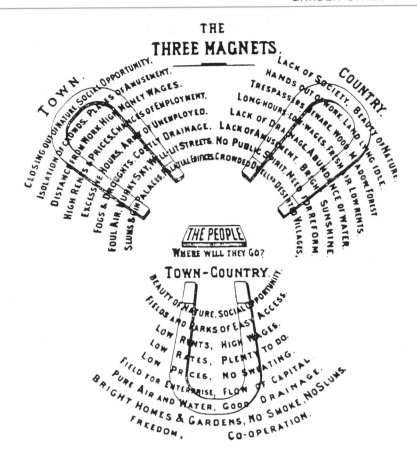

Figure 1

but this wondrous harvest of the clouds is seldom properly in-gathered, so that, in times of drought, there is frequently, even for drinking purposes, a most insufficient supply. Even the natural healthfulness of the country is largely lost for lack of proper drainage and other sanitary conditions, while, in parts almost deserted by the people, the few who remain are yet frequently huddled together as if in rivalry with the slums of our cities.

But neither the Town magnet nor the Country magnet represents the full plan and purpose of nature. Human society and the beauty of nature are meant to be enjoyed together. The two magnets must be made one. As man and woman by their varied gifts and faculties supplement each other, so should town and country. The town is the symbol of society – of mutual help and friendly co-operation, of fatherhood, motherhood, brotherhood, sisterhood, of wide relations between man and man – of broad, expanding sympathies – of science, art, culture, religion. And the country! The country is the symbol of God's love and care for man. All that we are and all that we have comes from it. Our bodies are formed of it; to it they return. We are fed by it, clothed by it, and by it are we warmed and sheltered. On its bosom we rest. Its beauty is the inspiration of art, of music, of poetry. Its forces propel all the wheels of industry. It is the source of all health, all wealth, all knowledge. But its fullness of joy and wisdom has not revealed itself to man. Nor can it ever, so long as this unholy, unnatural separation of society and nature endures. Town and country *must be married*, and out of this joyous union will spring a new hope, a new life, a new civilization. It is the purpose of this work to show how a first step can be taken in this direction by the construction of a Town–Country magnet; and I hope to convince the reader that this is practicable, here and now, and that on principles which are the very soundest, whether viewed from the ethical or the economic standpoint.

I will undertake, then, to show how in "Town–Country" equal, nay better, opportunities of social intercourse may be enjoyed than are enjoyed in any crowded city, while yet the beauties of nature may encompass and enfold each dweller therein; how higher wages are compatible with reduced rents and rates; how abundant opportunities for employment and bright prospects of advancement may be secured for all; how capital may be attracted and wealth created; how the most admirable sanitary conditions may be ensured; how beautiful homes and gardens may be seen on every hand; how the bounds of freedom may be widened, and yet all the best results of concert and co-operation gathered in by a happy people.

The construction of such a magnet, could it be effected, followed, as it would be, by the construction of many more, would certainly afford a solution of the burning question set before us by Sir John Gorst, "how to back the tide of migration of the people into the towns, and to get them back upon the land".

[. . .]

THE TOWN–COUNTRY MAGNET

The reader is asked to imagine an estate embracing an area of 6,000 acres, which is at present purely agricultural, and has been obtained by purchase in the open market at a cost of £40 an acre, or £240,000. The purchase money is supposed to have been raised on mortgage debentures, bearing interest at an average rate not exceeding 4 per cent. The estate is legally vested in the names of four gentlemen of responsible position and of undoubted probity and honour, who hold it in trust, first, as a security for the debenture-holders, and, secondly, in trust for the people of Garden City, the Town–Country magnet, which it is intended to build thereon. One essential feature of the plan is that all ground rents, which are to be based upon the annual value of the land, shall be paid to the trustees, who, after providing for interest and sinking fund, will hand the balance to the Central Council of the new municipality, to be employed by such Council in the creation and maintenance of all necessary public works – roads, schools, parks, etc. The objects of this land purchase may be stated in various ways, but it is sufficient here to say that some of the chief objects are these: To find for our industrial population work at wages of *higher purchasing power*, and to secure healthier surroundings and more regular employment. To enterprising manufacturers, co-operative societies, architects, engineers, builders, and mechanicians

of all kinds, as well as to many engaged in various professions, it is intended to offer a means of securing new and better employment for their capital and talents, while to the agriculturists at present on the estate as well as to those who may migrate thither, it is designed to open a new market for their produce close to their doors. Its object is, in short, to raise the standard of health and comfort of all true workers of whatever grade – the means by which these objects are to be achieved being a healthy, natural, and economic combination of town and country life, and this on land owned by the municipality.

Garden City, which is to be built near the centre of the 6,000 acres, covers an area of 1,000 acres, or a sixth part of the 6,000 acres, and might be of circular form, 1,240 yards (or nearly three-quarters of a mile) from centre to circumference. (Figure 2 is a ground plan of the whole municipal area, showing the town in the centre; and Figure 3, which represents one section or ward of the town, will be useful in following the description of the town itself – *a description which is, however, merely suggestive, and will probably be much departed from . . .*)

Six magnificent boulevards – each 120 feet wide – traverse the city from centre to circumference, dividing it into six equal parts or wards. In the centre is a circular space containing about five and a half acres, laid out as a beautiful and well-watered garden; and, surrounding this garden, each standing in its own ample grounds, are the larger public buildings – town hall, principal concert and lecture hall, theatre, library, museum, picture-gallery, and hospital.

The rest of the large space encircled by the "Crystal Palace" is a public park, containing 145 acres, which includes ample recreation grounds within very easy access of all the people.

Running all round the Central Park (except where it is intersected by the boulevards) is a wide glass arcade called the "Crystal Palace", opening on to the park. This building is in wet weather one of the favourite resorts of the people, whilst the knowledge that its bright shelter is ever close at hand tempts people into Central Park, even in the most doubtful of weathers. Here manufactured goods are exposed for sale, and here most of that class of shopping which requires the joy of deliberation and selection is done. The space enclosed by the Crystal Palace is, however, a good deal larger than is required for these purposes, and a considerable part of it is used as a Winter Garden – the whole forming a permanent exhibition of a most

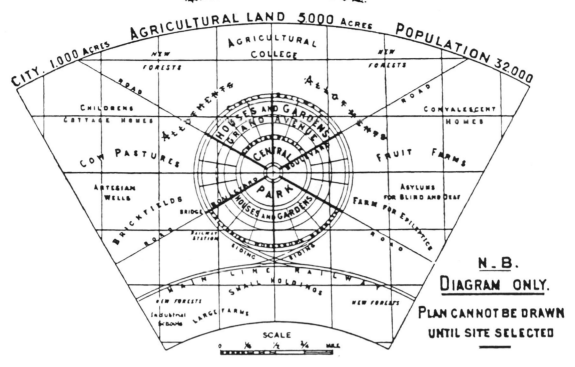

Figure 2

attractive character, whilst its circular form brings it near to every dweller in the town – the furthest removed inhabitant being within 600 yards.

Passing out of the Crystal Palace on our way to the outer ring of the town, we cross Fifth Avenue – lined, as are all the roads of the town, with trees – fronting which, and looking on to the Crystal Palace, we find a ring of very excellently built houses, each standing in its own ample grounds; and, as we continue our walk, we observe that the houses are for the most part built either in concentric rings, facing the various avenues (as the circular roads are termed), or fronting the boulevards and roads which all converge to the centre of the town. Asking the friend who accompanies us on our journey what the population of this little city may be, we are told about 30,000 in the city itself, and about 2,000 in the agricultural estate, and that there are in the town 5,500 building lots of an *average* size of 20 feet × 130 feet – the minimum space allotted for the purpose being 20 × 100. Noticing the very varied architecture and design which the houses and groups of houses display – some having common gardens

and co-operative kitchens – we learn that general observance of street line or harmonious departure from it are the chief points as to house building, over which the municipal authorities exercise control, for, though proper sanitary arrangements are strictly enforced, the fullest measure of individual taste and preference is encouraged.

Walking still toward the outskirts of the town, we come upon "Grand Avenue". This avenue is fully entitled to the name it bears, for it is 420 feet wide, and, forming a belt of green upwards of three miles long, divides that part of the town which lies outside Central Park into two belts. It really constitutes an additional park of 115 acres – a park which is within 240 yards of the furthest removed inhabitant. In this splendid avenue six sites, each of four acres, are occupied by public schools and their surrounding playgrounds and gardens, while other sites are reserved for churches, of such denominations as the religious beliefs of the people may determine, to be erected and maintained out of the funds of the worshippers and their friends. We observe that the houses fronting on Grand

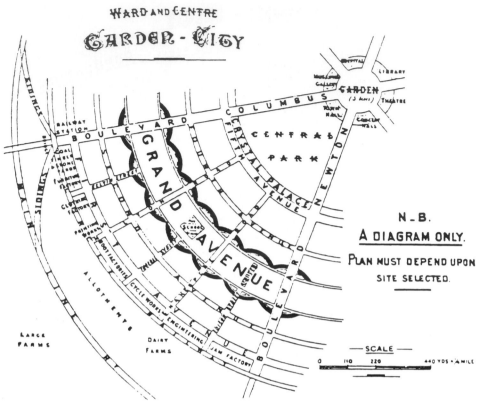

Figure 3

Avenue have departed (at least in one of the wards – that of which Figure 3 is a representation) – from the general plan of concentric rings, and, in order to ensure a longer line of frontage on Grand Avenue, are arranged in crescents – thus also to the eye yet further enlarging the already splendid width of Grand Avenue.

On the outer ring of the town are factories, warehouses, dairies, markets, coal yards, timber yards, etc., all fronting on the circle railway, which encompasses the whole town, and which has sidings connecting it with a main line of railway which passes through the estate. This arrangement enables goods to be loaded direct into trucks from the warehouses and workshops, and so sent by railway to distant markets, or to be taken direct from the trucks into the warehouses or factories; thus not only effecting a very great saving in regard to packing and cartage, and reducing to a minimum loss from breakage, but also, by reducing the traffic on the roads of the town, lessening to a very marked extent the cost of their maintenance. The smoke fiend is kept well within bounds in Garden City; for all machinery is driven by electric energy, with the result that the cost of electricity for lighting and other purposes is greatly reduced.

The refuse of the town is utilized on the agricultural portions of the estate, which are held by various individuals in large farms, small holdings, allotments, cow pastures, etc.; the natural competition of these various methods of agriculture, tested by the willingness of occupiers to offer the highest rent to the municipality, tending to bring about the best system of husbandry, or, what is more probable, the best systems adapted for various purposes. Thus it is easily conceivable that it may prove advantageous to grow wheat in very large fields, involving united action under a capitalist farmer, or by a body of co-operators; while the cultivation of vegetables, fruits, and flowers, which requires closer and more personal care, and more of the artistic and inventive faculty, may possibly be best dealt with by individuals, or by small groups of individuals having a common belief in the efficacy and value of certain dressings, methods of culture, or artificial and natural surroundings.

This plan, or, if the reader be pleased to so term it, this absence of plan, avoids the dangers of stagnation or dead level, and, though encouraging individual initiative, permits of the fullest co-operation, while the increased rents which follow from this form of competition are common or municipal property, and by far the larger part of them are expended in permanent improvements.

While the town proper, with its population engaged in various trades, callings, and professions, and with a store or depot in each ward, offers the most natural market to the people engaged on the agricultural estate, inasmuch as to the extent to which the townspeople demand their produce they escape altogether any railway rates and charges; yet the farmers and others are not by any means limited to the town as their only market, but have the fullest right to dispose of their produce to whomsoever they please. Here, as in every feature of the experiment, it will be seen that it is not the area of rights which is contracted, but the area of choice which is enlarged.

This principle of freedom holds good with regard to manufacturers and others who have established themselves in the town. These manage their affairs in their own way, subject, of course, to the general law of the land, and subject to the provision of sufficient space for workmen and reasonable sanitary conditions. Even in regard to such matters as water, lighting, and telephonic communication – which a municipality, if efficient and honest, is certainly the best and most natural body to supply – no rigid or absolute monopoly is sought; and if any private corporation or any body of individuals proved itself capable of supplying on more advantageous terms, either the whole town or a section of it, with these or any commodities the supply of which was taken up by the corporation, this would be allowed. No really sound system of *action* is in more need of artificial support than is any sound system of *thought*. The area of municipal and corporate action is probably destined to become greatly enlarged; but, if it is to be so, it will be because the people possess faith in such action, and that faith can be best shown by a wide extension of the area of freedom.

Dotted about the estate are seen various charitable and philanthropic institutions. These are not under the control of the municipality, but are supported and managed by various public-spirited people who have been invited by the municipality to establish these institutions in an open healthy district, and on land let to them at a pepper-corn rent, it occurring to the authorities that they can the better afford to be thus generous, as the spending power of these institutions greatly benefits the whole community. Besides, as those persons who migrate to the town are among its most energetic and resourceful members, it is but just and right that their more helpless brethren should be able to enjoy the benefits of an experiment which is designed for humanity at large.

"A Contemporary City"

from *The City of Tomorrow and its Planning* (1929)

Le Corbusier

Editors' Introduction

Le Corbusier (1887–1965) was one of the founding fathers of the Modernist movement and of what has come to be known as the International Style in architecture. Painter, architect, city planner, philosopher, author of revolutionary cultural manifestos, Le Corbusier exemplified the energy and efficiency of the Machine Age. His was the bold, nearly mystical rationality of a generation that was eager to accept the scientific spirit of the twentieth century on its own terms and to throw off all pre-existing ties – political, cultural, conceptual – with what it considered an exhausted, outmoded past.

Born Charles-Édouard Jeanneret, Le Corbusier grew up in the Swiss town of La Chaux-de-Fonds, noted for its watchmaking industry. He took his famous pseudonym after he moved to Paris to pursue a career in art and architecture. From the first, his designs for modern houses – he called them "machines for living" – were strikingly original, and many people were shocked by the spare cubist minimalism of his designs. The real shock, however, came in 1922 when Le Corbusier presented the public with his plan for "A Contemporary City of Three Million People." Laid out in a rigidly symmetrical grid pattern, the city consisted of neatly spaced rows of identical, strictly geometrical skyscrapers, as illustrated in Plate 37. This was not the city of the future, Le Corbusier insisted, but the city of today. It was to be built on the Right Bank, after demolishing several hundred acres of the existing urban fabric of Paris!

The "Contemporary City" proposal certainly caught the attention of the public, but it did not win Le Corbusier many actual urban planning commissions. Throughout the 1920s, 1930s, and 1940s, he sought out potential patrons wherever he could find them: the industrial capitalists of the Voisin automobile company, the communist rulers of the Soviet Union, and the fascist Vichy government of occupied France – mostly without success. Le Corbusier's real impact came not from cities he designed and built himself but from cities that were built by others incorporating the planning principles that he pioneered. Most notable among these was the notion of "the skyscraper in the park," an idea that is today ubiquitous. Whether in relatively complete examples like Brasilia and Chandigar, India (where new cities were built from scratch), or in partial examples such as the skyscraper parks and the high-rise housing blocks that have been built in cities worldwide, the Le Corbusier vision has truly transformed the global urban environment.

Le Corbusier's "Contemporary City" plan has often been contrasted to Frank Lloyd Wright's "Broadacre" (p. 331), and the comparison of a thoroughly centralized versus a thoroughly decentralized plan is indeed striking. Le Corbusier's boldness invites comparison with the original optimism of the post-World War II reconstruction and redevelopment efforts and even with the work of such visionary megastructuralists as Paolo Soleri. Some, however, have seen in the hyper-rationality of the pure Corbusian ideal an elitism and rigid class structure that runs counter to the democratic tradition. Lewis Mumford, Jane Jacobs, and Peter Hall may be counted as three of the severest critics. Allan Jacobs and Donald Appleyard's "Toward an Urban Design Manifesto" (p. 456) deliberately takes the form of a Le Corbusier pronouncement but rejects his program, opting instead for lively streets, participatory planning, and the integration of old buildings into the new urban fabric. Beneath all the sparkling clarity of Le Corbusier's urban designs are questions that must forever remain conjectural: How would democratic politics be practiced in a Corbusian city? What would social relationships be like amid the gleaming towers? Many of the "mega-urban regions" of Asia (p. 489) seem to rely on Corbusian principles, but would the "creative class" that Richard Florida (p. 129) writes about be comfortable in a Corbusian city? How would the long-distance virtual relationships envisioned by William J. Mitchell (p. 510) work, or not work, within a Cobusian environment? And what is it about the Corbusian skyscraper as a characteristic cultural form of modern Western urbanism, as described by Ian Buruma and Avishai Margalit (p. 136), that made the twin towers of the World Trade Center a target for Islamist attack on September 11, 2001?

Le Corbusier's writings include *The City of Tomorrow and its Planning* (New York: Dover, 1987) (translated by Frederich Etchells from *Urbanisme* [1929]), *Concerning Town Planning* (New Haven: Yale University Press, 1948) (translated by Clive Entwistle from *Propos d'Urbanisme* [1946]), and *L'Urbanisme des Trois Etablissements Humaines* (Paris: Editions de Minuit, 1959).

Excellent accounts of Le Corbusier's ideas may be found in Robert Fishman, *Urban Utopias in the Twentieth Century* (New York: Basic Books, 1977), Peter Hall, *Cities of Tomorrow* (Oxford: Blackwell, 1988), and Kenneth Frampton, *Le Corbusier: Architect of the Twentieth Century* (New York: Abrams, 2002, with photographs by Roberto Schezen). For a closer look at some of Le Corbusier's most important urban planning projects, consult Vikramaditya Prakash, *Chandigarh's Le Corbusier: The Struggle for Modernity in Postcolonial India* (Seattle: University of Washington Press, 2002) and Klaus-Peter Gast and Arthur Ruegg, *Le Corbusier: Paris-Chandigarh* (New York: Princeton Architectural Press, 2000). Jean-Louis Cohen, *Le Corbusier* (Köln: Taschen, 2005) contains excellent photographs of Le Corbusier's architectural work from all periods. For background on modernism as a movement, consult Richard Weston, *Modernism* (New York: Phaidon, 2001) and Christpher Wilk, *Modernism: Designing a New World* (London: Victoria and Albert Museum, 2006).

The existing congestion in the centre must be eliminated.

The use of technical analysis and architectural synthesis enabled me to draw up my scheme for a contemporary city of three million inhabitants. The result of my work was shown in November 1922 at the Salon d'Automne in Paris. It was greeted with a sort of stupor; the shock of surprise caused rage in some quarters and enthusiasm in others. The solution I put forward was a rough one and completely uncompromising. There were no notes to accompany the plans, and, alas! not everybody can read a plan. I should have had to be constantly on the spot in order to reply to the fundamental questions which spring from the very depths of human feelings. Such questions are of profound interest and cannot remain unanswered. When at a later date it became necessary that this book should be written, a book in which I could formulate the new principles of Town Planning, I resolutely decided *first of all* to find answers to these fundamental questions. I have used two kinds of argument: first, those essentially human ones which start from the mind or the heart or the physiology of our sensations as a basis; secondly, historical and statistical arguments.

Thus I could keep in touch with what is fundamental and at the same time be master of the environment in which all this takes place.

In this way I hope I shall have been able to help my reader to take a number of steps by means of which he can reach a sure and certain position. So that when I unroll my plans I can have the happy assurance that his astonishment will no longer be stupefaction nor his fears mere panic.

[. . .]

A CONTEMPORARY CITY OF THREE MILLION INHABITANTS

Proceeding in the manner of the investigator in his laboratory, I have avoided all special cases, and all that may be accidental, and I have assumed an ideal site to begin with. My object was not to overcome the existing state of things, but *by constructing a theoretically watertight formula to arrive at the fundamental principles of modern town planning*. Such fundamental principles, if they are genuine, can serve as the skeleton of any system of modern town planning; being as it were the rules according to which development will take place. We shall then be in a position to take a special case, no matter what: whether it be Paris, London, Berlin, New York or some small town. Then, as a result of what we have learnt, we can take control and decide in what direction the forthcoming battle is to be waged. For the desire to rebuild any great city in a modern way is to engage in a formidable battle. Can you imagine people engaging in a battle without knowing their objectives? Yet that is exactly what is happening. The authorities are compelled to do something, so they give the police white sleeves or set them on horseback, they invent sound signals and light signals, they propose to put bridges over streets or moving pavements under the streets; more garden cities are suggested, or it is decided to suppress the tramways, and so on. And these decisions are reached in a sort of frantic haste in order, as it were, to hold a wild beast at bay. That beast is the great city. It is infinitely more powerful than all these devices. And it is just beginning to wake. What will to-morrow bring forth to cope with it?

We must have some rule of conduct.

We must have fundamental principles for modern town planning.

Site

A level site is the ideal site [for the contemporary city (Figure 1)]. In all those places where traffic becomes over-intensified the level site gives a chance of a normal solution to the problem. Where there is less traffic, differences in level matter less.

The river flows far away from the city. The river is a kind of liquid railway, a goods station and a sorting house. In a decent house the servants' stairs do not go through the drawing room – even if the maid is charming (or if the little boats delight the loiterer leaning on a bridge).

Population

This consists of the citizens proper; of suburban dwellers; and of those of a mixed kind.

(a) Citizens are of the city: those who work and live in it.
(b) Suburban dwellers are those who work in the outer industrial zone and who do not come into the city: they live in garden cities.

(c) The mixed sort are those who work in the business parts of the city but bring up their families in garden cities.

To classify these divisions (and so make possible the transmutation of these recognized types) is to attack the most important problem in town planning, for such a classification would define the areas to be allotted to these three sections and the delimitation of their boundaries. This would enable us to formulate and resolve the following problems:

1 The *City*, as a business and residential centre.
2 The *Industrial City* in relation to the *Garden Cities* (i.e. the question of transport).
3 The *Garden Cities* and the *daily transport* of the workers.

Our first requirement will be an organ that is compact, rapid, lively and concentrated: this is the City with its well organized centre. Our second requirement will be another organ, supple, extensive and elastic; this is the *Garden City* on the periphery. Lying between these two organs, we must *require the legal establishment* of that absolute necessity, a protective zone which allows of extension, *a reserved zone* of woods and fields, a fresh-air reserve.

Density of population

The more dense the population of a city is the less are the distances that have to be covered. The moral, therefore, is that we must *increase the density of the centres of our cities, where business affairs are carried on.*

Lungs

Work in our modern world becomes more intensified day by day, and its demands affect our nervous system in a way that grows more and more dangerous. Modern toil demands quiet and fresh air, not stale air.

The towns of to-day can only increase in density at the expense of the open spaces which are the lungs of a city.

We must *increase the open spaces and diminish the distances to be covered.* Therefore the centre of the city must be constructed *vertically*.

The city's residential quarters must no longer be built along "corridor-streets", full of noise and dust and deprived of light.

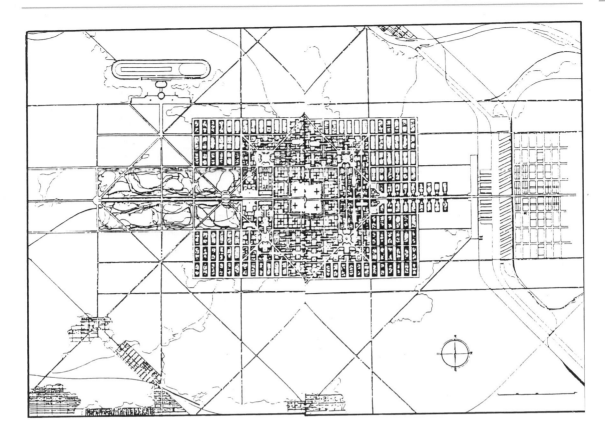

Figure 1

It is a simple matter to build urban dwellings away from the streets, without small internal courtyards and with the windows looking on to large parks; and this whether our housing schemes are of the type with "set-backs" or built on the "cellular" principle.

The street

The street of to-day is still the old bare ground which has been paved over, and under which a few tube railways have been run.

The modern street in the true sense of the word is a new type of organism, a sort of stretched-out workshop, a home for many complicated and delicate organs, such as gas, water and electric mains. It is contrary to all economy, to all security, and to all sense to bury these important service mains. They ought to be accessible throughout their length. The various storeys of this stretched-out workshop will each have their own particular functions. If this type of street, which I have called a "workshop", is to be realized,

it becomes as much a matter of construction as are the houses with which it is customary to flank it, and the bridges which carry it over valleys and across rivers.

The modern street should be a masterpiece of civil engineering and no longer a job for navvies.

The "corridor-street" should be tolerated no longer, for it poisons the houses that border it and leads to the construction of small internal courts or "wells".

Traffic

Traffic can be classified more easily than other things.

To-day traffic is not classified – it is like dynamite flung at hazard into the street, killing pedestrians. Even so, *traffic does not fulfil its function*. This sacrifice of the pedestrian leads nowhere.

If we classify traffic we get:

(a) Heavy goods traffic.
(b) Lighter goods traffic, i.e. vans, etc., which make short journeys in all directions.

(c) Fast traffic, which covers a large section of the town.

Three kinds of roads are needed, and in supermposed storeys:

(a) Below-ground there would be the street for heavy traffic. This storey of the houses would consist merely of concrete piles, and between them large open spaces which would form a sort of clearing-house where heavy goods traffic could load and unload.

(b) At the ground floor level of the buildings there would be the complicated and delicate network of the ordinary streets taking traffic in every desired direction.

(c) Running north and south, and east and west, and forming the two great axes of the city, there would be great *arterial roads for fast one-way traffic* built on immense reinforced concrete bridges 120 to 180 yards in width and approached every half-mile or so by subsidiary roads from ground level. These arterial roads could therefore be joined at any given point, so that even at the highest speeds the town can be traversed and the suburbs reached without having to negotiate any cross-roads.

The number of existing streets should be diminished by two-thirds. The number of crossings depends directly on the number of streets; and *cross-roads are an enemy to traffic*. The number of existing streets was fixed at a remote epoch in history. The perpetuation of the boundaries of properties has, almost without exception, preserved even the faintest tracks and footpaths of the old village and made streets of them, and sometimes even an avenue . . . The result is that we have cross-roads every fifty yards, even every twenty yards or ten yards. And this leads to the ridiculous traffic congestion we all know so well.

The distance between two bus stops or two tube stations gives us the necessary unit for the distance between streets, though this unit is conditional on the speed of vehicles and the walking capacity of pedestrians. So an average measure of about 400 yards would give the normal separation between streets, and make a standard for urban distances. My city is conceived on the gridiron system with streets every 400 yards, though occasionally these distances are subdivided to give streets every 200 yards.

This triple system of superimposed levels answers every need of motor traffic (lorries, private cars, taxis, buses) because it provides for rapid and *mobile* transit.

Traffic running on fixed rails is only justified if it is in the form of a convoy carrying an immense load; it then becomes a sort of extension of the underground system or of trains dealing with suburban traffic. *The tramway has no right to exist in the heart of the modern city.*

If the city thus consists of plots about 400 yards square, this will give us sections of about 40 acres in area, and the density of population will vary from 50,000 down to 6,000, according as the "lots" are developed for business or for residential purposes. The natural thing, therefore, would be to continue to apply our unit of distance as it exists in the Paris tubes to-day (namely, 400 yards) and to put a station in the middle of each plot.

Following the two great axes of the city, two "storeys" below the arterial roads for fast traffic, would run the tubes leading to the four furthest points of the garden city suburbs, and linking up with the metropolitan network . . . At a still lower level, and again following these two main axes, would run the one-way loop systems for suburban traffic, and below these again the four great main lines serving the provinces and running north, south, east and west. These main lines would end at the Central Station, or better still might be connected up by a loop system.

The station

There is only one station. The only place for the station is in the centre of the city. It is the natural place for it, and there is no reason for putting it anywhere else. The railway station is the hub of the wheel.

The station would be an essentially subterranean building. Its roof, which would be two storeys above the natural ground level of the city, would form the aerodrome for aero-taxis. This aerodrome (linked up with the main aerodrome in the protected zone) must be in close contact with the tubes, the suburban lines, the main lines, the main arteries and the administrative services connected with all these . . .

The plan of the city

The basic principles we must follow are these:

1 We must de-congest the centres of our cities.
2 We must augment their density.

3 We must increase the means for getting about.

4 We must increase parks and open spaces.

At the very centre we have the *station* with its landing stage for aero-taxis.

Running north and south, and east and west, we have the *main arteries* for fast traffic, forming elevated roadways 120 feet wide.

At the base of the sky-scrapers and all round them we have a great open space 2,400 yards by 1,500 yards, giving an area of 3,600,000 square yards, and occupied by gardens, parks and avenues. In these parks, at the foot of and round the sky-scrapers, would be the restaurants and cafes, the luxury shops, housed in buildings with receding terraces: here too would be the theatres, halls and so on; and here the parking places or garage shelters.

The sky-scrapers are designed purely for business purposes.

On the left we have the great public buildings, the museums, the municipal and administrative offices. Still further on the left we have the "Park" (which is available for further logical development of the heart of the city).

On the right, and traversed by one of the arms of the main arterial roads, we have the warehouses, and the industrial quarters with their goods stations.

All around the city is the *protected zone* of woods and green fields.

Further beyond are the *garden cities*, forming a wide encircling band.

Then, right in the midst of all these, we have the *Central Station*, made up of the following elements:

(a) The landing-platform; forming an aerodrome of 200,000 square yards in area.

(b) The entresol or mezzanine; at this level are the raised tracks for fast motor traffic: the only crossing being gyratory.

(c) The ground floor where are the entrance halls and booking offices for the tubes, suburban lines, main line and air traffic.

(d) The "basement": here are the tubes which serve the city and the main arteries.

(e) The "sub-basement": here are the suburban lines running on a one-way loop.

(f) The "sub-sub-basement": here are the main lines (going north, south, east and west).

The city

Here we have twenty-four sky-scrapers capable each of housing 10,000 to 50,000 employees; this is the business and hotel section, etc., and accounts for 400,000 to 600,000 inhabitants.

The residential blocks, of the two main types already mentioned, account for a further 600,000 inhabitants.

The garden cities give us a further 2,000,000 inhabitants, or more.

In the great central open space are the cafes, restaurants, luxury shops, halls of various kinds, a magnificent forum descending by stages down to the immense parks surrounding it, the whole arrangement providing a spectacle of order and vitality.

Density of population

(a) The sky-scraper: 1,200 inhabitants to the acre.

(b) The residential blocks with set-backs: 120 inhabitants to the acre. These are the luxury dwellings.

(c) The residential blocks on the "cellular" system, with a similar number of inhabitants.

This great density gives us our necessary shortening of distances and ensures rapid inter-communication.

Note. The average density to the acre of Paris in the heart of the town is 146, and of London 63; and of the over-crowded quarters of Paris 213, and of London 169.

Open spaces

Of the area (a), 95 per cent of the ground is open (squares, restaurants, theatres).

Of the area (b), 85 per cent of the ground is open (gardens, sports grounds).

Of the area (c), 48 per cent of the ground is open (gardens, sports grounds).

Educational and civic centres, universities, museums of art and industry, public services, county hall

The "Jardin anglais". (The city can extend here, if necessary.)

Sports grounds: Motor racing track, Racecourse, Stadium, Swimming baths, etc.

The protected zone (which will be the property of the city), with its aerodrome

A zone in which all building would be prohibited; reserved for the growth of the city as laid down by the municipality: it would consist of woods, fields, and sports grounds. The forming of a "protected zone" by continual purchase of small properties in the immediate vicinity of the city is one of the most essential and urgent tasks which a municipality can pursue. It would eventually represent a tenfold return on the capital invested.

Industrial quarters: types of buildings employed

For business: sky-scrapers sixty storeys high with no internal wells or courtyards . . .

Residential buildings with "set-backs", of six double storeys; again with no internal wells: the flats looking on either side on to immense parks.

Residential buildings on the "cellular" principle, with "hanging gardens", looking on to immense parks; again no internal wells. These are "service-flats" of the most modern kind.

Garden cities: their aesthetic, economy, perfection and modern outlook

A simple phrase suffices to express the necessities of tomorrow: WE MUST BUILD IN THE OPEN.

The lay-out must be of a purely geometrical kind, with all its many and delicate implications.

[. . .]

The city of to-day is a dying thing because it is not geometrical. To build in the open would be to replace our present haphazard arrangements, *which are all we have to-day*, by a *uniform* lay-out. Unless we do this *there is no salvation*.

The result of a true geometrical lay-out is *repetition*. The result of repetition is a *standard*, the perfect form (i.e. the creation of standard types). A geometrical lay-out means that mathematics play their part.

There is no first-rate human production but has geometry at its base. It is of the very essence of Architecture. To introduce uniformity into the building of the city we must *industrialize building*. Building is the one economic activity which has so far resisted industrialization. It has thus escaped the march of progress, with the result that the cost of building is still abnormally high.

The architect, from a professional point of view, has become a twisted sort of creature. He has grown to love irregular sites, claiming that they inspire him with original ideas for getting round them. Of course he is wrong. For nowadays the only building that can be undertaken must be either for the rich or built at a loss (as, for instance, in the case of municipal housing schemes), or else by jerry-building and so robbing the inhabitant of all amenities. A motor-car which is achieved by mass production is a masterpiece of comfort, precision, balance and good taste. A house built to order (on an "interesting" site) is a masterpiece of incongruity – a monstrous thing.

If the builder's yard were reorganized on the lines of standardization and mass production we might have gangs of workmen as keen and intelligent as mechanics.

The mechanic dates back only twenty years, yet already he forms the highest caste of the working world.

The mason dates . . . from time immemorial! He bangs away with feet and hammer. He smashes up everything round him, and the plant entrusted to him falls to pieces in a few months. The spirit of the mason must be disciplined by making him part of the severe and exact machinery of the industrialized builder's yard.

The cost of building would fall in the proportion of 10 to 2.

The wages of the labourers would fall into definite categories; to each according to his merits and service rendered.

The "interesting" or erratic site absorbs every creative faculty of the architect and wears him out. What results is equally erratic: lopsided abortions; a specialist's solution which can only please other specialists.

We must build *in the open*: both within the city and around it.

Then having worked through every necessary technical stage and using absolute ECONOMY, we shall be in a position to experience the intense joys of a creative art which is based on geometry.

THE CITY AND ITS AESTHETIC

(The plan of a city which is here presented is a direct consequence of purely geometric considerations.)

A new unit *on a large scale* (400 yards) inspires everything. Though the gridiron arrangement of the streets every 400 yards (sometimes only 200) is uniform (with a consequent ease in finding one's way about), no two streets are in any way alike. This is where, in a magnificent contrapuntal symphony, the forces of geometry come into play.

Suppose we are entering the city by way of the Great Park. Our fast car takes the special elevated motor track between the majestic sky-scrapers: as we approach nearer there is seen the repetition against the sky of the twenty-four sky-scrapers; to our left and right on the outskirts of each particular area are the municipal and administrative buildings; and enclosing the space are the museums and university buildings.

Then suddenly we find ourselves at the feet of the first sky-scrapers. But here we have, not the meagre shaft of sunlight which so faintly illumines the dismal streets of New York, but an immensity of space. The whole city is a Park. The terraces stretch out over lawns and into groves. Low buildings of a horizontal kind lead the eye on to the foliage of the trees. Where are now the trivial *Procuracies*? Here is the *city* with its crowds living in peace and pure air, where noise is smothered under the foliage of green trees. The chaos of New York is overcome. Here, bathed in light, stands the modern city [Figure 2].

Our car has left the elevated track and has dropped its speed of sixty miles an hour to run gently through the residential quarters. The "set-backs" permit of vast architectural perspectives. There are gardens, games and sports grounds. And sky everywhere, as far as the eye can see. The square silhouettes of the terraced roofs stand clear against the sky, bordered with the verdure of the hanging gardens. The uniformity of the units that compose the picture throw into relief the firm lines on which the far-flung masses are constructed. Their outlines softened by distance, the sky-scrapers raise immense geometrical facades all of glass, and in them is reflected the blue glory of the sky. An overwhelming sensation. Immense but radiant prisms.

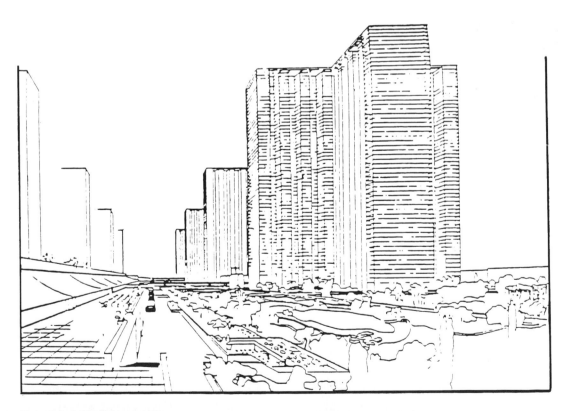

Figure 2 A Contemporary city

And in every direction we have a varying spectacle: our "gridiron" is based on a unit of 400 yards, but it is strangely modified by architectural devices! (The "set-backs" are in counterpoint, on a unit of 600 × 400.)

The traveller in his airplane, arriving from Constantinople or Pekin it may be, suddenly sees appearing through the wavering lines of rivers and patches of forests that clear imprint which marks a city which has grown in accordance with the spirit of man: the mark of the human brain at work.

As twilight falls the glass sky-scrapers seem to flame.

This is no dangerous futurism, a sort of literary dynamite flung violently at the spectator. It is a spectacle organized by an Architecture which uses plastic resources for the modulation of forms seen in light.

A city made for speed is made for success.

"Broadacre City: A New Community Plan"

Architectural Record (1935)

Frank Lloyd Wright

Editors' Introduction

For more than half a century, the question "Who is the greatest American architect?" could have only one answer: Frank Lloyd Wright (1867–1959). First with his revolutionary "prairie houses" that seemed to grow directly out of the Midwest landscape with their long, low cantilevered rooflines, and later with such masterpieces as the Imperial Hotel in Tokyo, the Guggenheim Museum in New York, and the breathtaking "Fallingwater" in Western Pennsylvania, Wright became the spokesman for "organic architecture" and a style of building that expressed "the nature of the materials."

To many, Wright's architecture and "the architecture of American democracy" were synonymous. As an unabashed egotist and a pioneer in the field of media celebrity, Wright encouraged the popular identification of himself with the American spirit. He cultivated an imperious image of plain-speaking, anti-collectivist democracy and sought personally to embody the notion of radical individualism. As an artistic genius, Wright despised the popular philistinism of his day and attributed the observable decline of American popular culture to "the mobocracy" and to the unprincipled bankers and politicians who served its interests. By the 1920s and 1930s, Wright had become a social revolutionary but not, characteristically, of the socialist Left. Rather, Wright called for a radical transformation of American society to restore earlier Emersonian and Jeffersonian virtues. The physical embodiment of that utopian vision was Broadacre City.

Wright unveiled his model of Broadacre City, illustrated in Plate 39, at Rockefeller Center, New York, in 1935. The article reprinted here represents his first and clearest statement of the revolutionary proposal whereby every citizen of the United States would be given a minimum of one acre of land per person, with the family homestead being the basis of civilization, and with government reduced to nothing more than a county architect who would be in charge of directing land allotments and the construction of basic community facilities. Many at the time thought the idea was totally outlandish, but Broadacre (and the small, efficient "Usonian" house) proved to be prophetic as sprawling suburban regions transformed the American landscape during the second half of the twentieth century.

Wright believed that two inventions – the telephone and the automobile – made the old cities "no longer modern," and he fervently looked forward to the day when dense, crowded conglomerations like New York and Chicago would wither and decay. In their place, Americans would reinhabit the rural landscape (and re-acquire the rural virtues of individual freedom and self-reliance) with a "city" of independent homesteads in which people would be isolated enough from one another to ensure family stability but connected enough, through modern telecommunications and transportation, to achieve a real sense of community. Borrowing an idea from the anarchist philosopher Kropotkin, Wright believed that the citizens of Broadacre should pursue a combination of manual and intellectual work every day, thus achieving a human wholeness that modern society and the modern city had destroyed. He also believed that a system of personal freedom and dignity through land ownership was the way to guarantee social harmony and avoid class struggle.

Broadacre City invites immediate comparison with the very different models of Ebenezer Howard's Garden City (p. 314) and with Le Corbusier's cities based on towers in a park (p. 322). Intriguingly, the overall population density of Broadacre, on the one hand, and the Garden City and Corbusian visions, on the other, were not all that different, depending on the actual acreage of the surrounding parkland or greenbelt. And both Wright's and Le Corbusier's plans are wedded to the automobile, one vision seeing a centralizing, the other a decentralizing, effect. But the most revealing comparisons are with Robert Fishman's description of the now-emerging "technoburbs" (p. 69) and Melvin Webber's prediction of a "post-urban age" (p. 473). One cannot help but wonder whether what seemed impossible in 1935 may actually be realized, with the help of computer-based telecommunications and the possibility of "telecommuting" to work over the Internet in the twenty-first century.

This selection is from *Architectural Record*, vol. 77 (April 1935). For more on Broadacre City see Robert Fishman's *Urban Utopias of the Twentieth Century* (New York: Basic Books, 1977). John Sergeant's *Frank Lloyd Wright's Usonian Houses: The Case for Organic Architecture* (New York: Whitney Library of Design, 1984) is also useful, and William Allin Storer's *A Frank Lloyd Wright Companion* (Chicago: University of Chicago Press, 1994) is an impressive and definitive reference book.

Three excellent biographies of Wright are Meryle Secrest, *Frank Lloyd Wright: A Biography* (Chicago: University of Chicago Press, 1998); Brendan Gill, *Many Masks: A Life of Frank Lloyd Wright* (New York: Da Capo Press, 1998); and Ada Louise Huxtable, *Frank Lloyd Wright* (New York: Viking, 2004). For good overviews of Wright's work see David Larkin and Bruce Brooks Pfeiffer (eds), *Frank Lloyd Wright: The Masterworks* (New York: Rizzoli, 1993) and Neil Levine, *The Architecture of Frank Lloyd Wright* (Princeton: Princeton University Press, 1996). But the very best sources on Wright are Wright himself, although his writing style is often quirky and hyperbolic. Of particular interest are *When Democracy Builds* (Chicago: University of Chicago Press, 1945), *Genius and the Mobocracy* (New York: Duell, Sloan & Pearce, 1949), and *The Living City* (New York: Horizon, 1958).

■ ■ ■ ■ ■ ■

Given the simple exercise of several inherently just rights of man, the freedom to decentralize, to re-distribute and to correlate the properties of the life of man on earth to his birthright – the ground itself – and Broadacre City becomes reality.

As I see Architecture, the best architect is he who will devise forms nearest organic as features of human growth by way of changes natural to that growth. Civilization is itself inevitably a form but not, if democracy is sanity, is it necessarily the fixation called "academic." All regimentation is a form of death which may sometimes serve life but more often imposes upon it. In Broadacres all is symmetrical but it is seldom obviously and never academically so.

Whatever forms issue are capable of normal growth without destruction of such pattern as they may have. Nor is there much obvious repetition in the new city. Where regiment and row serve the general harmony of arrangement both are present, but generally, both are absent except where planting and cultivation are naturally a process or walls afford a desired seclusion.

Rhythm is the substitute for such repetitions every-where. Wherever repetition (standardization) enters, it has been modified by inner rhythms either by art or by nature as it must, to be of any lasting human value.

The three major inventions already at work building Broadacres, whether the powers that over-built the old cities otherwise like it or not, are:

1 The motor car: general mobilization of the human being.
2 Radio, telephone and telegraph: electrical inter-communication becoming complete.
3 Standardized machine-shop production: machine invention plus scientific discovery.

The price of the major three to America has been the exploitation we see everywhere around us in waste and in ugly scaffolding that may now be thrown away. The price has not been so great if by way of popular government we are able to exercise the use of three inherent rights of any man:

1 His social right to a direct medium of exchange in place of gold as a commodity: some form of social credit.
2 His social right to his place on the ground as he has had it in the sun and air: land to be held only by use and improvements.
3 His social right to the ideas by which and for which he lives: public ownership of invention and scientific discoveries that concern the life of the people.

The only assumption made by Broadacres as ideal is that these three rights will be the citizen's so soon as the folly of endeavoring to cheat him of their democratic values becomes apparent to those who hold (feudal survivors or survivals), as it is becoming apparent to the thinking people who are held blindly abject or subject against their will.

The landlord is no happier than the tenant. The speculator can no longer win much at a game about played out. The present success-ideal, placing, as it does, premiums upon the wolf, the fox and the rat in human affairs and above all, upon the parasite, is growing more evident every day as a falsity just as injurious to the "successful" as to the victims of such success. Well – sociologically, Broadacres is release from all that fatal "success" which is, after all, only excess. So I have called it a new freedom for living in America. It has thrown the scaffolding aside. It sets up a new ideal of success.

In Broadacres, by elimination of cities and towns the present curse of petty and minor officialdom, government, has been reduced to one minor government for each county. The waste motion, the back and forth haul, that today makes so much idle business is gone. Distribution becomes automatic and direct, taking place mostly in the region of origin. Methods of distribution of everything are simple and direct. From the maker to the consumer by the most direct route.

Coal (one-third the tonnage of the haul of our railways) is eliminated by burning it at the mines and transferring that power, making it easier to take over the great railroad rights of way; to take off the cumbersome rolling stock and put the right of way into general service as the great arterial on which truck traffic is concentrated on lower side lanes, many lanes of speed traffic above and monorail speed trains at the center, continuously running. Because traffic may take off or take on at any given point, these arterials are traffic not dated but fluescent. And the great arterial as well as all the highways become great architecture, auto-

matically affording within their structure all necessary storage facilities of raw materials, the elimination of all unsightly piles of raw material.

In the hands of the state, but by way of the county, is all redistribution of land – a minimum of one acre going to the childless family and more to the larger family as effected by the state. The agent of the state in all matters of land allotment or improvement, or in matters affecting the harmony of the whole, is the architect. All building is subject to his sense of the whole as organic architecture. Here architecture is landscape and landscape takes on the character of architecture by way of the simple process of cultivation.

All public utilities are concentrated in the hands of the state and county government as are matters of administration, patrol, fire, post, banking, license and record, making politics a vital matter to everyone in the new city instead of the old case where hopeless indifference makes "politics" a grafter's profession.

In the buildings for Broadacres no distinction exists between much and little, more and less. Quality is in all, for all, alike. The thought entering into the first or last estate is of the best. What differs is only individuality and extent. There is nothing poor or mean in Broadacres.

Nor does Broadacres issue any dictum or see any finality in the matter either of pattern or style.

Organic character is style. Such style has myriad forms inherently good. Growth is possible to Broadacres as a fundamental form, not as mere accident of change but as integral pattern unfolding from within.

Here now may be seen the elemental units of our social structure [Figure 1]: the correlated farm, the factory – its smoke and gases eliminated by burning coal at places of origin, the decentralized school, the various conditions of residence, the home offices, safe traffic, simplified government. All common interests take place in a simple coordination wherein all are employed: *little* farms, *little* homes for industry, *little* factories, *little* schools, a *little* university going to the people mostly by way of their interest in the ground, *little* laboratories on their own ground for professional men. And the farm itself, notwithstanding its animals, becomes the most attractive unit of the city. The husbandry of animals at last is in decent association with them and with all else as well. True farm relief.

To build Broadacres as conceived would automatically end unemployment and all its evils forever. There would never be labor enough nor could under-consumption ever ensue. Whatever a man did would

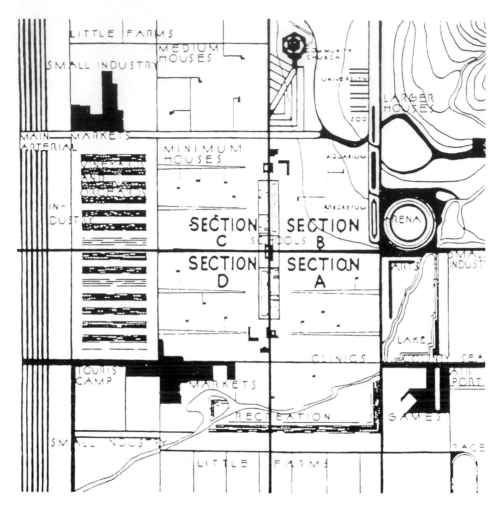

Figure 1

be done – obviously and directly – mostly by himself in his own interest under the most valuable inspiration and direction: under training, certainly, if necessary. Economic independence would be near, a subsistence certain; life varied and interesting.

Every kind of builder would be likely to have a jealous eye to the harmony of the whole within broad limits fixed by the county architect, an architect chosen by the county itself. Each county would thus naturally develop an individuality of its own. Architecture – in the broad sense – would thrive.

In an organic architecture the ground itself predetermines all features; the climate modifies them; available means limit them; function shapes them.

Form and function are one in Broadacres. But Broadacres is no finality! The model shows four square miles of a typical countryside developed on the acre as unit according to conditions in the temperate zone and accommodating some 1,400 families. It would swing north or swing south in type as conditions, climate and topography of the region changed.

In the model the emphasis has been placed upon diversity in unity, recognizing the necessity of cultivation as a need for formality in most of the planting. By a simple government subsidy certain specific acres or groups of acre units are, in every generation, planted to useful trees, meantime beautiful, giving privacy and various rural divisions. There are no rows of trees alongside the roads to shut out the view. Rows where they occur are perpendicular to the road or the trees are planted in groups. Useful trees like white pine, walnut, birch, beech, fir, would come to maturity as well as fruit and nut trees and they would come as a profitable crop meantime giving character, privacy and comfort to the whole city. The general park is a flowered meadow beside the stream and is bordered with ranks of trees, tiers gradually rising in height above the flowers at the ground level. A music-garden is sequestered from noise at one end. Much is made of general sports and festivals by way of the stadium, zoo, aquarium, arboretum and the arts.

The traffic problem has been given special attention as the more mobilization is made a comfort and a facility the sooner will Broadacres arrive. Every Broadacre citizen has his own car. Multiple-lane highways make travel safe and enjoyable. There are no grade crossings nor left turns on grade. The road system and construction is such that no signals nor any lamp-posts need be seen. No ditches are alongside the roads. No curbs either. An inlaid purfling over which the car cannot come without damage to itself takes its place to protect the pedestrian.

In the affair of air transport Broadacres rejects the present airplane and substitutes the self-contained mechanical unit that is sure to come: an aerator capable of rising straight up and by reversible rotors able to travel in any given direction under radio control at a maximum speed of, say, 200 miles an hour, and able to descend safely into the hexacomb from which it arose or anywhere else. By a doorstep if desired.

The only fixed transport trains kept on the arterial are the long-distance monorail cars traveling at a speed (already established in Germany) of 220 miles per hour. All other traffic is by motor car on the twelve lane levels or the triple truck lanes on the lower levels which have on both sides the advantage of delivery direct to warehousing or from warehouses to consumer. Local trucks may get to warehouse-storage on lower levels under the main arterial itself. A local truck road parallels the swifter lanes.

Houses in the new city are varied: make much of fireproof synthetic materials, factory-fabricated units adapted to free assembly and varied arrangement, but do not neglect the older nature-materials wherever they are desired and available. House-holders' utilities are nearly all planned in prefabricated utility stacks or units, simplifying construction and reducing building costs to a certainty. There is the professional's house with its laboratory, the minimum house with its workshop, the medium house ditto, the larger house and the house of machine-age luxury. We might speak of them

as a one-car house, a two-car house, a three-car house and a five-car house. Glass is extensively used as are roofless rooms. The roof is used often as a trellis or a garden. But where glass is extensively used it is usually for domestic purposes in the shadow of protecting overhangs.

Copper for roofs is indicated generally on the model as a permanent cover capable of being worked in many appropriate ways and giving a general harmonious color effect to the whole.

Electricity, oil and gas are the only popular fuels. Each land allotment has a pit near the public lighting fixture where access to the three and to water and sewer may be had without tearing up the pavements.

The school problem is solved by segregating a group of low buildings in the interior spaces of the city where the children can go without crossing traffic. The school building group includes galleries for loan collections from the museum, a concert and lecture hall, small gardens for the children in small groups and well-lighted cubicles for individual outdoor study: there is a small zoo, large pools and green playgrounds.

This group is at the very center of the model and contains at its center the higher school adapted to the segregation of the students into small groups.

This tract of four miles square, by way of such liberal general allotment determined by acreage and type of ground, including apartment buildings and hotel facilities, provides for about 1,400 families at, say, an average of five or more persons to the family.

To reiterate: the basis of the whole is general decentralization as an applied principle and architectural reintegration of all units into one fabric; free use of the ground held only by use and improvements; public utilities and government itself owned by the people of Broadacre City; privacy on one's own ground for all and a fair means of subsistence for all by way of their own work on their own ground or in their own laboratory or in common offices serving the life of the whole.

There are too many details involved in the model of Broadacres to permit complete explanation. Study of the model itself is necessary study. Most details are explained by way of collateral models of the various types of construction shown: highway construction, left turns, crossovers, underpasses and various houses and public buildings.

Anyone studying the model should bear in mind the thesis upon which the design has been built by the Taliesin Fellowship, built carefully not as a finality in any sense but as an interpretation of the changes inevitable to our growth as a people and a nation.

Individuality established on such terms must thrive. Unwholesome life would get no encouragement and the ghastly heritage left by over-crowding in overdone ultra-capitalistic centers would be likely to disappear in three or four generations. The old success ideals having no chance at all, new ones more natural to the best in man would be given a fresh opportunity to develop naturally.

"Towards Sustainable Development"

published as *Our Common Future* (1987)

World Commission on Environment and Development (The Brundtland Commission)

Editors' Introduction

The physical city is a man-made construct, but the relationship between the city and the surrounding natural environment has always helped to define the character and quality of urban life. If the very first cities were expressions of "hydraulic" civilizations based on the control of water for irrigation, and if the cities of the Industrial Revolution ushered in a new age of unprecedented environmental degradation, it was unpolluted nature – or at least the *idea* of unpolluted nature – that offered a continuing source of intellectual regeneration and moral comparison. The utopian visionaries of the nineteenth and twentieth centuries – Frederick Law Olmsted (p. 307), Ebenezer Howard (p. 314), Le Corbusier (p. 322), and Frank Lloyd Wright (p. 331) – created city plans that balanced the man-made urban areas with parks, public gardens, or surrounding rural zones. And in recent years the forces of nature ecology, green urbanism, and sustainability have grown in importance and become dominant forces in urban policy and planning.

In 1968, Stanford University biologist Paul Ehlich published *The Population Bomb*, predicting global overcrowding and persistent starvation in the underdeveloped nations by the 1980s. In 1970, the first Earth Day was celebrated in San Francisco. And in 1972, the Club of Rome published its influential report on *The Limits to Growth*, arguing that the world's governments and economic powers needed to begin cutting back on overproduction and overconsumption in the face of uneven development, exploding population growth, and declining resources. All these developments suggested a growing shift toward environmentalism as the new reform paradigm at the very time that capitalism was globalizing and the ideological certainties of the 1960s Left were beginning to lose favor. Then, in 1987, came *Our Common Future*, the report of the United Nations-sponsored World Commission on Environment and Development (WCED) and the clarion call for "sustainable development."

The WCED report is commonly called the Brundtland Report, so named after the chairperson of the Commission, Gro Brundtland of Norway, the only prime minister of a major country to have previously served as environment minister. In its "Call to Action," the Brundtland Report argued that during the twentieth century "the relationship between the human world and the planet that sustains it has undergone a profound change" and that the increased "rate of change is outstripping the ability of . . . our current capabilities." As a result, the world needs to embrace the concept of sustainable development, defined as "development that meets the needs of the present without compromising the ability of future generations to meet their own needs." Such sustainable development, the Report continues, "contains within it . . . the concept of 'needs,' in particular the needs of the world's poor, to which overriding priority should be given."

Some skeptics and conservatives dismissed the Brundtland Report as just one more attack on free-market capitalism in favor of massive government regulation of the economy. And even some radical environmentalists claimed that "sustainable development" was an oxymoron favored by big business interests to portray capitalism as ecologically benign. But the momentum behind the environmental vision of sustainability sparked by the Brundtland Report continued to grow with the Rio Declaration on Environment and Development of 1992, the Kyoto Protocol on global climate change of 1997–99, and the World Summit on Sustainable Development in Johannesburg in 2002. Today, the idea of sustainability has been broadly adopted as one of the key concepts behind urban development projects worldwide, but especially in Europe and North America and most particularly in the work of the New Urbanists such as Andrés Duany and Elizabeth Plater-Zyberk (p. 192) and Peter Calthorpe and William Fulton (p. 342).

For descriptions in this volume of sustainable development applied to urban contexts, consult Timothy Beatley, "Planning for Sustainability in European Cities" (p. 411) and Stephen Wheeler, "Planning Sustainable and Livable Cities" (p. 499). For a more expanded range of views on urban sustainability, consult Stephen Wheeler and Timothy Beatley (eds), *The Sustainable Urban Development Reader* (London and New York: Routledge, 2004). The literature on environmentalism in general is huge and ubiquitous. For the best dissenting view, consult Bjorn Lumborg, *The Skeptical Environmentalist: Measuring the Real State of the World* (Cambridge: Cambridge University Press, 2001).

■ ■ ■ ■ ■ ■

A CALL FOR ACTION

Over the course of this century, the relationship between the human world and the planet that sustains it has undergone a profound change.

When the century began, neither human numbers nor technology had the power radically to alter planetary systems. As the century closes, not only do vastly increased human numbers and their activities have that power, but major, unintended changes are occurring in the atmosphere, in soils, in waters, among plants and animals, and in the relationships among all of these. The rate of change is outstripping the ability of scientific disciplines and our current capabilities to assess and advise. It is frustrating the attempts of political and economic institutions, which evolved in a different, more fragmented world, to adapt and cope. It deeply worries many people who are seeking ways to place those concerns on the political agendas.

The onus lies with no one group of nations. Developing countries face the obvious life-threatening challenges of desertification, deforestation, and pollution, and endure most of the poverty associated with environmental degradation. The entire human family of nations would suffer from the disappearance of rain forests in the tropics, the loss of plant and animal species, and changes in rainfall patterns. Industrial nations face the life-threatening challenges of toxic chemicals, toxic wastes, and acidification. All nations may suffer from the releases by industrialized countries of carbon dioxide and of gases that react with the ozone layer, and from any future war fought with the nuclear arsenals controlled by those nations. All nations will have a role to play in changing trends, and in righting an international economic system that increases rather than decreases inequality, that increases rather than decreases numbers of poor and hungry.

The next few decades are crucial. The time has come to break out of past patterns. Attempts to maintain social and ecological stability through old approaches to development and environmental protection will increase instability. Security must be sought through change. The Commission has noted a number of actions that must be taken to reduce risks to survival and to put future development on paths that are sustainable. Yet we are aware that such a reorientation on a continuing basis is simply beyond the reach of present decision-making structures and institutional arrangements, both national and international.

This Commission has been careful to base our recommendations on the realities of present institutions, on what can and must be accomplished today. But to keep options open for future generations, the present generation must begin now, and begin together.

To achieve the needed changes, we believe that an active follow-up of this report is imperative. It is with this in mind that we call for the UN General Assembly, upon due consideration, to transform this report into a UN Programme on Sustainable Development. Special follow-up conferences could be initiated at the regional level. Within an appropriate period after the presentation of this report to the General Assembly, an international conference could be convened to review progress made, and to promote follow-up arrangements that will be needed to set benchmarks and to maintain human progress.

First and foremost, this Commission has been concerned with people – of all countries and all walks of life. And it is to people that we address our report. The changes in human attitudes that we call for depend on a vast campaign of education, debate, and public participation. This campaign must start now if sustainable human progress is to be achieved.

The members of the World Commission on Environment and Development came from 21 very different nations. In our discussions, we disagreed often on details and priorities. But despite our widely differing backgrounds and varying national and international responsibilities, we were able to agree to the lines along which change must be drawn.

We are unanimous in our conviction that security, wellbeing, and very survival of the planet depend on such changes, now.

A THREATENED FUTURE

The Earth is one but the world is not. We depend on one biosphere for sustaining our lives. Yet each community, each country, strives for survival and prosperity with little regard for its impact on others. Some consume the Earth's resources at a rate that would leave little for future generations. Others, many more in number, consume far too little and live with the prospect of hunger, squalor, disease, and early death.

Yet progress has been made. Throughout much of the world, children born today can expect to live longer and be better educated than their parents. In many parts, the newborn can also expect to attain a higher standard of living in a wider sense. Such progress provides hope as we contemplate the improvements still needed, and also as we face our failures to make this Earth a safer and sounder home for us and for those who are to come.

The failures that we need to correct arise both from poverty and from the short-sighted way in which we have often pursued prosperity. Many parts of the world are caught in a vicious downwards spiral: Poor people are forced to overuse environmental resources to survive from day to day, and their impoverishment of their environment further impoverishes them, making

their survival ever more difficult and uncertain. The prosperity attained in some parts of the world is often precarious, as it has been secured through farming, forestry, and industrial practices that bring profit and progress only over the short term.

Societies have faced such pressures in the past and, as many desolate ruins remind us, sometimes succumbed to them. But generally these pressures were local. Today the scale of our interventions in nature is increasing and the physical effects of our decisions spill across national frontiers. The growth in economic interaction between nations amplifies the wider consequences of national decisions. Economics and ecology bind us in ever-tightening networks. Today, many regions face risks of irreversible damage to the human environment that threaten the basis for human progress.

These deepening interconnections are the central justification for the establishment of this Commission. We traveled the world for nearly three years, listening. At special public hearings organized by the Commission, we heard from government leaders, scientists, and experts, from citizens' groups concerned about a wide range of environment and development issues, and from thousands of individuals – farmers, shanty-town residents, young people, industrialists, and indigenous and tribal peoples.

We found everywhere deep public concern for the environment, concern that has led not just to protests but often to changed behaviour. The challenge is to ensure that these new values are more adequately reflected in the principles and operations of political and economic structures.

We also found grounds for hope: that people can cooperate to build a future that is more prosperous, more just, and more secure; that a new era of economic growth can be attained, one based on policies that sustain and expand the Earth's resource base; and that the progress that some have known over the last century can be experienced by all in the years ahead. But for this to happen, we must understand better the symptoms of stress that confront us, we must identify the causes, and we must design new approaches to managing environmental resources and to sustaining human development.

SYMPTOMS AND CAUSES

Environmental stress has often been seen as the result of the growing demand on scarce resources and the pollution generated by the rising living standards of the relatively affluent. But poverty itself pollutes the environment, creating environmental stress in a different way. Those who are poor and hungry will often destroy their immediate environment in order to survive: They will cut down forests, their livestock will overgraze grasslands; they will overuse marginal land; and in growing numbers they will crowd into congested cities. The cumulative effect of these changes is so far-reaching as to make poverty itself a major global scourge.

On the other hand, where economic growth has led to improvements in living standards, it has sometimes been achieved in ways that are globally damaging in the longer term. Much of the improvement in the past has been based on the use of increasing amounts of raw materials, energy, chemicals, and synthetics and on the creation of pollution that is not adequately accounted for in figuring the costs of production processes. These trends have had unforeseen effects on the environment. Thus today's environmental challenges arise both from the lack of development and from the unintended consequences of some forms of economic growth.

[. . .]

Sustainable development is development that meets the needs of the present without compromising the ability of future generations to meet their own needs. It contains within it two key concepts:

- the concept of 'needs', in particular the essential needs of the world's poor, to which overriding priority should be given; and
- the idea of limitations imposed by the state of technology and social organization on the environment's ability to meet present and future needs.

Thus the goals of economic and social development must be defined in terms of sustainability in all countries – developed or developing, market-oriented or centrally planned. Interpretations will vary, but must share certain general features and must flow from a consensus on the basic concept of sustainable development and on a broad strategic framework for achieving it. Development involves a progressive transformation of economy and society. A development path that is sustainable in a physical sense could theoretically be pursued even in a rigid social and political setting. But physical sustainability cannot be secured unless development policies pay attention to such considerations as changes in access to resources and in the distribution of costs and benefits. Even the narrow notion of physical sustainability implies a concern for social equity between generations, a concern that must logically be extended to equity within each generation.

THE CONCEPT OF SUSTAINABLE DEVELOPMENT

The satisfaction of human needs and aspirations is the major objective of development. The essential needs of vast numbers of people in developing countries – for food, clothing, shelter, jobs – are not being met, and beyond their basic needs these people have legitimate aspirations for an improved quality of life. A world in which poverty and inequity are endemic will always be prone to ecological and other crises. Sustainable development requires meeting the basic needs of all and extending to all the opportunity to satisfy their aspirations for a better life.

Living standards that go beyond the basic minimum are sustainable only if consumption standards everywhere have regard for long-term sustainability. Yet many of us live beyond the world's ecological means, for instance in our patterns of energy use. Perceived needs are socially and culturally determined, and sustainable development requires the promotion of values that encourage consumption standards that are within the bounds of the ecologically possible and to which all can reasonably aspire.

Meeting essential needs depends in part on achieving full growth potential, and sustainable development clearly requires economic growth in places where such needs are not being met. Elsewhere, it can be consistent with economic growth, provided the content of growth reflects the broad principles of sustainability and non-exploitation of others. But growth by itself is not enough. High levels of productive activity and widespread poverty can coexist, and can endanger the environment. Hence sustainable development requires that societies meet human needs both by increasing productive potential and by ensuring equitable opportunities for all. An expansion in numbers can increase the pressure on resources and slow the rise in living standards in areas where deprivation is widespread. Though the issue is not merely one of population size but of the distribution of resources, sustainable development can only be pursued if demographic developments are in harmony with the changing productive potential of the ecosystem.

A society may in many ways compromise its ability to meet the essential needs of its people in the future – by overexploiting resources, for example. The direction of technological developments may solve some immediate problems but lead to even greater ones. Large sections of the population may be marginalized by ill-considered development.

Settled agriculture, the diversion of watercourses, the extraction of minerals, the emission of heat and noxious gases into the atmosphere, commercial forests, and genetic manipulation are all examples of human intervention in natural systems during the course of development. Until recently, such interventions were small in scale and their impact limited. Today's interventions are more drastic in scale and impact, and more threatening to life-support systems both locally and globally. This need not happen. At a minimum, sustainable development must not endanger the natural systems that support life on Earth: the atmosphere, the waters, the soils, and the living beings.

Growth has no set limits in terms of population or resource use beyond which lies ecological disaster. Different limits hold for the use of energy, materials, water, and land. Many of these will manifest themselves in the form of rising costs and diminishing returns, rather than in the form of any sudden loss of a resource base. The accumulation of knowledge and the development of technology can enhance the carrying capacity of the resource base. But ultimate limits there are, and sustainability requires that long before these are reached, the world must ensure equitable access to the constrained resource and reorient technological efforts to relieve the pressure.

Economic growth and development obviously involve changes in the physical ecosystem. Every ecosystem everywhere cannot be preserved intact. A forest may be depleted in one part of a water-shed and extended elsewhere, which is not a bad thing if the exploitation has been planned and the effects on soil erosion rates, water regimes, and genetic losses have been taken into account. In general, renewable resources like forests and fish stocks need not be depleted provided the rate of use is within the limits of regeneration and natural growth. But most renewable resources are part of a complex and interlinked ecosystem, and maximum sustainable yield must be defined after taking into account system-wide effects of exploitation.

As for nonrenewable resources, like fossil fuels and minerals, their use reduces the stock available for future generations. But this does not mean that such resources should not be used. In general the rate of depletion should take into account the criticality of that resource, the availability of technologies for minimizing depletion, and the likelihood of substitutes being available. Thus land should not be degraded beyond reasonable recovery. With minerals and fossil fuels, the rate of depletion and the emphasis on recycling and economy of use should be calibrated to ensure that the resource does not run out before acceptable substitutes are available. Sustainable development requires that the rate of depletion of nonrenewable resources should foreclose as few future options as possible.

Development tends to simplify ecosystems and to reduce their diversity of species. And species, once extinct, are not renewable. The loss of plant and animal species can greatly limit the options of future generations; so sustainable development requires the conservation of plant and animal species.

So-called free goods like air and water are also resources. The raw materials and energy of production processes are only partly converted to useful products. The rest comes out as wastes. Sustainable development requires that the adverse impacts on the quality of air, water, and other natural elements are minimized so as to sustain the ecosystem's overall integrity.

In essence, sustainable development is a process of change in which the exploitation of resources, the direction of investments, the orientation of technological development, and institutional change are all in harmony and enhance both current and future potential to meet human needs and aspirations.

"Designing the Region" and "Designing the Region is Designing the Neighborhood"

from *The Regional City: Planning for the End of Sprawl* (2001)

Peter Calthorpe and William Fulton

Editors' Introduction

What is the most appropriate scale for urban planning? The home? The street? The neighborhood? The city? Or the entire metropolitan region? In *The Regional City: Planning for the End of Sprawl*, Peter Calthorpe and William Fulton argue that every element of the planned environment interacts with every other and that planners embracing the New Urbanist conception should follow a holistic, ecological approach that harmonizes the intimate scale of the neighborhood with the metropolitan scale of the region.

Peter Calthorpe – principal of the architectural firm of Calthorpe Associates based in Berkeley, California – is an urban futurist rooted in both the realities of the present and the traditions of the past. An author, practicing architect, and former academic who taught at the University of California, Berkeley, Calthorpe is one of the founders of the Congress for the New Urbanism. William Fulton is president of the widely respected Solimar Research Group of Ventura, California, a firm that provides governments, foundations, and planning agencies with empirical research on land use, metropolitan growth, and related policy concerns. Founded in 1999, Solimar specializes in urban growth boundaries and what it calls "urban containment policy."

In *The Regional City*, Calthorpe and Fulton present a subtly nuanced exploration of the New Urbanism at its best, a unified vision that combines earlier ideas about neighborhoods as "Pedestrian Pockets" and "transit-oriented developments" in a sweeping regional conception of contemporary urban form.

As a leading proponent of ecology and environmentalism as applied to urban design, Calthorpe has long been a prophet of a new kind of twenty-first-century community: the intimate, human-scale model that is descended directly from Ebenezer Howard's Garden Cities (p. 314) and that quickly became an ubiquitous element of New Urbanist planning and design, first in America and later in Europe and elsewhere. Calthorpe began his career by writing and lecturing extensively on the necessity of ecologically sensitive design and energy-efficient building based on the application of passive solar techniques. For a time, he was a partner of Sim Van der Ryn, the California state architect in the administration of progressive governor Jerry Brown. Together, they published *Sustainable Communities* (1986), an influential volume that helped spread the ideas of solar energy, recycling, appropriate technology, and environmentalist approaches to urban planning and design.

In Doug Kelbaugh's *The Pedestrian Pocket Book* (1989), Calthorpe examined the "profound mismatch between the old suburban patterns . . . and the post-industrial culture in which we now find ourselves." The similarities between his solution and the one proposed by Howard in 1898 were striking. Both the Garden Cities and the Pedestrian Pockets were to be surrounded by greenbelts of permanent agricultural land. Both were relatively dense developments, allowing residents to walk to the urban center in a short period of time. Both combined residential, commercial, and workplace elements; and where the Garden City was served by a railroad connection, the Pedestrian Pocket avoided the typical suburban monoculture of the automobile by a system of light-rail transit connectors.

But although the Garden City is an important source of the New Urbanism style of community development, the real application of the Calthorpe plan is in the blossoming new ring of suburban development currently springing up around the old metropolitan cores, the area that journalist Joel Garreau has dubbed "Edge City" and which Robert Fishman calls the "technoburbs" (p. 69). In *The Next American Metropolis*, Calthorpe refined and matured the Pedestrian Pocket idea to fit the emerging realities of Edge City technoburbia with what he calls "transit-oriented developments" (TODs). And in *The Regional City*, he and Fulton present nothing less than "a new paradigm of community . . . that leads from the Edge City to the Regional City." This, they argue, offers a conception that presents a fuller, richer social, economic, and civic life than is possible in the cultural sterilities of Edge City as it has developed in the final decades of the twentieth century. Signaling a clear break with traditional modernism as exemplified by the theories of Le Corbusier (p. 322), Calthorpe and Fulton argue that the Regional City expresses a "set of principles rooted more in ecology than in mechanics," focusing on issues of human scale, diversity, and conservation rather than rationality and mechanical efficiency.

Calthorpe is the most practical of urban visionaries because his visions represent "a response to a transformation that has already expressed itself: the transformation from the industrial forms of segregation and centralization to the decentralized and integrated forms of the post industrial era." This, he writes, is the result of "a culture adjusting itself" to new realities. Moreover, Calthorpe and Fulton are comfortable recognizing that the creation of a new urban paradigm will require the clear and careful application of state power operating through local and regional planning agencies. They write that "the idea that a region can be 'designed' is central to the Regional City" and call for an active role on the part of central governments, especially in the provision of transit systems, open-space, and development financing.

Andrés Duany and Elizabeth Plater-Zyberk (p. 192), Jaime Correa, Steven Peterson and Barbara Littenberg, Mark Schimmenti, Daniel Solomon, and a number of other architects and planners are designing human-scale communities with design aspects similar to the ones Calthorpe and Fulton espouse. Because their architecture draws on traditional small town elements, these New Urbanist architects are sometimes referred to as neo-traditionalists, and some of their most famous designs – notably the town of Seaside and the Disney-sponsored community of Celebration in Florida – have been roundly criticized as artificial enclaves for the well-to-do that care more for nostalgia and middle-class amenity than the needs of working people and depressed inner-city neighborhoods.

This selection is from *The Regional City: Planning for the End of Sprawl* (Washington, DC: Island Press, 2001). Other books by Peter Calthorpe include *Sustainable Communities* (San Francisco: Sierra Club Books, 1986) with Sym Van der Ryn, and *The Next American Metropolis* (New York: Princeton Architectural Press, 1993). Other books by William Fulton include *The Reluctant Metropolis* (Baltimore: Johns Hopkins University Press, 2001) and *Guide to California Planning*, second edition (Pt. Arena: Solano Press, 2005).

Peter Katz, *The New Urbanism: Towards an Architecture of Community* (New York: McGraw-Hill, 1994) and Doug Kelbaugh, *Common Place: Toward Neighborhood and Regional Design* (Seattle: University of Washington Press, 1997) provide overviews of designs by Calthorpe and other New Urbanists. Kenneth B. Hall and Gerald A. Porterfield, *Community By Design: New Urbanism for Suburbs and Small Communities* (New York: McGraw-Hill Professional, 2001) and E. Talen, *New Urbanism and American Planning: The Conflict of Cultures* (London and New York: Routledge, 2005) offer detailed analyses of the New Urbanist movement. The 69-page *New Urbanism: Peter Calthorpe vs. Lars Lerup: Michigan Debates on Urbanism* (Ann Arbor: University of Michigan Press, 2005) is a lively and scintillating exchange of views with an afterword by editor Robert Fishman. See also James Howard Kunstler's *The Geography of Nowhere* (New York: Simon & Schuster, 1993) and *Home from Nowhere: Remaking Our Everyday World for the Twenty-first Century* (New York: Simon & Schuster, 1998) for a popular account of Calthorpe and other New Urbanists' work. Also of interest are Michael Leccese (ed.), *Charter of the New Urbanism* (New York: McGraw-Hill, 1999) and Todd W. Bressi (ed.), *The Seaside Debates: A Critique of the New Urbanism* (New York: Rizzoli, 2002).

DESIGNING THE REGION

At the heart of creating concrete visions for the Regional City is the notion that they can be "designed." We use the term "design" not in the typical sense of artistically configuring a physical form but to imply a process that synthesizes many disciplines. Regional design is an act that integrates multiple facets at once: the demands of the region's ecology, its economy, its history, its politics, its regulations, its culture, and its social structure. And its results are specific physical forms as well as abstract goals and policies – regional maps and neighborhood urban design standards as well as implementation strategies, governmental policies, and financing mechanisms.

Too often we plan and engineer rather than design. Engineering tends to optimize isolated elements without regard for the larger system, whereas planning tends to be ambiguous, leaving the critical details of place making to chance. If we merely plan and engineer, we forfeit the possibility of developing a "whole systems" approach or a "design" that recognizes the trade-offs between isolated efficiencies and integrated parts.

The engineering mentality often reduces complex, multifaceted problems to one measurable dimension. For example, traffic engineers optimize road size for auto capacity without considering the trade-off of neighborhood scale, walkability, or beauty. Civil engineers efficiently channelize our streams without considering recreational, ecological, or esthetic values. Commercial developers optimize the delivery of goods without balancing the social need of neighborhoods for local identity and meeting places. Again and again we sacrifice the synergy of the whole for the efficiency of the parts.

The idea that a region or even a neighborhood could or should be "designed" is central to the Regional City. We need to acknowledge that we can direct our growth and that such action can include complex trade-offs as well as unexpected synergies. The common impression is that our neighborhoods, towns, or regions evolve organically (and somewhat mysteriously). They are the product of invisible market forces or the summation of technical imperatives. There also is the illusion that these forces cannot and should not be tampered with. Planning failed in the past; therefore it will fail in the future.

The real illusion, of course, is that we cannot control the form of our communities. Historically, design played a large role in shaping our forms of settlement. The template that underlies much of our suburban growth was designed in the thirties by Frank Lloyd Wright with his Broadacre Cities plans and Clarence Stein's Greenbelt towns. These were then bastardized and codified by the HUD minimum property standards of the 1950s. The template for urban redevelopment was developed about the same time by Le Corbusier and a European group of architects called CIAM (Congrès Internationaux d'Architecture Moderne). Their vision of superblocks and high-rise development became the basis of our urban renewal programs of the 1960s.

The problem is not that our suburbs and cities are lacking design but that they are designed according to failed principles with flawed implementation. They are designed in accord with modernist principles and implemented by specialists. The modernist principles of specialization, standardization, and mass production in emulating our industrial economy had a severe effect on the character of our neighborhoods and regions.

At the neighborhood scale, specialization meant that each land use – residential, retail, commercial, or civic – was isolated and developed by "experts" who optimized their particular zones without any responsibility for the whole. Regional specialization meant that each area within the region could play an independent role: suburbs for the middle class and new businesses, cities for the poor and declining industries, and countryside for nature and agriculture.

As a complement to specialization, standardization led to the homogenization of our communities, a blindness to history and the demise of unique ecological systems. A "one size fits all" mentality of efficiency overrode the special qualities of place and community.

Mass production (in housing, transportation, offices, and so forth) upends the delicate balance between local enterprise, regional systems, and global networks. The logic of mass production moves relentlessly toward ever-increasing scales, which in turn reinforces the specialization and standardization of everyday life.

Against this modern alliance of specialization, standardization, and mass production stands a set of principles rooted more in ecology than in mechanics. They are the principles of diversity, conservation, and human scale. Diversity at each scale calls for more complex, differentiated communities shaped from the unique qualities of place and history. Conservation implies care for existing resources whether natural, social, or institutional. And the principle of human scale brings the individual back into a picture increasingly fashioned around remote and mechanistic concerns.

These alternative principles apply equally to the social, economic, and physical dimensions of communities. For example, the social implications of human scale may mean more police officers walking a beat rather than hovering overhead in a helicopter; the economic implications of human scale may mean economic policies that support small local business rather than major industries and corporations; and the physical implications of human scale may be realized in the form and detail of buildings as they relate to the street. Unlike the standard governmental categories of economic development, housing, education, and health services, each of these design principles incorporates physical design, social programs, and economic strategies. These principles, then, are the ones that we believe should form the foundation of a new regional and neighborhood design ethic.

Human scale

For several generations, the design of buildings, the planning of communities, and the growth of our institutions have exemplified the view that "bigger is better." Efficiency was correlated with large, centralized organizations and processes. Now the idea of decentralized networks of smaller working groups and more personalized institutions is gaining currency in both government and business. Efficiency is correlated with nimble, small working groups, not large hierarchical institutions.

Certainly, the reality of our time is a complex mix of both of these trends. For example, we have ever-larger retail outlets at the same time that Main Streets are making a comeback. Some businesses are growing larger and more centralized while the "new economy" is bursting with small-scale start-ups and intimate working groups. Housing production is diversifying home types at the same time that it consolidates into larger financing packages. Both directions are evolving at the same time, and the shape of our communities will have to accommodate this complex reality.

Yet people are reacting to an imbalance between these two forces. The building blocks of our communities – schools, local shopping areas, housing

subdivisions, apartment complexes, and office parks – have all grown into forms that defy human scale. And we are witnessing a reaction to this lack of scale in many ways. People uniformly long for an architecture that puts detail and identity back into what have too often become generic, if functional, buildings. They desire the character and scale of a walkable street, complete with shade trees and buildings that orient windows and entries their way. They idealize Main Street shopping areas and historic urban districts.

Human scale is a design principle that responds simultaneously to simple human desires and the emerging ethos of the new decentralized economics. The focus on human scale represents a shift away from top-down social programs, from characterless housing projects, and from more and more remote institutions. In its most concrete expression, human scale is the stoop of a townhouse or the front porch of a home rather than the stairwell of a high-rise or the garage door of a tract home. Human scale in economics means supporting individual entrepreneurs and local businesses. Human scale in community means a strong neighborhood focus and an environment that encourages everyday interaction.

Diversity

Diversity has multiple meanings and profound implications. It has the most challenging implications for the social, environmental, and economic dimensions of community planning. Perhaps its most obvious outcome is the creation of communities that are diverse in use and in population. As a planning axiom, it calls for a return to mixed-use neighborhoods that contain a broad range of uses as well as a broad range of housing types and people.

The four fundamental elements of community – civic places, commercial uses, housing opportunities, and natural systems – define the physical elements of diversity at any scale. As a physical principle, diversity in neighborhoods ensures that destinations are close at hand and that the shared institutions of community are integrated. It also implies an architecture rich in character and streetscapes that vary with place and use.

As a social principle, diversity is controversial and perhaps the most challenging of all. It implies creating neighborhoods that provide for a large range in age group, household type, income, and race. As already stated, neighborhoods have always (to a greater or lesser degree) been defined by commonalties even if energized by differences. But today we have reached an extreme: age, income, family size, and race are all divided into discrete market segments and locations that are built independently. Complete housing integration may be a distant goal, but inclusive neighborhoods that broaden the economic range, expand the mix of age and household types, and open the door to racial integration are feasible and desirable.

Diversity is a principle with significant economic implications. Gone are the days when economic-revitalization efforts focused on a single industry or a major governmental program. A more ecological understanding of industry clusters has emerged. This sensibility validates the notion that a range of complementary but differing enterprises (large and small, local, regional, and global) are important to maintaining a robust economy, and that now more than ever, the quality of life and the urbanism of a place, as well as the more traditional economic factors, play a significant role in the emerging economy.

Finally, diversity is a fundamental principle that can help to guide the preservation of local and regional ecologies. Clearly, understanding the complex nature of the existing or stressed habitats and watershed systems mandates a different approach to open-space planning. Active recreation, agriculture, and habitat preservation are often at odds. A broad range of open-space types, from the most active to the most passive, must be integrated in neighborhood and regional designs. Diversity in use, diversity in population, diversity in enterprise, and diversity in natural systems are fundamental to the Regional City.

Conservation

Conservation implies many things in community design beyond husbanding resources and protecting natural systems; it implies preserving and restoring the cultural, historic, and architectural assets of a place as well. Conservation calls for designing communities and buildings that require fewer resources – less energy, less land, less waste, and fewer materials – but it also implies caring for what we have and developing an ethic of reuse and repair – in both our physical and our social realms.

The principle of conservation and its complements, restoration and preservation, should be applied to the built environment as well as to the natural environment – not only to our historic building stock and neighborhood institutions, but also to human resources and human history. Communities should strive to conserve their cultural identity, physical history, and unique natural systems. Restoration and conservation are more than environmental themes; they are an approach to the way that we think about community at the regional and local levels.

Conserving resources has many obvious implications in community planning. Foremost are the quantities of farmlands and natural systems displaced by sprawling development and the quantity of auto travel required to support it. Even within more compact, walkable communities, conservation of resources can lead to new design strategies. The preservation of waterways and on-site water-treatment systems can add identity and natural amenities at the same time that they conserve water quality. Energy-conservation strategies in buildings often lead to environments that are climate responsive and unique to place.

Conserving the historic buildings and institutions of a neighborhood can preserve the icons of community identity. Restoring and enhancing the vernacular architecture of a place can simultaneously reduce energy costs, reestablish local history, and create jobs. Although the preservation movement has made great strides with landmark buildings, it is correct now in extending its agenda beyond building facades to the social fabric of neighborhoods and to the ecology of the communities that are the lifeblood of historic districts.

Conserving human resources is another implication of this fundamental principle. In too many of our communities, poverty, lack of education, and declining job opportunities lead to a tragic waste of human potential. As we have begun to see, communities are not viable when concentrations of poverty turn them into a wasteland of despair and crime. In this context, "conservation" takes on a larger meaning – the restoration and rehabilitation of human potential wherever it is being squandered and overlooked. There should be no natural or cultural environments that are disposable or marginalized. Conservation and restoration are practical undertakings that can be economically strengthening and socially enriching.

DESIGNING THE REGION IS DESIGNING THE NEIGHBORHOOD

What happens to regions or neighborhoods if they are "designed" according to these principles? An interesting set of parallel design strategies emerges at both the regional and the neighborhood levels. First and foremost, the region and its elements – the city, suburbs, and their natural environment – should be conceived as a unit, just as the neighborhood and its elements – housing, shops, open space, civic institutions, and businesses – should be designed as a unit. Treating each element separately is endemic to many of the problems that we now face. Just as a neighborhood needs to be developed as a whole system, the region must be treated as a human ecosystem, not a mechanical assembly.

Seen as this integrated whole, the region can be designed in much the same way as we would design a neighborhood. That the whole, the region, would be similar to its most basic pieces, its neighborhoods, is an important analogy. Both need protected natural systems, vibrant centers, human-scale circulation systems, a common civic realm, and integrated diversity. Developing such an architecture for the region creates the context for healthy neighborhoods, districts, and city centers. Developing such an architecture for the neighborhood creates the context for regions that are sustainable, integrated, and coherent. The two scales have parallel features that reinforce one another.

Major open-space corridors within the region, such as rivers, ridge lands, wetlands, or forests, can be seen as a "village green" at a megascale – the commons of the region. These natural commons establish an ecological identity as the basis of a region's character. Similarly, the natural systems and shared open spaces at the neighborhood scale are fundamental to its identity and character. A neighborhood's natural systems, like the region's, are as much a part of its commons as its civic institutions or commercial center.

Just as a neighborhood needs a vital center to serve as the crossroads of a local community, the region needs a vital central city to serve as its cultural heart and as a link to the global economy. In the Edge City metropolis, both types of centers are failing. In the suburbs, what were village centers of human proportions are overcome by remote discount centers and relentless commercial strips. In the central cities, poverty and disinvestment erode historic neighborhood communities. Both fall prey to specialized enterprises oriented to mass distribution rather than the local community. Like the commons, healthy centers, both urban and suburban, are fundamental to local and regional coherence.

Regional and neighborhood design has other parallels. Pedestrian scale within the neighborhood – walkable streets and nearby destinations – has a partner in transit systems at the regional scale. Transit can organize the region in much the same way as a street network orders a neighborhood. Transit lines focus growth and redevelopment in the region just as main streets can focus a neighborhood. Crossing local and metropolitan scales, transit supports the life of the pedestrian within each neighborhood and district by providing access to regional destinations. In a complementary fashion, pedestrian-friendly neighborhoods support transit by providing easy access for riders, not cars. The two scales, if designed as parallel strategies, reinforce each other.

As we have pointed out, diversity is a fundamental design principle for both the neighborhood and the region. A diverse population and job base within a region supports a resilient economy and a rich culture in much the same way that diverse uses and housing in a neighborhood support a complex and active community. The suburban trend to segregate development by age and income translates at the regional level into an increasing spatial and economic polarization . . . Both trends can be countered by policies that support inclusionary housing and mixed-use environments.

These parallels across scales are not merely coincidence. The fundamental nature of a culture and economy expresses itself at many scales simultaneously. Sprawl and our lack of regional structure is a manifestation of an older and quite different paradigm. Since World War II, our economy and culture have accelerated their movement toward the industrial qualities of mass production, standardization, and specialization. The massive suburbanization that marks this period is the direct expression of these qualities. As a counterpoint, the principles and concurrences just outlined define a new paradigm of community and growth, one that leads from the Edge City to the Regional City.

PART SIX

Urban planning theory and practice

INTRODUCTION TO PART SIX

Contemporary urban planning has come a long way from its origins in the Olmsted-inspired parks movement, the visionary plans of Ebenezer Howard and the Garden Cities that more or less followed his ideas, Daniel Burnham's monumental City Beautiful projects, the plans of eccentric Scottish biologist Patrick Geddes, Le Corbusier and his modernist followers, the brilliant Broadacre City vision of Frank Lloyd Wright, and a host of other imagined and implemented plans discussed in Part Five, "Urban Planning History and Visions." Today city and regional planning (or town and country planning as it is called in the UK) has matured into an important profession with its own body of theory and set of professional practices. This section focuses on the theory and practice of urban planning today.

Large local governments may employ dozens or even hundreds of professionally trained planners. Even most small and medium-sized cities and towns employ city planners and have explicit plans for their future development. In addition to professionals whose formal education is in city and regional (or town and country) planning, planning staffs are likely to include architects and urban designers, geographers, economists, civil engineers, transportation experts, environmental professionals, computer experts, and staff trained in negotiation and other communicative planning skills.

City plans are grounded in *analysis* of local conditions and contain a *vision* of an urban future the citizens, local elected officials, planning staff, and consultants consider desirable. Urban plans vary greatly in approach, content, sophistication, comprehensiveness, time frame, and format. Plans developed for the Los Angeles City Planning Commission or Curitiva, Brazil, run to many volumes built on mountains of data and sophisticated analysis. The town plan for a small town like J.B. Jackson's hypothetical Optimo City (p. 184) may contain a common-sense description of the town's situation and some practical suggestions worked out by the residents under the direction of a part-time planning consultant. The substance of city plans varies as widely as the planning culture of the cities that produce them. The plan for Optimo City might advance a narrow, business-as-usual vision for its future that envisions the city tearing down the courthouse and building a parking garage to attract more off-highway business. By contrast, green city plans in some of the most environmentally sensitive European cities that Timothy Beatley describes (p. 411) are filled with imaginative ideas for use of public bicycle depots, wind and solar power generation, community gardens, and recycling. The city and regional plans Peter Calthorpe, Andrés Duany and Elizabeth Plater-Zyberk have developed and implemented reflect their "New Urbanist" and "sustainable urban development" values.

The first two selections in this section pick up where Part Five leaves off. Sir Peter Hall's "The City of Theory" (p. 354) discusses the evolution and current status of twentieth-century urban planning theory, and Edward J. Kaiser and David R. Godschalk's "Twentieth Century Land Use Planning" (p. 366) describes current American land use planning practice. Both selections touch on planning in the first half of the twentieth century, but focus on how planning has evolved during the last half century and what it is like today.

In the first half of the twentieth century, urban planning was elitist and paid little attention to implementation. According to Hall, planning theory at that time was preoccupied with how to create stable cities "geared to a static world." During this "golden age," the planner was "free from political interference" and "serenely sure of his technical capacities." He (male) was also generally ineffective. Heavily influenced by the Garden Cities and City Beautiful movements, early planners worked out theoretically perfect physical plans in excruciating detail only to see their plans gather dust. Slowly a more conceptual, visionary, and flexible approach to creating city "general plans" emerged. By the 1950s, according to Kaiser and Godschalk (p. 366), there was a general consensus that urban general plans should be long-term, visionary documents, charting the desired physical form of the city. While an advance over earlier static architectural city planning, his ivory tower general planning approach was never very practical and was no longer defensible by the 1950s, as the pace of urban development and urban change accelerated.

In the last half century, urban planning theory has been buffeted by a whole series of conflicting approaches proposed by Marxists, advocacy planners, equity planners, systems analysts, pluralists, disjointed incrementalists, probabilistic planners, green urbanists, ecological designers, feminist planners, and communicative action theorists. In practice the general plan "trunk" of the urban land use planning "tree" has branched into management and policy as well as physical design. Planning no longer takes place in ivory towers. As planning has become more relevant it has also become more conflictual, as Paul Davidoff (p. 400) and John Forester (p. 387) describe. Citizens and decision makers did not care much about unrealistic static physical designs or utopian general plans developed in the first half of the twentieth century. But they care a great deal about plans that propose locating a hazardous waste disposal site near them, restricting the way in which they can use land they own (reducing its value), razing a historic church, developing open space, or building a highway in their neighborhood.

Cornell planning professor John Forester describes how planners actually work with neighborhood residents, local elected officials, interest groups, and private developers. Forester provides perhaps the best view we have of what current urban planning is really like as perceived by planners themselves. Based on interviews with dozens of practicing planners Forester describes planners' day-to-day activities, the nature of and limitations on their power, and the strategies they actually employ to get things done. Planners negotiate, mediate, resolve conflict, and serve as diplomats. Forester is interested in equity planning and applauds the efforts of many planners to redirect market forces and to empower people and communities poorly served by the private market. Planning practitioners who study Forester's theoretical writings on planning in the face of conflict have much to learn that can help them be more effective.

Planner/lawyer Paul Davidoff proposes an approach to urban planning that recognizes that different groups compete in the planning process. Davidoff envisioned a kind of planning practice in which city planners would act as *advocates* for poor people and disenfranchised groups. Davidoff and many planners whom he inspired saw "advocacy planning" as one way to bring about non-violent social change. Davidoff's concern with social justice is an enduring one. From the early efforts of nineteenth-century reformers advocating on behalf of slum residents in the new industrial slums, through the New Deal, New Frontier, and Great Society programs of the last century, some progressives have always made a connection between urban planning and social justice. A recent incarnation of this tradition is "equity planning," a subfield developed by Cleveland State University professor and former Cleveland city planning director Norman Krumholz.

Another enduring value in urban planning has been a concern to harmonize the built environment with the natural environment. By the middle of the nineteenth century, park planners like Joseph Paxton in England and Frederick Law Olmsted in the United States were working hard to bring nature into crowded cities. In the early twentieth century, Scottish biologist/planner Patrick Geddes had worked out an elaborate scheme for regional planning that reflected different ecosystems. Similar thinking informed Ian McHarg's theories in his classic *Design with Nature* (1969). Today "sustainable" urban planning as proposed by the Brundtland Commission (p. 337), and described by Timothy Beatley (p. 411) and Stephen Wheeler (p. 499), ecological design, and green urbanism have moved to the fore. The final selection in Part Six describes green urbanism in Europe and what planners worldwide can learn from it. Planning professor Timothy Beatley describes what environmentally conscious cities in Europe have actually done to retain their compact form, promote public transit, reduce auto dependency, substitute renewable energy sources for non-renewable energy sources, and support pedestrians. Beatley identifies core aspects of the European sustainable urban development agenda, provides specific examples of exemplary green practices that some European cities have implemented, and eloquently argues that green urbanism is possible and can produce livable cities that respect the natural environment of the Earth. Beatley's selection combines theory and practice. It is both realistic about urban environmental challenges and optimistic about what can be done. It is a fitting conclusion to this section on "Urban Planning Theory and Practice."

"The City of Theory"

from *Cities of Tomorrow: An Intellectual History
of Urban Planning and Design in the Twentieth Century,*
third edition (2001)

Peter Hall

Editors' Introduction

In this selection British geographer/planner Sir Peter Hall reviews the evolution of city planning theory in the United States and Europe during the last half century. Hall points out that before World War II, city planning was defined as the *craft* of physical planning. Professors who had been educated as architects, landscape architects, and other design professionals dominated teaching and writing. They taught students to prepare self-contained, end-state physical plans like Raymond Unwin's plan for the garden city of Letchworth, England: architectural drawings extended to city scale. Planners were viewed as a priviledged elite and were not expected to interact with the people they were planning for. Once a well-crafted, aesthetically pleasing, functional plan was complete on paper the early planning theorists believed that a city could be built just as a house is built from architectural drawings.

But, as Hall notes, few planners plan whole new towns from scratch. Rather they have to decide how to integrate new housing, streets, commercial and retail districts, industrial areas, parks and open space, and infrastructure into existing cities and emerging suburbs. City planning does not end the way house construction does with a finished product; it is an ongoing, messy process.

During the 1960s, systems analysts who had been educated as economists, computer scientists, quantitative geographers, and engineers developed a competing paradigm of what city planning should be. They argued that city planning should be a science, not a craft. They felt planners could and should use quantitative and exact methodologies to plan urban transportation, utility, or other urban systems on an ongoing basis. They were pioneers in using computers at a time when computers were new, costly, and hard to use. The systems theorists thought plans should be mathematical models rather than end-state architectural drawings. It may be helpful to imagine a planner from the old school saying, "Here is my plan," and unrolling a series of large, carefully drafted blueprints and design sketches confronting a systems planner saying, "Here's my plan," as she types an equation into a huge mainframe computer!

But planning for what kind of city and for whom? Normative theorists, shaken by 1960s racial and class conflict in cities, felt city planning was too important to leave to either elitist designers following the craft paradigm of planning or technocratic planners following the scientific paradigm. Liberal planners of the 1960s such as Paul Davidoff focused on planning outcomes and who the city was being built for. They saw planning as a vehicle to benefit poor and powerless people rather than to achieve design excellence or elegant mathematical modeling.

During the 1970s, Marxist theorists developed a coherent body of urban theory based on class conflict. Twentieth-century Marxists could not apply the categories Friedrich Engels applied to mid-nineteenth-century Manchester, England – the stark class conflict between an oppressed industrial proletariat and wealthy owners of the means of production that characterized mid-nineteenth century cities had become far more complicated.

But Marxist theorists felt updated class analysis was highly appropriate. Marxist urban theory dominated much academic discourse throughout the 1970s.

Hall ends his tour of planning theory with some critical comments on the current divorce between planning theory and practice. He argues that as graduate programs in city and regional planning have grown in number and size, and as a formal body of planning theory has developed, too often academics today merely debate each other's academic theories with little attention to actual practice or the needs of planning practitioners on the front lines. Finding academic planning theory irrelevant, city planning practitioners concern themselves only with the nuts and bolts of planning practice without the deeper understanding relevant theory could offer. Hall calls for an improved, reciprocal relationship between the two: theory that is informed by and relevant to planning practice and planning practice informed and improved by theory. John Forester's empirical research is a good example of how Peter Hall feels academic planning theorists might wed theory and practice. Forester interviewed dozens of practicing planners so that he was very knowledgeable about actual planning practice and then developed theory based on what they told him, which could in turn inform and improve planning practice.

Until assuming the Bartlett Professorship of City Planning at University College, London, in 1994, Peter Hall shuttled between the University of California Berkeley's Department of City and Regional Planning and the University of Reading's Geography Department. He has traveled widely and written prolifically on urban geography and city planning. In addition to his academic accomplishments, Peter Hall has worked with actual planning projects as varied as long-term plans for the London region and bringing high-speed bullet trains to California. His theory is informed by the experience of actual planning. Practicing planners as well as academics use his writings.

This selection is from *Cities of Tomorrow: An Intellectual History of Urban Planning and Design in the Twentieth Century*, third edition (Oxford: Blackwell, 2002). Other of Peter Hall's forty-plus books include *Urban and Regional Planning*, fourth edition (London and New York: Routledge, 2002), *Cities in Civilization* (New York: Pantheon, 1998), *Sociable Cities: The Legacy of Ebenezer Howard* (London: John Wiley & Sons, 1998) co-authored with Colin Ward, *The World Cities*, third edition (London: Weidenfeld and Nicolson; New York: St. Martin's, 1984), and *Great Planning Disasters* (Berkeley: University of California Press, 1982).

Section 2 of Eugenie Birch's *Urban and Regional Planning Reader* (London and New York: Routledge, 2007) contains additional writings on urban and regional planning theory. Other overviews of planning theory include Philip Allmendinger, *Planning Theory* (New York: Palgrave Macmillan, 2002), Susan Fainstein and Scott Campbell, *Readings in Urban Theory* (London: Blackwell, 2001), Nigel Taylor, *Urban Planning Theory Since 1945* (Thousand Oaks: Sage, 1998), Seymour Mandelbaum (ed.), *Explorations in Planning Theory* (Rutgers: CUPR Press, 1996), and John Friedmann, *Planning in the Public Domain: From Knowledge to Action* (Princeton: Princeton University Press, 1987).

PLANNING AND THE ACADEMY: PHILADELPHIA, MANCHESTER, CALIFORNIA, PARIS, 1955–1987

. . . about 1955 . . . city planning at last became legitimate; but in doing so, it began to sow the seeds of its own destruction. All too quickly, it split into two separate camps: the one, in the schools of planning, increasingly and exclusively obsessed with the theory of the subject; the other, in the offices of local authorities and consultants, concerned only with the everyday business of planning in the real world. That division was not at first evident; indeed, during the late 1950s and most of the 1960s, it seemed that at last a complete and satisfactory link had been forged between the world of theory and the world of practice. But all too soon, illusion was stripped aside: honeymoon was followed in quick succession during the 1970s by tiffs and temporary reconciliations, in the 1980s by divorce. And, in the process, planning lost much of its new-found legitimacy.

The prehistory of academic city planning 1930–1955

It was not that planning was innocent of academic influence before the 1950s. On the contrary: in virtually every urbanized nation, universities and polytechnics had created courses for the professional education of planners; professional bodies had come into existence to define and protect standards, and had forged links with the academic departments. Britain took an early lead when in 1909 . . . the soap magnate William Hesketh Lever, founder of Port Sunlight, won a libel action against a newspaper and used the proceeds to endow his local University of Liverpool with a Department of Civic Design. Stanley Adshead, the first professor, almost immediately created a new journal, the *Town Planning Review*, in which theory and good practice were to be firmly joined; its first editor was a young faculty recruit, Patrick Abercrombie, who was later to succeed Adshead in the chair first at Liverpool, then at Britain's second school of planning: University College London, founded in 1914. The Town Planning Institute – the Royal accolade was conferred only in 1959 – was founded in 1914 on the joint initiative of the Royal Institute of British Architects, the Institution of Civil Engineers and the Royal Institution of Chartered Surveyors; by the end of the 1930s, it had recognized seven schools whose examinations provided an entry to membership.

The United States was slower: though Harvard had established a planning course in 1909, neck and neck with Liverpool, it had no separate department until 1929. Nevertheless, by the 1930s America had schools also at MIT, Cornell, Columbia and Illinois, as well as courses taught in other departments at a great many universities across the country. And the American City Planning Institute, founded in 1917 as a breakaway from the National Conference on City Planning, ten years later became – mainly through the insistence of Thomas Adams – a full-fledged professional body on TPI lines, a status it retained when in 1938 it broadened to include regional planning and renamed itself the American Institute of Planners.

The important point about these, and other, initiatives was this: stemming as they did from professional needs, often through spin-offs from related professions like architecture and engineering, they were from the start heavily suffused with the professional styles of these design-based professions.

The job of the planners was to make plans, to develop codes to enforce these plans, and then to enforce those codes; relevant planning knowledge was what was needed for that job; planning education existed to convey that knowledge together with the necessary design skills. So, by 1950, the utopian age of planning . . . was over; planning was now institutionalized into comprehensive land-use planning. All this was strongly reflected in the curricula of the planning schools down to the mid-1950s, and often for years after that; and these in turn were reflected in the books and articles that academic planners wrote. Land-use planning, Keeble told his British audience in 1959 and Kent reminded the American counterpart in 1964, was a distinct and tightly bounded subject, quite different from social or economic planning. And these texts reflected the fact that "city planners early adopted the thoughtways and the analytical methods that engineers developed for the design of public works, and they then applied them to the design of cities."

The result, as Michael Batty has put it, was a subject that for the ordinary citizen was "somewhat mystical" or arcane, as law or medicine were, but that was – in sharp contrast to education for these older professions – not based on any consistent body of theory; rather, in it, "scatterings of social science bolstered the traditional architectural determinism." Planners acquired a synthetic ability not through abstract thinking, but by doing real jobs; in them, they used first creative intuition, then reflection. Though they might draw on bits and pieces of theory about the city – the Chicago school's sociological differentiation of the city, the land economists' theory of urban land rent differentials, the geographers' concepts of the natural region – these were employed simply as snippets of useful knowledge. In the important distinction later made by a number of writers, there was some theory in planning but there was no theory of planning. The whole process was very direct, based on a single-shot approach: survey (the Geddesian approach) was followed by analysis (an implicit learning approach), followed immediately by design.

True, as Abercrombie's classic text of 1933 argued, the making of the plan was only half the planner's job; the other half consisted of planning, that is implementation, but it was nowhere assumed that some kind of continuous learning process was needed. True, too, the 1947 Act provided for plans – and the surveys on which they were based – to be quinquennially updated; the assumption was still that the result would be a fixed land-use plan. And, a decade after that, though Keeble's equally classic text referred to the planning process, by this he simply meant the need for a spatial hierarchy of related plans from the regional to the local, and the need at each scale for survey before plan. Nowhere is found a discussion of implementation or updating.

Thus – apart from extremely generalized statements like Abercrombie's famous triad of 'beauty, health and convenience' – the goals were left implicit; the planner would develop them intuitively from his own values, which by definition were "expert" and apolitical.

So, in the classic British land-use planning system created by the 1947 Town and Country Planning Act, no repeated learning process was involved, since the planner would get it right first time:

> The process was therefore not characterized by explicit feedback as the search "homed in" on the best plan, for the notion that the planner had to learn about the nature of the problem was in direct conflict with his assumed infallibility as an expert, a professional . . . The assumed certainty of the process was such that possible links back to the reality in the form of new surveys were rarely if ever considered . . . This certainty, based on the infallibility of the expert, reinforced the apolitical, technical nature of the process. The political environment was regarded as totally passive, indeed subservient to the "advice" of the planners and in practice, this was largely the case.
>
> (Batty, 1979)

It was, as Batty calls it, the golden age of planning: the planner, free from political interference, serenely sure of his technical capacities, was left to get on with the job. And this was appropriate to the world outside, with which planning had to deal: a world of glacially slow change – stagnant population, depressed economy – in which major planning interventions would come only seldom and for a short time, as after a major war. Abercrombie, in the plan for the West Midlands he produced with Herbert Jackson in 1948, actually wrote that a major objective of the plan should be to slow down the rate of urban change, thus reducing the rate at which built structures became obsolescent: the ideal city would be a static, stable city:

> Let us assume . . . that a maximum population has been decided for a town, arrived at after consideration of all the factors appearing to be relevant . . . Allowance has been made for proper space for all conceivable purposes in the light of present facts and the town planner's experience and imagination. Accordingly, an envelope or green belt has been prescribed, outside which the land uses will be those involving little in the way of resident population. The

town planner is now in the happy position for the first time of knowing the limits of his problem. He is able to address himself to the design of the whole and the parts in the light of a basic overall figure for population. The process will be difficult enough in itself, but at least he starts with one figure to reassure him.

> (Abercrombie and Jackson, 1948)

American planning was never quite like that. Kent's text of 1964 on the urban general plan, though it deals with the same kind of land-use planning, reminds its students of end-directions which are continually adjusted as time passes. And, because the planner's basic understanding of the interrelationship between socio-economic forces and the physical environment was largely intuitive and speculative, Kent warned his student readers,

> In most cases it is not possible to know with any certainty what physical design measures should be taken to bring about a given social or economic objective, or what social and economic consequences will result from a given physical-design proposal. Therefore, the city council and the city-planning commission, rather than professional city planners, should make the final value judgements upon which the plan is based.
>
> (Kent, 1964)

But even Kent was certain that, despite all this, it was still possible for the planner to produce some kind of optimal land use plan; the problem of objectives was just shunted off.

The systems revolution

It was a happy, almost dream-like, world. But increasingly, during the 1950s, it did not correspond to reality. Everything began to get out of hand. In every industrial country, there was an unexpected baby boom, to which the demographers reacted with surprise, the planners with alarm; only its timing varied from one country to another, and everywhere it created instant demands for maternity wards and child-care clinics, only slightly delayed needs for schools and playgrounds. In every one, almost simultaneously, the great postwar economic boom got under way, bringing pressures for new investment in factories and offices. And, as boom generated affluence, these countries soon passed into the realms of high mass-consumption societies, with unprecedented demands for durable consumer goods: most notable among these, land-hungry homes and cars. The result everywhere – in America, in Britain, in the whole of western Europe – was that the pace of urban development and urban change began to accelerate to an almost superheated level. The old planning system, geared to a static world, was overwhelmed.

These demands in themselves would force the system to change; but, almost coincidentally, there were changes on the supply side too. In the mid-1950s there occurred an intellectual revolution in the whole cluster of urban and regional social studies, which provided planners with much of their borrowed intellectual baggage. A few geographers and industrial economists discovered the works of German theorists of location, such as Johann Heinrich von Thünen (1826) on agriculture, Alfred Weber (1909) on industry, Walter Christaller (1933) on central places, and August Lösch (1940) on the general theory of location; they began to summarize and analyse these works, and where necessary to translate them. In the United States, academics coming from a variety of disciplines began to find regularities in many distributions, including spatial ones. Geographers, beginning to espouse the tenets of logical positivism, suggested that their subject should cease to be concerned with descriptions of the detailed differentiation of the earth's surface, and should instead begin to develop general hypotheses about spatial distributions, which could then be rigorously tested against reality: the very approach which these German pioneers of location theory had adopted. These ideas, together with the relevant literature, were brilliantly synthesized by an American economist, Walter Isard, in a text that became immediately influential. Between 1953 and 1957, there occurred an almost instant revolution in human geography and the creation, by Isard, of a new academic discipline uniting the new geography with the German tradition of locational economics. And, with official blessing – as in the important report of Britain's Schuster Committee of 1950, which recommended a greater social science content in planning education – the new locational analysis began to enter the curricula of the planning schools.

The consequences for planning were momentous: with only a short timelag, "the discipline of physical planning changed more in the 10 years from 1960 to 1970, than in the previous 100, possibly even 1000 years" (Batty, 1979).

The subject changed from a kind of craft, based on personal knowledge of a rudimentary collection of concepts about the city, into an apparently scientific activity in which vast amounts of precise information were garnered and processed in such a way that the planner could devise very sensitive systems of guidance and control, the effects of which could be monitored and if necessary modified. More precisely, cities and regions were viewed as complex systems – they were, indeed, only a particular spatially based subset of a whole general class of systems – while planning was seen as a continuous process of control and monitoring of these systems, derived from the then new science of cybernetics developed by Norbert Wiener.

There was thus, in the language later used in the celebrated work of Thomas Kuhn, a "paradigm shift." It affected city planning as it affected many other related areas of planning and design. Particularly, its main early applications – already in the mid-1950s – concerned defence and aerospace; for these were the Cold War years, when the United States was engaging in a crash programme to build new and complex electronically controlled missile systems. Soon, from that field, spun off another application. Already in 1954, Robert Mitchell and Chester Rapkin – colleagues of Isard at the University of Pennsylvania – had published a book suggesting that urban traffic patterns were a direct and measurable function of the pattern of activities – and thus land uses – that generated them. Coupled with earlier work on spatial interaction patterns, and using for the first time the data-processing powers of the computer, this produced a new science of urban transportation planning, which for the first time claimed to be able scientifically to predict future urban-traffic patterns. First applied in the historic Detroit Metropolitan Area transportation study of 1955, further developed in the Chicago study of 1956, it soon became a standardized methodology employed in literally hundreds of such studies, first across the United States, then across the world.

Heavily engineering-based in its approach, it adopted a fairly standardized sequence. First, explicit goals and objectives were set for the performance of the system. Then, inventories were taken of the existing state of the system: both the traffic flows, and the activities that gave rise to them. From this, models were derived which sought to establish these relationships in precise mathematical form. Then, forecasts were made of the future state of the system, based on the relationships obtained from the models. From this, alternative solutions could be designed and evaluated in order to choose a preferred option. Finally, once implemented the network would be continually monitored and the system modified as necessary.

At first, these relationships were seen as operating in one direction: activities and land uses were given; from these, the traffic patterns were derived. So the resulting methodology and techniques were part of a new field, transportation planning, which came to exist on one side of traditional city planning. Soon, however, American regional scientists suggested a crucial modification: the locational patterns of activities – commercial, industrial, residential – were in turn influenced by the available transportation opportunities; these relationships, too, could be precisely modelled and used for prediction; therefore the relationship was two-way, and there was a need to develop an interactive system of land-use–transportation planning for entire metropolitan or subregional areas. Now, for the first time, the engineering-based approach invaded the professional territory of the traditional land-use planner. Spatial interaction models, especially the Garin–Lowry model – which, given basic data about employment and transportation links, could generate a resulting pattern of activities and land uses – became part of the planner's stock in trade. As put in one of the classic systems texts:

> In this general process of planning we particularise in order to deal with more specific issues: that is, a specific real world system or subsystem must be represented by a specific conceptual system or subsystem within the general conceptual system. Such a particular representation of a system is called a *model* . . . the use of models is a means whereby the high variety of the real world is reduced to a level of variety appropriate to the channel capacities of the human being.
>
> (Chadwick, 1971)

This involved more than a knowledge of computer applications – novel as that seemed to the average planner of the 1960s. It meant also a fundamentally different concept of planning. Instead of the old master-plan or blueprint approach, which assumed that the objectives were fixed from the start, the new concept was of planning as a process, "whereby programmes are adapted during their implementation as and when incoming information requires such changes". And this planning process was independent of the thing that was planned; as Melvin Webber put it, it was "a special way of deciding and acting" which involved a constantly recycled series of logical steps: goal-setting, forecasting of change in the outside world, assessment of chains of consequences of alternative courses of action, appraisal of costs and benefits as a basis for action strategies, and continuous monitoring. This was the approach of the new British textbooks of systems planning, which started to emerge at the end of the 1960s, and which were particularly associated with a group of younger British graduates, many teaching or studying at the University of Manchester. It was also the approach of a whole generation of subregional studies, made for fast-growing metropolitan areas in Britain during that heroic period of growth and change, 1965–75: Leicester–Leicestershire, Nottinghamshire–Derbyshire, Coventry–Warwickshire–Solihull, South Hampshire. All were heavily suffused with the new approach and the new techniques; in several, the same key individuals – McLoughlin in Leicester, Batty in Notts–Derby – played a directing or a crucial consulting role.

But the revolution was less complete – at least, in its early stages – than its supporters liked to argue: many of these "systems" plans had a distinctly blueprint tint, in that they soon resulted in all-too-concrete proposals for fixed investments like freeway systems. Underlying this, furthermore, were some curious metaphysical assumptions, which the new systems planners shared with their blueprint elders: the planning system was seen as active, the city system as purely passive; the political system was regarded as benign and receptive to the planner's expert advice. In practice, the systems planner was involved in two very different kinds of activity: as a social scientist, he or she was passively observing and analysing reality; as a designer, the same planner was acting on reality to change it – an activity inherently less certain, and also inherently subject to objectives that could only be set through a complex, often messy, set of dealings between professionals, politicians and public.

The core of this problem was a logical paradox: despite the claims of the systems planners, the urban planning system was different from (say) a weapons system. In this latter kind of system, to which the "systems approach" had originally and successfully been applied, the controls were inside the system; but here, the urban-regional system was inside its own

computer output of traffic patterns; in 1975, the same person was talking late into the night with community groups, in the attempt to organize against hostile forces in the world outside.

It was a remarkable inversion of roles. For what was wholly or partly lost, in that decade, was the claim to any unique and useful expertise, such as was possessed by the doctor or the lawyer. True, the planner could still offer specialized knowledge on planning laws and procedures, or on how to achieve a particular design solution; though often, given the nature of the context and the changed character of planning education, s/he might not have enough of either of these skills to be particularly useful. And, some critics were beginning to argue, this was because planning had extended so thinly over so wide an area that it became almost meaningless; in the title of Aaron Wildavsky's celebrated paper, "If Planning is Everything, Maybe it's Nothing".

The fact was that planning, as an academic discipline, had theorized about its own role to such an extent that it was denying its own claim to legitimacy. Planning, Faludi pointed out in his text of 1973, could be merely *functional*, in that the goals and objectives are taken as given; or *normative*, in that they are themselves the object of rational choice. The problem was whether planning was really capable of doing that latter job. As a result, by the mid-1970s planning had reached the stage of a "paradigm crisis"; it had been theoretically useful to distinguish the planning process as something separate from what is planned, yet this had meant a neglect of substantive theory, pushing it to the periphery of the whole subject. Consequently, new theory is needed which attempts to bridge current planning strategies and the urban physical and social systems to which strategies are applied.

The Marxist ascendancy

That became ever clearer in the following decade, when the logical positivists retreated from the intellectual field of battle and the Marxists took possession. As the whole world knows, the 1970s saw a remarkable resurgence – indeed a veritable explosion – of Marxist studies. This could not fail to affect the closely related worlds of urban geography, sociology, economics and planning. True, like the early neo-classical economists, Marx had been remarkably uninterested in questions of spatial location – even

though Engels had made illuminating comments on the spatial distribution of classes in mid-Victorian Manchester. The disciples now reverently sought to extract from the holy texts, drop by drop, a distillation that could be used to brew the missing theoretical potion. At last, by the mid-1970s, it was ready; then came a flood of new work. It originated in various places and in various disciplines: in England and the United States the geographers David Harvey and Doreen Massey helped to explain urban growth and change in terms of the circulation of capital; in Paris, Manuel Castells and Henri Lefebvre developed sociologically based theories. In the endless debates that followed among the Marxists themselves, a critical question concerned the role of the state. In France, Lokjine and others argued that it was mainly concerned, through such devices as macroeconomic planning and related infrastructure investment, directly to underpin and aid the direct productive investments of private capital. Castells, in contrast, argued that its main function had been to provide collective consumption – as in public housing, or schools, or transportation – to help guarantee the reproduction of the labour force and to dampen class conflict, essential for the maintenance of the system. Clearly, planning might play a very large role in both these state functions; hence, by the mid-1970s French Marxist urbanists were engaging in major studies of this role in the industrialization of such major industrial areas as Dieppe.

At the same time, a specifically Marxian view of planning emerged in the English-speaking world. To describe it adequately would require a course in Marxist theory. But, in inadequate summary, it states that the structure of the capitalist city itself, including its land-use and activity patterns, is the result of capital in pursuit of profit. Because capitalism is doomed to recurrent crises, which deepen in the current stage of late capitalism, capital calls upon the state, as its agent, to assist it by remedying disorganization in commodity production, and by aiding the reproduction of the labour force. It thus tries to achieve certain necessary objectives: to facilitate continued capital accumulation, by ensuring rational allocation of resources; by assisting the reproduction of the labour force through provision of social services, thus maintaining a delicate balance between labour and capital and preventing social disintegration; and by guaranteeing and legitimating capitalist social and property relations. As Dear and Scott put it: "In summary, planning is an historically-specific and socially-necessary response to the self-disorganizing tendencies of *privatized capitalist* social and property relations as these appear in urban space." In particular, it seeks to guarantee collective provision of necessary infrastructure and certain basic urban services, and to reduce negative externalities whereby certain activities of capital cause losses to other parts of the system.

But, since capitalism also wishes to circumscribe state planning as far as possible, there is an inbuilt contradiction: planning, because of this inherent inadequacy, always solves one problem only by creating another. Thus, say the Marxists, nineteenth-century clearances in Paris created a working-class housing problem; American zoning limited the powers of industrialists to locate at the most profitable locations. And planning can never do more than modify some parameters of the land development process; it cannot change its intrinsic logic, and so cannot remove the contradiction between private accumulation and collective action. Further, the capitalist class is by no means homogenous; different fractions of capital may have divergent, even contradictory interests, and complex alliances may be formed in consequence; thus, latter-day Marxist explanations come close to being pluralist, albeit with a strong structural element. But in the process, "the more that the State intervenes in the urban system, the greater is the likelihood that different social groups and fractions will contest the legitimacy of its decisions. *Urban life as a whole becomes progressively invaded by political controversies and dilemmas.*"

Because traditional non-Marxian planning theory has ignored this essential basis of planning, so Marxian commentators argue, it is by definition vacuous: it seeks to define what planning ideally ought to be, devoid of all context; its function has been to depoliticize planning as an activity, and thus to legitimate it. It seeks to achieve this by representing itself as the force which produces the various facets of real-world planning. But in fact, its various claims – to develop abstract concepts that rationally represent real-world processes, to legitimate its own activity, to explain material processes as the outcome of ideas, to present planning goals as derived from generally shared values, and to abstract planning activity in terms of metaphors drawn from other fields like engineering – all these are both very large and quite unjustified. The reality, Marxists argue, is precisely the opposite: viewed objectively, planning theory is nothing other

than a creation of the social forces that bring planning into existence.

It makes up a disturbing body of coherent criticism: yes, of course, planning cannot simply be an independent self-legitimating activity, as scientific inquiry may claim to be; yes, of course, it is a phenomenon that – like all phenomena – represents the circumstances of its time. As Scott and Roweis put it:

... there is a definite mismatch between the world of current planning theory, on the one hand, and the real world of practical planning intervention on the other hand. The one is the quintessence of order and reason in relation to the other which is full of disorder and unreason. Conventional theorists then set about resolving this mismatch between theory and reality by introducing the notion that planning theory is in any case not so much an attempt to explain the world as it is but as it ought to be. Planning theory then sets itself the task of rationalizing irrationalities, and seeks to materialize itself in social and historical reality (like Hegel's World Spirit) by bringing to bear upon the world a set of abstract, independent, and transcendent norms.

(Scott and Roweis, 1977)

It was powerful criticism. But it left in turn a glaringly open question, both for the unfortunate planner – whose legitimacy is now totally torn from him, like the epaulette from the shoulder of a disgraced officer – and, equally, for the Marxist critic: what, then, is planning theory about? Has it any normative or prescriptive content whatsoever? The answer, logically, would appear to be no. One of the critics, Philip Cooke, is uncompromising:

The main criticism that tends to have been made, justifiably, of planning is that it has remained stubbornly normative ... in this book it will be argued that [planning theorists] should identify mechanisms which cause changes in the nature of planning to be brought about, rather than assuming such changes to be either the creative idealizations of individual minds, or mere regularities in observable events.

(Cooke, 1983)

This is at least consistent: planning theory should avoid all prescription; it should stand right outside the planning process, and seek to analyse the subject

– including traditional theory – for what it is, the reflection of historical forces. Scott and Roweis, a decade earlier, seem to be saying exactly the same thing: planning theory cannot be normative, it cannot assume "transcendent operational norms". But then, they stand their logic on its own head, saying that "a viable theory of urban planning should not only tell us what planning is, but also what we can, and must, do as progressive planners".

This, of course, is sheer rhetoric. But it nicely displays the agony of the dilemma. Either theory is about unravelling the historical logic of capitalism, or it is about prescription for action. Since the planner-theorist – however sophisticated – could never hope to divert the course of capitalist evolution by more than a millimetre or a millisecond, the logic would seem to demand that s/he sticks firmly to the first and abjures the second. In other words, the Marxian logic is strangely quietist; it suggests that the planner retreats from planning altogether into the academic ivory tower.

Some were acutely conscious of the dilemma. John Forester tried to resolve it by basing a whole theory of planning action on the work of Jürgen Habermas. Habermas, perhaps the leading German social theorist of the post-World War Two era, had argued that latter-day capitalism justified its own legitimacy by spinning around itself a complex set of distortions in communication, designed to obscure and prevent any rational understanding of its own workings. Thus, he argued, individuals became powerless to understand how and why they act, and so were excluded from all power to influence their own lives,

> as they are harangued, pacified, mislead [sic], and ultimately persuaded that inequality, poverty, and ill-health are either problems for which the victim is responsible or problems so "political" and "complex" that they can have nothing to say about them. Habermas argues that democratic politics or planning requires the consent that grows from processes of collective criticism, not from silence or a party line.
>
> (Forester, 1980)

But, Forester argues, Habermas's own proposals for communicative action provide a way for planners to improve their own practice:

> By recognizing planning practice as norma-tively role-structured communication action which

distorts, covers up, or reveals to the public the prospects and possibilities they face, a critical theory of planning aids us practically and ethically as well. This is the contribution of critical theory to planning: pragmatics with vision – to reveal true alternatives, to correct false expectations, to counter cynicism, to foster inquiry, to spread political responsibility, engagement, and action. Critical planning practice, technically skilled and politically sensitive, is an organizing and democratizing practice.

> (Forester, 1980)

Fine. The problem is that – stripped of its Germanic philosophical basis, which is necessarily a huge over-simplification of a very dense analysis – the practical prescription all comes out as good old-fashioned democratic common sense, no more and no less than Davidoff's advocacy planning of fifteen years before: cultivate community networks, listen carefully to the people, involve the less-organized groups, educate the citizens in how to join in, supply information and make sure people know how to get it, develop skills in working with groups in conflict situations, emphasize the need to participate, compensate for external pressures. True, if in all this planners can sense that they have penetrated the mask of capitalism, that may help them to help others to act to change their environment and their lives; and, given the clear philosophical impasse of the late 1970s, such a massive metaphysical underpinning may be necessary.

The world outside the tower: practice retreats from theory

Meanwhile, if the theorists were retreating in one direction, the practitioners were certainly recipro-cating. Whether baffled or bored by the increasingly scholastic character of the academic debate, they lapsed into an increasingly untheoretical, unreflective, pragmatic, even visceral style of planning. That was not entirely new: planning had come under a cloud before, as during the 1950s, and had soon reappeared in a clear blue sky. What was new, strange, and seem-ingly unique about the 1980s was the divorce between the Marxist theoreticians of academe – essentially academic spectators, taking grandstand seats at what they saw as one of capitalism's last games – and the anti-theoretical, anti-strategic, anti-intellectual style of the players on the field down below. The 1950s were never like that; then, the academics were the coaches, down there with the team.

The picture is of course exaggerated. Many academics did still try to teach real-life planning through simulation of real-world problems. The Royal Town Planning Institute enjoined them to become ever more practice-minded. The practitioners had not all shut their eyes and ears to what comes out of the academy; some even returned there for refresher courses. And if all this was true in Britain, it was even more so of America, where the divorce had never been so evident. Yet the picture does describe a clear and unmistakable trend; and it was likely to be more than a cyclical one.

The reason is simple: as professional education of any kind becomes more fully absorbed by the academy, as its teachers become more thoroughly socialized within it, as careers are seen to depend on academic peer judgements, then its norms and values – theoretical, intellectual, detached – will become ever more pervasive; and the gap between teaching and practice will progressively widen. One key illustration: of the huge output of books and papers from the planning schools in the 1980s, there were many – often, those most highly regarded within the academic community – that were simply irrelevant, even completely incomprehensible, to the average practitioner.

Perhaps, it might be argued, that was the practitioner's fault; perhaps too we need fundamental science, with no apparent payoff, if we are later to enjoy its technological applications. The difficulty with that argument was to find convincing evidence that – not merely here, but in the social sciences generally – such payoff eventually comes. Hence the low esteem into which the social sciences had everywhere fallen, not least in Britain and the United States: hence too the diminished level of support for them, which – at any rate in Britain – had directly redounded on the planning schools. The relationship between planning and the academy had gone sour, and that is the major unresolved question that must now be addressed.

REFERENCES

Abercrombie, P. and Jackson, H. (1948) *West Midlands Plan.* Interim Confidential Edition. 5 vols. London: Ministry of Town and Country Planning.

Batty, M. (1979) "On Planning Processes", in: B. Goodall and A. Kirby (eds) *Resources and Planning.* Oxford: Pergamon.

Chadwick, G. (1971) *A Systems View of Planning: Towards a Theory of the Urban and Regional Planning Process.* Oxford: Pergamon.

Cooke, P.N. (1983) *Theories of Planning and Spatial Development.* London: Hutchinson.

Forester, J. (1980) "Critical Theory and Planning Practice", *Journal of the American Planning Association,* 46, 275–86.

Kent, T.J. (1964) *The Urban General Plan.* San Francisco: Chandler.

Scott, A.J. and Roweis, S.T. (1977) "Urban Planning in Theory and Practice: An Appraisal", *Environment and Planning,* 9, 1097–1119.

Webber, M.M. (1968–9) "Planning in an Environment of Change", *Town Planning Review,* 39, 179–95, 277–95.

"Twentieth Century Land Use Planning: A Stalwart Family Tree"

Journal of the American Planning Association (1995)

Edward J. Kaiser and David R. Godschalk

Editors' Introduction

Much of what city planning departments do is *physical planning* related to land use, transportation, capital improvements, and infrastructure. In the following selection, Edward J. Kaiser and David R. Godschalk describe how concepts of what land use plans should be like have evolved over time and what land use planning practice in the United States is like today.

Kaiser and Godschalk liken the development of land use planning to a tree. From disparate roots in planning theory and practice, a sturdy trunk based on the 1950s vision of a general plan for cities' long-term physical development has grown over time, with periodic branches, to a rich foliage of "hybrid" contemporary plans which typically blend aspects of design, policy, and management.

Kaiser and Godschalk begin their account of twentieth-century land use planning by describing the elitist, architecturally based, and often unrealistic plans that Peter Hall noted were common prior to World War II (p. 366). These have evolved into contemporary hybrid urban land use plans developed through participatory processes and which incorporate elements of policy and management that make them far more realistic than plans developed during planning's comfortable, but ineffective "golden age."

Kaiser and Godschalk trace the roots of their tree to the middle of the twentieth century. By 1950 there was a consensus that city plans should be focused on long-term physical development – what is called a "master plan" or "general plan."

In the 1950s, city general plans tended to be elitist, inspirational, long-range visions developed with little attention to implementation. In contrast, Kaiser and Godschalk argue that today, urban land use plans have become frameworks for community consensus on future growth supported by fiscally grounded actions to manage change. Plans are becoming more sensitive to the green planning issues raised by Timothy Beatley (p. 411) and sustainable urban development as proposed by the Brundtland Commission (p. 337) and discussed by Stephen Wheeler (p. 499).

Most modern city plans still contain maps showing projected long-range urban form for the city's land uses, transportation systems, community facilities, and other infrastructure. But today's plans do not consist of architecturally complete renderings of an entire static new town or section of a city like many pre-war plans. In addition to or in place of maps, the city plan's vision may be expressed in words and visual images.

At the beginning of the twenty-first century it appears that the "sturdy tree" of urban land use planning will continue to grow. Kaiser and Godschalk anticipate that the next generation of development plans will mature and adapt without abandoning the urban physical plan heritage. Edward J. Kaiser and David R. Godschalk are professors of city and regional planning at the University of North Carolina. They are co-authors (with Philip Berke) of *Urban*

Land Use Planning, fifth edition (Chicago: University of Illinois Press, 2006), a leading American text on land use planning.

Eugenie Birch (ed.), *The Urban and Regional Planning Reader* (London and New York: Routledge, 2007) is an anthology of classic and contemporary writings on urban and regional planning. Other books on urban planning in the United States include , Jay Stein (ed.), *Classic Readings in Urban Planning* (New York: McGraw-Hill, 2004), J. Barry Cullingworth and Roger Caves, *Planning in the USA: Policies, Issues, and Process*, second edition (London: Routledge, 2003), John M. Levy, *Contemporary Urban Planning*, sixth edition (New York: Prentice-Hall, 2001), Barbara Becker, Eric Damian Kelley, and Frank So, *Community Planning: An Introduction to the Comprehensive Plan* (Washington, DC: Milldale Press, 2000), International City Management Association (ICMA), *The Practice of Local Government Planning* (Washington, DC: ICMA, 1999), and Alexander Garvin, *The American City: What Works, What Doesn't* (New York: McGraw-Hill, 1995).

Leading books on European urban planning include Peter Hall, *Urban and Regional Planning*, fourth edition (London and New York: Routledge, 2002), J. Barry Cullingworth and Vincent Nadin, *Town and Country Planning in the UK*, fourteenth edition (London and New York: Routledge, 2006), and Patsy Healey and Stephen Graham (eds), *Managing Cities* (London: Wiley, 1995).

How a city's land is used defines its character, its potential for development, the role it can play within a regional economy and how it impacts the natural environment.
(Seattle Planning Commission, 1993)

During the twentieth century, community physical development plans have evolved from elite, City Beautiful designs to participatory, broad-based strategies for managing urban change. A review of land use planning's intellectual and practice history shows the continuous incorporation of new ideas and techniques. The traditional mapped land use design has been enriched with innovations from policy plans, land classification plans, and development management plans. Thanks to this flexible adaptation, local governments can use contemporary land use planning to build consensus and support decisions on controversial issues about space, development, and infrastructure. If this evolution persists, local plans should continue to be mainstays of community development policy into the twenty-first century.

Unlike the more rigid, rule-oriented modern architecture, contemporary local planning does not appear destined for deconstruction by a postmodern revolution. Though critics of comprehensive physical planning have regularly predicted its demise, the evidence demonstrates that spatial planning is alive and well in hundreds of United States communities. A 1994 tabulation found 2,742 local comprehensive plans prepared under state growth management regulations in twelve states. (See table 1.) This figure of course

significantly understates the overall nationwide total, which would include all those plans prepared in the other thirty-eight states and in the noncoastal areas of California and North Carolina. It is safe to assume that most, if not all, of these plans contain a mapped land use element. Not only do such plans help decision-makers to manage urban growth and change, they also provide a platform for the formation of community consensus about land use issues, now among the most controversial items on local government agendas.

This article looks back at the history of land use planning and forward to its future. It shows how planning ideas, growing from turn-of-the-century roots, culminated in a midcentury consensus on a general concept – the traditional land use design plan. That consensus was stretched as planning branched out to deal with public participation, environmental protection, growth management, fiscal responsibility, and effective implementation under turbulent conditions. To meet these new challenges, new types of plans arose: verbal policy plans, land classification plans, and growth management plans. These in turn became integrated into today's hybrid comprehensive plans, broadening and strengthening the traditional approach.

Future land use planning will continue to evolve in certain foreseeable directions, as well as in ways unforeseen. Among the foreseeable developments are even more active participation by interest groups, calling for planners' skills at consensus building and managing conflict; increased use of computers and

| State | Number of comprehensive plans | | | | Source |
	Cities/towns	Counties	Regions	Total	
California (coastal)	97	7	0	104	Coastal Commission
Florida	377	49	0	426	Department of Community Affairs
Georgia	298	94	0	392	Department of Community Affairs
Maine	270	0	0	270	Dept. of Economic and Community Development
Maryland	1	1	0	2	Planning Office
New Jersey	567	0	0	567	Community Affairs Department
North Carolina (coastal)	70	20	0	90	Division of Coastal Management
Oregon	241	36	1	278	Department of Local Community Development
Rhode Island	39	0	1	40	Department of Planning and Development
Vermont	235	0	10	245	Department of Housing and Community Affairs
Virginia	211	94	0	305	Department of Housing and Community Affairs
Washington	23	9	9	23	Office of Growth Management
Total	**2429**	**301**	**12**	**2742**	

Table 1 Local comprehensive plans in growth-managing states and coastal areas as of 1994
Source: Compiled from telephone survey of state sources

electronic media, calling for planners' skills in information management and communication; and continuing concerns over issues of diversity, sustainability, and quality of life, calling for planners' ability to analyze and seek creative solutions to complex and interdependent problems.

THE LAND USE PLANNING FAMILY TREE

We liken the evolution of the physical development plan to a family tree. The early genealogy is represented as the roots of the tree (figure 1). The general plan, constituting consensus practice at midcentury, is represented by the main trunk. Since the 1970s this traditional "land use design plan" has been joined by several branches – the verbal policy plan, the land classification plan, and the development management plan. These branches connect to the trunk although springing from different planning disciplines, in a way reminiscent of the complex structure of a Ficus tree. The branches combine into the contemporary, hybrid

comprehensive plan integrating design, policy, classification, and management, represented by the foliage at the top of the tree.

As we discuss each of these parts of the family tree, we show how plans respond both to social climate changes and to "idea genes" from the literature. We also draw conclusions about the survival of the tree and the prospects for new branches in the future. The focus of the article is the plan prepared by a local government – a county, municipality, or urban region – for the long-term development and use of the land.

ROOTS OF THE FAMILY TREE: THE FIRST 50 YEARS

New World city plans certainly existed before this century. They included L'Enfant's plan for Washington, William Penn's plan for Philadelphia, and General Oglethorpe's plan for Savannah. These plans, however, were blueprints for undeveloped sites, commissioned

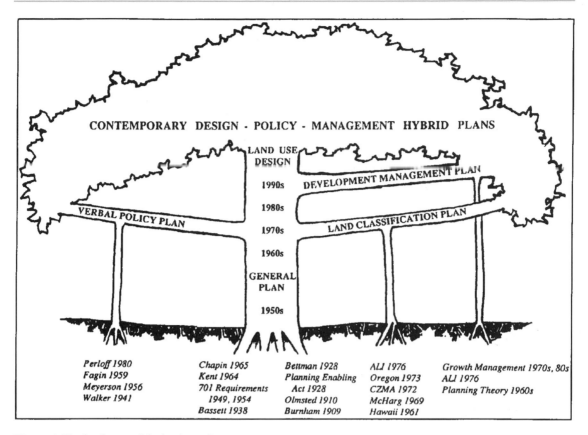

CONTEMPORARY DESIGN · POLICY · MANAGEMENT HYBRID PLANS

LAND USE DESIGN

DEVELOPMENT MANAGEMENT PLAN

1990s

1980s

VERBAL POLICY PLAN

LAND CLASSIFICATION PLAN

1970s

1960s

GENERAL PLAN

1950s

Perloff 1980	Chapin 1965	Bettman 1928	ALI 1976	Growth Management 1970s, 80s
Fagin 1959	Kent 1964	Planning Enabling	Oregon 1973	ALI 1976
Meyerson 1956	701 Requirements	Act 1928	CZMA 1972	Planning Theory 1960s
Walker 1941	1949, 1954	Olmsted 1910	McHarg 1969	
	Bassett 1938	Burnham 1909	Hawaii 1961	

Figure 1 The family tree of the land use plan

by unitary authorities with power to implement them unilaterally.

In this century, perhaps the most influential early city plan was Daniel Burnham's plan for Chicago, published by the Commercial Club of Chicago (a civic, not a government entity) in 1909. The archetypical plan-as-inspirational-vision, it focuses only on design of public spaces as a City Beautiful effort.

The City Beautiful approach was soon broadened to a more comprehensive view. At the 1911 National Conference on City Planning, Frederick Law Olmsted, Jr., son of the famous landscape architect and in his own right one of the fathers of planning, defined a city plan as encompassing all uses of land, private property, public sites, and transportation. Alfred Bettman, speaking at the 1928 National Conference of City Planning, envisioned the plan as a master design for the physical development of the city's territory, including "the general location and extent of new public improvements . . . and in the case of private develop-ments, the general distribution amongst various classes of land uses, such as residential, business, and industrial

uses . . . designed for . . . the future, twenty-five to fifty years" (Black 1968, 352–3). Together, Olmsted and Bettman anticipated the development of the midcentury land use plan.

Another early influence, the federal Standard City Planning Enabling Act of 1928, shaped enabling acts passed by many states. However, the Act left many planners and public officials confused about the difference between a master plan and a zoning ordinance, so that hundreds of communities adopted "zoning plans" without having created comprehensive plans as the basis for zoning (Black 1968, 353). Because the Act also did not make clear the importance of comprehensiveness or define the essential elements of physical development, no consensus about the essential content of the plan existed.

Ten years later, Edward Bassett's book, *The Master Plan* (1938), spelled out the plan's subject matter and format – supplementing the 1928 Act, and consistent with it. He argued that the plan should have seven elements, all relating to land areas (not buildings) and capable of being shown on a map: streets, parks, sites for public buildings, public reservations, routes for public utilities, pier-head and bulkhead lines (all public facilities), and zoning districts for private lands. Bassett's views were incorporated in many state enabling laws.

The physical plans of the first half of the century were drawn by and for independent commissions, reflecting the profession's roots in the Progressive Reform movement, with its distrust of politics. The 1928 Act reinforced that perspective by making the planning commission, not the legislative body, the principal client of the plan, and purposely isolating the commission from politics. Bassett's book reinforced the reliance on an independent commission. He conceived of the plan as a "plastic" map, kept within the purview of the planning commission, capable of quick and easy change. The commission, not the plan, was intended to be the adviser to the local legislative body and to city departments.

By the 1940s, both the separation of the planning function from city government and the plan's focus on physical development were being challenged. Robert Walker, in *The Planning Function in Local Government*, argued that the "scope of city planning is properly as broad as the scope of city government." The central planning agency might not necessarily do all the planning, but it would coordinate departmental planning in the light of general policy considerations – creating a comprehensive plan but one without a physical focus. That idea was not widely accepted. Walker also argued that the independent planning commission should be replaced by a department or bureau attached to the office of mayor or city manager. That argument did take hold, and by the 1960s planning in most communities was the responsibility of an agency within local government, though planning boards still advised elected officials on planning matters.

This evolution of ideas over 50 years resulted at midcentury in a consensus concept of a plan as focused on long-term physical development; this focus was a legacy of the physical design professions. Planning staff worked both for the local government executive officer and with an appointed citizen planning board, an arrangement that was a legacy of the Progressive insistence on the public interest as an antidote to governmental corruption. The plan addressed both public and private uses of the land, but did not deal in detail with implementation.

THE PLAN AFTER MIDCENTURY: NEW GROWTH INFLUENCES

Local development planning grew rapidly in the 1950s, for several reasons. First, governments had to contend with the postwar surge of population and urban growth, as well as a need for the capital investment in infrastructure and community facilities that had been postponed during the depression and war years. Second, municipal legislators and managers became more interested in planning as it shifted from being the responsibility of an independent commission to being a function within local government. Third, and very important, Section 701 of the Housing Act of 1954 required local governments to adopt a long-range general plan in order to qualify for federal grants for urban renewal, housing, and other programs, and it also made money available for such comprehensive planning. The 701 program's double-barreled combination of requirements and financial support led to more urban planning in the United States in the latter half of the 1950s than at any previous time in history.

At the same time, the plan concept was pruned and shaped by two planning educators. T. J. Kent, Jr. was a professor at the University of California at Berkeley, a planning commissioner, and a city councilman in the 1950s. His book, *The Urban General Plan* (1964), clarified the policy role of the plan. F. Stuart Chapin, Jr. was a TVA planner and planning director in Greensboro, NC in the 1940s, before joining the planning faculty at the University of North Carolina at Chapel Hill in 1949. His contribution was to codify the methodology of land use planning in the various editions of his book, *Urban Land Use Planning* (1957, 1965).

What should the plan look like? What should it be about? What is its purpose (besides the cynical purpose of qualifying for federal grants)? The 701 program, Kent, and Chapin all offered answers.

The "701" program comprehensive plan guidelines

In order to qualify for federal urban renewal aid and, later, for other grants – a local government had to prepare a general plan that consisted of plans for physical development, programs for redevelopment, and administrative and regulatory measures for controlling and guiding development. The 701 program specified what the content of a comprehensive development plan should include:

- A land use plan, indicating the locations and amounts of land to be used for residential, commercial, industrial, transportation, and public purposes
- A plan for circulation facilities
- A plan for public utilities
- A plan for community facilities

T.J. Kent's urban general plan

Kent's view of the plan's focus was similar to that of the 701 guidelines: long-range physical development in terms of land use, circulation, and community facilities. In addition, the plan might include sections on civic design and utilities, and special areas, such as historic preservation or redevelopment areas. It covered the entire geographical jurisdiction of the community, and was in that sense comprehensive. The plan was a vision of the future, but not a blueprint; a policy statement, but not a program of action; a formulation of goals, but not schedules, priorities, or cost estimates. It was to be inspirational, uninhibited by short-term practical considerations.

Kent (1964, 65–89) believed the plan should emphasize policy, serving the following functions:

- Policy determination – to provide a process by which a community would debate and decide on its policy
- Policy communication – to inform those concerned with development (officials, developers, citizens, the courts, and others) and educate them about future possibilities
- Policy effectuation – to serve as a general reference for officials deciding on specific projects
- Conveyance of advice – to furnish legislators with the counsel of their advisors in a coherent, unified form

The format of Kent's proposed plan included a unified, comprehensive, but general physical design for the future, covering the whole community and represented by maps. (See figure 2.) It also contained goals and policies (generalized guides to conduct, and the most important ingredients of the plan), as well as summaries of background conditions, trends, issues, problems, and assumptions. (See figure 3.) So that the plan would be suitable for public debate, it was to

be a complete, comprehensible document, containing factual data, assumptions, statements of issues, and goals, rather than merely conclusions and recommendations. The plan belonged to the legislative body and was intended to be consulted in decision-making during council meetings.

Kent recommended overall goals for the plan:

- Improve the physical environment of the community to make it more functional, beautiful, decent, healthful, interesting, and efficient
- Promote the overall public interest, rather than the interests of individuals or special groups within the community
- Effect political and technical coordination in community development
- Inject long-range considerations into the determination of short-range actions
- Bring professional and technical knowledge to bear on the making of political decisions about the physical development of the community

F. Stuart Chapin, Jr.'s urban land use plan

Chapin's ideas, though focusing more narrowly on the land use plan, were consistent with Kent's in both the 1957 and 1965 editions of *Urban Land Use Planning*, a widely used text and reference work for planners. Chapin's concept of the plan was of a generalized, but scaled, design for the future use of land, covering private land uses and public facilities, including the thoroughfare network.

Chapin conceived of the land use plan as the first step in preparing a general or comprehensive plan. Upon its completion, the land use plan served as a temporary general guide for decisions, until the comprehensive plan was developed. Later, the land use plan would become a cornerstone in the comprehensive plan, which also included plans for transportation, utilities, community facilities, and renewal, only the general rudiments of which are suggested in the land use plan. Purposes of the plan were to guide government decisions on public facilities, zoning, subdivision control, and urban renewal, and to inform private developers about the proposed future pattern of urban development.

The format of Chapin's land use plan included a statement of objectives, a description of existing conditions and future needs for space and services, and finally the mapped proposal for the future development

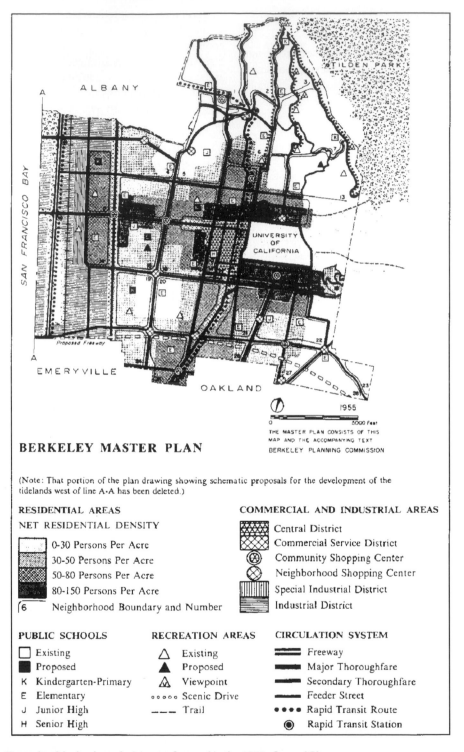

Figure 2 Example of the land use design map featured in the 1950s General Plan
Source: Kent 1991, 111

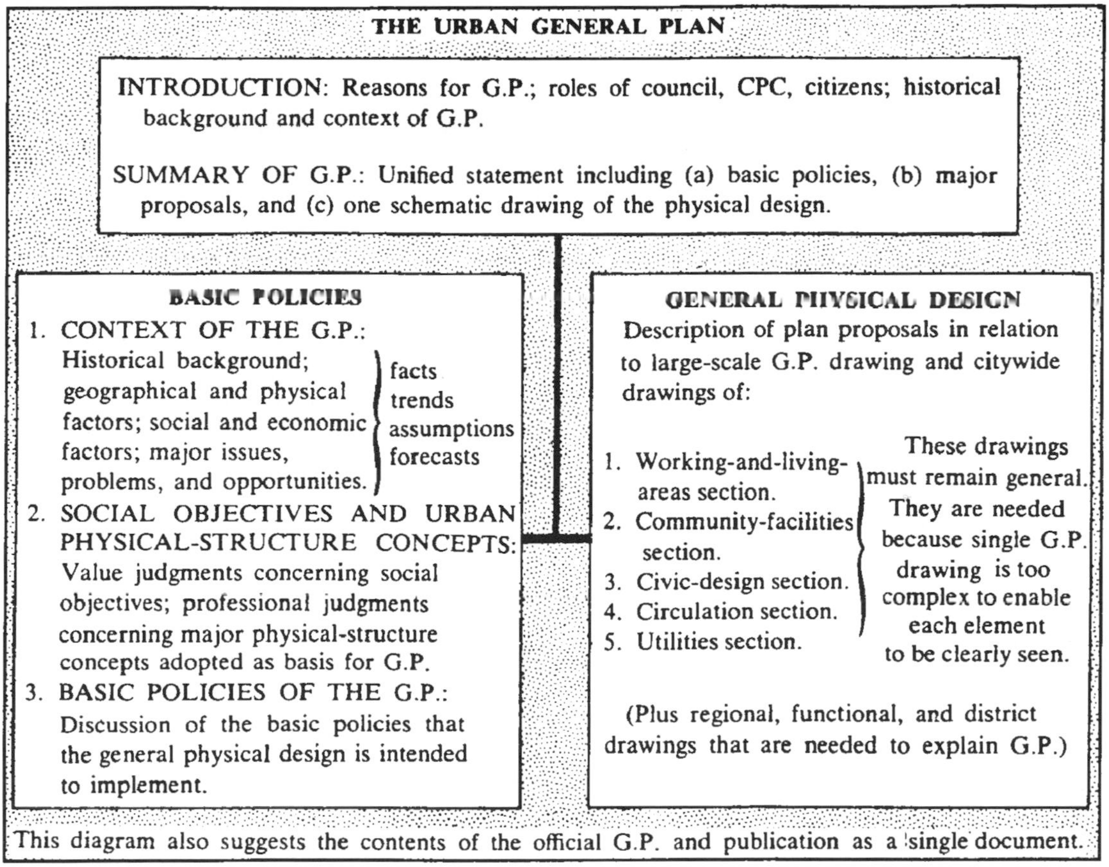

```
┌─────────────────────────────────────────────────────────────────────┐
│                      THE URBAN GENERAL PLAN                           │
│  ┌───────────────────────────────────────────────────────────────┐  │
│  │ INTRODUCTION: Reasons for G.P.; roles of council, CPC,        │  │
│  │ citizens; historical background and context of G.P.           │  │
│  │                                                                │  │
│  │ SUMMARY OF G.P.: Unified statement including (a) basic        │  │
│  │ policies, (b) major proposals, and (c) one schematic drawing  │  │
│  │ of the physical design.                                       │  │
│  └───────────────────────────────────────────────────────────────┘  │
└─────────────────────────────────────────────────────────────────────┘
```

BASIC POLICIES

1. CONTEXT OF THE G.P.:
   Historical background; geographical and physical factors; social and economic factors; major issues, problems, and opportunities. } facts, trends, assumptions, forecasts

2. SOCIAL OBJECTIVES AND URBAN PHYSICAL-STRUCTURE CONCEPTS:
   Value judgments concerning social objectives; professional judgments concerning major physical-structure concepts adopted as basis for G.P.

3. BASIC POLICIES OF THE G.P.:
   Discussion of the basic policies that the general physical design is intended to implement.

GENERAL PHYSICAL DESIGN
Description of plan proposals in relation to large-scale G.P. drawing and citywide drawings of:

1. Working-and-living-areas section.
2. Community-facilities section.
3. Civic-design section.
4. Circulation section.
5. Utilities section.

} These drawings must remain general. They are needed because single G.P. drawing is too complex to enable each element to be clearly seen.

(Plus regional, functional, and district drawings that are needed to explain G.P.)

This diagram also suggests the contents of the official G.P. and publication as a single document.

Figure 3 Components of the 1950s–1960s General Plan
Source: Kent 1964, 93

of the community, together with a program for implementing the plan (customarily including zoning, subdivision control, a housing code, a public works expenditure program, an urban renewal program, and other regulations and development measures).

The typical general plan of the 1950s and 1960s

Influenced by the 701 program, Kent's policy vision, and Chapin's methods, the plans of the 1950s and 1960s were based on a clear and straightforward concept: The plans' purposes were to determine, communicate, and effectuate comprehensive policy for the private and public physical development and redevelopment of the city. The subject matter was long-range physical development, including private uses of the land, circulation, and community facilities.

The standard format included a summary of existing and emerging conditions and needs; general goals; and a long-range urban form in map format, accompanied by consistent development policies. The coverage was comprehensive, in the sense of addressing both public and private development and covering the entire planning jurisdiction, but quite general. The tone was typically neither as "inspirational" as the Burnham plan for Chicago, nor as action-oriented as today's plans. Such was the well-defined trunk of the family tree in the 1950s and 1960s, in which today's contemporary plans have much of their origin.

CONTEMPORARY PLANS: INCORPORATING NEW BRANCHES

Planning concepts and practice have continued to evolve since midcentury, maturing in the process. By the 1970s, a number of new ideas had taken root. Referring back to the family tree in figure 1, we can see a trunk and several distinct branches:

- *The land use design*, a detailed mapping of future land use arrangements, is the most direct descendant of the 1950s plan. It still constitutes the trunk of the tree. However, today's version is more likely to be accompanied by action strategies, also mapped, and to include extensive policies.
- *The land classification plan*, a more general map of growth policy areas rather than a detailed land use pattern, is now also common, particularly for counties, metropolitan areas, and regions that want to encourage urban growth in designated development areas and to discourage it in conservation or rural areas. The roots of the land classification plan include McHarg's *Design With Nature* (1969), the 1976 American Law Institute (ALI) Model Land Development Code, the 1972 Coastal Zone Management Act, and the 1973 Oregon Land Use Law.
- *The verbal policy plan* de-emphasizes mapped policy or end-state visions and focuses on verbal action policy statements, usually quite detailed; sometimes called a strategic plan, it is rooted in Meyerson's middle-range bridge to comprehensive planning, Fagin's policies plan, and Perloff's strategies and policies general plan.
- *The development management plan* lays out a specific program of actions to guide development, such as a public investment program, a development code, and a program to extend infrastructure and services; and it assumes public sector initiative for influencing the location, type, and pace of growth. The roots of the development management plan are in the environmental movement, and the movements for state growth management and community growth control, as well as in ideas from Fagin (1959) and the ALI Code.

We looked for, but could not find, examples of land use plans that could be termed purely prototypical "strategic plans," in the sense of Bryson and Einsweiler. Hence, rather than identifying strategic planning as a separate branch on the family tree of the land use plan, we see the influence of strategic planning showing up across a range of contemporary plans. We tend to agree with the planners surveyed by Kaufman and Jacobs that strategic planning differs from good comprehensive planning more in emphasis (shorter range, more realistically targeted, more market-oriented) than in kind.

The land use design plan

The land use design plan is the most traditional of the four prototypes of contemporary plans and is the most direct descendent of the Kent–Chapin–701 plans of the 1950s and 1960s. It proposes a long-range future urban form as a pattern of retail, office, industrial, residential, and open spaces, and public land uses and a circulation system. Today's version, however, incorporates environmental processes, and sometimes agriculture and forestry, under the "open space" category of land use. Its land uses often include a "mixed use" category, honoring the neotraditional principle of closer mingling of residential, employment, and shopping areas. In addition, it may include a development strategy map, which is designed to bring about the future urban form and to link strategy to the community's financial capacity to provide infrastructure and services. The plans and strategies are often organized around strategic themes or around issues about growth, environment, economic development, transportation, or neighborhood/community scale change.

Like the other types of plans in vogue today, the land use design plan reflects recent societal issues, particularly the environmental crisis, the infrastructure crisis, and stresses on local government finance. Contemporary planners no longer view environmental factors as development constraints, but as valuable resources and processes to be conserved. They also may question assumptions about the desirability and inevitability of urban population and economic growth, particularly as such assumptions stimulate demand for expensive new roads, sewers, and schools. While at midcentury plans unquestioningly accommodated growth, today's plans cast the amount, pace, location, and costs of growth as policy choices to be determined in the planning process.

The 1990 Howard County (Maryland) General Plan, winner of an American Planning Association (APA) award in 1991 for outstanding comprehensive planning, exemplifies contemporary land use design. (See figure 4.) While clearly a direct descendent of the traditional general plan, the Howard County plan adds new types of goals, policies, and planning techniques. To enhance communication and public understanding, it is organized strategically around six themes/chapters

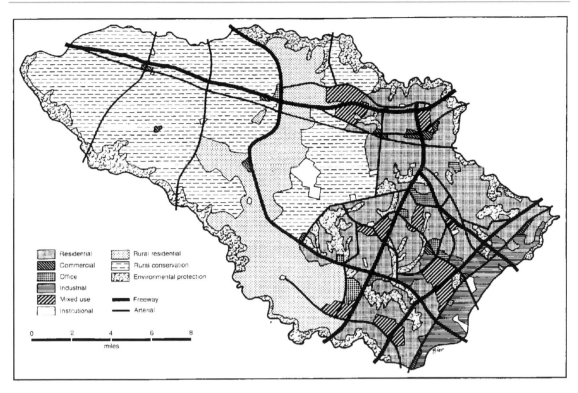

Figure 4 Howard County, Maryland, General Plan, Land Use 2010
Source: Adapted from Howard County 1990

(responsible regionalism, preservation of the rural area, balanced growth, working with nature, community enhancement, and phased growth), instead of the customary plan elements. Along with the traditional land use design, the plan includes a "policy map" (strategy map) for each theme and an overall policies map for the years 2000 and 2010. A planned service area boundary is used to contain urban growth within the eastern urbanized part of the county, home to the well-known Columbia New Town. The plan lays out specific next steps to be implemented over the next two years, and defines yardsticks for measuring success. An extensive public participation process for formulating the plan involved a 32-member General Plan Task Force, public opinion polling to discover citizen concerns, circulation of preplan issue papers on development impacts, and consideration of six alternative development scenarios.

The land classification plan

Land classification, or development priorities mapping, is a proactive effort by government to specify where

and under what conditions growth will occur. Often, it also regulates the pace or timing of growth. Land classification addresses environmental protection by designating "nondevelopment" areas in especially vulnerable locations. Like the land use design, the land classification plan is spatially specific and map-oriented. However, it is less specific about the pattern of land uses within areas specified for development, which results in a kind of silhouette of urban form. On the other hand, land classification is more specific about development strategy, including timing. Counties, metropolitan areas, and regional planning agencies are more likely than cities to use a land classification plan.

The land classification plan identifies areas where development will be encouraged (called urban, transition, or development areas) and areas where development will be discouraged (open space, rural, conservation, or critical environmental areas). For each designated area, policies about the type, timing, and density of allowable development, extension of infrastructure, and development incentives or constraints apply. The planning principle is to concentrate financial resources, utilities, and services within a limited, prespecified area suitable for development, and to relieve pressure on nondevelopment areas by withholding facilities that accommodate growth.

Ian McHarg's (1969) approach to land planning is an early example of the land classification concept. He divides planning regions into three categories: natural use, production, and urban. Natural use areas, those with valuable ecological functions, have the highest priority. Production areas, which include agriculture, forestry, and fishing uses, are next in priority. Urban areas have the lowest priority and are designated after allocating the land suitable to the two higher-priority uses. McHarg's approach in particular, and land classification generally, also reflect the emerging environmental consciousness of the 1960s and 1970s.

As early as 1961, Hawaii had incorporated the land classification approach into its state growth management system. The development framework plan of the Metropolitan Council of the Twin Cities Area defined "planning tiers," each intended for a different type and intensity of development. The concepts of the "urban service area," first used in 1958 in Lexington, Kentucky, and the "urban growth boundary," used throughout Oregon under its 1973 statewide planning act, classify land according to growth management policy. Typically, the size of an urban growth area is based on the amount of land necessary to accommodate development over a period of ten or twenty years.

Vision 2005: A Comprehensive Plan for Forsyth County, North Carolina exemplifies the contemporary approach to land classification plans. The plan, which won honorable mention from APA in 1989, employs a six category system of districts, plus a category for activity centers. It identifies both short- and long-range growth areas (4A and 4B in figure 5). Policies applicable to each district are detailed in the plan.

The verbal policy plan: shedding the maps

The verbal policy plan focuses on written statements of goals and policy, without mapping specific land use patterns or implementation strategy. Sometimes called a policy framework plan, a verbal policy plan is more easily prepared and flexible than other types of plans, particularly for incorporating nonphysical develop-

ment policy. Some claim that such a plan helps the planner to avoid relying too heavily on maps, which are difficult to keep up to date with the community's changes in policy. The verbal policy plan also avoids falsely representing general policy as applying to specific parcels of property. The skeptics, however, claim that verbal statements in the absence of maps provide too little spatial specificity to guide implementation decisions.

The verbal policy plan may be used at any level of government, but is especially common at the state level, whose scale is unsuited to land use maps. The plan usually contains goals, facts and projections, and general policies corresponding to its purposes – to understand current and emerging conditions and issues, to identify goals to be pursued and issues to be addressed, and to formulate general principles of action. Sometimes communities do a verbal policy plan as an interim plan or a first step in the planning process. Thus, verbal policies are included in most land use design plans, land classification plans, and development management plans.

The Calvert County, MD, Comprehensive Plan (Calvert County 1983), winner of a 1985 APA award, exemplifies the verbal policy plan. Its policies are concise, easy to grasp, and grouped in sections corresponding to the six divisions of county government responsible for implementation. It remains a policy plan, however, because it does not specify a program of specific actions for development management. Though the plan clearly addresses physical development and discusses specific spatial areas, it contains no land use map. (See figure 6 for an illustrative page from the Calvert County plan.)

The development management plan

The development management plan features a co-ordinated program of actions, supported by analyses and goals, for specific agencies of local government to undertake over a three-to-ten-year period. The program of actions usually specifies the content, geographic coverage, timing, assignment of responsibility, and coordination among the parts. Ideally, the plan includes most or all of the following components:

■ Description of existing and emerging development conditions, with particular attention to development processes, the political-institutional context,

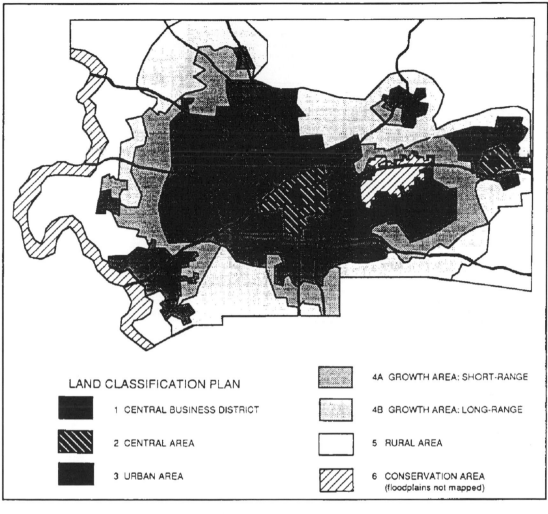

LAND CLASSIFICATION PLAN

◼ 1 CENTRAL BUSINESS DISTRICT

▨ 2 CENTRAL AREA

◼ 3 URBAN AREA

▦ 4A GROWTH AREA: SHORT-RANGE

▨ 4B GROWTH AREA: LONG-RANGE

☐ 5 RURAL AREA

▧ 6 CONSERVATION AREA
(floodplains not mapped)

Figure 5 Example of a land classification plan
Source: Adapted from Forsyth County City–County Planning Board 1988

and a critical review of the existing systems of development management

◼ Statement of goals and/or legislative intent, including management-oriented goals

◼ Program of actions – the heart of the plan – including:

1 Outline of a proposed development code, with: (a) procedures for reviewing development permits; (b) standards for the type of development, density, allowable impacts and/or performance standards; (c) site plan, site engineering, and construction practice requirements; (d) exactions and impact fee provisions and other incentives/disincentives; and (e) delineation of

districts where various development standards, procedures, exactions, fees, and incentives apply

2 Program for the expansion of urban infrastructure and community facilities and their service areas

3 Capital improvement program

4 Property acquisition program

5 Other components, depending on the community situation, for example, a preferential taxation program, an urban revitalization program for specific built-up neighborhoods, or a historic preservation program

◼ Official maps, indicating legislative intent, which may be incorporated into ordinances, with force

INDUSTRIAL DISTRICTS

Industrial Districts are intended to provide areas in the county which are suitable for the needs of industry. They should be located and designed to be compatible with the surrounding land uses, either due to existing natural features or through the application of standards.

RECOMMENDATIONS:

1. Identify general locations for potential industrial uses.
2. Permit retail sales as an accessory use in the Industrial District.

SINGLE-FAMILY RESIDENTIAL DISTRICTS

Single-Family Residential Districts are to be developed and promoted as neighborhoods free from any land usage which might adversely affect them.

RECOMMENDATIONS:

1. For new development, require buffering for controlling visual, noise, and activity impacts between residential and commercial uses.
2. Encourage single-family residential development to locate in the designated towns.
3. Allow duplexes, triplexes, and fourplexes as a conditional use in the "R-1" Residential Zone so long as the design is compatible with the single-family residential development.
4. Allow home occupations (professions and services, but not retail sales) by permitting the employment of one full-time equivalent individual not residing on the premises.

MULTIFAMILY RESIDENTIAL DISTRICTS

Multifamily Residential Districts provide for townhouses and multifamily apartment units. Areas designated in this category are those which are currently served or scheduled to be served by community or multi-use sewerage and water supply systems.

RECOMMENDATIONS:

1. Permit multifamily development in the Solomons, Prince Frederick, and Twin Beach Towns.
2. Require multifamily projects to provide adequate recreational facilities— equipment, structures, and play surfaces.
3. Evaluate the feasibility of increasing the dwelling unit density permitted in the multifamily Residential Zone (R-2)

Figure 6 An excerpt from a verbal policy plan
Source: Calvert County, Maryland 1983

of law – among them, goal-form maps (e.g., land classification plan or land use design); maps of zoning districts, overlay districts, and other special areas for which development types, densities, and other requirements vary; maps of urban services areas; maps showing scheduled capital improvements; or other maps related to development management standards and procedures

The development management plan is a distinct type, emphasizing a specific course of action, nor general policy. At its extreme the management plan actually incorporates implementation measures, so

that the plan becomes part of a regulative ordinance. Although the spatial specifications for regulations and other implementation measures are included, a land use map may not be.

One point of origin for development management plans is Henry Fagin's concept of the "policies plan," whose purpose was to coordinate the actions of line departments and provide a basis for evaluating their results, as well as to formulate, communicate, and implement policy (the traditional purpose). Such a plan's subject matter was as broad as the responsibilities of the local government, including but not limited to physical development. The format included a "state of the community" message, a physical plan, a financial plan, implementation measures, and detailed sections for each department of the government.

A more recent point of origin is *A Model Land Development Code* (American Law Institute 1976), intended to replace the 1928 Model Planning Enabling Act as a model for local planning and development management. The model plan consciously retains an emphasis on physical development (unlike Fagin's broader concept), but stresses a short-term program of action, rather than a long term, mapped goal form. The ALI model plan contains a statement of conditions and problems; objectives, policies, and standards; and a short-term (from one to five years) program of specific public actions. It may also include land acquisition requirements, displacement impacts, development regulations, program costs and fund sources, and environmental, social, and economic consequences. More than other plan types, the development management plan is a "course of action" initiated by government to control the location and timing of development.

The Sanibel, Florida, Comprehensive Land Use Plan (1981) exemplifies the development management plan. The plan outlines the standards and procedures of regulations (i.e., the means of implementation), as well as the analyses, goals, and statements of intent normally presented in a plan. Thus, when the local legislature adopts the plan, it also adopts an ordinance for its implementation. Plan and implementation are merged into one instrument, as can be seen in the content of its articles:

Article 1: Preamble: including purposes and objectives, assumptions, coordination with surrounding areas, and implementation

Article 2: Elements of the Plan: Safety, Human Support Systems, Protection of Natural Environmental, Economic and Scenic Resources, Intergovernmental Coordination, and Land Use

Article 3: Development Regulations: Definitions, Maps, Requirements, Permitted Uses, Subdivisions, Mobile Home and Recreation Vehicles, Flood and Storm Proofing, Site Preparation, and Environmental Performance Standards

Article 4: Administrative Regulations (i.e., procedures): Standards, Short Form Permits, Development Permits, Completion Permits, Amendments to the Plan, and Notice, Hearing and Decision Procedures on Amendments

Figure 7 shows the Sanibel plan's map of permitted uses, which is more like a zoning plan than a land use design plan, because it shows where regulations apply, and boundaries are exact.

THE CONTEMPORARY HYBRID PLAN: INTEGRATING DESIGN, POLICY, AND MANAGEMENT

The rationality of practice has integrated the useful parts of each of the separate prototypes reviewed here into contemporary hybrid plans that not only map and classify land use in both specific and general ways, but also propose policies and management measures. For example, Gresham, Oregon, combined land use design (specifying residential, commercial, and industrial areas, and community facilities and public lands) with an overlay of land classification districts (developed, developing, rural, and conservation), and also included standards and procedures for issuing development permits (i.e., a development code). Prepared with considerable participation by citizens and interest groups, such plans usually reflect animated political debates about the costs and benefits of land use alternatives.

The states that manage growth have created new land use governance systems whose influence has broadened the conceptual arsenals of local planners. DeGrove identifies the common elements of these systems:

■ Consistency – intergovernmentally and internally (i.e., between plan and regulations)
■ Concurrency – between infrastructure and new development
■ Compactness – of new growth, to limit urban sprawl affordability – of new housing
■ Economic development, or "managing to grow"
■ Sustainability – of natural systems

DeGrove attributes the changes in planning under growth management systems to new hard-nosed concerns for measurable implementation and realistic funding mechanisms. For example, Florida local governments must adopt detailed capital improvement programs as part of their comprehensive plans, and substantial state grants may be withheld if their plans do not meet consistency and concurrency requirements.

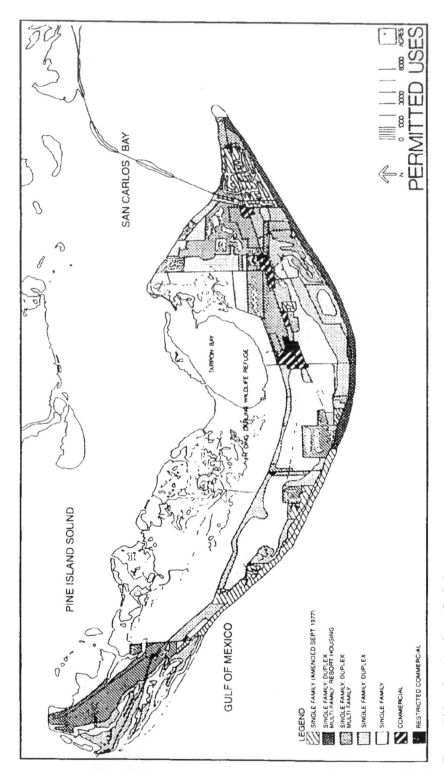

Figure 7 Map of permitted uses, Sanibel
Source: City of Sanibel 1981

Another important influence on contemporary plans is the renewed attention to community design. The neotraditional and transit-oriented design movements have inspired a number of proposals for mixed-use villages in land use plans. *Toward a Sustainable Seattle: A Plan for Managing Growth* (1994) exemplifies a city approach to the contemporary hybrid plan. Submitted as the Mayor's recommended comprehensive plan, it attempted to muster political support for its proposals. Three core values – social equity, environmental stewardship, and economic security and opportunity – underlie the plan's overall goal of sustainability. This goal is to be achieved by integrating plans for land use and transportation, healthy and affordable housing, and careful capital investment in a civic compact based on a shared vision. Citywide population and job growth targets, midway between growth completely by regional sprawl and growth completely by infill, are set forth within a 20-year time frame. The plan is designed to meet the requirements of the Washington State Growth Management Act.

The land use element designates urban center villages, hub urban villages, residential urban villages, neighborhood villages, and manufacturing/industrial centers, each with specific design guidelines (figure 8). The city's capacity for growth is identified, and then allocated according to the urban village strategy. Future development is directed to mixed-use neighborhoods, some of which are already established; existing single-family areas are protected. Growth is shaped to build community, promote pedestrian and transit use, protect natural amenities and existing residential and employment areas, and ensure diversity of people and activities. Detailed land use policies carry out the plan.

Loudoun County Choices and Changes: General Plan (1991), which won APA's 1994 award for comprehensive planning in small jurisdictions, exemplifies a county approach to the contemporary hybrid plan. Its goals are grouped into three categories:

1 Natural and cultural resources goals seek to protect fragile resources by limiting development or mitigating disturbances, while at the same time not unduly diminishing land values.
2 Growth management: goals seek to accommodate and manage the county's fair share of regional growth, guiding development into the urbanized eastern part of the county or existing western towns and their urban growth areas, and conserving agriculture and open space areas in the west. (See figure 9.)
3 Community design goals seek to concentrate growth in compact, urban nodes to create mixed-use communities with strong visual identities, human-scale street networks, and a range of housing and employment opportunities utilizing neo-traditional design concepts (illustrated in figure 9).

Three time horizons are addressed: the "ultimate" vision through 2040, the 20-year, long-range development pattern, and the five-year, short-range development pattern. The plan uses the concept of community character areas as an organizing framework for land use management. Policies are proposed for the overall county, as well as for the eastern urban growth areas, town urban growth areas, rural areas, and existing rural village areas. Implementation tools include capital facility and transportation proffers by developers, density transfers, community design guidelines, annexation guidelines, and an action schedule of next steps.

SUMMARY OF THE CONTEMPORARY SITUATION

Since midcentury, the nature of the plan has shifted from an elitist, inspirational, long-range vision that was based on fiscally innocent implementation advice, to a framework for community consensus on future growth that is supported by fiscally grounded actions to manage change. Subject matter has expanded to include the natural as well as the built environment. Format has shifted from simple policy statements and a single large-scale map of future land use, circulation, and community facilities, to a more complex combination of text, data, maps, and timetables. In a number of states plans are required by state law, and their content is specified by state agencies. Table 2 compares the general plan of the 1950s–1960s with the four contemporary prototype plans and the new 1990s hybrid design-policy management plan, which combines aspects of the prototype plans.

Today's prototype land use design continues to emphasize long-range urban form for land uses, community facilities, and transportation systems as shown by a map; but the design is also expressed in general policies. Land use design is still a common form of development plan, especially in municipalities.

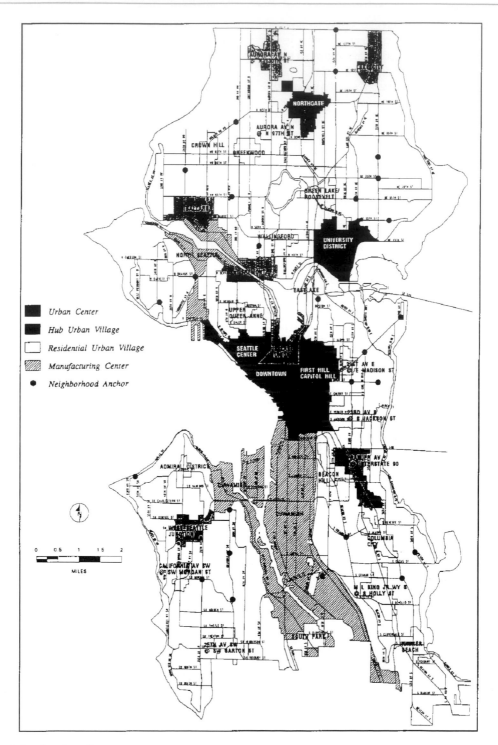

Figure 8 Seattle urban villages strategy
Source: Seattle Planning Department 1993

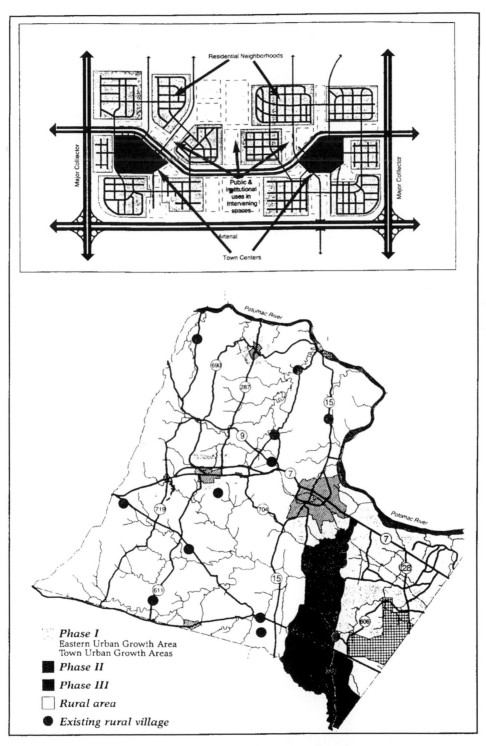

Figure 9 Neotraditional community schematic and generalized policy planning areas, Loudoun County, Virginia General Plan
Source: *Planning* 60, 3:10 (1994)

| | | Contemporary prototype plans | | | | |
Features of plans	1950s general plan	Land use design	Land classification plan	Verbal policy plan	Development management plan	1990s Hybrid design policy management plan
Land use maps	Detailed	Detailed	General	No	By growth areas	General *and* area specific
Nature of recommendations	General community goals	Land use policies & objectives	Growth locations & incentives	Variety of community policies	Specific management actions	Policy *and* actions
Time horizon	Long range	Long range	Long range	Intermediate range	Short range	Short *and* long range
Link to implementation	Very weak	Weak to moderate	Moderate	Moderate	Strong	Moderate to strong
Public participation	Pro-forma	Active	Moderate	Moderate	Active	Active
Capital improvements	Advisory	Recommended	Recommended	Recommended	Required	Recommended to required
Land use/transportation linkage	Moderate	Strong	Weak	Varies	Strong	Strong
Environmental protection	Weak	Moderate	Strong	Varies	Varies	Strong
Social policy linkage	Weak	Weak	Weak	Moderate to strong	Weak	Moderate

Table 2 Comparison of plan types

The land classification plan also still emphasizes mapping, but of development policy rather than policy about a pattern of urban land uses. Land classification is more specific about development management and environmental protection, but less specific about transportation, community facilities, and the internal arrangement of the future urban form. County and regional governments are more likely than are municipalities to use land classification plans.

The verbal policies plan eschews the spatial specificity of land use design and land classification plans and focuses less on physical development issues. It is more suited to regions and states, or may serve as an interim plan for a city or county while another type of plan is being prepared.

The development management plan represents the greatest shift from the traditional land use plan. It embodies a short-to-intermediate-range program of governmental actions for ongoing growth management rather than for long-range comprehensive planning.

In practice, these four types of plans are not mutually exclusive. Communities often combine aspects of them into a hybrid general plan that has policy sections covering environmental/social/economic/housing/infrastructure concerns, land classification maps defining spatial growth policy, land use design maps specifying locations of particular land uses, and development management programs laying out standards and procedures for guiding and paying for growth. Regardless of the type of plan used, the most progressive planning programs today regard the plan as but one part of a coordinated growth management program, rather than, as in the 1950s, the main planning product. Such a program incorporates a capital improvement program, land use controls, small area plans, functional plans, and other devices, as well as a general plan.

THE ENDURING LAND USE FAMILY TREE AND ITS FUTURE BRANCHES

For the first 50 years of this century, planning responded to concerns about progressive governmental reform, the City Beautiful, and the "City Efficient." Plans were advisory, specifying a future urban form, and were developed by and for an independent commission. By midcentury this type of plan, growing out of the design

tradition, had become widespread in local practice. During the 1950s and 1960s the 701 program, T. J. Kent, and F. Stuart Chapin, Jr. further articulated the plan's content and methodology. Over the last 30 years, environmental and infrastructure issues have pushed planning toward growth management. As citizen activists and interest groups have taken more of a role, land use politics have become more heated. Planning theorists, too, have questioned the mid-century approach to planning, and have proposed changes in focus, process, subject matter, and format, sometimes challenging even the core idea of rational planning. As a result, practice has changed, though not to a monolithic extent and without entirely abandoning the traditional concept of a plan. Instead, at least four distinct types of plans have evolved, all descending more or less from the midcentury model, but advocating very different concepts of what a plan should be. With a kind of self-correcting common sense, the plans of the 1990s have subsequently incorporated the useful parts of each of these prototypes to create today's hybrid design/policy/management plans.

To return to our analogy of the plan's family tree: Roots for the physical development plan became well established during the first half of this century. By 1950, a sturdy trunk concept had developed. Since then, new roots and branches have appeared – land classification plans, verbal policy plans, and development management plans. Meanwhile development of the main trunk of the tree – the land use design – has continued. Fortunately, the basic gene pool has been able to combine with new genes in order to survive as a more complex organism – the 1990s design-policy management hybrid plan. The present family tree of planning reflects both its heredity and its environment.

The next generation of physical development plans also should mature and adapt without abandoning their heritage. We expect that by the year 2000, plans will be more participatory, more electronically based, and concerned with increasingly complex issues. An increase in participation seems certain, bolstered by interest groups' as well as governments' use of expert systems and computer databases. A much broader consideration of alternative plans and scenarios, as well as a more flexible and responsive process of plan amendment, will become possible. These changes will call upon planners to use new skills of consensus building and conflict management, as more groups articulate their positions on planning matters, and government plans and interest group plans compete, each backed by experts.

With the advent of the "information highway," plans are more likely to be drafted, communicated, and debated through electronic networks and virtual reality images. The appearance of plans on CD ROM and cable networks will allow more popular access and input, and better understanding of plans' three-dimensional consequences. It will be more important than ever for planners to compile information accurately and ensure it is fairly communicated. They will need to compile, analyze, and manage complex databases, as well as to translate abstract data into understandable impacts and images.

Plans will continue to be affected by dominant issues of the times: aging infrastructure and limited public capital, central city decline and suburban growth, ethnic and racial diversity, economic and environmental sustainability, global competition and interdependence, and land use/transportation/air quality spillovers. Many of these are unresolved issues from the last 30 years, now grown more complex and interrelated. Some are addressed by new programs like the Intermodal Surface Transportation Efficiency Act (ISTEA) and HUD's Empowerment Zones and Enterprise Communities. To cope with others, planners must develop new concepts and create new techniques.

One of the most troubling new issues is an attempt by conservative politicians (see the Private Property Protection Act of 1995 passed by the US House of Representatives) and "wise use" groups to reverse the precedence of the public interest over individual private property rights. These groups challenge the use of federal, state, and local regulations to implement land use plans and protect environmental resources when the result is any reduction in the economic value of affected private property. Should their challenge succeed and become widely adopted in federal and state law, growth management plans based on regulations could become toothless. Serious thinking by land use lawyers and planners would be urgently needed to create workable new implementation techniques, setting in motion yet another planning evolution.

We are optimistic, however, about the future of land use planning. Like democracy, it is not a perfect institution but works better than its alternatives. Because land use planning has adapted effectively to this century's turbulence and become stronger in the process, we believe that the twenty-first century will see it continuing as a mainstay of strategies to manage community change.

"Planning in the Face of Conflict"

Journal of the American Planning Association (1987)

John Forester

Editors' Introduction

While good city planning needs to be inspired by a vision of the end results, and should be informed by theory, in democracies planning is never achieved without conflict. Planners, citizens, local elected officials, developers, and others invariably have different views on what a city *should* be like and *how* to build it. Passions run high at important city planning commission meetings.

Cornell planning professor John Forester got down in the trenches with practicing city planners and others involved in city development to study what the practice of city planning is really like in the face of conflict. The following selection summarizes what he learned about the process and his ideas on how planners can be effective in the face of conflict. It is valuable for the actual lessons Forester learned. But equally important Forester helps point the way out of the impasse Peter Hall describes in much academic planning theory today (p. 354). Unlike some ivory tower theorists, Forester listened carefully to practicing city planners and learned from them. His work synthesizes what he learned and develops theoretical concepts that are highly relevant to planning practice.

Forester found that city planners guide both developers and neighborhood residents through the complexities of the planning process using a variety of strategies. Planners have to be attentive to timing. Successful planners handle conflicts through both formal and informal channels. They must respond to complex and contradictory duties – tugged this way by local politicians, that way by legal mandates, and yet another way by citizen demands. Through all of this, successful city planners must be true to professional norms and hold fast to their own visions of high-quality city development. City planners who retreat to their planning offices to draw beautiful plans or create elegant computerized models of how they believe cities should develop without confronting the competing interest groups and conflicting ideas that all serious urban planning entails are doomed to fail. Planners need to wed professional expertise to practical skills for managing conflict among competing groups.

There are many lessons in Forester's work. People who want to be effective translating city plans into action need to expect opposition and should not be surprised or worn down by what often seems an endless and frustrating process. They need to be aware of their own power and also its limitations. They have to be sensitive to and understand the interests of the many different actors in the city development process.

Forester argues that city planners can self-consciously follow any of a number of strategies to keep projects on track and achieve success – as rule enforcers, negotiators and mediators, resource people, or shuttle diplomats.

Compare Forester's insights with David Harvey's description of social conflict around urban spatial issues (p. 225). Consider the kind of conflicts you would expect if you were trying to implement the different types of plans that Edward J. Kaiser and David R. Godschalk describe (p. 366).

John Forester is a professor of city and regional planning at Cornell University. He is interested in the ways planners shape participatory processes and manage public disputes in diverse settings, planning ethics, and mediation. He is a mediator for the Community Dispute Resolution Center of Tompkins County and has consulted on mediating urban planning disputes for the Consensus Building Institute.

Forester is the author of *The Deliberative Practitioner: Encouraging Participatory Planning Processes* (Cambridge: MIT Press, 1999) and *Critical Theory, Public Choice, and Planning Practice: Towards a Critical Pragmatism* (Buffalo: State University of New York Press, 1993). He teamed up with former Cleveland city planning director Norman Krumholz to write an account of Krumholz's experience implementing socially responsible equity planning in *Making Equity Planning Work* (Philadelphia: Temple University Press, 1992). He edited *Critical Theory and Public Life* (Cambridge: MIT Press, 1987), a book of essays about Jürgen Habermas's critical communications theory of society. He is the co-author, with Raphael Fischler and Deborah Shmueli, of *Israeli Planners and Designers: Profiles of Community Builders* (Albany: SUNY Press, 2001). He is working on a book, tentatively entitled *The Drama of Mediation: Exploring The Challenges of Participatory Planning Processes*. Another important article on how urban planners manage conflict is Forester's "Planning in the Face of Power," *Journal of the American Planning Association* 48(1) (1982).

Other books on what city planning is actually like include Allan Jacobs, *Making City Planning Work* (Chicago: American Society of Planning Officials, 1976) and Bruce W. McClendon (ed.), *Planners on Planning* (San Francisco: Jossey-Bass, 1996). Books on resolving urban planning conflicts include Patrick Field and Lawrence Susskind, *Dealing with an Angry Public: The Mutual Gains Approach To Resolving Disputes* (New York: Free Press, 1996) and Lawrence Susskind, Sarah McKearnan, and Jennifer Thomas-Larmer, *The Consensus Building Handbook: A Comprehensive Guide to Reaching Agreement* (Thousand Oaks: Sage, 1999).

▪▪▪▪▪▪

In the face of local land-use conflicts, how can planners mediate between conflicting parties and at the same time negotiate as interested parties themselves? To address that question, this article explores planners' strategies to deal with conflicts that arise in local processes of zoning appeals, subdivision approvals, special permit applications, and design reviews.

Local planners often have complex and contradictory duties. They may seek to serve political officials, legal mandates, professional visions, and the specific requests of citizens' groups, all at the same time. They typically work in situations of uncertainty, of great imbalances of power, and of multiple, ambiguous, and conflicting political goals. Many local planners, therefore, may seek ways both to negotiate effectively, as they try to satisfy particular interests, and to mediate practically, as they try to resolve conflicts through a semblance of a participatory planning process.

But these tasks – negotiating and mediating – appear to conflict in two fundamental ways. First, the negotiator's interest in the subject threatens the independence and the presumed neutrality of a mediating role. Second, although a negotiating role may allow planners to protect less powerful interests, a mediating role threatens to undercut this possibility and thus to leave existing inequalities of power intact. How can local planners deal with these problems? I discuss their strategies in detail below.

This article first presents local planners' own accounts of the challenges they face as simultaneous negotiators and mediators in local land-use permitting processes. Planning directors and staff in New England cities and towns, urban and suburban, shared their viewpoints with me during extensive open-ended interviews. The evidence reported here, therefore, is qualitative, and the argument that follows seeks not generalizability but strong plausibility across a range of planning settings.

The article next explores a repertoire of mediated negotiation strategies that planners use as they deal with local land-use permitting conflicts. It assesses the emotional complexity of mediating roles and asks: What skills are called for? Why do planners often seem reticent to adopt face-to-face mediating roles?

Finally, the article turns to the implications of these discussions. How might local planning organizations encourage both effective negotiation and equitable, efficient mediation? How might mediated-negotiation strategies empower the relatively powerless instead of simply perpetuating existing inequalities of power?

ELEMENTS OF LOCAL LAND-USE CONFLICTS

Consider first the settings in which planners face local permitting conflicts. Private developers typically propose projects. Formal municipal boards – typically planning boards and boards of zoning appeals – have decision-making authority to grant variances, special permits, or design approvals. Affected residents often have a say – but sometimes little influence – in formal public hearings before these boards. Planning staff report to these boards with analyses of specific proposals. When the reports are positive, they often recommend conditions to attach to a permit or suggest design changes to improve the final project. When the reports are negative, there are arguments to be made, reasons to be given.

Some municipalities have elected permit-granting boards; some have appointed boards. Some municipal ordinances mandate design review; others do not. Some local by-laws call for more than one planning board hearing on "substantial" projects, but others do not. Nevertheless, for several reasons, planners' roles in these different settings may be more similar than dissimilar.

Common planning responsibilities

First, planners must help both developers and neighborhood residents to navigate a potentially complex review process; clarity and predictability are valued goods. Second, the planners need to be concerned with timing. When a developer or neighborhood resident is told about an issue may be even more important than the issue itself. Third, planners typically need to deal with conflicts between project developers and affected neighborhood residents that usually concern several issues at once: scale, the income of tenants, new traffic, existing congestion, the character of a street, and so on. Such conflicts simultaneously involve questions of design, social policy, safety, transportation, and neighborhood character as well. Fourth, how much planners can do in the face of such conflicts depends not only upon their formal responsibilities, but also upon their informal initiatives. A zoning by-law, for example, can specify a time by which a planning board is to hold a public hearing, but it usually will not tell a planner how much information to give a developer or a neighbor, when to hold informal meetings with either or both, how to do it, just whom to invite, or how to negotiate with either party. So within the formal guidelines of zoning appeals, special permit applications, site plan and design reviews, planning staff can exercise substantial discretion and exert important influence as a result.

Planners' influence

The complexity of permitting processes is a source of influence for planning staff. Complexity creates uncertainties for everyone involved. Some planners eagerly use the resulting leverage, as an associate planning director explains, beginning with a truism but then elaborating:

> Time is money for developers. Once the money is in, the clock is ticking. Here we have some influence. We may not be able to stop a project that we have problems with, but we can look at things in more or less detail, and slow them down. Getting back to [the developers] can take two days or two months, but we try to make it clear, "We're people you can get along with." So many developers will say, "Let's get along with these people and listen to their concerns . . ."

He continues,

> But we have influence in other ways too. There are various ways to interpret the ordinance, for example. Or I can influence the building commissioner. He used to work in this office and we have a good relationship . . . his staff may call us about a project they're looking over and ask, "Hey, do you want this project or not?"

Planners think strategically about timing not only to discourage certain projects but to encourage or capture others. The associate director explains,

> On another project, we waited before pushing for changes. We wanted to let the developer get fully committed to it; then we'd push. If we'd pushed earlier, he might have walked away . . .

A director in another municipality echoes the point:

> Take an initial meeting with the developer, the mayor, and me. Depending on the benefits involved – fiscal or physical – the mayor might kick me under the table; "Not now," he's telling me. He doesn't want to discourage the project . . . and so I'll be able to work on the problems later . . .

For the astute, it seems, the complexity of the planning process creates more opportunities than headaches.

For the novice, no doubt, the balance shifts the other way.

But isn't everything, in the last analysis, all written down in publicly available documents for everyone to see? Hardly. Could all the procedures ever be made entirely clear? Consider the experience of an architect planner who grappled with these problems in several planning positions. The following conversation took place toward the end of my interview with this planner. The planner pulled a diagram from a folder and said, "Here's the new flow-chart I just drew up that shows how our design review process works. If you have any questions, let's talk. I think it's still pretty cryptic."

"If you think it's cryptic," interjected the zoning appeals planner, who was standing nearby and had overheard this, "just think what developers and neighborhood people will think!"

Both planners shook their heads and laughed, since the problem was all too plain: the arrows on the design review flow chart seemed to run everywhere. The chart was no doubt correct, but it did look complicated.

I recalled my first interview with the zoning appeals planner. Probing with a deliberately leading question, I had asked, "But what influence can you have in the process if everything's written down as public information, if it's all clear there on the page?"

The zoning appeals planner had grinned: "But that's just it! The process is not clear! And that's where I come in . . ." The architect-planner developed the point further:

> Where I worked before, the planning director wanted to adopt a new "policy and procedures" document that would have every last item defined. We were going to get it all clear. The whole staff spent a lot of time writing that, trying to get all the elements and subsections and so on clearly defined . . . But it was chaos. Once we had the document, everyone fought about what each item meant . . .

So clarity, apparently, has its limits!

Different actors, different strategies

Planning staff point almost poignantly to the different issues that arise as they work with developers and neighborhood residents. The candor of one planning director is worth quoting at length:

It's easy to sit down with developers or their lawyers. They're a known quantity. They want to meet. There's a common language – say, of zoning – and they know it, along with the technical issues. And they speak with one voice (although that's not to say that we don't play off the architect and the developer at times – we'll push the developer, for example, and the architect is happy because he agrees with us) . . .

But then there's the community. With the neighbors, there's no consistency. One week one group comes in, and the next week it's another. It's hard if there's no consistent view. One group's worried about traffic; the other group's not worried about traffic but about shadows. There isn't one point of view there. They also don't know the process (though there are cases where there are too many experts).

So at the staff level (as opposed to planning board meetings) we usually don't deal with both developers and neighbors simultaneously.

Although these comments may distress advocates of neighborhood power, they say much about the practical situation in which the director finds himself.

All people may be created equal, but when they walk into the planning department, they are simply not all the same. This director suggests that getting all the involved parties together around the table in the planners' conference room is not an obviously good idea, for several reasons. (It is, however, an idea we shall consider more closely below.)

First, the director suggests, planners generally know what to expect from developers; the developers' interests are often clearer than the neighbors', and project proponents may actually want to meet with the staff. Neighborhood residents may be less likely to treat planners as potential allies; after all, the planners are not the decision makers, and the decision makers can often easily ignore the planners' recommendations. Because developers may cultivate good relations with planning staff (this is in part their business, after all), while neighborhood groups do not, local planning staff may find meetings with developers relatively cordial and familiar, but meetings with neighborhood activists more guarded and uncertain.

Second, the planning director suggests that planners and developers often share a common professional language. They can pinpoint technical and regulatory issues and know that both sides understand what is being said. But on any given project, he implies, he may need to teach the special terms of the local zoning code to affected neighbors before they can really get to the issues at hand.

The planning director makes a third point. Developers speak with one voice; neighbors do not. When planners listen to developers talk, they know whom they're listening to, and they know what they're likely to hear repeated, elaborated, defended, or qualified next week. When planners listen to neighborhood residents, though, this director suggests, they can't be so sure how strongly to trust what they hear. "Who really speaks for the neighborhood?" the director wonders.

Planners must make practical judgments about who represents affected residents and about how to interpret their concerns. This director implies, therefore, that until planners find a way to identify "the neighborhood's voice," the problems of conducting joint mediated negotiations between developers and neighbors are likely to seem insurmountable. We return to this issue of representation below.

Inequalities of information, expertise, and financing

What about imbalances of power? Developers, typically, initiate site developments. Planners respond. Neighbors, if they are involved at all, then try to respond to both. Developers have financing and capital to invest; neighbors have voluntary associations and not capital, but lungs. Developers hire expertise; neighborhood groups borrow it. Developers typically have economic resources; neighbors often have time, but not always the staying power to turn that time into real negotiating power.

Where power relations are unbalanced, must mediated negotiation simply lead to coopting the weaker party? No, because, as we shall see below, mediated negotiation is not a gimmick or a recipe; it is a practical and political strategy to be applied in ways that address the specific relations of power at hand.

When either developers or neighborhood groups are so strong that they need not negotiate, mediated negotiation is irrelevant, and other political strategies are more appropriate. But when both developers and neighbors want to negotiate, planners can act both as mediators, assisting the negotiations, and as interested negotiators themselves. But how is this possible? What strategies can planners use?

PLANNERS' STRATEGIES: SIX WAYS TO MEDIATE LOCAL LAND-USE CONFLICTS

Consider the following six mediated-negotiation strategies that planning staff can utilize in the face of local land-use conflicts. They are *mediated* strategies because planners employ them to assure that the interests of the major parties legitimately come into play. They are *negotiation* strategies because (except for the first) they focus attention on the informal negotiations that may produce viable agreements even before formal decision-making boards meet.

Strategy one: The facts! The rules! (The planner as regulator)

The first strategy is a traditional response, pristine in its simplicity, but obviously more complex in practice. A young planner who handles zoning appeals and design review says:

> I see my role often as a fact finder so that the planning board can evaluate this project and form a recommendation; whether it's design review, special permits, or variances, you still need lots of facts . . .

Here of course is the clearest echo of the planner as technician and bureaucrat; the planner processes information and someone else takes responsibility for making decisions. But the echo quickly fades. A moment later, this planner continues,

> Our role is to listen to the neighbors, to be able to say to the board, "Okay, this project meets the technical requirements but there will be impacts . . ." The relief will usually then be granted, but with conditions . . . We'll ask for as much in the way of conditions as we think necessary for the legitimate protection of the neighborhood. The question is, is there a legitimate basis for complaint? And it's not just a matter of complaint, but of the merits.

This planner's role is much more complex than that of fact finder; it is virtually judicial in character. He implies, essentially, "I'm not just a bureaucrat, I'm a professional. I need to think not only about the technical requirements, but about what's legitimate protection for the neighbors. Now I have to think about the merits!" Thinking about the merits, though, does not yet mean thinking about politics, the feelings of other agencies, the chaos at community meetings – it means making professional judgments and then recommending to the planning board the conditions that should be attached to the permits.

Consider now a slightly more complex strategy.

Strategy two: Pre-mediate and negotiate – representing concerns

When developers meet with planners to discuss project proposals, neighborhood representatives rarely join them. Yet planners might nevertheless speak *for* neighborhood concerns as well as *about* them. A planning director in a municipality where neighborhood groups are well-organized, vocal, and influential notes,

> We temper our recommendations to developers. While we might accept A, the neighbors want D, and so we'll tell the developers to think about something in the middle – if they can make it work.

Here, the planner anticipates the concerns of affected residents and changes the informal staff recommendation accordingly to search for an acceptable compromise with the developers. He explains,

> What we do is premediate rather than mediate after the fact. We project people's concerns and then raise them; so we do more before [explicit conflict arises] . . . The only other way we step in and mediate, later, is when we support changes to be made in a project, changes that consider the neighbors' views; but that's later, after the public hearing . . .

Unlike the planner-regulator quoted above, this planning director relies on far more than his professional judgment when he meets with a developer. He will negotiate to reach project outcomes that satisfy local statutes, professional standards, and the interests of affected residents as well. His calculation is not only judicial, but explicitly political. He anticipates the concerns of interested community members. So he seeks to represent neighborhood interests – without neighborhood representatives.

Such premediation – articulating others' concerns well before they can erupt into overt conflict – involves a host of political, strategic, and ethical issues. What relationships does the planner have with neighborhood groups? In what senses can the planner "know what the community wants"? To which "key actors" might the planner "steer" the developer? How much information and how much advice should the planner give, or withhold?

Such questions arise whether or not project developers ever meet with neighbors. In many cases, where "neighborhoods" are sprawling residential areas, and where "the interests of the neighbors" seem most difficult to represent through actual neighborhood representatives, the planners' premediation may be the only mediation that takes place.

Strategy three: Let them meet – the planner as a resource

The planner's influence might be used in still other ways. The director continues:

> Regardless of how our first meeting with a developer goes, we recommend to them that they meet with neighbors and the neighbors' representatives [on the permit-granting board]. We usually can give the developer a good inkling about what to expect both professionally and politically. The same elected representative might say that a project is "okay" professionally, but not "okay" for them in their elected capacity. We try to encourage back and forth meetings . . .

The director, then, regularly takes the pulse of neighborhood groups and elected representatives. Working in city hall has its advantages: "We'll discuss a project with the representatives; we see them so much here, just in the halls, and they ask us to let them know what's happening in their parts of the city." So the director listens to the developers, listens to the neighbors, and "encourages back and forth meetings."

A planning director who seldom met jointly with neighbors and developers had an acute sense of other strategies he used:

> We . . . urge the applicants, the developers, to deal directly with the neighborhood for several reasons: First, if the neighbors are confronted at a hearing with glossy plans, they'll think it's all a *fait accompli*;

so they'll just adopt the "guns blazing, full charge ahead" strategy, since they think it'll just be a "yea" or "nay" decision. Second, we tell them to talk to the neighbors since if they can come up with something that the neighbors will "okay," it'll be easier at the board of appeals. Third, we try to get them to meet one on one, or maybe as a group, but in as deinstitutionalized a way as possible, informally. We try to get the developers to sell their case that way; it'll get a much better hearing than at the big formal public hearings.

But why should planners be reluctant to convene joint negotiating sessions between developers and neighbors, yet still be willing to encourage both parties to meet on their own? Why don't these planners embrace opportunities to mediate local land-use conflicts face to face? One planner could hardly imagine such a mediating role:

> Work as a neutral between developers and neighbors? I don't know how I'd approach it. I'd just answer questions, suggest what could be done, and so on. That's what our role should be – although we should reach compromises between developers and neighbors. But we have to work within the rules – that's my reference point – to say what the rules of the game are; that's the job.

This planner's image of a "neutral" between disputing parties is less that of a mediator facilitating agreement than it is of a referee in a boxing match. The referee assures that the rules are followed, but the antagonists might still kill each other. No wonder planners might find this image of mediation unattractive!

A senior planner envisions further complications:

> If I could be assured I could be wholly independent, then I could mediate – but I still have to pay my bills . . . The planning department always has some vested interests, as much as we try to stay objective, independent . . . I work for a mayor, for the elected representatives, for 14 committees . . . So there's always the question of compromise on my part: if the mayor says, "Tell me how to make this project work," for example. It took me a long time before I was able to say, "I'm going to have to say no." We have a very strong mayor . . .

Strategy four: Perform shuttle diplomacy – probe and advise both sides

A planning director proposes another way to facilitate developer–neighbor negotiations:

> I feel more comfortable in shuttle diplomacy, if you will; trying to get the neighbors' concerns on the table, to get the developers to deal with them . . . I'd rather bounce ideas off each side individually than be caught in the middle if they're both there. If both sides are there, I'm less likely to give my own ideas than if I'm alone with each of them.

Shuttle diplomacy, this director suggests, allows planners to address the concerns of each party in a professionally effective way. He explains:

> If I'm with the developer, I feel I can make a much more extreme proposal – "knock off three stories" – but I wouldn't dare say that if neighbors were there. The neighbors would be likely to pick up and run with it, and it could damage the negotiations rather than help them . . . I'm willing to back off on an issue if the developer has a good argument, but the neighborhood might not, and then they might use my point as a club to hit the developer with: "Well, the planning director suggested that; it must be a good idea" . . . and then I can't unsay it . . .

This planning director is as concerned about how his suggestions, proposals, queries, and arguments will be understood and used as he is about what ought to be altered in the project at hand. He recognizes clearly that when he talks he acts politically and inevitably fuels one argument or another. He not only conveys information in talking, but he acts practically, influentially. He focuses attention on specific problems, shapes future agendas, legitimates a point of view, and suggests lines of further argument.

The director continued,

> I might not want to concede to a developer that there won't be a traffic problem, because I want to push him to relieve a problem or a perceived problem . . . but I could say to the neighbors aside, "Look, this will be no big deal; it'll be five trips, not fifty." I can say that in a private meeting, but in a public meeting if I say it to a neighborhood representative I'm insulting him, even if the developer

snickers silently . . . So I lose my ability to be frank with both sides if we're all together. Not that this should be completely shuttle diplomacy, but it has its place.

These comments suggest that planning staff can certainly mediate conflicts in local permitting processes, if not in ways that mediators are thought typically to act. The planners may not be independent third parties who assist developers and neighbors in face-to-face meetings to reach development agreements – but they might still mediate such conflicts as "shuttle diplomats."

Strategy five: Active and interested mediation – thriving as a non-neutral

We can consider a case that involves not a zoning appeal but a rezoning proposal. One planner, who had earlier worked as a community organizer, had convened a working group of five community representatives and five local business representatives to draft a rezoning proposal for a large stretch of the major arterial street in their municipality. She considers her work on that project a kind of mediation and reflects about how she as a planner acts as a mediator, dealing with substantive and affective issues alike:

> Am I in a position of having to think about everyone's interest and yet being trusted by no one? Sure, all the time. But I've been in this job for seven years, and I have a reputation that's good, fortunately . . . Trust is an issue of your integrity and planning process. I talk to people a lot; communication is a big part of it . . . My approach is to let people let off steam – let them say negative things about other people to me, and then in a different conversation at another time, I'll be sure to say something positive about that person – to try to let them feel that they can say whatever they want to me, and to try to confront them with the fact that the other person isn't just out to ruin the process. But I'd do that in another conversation; I let them let off steam if they're angry.

This planner is well aware that distrust on all sides is an abiding issue, so she tries to build trust as she works. She works to assure others that she will listen to them and more; that she will acknowledge and respect their thoughts and feelings, whatever they have to say. She

pays attention first to the person, then to the words. Then, as she establishes trust with her committee members and with others, she can also make sure, carefully, that real evidence is not ignored.

She realizes that anger makes its own demands, so she responds with an interested patience. She seeks throughout to mediate the conflicting interests of the groups with whom she's working:

> I also make a point to tell each side the other's concerns – categorically, not with names, but all the other sides' concerns . . . Why's that important? I like to let people anticipate the arguments and prepare a defense, either to stand or fall on its own merits. For people to be surprised is unfortunate. It's better to let people know what's coming so they can build a case. They can hear an objection, if you can retain credibility, and absorb it; but in another setting they might not be able to hear it . . . If they hear an objection first as a surprise, you're likely to get blamed for it. If concerns are raised in an emotional setting, people concentrate more on the emotion than on the substance. This is a concern of mine. In emotional settings, lots gets thrown out, and lots is peripheral, but possibly also central later . . .

This planner is keenly aware that emotion and substance are interwoven, and that planners who focus only upon substance and try to ignore or wish away emotion do so at their own practical peril. Yet she is saying even more.

She knows that in some settings disputing groups can hear objections, understand the points at stake, and address them, while in other settings those points may be lost. She tries to present each side's concerns to the other so that they can be understood and addressed. Anticipating issues is central; learning of important objections late in the process will be mostly emotionally and financially, and planning staff are likely to share the blame. "Why didn't you tell us sooner . . .?" the refrain is likely to sound.

Consider next, then, this planner-mediator's thoughts about the sort of mediation role she is performing. She continues,

> But what I do is different from the independent mediator model. In a job like mine, you have an on-going relationship with parties in the city. You have more information than a mediator does about

the history of various individuals, about participating organizations, about the political history of city agencies, and so on. You also have a vested interest in what happens. You want the process to be credible. You want the product to be successful; in my case I want the city council to adopt the committee's proposal. And you're invested . . . both professionally and emotionally. And then you have an opinion about particular proposals; you're a professional, you should have one, you should be able to look at a proposal and have an opinion.

Thus, she suggests, mediation has its place in local land-use conflicts, but the "rules of the game" will not be those that labor mediators follow. Indeed, planners who now mediate local land-use conflicts are not waiting for someone else to write the rules of the game, they are writing them themselves.

Strategy six: Split the job – you mediate, I'll negotiate

Consider finally a planning strategy that promotes face-to-face mediation with planning staff at the table – but as negotiators or advisors, not as mediators. A planning director explains:

> There's another way we deal with these conflicts; we might involve a local planning board member. For example, if there's a sophisticated neighborhood group that's well organized, we've brought in an architect from the board who's as good with words as he is with his pencil . . . The chair of the board might ask the board member to be a liaison to the neighborhood, say, and sometimes he'll talk just to the neighbors, sometimes with both . . .

Here the "process manager" comes from the planning board with highly developed "communications skills." How does the planner feel in these situations?

> It's more comfortable from my point of view, and the citizens', to have a board member in the convening role. I'm still a hired hand. It seems more appropriate in a negotiating situation to have a citizen in that role and not an employee . . . Since they've come from the neighborhoods, a board member is in a better position to bring neighbors and developers together – if they behave properly. Some board members are good communicators;

some are more dynamic than others in pressing for specific solutions.

This planner identifies so strongly with the professional and political mandate of his position that he cannot imagine a role as neutral convener or mediator of neighborhood–developer negotiations. But that does not prevent mediation; it means rather that the planner retains a substantively interested posture while another party, here a planning board member, convenes informal, but organized, project negotiations between developers and neighbors. This planner's example makes the point:

> Take the example of the Mayfair Hospital site. The hospital was going to close, and the neighbors and the planning board were concerned about what might happen with the site. So Jan from the planning board got involved with the hospital and the neighborhood to look at the possibilities. Both the neighbors and the hospital set up re-use committees, and Jan and I went to the meetings. There was widespread agreement that the best use of the site would be residential – the neighbors definitely preferred that to an institutional use – but then there was a lot of haggling over scale, density, and so on. Ultimately, a special zoning district was proposed that included the site; the neighbors supported it, and it went to [the elected representatives] where they voted to rezone the several acres involved . . .

When local planners feel they cannot mediate disputes themselves, then one strategy may be to search for informal, most likely volunteer, mediators. These ad hoc mediators might be "borrowed" from respected local institutions, and their facilitation of meetings between disputing parties might allow planning staff to participate as professionally interested parties concerned with the site in question.

Table 1 summarizes the six approaches presented. Together, these approaches form a repertoire of strategies that land-use planners can use to encourage mediated negotiations in the face of conflicts in local zoning, special permit, and design review processes. To refine these strategies, local planning staff can build upon several basic theories and techniques of conflict resolution. Consider now the distinctive competences and sensitivities required by these strategies.

Table 1 A repertoire of mediated-negotiation strategies used by local land-use planners in permitting processes

THE EMOTIONAL COMPLEXITY OF MEDIATED-NEGOTIATION STRATEGIES

More than a lack of independence keeps planners from easily adopting roles as mediators. The emotional complexity of the mediating role makes quite different demands upon planners than those that they have traditionally been prepared to meet. The community-organizer-turned-planner makes the point brilliantly:

In the middle, you get all the flak. You're the release valve. You're seen as having some power, and you do have some . . .Look, if you have a financial interest in a project, or an emotional one, you want the person in the middle to care about your point of view, and if you don't think they do, you'll be angry!

["So when planners try to be professional by appearing detached, objective, does it get people angry at them?" I asked.]

Sure!

This comment cuts to the heart of planners' professional identities. Must "professional," "objective," and "detached" be synonymous? If so, this planner suggests, then planners' own striving for an independent professionalism will fuel the anger, resentment, and suspicion of the same people those planners presume to serve!

Thus we can understand the caution with which a planner speaks of his way of handling emotional participants in public hearings:

How do I deal with people's anger? I try to keep cool, but occasionally I get irritated. But that's how we're expected to behave, to be rational. It's all right for citizens to be irrational, but not the staff!

How does one keep cool, be rational, and still respond to the claims of an emotional public at formal hearings? This planner elaborates:

It's one thing to begin the discussion of a project [to present our analysis] and anticipate problems. But it's another thing to *rebut* a neighborhood resident in public in a gentle way . . . Part of the problem is that if you antagonize people it'll haunt you in the future . . . We're here for the long haul, and we have to try to maintain our credibility . . .

The planner's problem here is precisely *not* the facts of the case: the facts themselves may be clear enough. But how should the planner present the analysis that he feels must be made and how should he decide which arguments to make and which to hold back at a given time?

The biggest problem I have in the board meetings is when to respond and when to keep quiet. In a hearing, for example, I can't possibly respond to all the accusations and issues that come up. So I have to pick a direction, to deal with a generally felt concern. It's just not effective to enter into a debate on each point in turn; it's better to clarify things, to explain what's misunderstood . . .

This planner does much more than simply recapitulate facts. He tries to avoid an adversarial posture, even when he feels the situation is quite conflictual. He listens as much to the individuals and their concern as he does to each point. He knows that points and demands and positions may change as issues are clarified, but that if he cannot respond to people's concerns, he's in some trouble. Because he and his staff are there "for the long haul," he wants to be able to work with neighbors, community leaders, and elected representatives alike not just now but in the future as well. How he relates to the parties involved in local disputes, he suggests, is as important as what he has to say.

Another planner points to the skills involved:

Whom would I try to hire to deal with such conflicts? I'd look for someone who's a careful listener, someone who's good at explaining a position coherently, succinctly, quietly, in a calm tone . . . someone who could hear a point, understand it whether he or she agreed with it or not, and then verbalize a clear, concise response. Most people though – myself included – try to jump the gun and answer before it's appropriate. So I want someone who's able to stay cool and stay on the issues . . .

A community development director first mentions "a good listener" and then elaborates:

[To deal with these conflicting situations I'd want to hire staff] who won't say, "I know best," who won't get people's backs up just by their style. I'd want someone with some openness, with a sense of how things work who won't accept everything, but who won't offend people. They have to have critical judgment – to leave doors open, to give people a sense of involvement and a sense of the feasible – [someone who] can't be convinced of something that's not likely to work, just for the sake of getting agreement . . .

This planning director also points to the balance necessary between what planners say and how they say it. The "how" counts; he doesn't want staff who will "get people's backs up," "offend people," and not communicate an openness to others' concerns. Nor does he want someone who will sacrifice project viability for the temporary comfort of agreement. He asks for substantive judgment and the skills to manage a process.

Referring to the demands of working and negotiating with developers as they navigate the approval process, the director stresses the role of diplomacy:

We [planners] have access to information, to resources, to skills . . . so developers usually want to work with us. They have certain problems getting through the process . . . so we'll go to them and ask, "What do you want?" and we'll start a process of meetings . . . It's diplomacy; that's the real work. You have to have the technical skill . . . but that's the first 25 percent. The next 75 percent is diplomacy, working through the process.

Percentages aside, the point remains. To the extent that planning practitioners and educators focus predominantly upon facts, rules, likely consequences, and mitigation measures, they may fail to attend to the pressing emotional and communicative dimensions of local land-use conflicts. Because the planning profession has not traditionally embraced the diplo-

mat's skills, it should surprise no one that practicing planners envision mediating roles with more reticence than relish.

In the next section, we turn to administrative and political questions. What, initially, can be done in planning organizations to improve planners' abilities to mediate local land-use negotiations successfully? What about imbalances of power?

ADMINISTRATIVE IMPLICATIONS FOR PLANNING ORGANIZATIONS

What does this analysis imply for policymakers and planners who wish to build options for mediation into local review processes? Mediation may offer several opportunities, under conditions of interdependent power: a shift from adversarial to collaborative problem-solving; voluntary development controls and agreements; improved city–developer–neighborhood relationships enabling early and effective reviews of future projects; more effective neighborhood voice; and joint gains ("both gain" outcomes) for the municipality, neighbors, and developers alike. Such opportunities present themselves *only* when no single party is so dominant that it need not negotiate at all, that it is likely simply to get what it wants in any case.

Planners already use the strategies reviewed in diverse settings. Which strategy a planner uses, and at which times, depends largely on practical judgment: What skills does the planner have? How willing are developers or neighbors, or other agency staff, to meet jointly? Does enough time exist to allow early, joint meetings? Are the practical and political alternatives of any one party so attractive that they see no point in mediated negotiations?

No strategy is likely to be desirable in all circumstances, so no one approach will provide the model to formalize into new zoning or permitting procedures. But to say that we should not formalize these strategies does not mean that we cannot regularly use them. How, then, can planners apply the mediated-negotiation strategies in local zoning, permitting, and design review processes?

First, planning staff must distinguish clearly the two complementary but distinct mandates they typically must serve: to press professionally, and thus to negotiate, for particular substantive goals (design quality or affordable housing, for example), and to enable a participatory process that gives voice to affected parties; thus, like mediators, to facilitate negotiations between disputants.

Second, planning staff need to adopt, administratively if not formally, a goal of supplementing (not substituting for) formal permitting processes with mediated negotiations: attempting to craft workable and voluntary tentative agreements before formal hearing dates.

Third, planning staff should examine each of the strategies reviewed here. They need to determine how each could work, given the size of their agency, their zoning and related by-laws, the political and institutional history of elected officials, neighborhood groups, and other agencies. Planning staff must ask which skills and competencies they need to develop to employ each of these strategies appropriately.

Fourth, planning staff must be able to show others – developers, neighborhood groups, public works department staff, elected and appointed officials – how and when mediated negotiations can lead to "both gain" outcomes and so improve the local land-use planning and development process. Planners also have to be clear about what mediated negotiation will not do: it will not solve problems of radically unbalanced power, for example. It can, however, refine an adversarial process into a partially collaborative one. It will not solve problems of basic rights, but it can often expand the range of affected parties' interests that developers will take into account. Mediated negotiations will neither necessarily co-opt project opponents (as skeptical neighborhood residents might suspect) nor stall proposals and projects (as skeptical developers and builders might suspect). Yet when each side can effectively threaten the other, when each side's interests depend upon the other's actions, then mediated negotiations may enable voluntary agreements, incorporate measures of control on both sides, allow "both gain" trades to be achieved, and do so more efficiently for all sides than pursuing alternative strategies (e.g. going to court or, sometimes, community organizing).

Fifth, planners need administratively to create an organized process to match incoming projects with one or more of the mediated-negotiation strategies and to review their progress as they go along. With staff training in negotiation and mediation principles and techniques, planning departments would be better able to carry out these strategies effectively once they have organized administratively to promote them.

DEALING WITH POWER IMBALANCES: CAN THE SIX STRATEGIES MAKE A DIFFERENCE?

The six strategies we have considered are hardly "neutral." Planners who adopt them inevitably either perpetuate or challenge existing inequalities of information, expertise, political access, and opportunity. Consider each approach, briefly, in turn.

To provide only the facts, or information about procedures, to whomever asks for them seems to treat everyone equally. Yet where severe inequalities exist, to treat the strong and the weak alike only ensures that the strong remain strong, the weak remain weak. The planner who pretends to act as a neutral regulator may sound egalitarian but nevertheless act, ironically, to perpetuate and ignore existing inequalities.

The premediation strategy can involve substantial discretion on the part of the planning staff. If the staff fail to put the interests of weaker parties "on the negotiating table," then here, too, inequalities will be perpetuated, not mitigated. If the staff do defend neighborhood interests in the development negotiations, they may challenge existing inequalities. But which "neighborhood interests" should the planning staff identify? How should neighborhoods – especially weakly organized ones – be represented? These questions are both practical and theoretical and they have no purely technical, "recipe"-like answers.

At first glance, the strategy of letting developers and neighbors meet without an active staff presence seems only to reproduce the initial strengths of the parties. Yet depending on how the planning staff intervene, one party or another may be strengthened or weakened. At times planners have helped developers anticipate and ultimately evade the concerns of citizens who opposed projects. Yet planners may also provide expertise, access, information, and so on to strengthen weaker citizens' positions.

The same discretion exists for planning staff who act as shuttle diplomats. Here a planner may counsel weaker parties to help them both before and during actual negotiations by identifying concerns that might effectively be raised, experts or other influentials who might be called upon, prenegotiation strategies and tactics to be employed, and so on. The shuttle diplomat need not appear neutral to all parties but he or she does need to appear useful to, or needed by, those parties. Planners who act as "interested mediators" face many of the same problems and opportunities that shuttle diplomats confront. In addition, though, the activist mediator may risk being perceived by planning board members, officials, or elected representatives as making deals that preempt their own formal authority. Thus the invisibility of the shuttle diplomat has its advantages; the planners can give counsel discreetly, suggesting packages and "deals" but avoiding the glare – and the heat – of the limelight.

Finally, the strategy of separating mediation and negotiation functions also involves substantial staff discretion. Here, too, the ways that mediators and negotiators consider the interests and enable the voice of weaker parties will affect existing power imbalances.

Because negotiations always involve questions of relative power, they depend heavily upon the parties' *prenegotiation* work of marshalling resources, developing options, and organizing support. Thus politically astute planners need both organizing and mediated-negotiation skills if conflicts are to be addressed without pretending that structural power imbalances just do not exist. Finally, note that a planner who explicitly calls everyone's attention to class-based power imbalances, for example, may not obviously do better in any practical sense of the word than an activist mediator who knows the same thing and acts on it in just the same ways without explicitly framing the planning negotiations in those terms.

CONCLUSION

The repertoire of mediated-negotiation strategies inevitably requires that planners exercise practical judgment, both politically and ethically. These judgments involve who is and who is not invited to meetings; where, when, and which meetings are held; what issues should and should not appear on agendas; whose concerns are and are not acknowledged; how interventionist the planner's role is; and so on.

In local planning processes, then, planners often have the administrative discretion not only to mediate among conflicting parties, but to negotiate as interested parties themselves. Planning staff can routinely engage in the complementary tasks of supporting organizing efforts, negotiating, and mediating. In these ways, local planners can use a range of mediated-negotiation strategies to address practically existing power imbalances of access, information, class, and expertise that perpetually threaten the quality of local planning outcomes.

Mediated negotiations in local permitting processes will, of course, not resolve the structural problems of our society. Yet when local conflicts involve multiple issues, when differences in interests can be exploited by trading to achieve joint gains, and when diverse interests rather than fundamental rights are at stake, mediated-negotiation strategies for planners make good sense, politically, ethically, and practically.

"Advocacy and Pluralism in Planning"

Journal of the American Institute of Planners (1965)

Paul Davidoff

Editors' Introduction

Each year at the annual meeting of the Association of Collegiate Schools of Planning (ACSP), North American professors of urban planning present the Paul Davidoff Award to a city planning scholar whose work exemplifies the practice and ideals of the author of this selection, lawyer/planner/professor/activist Paul Davidoff. It is an honor to receive the Davidoff award, because Davidoff exemplified professional commitment to vigorous advocacy on behalf of the less fortunate members of society.

During the 1960s, Davidoff, a lawyer and city planner, taught city planning students at Hunter College and simultaneously fought successfully to get racially integrated low-income housing built in exclusive white suburbs. This experience as an advocate for low-income minorities shaped his view of what city planning should be like.

Unlike the "advocacy planning" Davidoff proposes, most city planning is performed by a local government planning department working under the direction of a planning commission that develops plans it feels will best serve the welfare of the whole community, not of individual interest groups such as organizations of homeless people, merchants, environmentalists, or bicycle enthusiasts. While city planning commissions may explore many alternatives and consider conflicting interest group demands before finalizing plans, they end up with a single unitary plan.

Davidoff's vision for how planning might be structured is quite different. He argues that different groups in society have different interests. If they were recognized, these interests would result in plans that are different from each other in fundamental ways. Business elites and other articulate, wealthy, and powerful groups have the skill and resources to shape city plans to serve their interests. But what about the poor and powerless? Davidoff argued that there should be planners acting as *advocates* articulating the interests of these and other groups, much as a lawyer represents a client. For example, one planner might develop and advocate for a plan that would meet the needs of poor West Indian residents of London's Brixton neighborhood. Another planner might develop a different plan representing the point of view of shopkeepers in the same area. And yet another might work with Brixton environmentalists to develop and advocate for a plan for the Brixton area incorporating the kind of sustainable urban development urged by the Brundtland Commission (p. 337), Timothy Beatley (p. 411), and Stephen Wheeler (p. 499). Confronted with these different visions (and the empirical data which makes the most compelling case for them), the local planning commission could weigh the merits of the competing plans much as a court weighs evidence and conflicting characterizations of a legal case by competing lawyers. Davidoff believed the plan that would emerge from such a process would be better than a plan prepared by planning department staff without the interplay of competing advocacy planners. And, Davidoff reasoned, the needs of the poor and powerless would be better met in city plans if – a big if – they were adequately represented by advocacy planners speaking on their behalf.

Davidoff's view of planning profoundly influenced activist planners of the 1960s and 1970s, many of whom defined themselves as advocacy planners and developed plans to meet the needs and interests of under-represented groups, with some notable successes. "Equity planners" continue this tradition today.

Read the accounts of the evolution of urban planning theory and practice by Edward J. Kaiser and David R. Godschalk (p. 366) and Peter Hall (p. 354) to better understand the context within which Davidoff developed his critique. Compare Davidoff's humanistic, grassroots, pluralistic approach to city planning with Le Corbusier's brilliant, but elitist vision of an elite cadre of CIAM architects imposing the forms they felt modern machine culture demanded on the fabric of cities (p. 322). Compare Davidoff's views with John Forester's comments on how planners working within the system can use their influence to empower stake-holders in the planning process (p. 387). Compare the advocacy planning approach to strategies to empower communities to reach the highest possible level on the "ladder" of citizen participation that Sherry Arnstein developed (p. 283).

While he was the planning director of Cleveland, Ohio, Norman Krumholz worked hard to make city planning responsive to the needs of the poor and powerless. As a planning professor at Cleveland State University he continues to develop the theory and practice of equity planning. Krumholtz describes his experience practicing equity planning as the planning director of Cleveland, Ohio, in *Making Equity Planning Work* (Philadelphia: Temple University Press, 1990) co-authored with John Forester, and the experience of other equity planners in *Reinventing Cities: Equity Planners Tell Their Stories* (Philadelphia: Temple University Press, 1994) co-authored with Pierre Clavel. Chester Hartman's *Cities for Sale: The Transformation of San Francisco* (Berkeley: University of California Press, 2002) describes how advocacy planners and lawyers fought to make urban renewal more responsive to residents of a low income area. For an application of advocacy planning to women, see Jacqueline Levitt, "Feminist Advocacy Planning in the 1980s" in Barry Checkoway (ed.), *Strategic Perspectives in Planning Practice* (Lexington: Lexington Books, 1986).

The present can become an epoch in which the dreams of the past for an enlightened and just democracy are turned into a reality. The massing of voices protesting racial discrimination have roused this nation to the need to rectify racial and other social injustices. The adoption by Congress of a host of welfare measures and the Supreme Court's specification of the meaning of equal protection by law both reveal the response to protest and open the way for the vast changes still required.

The just demand for political and social equality on the part of the Negro and the impoverished requires the public to establish the bases for a society affording equal opportunity to all citizens. The compelling need for intelligent planning, for specification of new social goals and the means for achieving them, is manifest. The society of the future will be an urban one, and city planners will help to give it shape and content.

The prospect for future planning is that of a practice which openly invites political and social values to be examined and debated. Acceptance of this position means rejection of prescriptions for planning which would have the planner act solely as a technician. It has been argued that technical studies to enlarge the information available to decision makers must take precedence over statements of goals and ideals:

We have suggested that, at least in part, the city planner is better advised to start from research into the functional aspects of cities than from his own estimation of the values which he is attempting to maximize. This suggestion springs from a conviction that at this juncture the implications of many planning decisions are poorly understood, and that no certain means are at hand by which values can be measured, ranked, and translated into the design of a metropolitan system.

While acknowledging the need for humility and openness in the adoption of social goals, this statement amounts to an attempt to eliminate, or sharply reduce, the unique contribution planning can make: understanding the functional aspects of the city and recommending appropriate future action to improve the urban condition.

Another argument that attempts to reduce the importance of attitudes and values in planning and other policy sciences is that the major public questions are themselves matters of choice between technical methods of solution. Dahl and Lindblom put forth this position at the beginning of their important textbook *Politics, Economics, and Welfare*:

In economic organization and reform, the "great issues" are no longer the great issues, if they ever were. It has become increasingly difficult for thoughtful men to find meaningful alternatives posed in the traditional choices between socialism and capitalism, planning and the free market, regulation and laissez faire, for they find their actual choices neither so simple nor so grand. Not so simple, because economic organization poses knotty problems that can only be solved by painstaking attention to technical details – how else, for example, can inflation be controlled? Nor so grand, because, at least in the Western world, most people neither can nor wish to experiment with the whole pattern of socio-economic organization to attain goals more easily won. If, for example, taxation will serve the purpose, why "abolish the wages system" to ameliorate income inequality?

These words were written in the early 1950s and express the spirit of that decade more than that of the 1960s. They suggest that the major battles have been fought. But the "great issues" in economic organization, those revolving around the central issue of the nature of distributive justice, have yet to be settled. The world is still in turmoil over the way in which the resources of nations are to be distributed. The justice of the present social allocation of wealth, knowledge, skill, and other social goods is clearly in debate. Solutions to questions about the share of wealth and other social commodities that should go to different classes cannot be technically derived; they must arise from social attitudes.

Appropriate planning action cannot be prescribed from a position of value neutrality, for prescriptions are based on desired objectives. One conclusion drawn from this assertion is that "values are inescapable elements of any rational decision-making process" and that values held by the planner should be made clear. The implications of that conclusion for planning have been described elsewhere and will not be considered in this article. Here I will say that the planner should do more than explicate the values underlying his prescriptions

for courses of action; he should affirm them; he should be an advocate for what he deems proper.

Determinations of what serves the public interest, in a society containing many diverse interest groups, are almost always of a highly contentious nature. In performing its role of prescribing courses of action leading to future desired states, the planning profession must engage itself thoroughly and openly in the contention surrounding political determination. Moreover, planners should be able to engage in the political process as advocates of the interests both of government and of such other groups, organizations, or individuals who are concerned with proposing policies for the future development of the community.

The recommendation that city planners represent and plead the plans of many interest groups is founded upon the need to establish an effective urban democracy, one in which citizens may be able to play an active role in the process of deciding public policy. Appropriate policy in democracy is determined through a process of political debate. The right course of action is always a matter of choice, never of fact. In a bureaucratic age great care must be taken that choices remain in the area of public view and participation.

Urban politics, in an era of increasing government activity in planning and welfare, must balance the demands for ever-increasing central bureaucratic control against the demands for increased concern for the unique requirements of local, specialized interests. The welfare of all and the welfare of minorities are both deserving of support; planning must be so structured and so practiced as to account for this unavoidable bifurcation of the public interest.

The idealized political process in a democracy serves the search for truth in much the same manner as due process in law. Fair notice and hearings, production of supporting evidence, cross-examination, reasoned decision are all means employed to arrive at relative truth: a just decision. Due process and two- (or more) party political contention both rely heavily upon strong advocacy by a professional. The advocate represents an individual, group, or organization. He affirms their position in language understandable to his client and to the decision makers he seeks to convince.

If the planning process is to encourage democratic urban government then it must operate so as to include rather than exclude citizens from participating in the process. "Inclusion" means not only permitting the citizen to be heard. It also means that he be able to become well informed about the underlying reasons for planning proposals, and be able to respond to them in the technical language of professional planners.

A practice that has discouraged full participation by citizens in plan making in the past has been based on what might be called the *"unitary plan."* This is the idea that only one agency in a community should prepare a comprehensive plan; that agency is the city planning commission or department. Why is it that no other organization within a community prepares a plan? Why is only one agency concerned with establishing both general and specific goals for community development, and with proposing the strategies and costs required to effect the goals? Why are there not plural plans?

If the social, economic, and political ramifications of a plan are politically contentious, then why is it that those in opposition to the agency plan do not prepare one of their own? It is interesting to observe that "rational" theories of planning have called for consideration of alternative courses of action by planning agencies. As a matter of rationality it has been argued that all of the alternative choices open as means to the ends ought be examined. But those, including myself, who have recommended agency consideration of alternatives have placed upon the agency planner the burden of inventing "a few representative alternatives." The agency planner has been given the duty of constructing a model of the political spectrum, and charged with sorting out what he conceives to be worthy alternatives. This duty has placed too great a burden on the agency planner, and has failed to provide for the formulation of alternatives by the interest groups who will eventually be affected by the completed plans.

Whereas in a large part of our national and local political practice contention is viewed as healthy, in city planning where a large proportion of the professionals are public employees, contentious criticism has not always been viewed as legitimate. Further, where only government prepares plans, and no minority plans are developed, pressure is often applied to bring all professionals to work for the ends espoused by a public agency. For example, last year a Federal official complained to a meeting of planning professors that the academic planners were not giving enough support to Federal programs. He assumed that every planner should be on the side of the Federal renewal program. Of course government administrators will seek to gain the support of professionals outside of government, but such support should not be expected as a matter of loyalty. In a democratic system opposition to a public agency should be just as normal and appropriate as support. The agency, despite the fact that it is concerned with planning, may be serving undesired ends.

In presenting a plea for plural planning I do not mean to minimize the importance of the obligation of the public planning agency. It must decide upon appropriate future courses of action for the community. But being isolated as the only plan maker in the community, public agencies as well as the public itself may have suffered from incomplete and shallow analysis of potential directions. Lively political dispute aided by plural plans could do much to improve the level of rationality in the process of preparing the public plan.

The advocacy of alternative plans by interest groups outside of government would stimulate city planning in a number of ways. First, it would serve as a means of better informing the public of the alternative choices open, *alternatives strongly supported by their proponents*. In current practice those few agencies which have portrayed alternatives have not been equally enthusiastic about each. A standard reaction to rationalists' prescription for consideration of alternative courses of action has been "it can't be done; how can you expect planners to present alternatives which they don't approve?" The appropriate answer to that question has been that planners, like lawyers, may have a professional obligation to defend positions they oppose. However, in a system of plural planning, the public agency would be relieved of at least some of the burden of presenting alternatives. In plural planning the alternatives would be presented by interest groups differing with the public agency's plan. Such alternatives would represent the deep-seated convictions of their proponents and not just the mental exercises of rational planners seeking to portray the range of choice.

A second way in which advocacy and plural planning would improve planning practice would be in forcing the public agency to compete with other planning groups to win political support. In the absence of opposition or alternative plans presented by interest groups the public agencies have had little incentive to improve the quality of their work or the rate of production of plans. The political consumer has been offered a yes–no ballot in regard to the comprehensive plan; either the public agency's plan was to be adopted or no plan would be adopted.

A third improvement in planning practice which might follow from plural planning would be to force those who have been critical of "establishment" plans to produce superior plans, rather than only to carry out the very essential obligation of criticizing plans deemed improper.

THE PLANNER AS ADVOCATE

Where plural planning is practiced, advocacy becomes the means of professional support for competing claims about how the community should develop. Pluralism in support of political contention describes the process; advocacy describes the role performed by the professional in the process. Where unitary planning prevails, advocacy is not of paramount importance, for there is little or no competition for the plan prepared by the public agency. The concept of advocacy as taken from legal practice implies the opposition of at least two contending viewpoints in an adversary proceeding.

The legal advocate must plead for his own and his client's sense of legal propriety or justice. The planner as advocate would plead for his own and his client's view of the good society. The advocate planner would be more than a provider of information, an analyst of current trends, a simulator of future conditions, and a detailer of means. In addition to carrying out these necessary parts of planning, he would be a *proponent* of specific substantive solutions.

The advocate planner would be responsible to his client and would seek to express his client's views. This does not mean that the planner could not seek to persuade his client. In some situations persuasion might not be necessary, for the planner would have sought out an employer with whom he shared common views about desired social conditions and the means toward them. In fact one of the benefits of advocate planning is the possibility it creates for a planner to find employment with agencies holding values close to his own. Today the agency planner may be dismayed by the positions affirmed by his agency, but there may be no alternative employer.

The advocate planner would be above all a planner. He would be responsible to his client for preparing plans and for all of the other elements comprising the planning process. Whether working for the public agency or for some private organization, the planner would have to prepare plans that take account of the arguments made in other plans. Thus the advocate's plan might have some of the characteristics of a legal brief. It would be a document presenting the facts and reasons for supporting one set of proposals, and facts and reasons indicating the inferiority of counter-proposals. The adversary nature of plural planning might, then, have the beneficial effect of upsetting the tradition of writing plan proposals in terminology which makes them appear self-evident.

A troublesome issue in contemporary planning is that of finding techniques for evaluating alternative plans. Technical devices such as cost–benefit analysis by themselves are of little assistance without the use of means for appraising the values underlying plans. Advocate planning, by making more apparent the values underlying plans, and by making definitions of social costs and benefits more explicit, should greatly assist the process of plan evaluation. Further, it would become clear (as it is not at present) that there are no neutral grounds for evaluating a plan; there are as many evaluative systems as there are value systems.

The adversary nature of plural planning might also have a good effect on the uses of information and research in planning. One of the tasks of the advocate planner in discussing the plans prepared in opposition to his would be to point out the nature of the bias underlying information presented in other plans. In this way, as critic of opposition plans, he would be performing a task similar to the legal technique of cross-examination. While painful to the planner whose bias is exposed (and no planner can be entirely free of bias) the net effect of confrontation between advocates of alternative plans would be more careful and precise research.

Not all the work of an advocate planner would be of an adversary nature. Much of it would be educational. The advocate would have the job of informing other groups, including public agencies, of the conditions, problems, and outlook of the group he represented. Another major educational job would be that of informing his clients of their rights under planning and renewal laws, about the general operations of city government, and of particular programs likely to affect them.

The advocate planner would devote much attention to assisting the client organization to clarify its ideas and to give expression to them. In order to make his client more powerful politically the advocate might also become engaged in expanding the size and scope of his client organization. But the advocate's most important function would be to carry out the planning process for the organization and to argue persuasively in favor of its planning proposals.

Advocacy in planning has already begun to emerge as planning and renewal affect the lives of more and more people. The critics of urban renewal have forced response from the renewal agencies, and the ongoing debate has stimulated needed self-evaluation by public agencies. Much work along the lines of advocate planning has already taken place, but little of it by professional planners. More often the work has been conducted by trained community organizers or by student groups. In at least one instance, however, a planner's professional aid led to the development of an alternative renewal approach, one which will result in the dislocation of far fewer families than originally contemplated.

Pluralism and advocacy are means for stimulating consideration of future conditions by all groups in society. But there is one social group which at present is particularly in need of the assistance of planners. This group includes organizations representing low-income families. At a time when concern for the condition of the poor finds institutionalization in community action programs, it would be appropriate for planners concerned with such groups to find means to plan with them. The plans prepared for these groups would seek to combat poverty and would propose programs affording new and better opportunities to the members of the organization and to families similarly situated. The difficulty in providing adequate planning assistance to organizations representing low-income families may in part be overcome by funds allocated to local antipoverty councils. But these councils are not the only representatives of the poor; other organizations exist and seek help. How can this type of assistance be financed? This question will be examined below, when attention is turned to the means for institutionalizing plural planning.

THE STRUCTURE OF PLANNING

Planning by special interest groups

The local planning process typically includes one or more "citizens'" organizations concerned with the nature of planning in the community. The Workable Program requirement for "citizen participation" has enforced this tradition and brought it to most large communities. The difficulty with current citizen participation programs is that citizens are more often *reacting* to agency programs than proposing their concepts of appropriate goals and future action.

The fact that citizens' organizations have not played a positive role in formulating plans is to some extent a result of both the enlarged role in society played by government bureaucracies and the historic weakness of municipal party politics. There is something very shameful to our society in the necessity to have organized "citizen participation." Such participation should be the norm in an enlightened democracy. The formalization of citizen participation as a required practice in localities is similar in many respects to totalitarian shows of loyalty to the state by citizen parades.

Will a private group interested in preparing a recommendation for community development be required to carry out its own survey and analysis of the community? The answer would depend upon the quality of the work prepared by the public agency, work which should be public information. In some instances the public agency may not have surveyed or analyzed aspects the private group thinks important; or the public agency's work may reveal strong biases unacceptable to the private group. In any event, the production of a useful plan proposal will require much information concerning the present and predicted conditions in the community. There will be some costs associated with gathering that information, even if it is taken from the public agency. The major cost involved in the preparation of a plan by a private agency would probably be the employment of one or more professional planners.

What organizations might be expected to engage in the plural planning process? The first type that comes to mind are the political parties; but this is clearly an aspirational thought. There is very little evidence that local political organizations have the interest, ability, or concern to establish well-developed programs for their communities. Not all the fault, though, should be placed upon the professional politicians, for the registered members of political parties have not demanded very much, if anything, from them as agents.

Despite the unreality of the wish, the desirability for active participation in the process of planning by the political parties is strong. In an ideal situation local parties would establish political platforms which would contain master plans for community growth and both the majority and minority parties in the legislative branch of government would use such plans as one basis for appraising individual legislative proposals. Further, the local administration would use its planning agency to carry out the plans it proposed to the electorate. This dream will not turn to reality for a long time. In the interim other interest groups must be sought to fill the gap caused by the present inability of political organizations.

The second set of organizations which might be interested in preparing plans for community development are those that represent special interest groups having established views in regard to proper public policy. Such organizations as chambers of commerce, real estate boards, labor organizations, pro- and anti-civil rights groups, and anti-poverty councils come to mind. Groups of this nature have often played parts in the development of community plans, but only in a very few instances have they proposed their own plans.

It must be recognized that there is strong reason operating against commitment to a plan by these organizations. In fact it is the same reason that in part limits the interests of politicians and which limits the potential for planning in our society. The expressed commitment to a particular plan may make it difficult for groups to find means for accommodating their various interests. In other terms, it may be simpler for professionals, politicians, or lobbyists to make deals if they have not laid their cards on the table.

There is a third set of organizations that might be looked to as proponents of plans and to whom the foregoing comments might not apply. These are the ad hoc protest associations which may form in opposition to some proposed policy. An example of such a group is a neighborhood association formed to combat a renewal plan, a zoning change, or the proposed location of a public facility. Such organizations may seek to develop alternative plans, plans which would, if effected, better serve their interests.

From the point of view of effective and rational planning it might be desirable to commence plural planning at the level of city-wide organizations, but a more realistic view is that it will start at the neighborhood level. Certain advantages of this outcome should be noted. Mention was made earlier of tension in government between centralizing and decentralizing forces. The contention aroused by conflict between the central planning agency and the neighborhood organization may indeed be healthy, leading to clearer

definition of welfare policies and their relation to the rights of individuals or minority groups.

Who will pay for plural planning? Some organizations have the resources to sponsor the development of a plan. Many groups lack the means. The plight of the relatively indigent association seeking to propose a plan might be analogous to that of the indigent client in search of legal aid. If the idea of plural planning makes sense, then support may be found from foundations or from government. In the beginning it is more likely that some foundation might be willing to experiment with plural planning as a means of making city planning more effective and more democratic. Or the Federal Government might see plural planning, if carried out by local anti-poverty councils, as a strong means of generating local interest in community affairs.

Federal sponsorship of plural planning might be seen as a more effective tool for stimulating involvement of the citizen in the future of his community than are the present types of citizen participation programs. Federal support could only be expected if plural planning were seen, not as a means of combating renewal plans, but as an incentive to local renewal agencies to prepare better plans.

The public planning agency

A major drawback to effective democratic planning practice is the continuation of that non-responsible vestigial institution, the planning commission. If it is agreed that the establishment of both general policies and implementation policies are questions affecting the public interest and that public interest questions should be decided in accord with established democratic practices for decision making, then it is indeed difficult to find convincing reasons for continuing to permit independent commissions to make planning decisions. At an earlier stage in planning the strong arguments of John T. Howard and others in support of commissions may have been persuasive. But it is now more than a decade since Howard made his defense against Robert Walker's position favoring planning as a staff function under the mayor. With the increasing effect planning decisions have upon the lives of citizens the Walker proposal assumes great urgency.

Aside from important questions regarding the propriety of independent agencies which are far removed from public control determining public policy, the failure to place planning decision choices in the hands of elected officials has weakened the ability of professional planners to have their proposals effected. Separating planning from local politics has made it difficult for independent commissions to garner influential political support. The commissions are not responsible directly to the electorate and in turn the electorate is, at best, often indifferent to the planning commission.

During the last decade, in many cities power to alter community development has slipped out of the hands of city planning commissions, assuming they ever held it, and has been transferred to development co-ordinators. This has weakened the professional planner. Perhaps planners unknowingly contributed to this by their refusal to take concerted action in opposition to the perpetuation of commissions.

Planning commissions are products of the conservative reform movement of the early part of this century. The movement was essentially anti-populist and pro-aristocracy. Politics was viewed as dirty business. The commissions are relics of a not-too-distant past when it was believed that if men of good will discussed a problem thoroughly, certainly the right solution would be forthcoming. We know today, and perhaps it was always known, that there are no right solutions. Proper policy is that which the decision-making unit declares to be proper.

Planning commissions are responsible to no constituency. The members of the commissions, except for their chairman, are seldom known to the public. In general the individual members fail to expose their personal views about policy and prefer to immerse them in group decision. If the members wrote concurring and dissenting opinions, then at least the commissions might stimulate thought about planning issues. It is difficult to comprehend why this aristocratic and undemocratic form of decision making should be continued. The public planning function should be carried out in the executive or legislative office and perhaps in both. There has been some question about which of these branches of government would provide the best home, but there is much reason to believe that both branches would be made more cognizant of planning issues if they were each informed by their own planning staffs. To carry this division further, it would probably be advisable to establish minority and majority planning staffs in the legislative branch.

At the root of my last suggestion is the belief that there is or should be a Republican and Democratic way of viewing city development; that there should be

conservative and liberal plans, plans to support the private market, and plans to support greater government control. There are many possible roads for a community to travel and many plans should show them. Explication is required of many alternative futures presented by those sympathetic to the construction of each such future. As indicated earlier, such alternatives are not presented to the public now. Those few reports which do include alternative futures do not speak in terms of interest to the average citizen. They are filled with professional jargon and present sham alternatives. These plans have expressed technical land use alternatives rather than social, economic, or political value alternatives. Both the traditional unitary plans and the new ones that present technical alternatives have limited the public's exposure to the future states that might be achieved. Instead of arousing healthy political contention as diverse comprehensive plans might, these plans have deflated interest.

The independent planning commission and unitary plan practice certainly should not co-exist. Separately they dull the possibility for enlightened political debate; in combination they have made it yet more difficult. But when still another hoary concept of city planning is added to them, such debate becomes practically impossible. This third of a trinity of worn-out notions is that city planning should focus only upon the physical aspects of city development.

AN INCLUSIVE DEFINITION OF THE SCOPE OF PLANNING

The view that equates physical planning with city planning is myopic. It may have had some historic justification, but it is clearly out of place at a time when it is necessary to integrate knowledge and techniques in order to wrestle effectively with the myriad of problems afflicting urban populations.

The city planning profession's historic concern with the physical environment has warped its ability to see physical structures and land as servants to those who use them. Physical relations and conditions have no meaning or quality apart from the way they serve their users. But this is forgotten every time a physical condition is described as good or bad without relation to a specified group of users. High density, low density, green belts, mixed uses, cluster developments, centralized or decentralized business centers are per se neither good nor bad. They describe physical relations or conditions, but take on value only when seen in terms of their social, economic, psychological, physiological, or aesthetic effects upon different users.

The profession's experience with renewal over the past decade has shown the high costs of exclusive concern with physical conditions. It has been found that the allocation of funds for removal of physical blight may not necessarily improve the overall physical condition of a community and may engender such harsh social repercussions as to severely damage both social and economic institutions. Another example of the deficiencies of the physical bias is the assumption of city planners that they could deal with the capital budget as if the physical attributes of a facility could be understood apart from the philosophy and practice of the service conducted within the physical structure. This assumption is open to question. The size, shape, and location of a facility greatly interact with the purpose of the activity the facility houses. Clear examples of this can be seen in public education and in the provision of low cost housing. The racial and other socioeconomic consequences of "physical decisions" such as location of schools and housing projects have been immense, but city planners, while acknowledging the existence of such consequences, have not sought or trained themselves to understand socioeconomic problems, their causes or solutions.

The city planning profession's limited scope has tended to bias strongly many of its recommendations toward perpetuation of existing social and economic practices. Here I am not opposing the outcomes, but the way in which they are developed. Relative ignorance of social and economic methods of analysis has caused planners to propose solutions in the absence of sufficient knowledge of the costs and benefits of proposals upon different sections of the population.

Large expenditures have been made on planning studies of regional transportation needs, for example, but these studies have been conducted in a manner suggesting that different social and economic classes of the population did not have different needs and different abilities to meet them. In the field of housing, to take another example, planners have been hesitant to question the consequences of locating public housing in slum areas. In the field of industrial development, planners have seldom examined the types of jobs the community needs; it has been assumed that one job was about as useful as another. But this may not

be the case where a significant sector of the population finds it difficult to get employment.

"Who gets what, when, where, why, and how" are the basic political questions which need to be raised about every allocation of public resources. The questions cannot be answered adequately if land use criteria are the sole or major standards for judgment.

The need to see an element of city development, land use, in broad perspective applies equally well to every other element, such as health, welfare, and recreation. The governing of a city requires an adequate plan for its future. Such a plan loses guiding force and rational basis to the degree that it deals with less than the whole that is of concern to the public.

The implications of the foregoing comments for the practice of city planning are these. First, state planning enabling legislation should be amended to permit planning departments to study and to prepare plans related to any area of public concern. Second, planning education must be redirected so as to provide channels of specialization in different parts of public planning and a core focused upon the planning process. Third, the professional planning association should enlarge its scope so as to not exclude city planners not specializing in physical planning.

A year ago at the AIP convention it was suggested that the AIP Constitution be amended to permit city planning to enlarge its scope to all matters of public concern. Members of the Institute in agreement with this proposal should seek to develop support for it at both the chapter and national level. The Constitution at present states that the Institute's "particular sphere of activity shall be the planning of the unified development of urban communities and their environs and of states, regions and the nation *as expressed through determination of the comprehensive arrangement of land and land occupancy and regulation thereof.*" It is time that the AIP delete the words in my italics from its Constitution. The planner limited to such concerns is not a city planner, he is a land planner or a physical planner. A city is its people, their practices, and their political, social, cultural and economic institutions as well as other things. The city planner must comprehend and deal with all these factors.

The new city planner will be concerned with physical planning, economic planning, and social planning. The scope of his work will be no wider than that presently demanded of a mayor or a city councilman. Thus, we cannot argue against an enlarged planning function on grounds that it is too large to handle. The mayor needs assistance; in particular he needs the assistance of a planner, one trained to examine needs and aspirations in terms of both short- and long-term perspectives. In observing the early stages of development of Community Action Programs, it is apparent that our cities are in desperate need of the type of assistance trained planners could offer. Our cities require for their social and economic programs the type of long-range thought and information that have been brought forward in the realm of physical planning. Potential resources must be examined and priorities set.

What I have just proposed does not imply the termination of physical planning, but it does mean that physical planning be seen as part of city planning. Uninhibited by limitations on his work, the city planner will be able to add his expertise to the task of coordinating the operating and capital budgets and to the job of relating effects of each city program upon the others and upon the social, political, and economic resources of the community.

An expanded scope reaching all matters of public concern will make planning not only a more effective administrative tool of local government but it will also bring planning practice closer to the issues of real concern to the citizens. A system of plural city planning probably has a much greater chance for operational success where the focus is on live social and economic questions instead of rather esoteric issues relating to physical norms.

THE EDUCATION OF PLANNERS

Widening the scope of planning to include all areas of concern to government would suggest that city planners must possess a broader knowledge of the structure and forces affecting urban development. In general this would be true. But at present many city planners are specialists in only one or more of the functions of city government. Broadening the scope of planning would require some additional planners who specialize in one or more of the services entailed by the new focus.

A prime purpose of city planning is the coordination of many separate functions. This coordination calls for men holding general knowledge of the many elements comprising the urban community. Educating a man for performing the coordinative role is a difficult job, one not well satisfied by the present tradition of two years

of graduate study. Training of urban planners with the skills called for in this article may require both longer graduate study and development of a liberal arts undergraduate program affording an opportunity for holistic understanding of both urban conditions and techniques for analyzing and solving urban problems.

The practice of plural planning requires educating planners who would be able to engage as professional advocates in the contentious work of forming social policy. The person able to do this would be one deeply committed to both the process of planning and to particular substantive ideas. Recognizing that ideological commitments will separate planners, there is tremendous need to train professionals who are competent to express their social objectives.

The great advances in analytic skills, demonstrated in the recent May issue of this journal [*Journal of the American Institute of Planners*] dedicated to techniques of simulating urban growth processes, portend a time when planners and the public will be better able to predict the consequences of proposed courses of action. But these advances will be of little social advantage if the proposals themselves do not have substance. The contemporary thoughts of planners about the nature of man in society are often mundane, unexciting or gimmicky. When asked to point out to students the planners who have a developed sense of history and philosophy concerning man's situation in the urban world one is hard put to come up with a name. Sometimes Goodman or Mumford might be mentioned. But planners seldom go deeper than acknowledging the goodness of green space and the soundness of proximity of linked activities. We cope with the problems of the alienated man with a recommendation for reducing the time of the journey to work.

CONCLUSION

The urban community is a system comprised of interrelated elements, but little is known about how the elements do, will, or should interrelate. The type of knowledge required by the new comprehensive city planner demands that the planning profession be comprised of groups of men well versed in contemporary philosophy, social work, law, the social sciences, and civic design. Not every planner must be knowledgeable in all these areas, but each planner must have a deep understanding of one or more of these areas and he must be able to give persuasive expression to his understanding. As a profession charged with making urban life more beautiful, exciting, and creative, and more just, we have had little to say.

Our task is to train a future generation of planners to go well beyond us in its ability to prescribe the future urban life.

"Planning for Sustainability in European Cities: A Review of Practice in Leading Cities"

from *The Sustainable Urban Development Reader* (2003)

Timothy Beatley

Editors' Introduction

The urbanization of the human population described by Kingsley Davis (p. 17) and industrialization without regard for the environmental consequences in both capitalist and communist countries, raise serious concerns about planet Earth's capacity to sustain urban life, as the Bruntland Commission concluded (p. 337). Green urbanism and sustainable urban development are alternatives proposed to align human development with natural processes and assure that natural resources will be available to subsequent generations.

Timothy Beatley, a professor of urban planning from the University of Virginia, has studied green urbanism and planning for sustainability in European cities. In this selection he succinctly summarizes promising sustainability practices European cities are engaged in, as well as offering a critique of their limitations.

Most European cities take sustainable urban development more seriously than cities in the US and elsewhere in the world. Vigorous green politics, participation in EU-sponsored information sharing, and hundreds of exemplary local projects are evidence of their concern. Because European cities have pioneered new policies, the many successful sustainability practices they have adopted and some shortcomings they have experienced offer important lessons to other cities worldwide.

Hard-headed decision-makers often dismiss alternative green visions as hopelessly unrealistic. In their view it is easy to dream of cities where renewable solar and wind power rather than non-renewable resources like coal, oil, and natural gas produce much of the communities' energy needs. It is pleasant to fantasize about communities where people walk and bicycle to work rather than drive cars, recycle sewage sludge into biogas for their buses, and grow many of their vegetables in urban gardens. But, these critics of green urbanism say, alternative energy sources can never really produce nearly enough energy to run cities. Bicycle and pedestrian paths are nice, but people need cars and super highways to get around. And agribusiness, not urban farms, is the only way to grow enough food cheaply enough to feed the earth's 6 billion-plus inhabitants. Beatley disagrees. He produces hard evidence from Europe that compact, walkable, energy-efficient, green communities can be created; cities that are sustainable, livable, and also economically viable.

Fundamental to many European sustainability practices is the way European cities have limited sprawl and encouraged compact development. European cities have much higher average densities than American cities. They are generally more compact than their American counterparts because their citizens accept much higher density development than suburb-loving, auto-dependent Americans. Americans who have chosen to live in low-density suburbs travel to Paris or Amsterdam to enjoy the energy and street life!

European cities have achieved relatively high average densities mainly by restricting sprawl, building new areas adjacent to the existing city core at relatively high densities, and fostering urban development and industrial reuse.

Beatley points out that this higher density makes possible more efficient public transit and energy systems and facilitates pedestrian spaces.

European cities generally invest much more per capita in public transportation systems such as high-speed rail lines (bullet trains), subways, and buses than cities in the United States. They work hard to reduce auto dependency through imaginative incentive systems to reduce driving, car-sharing programs, and promotion of bicycle use. European transit systems are consistent with land use plans and well integrated with each other. This makes it possible for households to get around with one car or no car at all. If there is a really good public transit system serving a compact city, as is the case in many European cities, people will use buses rather than cars for many trips. This in turn generates revenue to keep the system viable. In contrast, if there is a terrible bus system that functions poorly as the transit system of last resort, it will not attract riders who have other alternatives. Without enough revenue the system will deteriorate further. In many European countries gasoline taxes are two or three times as high as in the United States. This is a good example of how Wilbur Thompson's ideas about pricing goods to achieve urban policies can be put into practice (p. 266). High gas taxes discourage driving and contribute to more compact cities, better and more used public transportation, and less air pollution.

Amsterdam and other European cities work hard to make bicycling possible. Some even provide free public bicycle depots. This bundle of carrots and sticks makes for much more environmentally friendly transit in Europe than in North America.

Not only are European cities generally better served by public transit, they are often more pedestrian-friendly than sprawling, low-density cities in the United States and elsewhere in the world. Greater density makes it easier for people to get around on foot. Conscious policies to promote attractive, exciting pedestrian areas reinforce walking as an alternative. European cities demonstrate that Stephen Wheeler's goals of sustainability and livability (p. 499) can go hand in hand.

European cities have made impressive strides in increasing energy efficiency and reducing waste. Stockholm, for example, has reorganized its government departments so that the city offices dealing with waste, water, and energy are all grouped together. Sewage sludge in Stockholm is used for fertilizer and – hard-headed critics notwithstanding – to produce biogas to fuel the cities' buses and a local power plant.

Today green politics has established itself as a worldwide political movement. Green parties have a political agenda to protect the environment and manage Earth's resources responsibly. In Europe green parties have been successful in electing enough representatives to be political forces in the governing coalitions of a number of countries. In the United States some local green party candidates have been elected and green politics is stimulating new thinking about the nature of urban development.

Green urbanism is not a new idea. Before World War I, eccentric Scottish biologist Patrick Geddes had classified the environmental needs of different ecological systems and developed a systematic approach to building cities that respect natural systems. Ian McHarg, another Scot, wrote a seminal book titled *Design with Nature* that inspired the environmentally conscious generation of the 1960s. Today there is a movement in ecological design among architects, and respect for the natural environment is a cornerstone of the New Urbanism movement described by Andrés Duany and Elizabeth Plater-Zyberk (p. 192) and Peter Calthorpe and William Fulton (p. 342).

Timothy Beatley is the Theresa Heinz Professor of Urban and Environmental Planning at the University of Virginia, Charlottesville. The practices reviewed in this selection are discussed in greater detail in Beatley, *Green Urbanism: Learning from European Cities* (Washington, DC: Island Press, 2000). Professor Beatley's other books include *The Ecology of Place* (Washington, DC: Island Press, 1997) and *An Introduction to Coastal Zone Management* (Washington, DC: Island Press, 1993).

Key classic and contemporary writings on sustainable urban development and green urbanism are contained in *The Sustainable Urban Development Reader* (London and New York: Routledge, 2004), which Beatley co-edited with Stephen Wheeler.

Other readings on the practice of sustainable urban development include Gwendolyn Hallsmith, *The Key to Sustainable Cities: Meeting Human Needs, Transforming Community Systems* (Gabriola Island, BC: New Society Publishers, 2003), Judy Corbett, Michael Corbett, and Robert Thayer, *Designing Sustainable Communities: Learning from Village Homes* (Washington, DC: Island Press, 2000), Stacy Mitchell and Mark Roseland, *Toward Sustainable Communities: Resources for Citizens and Their Governments* (Gabriola Island, BC: New Society

Publishers, 1998), and Alex Wilson, Jen Uncapher, Lisa McManigal, and L. Hunter Lovins, *Green Development: Integrating Ecology and Real Estate* (New York: John Wiley, 1998).

For more on green politics in Europe see John S. Dryzek (ed.), *Green States and Social Movements: Environmentalism in the United States, United Kingdom, Germany and Norway* (London and New York: Oxford University Press, 2002), Michael Dobson, *Green Political Thought*, third edition (London and New York: Routledge), and Michael O'Neill, *Green Parties and Political Change in Contemporary Europe: New Politics, Old Predicaments* (Aldershot and Brookfield: Ashgate, 1997). For one view on green politics in America, see John Rensenbrink, *Against All Odds: The Green Transformation of American Politics* (Raymond: Leopold Press, 1999).

Classic statements that anticipate the green urbanism movement are Patrick Geddes, *Cities in Evolution* (London. Williams & Norgate, 1015), reprinted in Richard LeGates and Frederic Stout, *Early Urban Planning 1870–1940* (London: Routledge/Thoemmes, 1998) and Ian McHarg, *Design with Nature* (Garden City: Doubleday & Company, 1969).

■ ■ ■ ■ ■ ■

INTRODUCTION: LEARNING FROM EUROPEAN CITIES

In few other parts of the world is there as much interest in sustainability as in Europe, especially northern and northwestern Europe, and as much tangible evidence of applying this concept to cities and urban development. For approximately the last six years this author has been researching innovative urban sustainability practice in European cities. The findings from the first phase of this work are presented in the book *Green Urbanism: Learning from European Cities* (Island Press, 2000). What follows is a summary of some of the key themes and most promising ideas and strategies found in the 30 or so cities, in 11 countries, described in this book, as well as more recent case studies and field work.

An initial observation from this work is just how important sustainability is at the municipal level in Europe, especially evident in the cities chosen. "Sustainable cities" resonates well and has important political meaning and significance in these cities, and on the European urban scene generally. One measure of this is the success of the Sustainable Cities and Towns campaign, an EU-funded informal network of communities pursuing sustainability begun in 1994. Participating cities have signed the so-called Aalborg Charter (from Aalborg, Denmark, the site of the first campaign conference), and more than 1800 cities and towns have done so. Among the activities of this organization are the publication of a newsletter, networking between cities, and initiation of conferences and workshops. The organization has also created the annual European Sustainable City award (with the first of these awards issued in 1996), and it is clear that they have been coveted and highly valued by politicians and city officials.

Many European cities have also gone through, or are currently going through, some form of local Agenda 21 process (including many of the same cities that have signed the Aalborg charter), and this is another important indicator of the relevance of local sustainability. Indeed, in the countries studied, high percentages of municipal governments are participating (for instance, in Sweden 100 percent of all local governments are at some stage in the local Agenda 21 process). Often these programs represent tremendous local efforts to engage the community in a dialogue about sustainability, and typically involve the creation of a local sustainability forum, sustainability indicators, local state-of-the-environment reports, and the preparation of comprehensive local sustainability action plans. European cities and towns demonstrate serious commitment to environmental and sustainability values and what follows are a few of the more important ways in which these concerns are being addressed.

Compact cities and regions

Urban form and land use patterns are primary determinants of urban sustainability. While European cities have been experiencing considerable decentralization pressures, they are typically much more compact and dense than American cities. Peter Newman and Jeffrey Kenworthy have monitored and tracked average density in a number of cities throughout the world.

Western European cities like Amsterdam and Paris have substantially higher densities, as measured in persons per hectare, than typical American cities. Overall or whole-city densities for European cities are typically in the 40–60 persons per hectare range; American cities are much lower, commonly under 20 persons per hectare. Even American cities that we tend to think of as particularly dense, for example New York, are comparatively less dense when the entire metropolitan wide pattern is considered. Density and compactness directly translate into much lower energy use, per capita, and lower carbon emissions, air and water pollution, and other resource demands compared with less dense, less compact cities.

Many of these European examples, moreover, show that compactness and density need not translate to skyscrapers and excessive high-rise. Density and compactness in cities like Amsterdam happens through a building pattern of predominately low-rise structures. While many sustainability proponents advocate the need for the green high-rise development (e.g. see Ken Yeang's designs for bio-climatic skyscrapers), these European cities demonstrate convincingly that tremendous compactness and density can be accomplished at a clearly human scale. The European model is appealing to many precisely because of its more traditional form of density and compactness, and many believe its more human scale.

These characteristics of urban form make many other dimensions of local sustainability more feasible, of course (e.g. public transit, walkable places, energy efficiency). There are many factors that explain this urban form, including an historic pattern of compact villages and cities, a limited land base in many countries, and different cultural attitudes about land. Nevertheless in the cities studied there are conscious policies aimed at strengthening a tight urban core. Indeed, the major new growth areas in almost every city studied are situated in locations within or adjacent to existing developed areas, and are designed generally at relatively high densities.

Exemplary and for the most part effective efforts at maintaining the traditional tight urban form can be seen in many cities. Cities like Amsterdam are actively promoting urban redevelopment and industrial reuse (e.g. through its eastern docklands redevelopment). Berlin's plan calls for most future growth to be accommodated with its urbanized area through a variety of infill and reurbanization strategies. Freiburg, Germany, has been able to effectively steer relatively compact,

high-density new growth along the main corridors of its tram system, as well as to protect existing housing supply in the center (there is now a prohibition on the conversion of housing to offices and other uses).

European cities are utilizing a variety of planning strategies to promote compactness and to maintain a tight urban form. These include strict limits on building outside of designated development areas, a strong role for municipal governments in designing and developing new growth areas, extensive public acquisition and ownership of land (especially in Scandinavian cities like Stockholm), and a willingness to make significant transportation and other infrastructure investments that facilitate and support compactness.

GREEN URBANISM: COMPACT AND ECOLOGICAL URBAN FORM

Growth areas and redevelopment districts in these European cities are incorporating a wide range of ecological design and planning concepts, from solar energy to natural drainage to community gardens, and effectively demonstrate that *ecological* and *urban* can go together. Good examples of this compact green growth can be seen in the new development districts planned for or recently completed in Utrecht (Leidsche Rijn), Frieburg (Rieselfeld), Amsterdam (e.g. IJburg), Copenhagen (Orestad), Helsinki (Viikki), and Stockholm (Hammerby Sjostad).

Leidsche Rijn, for example, is an innovative new growth district in the Dutch city of Utrecht. In addition to incorporating a mixed-use design, and a balance of jobs and housing (30,000 dwelling units and 30,000 new jobs), it will include a number of ecological design features. Much of the area will be heated through district heating supplied from the waste energy of a nearby power plant, a double-water system which will provide recycled water for non-potable uses, and a storm water management through a system of natural swales (what the Dutch call "wadies"). Higher-density uses will be clustered around several new train stations and bicycle-only and bicycle/pedestrian-only bridges will provide fast, direct connections to the city center. Homes and buildings will meet a low-energy standard and only certified sustainably harvested wood will be allowed.

European cities also provide excellent and generally successful examples of redevelopment and adaptive reuse of older, deteriorated areas within the center-city. Good examples include Amsterdam's eastern docklands, where 8000 new homes have been accommodated on recycled land. In *Java-eiland*, one major piece of this project, an overall plan (prepared by urban designer Sjoerd Soeters) lays out broad density, massing, and circulation for the district. Diversity and distinctiveness in actual design of the buildings, however, was encouraged through a restriction on the number of buildings that could be designed by a single architect. The result is a stimulating community where buildings have been created by scores of different designers. This island district successfully balances connection to the past (a series of canals and building scale reminiscent of historic Amsterdam) with unique modern design (each of the pedestrian bridges crossing the canals offers a distinctive look and design). *Java-eiland* demonstrates that city building can occur in ways that create interesting and organically evolved places, and which also acknowledge and respect history and context, overcoming sameness.

European cities on the whole (and especially the cities examined in this study) have been able to maintain and strengthen their center cities and urban cores. In no small part this is a function of historic density and compactness, but they are also the result of numerous efforts to maintain and enhance the quality and attractiveness of the city-center. In the cities studied, the center has remained a mixed-use zone, with a significant residential population. *Groningen*, for instance, has undertaken a host of actions to improve its center including the creation of new pedestrian-only shopping areas (creating a system of two linked circles of pedestrian areas), and installation of (yellow) brick surfaces and new street furniture in walking areas, among other actions.

Committed to a policy of compact urban form, Groningen has also made a strong effort to keep all major new public buildings and public attractions close-in. As one example, a new modern art museum has been sited and designed to provide an important pedestrian link between the city's main train station and the town center.

SUSTAINABLE MOBILITY

Achieving a more sustainable mix of mobility options is a major challenge, and in almost all of the cities studied in *Green Urbanism* a very high level of priority

is given to building and maintaining a relatively fast, comfortable and reliable systems of public transport.

There are impressive examples of cities that have been working hard to expand and enhance transit, in the face of rising auto use in many areas. Zurich implements an aggressive set of measures to give priority to its transit on streets. Trams and buses travel on protected, dedicated lanes. A traffic control system gives trams and buses green lights at intersections and numerous changes and improvements have been made to reduce the interference of autos with transit movement (e.g. bans on left turns on tram line roads; prohibiting stopping or parking in certain areas; building pedestrian islands; etc.) A single ticket is good for all modes of transit in the city (including buses, trams, and a new underground regional metro system). The frequency of service is high and there are few areas in the canton that are not within a few hundred meters of a station of stop. Cities like Freiburg and Copenhagen have made similar strides.

In these European cities transit modes are integrated to an impressive degree. This means coordination of investments and routes so that transit modes complement each other. In most of the cities studied, for instance, regional and national trains systems are fully integrated with local routes. It is easy, as well, to shift from one mode to another. Local transit centers are viewed in these cities as multi-modal, mixed-use centers of activity. Arnhem's new central train station in the Netherlands is a case in point. It integrates in a single location high-speed and conventional train service, local transit, bicycle parking, rental, and repair, as well as shops, offices and housing. These uses are all within a few hundred meters of the city center.

The ease of traveling throughout Europe is aided tremendously by the commitment on this continent to high-speed rail. Cross-national movement by high-speed train is increasingly comfortable and easy, and investments in dedicated tracks and infrastructure reflect impressive forward thinking on this issue. And increasingly it is not just the northern and northwestern European nations leading the way. Major new high-speed rail systems are under construction in Italy and Spain for instance. Overall, plans are on the books to double the length of dedicated high-speed rail track in Europe over the next eight years. And, the newest generation of trains will travel faster – on average 300 kph or higher.

Importantly, investments in transit complement, and are coordinated with, important land use decisions.

Virtually all the major new growth areas identified in this study have good public transit service as a basic, underlying design assumption. The cities studied here do not wait until after the housing is built, but rather the lines and investments occur contemporaneously with the projects. The new community growth area *Rieselfeld* in Freiburg, for instance, has a new tram line even before the project has been fully built. In Amsterdam, as a further example, at the new neighborhood of *Nieuw Sloten*, tram service began when the first homes were built. In the new ecological housing district *Kronsberg*, in Hannover, three new tram stops ensure that no resident is further than 600 meters away from a station. There is an recognition in these cities of the importance of providing new residents with options, and establishing mobility patterns early.

Car sharing has become a viable and increasingly popular option in Europe cities. Here, by joining a car sharing company or organization residents have access to neighborhood-based cars, on an hourly or per-kilometer cost. There are now some 100,000 members served by car sharing companies or organizations in 500 European cities. Some of the newest car sharing companies, such as *GreenWheels* in the Netherlands, are also pursuing creative strategies for enticing new customers. This company has been developing strategic alliances, for example with the national train company, to provide packages of benefits at reduced prices. One of the key issues for the success of car sharing is the availability of convenient spaces, and a number of cities, including Amsterdam and Utrecht, have been setting-aside spaces for this purpose. In cities such as Hannover, Germany, the car sharing organization there (a non-profit called Okostadt) has strategically placed cars at the stations of the Stadtbahn, or city tram, furthering enhancing their accessibility.

Thinking beyond the automobile

Many of these cities are in the vanguard of new mobility ideas and concepts and are working hard to incorporate them into new development areas. Amsterdam, for example, has taken an important strategy in developing *Jjburg*. It is working to develop a comprehensive mobility package that all new residents will be offered and which includes, among other things, a free transit pass (for certain specified period) and discounted membership in local car

sharing companies. Minimizing from the beginning the reliance on automobiles, and giving residents more mobility options, are the goals. Eventually this new area will be served both by an extension of the city's underground metro and fast tram.

An increasing number of carfree housing estates are also being developed in these cities, as a further reflection of the commitment to minimizing auto-dependence. The *GWL-Terrein* project, also in Amsterdam, built on the city's old waterworks site, incorporates only very limited peripheral parking. An on-site car sharing company, in combination with good tram service, are part of what makes this concept work there. The interior of the project incorporates extensive gardens (and 120 community gardens available to residents) and pedestrian environment, with key-lock access for fire and emergency vehicles.

Another carfree experiment is the new ecological district *Vauban*, in Freiburg. Built on the former site of an army barracks, this project is unique because it gives new residents the opportunity to declare their intentions to be carfree, and rewards them financially for doing so. Specifically, if residents choose to have a car, they must pay approximately $13,000 for the cost of a space in the nearby parking garage (a bit less than one-tenth the cost of the housing units). In this way there is a strong financial incentive to choose to be carfree and so far about half the residents have taken the carfree path. Projects like *Vauban* challenge new residents to think and act more sustainably and reward them for doing so.

Bicycles are an impressive mobility option in almost all of the cities studied in *Green Urbanism*, and many of these cities have taken tremendous efforts to expand bicycle facilities and to promote bicycle use. Berlin has 800 km of bike lanes, and Vienna has more than doubled its bicycle network since the late 1980s. Copenhagen now has a policy of installing bike lanes along all major streets, and bicycle use in that city has risen substantially. Few have gone as far, of course, as the Dutch cities, with cities like Groningen, where more than half of the daily trips are made on bicycles. In virtually all new growth areas in the Dutch cities, as well as many Scandinavian and German cities, bicycle mobility is an essential design feature, including providing important connections to existing city bicycle networks.

A number of actions have been taken by these cities to promote bicycle use. These include separated bike lanes with separate signaling, separate signaling and priority at intersections, signage and provision of extensive bicycle parking facilities (e.g. especially at train stations, public buildings), and minimum bicycle storage and parking standards for new development. Many cities are gradually converting spaces for auto parking to spaces to bicycles. Utrecht has discovered that it can fit 6–10 bicycles in the same space it takes to park one automobile. Tilburg, in the Netherlands, has recently built an underground valet bicycle parking facility in the heart of that city's shopping district. Freiburg's mobility center combines two levels of bicycle parking, with car-sharing cars on the ground level, a café, travel agency, and office of the Deutsche Bahn (and the structure has a green roof and a photovoltaic array generating electricity!).

These cities are also innovating in the area of public bikes. The most impressive program is Copenhagen's "City Bikes," which now makes available more than 2,000 public bicycles throughout the center of the city. The bikes are brightly painted (companies sponsor and purchase the bikes in exchange for the chance to advertise on their wheels and frames), and can be used by simply inserting a coin as a deposit. The bikes are geared in such a way that the pedaling is difficult enough to discourage their theft. The program has been a success, and the number of bikes has been expanding. These sustainable European cities have discovered that bicycles are an important and legitimate alternative mode of transport to the car and with modest planning and investments substantial ridership can be achieved.

BUILDING PEDESTRIAN CITIES; EXPANDING THE PUBLIC REALM

European cities represent, as well, exemplary efforts at creating walkable, pedestrian urban environments. Relatively compact, dense, and mixed-use urban environments make cities much more walkable, of course. And most European cities and regions benefit from having a compact historic core, designed and evolved around walking and face-to-face commerce. The vitality, beauty, and attraction of European cities is in no small part a function of the impressive public and pedestrian spaces. Cities like Barcelona and Venice remain positive and compelling models of pedestrian urban society. The uses of these spaces are varied and many: they are outdoor stages, the "living rooms" in which citizens socialize, interact, and come

together, places where political events occur and democracy plays itself out. These areas are now the social heart of these communities – places where children play, casual conversations and unexpected meetings take place, and people come to watch and be seen.

The overall land use pattern in these cities, and the priority given to maintaining their compact form, certainly make a walking culture more feasible. What is especially impressive, however, is the continued attention given to this issue and the continued expanding of pedestrian areas and the strengthening of the public and pedestrian realm. Cities like Copenhagen have set the stage, beginning in the early 1960s, gradually taking back their urban centers from cars. That city pedestrianized the Stroget, one of its main downtown streets, in 1962. Copenhagen continues this pedestrianizing in a gradual way each year. The city has adopted the policy of converting 2–3 percent of its downtown parking to pedestrian space each year, to dramatic effect over a 20–40 year period. Today the amount of pedestrian space is tremendous. Eighteen pedestrian squares have been created in Copenhagen where there was once auto parking – some 100,000 square meters in all. Had proponents of public space in Copenhagen attempted to convert this amount of space all at once it would have been very politically difficult to do so.

Many other cities have followed suit, especially Dutch and German cities, but examples can be found throughout Europe. Cities like Vienna and Groningen have pedestrianized much of their centers, creating delightful, highly functional public spaces. Groningen's compact city policy ensures that major new public buildings and facilities are kept in the center, and accessible through walking – it is a compact city of "short distances." In cities like Leiden, emphasis has been given to installing new pedestrian bridges over canals connecting major streets, and every new residential area is designed to include a grocery, post office, and other shops within an easy walk. The greater mixing of uses means that residents of these cities typically have many shops, services, cafés within a walkable range.

The experience of these European cities in pedestrianizing much of their urban centers has been a positive one, both economically and in terms of quality of life. The spaces created commonly contain fountains, sculptures and public art, extensive seating and, of course, many reasons for being there – restaurants, cafés, shops. Each city has its own unique history and features that can be used to strengthen the unique character of its pedestrian environment. Freiburg's "backle," or urban streams that run through the streets of its old center, as well as its pebble mosaics are delightful and special and this city has done an excellent job expanding and adding onto to these unique qualities of place.

Good public transit appears a major factor strengthening the pedestrian realm in these cities, as well as commitments to bicycles, as in the case of Copenhagen. Extensive efforts to calm urban traffic, to restrict auto access, and to raise the cost of parking and auto mobility are also important elements. A number of European cities have experimented with or are anticipating some form of road pricing. The City of London is the most recent notable example, now charging a fee of five pounds for cars wishing to enter central London (and already resulting in a significant reduction in car traffic there). These European experiences support that a pedestrian culture and community life is indeed possible, even where the climate may be harsh and that these spaces serve an incredible range of social, cultural, and economic functions.

GREENING THE URBAN ENVIRONMENT

Ensuring that compact cities are also green cities is a major challenge, and there are a number of impressive greening initiatives among the study cities. First, in many of these cities there is an extensive greenbelt and regional open space structure, with a considerable amount of natural land actually owned by the cities. Extensive tracts of forest and open lands are owned by cities such as Vienna, Berlin, and Graz, among others. Cities such as Helsinki and Copenhagen are spatially structured so that large wedges of green nearly penetrate the center for these cities. Helsinki's large *Keskuspuisto* central park extends in an almost unbroken wedge from the center to an area of old growth forest to the north of city. It is 1000 hectares is size and 11 km long.

In Hannover an extensive system of protected greenspaces exists, including the *Eilenriede*, a 650 hectare dense forest located in the center of the city. Hannover has also recently completed a 80-kilometre long *green ring* (der grune ring) which circles the city, providing a continuous hiking and biking route, and exposing residents to a variety of landscape types, from hilly Borde to the river valleys of the Leineaue river.

There is a trend in the direction of creating and strengthening ecological networks within and between urban centers. This is perhaps most clearly evident in Dutch cities, where extensive attention to ecological networks has occurred at the national and provincial levels. Under the national government's innovative Nature Policy Plan, a national ecological network has been established consisting of core areas, nature development areas, and corridors, which must be more specifically elaborated and delineated at the provincial level. Cities in turn are attempting to tie into this network and build upon it. At a municipal level, such networks can consist of ecological waterways (e.g. canals), tree corridors, and green connections between parks and open space systems. Dutch cities like Groningen, Amsterdam, and Utrecht have full time urban ecology staff, and are working to create and restore these important ecological connections and corridors.

Many examples exist of efforts to mandate or subsidize the greening of existing urban areas. There is a continuing trend, for instance, towards installation of ecological or green rooftops, especially in German, Austrian, and Dutch cities. Linz, Austria, for instance, has one of the most extensive green roof programs in Europe. Under this program, the city frequently requires building plans to compensate for the loss of green space taken by a building. Creation of green roofs has frequently been the response. Also since the late 1980s the city has subsidized the installation of green roofs – specifically, it will pay up to 35 percent of the costs. The program has been quite successful and there are now some 300 green roofs scattered around the city. They have been incorporated into many different types of buildings including a hospital, a kindergarten, a hotel, a school, a concert hall, and even the roof of a gas station. Green roofs have been shown to provide a number of important environmental benefits, and to accommodate a surprising amount of biological diversity. Many other innovative urban greening strategies can be found in these cities from green streets, to green bridges, to urban stream daylighting.

RENEWABLE ENERGY AND CLOSED-LOOP CITIES

A number of the cities have taken action to promote more closed-loop urban metabolism, in which, as in nature, wastes represent inputs or "food," for other urban processes. The city of Stockholm has made some of the most impressive progress in this area, and has even administratively reorganized its governmental structure so that the departments of waste, water, and energy are grouped within an eco-cycles division. A number of actions in support of ecocycle balancing have already occurred. These include, for instance: the conversion of sewage sludge to fertilizer and its use in food production, and the generation of biogas from sludge. The biogas is used to fuel public vehicles in the city, and to fuel a combined heat and power plant. In this way, wastes are returned to residents in the form of district heating. Another powerful example of the closed-loop concept can be seen in Rotterdam, [where] the Roca3 power plant supplies [sic] district heating and carbon dioxide to 120 greenhouses in the area. A waste product becomes a useful input, and in this case prevents some 130,000 metric tonnes of carbon emissions annually.

Energy is very much on the planning agenda, and these exemplary cities are taking a host of serious measures to conserve energy and to promote renewable sources. The heavy use of combined heat and power (CHP) generation, and district heating, especially in northern European cities, is one reason for typically lower per capita levels of CO_2 production here. Helsinki, for instance, has one of the most extensive district heating systems: more than 91 percent of the city's buildings are connected to it. The result is a substantial increase in fuel efficiency, and significant reductions in pollution emissions. District heating and decentralized combined heat and power plants are now commonly integrated into new housing districts in these cities. In *Kronsberg*, in Hannover, for instance, heat is provided by two CHP plants, one of which, serving about 600 housing units and a small school, is actually located in the basement of a building of flats.

Many cities, including Heidelberg and Freiburg, have set ambitious maximum energy consumption standards for new construction projects. Heidelberg has recently sponsored a low-energy social housing project, to demonstrate the feasibility of very low-energy designs (specifically a standard of 47 kwh/m^2

per year). The Dutch are promoting the concept of energy-balanced housing – housing that will over the course of a year produce as much energy as it uses – and the first two of these units have been completed in the *Nieuwland* district in Amersfoort.

Many cities such as Heidelberg have undertaken programs to evaluate and reduce energy consumption in schools and other public buildings. Incentive programs have been established which allow schools to keep a certain percentage of the savings from energy conservation and retrofitting investments. Heidelberg has engaged in an innovative system of performance contracts, in which private retrofitting companies get to keep a certain share of the conservation benefits.

There is an explosion of interest in solar and other renewable energy sources in these cities (and countries). Cities like Freiburg and Berlin have been competing for the label "solar city," with each providing significant subsidies for solar installations. In the Netherlands, major new development areas, such as *Nieuwland* in Amersfoort and *Nieuw Sloten* in Amsterdam, are incorporating solar energy, both passive and active, into their designs. In Nieuwland, described as a "solar suburb," there are more than 900 homes with rooftop photovoltaics, 1100 homes with thermal solar units, and a number of major public buildings producing power from solar (including several schools, a major sports hall, and a childcare facility). What is particularly exciting is to see the effective integration of solar into the architectural design of homes, schools and other buildings.

The degree of public and governmental support in these European cities, financial and technical, for renewable energy developments is truly impressive. Reflecting a generally overall level of concern for global warming issues and energy self-sufficiency, significant production subsidies and consumer subsidies have both been given. The degree of creativity in incorporating renewable energy ideas and technologies in many of these cities is also quite impressive. Oslo's new international airport, for example, provides heating through a bark/wood bio-energy district heating system. This system provides heat for buildings through 8 km of pipes, as well as the airport's de-icing system. The moist bark fuel is a local product, and costs only one-third as much as fuel oil. In Sundsvall, Sweden, snow is collected, stored, and used as a major cooling source for the city's main hospital. In Copenhagen, twenty 2 MW wind turbines have been installed offshore which will together generate enough energy for about 30,000 homes.

Green cities, green governance

Many of these cities are taking a hard look at ways their own operations and management can become more environmentally responsible. As a first step, many local governments have undertaken some form of internal environmental audit. Variously called green audits or environmental audits, they represent attempts to study comprehensively the environmental implications of a city's policies and governance structure. A number of local governments are now going through the process of becoming certified (the London borough of Sutton being the first) under the EU's Eco-Management and Audit Scheme (EMAS), an environmental management system more commonly applied to private companies. Several German cities are preparing environmental budgets, under a pilot program. The cities of Den Haag and London have calculated their ecological footprints and are using these measures as policy guideposts. Albertslund, Denmark, has developed an innovative system of "green accounts," used to track and evaluate key environmental trends at city and district levels, and many of the study cities have developed sustainability indicators (e.g. Leicester, London and Den Haag). Cities like Lahti, Helsinki, and Bologna have gone through extensive in-house education and involvement of city personnel, often as part of the local Agenda 21 process, in examining environmental impacts and in identifying ways that personnel and city departments can reduce waste, energy, and environmental impacts.

Municipal governments have taken a variety of measures to reduce the environmental impacts of their actions. A number of communities have adopted environmental purchasing and procurement policies. Cities like Alberstlund have adopted policies mandating that only organic food can be served in schools and child care facilities, and restricting use of pesticides in public parks and grounds. Other cities are aggressively promoting the development of environmental vehicles. Stockholm's environmental vehicles program is one of the largest (a pilot program under the EU-funded initiative ZEUS), with over 300 vehicles. A number of cities have sought to modify the mobility patterns of employees, for instance by creating financial incentives for the use of transit or bicycles. Cities like Saarbrucken, Germany, have made great strides in reducing energy, waste, and resource consumption in public buildings.

Communities have also engaged in extensive public involvement and outreach on sustainability matters. A variety of creative approaches have been taken. Leicester, for instance, has developed alliances with the local media and has sponsored a series of educational campaigns on particular community issues. As a further example, it has established (with its NGO partner Environ) an environmental center and cyber-cafe called the Ark, as well as a demonstration ecological home. Officials in these exemplary cities often express the belief that it is essential to set a positive example for the community and that before they could ask citizens to change their behaviors and lifestyles, the municipal government must have its environmental house in order.

Understanding European cities: some concluding thoughts

To be sure, many European cities are facing some serious problems and trends working against sustainability, in particular a dramatic rise in automobile ownership and use, and a continuing pattern of deconcentration of people and commerce. And, with their relatively affluent populations consuming substantial amounts of resources, European cities exert a tremendous ecological footprint on the world. Yet, these most exemplary cities provide both tangible examples of sustainable practice, and inspiration that progress can be made in the face of these difficult pressures.

The lessons are several. These cities demonstrate the critical role that municipalities can and must play in addressing serious global environment problems, including reliance on fossil-fuels and global climate change. Innovations in the urban environment offer tremendous potential for dramatically reducing our ecological impacts (European cities produce about half the per-capita carbon emissions of American cities), while at the same time enhancing our quality of life (e.g. by expanding personal mobility options with bicycles and transit).

Many, indeed most, of the ideas, initiatives, strategies undertaken in these innovative cities serve, in addition to reducing ecological footprints, to enhance livability and quality of life. Taking back space from the auto and converting it to pedestrian and public space does much to enhance the desirability of these cities.

Investments in public transit reduce dramatically energy consumption, CO2 emissions, and urban air quality problems, but at the same time provide tremendous levels of independence and mobility to the youngest and oldest members of society. Making bicycling riding safer and easier helps the environment, but also provides a badly needed form of physical exercise.

These experiences demonstrate clearly that it is possible to apply virtually every green or ecological strategy or technique, from solar and wind energy to greywater recycling, in very urban, very compact settings. Green Urbanism is not an oxymoron. Moreover, the lesson of these European cities is that municipal governments can do much to help bring these ideas about, from making parking spaces available for car sharing companies to providing density bonuses for green rooftops, to producing or purchasing green power.

There are also process lessons here. Key among them is an understanding of the great power of partnerships and collaboration between different parties with an interest in sustainability. While not always easy, success at achieving sustainability will depend on them. This means getting different departments to talk to each other and to work together (as in Stockholm), and getting different public and private actors to join together in common initiatives that demonstrate that green urban ideas are possible and desirable.

It is important to recognize, to be sure, the differences in governmental structure. The economic and planning frameworks in place in these countries (compared with, say, the United States) often facilitate many of the exemplary urban sustainability projects described here. The role of economic incentives and the economic incentive structures is critical and undeniable. High prices at the gas pump (typically $4–5 per gallon in Western Europe) have been a conscious policy decision in European countries, and in countries like Germany, have provided essential funding for public transit. Such high prices, relative to countries like the United States, undoubtedly help to encourage more compact land use and personal choices in favor of more sustainable modes of mobility. Also, carbon taxes in countries like Denmark help to substantially level the economic playing field between conventional fossil-fuel energy and more sustainable, renewable forms of energy. Higher energy prices generally help to promote greater conservation and energy efficiency improvements. The important role of adjusting incentives and economic signals is itself a key lesson from the European scene. Rather than seen as a pre-existing background condition, raising gasoline and energy taxes can be seen as an example of an important strategic societal and political choice.

There are other political, social, and cultural conditions, to be sure, that favor many of the exemplary ideas discussed here. Parliamentary governmental structures that give relative voice and power to green party and other social and environmental views (with local representation of these views as well) have been important. Historically stronger planning and land use control systems are helpful also, as well as generally stronger and more proactive roles afforded to government. Many of the important (more activist) urban sustainability activities undertaken in these European cities – as market stimulators, promoters of innovation, and financial underwriters for innovative urban sustainability practices and projects – are common and accepted roles for local governments to play.

But there are also certainly many underlying value differences (compared with the U.S.) that further explain good practice. Prevailing European views of land are less imbued with a sense of personal use and freedom, and there is little expectation, for instance, on the part of a rural landowner or farmer that his or her land will eventually be convertible to urban development.

There are also a number of more regionally unique cultural values and differences, each with significant planning and land use shaping implications. A stronger desire to live within a city or town center clearly exists throughout much of Europe, borne undoubtedly from an older, more developed urban culture. Importance given to strolling, spending time in public places, and to the values of the public realm more generally, in countries like Spain and Italy, certainly help explain the success of pedestrian spaces in these countries. Pace of life, cultural organization of the day, and the number of hours in the work week are also clearly important. In Italy, public and pedestrian spaces are used in part because there is time to use them – the culture organizes its day so as to support the early evening stroll, after the shops close but before the evening meal. To many observers of the European scene there are also lessons to emulate – suggestions and ideas for humanizing cities and strengthening their livability and sociability, as well as their sustainability. The lessons are many and profound on many levels.

Plate 33 Central Park, New York, 1863. Frederick Law Olmsted and his partner Calvert Vaux conceived and executed a park, which was not only a masterpiece of design excellence, but also articulated a philosophy about what urban parks were for. Central Park provided the illusion of nature in the city. It was an oasis of calm and an intended meeting place for different classes. Central Park provided areas for quiet contemplation, boating, strolling, riding, baseball, Sunday school picnics, and countless other activities. (Courtesy of the Museum of the City of New York.)

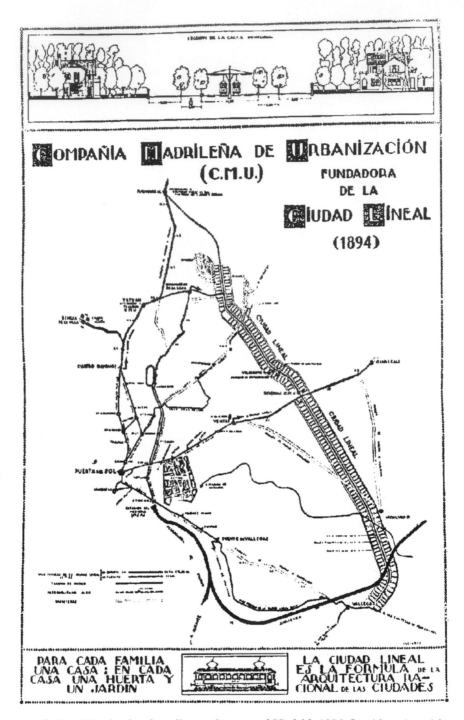

Plate 34 Arturo Soria y Mata's plan for a linear city around Madrid, 1894. Spanish engineer/planner Arturo Soria y Mata envisaged the liberating force of new transportation technology. He conceived of "linear cities" built along electric streetcar lines which would provide for access to nature, large and relatively inexpensive lots, quick transportation, and efficient provision of infrastructure. Portions of the linear city around Madrid pictured here were built. (From Arturo Soria y Mata, *The Linear City*.)

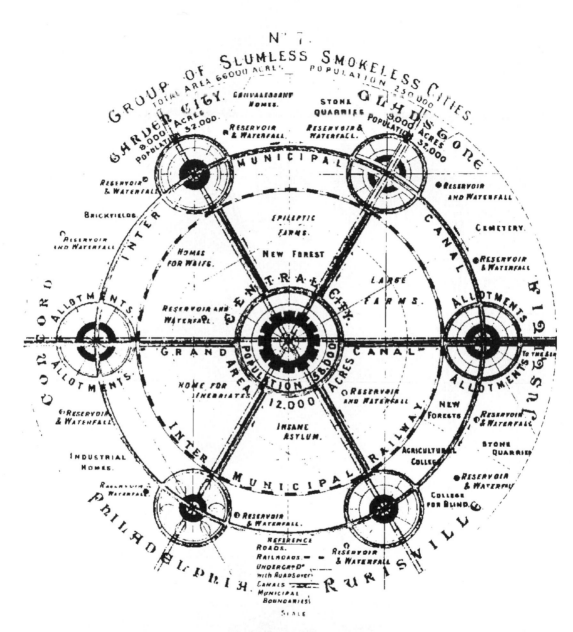

Plate 35 Ebenezer Howard's plan for a Garden City, 1898. From Ebenezer Howard, *To-morrow: A Peaceful Path for Real Reform* (London: Swan Sonnenschein, 1898). Reacting to the squalor of the nineteenth-century city, Ebenezer Howard proposed self-sufficient "Garden Cities" of about 32,000 people, carefully planned and surrounded by a permanent green belt. The Garden City movement spread worldwide and continues to inspire city planners.

Plate 36 Plan for Welwyn Garden City, 1909. Welwyn, the second of Britain's Garden Cities, closely followed Howard's vision. Howard lived to see Welwyn built and spent the last years of his life there.

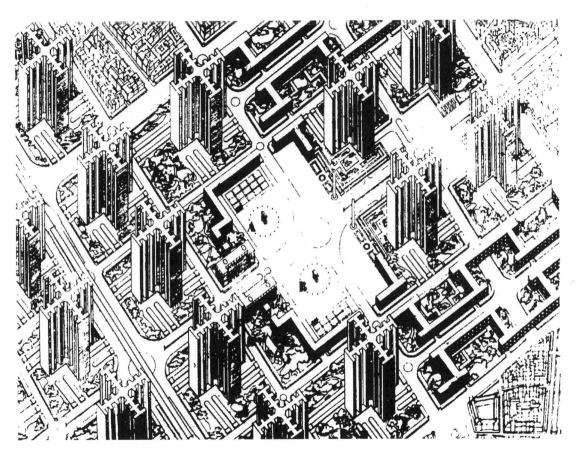

Plate 37 Le Corbusier's "Plan Voisin" for a hypothetical city of three million people, 1925. Visionary modernist Le Corbusier planned huge new cities of steel and concrete dominated by highways and large modern buildings in park-like settings. This 1925 plan contemplates an entire new city of three million people built on these principles.

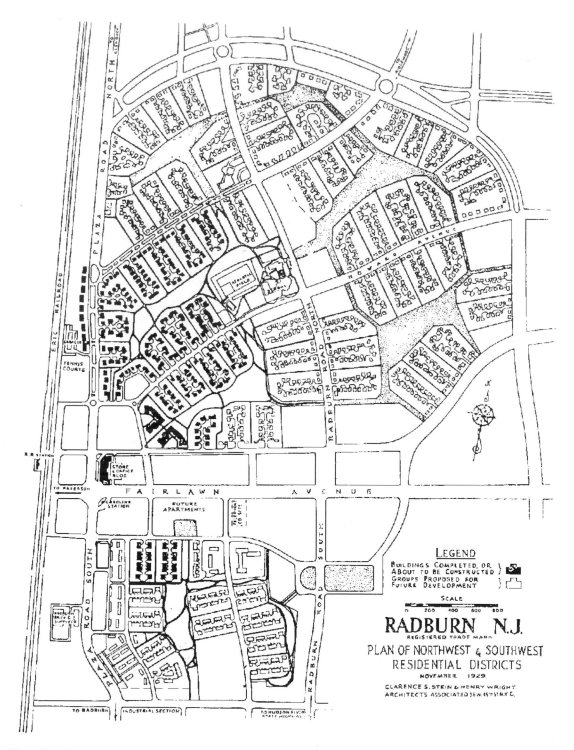

Plate 38 Plan for Radburn, New Jersey, 1929. Architects Clarence Stein and Henry Wright responded to the automobile by designing cities with a separation between pedestrian and traffic arteries, superblocks, and separate city areas with strong neighborhood identity. Their plan for Radburn, New Jersey, has inspired generations of planners.

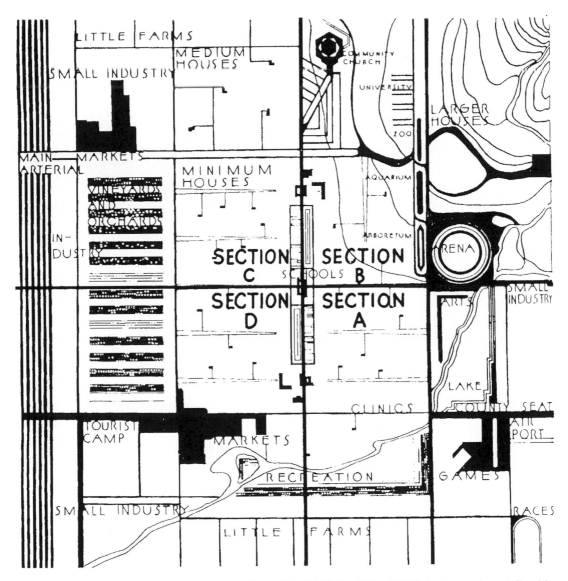

Plate 39 Frank Lloyd Wright's plan for Broadacre City, 1935. Frank Lloyd Wright's extreme decentralist vision of the modern city contemplated a new form of Jeffersonian democracy with decentralized households living with at least one acre of land for each person.

Plate 40 **Paseo del Rio, San Antonio, Texas.** Good city planning can turn the most mundane landscape into a good urban space. Through adopting a bold vision and seeing it through to completion, San Antonio, Texas, turned a blighted riverfront into a magnificent area of parkland, water-oriented activities, and retail shopping. (Photograph by Alexander Garvin.)

Plate 41 Quincy Market, Boston, Massachusetts. Suburban malls and shopping centers may be "the new downtown" in many metropolitan areas, but some cities like Boston, Massachusetts are rebuilding their urban core into attractive shopping and entertainment areas. Boston's Quincy Market has been transformed from a seedy and economically marginal market area into a bustling and successful commercial area. (Photograph by Alexander Garvin.)

Plate 42 Peter Calthorpe's plan for a transit-oriented development: "The Crossings," Mountain View, California. Architects and planners are paying increasing attention to developing commercial properties and housing in relation to transit modes. Commuters in this new California development can walk to a light rail line that connects to work sites and shopping. (Image courtesy of Peter Calthorpe Associates.)

PART SEVEN

Perspectives on urban design

INTRODUCTION TO PART SEVEN

This section focuses on urban *design* – the way in which humans actually shape the built environment. Urban designers are usually trained as architects with further training in urban design, city and regional planning, or both. They focus on the design of sites larger than individual buildings – neighborhoods, park systems, highway corridors, or even entire new towns. Professionals from the related field of landscape architecture are educated to manage the relationship between the natural environment and the built environment. As in other areas where academics study cities or professionals work to build cities, material from many disciplines and professional fields is relevant to urban design. The best designers are also social scientists who study how people use the environments they are designing.

The selections in this section by architect Camillo Sitte, urban designer Kevin Lynch, sociologist William Whyte and city planners Allan Jacobs and Donald Appleyard are examples of the way different approaches and perspectives can enrich urban design. Urban designers may disagree on what makes for a good design, but they share a belief in the value of design itself. They believe professionals should consciously think about physical relationships in the creation of urban space. Design is often expressed through drawings – generated by hand or computer – but design ideas may be expressed in many other ways.

Good urban design usually begins with intensive observation. Camillo Sitte (p. 427) observed squares and plazas in cities all over Europe in order to develop his principles for designing cities. Kevin Lynch and his students (p. 438) surveyed people to see how they perceived the cities where they lived and asked them to draw maps to help understand what parts were and were not clear to them. William Whyte (p. 448) and his students spent hundreds of hours watching people use New York City parks and plazas – filming them, analyzing the films, and quantifying what they observed about how people used parks and plazas in order to specify principles for good park design. Allan Jacobs and Donald Appleyard's "urban design manifesto" (p. 456) grew out of many hours surveying, sketching, counting, mapping, photographing, measuring, and simply walking and looking at San Francisco.

Urban designers often criticize how badly the built environment fits human needs. Ugly, impersonal, dirty, dangerous, dysfunctional, race- and gender-segregated areas dominate many large cities today. Urban design is never value-free. Designers may have different priorities with respect to the value of improving traffic flow versus making pedestrian friendly streets, economically revitalizing an area versus retaining historic buildings, or protecting the natural environment versus keeping the city competitive in the global economy. The design of the built environment may not *determine* human behavior, but implicit in the following selections is the notion that bad design can numb the human spirit and good design can have powerful, positive influences on human beings.

The section begins with Austrian architect Camillo Sitte's theories on "the art of building cities," written at the end of the nineteenth century. Sitte is often viewed as the father of modern urban design. Dissatisfied with the impersonality of building projects in his own city of Vienna, Sitte set out to rediscover the artistic principles that guided classical Greek and Roman city builders and their medieval and renaissance successors. Sitte emphasizes the aesthetic, artistic character of city design. Underlying Sitte's theories is a profound belief in the importance of public places as venues to celebrate civic life. Like Lewis Mumford (p. 85) Sitte values human-scale development, the retention of historical elements, the joys of irregularity, and cities as theaters for the display of culture and civilization. Sitte's writings ran counter to the grand building schemes the new industrial wealth in Europe was creating during his own day. But he touched a responsive chord, and he quickly developed followers throughout the world during his lifetime. In the 1930s, modernists, enamored of industrial technology and efficiency, dismissed Sitte. But in today's postmodern era his ideas are enjoying a resurgence of interest. Sitte's perspective – the aesthetic character of cities – did not deal with many of the difficult and important issues raised in prior sections of this Reader. Issues of class and race, economics and governance, and how to implement city plans are nowhere discussed in his work. But within the sphere of aesthetics and design he made an important contribution.

Seventy years after *The Art of Building Cities* was published another slim volume on urban design appeared. Massachusetts Institute of Technology planning professor Kevin Lynch's *The Image of the City* quickly established itself as the foundation of contemporary urban design. Lynch asked basic questions: How do people perceive the built environment? What are the underlying elements common to human perception of the city? Armed with a better theoretical understanding of how people perceive the city image, what can urban designers actually do to design cities better? The selection by Lynch reprinted here describes the city image and the characteristics of five elements of urban form that Lynch considers to be fundamental. It is rich in suggestions about how these findings can shape better urban design. Urban planners like Allan Jacobs and Donald Appleyard have put Lynch's ideas into practice all over the world.

Sociologist William Whyte's writing on the design of spaces summarizes ideas he developed studying the way in which New Yorkers use urban parks and plazas. Whyte was a sharp observer and a fine writer. His writings are exemplary of the way in which social scientists can link understanding of human behavior to good urban design. Whyte's work also illustrates how urban research can lead directly to changes in city policy. The New York City Planning Department and organizations in New York involved in planning and managing parks have incorporated Whyte's ideas directly into policy. Other cities have followed his lead.

Urban planners Allan Jacobs and Donald Appleyard also illustrate the connection between theory and practice. Jacobs is an emeritus professor of city planning at the University of California, Berkeley, where Appleyard also taught until his death. Both were students of Kevin Lynch. Jacobs served as the director of the San Francisco City Planning Department, and Appleyard worked with him on notable studies of street livability and urban design. Under Jacobs's leadership, the San Francisco City Planning Department produced an award-winning urban design plan, which draws heavily on Lynch's ideas and the insights of Appleyard's studies. That work has profoundly shaped the development of San Francisco over the last quarter century.

Designing the urban environment requires sensitivity to human needs and the natural environment. Urban planners, architects, landscape architects, and other design professions must be sensitive to biological and other natural systems, physical form and function, aesthetics, economics, transportation systems, and all the myriad ways in which people use urban space. Each of the major subdivisions of this anthology can contribute to sensitive urban design.

"Author's Introduction," "The Relationship Between Buildings, Monuments, and Public Squares," and "The Enclosed Character of the Public Square"

from *The Art of Building Cities* (1889)

Camillo Sitte

Editors' Introduction

Camillo Sitte (1843–1903), an Austrian architect, felt that much was lost in the transformation of his native city of Vienna in the latter part of the nineteenth century. Sitte witnessed the old city walls – no longer useful against modern artillery – torn down, a ring road (Ringstrasse) with new electric streetcars built to encircle the city, and old areas in the city leveled for monumental boulevards and impressive new buildings. Sitte felt nostalgia for the oddly shaped cathedral squares and narrow streets of Vienna, Salzburg, and other European cities that had evolved over time. He mourned the loss of structures built to human scale and public spaces embellished with statues, fountains, and other municipal art that adorned cites in classical Greece and Rome, the Middle Ages, and the Renaissance.

Sitte knew instinctively that he enjoyed qualities in cities he had visited that had been built before the modern era. He also admired the urban form of Greek and Roman cities of classical antiquity discernible in ruins and fragments of existing cities. But what exactly were these qualities? And how might modern-day designers, architects, and city planners incorporate the principles that Sitte and so many others enjoyed into new building projects? To answer these questions, Sitte embarked on a careful study of the built environment of notable European cities. Armed with a sketchbook he visited Athens, Rome, Florence, Venice, Paris, Pisa, Salzburg, Rothenburg on the Tauber, Dresden, and dozens of other European cities. Everywhere he went, Sitte carefully sketched the physical form of squares and plazas, the outline of cathedrals and public buildings, the location of statues and fountains. He thought about scale and building materials, views and elevations, the integration of ornamental features with functional buildings. He imagined what civic life in these urban spaces must have been like at the time of Pericles and Julius Caesar, of Lorenzo the Magnficent, and Louis XI of France. Sitte reflected on how the architects and city planners designed aesthetically pleasing spaces that reinforced civic culture. The result was a masterful little book that set in motion the modern study of urban design. Sitte produced a volume that was passionate in its advocacy of human-scale building and consideration of "artistic principles" in city building.

Sitte celebrated public space and was particularly enamoured of public squares and plazas. He applauded the practice in ancient Greece and Rome and during the Italian Renaissance of concentrating outstanding buildings around a single public square or plaza and ornamenting this center of community life with fountains, monuments, and statues.

Sitte had limited influence on the rebuilding of his native Vienna, but enormous and continuing impact elsewhere. "Sitte Schülen" (Sitte schools) sprang up all over Europe as young architects and planners read his book and discussed how to implement his ideas. *The Art of Building Cities* was translated into other languages. In the United States it was the Bible of the turn-of-the-century municipal arts movement. Dozens of local committees inspired by Sitte and the American writer Charles Mulford Robinson formed to "embellish and adorn" American cities.

Sitte fell out of favor in the interwar period when Le Corbusier and the insurgent young architects of the Congrès International de Architecture Moderne (CIAM) developed plans to raze and rebuild what they saw as the obsolete cities using modern materials, monumental scale, and designs inspired by industrial society (p. 322). Today there is a renewed interest in human-scale postmodernist designs, and "New Urbanist" architects and planners like Andrés Duany and Elizabeth Plater-Zyberk (p. 192) and Peter Calthorpe and William Fulton (p. 342) are rediscovering Sitte's writings and find much of value in the principles he developed more than a century ago.

Note the connections between Sitte's celebration of plazas and public squares in classical Greece and Rome and in medieval and renaissance European cities with Lewis Mumford's notion that cities, above all, should be theaters in which humans can display their culture (p. 85) and William Whyte's views on how urban parks and plazas contribute to urban life (p. 448).

Camillo Sitte's treatise is available in many languages and editions. The most recent English language edition is Camillo Sitte, *The Art of Building Cities: City Building According to its Artistic Fundamentals* (Westport: Hyperion Press, 1979).

An exhaustive study of the form of European (and non-European) cities before the Industrial Revolution is A.E.J. Morris, *The History of Urban Form Before the Industrial Revolutions* (New York: John Wiley, 1994). Another study of public spaces in European cities is Paul Zucker, *Town and Square from the Agora to the Village Green*, second edition (Cambridge: MIT Press, 1959). Erwin Anton Gutkind's eight-volume *International History of City Development* (New York: Free Press, 1964) has hundreds of illustrations of public spaces in European cities.

Spiro Kostoff's monumental studies of urban form, *The City Shaped* (New York: Little Brown, 1991) and *The City Assembled* (New York: Little Brown, 1992), contain illustrations of many of the features Sitte discusses. An account of the transformation of Vienna at the time that Sitte lived and wrote is contained in Carl Shorske, *Fin-De-Siècle Vienna: Politics and Culture* (New York: Vintage, 1981).

A biography and study of Sitte's work is George and Christiane Collins, *Camillo Sitte: The Birth of Modern City Planning* (New York: Rizzoli, 1981).

■ ■ ■ ■ ■ ■

AUTHOR'S INTRODUCTION

Memory of travel is the stuff of our fairest dreams. Splendid cities, plazas, monuments, and landscapes thus pass before our eyes, and we enjoy again the charming and impressive spectacles that we have formerly experienced. If we could but stop again at those places where beauty never satiates, we could bear many dreary hours with a light heart and pursue life's long struggle with new energies. Assuredly the imperturbable lightheartedness of the South, on the Hellenic coast, in lower Italy and other favored climes, is above all a gift of nature. And the old cities of these countries, built after the beauty of nature itself, continue to augment nature's gentle and irresistible influence upon the soul of man. Only the person who has never understood the beauty of an ancient city could contradict this assertion.

Let him go ramble on the ruins of Pompeii to convince himself of it. If, after a day of patient investigation there, he walks across the bare Forum, he will be drawn, in spite of himself, to the summit of the monumental staircase toward the terrace of Jupiter's temple. On this platform, which dominates the entire place, he will sense, rising within him, waves of harmony like the pure, full tones of sublime music. Under this influence he will truly understand the words of Aristotle, who thus summarized all principles of city building: "A city should be built to give its inhabitants security and happiness."

The science of the technician will not suffice to accomplish this. We need, in addition, the talent of the artist. Thus it was in ancient times, in the Middle Ages, and in the Renaissance, wherever fine arts were held in esteem. It is only in our mathematical century that the construction and extension of cities has become a purely technical matter. Perhaps, then, it is not beside the point to recall that these problems have diverse aspects, and that he who has been given the least attention in our time is perhaps not the least important.

The object of this study, then, is clear. It is not our purpose to republish ancient and trite ideas, nor to reopen sterile complaints against the already proverbial banality of modern streets. It is useless to hurl general condemnations and to put everything that has been done in our time and place once more to the pillory. That kind of purely negative effort should be left to the critic who is never satisfied and who can only contradict. Those who have enough enthusiasm and faith in good causes should be convinced that our own era can create works of beauty and worth. We shall examine the plan of a number of cities, but neither as historian nor as critic. We wish to seek out, as technician and artist, the elements of composition which formerly produced such harmonious effects, and those which today produce only loose and dull results. Perhaps this study will permit us to find the means of satisfying the three principal requirements of practical city building: to rid the modern system of blocks and regularly aligned houses; to save as much as possible of that which remains from ancient cities; and in our creation to approach more closely the ideal of the ancient models.

This standpoint of practical art will lead us to consider especially the cities of the Middle Ages and the Renaissance. We shall be content, on recalling examples from Greek and Roman conceptions, either to explain the creations of following epochs, or to support the ideas that we propose to develop. For the principal architectural elements of cities have greatly changed since antiquity. Public squares (Forum, market, etc.) are used in our times not so much for great popular festivals or for the daily needs of our life. The sole reason for their existence is to provide more air and light, and to break the monotony of oceans of houses. At times they also enhance a monumental edifice by freeing its walls. It was quite different in ancient times. Public squares, or plazas, were then of prime necessity, for they were theaters for the principal scenes of public life, which today take place in enclosed halls. Under the open sky, on the agora, the council of the ancient Greeks gathered.

The market place, a further center of activity for our ancestors, has persisted, it is true, to the present time, but more and more it is being replaced by vast enclosed halls. And how many other scenes of public life have totally disappeared? Sacrifices before the temples, games, and theatrical presentations of all kinds. The temples themselves were scarcely covered, and the principal part of dwellings, around which were grouped large and small rooms, consisted of an open court. In a word, the distinction between the public square and other structures was so slight that it is amazing to our modern minds, accustomed to a very different state of things.

A review of the writings of the period proves to us that the ancients themselves sensed this similarity. Thus Vitruvius does not discuss the Forum in connection with the placement of public buildings or the arrangement of streets in his account of Dinocrates and his plan of Alexandria. But he does mention it in the same chapter which discusses the Basilica, and in the same book (1, 5) he deals with the theaters, palaces, the circus, and the baths. That is to say, all gathering places under the open sky constituted architectural works. The ancient Forum corresponds exactly to this definition, and Vitruvius logically places it in this group. This close relationship between the Forum and a public hall enhanced architecturally by statues and paintings is brought out clearly by the Latin writer's description, and more clearly still by an examination of the Forum of Pompeii. Vitruvius writes again on this subject:

The Greeks arrange their market places in the form of a square and surround them by vast double column supporting stone or marble architraves above which run the promenades. In Italian cities the Forum takes another aspect, for from time immemorial it has been the theater of gladiatorial combats. The columns, therefore, must be less densely grouped. They shelter the stalls of the silversmiths, and their upper floors have projections in the form of balconies which are advantageously placed for frequent use and for public revenue.

This description illustrates well the correspondence between theater and Forum. This relationship appears still more striking when we examine the plan of the Forum of Pompeii (Figure 1). The square is surrounded on all sides by public buildings. The temple of Jupiter alone rises in isolation. And the two-story colonnade

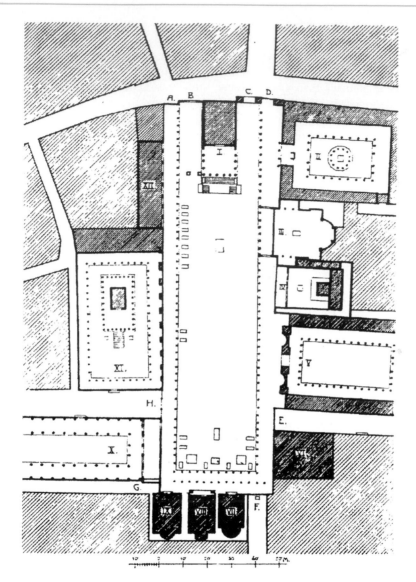

Figure 1 Forum of Pompeii
I Temple of Jupiter, II Enclosed Market, III Temple of Household Gods, IV Temple of Vespasianus, V Eumachia, VI Comitium, VII–IX Public Buildings, X Basilica, XI Temple of Apollo, XII Market Hall

which surrounds the entire space is interrupted only by the peristyle of the temple of the household gods, which makes a greater projection than the other buildings. The center of the Forum remains free, but its periphery is occupied by numerous monuments, the pedestals of which, covered with inscriptions, are still visible.

What a grandiose impression this place must have made! To our modern point of view its effect is like that of a great concert hall without ceiling. In every direction the eye fell upon edifices which in no respect resembled our files of modern houses, and there were

far fewer streets opening directly on the plaza. Streets ran behind buildings III, IV, and V, but they did not extend as far as the Forum. Streets C, D, E, and F were closed by grilles, and even those on the north side passed under the monumental portals, A and B.

Forum Romanum (Figure 2) was conceived according to the same principles. It is surrounded, of course, by buildings more varied in type but all monumental. The streets which open onto it were arranged to avoid too frequent openings in the frame of the plaza. Monuments are located around its sides rather than in

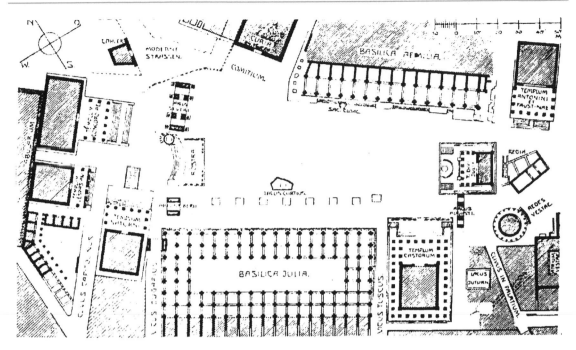

Figure 2 Forum Romanum

its center. In brief, the place of the forum in cities corresponds to that of the principal room of a house. It is to the city, so to speak, the principal hall, as well arranged as it is richly furnished. There stand assembled in immense bulk the columns, the statues, the monuments, and everything that can contribute to the splendor of the place. The art treasures of some of them were said to be numbered in hundreds and thousands. As they did not encumber the midst of the plaza, but were always located at the periphery, it was possible to encompass them all with a single glance; and the spectacle must have been imposing. This concentration of plastic and architectural masterpieces at a single point was a stroke of genius. Aristotle had taught it. He advocated grouping the temples of the gods with public buildings. Pausanias wrote similarly, "A city without public edifices and squares is not worthy of its name."

The market place of Athens is arranged in its principal features according to the same rules, as well as may be judged from the restoration projects. They are applied on a still grander scale in the consecrated cities of Hellenic antiquity (Olympia, Delphi, Eleusis) (Figure 3). Masterpieces of architecture, painting, and sculpture are found there in a superb and imposing union capable of rivaling the most powerful tragedies and the most majestic symphonies. The Acropolis of

Athens (Figure 4) is the most finished creation of this character. A high plateau surrounded by high walls is the base of it. The lower entrance portal, the enormous flight of steps, and monumental vestibules constitute the first phrase of this symphony in marble, gold, ivory, bronze, and color. The interior temples and monuments are the stone myths of the Greek people. The highest poetry and thought are embodied in them. It is truly the center of a considerable city, an expression of the feelings of a great people. It is no longer a simple square in the ordinary sense of the term, but the work of several centuries grown to the maturity of pure art.

It is impossible to establish a higher aim in this style, and it is difficult to imitate successfully this splendid model, but it should always remain before our eyes in all our works as the most sublime ideal to attain. In the progress of our study we shall see that the principles which have inspired such building are not entirely lost, but that they remain to us.

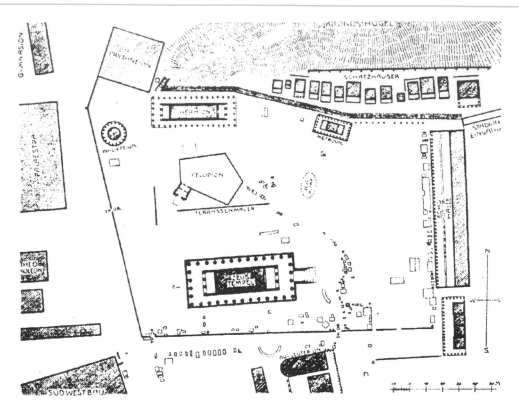

Figure 3 The Temple of Zeus and Plaza of Olympia

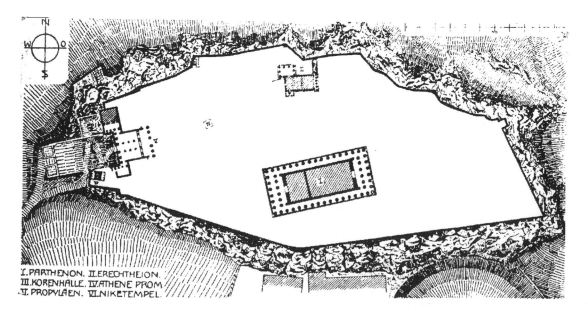

Figure 4 The Acropolis of Athens in the age of Pericles

THE RELATIONSHIP BETWEEN BUILDINGS, MONUMENTS, AND PUBLIC SQUARES

In the South of Europe, and especially in Italy, where ancient cities and ancient public customs have remained alive for ages, even to the present in some places, public squares still follow the type of the ancient forum. They have preserved their role in public life. Their natural relationships with the buildings which enclose them may still be readily discerned. The distinction between the forum, or agora, and the market place also remains. As before, we find the tendency to concentrate outstanding buildings at a single place, and to ornament this center of community life with fountains, monuments, and statues which can bring back historical memories and which, during the Middle Ages and the Renaissance, constituted the glory and pride of each city.

It was there that traffic was most intense. That is where public festivals and theatrical presentations were held. There it was that official ceremonies were conducted and laws promulgated. In Italy, according to varying circumstances, two or three public places, rarely a single one, served these practical purposes.

The existence of two powers, temporal and spiritual, required two distinct centers: one, the cathedral square (Figure 5) dominated by the campanile, the baptistry, and the palace of the bishop; the other, the Signoria, or manor place, which is a kind of vestibule to a royal residence. It is enclosed by houses of the country's great and adorned with monuments. Sometimes we see there a loggia, or open gallery, used by a military guard, or a high terrace from which laws and public statements were promulgated. The Signoria of Florence (Figure 6) is the finest example of this. The market square, rarely lacking even in cities of northern Europe, is the meeting place of the citizens. There stand the City Hall and the more or less richly decorated traditional fountain, the sole vestige of the past that has been conserved since the lively activity of merchants and traders has been moved within to iron cages and glass market places.

The important function of the public square in the community life of past ages is evident. The period of the Renaissance saw the birth of masterpieces in the manner of the Acropolis of Athens, where everything concurred to produce a finished artistic effect. The cathedral place at Pisa, an Acropolis of Pisa (Figure 5), is the proof of this. It includes everything that the people of the City have been able to create in building religious edifices of unparalleled richness and grandeur. The splendid cathedral, the campanile, the baptistry, the incomparable Campo Santo are not depreciated by profane or banal surroundings of any kind. The effect produced by such a place, removed from the world of baseness while rich in the noblest works of the human spirit, is overpowering. Even those with a poorly developed sensitiveness to art are unable to escape the power of this impression. There is nothing there to distract our thoughts or to intrude our daily affairs. The esthetic enjoyment of those who look upon

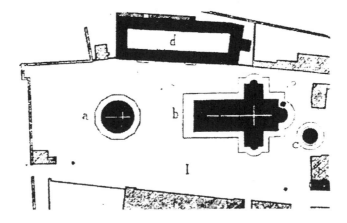

Figure 5 Pisa: Cathedral Square
Key
a. Saint-Jean b. Cathedral c. Campanile d. Campo Santo (Cemetery)

Figure 6 Florence: Piazza of the Signoria

the noble facade of the Cathedral is not spoiled by the sight of a modern haberdashery, by the cries of drivers and porters, or by the tumult of a cafe. Peace reigns over the place. It is thus possible to give full attention to the artwork assembled there.

This situation is almost unique, although that of Saint Francis of Assisi and the arrangement of the Certosa de Pavia closely approach it. In general, the modern period does not encourage the formation of such perfect groupings. Cities, even in the fatherland of art, undergo the fate of palaces and dwellings. They no longer have distinct character. They present a mixture of motifs borrowed as much from the architecture of the north as from that of the southern countries. Ideas and tastes have been mingled as the people themselves have been interchanged. Local characteristics are gradually disappearing. The market place alone, with its City Hall and fountain, has here and there remained intact.

In passing we should like to remark that our intention is not to suggest a sterile imitation of the beauties spoken of as "picturesque" in the ancient cities for our present needs. The proverb, "Necessity breaks even iron," is fully applicable here. Changes made necessary by hygiene or other requirements must be carried out, even if the picturesque suffers from it. But that does not prevent us from examining the work of our forebears at close range to determine how much of it may be adapted to modern conditions. In this way alone can we resolve the esthetic part of the practical problem of city building, and determine what can be saved from the heritage of our ancestors.

Before determining the question in a positive manner, we state the principle that during the Middle Ages and Renaissance public squares were often used for practical purposes, and that they formed an entirety with the buildings which enclosed them. Today they serve at best as places for stationing vehicles, and they have no relation to the buildings which dominate them. Our parliament buildings have no agora enclosed by columns. Our universities and cathedrals have lost their atmosphere of peace. Surging throngs no longer circulate on market days before our City Halls. In brief, activity is lacking precisely in those places where, in ancient times, it was most intense – near public structures. Thus, to a great extent, we have lost that which contributed to the splendor of public squares.

And the fabric of their very splendor, the numerous statues, is almost entirely lacking today. What have we to compare to the richness of ancient forums and to works of majestic style like the Signoria of Florence and its Loggia dei Lanzi?

A few years ago there flourished at Vienna a remarkable school of sculpture whose works of merit cannot be scorned. They were generally used to adorn buildings. In only a few exceptional cases were their works used in public squares. Statues adorn the two museums, the palace of Parliament, the two Court theaters, the City Hall, the new university, the Votive Church. But there is no interest in adorning public open spaces. And that is true not only in Vienna, but nearly everywhere.

Buildings lay claim to so many statues that commissions are needed to find new subjects to be represented. It is often necessary to wait for years to find a suitable place for a statue although many appropriate places remain empty in the meantime. After long efforts we have reconciled ourselves to modern public squares as vast as they are deserted, and the monument, without a place of refuge, becomes stranded on some small and ancient space. That is even more strange, yet true. After much groping about, this fortunate result occurs, for it is thus that a work of art derives its value and produces a more powerful impression. Indifferent artists who neglect to provide for such effects must bear the entire responsibility of it.

The story of Michelangelo's *David* at Florence shows how mistakes of this kind are perpetrated in modern times. This gigantic marble statue stands close to the walls of the Palazzo Vecchio, to the left of its principal entrance, in the exact place chosen by Michelangelo. The idea of erecting a statue on this place of ordinary appearance would have appeared to moderns as absurd if not insane. Michelangelo chose it, however, and without doubt deliberately; for all those who have seen the masterpiece in this place testify to the extraordinary impression that it makes. In contrast to the relative scantiness of the place, affording an easy comparison with human stature, the enormous statue seems to swell even beyond its actual dimensions. The sombre and uniform, but powerful, walls of the palace provide a background on which we could not wish to improve to make all the lines of the figure stand out.

Today the David is moved into one of the academy's halls under a glass cupola in the midst of plaster reproductions, photographs, and engravings. It serves as a model for study and an object of research for historians and critics. A special mental preparation is needed now to resist the morbid influences of an art prison that we call a museum, and to have the ability to enjoy the imposing work. Moreover, the spirit of the times, which believed that it was perfecting art, and which was still not satisfied with this innovation, had a bronze cast made of the David in its original grandeur and put it up on a vast plaza (naturally in its mathematical center) far from Florence at the Via dei Colli. It has a superb horizon before it; behind it, cafes; on one side, a carriage station, a corso; and from all sides the murmurs of Baedeker readers ascend to it. In this setting the statue produces no effect at all. The opinion that its dimensions do not exceed human stature is often heard. Michelangelo thus understood best the

kind of placement that would be suitable for his work, and, in general, the ancients were abler than we are in these matters.

The fundamental difference between the procedures of former times and those of today rests in the fact that we constantly seek the largest possible space for each little statue. Thus we diminish the effect that it could produce, instead of augmenting it with the assistance of a neutral background such as painters have used in their portraits.

This explains why the ancients erected their monuments by the sides of public places, as is shown in the view of the Signoria of Florence. In this way, the number of statues could increase indefinitely without obstructing the circulation of traffic, and each of them had a fortunate background. Contrary to this, we hold the middle of a public place as the sole spot worthy to receive a monument. Thus no esplanade, however magnificent, can have more than one. If by misfortune it is irregular and if its center cannot be located geometrically we become confused and allow the space to remain empty for eternity.

THE ENCLOSED CHARACTER OF THE PUBLIC SQUARE

The old practice of setting churches and palaces back against other buildings brings to mind the ancient forum and its unbroken frame of public buildings. In examining the public squares that came into being during the Middle Ages and the Renaissance, especially in Italy, it is seen that this pattern has been retained for ages by tradition. The old plazas produce a collective harmonious effect because they are uniformly enclosed. In fact, the public square owes its name to this characteristic in an expanse at the center of a city. It is true that we now use the term to indicate any parcel of land bounded by four streets on which all construction has been renounced.

That can satisfy the public health officer and the technician, but for the artist these few acres of ground are not yet a public square. Many things must be done to embellish the area to give it character and importance. For just as there are furnished and unfurnished rooms, we could speak of complete and incomplete squares. The essential thing of both room and square is the quality of enclosed space. It is the most essential condition of any artistic effect, although it is ignored by those who are now elaborating on city plans.

The ancients, on the contrary, employed the most diverse methods of fulfilling this condition under the most diverse circumstances. They were, it is true, supported by tradition and favored by the usual narrowness of streets and less active traffic movement. But it is precisely in cases where these aids were lacking that their talent and artistic feeling is displayed most conspicuously.

A few examples will assist in accounting for this. The following is the simplest. Directly facing a monumental building a large gap was made in a mass of masonry, and the square thus created, completely surrounded by buildings, produced a happy effect. Such is the Piazza S. Giovanni at Brescia. Often a second street opens on to a small square, in which case care is taken to avoid an excessive breach in the border, so that the principal building will remain well enclosed. The methods used by the ancients to accomplish this were so greatly varied that chance alone could not have guided them. Undoubtedly they were often assisted by circumstance, but they also knew how to use circumstances admirably.

Today in such cases all obstructions would be taken down and large breaches in the border of the public place would be opened, as is done when we decide to "modernize" a city. Ancient streets would be found to open on the square in a manner precisely contrary to the methods of modern city builders, and mere chance would not account for this. Today the practice is to join two streets that intersect at right angles at each corner of the square, probably to enlarge as much as possible the opening made in the enclosure and to destroy every impression of cohesion. Formerly the procedure was entirely different. There was an effort to have only one street at each angle of the square. If a second artery was needed in a direction at right angles to the first it was designed to terminate at a sufficient distance from the square to remain out of view from the square. And better still, the three or four streets which came in at the corners each ran in a different direction. This interesting arrangement was reproduced so frequently, and more or less completely, that it can be considered as one of the conscious or subconscious principles of ancient city building.

Careful study shows that there are many advantages to an arrangement of street openings in the form of turbine arms. From any part of the square there is but one exit on the streets opening into it, and the enclosure of buildings is not broken. It even seems to enclose the square completely, for the buildings set at an angle conceal each other, thanks to perspective, and unsightly impressions which might be made by openings are avoided. The secret of this is in having streets enter the square at right angles to the visual lines instead of parallel to them. Joiners and carpenters have followed this principle since the Middle Ages when, with subtle art, they sought to make joints of wood and stone inconspicuous if not invisible.

The Cathedral Square at Ravenna shows the purest type of the arrangement just described. The square of

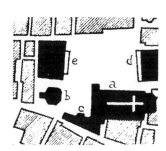

Figure 7 Pistoia: Cathedral Square
Key
a. Cathedral
c. Residence
e. Palais du Podestat
b. Baptistery
d. Palais de la Commune

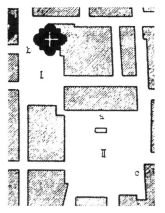

Figure 8 Parma
Key
a. Pal. del Commune
c. Pal. della Podesteria
I. Piazza d. Steccata
b. Madonna della Steccata
II. Piazza Grande

Pistoia (Figure 7) is in the same manner; as is . . . the Piazza Grande at Parma (Figure 8).

The ancients had recourse to still other means of closing in their squares. Often they broke the infinite perspective of a street by a monumental portal or by several arcades of which the size and number were determined by the intensity of traffic circulation. This splendid architectural pattern has almost entirely disappeared, or, more accurately, it has been suppressed. Again Florence gives us one of the best examples in the portico of the Uffizi with its view of the Arno in the distance. Every Italian city of average importance has its portico, and this is also true north of the Alps. We mention only the Langasser Thor at Danzig, the entrance portal of the City Hall and Chancellery at Bruges, the Kerkboog at Nimeguen, the great Bell Tower at Rouen, the monumental Portals of Nancy, and the windows of the Louvre.

More or less ornate portals like those that simply but effectively frame the Piazza dei Signori at Verona (Figure 9) are to be found in all the royal residences, in the chateaux and city halls, and they are used as much for vehicular traffic as by pedestrians. While ancient architects used this pattern wherever possible with infinite variations, our modern builders seem to ignore its existence. Let us recall, to demonstrate again the persistence of ancient traditions, that at Pompeii, too, there is an Arc de Triomphe at the entrance to the Forum.

Columns were used with porticos to form enclosures for public squares. Saint Peter's in Rome is the best example of this . . .

Arcades were used to embellish monumental buildings more frequently in former times than at present, either on the higher stories, as in the City Halls of Halle (1548) and Cologne (1568), or on the ground level . . .

All of these above-mentioned architectural forms in former times made up a complete system of enclosing public squares. Today there is a contrary tendency to open them on all sides. It is easy to describe the results that have come about. It has tended to destroy completely the old public squares. Wherever these openings have been made the cohesive effect of the square has been completely nullified.

Figure 9 Verona: Piazza dei Signori

"The City Image and its Elements"

from *The Image of the City* (1960)

Kevin Lynch

Editors' Introduction

Kevin Lynch (1918–1989) is the towering figure of twentieth-century urban design. *The Image of the City*, from which this selection is taken, is the most widely read urban design book of all time. Lynch was a professor of urban studies and planning at the Massachusetts Institute of Technology where he taught courses in urban design.

As a young student Kevin Lynch apprenticed himself to Frank Lloyd Wright, the brilliant and opinionated designer who was eventually recognized as one of America's greatest architects. Wright utilized natural materials, skylights and walls of windows to embrace the natural environment. Drawing widely on material from psychology and the humanities, Lynch sought to understand how people perceive their environments and how design professionals can respond to the deepest human needs. Lynch's rambling, profoundly humane writings weave together a unique blend of theory and practical design suggestions drawn from his voluminous reading in history, anthropology, architecture, art, literature, and a host of other areas.

This influential chapter on "The City Image and its Elements" presents Lynch's best-known concepts on how people perceive cities. Lynch argues that people perceive cities as consisting of underlying city form "elements" such as *paths* (along which movement flows) and *edges* (which differentiate one part of the urban fabric from another). If they understand how people perceive these elements and design to make cities more imageable, Lynch argues, urban designers can create more psychologically satisfying urban environments.

Urban designers throughout the world today sketch out the elements of cities or parts of cities they are designing as paths, edges, nodes, landmarks, and districts – the underlying elements of city form that Lynch identified – and draw on his theories and practical suggestions to strengthen the city image. Planners in cities as diverse as San Francisco, Cairo, Havana, and Ciudad Guyana in Venezuela, have used Lynch's concepts to inform their urban planning and design strategies.

Compare Lynch's ideas about what people find psychologically satisfying and aesthetically appealing about cities with Camillo Sitte's ideas (p. 427). Contrast his practical suggestions with William Whyte's applied principles and standards for park and plaza design (p. 448), and Frederick Law Olmsted's vision of urban parks (p. 307).

In addition to *The Image of the City* (Cambridge: MIT Press, 1960), Lynch's many books include a textbook on site design co-authored with Gary Hack, *Site Planning*, third edition (Cambridge: MIT Press, 1984), a book on historic preservation, *What Time Is This Place* (Cambridge: MIT Press, 1979), a book on regional planning, *Managing the Sense of a Region* (Cambridge: MIT Press, 1976), and his magnum opus, *Good City Form* (Cambridge: MIT Press, 1991). Other of Lynch's writings are contained in Kevin Lynch, Tridib Banerjee, and Michael Southworth (eds), *City Sense and City Design: Writings and Projects of Kevin Lynch* (Cambridge: MIT Press, 1995).

Classic and contemporary writings on urban design are collected in Michael Larice and Elizabeth Macdonald, *The Urban Design Reader* (London and New York: Routledge, 2006).

Other books on the way in which people perceive urban space include Anthony Hiss, *The Experience of Place* (New York: Knopf, 1990) and Robert Sommer, *Personal Space* (Englewood Cliffs: Prentice-Hall, 1969). For more on urban design see Edmund Bacon, *The Design of Cities* (New York: Viking Press, 1976), Spiro Kostoff, *The City Shaped* (New York: Little Brown, 1991) and *The City Assembled* (New York: Little Brown, 1992), Mike Greenberg, *The Poetics of Cities: Designing Neighborhoods that Work* (Columbus: Ohio State University Press, 1995), and Doug Kelbaugh, *Common Place: Toward Neighborhood and Regional Design* (Seattle: University of Washington Press, 1997).

■ ■ ■ ■ ■ ■

There seems to be a public image of any given city which is the overlap of many individual images. Or perhaps there is a series of public images, each held by some significant number of citizens. Such group images are necessary if an individual is to operate successfully within his environment and to cooperate with his fellows. Each individual picture is unique, with some content that is rarely or never communicated, yet it approximates the public image, which, in different environments, is more or less compelling, more or less embracing.

This analysis limits itself to the effects of physical, perceptible objects. There are other influences on imageability, such as the social meaning of an area, its function, its history, or even its name. These will be glossed over, since the objective here is to uncover the role of form itself. It is taken for granted that in actual design form should be used to reinforce meaning, and not to negate it.

The contents of the city images so far studied, which are referable to physical forms, can conveniently be classified into five types of elements: paths, edges, districts, nodes, and landmarks.

Indeed, these elements may be of more general application, since they seem to reappear in many types of environmental images. . . . These elements may be defined as follows:

1 *Paths*. Paths are the channels along which the observer customarily, occasionally, or potentially moves. They may be streets, walkways, transit lines, canals, railroads. For many people, these are the predominant elements in their image. People observe the city while moving through it, and along these paths the other environmental elements are arranged and related.

2 *Edges*. Edges are the linear elements not used or considered as paths by the observer. They are the boundaries between two phases, linear breaks in con-

tinuity: shores, railroad cuts, edges of development, walls. They are lateral references rather than coordinate axes. Such edges may be barriers, more or less penetrable, which close one region off from another; or they may be seams, lines along which two regions are related and joined together. These edge elements, although probably not as dominant as paths, are for many people important organizing features, particularly in the role of holding together generalized areas, as in the outline of a city by water or wall.

3 *Districts*. Districts are the medium-to-large sections of the city, conceived of as having two-dimensional extent, which the observer mentally enters "inside of," and which are recognizable as having some common, identifying character. Always identifiable from the inside, they are also used for exterior reference if visible from the outside. Most people structure their city to some extent in this way, with individual differences as to whether paths or districts are the dominant elements. It seems to depend not only upon the individual but also upon the given city.

4 *Nodes*. Nodes are points, the strategic spots in a city into which an observer can enter, and which are the intensive foci to and from which he is traveling. They may be primarily junctions, places of a break in transportation, a crossing or convergence of paths, moments of shift from one structure to another. Or the nodes may be simply concentrations, which gain their importance from being the condensation of some use or physical character, as a street-corner hangout or an enclosed square. Some of these concentration nodes are the focus and epitome of a district, over which their influence radiates and of which they stand as a symbol. They may be called cores. Many nodes, of course, partake of the nature of both junctions and concentrations. The concept of node is related to the concept of path, since junctions are typically the convergence of paths, events on the journey. It is similarly

related to the concept of district, since cores are typically the intensive foci of districts, their polarizing center. In any event, some nodal points are to be found in almost every image, and in certain cases they may be the dominant feature.

5 *Landmarks.* Landmarks are another type of point-reference, but in this case the observer does not enter within them, they are external. They are usually a rather simply defined physical object: building, sign, store, or mountain. Their use involves the singling out of one element from a host of possibilities. Some landmarks are distant ones, typically seen from many angles and distances, over the tops of smaller elements, and used as radial references. They may be within the city or at such a distance that for all practical purposes they symbolize a constant direction. Such are isolated towers, golden domes, great hills. Even a mobile point, like the sun, whose motion is sufficiently slow and regular, may be employed. Other landmarks are primarily local, being visible only in restricted localities and from certain approaches. These are the innumerable signs, storefronts, trees, doorknobs, and other urban detail, which fill in the image of most observers. They are frequently used clues of identity and even of structure, and seem to be increasingly relied upon as a journey becomes more and more familiar.

The image of a given physical reality may occasionally shift its type with different circumstances of viewing. Thus an expressway may be a path for the driver, and edge for the pedestrian. Or a central area may be a district when a city is organized on a medium scale, and a node when the entire metropolitan area is considered. But the categories seem to have stability for a given observer when he is operating at a given level.

None of the element types isolated above exist in isolation in the real case. Districts are structured with nodes, defined by edges, penetrated by paths, and sprinkled with landmarks. Elements regularly overlap and pierce one another. If this analysis begins with the differentiation of the data into categories, it must end with their reintegration into the whole image . . .

PATHS

For most people interviewed, paths were the predominant city elements, although their importance varied according to the degree of familiarity with the city. People with least knowledge of Boston tended to think of the city in terms of topography, large regions, generalized characteristics, and broad directional relationships. Subjects who knew the city better had usually mastered part of the path structure; these people thought more in terms of specific paths and their interrelationships. A tendency also appeared for the people who knew the city best of all to rely more upon small landmarks and less upon either regions or paths.

The potential drama and identification in the highway system should not be underestimated. One Jersey City subject, who can find little worth describing in her surroundings, suddenly lit up when she described the Holland Tunnel. Another recounted her pleasure:

> You cross Baldwin Avenue, you see all of New York in front of you, you see the terrific drop of land [the Palisades] . . . and here's this open panorama of lower Jersey City in front of you and you're going down hill, and there you know: there's the tunnel, there's the Hudson River and everything. . . . I always look to the right to see if I can see the . . . Statue of Liberty. . . . Then I always look up to see the Empire State Building, see how the weather is. . . . I have a real feeling of happiness because I'm going someplace, and I love to go places.

Particular paths may become important features in a number of ways. Customary travel will of course be one of the strongest influences, so that major access lines, such as Boylston Street, Storrow Drive, or Tremont Street in Boston, Hudson Boulevard in Jersey City, or the freeways in Los Angeles, are all key image features. . . .

Concentration of special use or activity along a street may give it prominence in the minds of observers. Washington Street is the outstanding Boston example: subjects consistently associated it with shopping and theatres. . . . People seemed to be sensitive to variations in the amount of activity they encountered, and sometimes guided themselves largely by following the main stream of traffic. Los Angeles' Broadway was recognized by its crowds and its street cars; Washington Street in Boston was marked by a torrent of pedestrians. . . .

Characteristic spatial qualities were able to strengthen the image of particular paths. In the simplest sense, streets that suggest extremes of either width or narrowness attracted attention. Cambridge Street,

Commonwealth Avenue, and Atlantic Avenue are all well known in Boston, and all were singled out for their great width. . . . Spatial qualities of width and narrowness derived part of their importance from the common association of main streets with width and side streets with narrowness. Looking for, and trusting to the "main" (i.e., wide) street becomes automatic, and in Boston the real pattern usually supports this assumption. Some of the orientation difficulties in Boston's financial district, or the anonymity of the Los Angeles grid, may be due to this lack of spatial dominance.

Special façade characteristics were also important for path identity. Beacon Street and Commonwealth Avenue were distinctive partly because of the building façades that line them. . . .

Proximity to special features of the city could also endow a path with increased importance. In this case the path would be acting secondarily as an edge. Atlantic Avenue derived much importance from its relation to the wharves and the harbor, Storrow Drive from its location along the Charles River.

[. . .]

Where major paths lacked identity, or were easily confused one for the other, the entire city image was in difficulty. . . . Many of the paths in Jersey City were difficult to find, both in reality and in memory.

That the paths, once identifiable, have continuity as well, is an obvious functional necessity. People regularly depended upon this quality.

[. . .]

People tended to think of path destinations and origin points: they liked to know where paths came from and where they led. Paths with clear and well-known origins and destinations had stronger identities, helped tie the city together, and gave the observer a sense of his bearings whenever he crossed them. . . .

[. . .]

A few important paths may be imaged together as a simple structure, despite any minor irregularities, as long as they have a consistent general relationship to one another. The Boston street system is not conducive to this kind of image, except perhaps for the basic parallelism of Washington and Tremont Streets. But the Boston subway system, whatever its involutions in true scale, seemed fairly easy to visualize as two parallel lines cut at the center by the Cambridge–Dorchester line, although the parallel lines might be confused one with the other, particularly since both go to North Station. The freeway system in

Los Angeles seemed to be imaged as a complete structure. . . .

[. . .]

A large number of paths may be seen as a total network, when repeating relationships are sufficiently regular and predictable.

The Los Angeles grid is a good example. Almost every subject could easily put down some twenty major paths in correct relation to each other. At the same time, this very regularity made it difficult for them to distinguish one path from another.

Boston's Back Bay is an interesting path network. Its regularity is remarkable in contrast to the rest of the central city, an effect that would not occur in most American cities. But this is not a featureless regularity. The longitudinal streets were sharply differentiated from the cross streets in everyone's mind, much as they are in Manhattan. The long streets all have individual character – Beacon Street, Marlboro Street, Commonwealth Avenue, Newbury Street, each one is different – while the cross streets act as measuring devices. The relative width of the streets, the block lengths, the building frontages, the naming system, the relative length and number of the two kinds of streets, their functional importance, all tend to reinforce this differentiation. Thus a regular pattern is given form and character. The alphabet formula for naming the cross streets was frequently used as a location device, much as the numbers are used in Los Angeles.

[. . .]

The frequent reduction of the South End to a geometrical system was typical of the constant tendency of the subjects to impose regularity on their surroundings. Unless obvious evidence refuted it, they tried to organize paths into geometrical networks, disregarding curves and non-perpendicular intersections. The lower area of Jersey City was frequently drawn as a grid, even though it is one only in part. Subjects absorbed all of central Los Angeles into a repeating network, without being disturbed by the distortion at the eastern edge. Several subjects even insisted on reducing the street maze of Boston's financial district to a checkerboard! . . .

EDGES

Edges are the linear elements not considered as paths: they are usually, but not quite always, the boundaries between two kinds of areas. They act as lateral references. They are strong in Boston and Jersey City but weaker in Los Angeles. Those edges seem strongest which are not only visually prominent, but also continuous in form and impenetrable to cross movement. The Charles River in Boston is the best example and has all of these qualities.

The importance of the peninsular definition of Boston has already been mentioned. It must have been much more important in the 18th century, when the city was a true and very striking peninsula. Since then the shore lines have been erased or changed, but the picture persists. One change, at least, has strengthened the image: the Charles River edge, once a swampy backwater, is now well defined and developed. It was frequently described, and sometimes drawn in great detail. Everyone remembered the wide open space, the curving line, the bordering highways, the boats, the Esplanade, the Shell.

The water edge on the other side, the harborfront, was also generally known, and remembered for its special activity. But the sense of water was less clear, since it was obscured by many structures, and since the life has gone out of the old harbor activities. . . .

[. . .]

In Jersey City, the waterfront was also a strong edge, but a rather forbidding one. It was a no-man's land, a region beyond the barbed wire. Edges, whether of railroads, topography, throughways, or district boundaries, are a very typical feature of this environment and tend to fragment it. Some of the most unpleasant edges, such as the bank of the Hackensack River with its burning dump areas, seemed to be mentally erased.

[. . .]

While continuity and visibility are crucial, strong edges are not necessarily impenetrable. Many edges are uniting seams, rather than isolating barriers, and it is interesting to see the differences in effect. Boston's Central Artery seems to divide absolutely, to isolate. Wide Cambridge Street divides two regions sharply but keeps them in some visual relation. Beacon Street, the visible boundary of Beacon Hill along the Common, acts not as a barrier but as a seam along which the two major areas are clearly joined together. Charles Street at the foot of Beacon Hill both divides and unites, leaving the lower area in uncertain relation to the hill above. Charles Street carries heavy traffic but also contains the local service stores and special activities associated with the Hill. It pulls the residents together by attracting them to itself. It acts ambiguously either

as linear node, edge, or path for various people at various times.

Edges are often paths as well. Where this was so, and where the ordinary observer was not shut off from moving on the path . . . then the circulation image seemed to be the dominant one. The element was usually pictured as a path, reinforced by boundary characteristics.

[. . .]

It is difficult to think of Chicago without picturing Lake Michigan. It would be interesting to see how many Chicagoans would begin to draw a map of their city by putting down something other than the line of the lake shore. Here is a magnificent example of a visible edge, gigantic in scale, that exposes an entire metropolis to view. Great buildings, parks, and tiny private beaches all come down to the water's edge, which throughout most of its length is accessible and visible to all. The contrast, the differentiation of events along the line, and the lateral breadth are all very strong. The effect is reinforced by the concentration of paths and activities along its extent. The scale is perhaps unrelievedly large and coarse, and too much open space is at times interposed between city and water, as at the Loop. Yet the façade of Chicago on the Lake is an unforgettable sight.

DISTRICTS

Districts are the relatively large city areas which the observer can mentally go inside of, and which have some common character. They can be recognized internally, and occasionally can be used as external reference as a person goes by or toward them. Many persons interviewed took care to point out that Boston, while confusing in its path pattern even to the experienced inhabitant, has, in the number and vividness of its differentiated districts, a quality that quite makes up for it. As one person put it: Each part of Boston is different from the other. You can tell pretty much what area you're in.

Jersey City has its districts too, but they are primarily ethnic or class districts with little physical distinction. Los Angeles is markedly lacking in strong regions, except for the Civic Center area. The best that can be found are the linear, street-front districts of Skid Row or the financial area. . . .

Subjects, when asked which city they felt to be a well-oriented one, mentioned several, but New York (meaning Manhattan) was unanimously cited. And this city was cited not so much for its grid, which Los Angeles has as well, but because it has a number of well-defined characteristic districts, set in an ordered frame of rivers and streets. Two Los Angeles subjects even referred to Manhattan as being "small" in comparison to their central area! Concepts of size may depend in part on how well a structure can be grasped.

In some Boston interviews, the districts were the basic elements of the city image. One subject, for example, when asked to go from Faneuil Hall to Symphony Hall, replied at once by labeling the trip as going from North End to Back Bay. But even where they were not actively used for orientation, districts were still an important and satisfying part of the experience of living in the city. Recognition of distinct districts in Boston seemed to vary somewhat as acquaintance with the city increased. People most familiar with Boston tended to recognize regions but to rely more heavily for organization and orientation on smaller elements. A few people extremely familiar with Boston were unable to generalize detailed perceptions into districts: conscious of minor differences in all parts of the city, they did not form regional groups of elements.

The physical characteristics that determine districts are thematic continuities which may consist of an endless variety of components: texture, space, form, detail, symbol, building type, use, activity, inhabitants, degree of maintenance, topography. In a closely built city such as Boston, homogeneities of façade – material, modeling, ornament, color, skyline, especially fenestration – were all basic clues in identifying major districts. Beacon Hill and Commonwealth Avenue are both examples. The clues were not only visual ones: noise was important as well. At times, indeed, confusion itself might be a clue, as it was for the woman who remarked that she knows she is in the North End as soon as she feels she is getting lost.

Usually, the typical features were imaged and recognized in a characteristic cluster, the thematic unit. The Beacon Hill image, for example, included steep narrow streets; old brick row houses of intimate scale; inset, highly maintained, white doorways; black trim; cobblestones and brick walks; quiet; and upper-class pedestrians. The resulting thematic unit was distinctive by contrast to the rest of the city and could be recognized immediately. . . .

A certain reinforcement of clues is needed to produce a strong image. All too often, there are a few

distinctive signs, but not enough for a full thematic unit. Then the region may be recognizable to someone familiar with the city, but it lacks any visual strength or impact. Such, for example, is Little Tokyo in Los Angeles, recognizable by its population and the lettering on its signs but otherwise indistinguishable from the general matrix. Although it is a strong ethnic concentration, probably known to many people, it appeared as only a subsidiary portion of the city image.

Yet social connotations are quite significant in building regions. A series of street interviews indicated the class overtones that many people associate with different districts. Most of the Jersey City regions were class or ethnic areas, discernible only with difficulty for the outsider. Both Jersey City and Boston have shown the exaggerated attention paid to upper-class districts, and the resulting magnification of the importance of elements in those areas. District names also help to give identity to districts even when the thematic unit does not establish a striking contrast with other parts of the city, and traditional associations can play a similar role.

When the main requirement has been satisfied, and a thematic unit that contrasts with the rest of the city has been constituted, the degree of internal homogeneity is less significant, especially if discordant elements occur in a predictable pattern. Small stores on street corners establish a rhythm on Beacon Hill that one subject perceived as part of her image. These stores in no way weakened her non-commercial image of Beacon Hill but merely added to it. Subjects could pass over a surprising amount of local disagreement with the characteristic features of a region.

Districts have various kinds of boundaries. Some are hard, definite, precise. Such is the boundary of the Back Bay at the Charles River or at the Public Garden. All agreed on this exact location. Other boundaries may be soft or uncertain, such as the limit between downtown shopping and the office district, to whose existence and approximate location most people would testify. Still other regions have no boundaries at all, as did the South End for many of our subjects. . . .

These edges seem to play a secondary role: they may set limits to a district, and may reinforce its identity, but they apparently have less to do with constituting it. Edges may augment the tendency of districts to fragment the city in a disorganizing way. A few people sensed disorganization as one result of the large number of identifiable districts in Boston: strong edges, by hindering transitions from one district to another, may add to the impression of disorganization.

That type of district which has a strong core, surrounded by a thematic gradient which gradually dwindles away, is not uncommon. Sometimes, indeed, a strong node may create a sort of district in a broader homogeneous zone, simply by "radiation," that is, by the sense of proximity to the nodal point. These are primarily reference areas, with little perceptual content, but they are useful organizing concepts, nevertheless.

Some well-known Boston districts were unstructured in the public image. The West End and North End were internally undifferentiated for many people who recognized these regions. Even more often, thematically vivid districts such as the market area seemed confusingly shapeless, both externally and internally. The physical sensations of the market activity are unforgettable. Faneuil Hall and its associations reinforce them. Yet the area is shapeless and sprawling, divided by the Central Artery, and hampered by the two activity centers which vie for dominance: Faneuil Hall and Haymarket Square. Dock Square is spatially chaotic. The connections to other areas are either obscure or disrupted by the Artery. Thus the market district simply floated in most images. Instead of fulfilling its potential role as a mosaic link at the head of the Boston peninsula, as does the Common farther down, the district, while distinctive, acted only as a chaotic barrier zone. Beacon Hill, on the other hand, was very highly structured, with internal subregions, a node at Louisburg Square, various landmarks, and a configuration of paths.

Again, some regions are introvert, turned in upon themselves with little reference to the city outside them, such as Boston's North End or Chinatown. Others may be extrovert, turned outward and connected to surrounding elements. The Common visibly touches neighboring regions, despite its inner path confusions. Bunker Hill in Los Angeles is an interesting example of a district of fairly strong character and historical association, on a very sharp topographical feature lying even closer to the city's heart than does Beacon Hill. Yet the city flows around this element, buries its topographic edges in office buildings, breaks off its path connections, and effectively causes it to fade or even disappear from the city image. Here is a striking opportunity for change in the urban landscape.

Some districts are single ones, standing alone in their zone. The Jersey City and Los Angeles regions are practically all of this kind, and the South End is a Boston example. Others may be linked together, such as Little Tokyo and the Civic Center in Los Angeles, or West End–Beacon Hill in Boston. In one part of central Boston, inclusive of the Back Bay, the Common, Beacon Hill, the downtown shopping district, and the financial and market areas, the regions are close enough together and sufficiently well joined to make a continuous mosaic of distinctive districts. Wherever one proceeds within these limits, one is in a recognizable area. The contrast and proximity of each area, moreover, heightens the thematic strength of each. The quality of Beacon Hill, for example, is sharpened by its nearness to Scollay Square, and to the downtown shopping district.

NODES

Nodes are the strategic foci into which the observer can enter, typically either junctions of paths, or concentrations of some characteristic. But although conceptually they are small points in the city image, they may in reality be large squares, or somewhat extended linear shapes, or even entire central districts when the city is being considered at a large enough level. Indeed, when conceiving the environment at a national or international level, then the whole city itself may become a node.

The junction, or place of a break in transportation, has compelling importance for the city observer. Because decisions must be made at junctions, people heighten their attention at such places and perceive nearby elements with more than normal clarity. This tendency was confirmed so repeatedly that elements located at junctions may automatically be assumed to derive special prominence from their location. The perceptual importance of such locations shows in another way as well. When subjects were asked where on a habitual trip they first felt a sense of arrival in downtown Boston, a large number of people singled out break-points of transportation as the key places. In a number of cases, the point was at the transition from a highway (Storrow Drive or the Central Artery) to a city street; in another case, the point was at the first railroad stop in Boston (Back Bay Station) even though the subject did not get off there. Inhabitants of Jersey City felt they had left their city when they had passed through the Tonnelle Avenue Circle. The transition from one transportation channel to another seems to mark the transition between major structural units.

[. . .]

The subway stations, strung along their invisible path systems, are strategic junction nodes. Some, like Park Street, Charles Street, Copley, and South Station, were quite important in the Boston map, and a few subjects would organize the rest of the city around them. . . .

Major railroad stations are almost always important city nodes, although their importance may be declining. Boston's South Station was one of the strongest in the city, since it is functionally vital for commuter, subway rider, and intercity traveler, and is visually impressive for its bulk fronting on the open space of Dewey Square. The same might have been said for airports, had our study areas included them. . . .

The other type of node, the thematic concentration, also appeared frequently. Pershing Square in Los Angeles was a strong example, being perhaps the sharpest point of the city image, characterized by highly typical space, planting, and activity. . . .

Louisburg Square is another thematic concentration, a well-known quiet residential open space, redolent of the upper-class themes of the Hill, with a highly recognizable fenced park. It is a purer example of a concentration than is the Jordan–Filene corner, since it is no transfer point at all, and was only remembered as being "somewhere inside" Beacon Hill. Its importance as a node was out of all proportion to its function.

Nodes may be both junctions and concentrations, as is Jersey City's Journal Square, which is an important bus and automobile transfer and is also a concentration of shopping. Thematic concentrations may be the focus of a region, as is the Jordan–Filene corner, and perhaps Louisburg Square. Others are not foci but are isolated special concentrations, such as Olvera Street in Los Angeles.

A strong physical form is not absolutely essential to the recognition of a node: witness Journal Square and Scollay Square. But where the space has some form, the impact is much stronger. The node becomes memorable.

[. . .]

Nodes, like districts, may be introvert or extrovert. Scollay Square is introverted, it gives little directional sense when one is in it or its environs. The principal

direction in its surroundings is toward or away from it; the principal locational sensation on arrival is simply "here I am." Boston's Dewey Square, on the other hand, is extroverted. General directions are explained, and connections are clear to the office district, the shopping district, and the waterfront. . . .

Many of these qualities may be summed up by the example of a famous Italian node: the Piazza San Marco in Venice. Highly differentiated, rich and intricate, it stands in sharp contrast to the general character of the city and to the narrow, twisting spaces of its immediate approaches. Yet it ties firmly to the major feature of the city, the Grand Canal, and has an oriented shape that clarifies the direction from which one enters. It is within itself highly differentiated and structured: into two spaces (Piazza and Piazzetta) and with many distinctive landmarks (Duomo, Palazzo Ducale, Campanile, Libreria). Inside, one feels always in clear relation to it, precisely micro-located, as it were. So distinctive is this space that many people who have never been to Venice will recognize its photograph immediately.

LANDMARKS

Landmarks, the point references considered to be external to the observer, are simple physical elements which may vary widely in scale. There seemed to be a tendency for those more familiar with a city to rely increasingly on systems of landmarks for their guides – to enjoy uniqueness and specialization, in place of the continuities used earlier.

Since the use of landmarks involves the singling out of one element from a host of possibilities, the key physical characteristic of this class is singularity, some aspect that is unique or memorable in the context.

Landmarks become more easily identifiable, more likely to be chosen as significant, if they have a clear form; if they contrast with their background; and if there is some prominence of spatial location. Figure–background contrast seems to be the principal factor. The background against which an element stands out need not be limited to immediate surroundings: the grasshopper weathervane of Faneuil Hall, the gold dome of the State House, or the peak of the Los Angeles City Hall are landmarks that are unique against the background of the entire city.

In another sense, subjects might single out landmarks for their cleanliness in a dirty city (the Christian

Science buildings in Boston) or for their newness in an old city (the chapel on Arch Street). The Jersey City Medical Center was as well known for its little lawn and flowers as for its great size. The old Hall of Records in the Los Angeles Civic Center is a narrow, dirty structure, set at an angle to the orientation of all the other civic buildings, and with an entirely different scale of fenestration and detail. Despite its minor functional or symbolic importance, this contrast of siting, age, and scale makes it a relatively well-identified image, sometimes pleasant, sometimes irritating. It was several times reported to be "pie-shaped," although it is perfectly rectangular. This is evidently an illusion of the angled siting.

Spatial prominence can establish elements as landmarks in either of two ways: by making the element visible from many locations (the John Hancock Building in Boston, the Richfield Oil Building in Los Angeles), or by setting up a local contrast with nearby elements, i.e., a variation in setback and height. In Los Angeles, on 7th Street at the corner of Flower Street, is an old, two-story gray wooden building, set back some ten feet from the building line, containing a few minor shops. This took the attention and fancy of a surprising number of people. One even anthropomorphized it as the "little gray lady." The spatial setback and the intimate scale is a very noticeable and delightful event, in contrast to the great masses that occupy the rest of the frontage.

[. . .]

Distant landmarks, prominent points visible from many positions, were often well known, but only people unfamiliar with Boston seemed to use them to any great extent in organizing the city and selecting routes for trips. It is the novice who guides himself by reference to the John Hancock Building and the Custom House.

Few people had an accurate sense of where these distant landmarks were and how to make one's way to the base of either building. Most of Boston's distant landmarks, in fact, were "bottomless"; they had a peculiar floating quality. The John Hancock Building, the Custom House, and the Court House are all dominant on the general skyline, but the location and identity of their base is by no means as significant as that of their top.

The gold dome of Boston's State House seems to be one of the few exceptions to this elusiveness. Its unique shape and function, its location at the hill crest and its exposure to the Common, the visibility from long distances of its bright gold dome, all make it a key sign for central Boston. It has the satisfying qualities of recognizability at many levels of reference, and of coincidence of symbolic with visual importance.

People who used distant landmarks did so only for very general directional orientation, or, more frequently, in symbolic ways. For one person, the Custom House lent unity to Atlantic Avenue because it can be seen from almost any place on that street. For another, the Custom House set up a rhythm in the financial district, for it can be seen intermittently at many places in that area.

The Duomo of Florence is a prime example of a distant landmark: visible from near and far, by day or night; unmistakable; dominant by size and contour; closely related to the city's traditions; coincident with the religious and transit center; paired with its campanile in such a way that the direction of view can be gauged from a distance. It is difficult to conceive of the city without having this great edifice come to mind.

But local landmarks, visible only in restricted localities, were much more frequently employed in the three cities studied. They ran the full range of objects available. The number of local elements that become landmarks appears to depend as much upon how familiar the observer is with his surroundings as upon the elements themselves. Unfamiliar subjects usually mentioned only a few landmarks in office interviews, although they managed to find many more when they went on field trips. Sounds and smells sometimes reinforced visual landmarks, although they did not seem to constitute landmarks by themselves.

[. . .]

Element interrelations

These elements are simply the raw material of the environmental image at the city scale. They must be patterned together to provide a satisfying form. The preceding discussions have gone as far as groups of similar elements (nets of paths, clusters of landmarks, mosaics of regions). The next logical step is to consider the interaction of pairs of unlike elements.

Such pairs may reinforce one another, resonate so that they enhance each other's power: or they may conflict and destroy themselves. A great landmark may dwarf and throw out of scale a small region at its base.

Properly located, another landmark may fix and strengthen a core; placed off center, it may only mislead, as does the John Hancock Building in relation to Boston's Copley Square. . . .

[. . .]

We are continuously engaged in the attempt to organize our surroundings, to structure and identify them. Various environments are more or less amenable to such treatment. When reshaping cities it should be possible to give them a form which facilitates these organizing efforts rather than frustrates them.

S
E
V
E
N

"The Design of Spaces"

from *City: Rediscovering the Center* (1988)

William H. Whyte

Editors' Introduction

Puzzled by why some of New York's parks and plazas were well used while others were almost empty, the New York City Planning Commission asked sociologist William Whyte (1918–1999) to study park and plaza use and help draft a comprehensive design plan for the city.

Whyte's lucid writing on planning and design gave him great credibility. Hunter College appointed him a Distinguished Professor, and the National Geographic Society gave him the first domestic "expedition grant" they had ever made.

Whyte worked with bright young designers and planners at the New York City Planning Department, Hunter College sociology students, and other talented people he drew to "The Street Life Project." This team produced an exceptional study of how people use urban space and a set of urban design guidelines for New York that have been widely praised and used in New York and many other cities.

The Street Life Project is an excellent example of how to do urban research. Whyte formed hypotheses about how people use urban space. Then he tested each hypothesis by filming people using different plazas and parks in New York City and carefully analyzing the films. His results were often startling. Some initial hypotheses were validated, but Whyte found that he had to reject or modify many others that had seemed intuitively obvious. For example, Whyte hypothesized that the number of people using a plaza would be related to the *amount* of space or its *shape*. Big parks should have more people than little ones. A long skinny strip park should have fewer people than a rectangular one. But Chart 1 in Figure 1 shows that is not the case. New Yorkers use tiny Greenacre Park much more than the much larger J.C. Penny Park. One of New York's most popular parks is just a long, narrow indentation in a building. Whyte eventually concluded that the amount of *sitting* space in a park or plaza was much more important than either the total space or its shape.

Most writing on urban design ignores gender differences or is written from a male perspective with a separate section on design implications for women. Whyte is one of the few authors to notice gender and to weave its significance into the fabric of his study. He noticed that women are more discriminating than men as to where they will sit and are more sensitive to annoyances. He concluded that if a plaza has a high proportion of women, it is probably a good and well-managed one.

William Whyte's ideas have had a wide impact. The New York City Planning Commission held hearings on his recommendations and, after much debate, adopted many of his suggestions as requirements or guidelines for new development. A public–private partnership inspired by Whyte's ideas tore down walls isolating Bryant Park from the street, put in sitting space and food, and transformed a dangerous, little-used park into one bustling with life (while at the same time evicting many homeless park "residents" and raising troubling questions about who should control public space and prescribe how it is to be used).

Note the similarity between Whyte's description of Seagram's Plaza as "the best of stages" and Lewis Mumford's emphasis on the city as theater (p. 85). Mike Davis (p. 178) found designers in Los Angeles consciously design public spaces to keep *people away*. As Whyte comments, "it takes real work to create a lousy place." Note the importance of good public spaces in Camillo Sitte's work on *The Art of Building Cities* (p. 427) and in H.D.F. Kitto's description of the Greek polis (p. 35).

This selection is from *City: Rediscovering the Center* (New York: Anchor Books, 1988). In other chapters of *City*, Whyte explores water, wind, trees, light, steps and entrances, undesirables, walls, sun and shadows, and many other factors. Whyte produced a delightful film titled *Public Spaces/Human Places* based on his research (available from Direct Cinema Limited in Los Angeles). The original report of Whyte's classic street life project has recently been reprinted as *The Social Life of Small Urban Spaces* (New York: Project for Public Spaces Inc., 2001). An anthology of William Whyte's most important writings is Albert LaFarge (ed.), *The Essential William Whyte* (New York: Fordham University Press, 2000). Whyte's other main book on planning is *The Last Landscape* (Garden City: Doubleday, 1968). His best-known book is *The Organization Man* (New York: Simon & Schuster, 1956) – a study of corporate culture that sold more than two million copies.

There are many books concerning human aspects of design. Oscar Newman's influential *Defensible Space* (New York: Macmillan, 1972) is a study of the way in which low-rent public housing project residents use space, with suggestions to architects and planners on how to meet their security concerns. Clare Cooper and Wendy Sarkissien's *Housing as if People Mattered* (Berkeley: University of California Press, 1986) provides practical suggestions for designing housing responsive to the needs of all its residents, particularly working women and children. Allan Jacobs's *Looking at Cities* (Cambridge: MIT Press, 1985) provides a stimulating discussion of how close observation like Whyte undertook can inform city planning, and how to do it.

■ ■ ■ ■ ■ ■

. . . Since 1961 New York City had been giving incentive bonuses to developers who would provide plazas . . . Every new office building qualified for the bonus by providing a plaza or comparable space; in total, by 1972 some twenty acres of the world's most expensive open space.

Some plazas attracted lots of people . . .

But on most plazas there were few people. In the middle of the lunch hour on a beautiful day the number of people sitting on plazas averaged four per thousand square feet of space – an extraordinarily low figure for so dense a center . . .

. . . The city was being had. For the millions of dollars of extra floor space it was handing out to developers, it had every right to demand much better spaces in return.

I put the question to the chairman of the city planning commission, Donald Elliott . . . He felt tougher zoning was in order. If we could find out why the good places worked and the bad ones didn't and come up with tight guidelines, there could be a new code . . .

We set to work. We began studying a cross section of spaces – in all, sixteen plazas, three small parks, and a number of odds and ends of space . . .

[. . .]

We started by charting how people used plazas. We mounted time-lapse cameras at spots overlooking the plaza . . . and recorded the dawn-to-dusk patterns. We made periodic circuits of the plazas and noted on sighting maps where people were sitting, their gender, and whether they were alone or with others . . . We also interviewed people and found where they worked, how frequently they used the plaza, and what they thought of it. But mostly we watched what they did.

Most of them were young office workers from nearby buildings. Often there would be relatively few from the plaza's own building. As some secretaries confided, they would just as soon put a little distance between themselves and the boss come lunchtime. In most cases the plaza users came from a building within a three-block radius. Small parks, such as Paley and Greenacre, had a somewhat more varied mix of people – with more upper-income older people – but even here office workers predominated.

This uncomplicated demography underscores an elemental point about good spaces: supply creates demand. A good new space builds a new constituency. It gets people into new habits – such as alfresco lunches – and induces them to use new paths . . .

The best-used plazas are sociable places, with a higher proportion of couples and groups than you will find in less-used places. At the plazas in New York, the proportion of people in twos or more runs about 50–62 percent; in the least-used, 25–30 percent. A high proportion is an index of selectivity. If people go to a place in a group or rendezvous there, it is most often because they decided to beforehand. Nor are these places less congenial to the individual. In absolute numbers, they attract more individuals than do the less-used spaces. If you are alone, a lively place can be the best place to be.

The best-used places also tend to have a higher than average proportion of women. The male–female ratio of a plaza reflects the composition of the work force and this varies from area to area. In midtown New York it runs about 60 percent male, 40 percent female. Women are more discriminating than men as to where they will sit, they are more sensitive to annoyances, and they spend more time casing a place. They are also more likely to dust off a ledge with their handkerchief.

The male–female ratio is one to watch. If a plaza has a markedly low proportion of women, something is wrong. Conversely, if it has a high proportion, the plaza is probably a good and well-managed one and has been chosen as such.

The rhythms of plaza life are much alike from place to place. In the morning hours, patronage will be sporadic . . .

Around noon the main clientele begins to arrive. Soon activity will be near peak and will stay there until a little before two . . .

Some 80 percent of the people activity on plazas comes during the lunchtime, and there is very little of any kind after five-thirty . . .

During the lunch period, people will distribute themselves over space with considerable consistency, with some sectors getting heavy use day in and day out, others much less so. We also found that off-peak use often gives the best clues to people's preferences. When a place is jammed, people sit where they can; this may or may not be where they most want to. After the main crowd has left, however, the choices can be significant. Some parts of the plaza become empty; others continue to be use . . .

Men show a tendency to take the front row seats and if there is a kind of gate they will be the guardians of it. Women tend to favor places slightly secluded. If there are double-sided ledges parallel to the street, the inner side will usually have a higher proportion of women; the outer, of men.

Of the men up front the most conspicuous are the girl watchers. As I have noted, they put on such a show of girl watching as to indicate that their real interest is not so much the girls as the show. It is all machismo. Even in the Wall Street area, where girl watchers are especially demonstrative you will hardly ever see one attempt to pick up a girl.

Plazas are not ideal places for striking up acquaintances. Much better is a very crowded street with lots of eating and quaffing going on. An outstanding example is the central runway of the South Street Seaport. At lunch sometimes, one can hardly move for the crush. As in musical chairs, this can lead to interesting combinations. On most plazas, however, there isn't much mixing. If there are, say, two smashing blondes on a ledge, the men nearby will usually put on an elaborate show of disregard. Look closely, however, and you will see them giving away the pose with covert glances.

Lovers are to be found on plazas, but not where you would expect them. When we first started interviewing, people would tell us to be sure to see the lovers in the rear places. But they weren't usually there. They would be out front. The most fervent embracing we've recorded on film has taken place in the most visible of locations, with the couple oblivious of the crowd. (In a long clutch, however, I have noted that one of the lovers may sneak a glance at a wristwatch.)

Certain locations become rendezvous points for groups of various kinds. The south wall of the Chase Manhattan Plaza was, for a while, a gathering point for camera bugs, the kind who are always buying new lenses and talking about them. Patterns of this sort may last no more than a season – or persist for years. A black civic leader in Cincinnati told me that when he wants to make contact, casually, with someone, he usually knows just where to look at Fountain Square . . .

Standing patterns on the plazas are fairly regular. When people stop to talk they will generally do so athwart one of the main traffic flows, as they do on streets. They also show an inclination to station themselves near objects, such as a flagpole or a piece of sculpture. They like well-defined places, such as steps or the border of a pool. What they rarely choose is the middle of a large space.

There are a number of explanations. The preference might be ascribed to some primeval instinct: you have a full view of all comers but your rear is covered. But this doesn't explain the inclination men have for lining up at the curb. Typically, they face inward, with their backs exposed to the vehicle traffic of the street.

Whatever their cause, people's movements are one of the great spectacles of a plaza. You do not see this in architectural photographs, which are usually devoid of human beings and are taken from a perspective that few people share. It is a misleading one. Looking down on a bare plaza, one sees a display of geometry, done almost in monochrome. Down at eye level the scene comes alive with movement and color – people walking quickly, walking slowly, skipping up steps, weaving in and out on crossing patterns, accelerating and retarding to match the moves of others. Even if the paving and the walls are gray, there will be vivid splashes of color – in winter especially, thanks to women's fondness for red coats and colored umbrellas.

There is a beauty that is beguiling to watch, and one senses that the players are quite aware of this themselves. You can see this in the way they arrange themselves on ledges and steps. They often do so with a grace that they must appreciate themselves. With its brown-gray setting, Seagram is the best of stages – in the rain, too, when an umbrella or two puts spots of color in the right places, like Corot's red dots.

Let us turn to the factors that make for such places. The most basic one is so obvious it is often overlooked: people. To draw them, a space should tap a strong flow of them. This means location, and, as the old adage has it, location and location. The space should be in the heart of downtown, close to the 100 percent corner – preferably right on top of it.

Because land is cheaper further out, there is a temptation to pick sites away from the center. There may also be some land for the asking – some underused spaces, for example, left over from an ill-advised civic center campus of urban renewal days. They will be poor bargains. A space that is only a few blocks too far might as well be ten blocks for all the people who will venture to walk to it.

People *ought* to walk to it, perhaps; the exercise would do them good. But they don't. Even within the core of downtown the effective radius of a good place is about three blocks. About 80 percent of the users will have come from a place within that area. This does indicate a laziness on the part of pedestrians and this may change a bit, just as the insistence on close-in parking may. But there is a good side to the constrained radius. Since usage is so highly localized, the addition of other good open spaces will not saturate demand. They will increase it.

Given a fine location, it is difficult to design a space that will not attract people. What is remarkable is how often this has been accomplished. Our initial study made it clear that while location is a prerequisite for

success, it in no way assures it. Some of the worst plazas are in the best spots . . .

All of the plazas and small parks that we studied had good locations; most were on the major avenues, some on attractive side-streets. All were close to bus-stops or subway stations and had strong pedestrian flows on the sidewalks beside them. Yet when we rated them according to the number of people sitting at peak time, there was a wide range: from 160 people at 77 Water Street to 17 at 280 Park Avenue (Figure 1).

How come? The first factor we studied was the sun. We thought it might well be the critical one, and

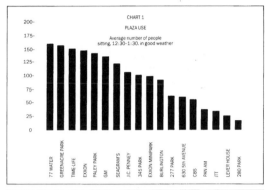

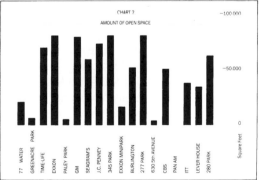

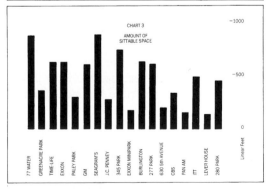

Figure 1

our first time-lapse studies seemed to bear this out. Subsequent studies did not. As I will note later they show that sun was important but did not explain the differences in popularity of plazas.

Nor did aesthetics . . . The elegance and purity of a complex's design, we had to conclude, had little relationship to the usage of the spaces around it.

[. . .]

Another factor we considered was the shape of spaces. Members of the commission's urban design group believed this was very important and hoped our findings would support tight criteria for proportions and placement. They were particularly anxious to rule out strip plazas: long, narrow spaces that were little more than enlarged sidewalks, and empty of people more times than not . . .

Our data did not support such criteria. While it was true that most strip plazas were little used, it did not follow that their shape was the reason. Some squarish plazas were little used too, and, conversely, several of the most heavily used spaces were in fact long, narrow strips. One of the five most popular sitting places in New York is essentially an indentation in a building, long and narrow. Our research did not prove shape unimportant or designers' instincts misguided. As with the sun, however, it proved that other factors were more critical.

If not the shape of the space, what about the *amount* of it? Some conservationists believed this would be the key factor. In their view, people seek open space as a relief from overcrowding and it would follow that places with the greatest sense of space and light and air would draw the best. If we ranked plazas by the amount of space they provided, there surely would be a positive correlation between space and people.

Once again we found no clear relationship. Several of the smallest spaces had the largest number of people, and several of the largest spaces had the least number of people . . .

What about the amount of *sittable* space? Here we began to get close. As we tallied the number of linear feet of sitting space, we could see that the plazas with the most tended to be among the most popular . . .

. . . No matter how many other variables we checked, one basic point kept coming through. We at last recognized that it was the major one.

People tend to sit most where there are places to sit.

This may not strike the reader as an intellectual bombshell, and now that I look back on our study I wonder why it was not more apparent to us from the beginning . . . Whatever the attractions of a space, it cannot induce people to come and sit if there is no place to sit.

INTEGRAL SEATING

The most basic kind of seating is the kind that is built into a place, such as steps and ledges. Half the battle is seeing to it that these features are usable by people. And there is a battle. Another force has been diligently at work finding ways to deny these spaces. Here are some of the ways:

> Horizontal metal strip with sawtooth points.
> Jagged rocks set in concrete (Southbridge House, New York City).
> Spikes imbedded in ledges (Peachtree Plaza Hotel).
> Railing placed to hit you in small of back (GM Plaza, New York City).
> Canted ledges of slippery marble (Celanese Building, New York City).

It takes real work to create a lousy place. In addition to spikes and metal objects, there are steps to be made steep, additional surveillance cameras to be mounted, walls to be raised high. Just not doing such things can produce a lot of sitting space.

It won't be the most comfortable kind but it will have the great advantage of enlarging choice. The more sittable the inherent features are made, the more freedom people have to sit up front, in the back, to the side, in the sun, or out of it. This means designing ledges and parapets and other flat surfaces so they can do double duty as seating, tables, and shelves. Since most building sites have some slope in them, there are bound to be opportunities for such features, and it is no more trouble to leave them sittable than not.

[. . .]

SITTING HEIGHT

One guideline we thought would be easy to establish was for sitting heights. It seemed obvious enough that somewhere around sixteen to seventeen inches would probably be the optimum. But how much higher or lower could a surface be and still be sittable? Thanks to slopes, several of the most popular ledges provided a range of continuously variable heights. The front ledge at Seagram, for example started at seven inches at one corner and rose to forty-four inches at the other. Here was an opportunity for a definitive study, we thought; by recording over time how many people sat at what heights, we would get a statistical measure of preferences.

We didn't . . . We had to conclude that people will sit almost anywhere between a height of one foot and three, and this was the range that was to be specified in the zoning. People will sit lower or higher, of course, but there are apt to be special conditions – a wall too high for most adults to mount but just right for teenagers.

A dimension that is truly important is the human backside. It is a dimension many architects ignore. Rarely will you find a ledge or bench that is deep enough to be sittable on both sides . . . Most frustrating are the ledges just deep enough to tempt people to sit on both sides, but too shallow to let them do so comfortably. At peak times people may sit on both sides but they won't be comfortable doing it. They will be sitting on the forward edge, awkwardly.

Thus to another of our startling findings: ledges and spaces two backsides deep seat more people than those that are not as deep . . .

[. . .]

Steps work for the same reason. They afford an infinity of possible groupings, and the excellent sight lines make all the seating great for watching the theatre of the street . . .

[. . .]

Circulation and sitting, in sum, are not antithetical but complementary. I stress this because a good many planners think that the two should be kept separate. More to the point, so do some zoning codes. New York's called for "pedestrian circulation areas" separate from "activity areas" for sitting. People ignore such boundaries.

We felt that pedestrian circulation through and within plazas should be encouraged. Plazas that are sunken or elevated tend to attract low flows, and for that reason the zoning specifies that plazas be not more than three feet below street level or above it. The easier the flow between street and plaza the more likely they are to come in and tarry and sit.

This is true of the handicapped also. If a place is planned with their needs in mind, the place is apt to function more easily for everyone. Drinking fountains that are low enough for wheelchair users are low

enough for children. Walkways that are made easier for the handicapped by ramps, handrails, and steps of gentle pitch are easier for all. The guidelines make such amenities mandatory . . . For the benefit of the handicapped, it is required that at least 5 percent of the seating spaces have backrests. These are not segregated for the handicapped. No facilities are segregated. The idea is to make all of the place useful for everyone.

BENCHES

Benches are design artifacts the purpose of which is to punctuate architectural photographs. They are most often sited in modular form, spaced equidistant from one another in a symmetry that is pleasing in plan view. They are not very good, however, for sitting. There are usually too few of them; they are too short and too narrow; they are isolated from other benches and from what action there is to look at.

[. . .]

Watch how benches fill up. The first arrival will usually take the end of a bench, not the middle. The next arrival will take the end of another bench. Subsequent arrivals head for whatever end spots are not taken. Only when there are few other places left will people sit in the middle of the bench, and some will elect to stand.

Since it's the ends of the benches that do most of the work, it could be argued that benches ought to be shortened so they're all end and no middle. But the unused middles are functional for not being used. They provide buffer space. They also provide choice, and if it is the least popular choice, that does not negate its utility.

[. . .]

CHAIRS

We come now to a wonderful invention: the movable chair. Having a back, it is comfortable, and even more so if it has armrests as well. But the big asset is movability. Chairs enlarge choice: to move into the sun, out of it; to move closer to someone, further away from another.

The possibility of choice is as important as the exercise of it. If you know you can move if you want to, you can feel all the more comfortable staying put. This is why, perhaps, people so often approach a chair and then, before sitting on it, move the chair a few inches this way or that, finally ending up with the chair just about where it was in the first place. These moves are functional. They are a declaration of one's free will to oneself, and rather satisfying. In this one small matter you are the master of your fate.

Small moves can say things to other people. If a newcomer chooses a chair next to a couple in a crowded situation, he may make several moves with the chair. He is conveying a message: Sorry about the closeness, but it can't be helped and I am going to respect your privacy as you will mine. A reciprocal shift of a chair may signal acknowledgment.

Chair arranging by groups is a ritual worth watching. In a group of three or four women, one may be dominant and direct the sitting, including the fetching of an extra chair. More times, the members of the group work it out themselves, often with false starts and second choices. The chair arranging can take quite a bit of time on occasion – it is itself a form of recreation – but people enjoy it. Watching these exercises in civility is one of the pleasures of a good place.

Fixed individual seats deny choice. They may be good to look at, and in the form of stools, metal love seats, granite cubes, and the like, they make interesting decorative elements. That is their primary function. For sitting, however, they are inflexible and socially uncomfortable.

[. . .]

Where space is at a premium – in theatres, stadia – fixed seats are a necessity. In open spaces, however, they are uncalled for; there is so much space around them that the compression makes for awkward sitting . . . On one campus a group of metal love seats was cemented to the paving with epoxy glue; in short order they were wrenched out of position by students. The designer is unrepentant. His love seats have won several design awards.

[. . .]

A salute to grass is in order. It is a wonderfully adaptable substance, and while it is not the most comfortable seating, it is fine for napping, sunbathing, picnicking, and Frisbee throwing. Like movable chairs, it also has the great advantage of offering people the widest possible choice of sitting arrangements. There are an infinity of possible groupings, but you will note that the most frequent has people self-positioned at oblique angles from each other.

Grass offers a psychological benefit as well. A patch of green is a refreshing counter to granite and concrete, and when people are asked what they would like to see in a park, trees and grass usually are at the top of the list . . .

RELATIONSHIP TO THE STREET

Let us turn to a more difficult consideration. With the kind of amenities we have been discussing, there are second chances. If the designers have goofed on seating, more and better seating can be provided. If they have been too stingy with trees, more trees can be planted. If there is no food, a food cart can be put in – possibly a small pavilion or gazebo. If there is no water feature, a benefactor might be persuaded to donate a small pool or fountain. Thanks to such retrofitting, spaces regarded as hopeless dogs have been given new life.

What is most difficult to change, however, is what is most important: the location of the space and its relationship to the street. The real estate people are right about location, location, location. For a space to function truly well it must be central to the constituency it is to serve – and if not in physical distance, in visual accessibility . . .

The street functions as part of the plaza or square; indeed, it is often hard to tell where the street leaves off and the plaza begins. The social life of the spaces flows back and forth between them. And the most vital space of all is often the street corner. Watch one long enough and you will see how important it is to the life of the large spaces. There will be people in 100 percent conversations or prolonged goodbyes. If there is a food vendor at the corner, like Gus at Seagram, people will be clustered around him, and there will be a brisk traffic between corner and plaza.

It is a great show, and one of the best ways to make the most of it is, simply, not to wall off the plaza from it. Frederick Law Olmsted spoke of an "interior park" and an "outer park," and he argued that the latter – the surrounding streets – was vital to the enjoyment of the former. He thought it an abomination to separate the two with walls or, worse yet, with a spiked iron fence. "In expression and association," he said, "it is in the most distinct contradiction and discord with all the sentiment of a park. It belongs to a jail or to the residence of a despot who dreads assassination."

But walls are still being put up, usually in the mistaken notion that they will make the space feel safer. They do not . . . they make a space feel isolated and gloomy. Lesser defensive measures can work almost as much damage. The front rows of a space – whether ledges or steps or benches – are the best of sitting places, yet they are often modified against human use. At the General Motors Building on Fifth Avenue, the front ledges face out on one of the greatest of promenades. But you cannot sit on the ledges for more than a minute or so. There is a fussy little railing that catches you right in the small of your back. I do not think it was deliberately planned to do so. But it does and you cannot sit for more than a few moments before your back hurts. Another two inches of clearance for the railing and you would be comfortable. But day after day, year after year, one of the great front rows goes scarcely used, for want of two inches. Canted ledges, especially ones of polished marble, are another nullifying feature. You can almost sit on them if you keep pressing down on your heel hard enough.

[. . .]

A good space beckons people in, and the progression from street to interior is critical in this respect. Ideally, the transition should be such that it's hard to tell where one ends and the other begins. You shouldn't have to make a considered decision to enter; it should be almost instinctive . . .

[. . .]

SEVEN

"Toward an Urban Design Manifesto"

Journal of the American Institute of Planners (1987)

Allan Jacobs and Donald Appleyard

Editors' Introduction

Allan Jacobs and Donald Appleyard deplore many of the same aspects of Los Angeles, London, New York, and other large cities that Mike Davis (p. 178), Ali Madanipour (p. 158), Edward Soja (p. 166), and David Harvey (p. 225) criticize: vast anonymous areas developed by giant public and private developers; dangerous, polluted, noisy, anonymous living environments; fortress-like buildings which present windowless façades to the street; and pervasive semiotics that tell "outsiders" they are not welcome in subtle and not-so-subtle ways. But "Toward an Urban Design Manifesto" moves beyond observation and critique to set out goals for urban life and advance ideas for how the urban fabric of cities might be designed to encourage livable urban environments.

Allan Jacobs is an emeritus professor of city and regional planning at the University of California, Berkeley. Donald Appleyard also taught at Berkeley until his death shortly before this selection was published. Both Jacobs and Appleyard worked closely with Kevin Lynch, whose *Image of the City* (p. 438) and other writings strongly influenced their work.

Jacobs alternated between careers as a practicing city planner (in Pittsburgh, Philadelphia, New Delhi, and San Francisco) and teaching urban planning and urban design at the University of Pennsylvania and the University of California, Berkeley. While he was San Francisco's City Planning Director, Jacobs enlisted Appleyard to work on studies of street livability in San Francisco and to help develop an award-winning citywide urban design plan reflecting Lynch's ideas.

Jacobs and Appleyard title their piece a "manifesto" and model it on the celebrated Charter of Athens adopted by the International Congress of Modern Architecture (CIAM) – the organization that advanced ideas for building contemporary cities based on Le Corbusier's principles (p. 322).

Jacobs and Appleyard do not like the vast clearance projects, highways, and high-rise buildings surrounded by enormous open space that have resulted from the CIAM's design ideology. They acknowledge that the Garden City ideas of Ebenezer Howard (p. 314) have produced some pleasant communities, but dismiss them as more like suburbs than true *cities*. Their manifesto suggests an approach that is more subtle and humane than the CIAM's and more truly urban than Howard's.

Jacobs and Appleyard's manifesto is grounded in both a command of academic theory and their own practical experience in city design. In this manifesto they propose urban development at densities higher than Howard proposed for Garden City designs – high enough to qualify as truly urban. But they do not endorse urban densities nearly as great as the CIAM theorists do, particularly in megastructures surrounded by open space. (Le Corbusier deliberately shocked the architectural establishment by producing a plan to tear down much of central Paris and replace it with modern concrete and steel high-rise buildings!)

While Jacobs and Appleyard favor *reasonable* standards for decibel levels and street widths, they oppose excessive engineering standards that destroy the texture of urban life. Like Jane Jacobs (p. 98), William Whyte

(p. 448), and Lewis Mumford (p. 85) they relish some of the disorder that makes urban life enjoyable, including noise, smell, and jumbled uses that some engineers and many modernist architects wanted to separate into orderly zones. Jacobs and Appleyard value pedestrians and public space. Unlike the elitist CIAM theorists, they argue that participatory planning of the kind Sherry Arnstein (p. 233) and John Forester (p. 387) describe is essential.

In *Making City Planning Work* (Chicago: Planner's Press, 1976), Jacobs alternates chapters describing the practical aspects of a city planning director's job with case studies on successful and not so successful projects he undertook during his tenure as San Francisco's city planning director. Jacobs's *Looking at Cities* (Cambridge: MIT Press, 1985) grew out of a class he taught at Berkeley in which students took him to an unfamiliar neighborhood, left him to observe it carefully, and then compared what he found out from observation with what they learnt by examining data and city planning reports on the same neighborhood. *Looking at Cities* reminds professionals to open their eyes and experience the areas they are planning. It outlines a methodology for reading clues in the built environment that can improve urban planning practice. Jacobs's most recent books are *Great Streets* (Cambridge: MIT Press, 1995) and *The Boulevard Book: History, Evolution, Design of Multiway Boulevards* (Cambridge: MIT Press, 2001), the latter co-authored with Elizabeth MacDonald and Yodan Rofe. Both of these recent books reflect many of the values expressed in this selection.

Donald Appleyard's *The View from the Road*, co-authored with Kevin Lynch and John Myer (Cambridge: MIT Press, 1963) and *Livable Streets* (Berkeley: University of California Press, 1981) show how ideas can be translated into action in street design.

Elizabeth McDonald and Michael Larice's *Urban Design Reader* (London and New York: Routledge, 2006) contains classic and contemporary selections on urban design. Other books on urban design Include Jonathon Barnett, *An Introduction to Urban Design* (New York: Harper & Row, 1982), Edmund N. Bacon, *Design of Cities* (New York: Penguin, 1976), Vincent Scully, *American Architecture and Urbanism* (New York: Praeger, 1969), and Gordon Cullen, *Townscape* (New York: Penguin, 1976).

We think it's time for a new urban design manifesto. Almost 50 years have passed since Le Corbusier and the International Congress of Modern Architecture (CIAM) produced the Charter of Athens, and it is more than 20 years since the first Urban Design Conference, still in the CIAM tradition, was held (at Harvard in 1957). Since then the precepts of CIAM have been attacked by sociologists, pecently by architects them-selves. But it is still a strong influence, and we will take it as our starting point. Make no mistake: the charter was, simply, a manifesto – a public declaration that spelled out the ills of industrial cities as they existed in the 1930s and laid down physical requirements necessary to establish healthy, humane, and beautiful urban environments for people. It could not help but deal with social, economic, and political phenomena, but its basic subject matter was the physical design of cities. Its authors were (mostly) socially concerned architects, determined that their art and craft be responsive to social realities as well as to improving the lot of man. It would be a mistake to write them off as simply elitist designers and physical determinists.

So the charter decried the medium-size (up to six storys) high-density buildings with high land coverage that were associated so closely with slums. Similarly, buildings that faced streets were found to be detri-mental to healthy living. The seemingly limitless horizontal expansion of urban areas devoured the countryside, and suburbs were viewed as symbols of terrible waste. Solutions could be found in the demo-lition of unsanitary housing, the provision of green areas in every residential district, and new high-rise, high-density buildings set in open space. Housing was to be removed from its traditional relationship facing streets, and the whole circulation system was to be revised to meet the needs of emerging mechanization (the automobile). Work areas should be close to but separate from residential areas. To achieve the new city, large land holdings, preferably owned by the public, should replace multiple small parcels (so that projects could be properly designed and developed).

Now thousands of housing estates and redevel-opment projects in socialist and capitalist countries the world over, whether built on previously undeveloped land or developed as replacements for old urban areas, attest to the acceptance of the charter's dictums. The design notions it embraced have become part of a world design language, not just the intellectual property of an enlightened few, even though the principles have been devalued in many developments.

Of course, the Charter of Athens has not been the only major urban philosophy of this century to influence the development of urban areas. Ebenezer Howard, too, was responding to the ills of the nineteenth-century industrial city, and the Garden City movement has been at least as powerful as the Charter of Athens. New towns policies, where they exist, are rooted in Howard's thought. But you don't have to look to new towns to see the influence of Howard, Olmsted, Wright, and Stein. The superblock notion, if nothing else, pervades large housing projects around the world, in central cities as well as suburbs. The notion of buildings in a park is as common to garden city designs as it is to charter-inspired development. Indeed, the two movements have a great deal in common: superblocks, separate paths for people and cars, interior common spaces, housing divorced from streets, and central ownership of land. The garden city-inspired communities place greater emphasis on private outdoor space. The most significant difference, at least as they have evolved, is in density and building type: the garden city people preferred to accommodate people in row houses, garden apartments, and maisonettes, while Corbusier and the CIAM designers went for high-rise buildings and, inevitably, people living in flats and at significantly higher densities.

We are less than enthralled with what either the Charter of Athens or the Garden City movement has produced in the way of urban environments. The emphasis of CIAM was on buildings and what goes on within buildings that happen to sit in space, not on the public life that takes place constantly in public spaces. The orientation is often inward. Buildings tend to be islands, big or small. They could be placed anywhere. From the outside perspective, the building, like the work of art it was intended to be, sits where it can be seen and admired in full. And because it is large it is best seen from a distance (at a scale consistent with a moving auto). Diversity, spontaneity, and surprise are absent, at least for the person on foot. On the other hand, we find little joy or magic or spirit in the charter cities. They are not urban, to us, except according to some definition one might find in a census. Most garden cities, safe and healthy and even gracious as

they may be, remind us more of suburbs than of cities. But they weren't trying to be cities. The emphasis has always been on "garden" as much as or more than on "city."

Both movements represent overly strong design reactions to the physical decay and social inequities of industrial cities. In responding so strongly, albeit understandably, to crowded, lightless, airless, "utilitiless," congested buildings and cities that housed so many people, the utopians did not inquire what was good about those places, either socially or physically. Did not those physical environments reflect (and maybe even foster) values that were likely to be meaningful to people individually and collectively, such as publicness and community? Without knowing it, maybe these strong reactions to urban ills ended up by throwing the baby out with the bathwater.

In the meantime we have had a lot of experience with city building and rebuilding. New spokespeople with new urban visions have emerged. As more CIAM-style buildings were built people became more disenchanted. Many began to look through picturesque lenses back to the old preindustrial cities. From a concentration on the city as a kind of sculpture garden, the townscape movement, led by the *Architectural Review*, emphasized "urban experience." This phenomenological view of the city was espoused by Rasmussen, Kepes, and ultimately Kevin Lynch and Jane Jacobs. It identified a whole new vocabulary of urban form – one that depended on the sights, sounds, feels, and smells of the city, its materials and textures, floor surfaces, facades, style, signs, lights, seating, trees, sun, and shade all potential amenities for the attentive observer and user. This has permanently humanized the vocabulary of urban design, and we enthusiastically subscribe to most of its tenets, though some in the townscape movement ignored the social meanings and implications of what they were doing.

The 1960s saw the birth of community design and an active concern for the social groups affected, usually negatively, by urban design. Designers were the "soft cops," and many professionals left the design field for social or planning vocations, finding the physical environment to have no redeeming social value. But at the beginning of the 1980s the mood in the design professions is conservative. There is a withdrawal from social engagement back to formalism. Supported by semiology and other abstract themes, much of architecture has become a dilettantish and narcissistic pursuit, a chic component of the high art consumer culture, increasingly remote from most people's everyday lives, finding its ultimate manifestation in the art gallery and the art book. City planning is too immersed in the administration and survival of housing, environmental, and energy programs and in responding to budget cuts and community demands to have any clear sense of direction with regard to city form.

While all these professional ideologies have been working themselves out, massive economic, technological, and social changes have taken place in our cities. The scale of capitalism has continued to increase, as has the scale of bureaucracy, and the automobile has virtually destroyed cities as they once were.

In formulating a new manifesto, we react against other phenomena than did the leaders of CIAM 50 years ago. The automobile cities of California and the Southwest present utterly different problems from those of nineteenth-century European cities, as do the CIAM-influenced housing developments around European, Latin American, and Russian cities and the rash of squatter settlements around the fast-growing cities of the Third World. What are these problems?

PROBLEMS FOR MODERN URBAN DESIGN

Poor living environments

While housing conditions in most advanced countries have improved in terms of such fundamentals as light, air, and space, the surroundings of homes are still frequently dangerous, polluted, noisy, anonymous wastelands. Travel around such cities has become more and more fatiguing and stressful.

Giantism and loss of control

The urban environment is increasingly in the hands of the large-scale developers and public agencies. The elements of the city grow inexorably in size, massive transportation systems are segregated for single travel modes, and vast districts and complexes are created that make people feel irrelevant.

People, therefore, have less sense of control over their homes, neighborhoods, and cities than when they lived in slower-growing locally based communities. Such giantism can be found as readily in the housing

projects of socialist cities as in the office buildings and commercial developments of capitalist cities.

Large-scale privatization and the loss of public life

Cities, especially American cities, have become privatized, partly because of the consumer society's emphasis on the individual and the private sector, creating Galbraith's "private affluence and public squalor," but escalated greatly by the spread of the automobile. Crime in the streets is both a cause and a consequence of this trend, which has resulted in a new form of city: one of closed, defended islands with blank and windowless facades surrounded by wastelands of parking lots and fast-moving traffic. As public transit systems have declined, the number of places in American cities where people of different social groups actually meet each other has dwindled. The public environment of many American cities has become an empty desert, leaving public life dependent for its survival solely on planned formal occasions, mostly in protected internal locations.

Centrifugal fragmentation

Advanced industrial societies took work out of the home, and then out of the neighborhood, while the automobile and the growing scale of commerce have taken shopping out of the local community. Fear has led social groups to flee from each other into homogeneous social enclaves. Communities themselves have become lower in density and increasingly homogeneous. Thus the city has spread out and separated to form extensive monocultures and specialized destinations reachable often only by long journeys – a fragile and extravagant urban system dependent on cheap, available gasoline, and an effective contributor to the isolation of social groups from each other.

Destruction of valued places

The quest for profit and prestige and the relentless exploitation of places that attract the public have led to the destruction of much of our heritage, of historic places that no longer turn a profit, of natural amenities that become overused. In many cases, as in San

Francisco, the very value of the place threatens its destruction as hungry tourists and entrepreneurs flock to see and profit from it.

Placelessness

Cities are becoming meaningless places beyond their citizens' grasp. We no longer know the origins of the world around us. We rarely know where the materials and products come from, who owns what, who is behind what, what was intended. We live in cities where things happen without warning and without our participation. It is an alien world for most people. It is little surprise that most withdraw from community involvement to enjoy their own private and limited worlds.

Injustice

Cities are symbols of inequality. In most cities the discrepancy between the environments of the rich and the environments of the poor is striking. In many instances the environments of the rich, by occupying and dominating the prevailing patterns of transportation and access, make the environments of the poor relatively worse. This discrepancy may be less visible in the low-density modern city, where the display of affluence is more hidden than in the old city; but the discrepancy remains.

Rootless professionalism

Finally, design professionals today are often part of the problem. In too many cases, we design for places and people we do not know and grant them very little power or acknowledgment. Too many professionals are more part of a universal professional culture than part of the local cultures for whom we produce our plans and products. We carry our "bag of tricks" around the world and bring them out wherever we land. This floating professional culture has only the most superficial conception of particular place. Rootless, it is more susceptible to changes in professional fashion and theory than to local events. There is too little inquiry, too much proposing. Quick surveys are made, instant solutions devised, and the rest of the time is spent persuading the clients. Limits on time and budgets drive us on, but so do lack of understanding and the

placeless culture. Moreover, we designers are often unconscious of our own roots, which influence our preferences in hidden ways.

At the same time, the planning profession's retreat into trendism, under the positivist influence of social science, has left it virtually unable to resist the social pressures of capitalist economy and consumer sovereignty. Planners have lost their beliefs. Although we believe citizen participation is essential to urban planning, professionals also must have a sense of what we believe is right, even though we may be vetoed.

GOALS FOR URBAN LIFE

We propose, therefore, a number of goals that we deem essential for the future of a good urban environment: livability; identity and control; access to opportunity, imagination, and joy; authenticity and meaning; open communities and public life; self-reliance; and justice.

Livability

A city should be a place where everyone can live in relative comfort. Most people want a kind of sanctuary for their living environment, a place where they can bring up children, have privacy, sleep, eat, relax, and restore themselves. This means a well-managed environment relatively devoid of nuisance, overcrowding, noise, danger, air pollution, dirt, trash, and other unwelcome intrusions.

Identity and control

People should feel that some part of the environment belongs to them, individually and collectively, some part for which they care and are responsible, whether they own it or not. The urban environment should be an environment that encourages people to express themselves, to become involved, to decide what they want and act on it. Like a seminar where everybody has something to contribute to communal discussion, the urban environment should encourage participation. Urbanites may not always want this. Many like the anonymity of the city, but we are not convinced that the freedom of anonymity is a desirable freedom. It would be much better if people were sure enough of themselves to stand up and be counted. Environments should therefore be designed for those who use them or are affected by them, rather than for those who own them. This should reduce alienation and anonymity (even if people want them); it should increase people's sense of identity and rootedness and encourage more care and responsibility for the physical environment of cities.

Respect for the existing environment, both nature and city, is one fundamental difference we have with the CIAM movement. Urban design has too often assumed that new is better than old. But the new is justified only if it is better than what exists. Conservation encourages identity and control and, usually, a better sense of community, since old environments are more usually part of a common heritage.

Access to opportunity, imagination, and joy

People should find the city a place where they can break from traditional molds, extend their experience, meet new people, learn other viewpoints, have fun. At a functional level, people should have access to alternative housing and job choices; at another level, they should find the city an enlightening cultural experience. A city should have magical places where fantasy is possible, a counter to and an escape from the mundaneness of everyday work and living. Architects and planners take cities and themselves too seriously; the result too often is deadliness and boredom, no imagination, no humor, alienating places. But people need an escape from the seriousness and meaning of the everyday. The city has always been a place of excitement; it is theater, a stage upon which citizens can display themselves and see others. It has magic, or should have, and that depends on a certain sensuous, hedonistic mood, on signs, on night lights, on fantasy, color, and other imagery. There can be parts of the city where belief can be suspended, just as in the experience of fiction. It may be that such places have to be framed so that people know how to act. Until now such fantasy and experiment have been attempted mostly by commercial facilities, at rather low levels of quality and aspiration, seldom deeply experimental. One should not have to travel as far as the Himalayas or the South Sea Islands to stretch one's experience. Such challenges could be nearer home. There should be a place for community utopias; for historic, natural, and anthropological evocations of the modern city, for encounters with the truly exotic.

Authenticity and meaning

People should be able to understand their city (or other people's cities), its basic layout, public functions, and institutions; they should be aware of its opportunities. An authentic city is one where the origins of things and places are clear. All this means an urban environment should reveal its significant meanings; it should not be dominated only by one type of group, the powerful; neither should publicly important places be hidden. The city should symbolize the moral issues of society and educate its citizens to an awareness of them.

That does not mean everything has to be laid out as on a supermarket shelf. A city should present itself as a readable story, in an engaging and, if necessary, provocative way, for people are indifferent to the obvious, overwhelmed by complexity. A city's offerings should be revealed or they will be missed. This can affect the forms of the city, its signage, and other public information and education programs.

Livability, identity, authenticity, and opportunity are characteristics of the urban environment that should serve the individual and small social unit, but the city has to serve some higher social goals as well. It is these we especially wish to emphasize here.

Community and public life

Cities should encourage participation of their citizens in community and public life. In the face of giantism and fragmentation, public life, especially life in public places, has been seriously eroded. The neighborhood movement, by bringing thousands, probably millions of people out of their closed private lives into active participation in their local communities, has begun to counter that trend, but this movement has had its limitations. It can be purely defensive, parochial, and self-serving. A city should be more than a warring collection of interest groups, classes, and neighborhoods; it should breed a commitment to a larger whole, to tolerance, justice, law, and democracy. The structure of the city should invite and encourage public life, not only through its institutions, but directly and symbolically through its public spaces. The public environment, unlike the neighborhood, by definition should be open to all members of the community. It is where people of different kinds meet. No one should be excluded unless they threaten the balance of that life.

Urban self-reliance

Increasingly cities will have to become more self-sustaining in their uses of energy and other scarce resources. "Soft energy paths" in particular not only will reduce dependence and exploitation across regions and countries but also will help reestablish a stronger sense of local and regional identity, authenticity, and meaning.

An environment for all

Good environments should be accessible to all. Every citizen is entitled to some minimal level of environmental livability and minimal levels of identity, control, and opportunity. Good urban design must be for the poor as well as the rich. Indeed, it is more needed by the poor.

We look toward a society that is truly pluralistic, one where power is more evenly distributed among social groups than it is today in virtually any country, but where the different values and cultures of interest- and place-based groups are acknowledged and negotiated in a just public arena.

These goals for the urban environment are both individual and collective, and as such they are frequently in conflict. The more a city promises for the individual, the less it seems to have a public life; the more the city is built for public entities, the less the individual seems to count. The good urban environment is one that somehow balances these goals, allowing individual and group identity while maintaining a public concern, encouraging pleasure while maintaining responsibility, remaining open to outsiders while sustaining a strong sense of localism.

AN URBAN FABRIC FOR AN URBAN LIFE

We have some ideas, at least, for how the fabric or texture of cities might be conserved or created to encourage a livable urban environment. We emphasize the structural qualities of the good urban environment – qualities we hope will be successful in creating urban experiences that are consonant with our goals.

Do not misread this. We are not describing all the qualities of a city. We are not dealing with major transportation systems, open space, the natural environment, the structure of the large-scale city, or even the structure of neighborhoods, but only the grain of the good city.

There are five physical characteristics that must be present if there is to be a positive response to the goals and values we believe are central to urban life. They must be designed, they must exist, as prerequisites of a sound urban environment. All five must be present, not just one or two. There are other physical characteristics that are important, but these five are essential: livable streets and neighborhoods; some minimum density of residential development as well as intensity of land use; an integration of activities – living, working, shopping – in some reasonable proximity to each other; a manmade environment, particularly buildings, that defines public space (as opposed to buildings that, for the most part, sit in space); and many, many separate, distinct buildings with complex arrangements and relationships (as opposed to few, large buildings).

Let us explain, keeping in mind that all five of the characteristics must be present. People, we have said, should be able to live in reasonable (though not excessive) safety, cleanliness, and security. That means livable streets and neighborhoods: with adequate sunlight, clean air, trees, vegetation, gardens, open space, pleasantly scaled and designed buildings; without offensive noise; with cleanliness and physical safety. Many of these characteristics can be designed into the physical fabric of the city.

The reader will say, "Well of course, but what does that mean?" Usually it has meant specific standards and requirements, such as sun angles, decibel levels, lane widths, and distances between buildings. Many researchers have been trying to define the qualities of a livable environment. It depends on a wide array of attributes, some structural, some quite small details. There is no single right answer. We applaud these efforts and have participated in them ourselves. Nevertheless, desires for livability and individual comfort by themselves have led to fragmentation of the city. Livability standards, whether for urban or for suburban developments, have often been excessive.

Our approach to the details of this inclusive physical characteristic would center on the words "reasonable, though not excessive . . ." Too often, for example, the requirement of adequate sunlight has resulted in buildings and people inordinately far from each other, beyond what demonstrable need for light would dictate. Safety concerns have been the justifications for ever wider streets and wide, sweeping curves rather than narrow ways and sharp corners. Buildings are removed from streets because of noise considerations when there might be other ways to deal with this concern. So although livable streets and neighborhoods are a primary requirement for any good urban fabric – whether for existing, denser cities or for new development – the quest for livable neighborhoods, if pursued obsessively, can destroy the urban qualities we seek to achieve.

A *minimum density* is needed. By density we mean the number of people (sometimes expressed in terms of housing units) living on an area of land, or the number of people using an area of land.

Cities are not farms. A city is people living and working and doing the things they do in relatively close proximity to each other.

We are impressed with the importance of density as a perceived phenomenon and therefore relative to the beholder and agree that, for many purposes, perceived density is more important than an "objective" measurement of people per unit of land. We agree, too, that physical phenomena can be manipulated so as to render perceptions of greater or lesser density. Nevertheless, a narrow, winding street, with a lot of signs and a small enclosed open space at the end, with no people, does not make a city. Cities are more than stage sets. Some minimum number of people living and using a given area of land is required if there is to be human exchange, public life and action, diversity and community.

Density of people alone will account for the presence or absence of certain uses and services we find important to urban life. We suspect, for example, that the number and diversity of small stores and services – for instance, groceries, bars, bakeries, laundries and cleaners, coffee shops, secondhand stores, and the like – to be found in a city or area is in part a function of density. That is, that such businesses are more likely to exist, and in greater variety, in an area where people live in greater proximity to each other ("higher" density). The viability of mass transit, we know, depends partly on the density of residential areas and partly on the size and intensity of activity at commercial and service destinations. And more use of transit, in turn, reduces parking demands and permits increases in density. There must be a critical mass of people, and they must spend a lot of their time in reasonably close proximity to each other, including when they are at home, if there is to be an urban life. The goal of local control and community identity is associated with density as well. The notion of an optimum density is elusive and is easily confused with the health and livability of urban areas, with lifestyles, with housing types, with the size of area being considered (the building site or the neighborhood or the city), and with the economics of development. A density that might be best for child rearing might be less than adequate to support public transit. Most recently, energy efficiency has emerged as a concern associated with density, the notion being that conservation will demand more compact living arrangements.

Our conclusion, based largely on our experience and on the literature, is that a minimum net density (people or living units divided by the size of the building site, excluding public streets) of about 15 dwelling units (30–60 people) per acre of land is necessary to support city life. By way of illustration, that is the density produced with generous town houses (or row houses). It would permit parcel sizes up to 25 feet wide by about 115 feet deep. But other building types and lot sizes also would produce that density. Some areas could be developed with lower densities, but not very many. We don't think you get cities at 6 dwellings to the acre, let alone on half-acre lots. On the other hand, it is possible to go as high as 48 dwelling units per acre (96 to 192 people) for a very large part of the city and still provide for a spacious and gracious urban life. Much of San Francisco, for example, is developed with three story buildings (one unit per floor) above a parking story, on parcels that measure 25 feet by 100 or 125 feet. At those densities, with that kind of housing, there can be private or shared gardens for most people, no common hallways are required, and people can have direct access to the ground. Public streets and walks adequate to handle pedestrian and vehicular traffic generated by these densities can be accommodated in rights-of-way that are 50 feet wide or less. Higher densities, for parts of the city, to suit particular needs and lifestyles, would be both possible and desirable. We are not sure what the upper limits would be but suspect that as the numbers get much higher than 200 people per net residential acre, for larger parts of the city, the concessions to less desirable living environments mount rapidly.

Beyond residential density, there must be a minimum intensity of people using an area for it to be urban, as we are defining that word. We aren't sure what the numbers are or even how best to measure this kind of intensity. We are speaking here, particularly, of the public or "meeting" areas of our city. We are confident that our lowest residential densities will provide most meeting areas with life and human exchange, but are not sure if they will generate enough activity for the most intense central districts.

There must be an *integration of activities* – living, working, and shopping as well as public, spiritual, and recreational activities – reasonably near each other.

The best urban places have some mixtures of uses. The mixture responds to the values of publicness and diversity that encourage local community identity. Excitement, spirit, sense, stimulation, and exchange are more likely when there is a mixture of activities than when there is not. There are many examples that we all know. It is the mix, not just the density of people and uses, that brings life to an area, the life of people going about a full range of normal activities without having to get into an automobile.

We are not saying that every area of the city should have a full mix of all uses. That would be impossible. The ultimate in mixture would be for each building to have a range of uses from living, to working, to shopping, to recreation. We are not calling for a return to the medieval city. There is a lot to be said for the notion of "living sanctuaries," which consist almost wholly of housing. But we think these should be relatively small, of a few blocks, and they should be close and easily accessible (by foot) to areas where people meet to shop or work or recreate or do public business. And except for a few of the most intensely developed office blocks of a central business district or a heavy industrial area, the meeting areas should have housing within them. Stores should be mixed with offices. If we envision the urban landscape as a fabric, then it would be a salt-and-pepper fabric of many colors, each color for a separate use or a combination. Of course, some areas would be much more heavily one color than another, and some would be an even mix of colors. Some areas, if you squinted your eyes, or if you got so close as to see only a small part of the fabric, would read as one color, a red or a brown or a green. But by and large there would be few if any distinct patterns, where one color stopped and another started. It would not be patchwork quilt, or an even-colored fabric. The fabric would be mixed.

In an urban environment, *buildings* (and other objects that people place in the environment) *should be arranged in such a way as to define and even enclose public space, rather than sit in space*. It is not enough to have high densities and an integration of activities to have cities. A tall enough building with enough people living (or even working) in it, sited on a large parcel, can easily produce the densities we have talked about and can have internally mixed uses, like most "mixed use" projects. But that building and its neighbors will be unrelated objects sitting in space if they are far enough apart, and the mixed uses might be only privately available. In large measure that is what the Charter of Athens, the garden cities, and standard suburban development produce.

Buildings close to each other along a street, regardless of whether the street is straight, or curved, or angled, tend to define space if the street is not too wide in relation to the buildings. The same is true of a plaza or a square. As the spaces between buildings become larger (in relation to the size of the buildings, up to a point), the buildings tend more and more to sit in space. They become focal points for few or many people, depending on their size and activity. Except where they are monuments or centers for public activities (a stadium or meeting hall), where they represent public gathering spots, buildings in space tend to be private and inwardly oriented. People come to them and go from them in any direction. That is not so for the defined outdoor environment. Avoiding the temptation to ascribe all kinds of psychological values to defined spaces (such as intimacy, belonging, protection – values that are difficult to prove and that may differ for different people), it is enough to observe that spaces surrounded by buildings are more likely to bring people together and thereby promote public interaction. The space can be linear (like streets) or in the form of plazas of myriad shapes. Moreover, interest and interplay among uses is enhanced. To be sure, such arrangements direct people and limit their freedom – they cannot move in just any direction from any point – but presumably there are enough choices (even avenues of escape) left open, and the gain is in greater potential for sense stimulation, excitement, surprise, and focus. Over and over again we seek out and return to defined ways and spaces as symbolic of urban life emphasizing the public space more than the private building.

It is important for us to emphasize *public places* and a *public* way system. We have observed that the central value of urban life is that of publicness, of people from different groups meeting each other and of people acting in concert, albeit with debate. The most important public places must be for *pedestrians*, for no public life can take place between people in automobiles. Most public space has been taken over by the automobile, for travel or parking. We must fight to restore more for the pedestrian. Pedestrian malls are not simply to benefit the local merchants. They have an essential public value. People of different kinds meet each other directly. The level of communication may be only visual, but that itself is educational and can encourage tolerance. The revival of street activities, street vending, and street theater in American cities may be the precursor of a more flourishing public environment, if the automobile can be held back.

There also must be symbolic, public meeting places, accessible to all and publicly controlled. Further, in order to communicate, to get from place to place, to interact, to exchange ideas and goods, there must be a healthy public circulation system. It cannot be privately controlled. Public circulation systems should be seen as significant cultural settings where the city's finest products and artifacts can be displayed, as in the piazzas of medieval and renaissance cities.

Finally, *many different buildings and spaces with complex arrangements and relationships* are required. The often elusive notion of human scale is associated with this requirement – a notion that is not just an architect's concept but one that other people understand as well.

Diversity, the possibility of intimacy and confrontation with the unexpected, stimulation, are all more likely with many buildings than with few taking up the same ground areas.

For a long time we have been led to believe that large land holdings were necessary to design healthy, efficient, aesthetically pleasing urban environments. The slums of the industrial city were associated, at least in part, with all those small, overbuilt parcels. Socialist and capitalist ideologies alike called for land assembly to permit integrated, socially and economically useful developments. What the socialist countries would do via public ownership the capitalists would achieve through redevelopment and new fiscal mechanisms that rewarded large holdings. Architects of both ideological persuasions promulgated or were easily convinced of the wisdom of land assembly. It's not hard to figure out why. The results, whether by big business or big government, are more often than not inward-oriented, easily controlled or controllable, sterile, large-building projects, with fewer entrances, fewer windows, less diversity, less innovation, and less individual expression than the urban fabric that existed previously or that can be achieved with many actors and many buildings. Attempts to break up facades or otherwise to articulate separate activities in large

buildings are seldom as successful as when smaller properties are developed singly.

Health, safety, and efficiency can be achieved with many smaller buildings, individually designed and developed. Reasonable public controls can see to that. And, of course, smaller buildings are a lot more likely if parcel sizes are small than if they are large. With smaller buildings and parcels, more entrances must be located on the public spaces, more windows and a finer scale of design diversity emerge. A more public, lively city is produced. It implies more, smaller groups getting pieces of the public action, taking part, having a stake. Other stipulations may be necessary to keep public frontages alive, free from the deadening effects of offices and banks, but small buildings will help this more than large ones. There need to be large buildings, too, covering large areas of land, but they will be the exception, not the rule, and should not be in the centers of public activity.

ALL THESE QUALITIES . . . AND OTHERS

A good city must have all those qualities. Density without livability could return us to the slums of the nineteenth century. Public places without small-scale, fine-grain development would give us vast, overscale cities. As an urban fabric, however, those qualities stand a good chance of meeting many of the goals we outlined. They directly attend to the issue of livability though they are aimed especially at encouraging public places and a public life. Their effects on personal and group identity are less clear, though the small-scale city is more likely to support identity than the large-scale city. Opportunity and imagination should be encouraged by a diverse and densely settled urban structure. This structure also should create a setting that is more meaningful to the individual inhabitant and small group than the giant environments now being produced. There is no guarantee that this urban structure will be a more just one than those presently existing. In supporting the small against the large, however, more justice for the powerless may be encouraged.

Still, an urban fabric of this kind cannot by itself meet all these goals. Other physical characteristics are important to the design of urban environments. Open space, to provide access to nature as well as relief from the built environment, is one. So are definitions, boundaries if you will, that give location and identity to neighborhoods (or districts) and to the city itself.

There are other characteristics as well: public buildings, educational environments, places set aside for nurturing the spirit, and more. We still have work to do.

MANY PARTICIPANTS

While we have concentrated on defining physical characteristics of a good city fabric, the process of creating it is crucial. As important as many buildings and spaces are many participants in the building process. It is through this involvement in the creation and management of their city that citizens are most likely to identify with it and, conversely, to enhance their own sense of identity and control.

AN ESSENTIAL BEGINNING

The five characteristics we have noted are essential to achieving the values central to urban life. They need much further definition and testing. We have to know more about what configurations create public space: about maximum densities, about how small a community can be and still be urban (some very small Swiss villages fit the bill, and everyone knows some favorite examples), about what is perceived as big and what small under different circumstances, about landscape material as a space definer, and a lot more. When we know more we will be still further along toward a new urban design manifesto.

We know that any ideal community, including the kind that can come from this manifesto, will not always be comfortable for every person. Some people don't like cities and aren't about to. Those who do will not be enthralled with all of what we propose.

Our urban vision is rooted partly in the realities of earlier, older urban places that many people, including many utopian designers, have rejected, often for good reasons. So our utopia will not satisfy all people. That's all right. We like cities. Given a choice of the kind of community we would *like* to live in – the sort of choice earlier city dwellers seldom had – we would choose to live in an urban, public community that embraces the goals and displays the physical characteristics we have outlined. Moreover, we think it responds to what people want and that it will promote the good urban life.

PART EIGHT

The future of the city

INTRODUCTION TO PART EIGHT

The desire to peer into the future is a human trait as old as the biblical prophets and the oracle at Delphi. And the desire to project *urban* futures is at least as old as Plato's description of the ideal city-state in *The Republic*. But the pace of futurist predictions seems to quicken at times when great cultural and historic shifts are taking place. Such was the case during the Industrial Revolution of the nineteenth century when fantasists like Jules Verne and political idealists like Edward Bellamy (author of *Looking Backward, 2000–1887*) captured the popular imagination, and when the utopian visionaries discussed in Part Five of this volume helped to establish the theoretical basis of urban planning practice. Such is the case today as the realization becomes every day more clear that the advanced economies of the world are entering a new global, information-based, post-industrial stage of development that promises new forms of urban civilization and human community.

Clearly, a new urban paradigm is now emerging. But in order to predict the shape of the emerging postmodern city, one must have a clear sense of the probable direction of the world urbanization process described by Kingsley Davis in "The Urbanization of the Human Population," the essay with which the "Evolution of Cities" section of this volume begins (p. 17). Will the percentage of the world's total population living in cities, now at about 50 percent, continue to increase in the decades and centuries to come? Will the "townward drift," as Frederick Law Olmsted (p. 307) called it, continue? To what eventual level? 80 percent? 90? 100 percent? Or has urbanization reached its peak, ready to stabilize at more or less the present level? Might the S-curve on the chart representing world urbanization become a bell curve, with the percentage of the human population living in cities going into a long, gradual decline until only a small number of the total population remains urbanized? This last possibility has spawned an intriguing, if highly conjectural body of literature. Some, pondering the possible effects of modern transportation and tele-communications technologies, see a gradual withering away of cities. Others, particularly environmentalists, talk about urbanization reaching its natural limits and beginning to reverse in an age of ecological and economic constraints, or suggest that without the kind of sustainable development called for in the Brundtland Report (p. 337), unrestrained growth is sure to result in widespread economic and environmental collapse.

Today, the variety of possible futures from which to choose is extraordinarily diverse. Each option mirrors some of our deepest hopes and fears. Marshall McLuhan, the 1960s guru of communications theory, suggested that the whole world would one day become a "global village," with every member of humanity interacting with every other in a real-time simulacrum of the neolithic community. Some of the more radical members of the environmentalist movement have gone "back to nature" by establishing rural communes along the fringes of urbanized civilization, while other equally radical social activists have established "urban kibbutzim" in the very hearts of the inner cities. Looking at the new global economy, some like Saskia Sassen (p. 197) see a future urban society divided between the affluent elite and the increasingly disenfranchised corporate service class, but Richard Florida (p. 129) and others see an urban world dominated and enriched by an emerging "creative class." Techno-optimists see Earth's future in space colonization projects while techno-pessimists envision post-apocalypse cities like the ones depicted in *Terminator, Bladerunner, The Matrix*, and other popular science fiction films. And some Islamic fundamentalists see all modern cities as essentially Western – and therefore morally corrupt – cities that must be destroyed.

One possible urban future is post-urbanism. Melvin Webber, author of "The Post-City Age" (p. 473), argues that certain technological developments will result in an end to traditional cities and the emergence of a post-urban period of human development. Because he wrote his prophetic essay in 1968, the technologies he referred to were air travel and telephones. Today, computers and telecommunications make his predictions even more plausible. Even if we reject the basic idea of a reversal of the course of urbanization, Webber's insight into the way the new technological world may leave some people informationally rich and others informationally poor – the "digital divide" – is a troubling reality that cannot be ignored.

Webber is something of a utopian visionary engaged in brilliant, if somewhat mystical, speculation about humankind's future. It is perhaps more useful, when considering the future of the city, to come down to earth, to shorten the futurist perspective to the near term, and to project immediate futures based on present and observable trends. Among the most important such trends today are the parallel emergence of (1) a global, post-industrial, telecommunications-based economy; (2) a new, far-flung suburban ring of development that journalist Joel Garreau has dubbed "Edge City" and that Robert Fishman has analyzed as "technoburbia" (p. 69); (3) the growth of "global" cities and mega-urban regions, principally in Asia; (4) transnational environmental concerns such as global warming and atmospheric pollution; and (5) the persistence, even intensification, of classic social conflicts along racial, class, and cultural lines that have always been a feature of urban life. These influences are likely to determine the course of early twenty-first-century urban development.

Manuel Castells (p. 478) is a critical analyst of urban life who has carefully analyzed the near-term futures of urban society – both for the United States and Western Europe – and suggested possible directions for urban development strategies in ways that combine the influences of new technologies and old social inequalities based on uneven access to economic empowerment. No one thinks that solving the problem of inner-city poverty and race-and-class-based segregation will be an easy task, nor that one solution is the only one worth considering. Clearly, there are many contentious issues of politics and urban governance, and even deeper concerns at the level of social structure and cultural development that will affect the future of cities. But as the neo-suburban technoburbs and the New Urbanist TODs (Transit-Oriented Developments) beckon, and as local economies and national sovereignties give way to global interdependence, the ongoing and sometimes explosive violence of racial, class, and religious discrimination is still there at the core of many cities, threatening to destroy urban civilization and to undermine the human community. Even in an age of international terrorism, defusing those possible sources of internal conflict remains part of the unfinished business of the city of the future.

In "European Cities, the Informational Society, and the Global Economy", Castells surveys the urban future of Western Europe. Recognizing that racial and class conflicts – in this case exacerbated by immigration from the underdeveloped world – will be a troubling feature of future urban development, Castells worries about the effects of economic globalism and telecommunications technologies on the democratic values of equality and social justice. He argues that technology-based cities represent "a new industrial space," located in centers throughout Europe, Asia, and America and tied to a global "informational" economy, that are fundamentally different from anything that has ever come before. A two-tier economy and a widening gulf between the educated elites and the ghettoized, marginalized urban populations is intensifying rather than diminishing in the cities of the information economy and the global marketplace. The new urban order, abetted by new economic realities, seems cut off from its past, from history and cultural tradition as sources of communal meaning and individual identity.

In *The Rise of the Network Society* (1998), Castells argued that the implicit tendency of the work-styles of the post-industrial economy is to detach themselves from traditional cultures, values, and communities. Information flows through networks and across vast distances, and "the historical emergence of the space of flows," he writes, supersedes "the meaning of the space of places." For Castells, then, the real challenge of the new informational cities of Western Europe is to reconcile the "new techno-economic paradigm" and "place-based social meaning" in a way that will avoid what he calls "urban schizophrenia." "At the cultural level," he writes, "local societies, territorially defined, must preserve their identities, and build upon their historical roots, regardless of their economic and functional dependence upon the space of flows."

In addition to computerized telecommunications, global economic integration is one of the leading forces propelling the systematic changes that are transforming cities worldwide. This is nowhere more true than in Asia, one of the areas of the world that has become a major focus of burgeoning globalized economic activity based on the dramatic modernization campaigns of central governments and the availability of inexpensive labor for manufacturing operations. In *Beyond Metropolis* (p. 489), Aprodicio Laquian notes that "Asia currently has twelve of the world's largest urban agglomerations. By 2015, this number may increase to fourteen."

According to Laquian, most of these new cities are actually "mega-urban regions" extending out from the core of the major urban centers of the entire global quadrant extending from Japan and Korea and China . . . through the Philippines, Indonesia, and Malaysia, Thailand, Vietnam, and Myanmar . . . to Bangladesh, India, and Pakistan. Some of these Asian cities are ancient capitals. Others date from the recent colonial past. But almost all of them have experienced extraordinary regional population growth over the course of the past twenty to forty years and are predicted to grow even larger in the decades to come. Interestingly, it is not just these well-known cities themselves – Tokyo, Shanghai, Manila, Jakarta, Dhaka, Mumbai, Karachi – that are growing. In some cases, the population growth of these central cities has actually begun to stabilize. But the "mega-urban" regions surrounding and extending from these cities are growing in ways that will mark almost all of them as "global cities" (that is, city regions of ten million or more) within the next ten years. In case after case, Laquian describes how national governments have responded to what might be called "natural" urban–regional development – cities have always had hinterlands, after all – by putting in place regional administrative structures to carry out the planning and development needs of the new "mega-urban" regions. Shanghai, the city depicted on the cover of this book, is a city of about thirteen million, but it is the "head of the dragon" of the entire Yangtze River delta region and is now "the dominant core of a megalapolitan region covering about 100,000 square kilometers with a population of 72.7 million."

If Castells is the great theoretician of the techno-urbanist economy of the future, and Laquian the chronicler of mega-regional development in the global cities of Asia, Stephen Wheeler, in "Planning Sustainable and Livable Cities" (p. 499), suggests that advanced telecommunications technologies and regional administrative structures are just new developments in the infrastructures of production, management, and control that have always been a part of urban life. Increasingly, he argues, another factor will greatly influence the future of cities: the natural environment, now seen not just as a regional rural hinterland but as a global context in which cities can either flourish or die.

The contemporary concept of urban sustainability has a long pre-history. By the middle of the nineteenth century, park planners like Joseph Paxton in England and Frederick Law Olmsted in the United States were working hard to bring nature into crowded cities. In the early twentieth century, Patrick Geddes had worked out an elaborate scheme for regional planning that reflected different ecosystems. And utopian planners like Ebenezer Howard (p. 314) proposed new cities greenbelted by open countryside. Similar thinking informed Ian McHarg's theories in his classic *Design with Nature* (1965).

Today – especially since the publication of the Brundtland Report – issues of sustainability, including "sustainable" urban planning, have moved to the forefront of global consciousness. Urban sustainability – the ability of cities to grow and develop without overwhelming the environmental carrying capacity of the Earth – have become an important feature of contemporary thinking about the future of cities. In "Planning Sustainable and Livable Cities," Wheeler summarizes the current sustainable urban planning literature and provides his own definition of sustainable urban development and the operating principles needed to achieve it. Wheeler illustrates his ideas with actual examples of good sustainable planning practice and thereby points the way toward desirable urban futures that can actually be realized. Wheeler's vision of a sustainable urban future is not narrowly ecological, in the sense of placing natural above human values. "In the long run," he writes, "sustainable development will require systemic cultural change that builds democracy and social capital."

But can social capital, or even the accepted norms of urban social life, withstand the underlying implications of the telecommunications revolution? William Mitchell thinks not and, in "The Teleserviced City" (p. 510) from his imaginative book *E-topia* (2001), suggests that the future of urban social life will be transformed in very fundamental ways by the very nature of digital electronics. Not only will urban space and urban economic relationships be transformed by computers and the daily online transactions that already characterize our lives, but urban communities and even individual urban personalities, human identity itself, are on the verge of a major new historic and cultural revolution, equal in effect to the Neolithic, Urban, and Industrial Revolutions that V. Gordon Childe (p. 27) described. Mitchell's vision of the inevitable future is not without promise, but many will find it deeply challenging. And it is how we will adapt to the challenges of the future – population growth, the new globalism, the emerging technologies, environmental constraints, and the ongoing struggles for social justice – that will determine how in the years to come cities and citizens will achieve, or fail to achieve, a meaningful sense of personal identity and social community, and thereby give ongoing justification or final refutation to the validity of the urban project.

In the end, a consideration of the many possibilities of the urban future leaves more questions than answers. As we move together into that future, one of our best guides will be "The Urban Future" (p. 517), the concluding chapter of Joel Kotkin's *The City: A Global History* (2005). Kotkin is an economist, but he is also an interdisciplinary generalist. His broad overview of the big themes and major trends of urban history provides a reasoned and judicious metric on which to measure our urban expectations.

Kotkin sees a continuing crisis of development in the "megacities" described by Aprodicio Laquian and others. Some, like Singapore, Kuala Lumpur and Shanghai, have successfully "integrated themselves into the global economy," while others, like Bangkok, Manila and Cairo, have found that massive growth is "often more a burden than an advantage." Kotkin also sees emerging limits to the economic supremacy of the most successful global cities. New York, London, Tokyo, and the others will not collapse, but the "destruction of distance" made possible by modern telecommunications will allow many smaller cities to become important, specialized centers of the world economy. He notes that the mega-retailer Wal-Mart, for example, has been able to operate quite effectively from Bentonville, Arkansas!

Kotkin notes the rise of the "ephemeral city," with populations of "creative class" residents like those described by Richard Florida producing services and experiences, not industrial products, but wonders if "this form of culturally based growth may not be self-sustaining." Does a thriving arts culture create economic success? Or is it the other way around? Kotkin also notes the persistence of urban crime as an ongoing security challenge for urban residents and points to the terrorist threat from radical fundamentalist Islam as "the most lethal threat to the security of cities globally." But it is the city as a "sacred place," a place where the diurnal rituals of human society play out on an everyday basis, that provides Kotkin with his final hope for the urban future. "It is in the city," he writes, "this ancient confluence of the sacred, safe, and busy, where humanity's future will be shaped for centuries to come."

"The Post-City Age"

Daedalus (1968)

Melvin M. Webber

Editors' Introduction

In the introduction to Kingsley Davis's "The Urbanization of the Human Population" (p. 17), we asked what the future of urbanization might be. Would the urban population of the Earth stabilize at 50 percent or 70 percent? Might it increase to 100 percent? Or might the S-curve of historic urbanization become a standard bell curve in the future, charting a steady decline in the proportion of the total human population that lives in cities? As long ago as 1968, Melvin Webber thought through the possibility of a gradual decline in urbanization in his brilliant and prophetic essay "The Post-City Age." Prophetic? Maybe, maybe not. Major urban centers have not decreased in density and population during the last thirty years. To the contrary, major global cities – as well as newly emerging mega-cities in Asia, Africa, and Latin America – have grown tremendously in the decades since Webber made his bold and counter-intuitive prediction. Admitting as much in a later essay entitled "Tenacious Cities," Webber wrote, "despite growing ease of interaction over distance and the eroding requirements for propinquity . . . metropolitan areas have not disappeared. Indeed, they continue to grow."

Still, Webber was prescient in pointing to two major technological developments – air transportation and the telephone – that even in 1968 were eliminating the traditional space-time constraints on human interaction and bringing about stunning cultural changes in the nature of an increasingly global human civilization.

Webber argues that the widespread availability of commercial air transportation and global telephonic communication permits an educated and affluent class of people to live anywhere – in suburbs, in rural districts, on mountain-tops, for that matter – and still be thoroughly "urban," participating fully in intellectual, professional, and economic life. At the same time, it is precisely the poor populations that are trapped in inner cities and deprived of access to technology that are becoming increasingly "rural" in the sense that they are non-participants in global community affairs.

Are these insights still applicable, perhaps even more applicable, now that networks of instantaneous global telecommunications exist to serve the needs of technologically advanced individuals and populations? Is the society that Webber foresaw in 1968 something like the contemporary "technoburbs" (p. 69) that Robert Fishman describes? Might not computers and modern telecommunications make the Broadacre City that Frank Lloyd Wright prophesied (p. 331) even more possible today and in the future? And might Webber's vision of a "post-city age" prefigure the kind of "teleserviced city" that William Mitchell describes in *E-topia* (p. 510)?

Melvin Webber is professor emeritus of urban planning at the University of California, Berkeley, where he was the director of the Institute of Urban and Regional Development and the University of California Transportation Center. He is the author of *Explorations into Urban Structure* (Philadelphia: University of Pennsylvania Press, 1964) and has written extensively on issues of urban transit and cyberspace citizenship. He was a winner of the Distinguished Planning Education Award of the American Institute of Planners.

THE POST-CITY AGE

We are passing through a revolution that is unhitching the social processes of urbanization from the locationally fixed city and region. Reflecting the current explosion in science and technology, employment is shifting from the production of goods to services; increasing ease of transportation and communication is dissolving the spatial barriers to social intercourse; and Americans are forming social communities comprised of spatially dispersed members. A new kind of large scale urban society is emerging that is increasingly independent of the city. In turn, the problems of the city place generated by early industrialization are being supplanted by a new array different in kind. With but a few remaining exceptions (the new air pollution is a notable one), the recent difficulties are not place type problems at all. Rather, they are the transitional problems of a rapidly developing society-economy-and-polity whose turf is the nation. Paradoxically, just at the time in history when policy-makers and the world press are discovering the city, "the age of the city seems to be at an end."

Our failure to draw the rather simple conceptual distinction between the spatially defined city or metropolitan area and the social systems that are localized there clouds current discussions about the "crisis of our cities." The confusion stems largely from the deficiencies of our language and from the anachronistic thoughtways we have carried over from the passing era. We still have no adequate descriptive terms for the emerging social order, and so we use, perforce, old labels that are no longer fitting. Because we have named them so, we suppose that the problems manifested inside cities are, therefore and somehow, "city problems." Because societies in the past had been spatially and locally structured, and because urban societies used to be exclusively city-based, we seem still to assume that territorial is a necessary attribute of social systems.

The error has been a serious one, leading us to seek local solutions to problems whose causes are not of local origin and hence are not susceptible to municipal treatment. We have been tempted to apply city-building instruments to correct social disorders, and we have then been surprised to find that they do not work. (Our experience with therapeutic public housing, which was supposed to cure "social pathologies," and urban renewal, which was supposed to improve the lives of the poor, may be our most spectacular failures.)

We have lavished large investments on public facilities, but neglected the quality and the distribution of the social services. And we have defended and reinforced home-rule prerogatives of local and state governments with elaborate rhetoric and protective legislation.

Neither crime-in-the-streets, poverty, unemployment, broken families, race riots, drug addiction, mental illness, juvenile delinquency, nor any of the commonly noted "social pathologies" marking the contemporary city can find its causes or its cure there. We cannot hope to invent local treatments for conditions whose origins are not local in character, nor can we expect territorially defined governments to deal effectively with problems whose causes are unrelated to territory or geography. The concepts and methods of civil engineering and city planning suited to the design of unitary physical facilities cannot be used to serve the design of social change in a pluralistic and mobile society. In the novel society now emerging – with its sophisticated and rapidly advancing science and technology, its complex social organization, and its internally integrated societal processes – the influence and significance of geographic distance and geographic place are declining rapidly.

This is, of course, a most remarkable change. Throughout virtually all of human history, social organization coincided with spatial organization. In preindustrial society, men interacted almost exclusively with geographic neighbors. Social communities, economies, and polities were structured about the place in which interaction was least constrained by the frictions of space. With the coming of large-scale industrialization during the latter half of the nineteenth century, the strictures of space were rapidly eroded, abetted by the new ease of travel and communication that the industrialization itself brought.

The initial counterparts of industrialization in the United States were, first, the concentration of the nation's population into large settlements and, then, the cultural urbanization of the population. Although these changes were causally linked, they had opposite spatial effects. After coming together at a common place, people entered larger societies tied to no specific place. Farming and village peoples from throughout the continent and the world migrated to the expanding cities, where they learned urban ways, acquired the occupational skills that industrialization demanded, and became integrated into the contemporary society.

In recent years, rising societal scale and improvements in transportation and communications systems have loosed a chain of effects robbing the city of its once unique function as an urbanizing instrument of society. Farmers and small-town residents, scattered throughout the continent, were once effectively removed from the cultural life of the nation. City folks visiting the rural areas used to be treated as strangers, whose styles of living and thinking were unfamiliar. News of the rest of the world was hard to get and then had little meaning for those who lived the local life. Country folk surely knew there was another world out there somewhere, but little understood it and were affected by it only indirectly. The powerful anti-urban traditions in early American thought and politics made the immigrant city dweller a suspicious character whose crude ways marked him as un-Christian (which he sometimes was) and certainly un-American. The more sophisticated urban upper classes – merchants, landowners, and professional men – were similarly suspect and hence rejected. In contrast, the small-town merchant and the farmer who lived closer to nature were the genuine Americans of pure heart who lived the simple, natural life. Because the contrasts between the rural and the urban ways-of-life were indeed sharp, antagonisms were real, and the differences became institutionalized in the conduct of politics. America was marked by a diversity of regional and class cultures whose followers interacted infrequently, if ever.

By now this is nearly gone. The vaudeville hicktown and hayseed characters have left the scene with the vaudeville act. Today's urbane farmer watches television documentaries, reads the national news magazines, and manages his acres from an office (maybe located in a downtown office building), as his hired hands ride their tractors while listening to the current world news broadcast from a transistor. Farming has long since ceased to be a handicraft art; it is among the most highly technologized industries and is tightly integrated into the international industrial complex.

During the latter half of the nineteenth century and the first third of the twentieth, the traditional territorial conception that distinguished urbanites and ruralites was probably valid: The typical rural folk lived outside the cities, and the typical urbanites lived inside. By now this pattern is nearly *reversed*. Urbanites no longer reside exclusively in metropolitan settlements, nor do ruralites live exclusively in the hinterlands. Increasingly, those who are least integrated into modern society – those who exhibit most of the attributes of rural folk – are concentrating within the highest-density portions of the large metropolitan centers. This profoundly important development is only now coming to our consciousness, yet it points up one of the major policy issues of the next decades.

Cultural diffusion is integrating immigrants, city residents, and hinterland peoples into a national urban society, but it has not touched all Americans evenly. At one extreme are the intellectual and business elites, whose habitat is the planet; at the other are the lower-class residents of city and farm who live in spatially and cognitively constrained worlds. Most of the rest of us, who comprise the large middle class, lie somewhere in-between, but in some facets of our lives we all seem to be moving from our ancestral localism toward the unbounded realms of the cosmopolites.

High educational attainments and highly specialized occupations mark the new cosmopolites. As frequent patrons of the airlines and the long-distance telephone lines, they are intimately involved in the communications networks that tie them to their spatially dispersed associates. They contribute to and consume the specialized journals of science, government, and industry, thus maintaining contact with information resources of relevance to their activities, whatever the geographic sources or their own locations. Even though some may be employed by corporations primarily engaged in manufacturing physical products, these men trade in information and ideas. They are the producers of the information and ideas that fuel the engines of societal development. For those who are tuned into the international communications circuits, cities have utility precisely because they are rich in information. The way such men use the city reveals its essential character most clearly, for to them the city is essentially a massive communications switchboard through which human interaction takes place.

Indeed, cities exist *only* because spatial agglomeration permits reduced costs of interaction. Men originally elected to locate in high-density settlements precisely because space was so costly to overcome. It is still cheaper to interact with persons who are nearby, and so men continue to locate in such settlements. Because there *are* concentrations of associates in city places, the new cosmopolites establish their offices there and then move about from city to city conducting their affairs. The biggest settlements attract the most long-distance telephone and airline traffic and have undergone the most dramatic growth during the era of city-building.

The recent expansion of Washington, DC is the most spectacular evidence of the changing character of metropolitan development. Unlike the older settlements whose growth was generated by expanding manufacturing activities during the nineteenth and early-twentieth centuries, Washington produces almost no goods whatsoever. Its primary products are information and intelligence, and its fantastic growth is a direct measure of the predominant roles that information and the national government have come to play in contemporary society.

This terribly important change has been subtly evolving for a long time, so gradually that it seems to have gone unnoticed. The preindustrial towns that served their adjacent farming hinterlands were essentially alike. Each supplied a standardized array of goods and services to its neighboring market area. The industrial cities that grew after the Civil War and during the early decades of this century were oriented to serving larger markets with the manufacturing products they were created to produce. As their market areas widened, as product specialization increased, and as the information content of goods expanded, establishments located in individual cities became integrated into the spatially extensive economies. By now, the large metropolitan centers that used to be primarily goods-producing loci have become interchange junctions within the international communications networks. Only in the limited geographical, physical sense is any modern metropolis a discrete, unitary, identifiable phenomenon. At most, it is a localized node within the integrating international networks, finding its significant identity as contributor to the workings of that larger system. As a result, the new cosmopolites belong to none of the world's metropolitan areas, although they use them. They belong, rather, to the national and international communities that merely maintain information exchanges at these metropolitan junctions.

Their capacity to interact intimately with others who are spatially removed depends, of course, upon a level of wealth adequate to cover the dollar costs of long-distance intercourse, as well as upon the cognitive capacities associated with highly skilled professional occupations. The intellectual and business elites are able to maintain continuing and close contact with their associates throughout the world because they are rich not only in information, but also in dollar income.

As the costs of long-distance interaction fall in proportion to the rise in incomes, more and more people are able and willing to pay the transportation and communication bills. As expense-account privileges are expanded, those costs are being reduced to zero for ever larger numbers of people. As levels of education and skill rise, more and more people are being tied into the spatially extensive communities that used to engage only a few.

Thus, the glue that once held the spatial settlement together is now dissolving, and the settlement is dispersing over ever widening terrains. At the same time, the pattern of settlement upon the continent is also shifting (moving toward long strips along the coasts, the Gulf, and the Great Lakes). These trends are likely to be accelerated dramatically by cost-reducing improvements in transportation and communications technologies now in the research-and-development stages. (The SST, COMSAT communications, high-speed ground transportation with speeds up to 500 mph, TV and computer-aided educational systems, no-toll long-distance telephone service, and real-time access to national computer-based information systems are likely to be powerful ones.) Technological improvements in transport and communications reduce the frictions of space and thereby ease long-distance intercourse. Our compact, physical city layouts directly mirror the more primitive technologies in use at the time these were built. In a similar way, the locational pattern of cities upon the continent reflects the technologies available at the time the settlements grew. If currently anticipated technological improvements prove workable, each of the metropolitan settlements will spread out in low-density patterns over far more extensive areas than even the most frightened future-mongers have yet predicted. The new settlement-form will little resemble the nineteenth-century city so firmly fixed in our images and ideologies. We can also expect the large junction points will no longer have the communications advantage they now enjoy, and smaller settlements will undergo a major spurt of growth in all sorts of now isolated places where the natural amenities are attractive.

Moreover, as ever larger percentages of the nation's youth go to college and thus enter the national and international cultures, attachments to places of residence will decline dramatically. This prospect, rather than the spatial dispersion of metropolitan areas, portends the functional demise of the city. The signs are already patently clear among those groups whose worlds are widest and least bounded by parochial constraints.

Consider the extreme cosmopolite, if only for purposes of illustrative cartooning. He might be engaged in scientific research, news reporting, or international business, professions exhibiting critical common traits. The astronomer, for example, maintains instantaneous contact with his colleagues around the world; indeed, he is a day-to-day collaborator with astronomers in all countries. His work demands that he share information and that he and his colleagues monitor stellar events jointly, as the earth's rotation brings men at different locales into prime viewing position. Because he is personally committed to their common enterprise, his social reference group is the society of astronomers. He assigns his loyalties to the community of astronomers, since their work and welfare matter most to him.

To be sure, as he plays out other roles – say, as citizen, parent, laboratory director, or grocery shopper – he is a member of many other communities, both interest-based and place-defined ones. But the striking thing about our astronomer, and the millions of people like him engaged in other professions, is how little of his attention and energy he devotes to the concerns of place-defined communities. Surely, as compared to his grandfather, whose life was largely bound up in the affairs of his locality, the astronomer, playwright, newsman, steel broker, or wheat dealer lives in a life-space that is not defined by territory and deals with problems that are not local in nature. For him, the city is but a convenient setting for the conduct of his professional work; it is not the basis for the social communities that he cares most about.

"European Cities, the Informational Society, and the Global Economy"

Journal of Economic and Social Geography (1993)

Manuel Castells

Editors' Introduction

The information revolution sweeping the world today has profound implications for the future of cities. The power of computers is increasing and the cost of computing is dropping at astonishing rates. Fax machines, modems, fiber-optic cable, communication satellites, a global electronic e-mail system, videoconferencing, the Internet, information highways, virtual reality, and multi-media – all are transforming traditional urban space and communities in ways that are still only partially understood.

Manuel Castells was born in Spain and educated as a sociologist in France. He has taught at many universities in Europe, North America, Latin America, and Asia, including the University of Paris and the University of California, Berkeley. He currently teaches at the Universidad Autonoma de Barcelona and the University of Southern California. As a young man, Castells fled Franco's authoritarian and intellectually stifling Spain for the freedom and intellectual excitement of Paris. He became a neo-Marxian and crafted sophisticated theories on the role of the capitalist state and grassroots urban protest movements. Later, Castells has turned his attention to the implications of high technology and the information revolution for community life and urban development.

Castells sees information technologies as the fundamental instrument of the new organizational logic transforming the world today. Accordingly he uses the term "informational" as an adjective to describe a type of city – just as "industrial" or "colonial" were used in the nineteenth century. In his magisterial three-volume *The Rise of the Network Society* (1998) Castells describes how information technology will restructure relationships between rich and poor regions, labor and capital, centralization and decentralization of services, governments and nongovernmental entities, the individual and society.

Castells argues that what he calls the "space of flows" will increasingly govern the actions of power-holding organizations rather than territorially based institutions operating in the "space of places." Industry and services will be organized worldwide around the operation of their information-generating units. He envisions powerful, secretive, multi-national institutions not tied to any particular place as the dominant institutions of the future.

In "European Cities, the Informational Society, and the Global Economy," Castells predicts how the basic elements of the new informational society will affect the cities of Western Europe. Among the main influences, he lists the inevitability of European Union political and economic integration, the cultural pressures of Third World immigration, social conflicts arising out of gentrification and the "polarized occupational structure," and the increasing importance of environmentalism and gender-based social restructuring. Thus, he argues that the "fundamental urban dualism of our time . . . opposed the cosmopolitanism of the elite, living on a daily connection to the whole world to the tribalism of local communities, retrenched in their spaces that they try to control as their last stand against the macro-forces that shape their lives out of their reach." Castells's vision may be compared to that of Richard Florida's "creative class" thesis (p. 129), and to that of Saskia Sassen (p. 197) who sees a globalized economy leading to highly polarized urban social systems.

Castells's analysis raises many questions. Will local and even national governments wither in importance with the information dominance of powerful multinational institutions? Will the information revolution lead to greater social inequality between rich and poor nations? Are "Dual Cities" inevitable where rich and powerful professional managers manipulate impoverished masses, both blocks and continents away? May his vision of cosmopolitan elites versus retrenched local communities be too dualistic, too either-or, when the lifestyles of the broad mass of middle-class urban residents may be somewhere between these extremes? Will mass urban social movements emerge to reassert popular power against such a techno-Orwellian future? And can they succeed against so powerful and subtle a global system? Thirty years ago, radicals could identify and work to overthrow a territorially based dictator like Francisco Franco. But who and where is the enemy in the emerging space of flows?

Many of the visionary writers in this anthology predicted that the information (and transportation) technology of their times would undermine old city patterns and make possible new ones. Ebenezer Howard argued that Garden Cities were possible because they could communicate by telegraph and send goods by railroad (p. 314). Frank Lloyd Wright foresaw instant communication and rapid transportation in Broadacre City by pneumatic tube and helicopter (p. 331). And Melvin Webber predicted a "post-city age" as a result of telephones and air transportation (p. 473).

Other books on information technology and cities by Castells are *The Rise of the Network Society*, three volumes (Oxford: Blackwell, 1998; second edition, 2000), *The Informational City: Information Technology, Economic Restructuring, and the Urban-Regional Process* (Oxford and Cambridge: Blackwell, 1991), an edited anthology *High Tech, Space, and Society* (Beverly Hills: Sage, 1985), and *Technopoles of the World* (London and New York: Routledge, 1994) jointly authored with Sir Peter Hall.

Other books by Castells include *The Power of Identity: The Information Age – Economy, Society, and Culture* (Oxford: Blackwell, 1997), *Dual City: Restructuring New York* (New York: Russell Sage Foundation, 1991) edited with John Mollenkopf, *The City and the Grassroots* (Berkeley: University of California Press, 1983), and *The Urban Question* (London: Edward Arnold, 1977). An excellent anthology of works from throughout his career is Ida Susser (ed.), *The Castells Reader on Cities and Social Theory* (Oxford: Blackwell, 2002).

■ ■ ■ ■ ■ ■

An old axiom in urban sociology considers space as a reflection of society. Yet life, and cities, are always too complex to be captured in axioms. Thus the close relationship between space and society, between cities and history, is more a matter of expression than of reflection. The social matrix expresses itself into the spatial pattern through a dialectical interaction that opposes social contradictions and conflicts as trends fighting each other in an endless supersession. The result is not the coherent spatial form of an overwhelming social logic be it the capitalist city, the preindustrial city or the ahistorical utopia but the tortured and disorderly, yet beautiful patchwork of human creation and suffering.

Cities are socially determined in their forms and in their processes. Some of their determinants are structural, linked to deep trends of social evolution that transcend geographic or social singularity. Others

are historically and culturally specific. And all are played out, and twisted, by social actors that impose their interests and their values, to project the city of their dreams and to fight the space of their nightmares.

Sociological analysis of urban evolution must start from the theoretical standpoint of considering the complexity of these interacting trends in a given time–space context. The last twenty years of urban sociology have witnessed an evolution of thinking (including my own) from structuralism to subjectivism, then to an attempt, however imperfect, at integrating both perspectives into a structural theory of urban change that, if a label rooted in an intellectual tradition is necessary, I would call Marxian, once history has freed the Marxian theoretical tradition from the terrorist tyranny of Marxism-Leninism.

I intend to apply this theoretical perspective to the understanding of the fundamental transformations that are taking place in West European cities at the end of the second millennium. In order to understand such transformations we have to refer to major social trends that are shaking up the foundations of our existence: the coming of a technological revolution centered on information technologies, the formation of a global economy, and transition to a new society, the informational society, that without ceasing to be capitalist or statist replaces the industrial society as the framework of social institutions.

But this analysis has to be at the same time general and structural (if we accept that a historical transformation is under way) and specific to a given social and cultural context, such as Western Europe (with all due acknowledgement to its internal differentiation).

In recent years, a new trademark has become popular in urban theory: capitalist restructuring. Indeed it is most relevant to pinpoint the fundamental shift in policies that both governments and corporations have introduced in the 1980s to steer capitalist economies out of their 1970s crises. Yet more often than not, the theory of capitalist restructuring has missed the specificity of the process of transformation in each area of the world, as well as the variation of the cultural and political factors that shape the process of economic restructuring, and ultimately determine its outcome.

Thus the deindustrialization processes of New York and London take place at the same time that a wave of industrialization of historic proportions occurs in China and in the Asian Pacific. The rise of the informal economy and of urban dualism takes place in Los Angeles, as well as in Madrid, Miami, Moscow, Bogota and Kuala Lumpur, but the social paths and social consequences of such similarly structural processes are so different as to induce a fundamental variegation of each resulting urban structure.

I will try therefore to analyze some structural trends underlying the current transformation of European cities, while accounting for the historical and social specificity of the processes emerging from such structural transformation.

THE THREAD OF THE NEW HISTORY

Urban life muddles through the pace of history. When this pace accelerates, cities and their people become confused, spaces turn threatening, and meaning escapes from experience. In such disconcerting yet magnificent times, knowledge becomes the only source to restore meaning, and thus meaningful action.

At the risk of schematism, and for the sake of clarity, I will summarize what seem to be the main trends that, together and in their interaction, provide the framework of social, economic and political life for European cities in this particular historical period.

First of all, we live in the midst of a fundamental technological revolution that is characterized by two features:

As all major technological revolutions in history, the effects are pervasive. They are not limited to industry, or to the media, or to telecommunications or transportation. New technologies, that have emerged in their applications in full strength since the mid 1970s, are transforming production and consumption, management and work, life and death, culture and warfare, communication and education, space and time. We have entered a new technological paradigm.

As the industrial revolution was based on energy (although it embraced many other technological fields) the current revolution is based upon information technologies, in the broadest sense of the concept, which includes genetic engineering (the decoding and reprogramming of the codes of living matter).

This technological informational revolution is the backbone (although not the determinant) of all other major structural transformations:

It provides the basic infrastructure for the formation of a functionally interrelated world economic system.

It becomes a crucial factor in competitiveness and productivity for countries throughout the world, ushering in a new international division of labour.

It allows for the simultaneous process of centralization of messages and decentralization of their reception, creating a new communications world made up at the same time of the global village and of the incommunicability of those communities that are switched off from the global network. Thus an asymmetrical space of communication flows emerges from the uneven appropriation of a global communication system.

It creates a new, intimate linkage between the productive forces of the economy and the cultural capacity of society. Because knowledge generation and information processing are at the roots of the new productivity, the ability of a society to accumulate knowledge and manipulate symbols translates into economic productivity and political military might, anchoring the sources of wealth and power in the informational capacity of each society.

While this technological revolution does not determine per se the emergence of a social system, it is an essential component of the new social structure that characterizes our world: the informational society. By this concept, I understand a social structure where the sources of economic productivity, cultural hegemony and political military power depend, fundamentally, on the capacity to retrieve, store, process and generate information and knowledge. Although information and knowledge have been critical for economic accumulation and political power throughout history, it is only under the current technological, social, and cultural parameters that they become directly productive forces. In other words, because of the interconnection of the whole world and the potential automation of most standard production and management functions, the generation and control of knowledge, information and technology is a necessary and sufficient condition to organize the overall social structure around the interests of the information holders. Information becomes the critical raw material of which all social processes and social organizations are made. Material production, as well as services, become subordinate to the handling of information in the system of production and in the organization of society. Empirically speaking, an ever growing majority of employment in Western European cities relates to information processing jobs. The growing proportion of employment in service activities is not the truly distinctive feature, because of the ambiguity of the notion of 'services' (e.g. in Third World cities a majority of the population also works in 'services', although these are indeed very different kinds of activities). What is truly fundamental is the growing quantitative size and qualitative importance of information processing activities in both goods production and services delivery. The contradictory but ineluctable emergence of the informational society shapes European cities as the onset of the industrial era marked forever the urban and rural spaces of the nineteenth century.

A third major structural trend of our epoch is the formation of a global economy. The global economy concept must be distinguished from the notion of a world economy, which reflects a very old historical reality for most European nations, and particularly for the Netherlands, which emerged as a nation through its role as one of the nodal centres of the sixteenth century's world economy. Capitalism has accumulated, since its beginnings, on a worldwide scale. This is not to say that the capitalist economy was a global economy. It is only now becoming such.

By global economy we mean an economy that works as a unit in real time on a planetary scale. It is an economy where capital flows, labour markets, commodity markets, information, raw materials, management, and organization are internationalized and fully interdependent throughout the planet, although in an asymmetrical form, characterized by the uneven integration into the global system of different areas of the planet. Major functions of the economic system are fully internationalized and interdependent on a daily basis. But many others are segmented and unevenly structured depending upon functions, countries, and regions. Thus the global economy embraces the whole planet, but not all regions and all people on the planet. In fact, only a minority of people are truly integrated into the global economy, although all the dominant economic and political centres on which people depend are indeed integrated into the global economic networks (with the possible exception of Bhutan . . .). With the disintegration of the Soviet empire, the last area of the planet that was not truly integrated into the global economy, it is restructuring itself in the most dramatic conditions to be able to reach out to the perceived avenues of prosperity of our economic model. (China already started its integration in the global capitalist economy in December 1979, while trying to preserve its statist political regime.)

This global economy increasingly concentrates wealth, technology, and power in 'the North', a vague geopolitical notion that replaces the obsolete West–East differentiation, and that roughly corresponds to the OECD countries. The East has disintegrated and is quickly becoming an economic appendage of the North. Or at least such is the avowed project of its new leaders. The 'South' is increasingly differentiated. East Asia is quickly escaping from the lands of poverty and underdevelopment to link up, in fact, with the rising sun of Japan, in a model of development that Japanese writers love to describe as 'the flying geese pattern', with Japan of course leading the way, and the other Asian nations taking off harmoniously under the technological guidance and economic support of Japan. China is at the crossroads of a potential process of substantial economic growth at a terrible human cost as hundreds of millions of peasants are uprooted without structures able to integrate them into the new urban industrial world. South and Southeast Asia struggle to survive the process of change, looking for a subordinate yet livable position in the new world order. Most of Africa, on the contrary, finds itself increasingly disconnected from the new global economy, reduced to piecemeal, secondary functions that see the continent deteriorate, with the world only waking up from time to time to the structural genocide taking place there when television images strike the moral consciousness of public opinion and affect the political interests of otherwise indifferent policy makers. Latin America, and many regions and cities around the world, struggle in the in-between land of being only partially integrated into the global economy, and then submitted to the tensions between the promise of full integration and the daily reality of a marginal existence.

In this troubled world, Western Europe has, in fact, become a fragile island of prosperity, peace, democracy, culture, science, welfare and civil rights. However, the selfish reflex of trying to preserve this heaven by erecting walls against the rest of the world may undermine the very fundamentals of European culture and democratic civilization, since the exclusion of the other is not separable from the suppression of civil liberties and a mobilization against alien cultures. Major European cities have become nodal centres of the new global economy, but they have also seen themselves transformed into magnets of attraction for millions of human beings from all around the world who want to share the peace, democracy, and prosperity of Europe in exchange for their hard labour and their commitment to a promised land. But the overcrowded and aged Western Europe of the late twentieth century docs not seem as open to the world as was the young, mostly empty America of the beginning of the century. Immigrants are not welcome as Europe tries to embark on a new stage of its common history, building the supranational Europe without renouncing its national identities. Yet the cultural isolation of the pan-European construction is inseparable from the affirmation of ethnic nationalism that will eventually turn not only against the 'alien immigrants' but against European foreigners as well. European cities will have to cope with this new global economic role while accommodating to a multi-ethnic society that emerges from the same roots that sustain the global economy.

The fourth fundamental process under way in European cities is the process of European integration, into what will amount in the twenty-first century to some form of confederation of the present nation states. This is an ineluctable process for at least fifteen countries (the current twelve EC countries plus Sweden, Austria and Switzerland) regardless of the fate of the symbolic Maastricht Treaty. If, as is generally accepted, the European Community is heading toward a common market, a common resident status for all its citizens, a common technology policy, a common currency, a common defense and a common foreign policy, all the basic prerogatives of the national state will be shifted to the European institutions by the end of the century. This will certainly be a tortuous path, with the nostalgics of the past, neofascists, neo-communists and fundamentalists of all kinds fighting the tide of European solidarity, fueling the fears of ignorance among people, building upon demagogy and opportunism. Yet however difficult the process, and with whatever substantial modifications to the current technocratic blueprints, Europe will come into existence: there are too many interests and too much political will at stake to see the project destroyed after having come this far.

The process of European integration will cause the internationalization of major political decision-making processes, and thus will trigger the fear of sub-ordination of specific social interests to supranational institutions. But most of these specific interests express themselves on a regional or local basis rather than at the national level. Thus we are witnessing the renewal of the role of regions and cities as locuses of autonomy and political decision. In particular, major cities throughout Europe constitute the nervous system of both the economy and political system of the continent. The more national states fade in their role, the more cities emerge as a driving force in the making of the new European society.

IDENTITY AND SOCIAL MOVEMENTS

The process of historical transition experienced by Europe's cities leads to an identity crisis in its cultures and among its people, that becomes another major element of the new urban experience. This identity crisis is the result of two above mentioned processes that, however contradictory among themselves, jointly contribute to shake up the foundations of European national and local cultures. On the one hand, the march to supranationality blurs national identities and makes people uncertain about the power holders of their destiny, thus pushing them into withdrawal, either individualistic (neolibertarianism) or collective (neonationalism). On the other hand, the arrival of millions of immigrants and the consolidation of multi-ethnic, multicultural societies in most West European countries, confronts Europe head on with the reality of a nonhomogeneous culture, precisely at the moment when national identity is most threatened. There follows a crisis of cultural identity (with the corollary of collective alienation) that will mark urban processes in Europe for years to come.

More to the point: major cities will concentrate the overwhelming proportion of immigrants and ethnic minority citizens (the immigrants' sons and daughters). Thus they will also be at the forefront of the waves of racism and xenophobia that will shake up the institutions of the new Europe even before they come into existence. As a reaction to the national identity crisis we observe the emergence of territorially defined identities at the level of the region, of the city, of the neighbourhood. European cities will be increasingly oriented toward their local culture, while increasingly distrustful of higher order cultural identities. The issue then is to know if cities can reach out to the whole world without surrendering to a localistic, quasi-tribal reaction that will create a fundamental divide between local cultures, European institutions, and the global economy.

European cities are also affected by the rise of the social movements of the informational society, and in particular by the two central movements of the informational society: the environmental movement, and the women's movement.

The environmental movement is at the origin of the rise of the ecological consciousness that has substantially affected urban policies and politics. The issue of sustainable development is indeed a fundamental theme of our civilization and a dominant topic on today's political agendas. Because major cities in Europe are at the same time nodal centres of economic growth and the living places for the most environmentally conscious segment of the population, the battles for integration will be fought in the streets and institutions of these major cities.

The structural process of transformation of women's condition, in dialectical interaction with the rise of the feminist movement, has completely changed the social fabric of cities. Labour markets have been massively feminized, resulting in a change in the conditions of work and management, of struggle and negotiation, and ultimately in the weakening of a labour movement that could not overcome its sexist tradition. This also points to the possibility of a new informational labour movement that because it will have to be based on women's rights and concerns, as well as on those of men, will be historically different from its predecessor.

At the same time, the transformation of households and the domestic division of labour is fundamentally changing the demands on collective consumption, and thus urban policy. For instance, childcare is becoming as important an issue as housing in today's cities. Transportation networks have to accommodate the demands of two workers in the family, instead of relying on the free driving service provided by the suburban housewife in the not so distant past.

Some of the new social movements, the most defensive, the most reactive, have taken and will be taking the form of territorially based countercultures, occupying a given space to cut themselves off from the outside world, hopeless of being able to transform the society they refuse. Because such movements are likely to occur in major cities that concentrate a young, educated population, as well as marginal cultures that accommodate themselves in the cracks of the institutions, we will be witnessing a constant struggle over the occupation of meaningful space in the main European cities, with business corporations trying to appropriate the beauty and tradition for their noble quarters, and urban countercultures making a stand on the use value of the city, while local residents try to get on with their living, refusing to be bent by the alien wind of the new history.

Beyond the territorial battles between social movements and elite interests, the new marginality, unrelated to such social movements, is spreading over the urban space. Drug addicts, drug dealers and drug victims populate the back alleys of European cities, creating the unpredictable, awakening our own psychic terrors, and tarnishing the shine of civilized prosperity at the daily coming of darkness. The 'black holes' of our society, those social conditions from where there is no return, take their territory too, making cities tremble at the fear of their unavowed misery.

The occupation of urban space by the new poverty and the new marginality takes two forms: the tolerated ghettoes where marginalized people are permitted to stay, out of sight of the mainstream society; the open presence in the core area of cities of 'street people', a risky strategy, but at the same time a survival technique since only there do they exist, and thus only there can they relate to society, either looking for a chance or provoking a final blow.

Because the informational society concentrates wealth and power, while polarizing social groups according to their skills, unless deliberate policies correct the structural tendencies we are also witnessing the emergence of a social dualism that could ultimately lead to the formation of a dual city, a fundamental concept that I will characterize below, when considering the spatial consequences of the structural trends and social processes that I have proposed as constituting the framework that underlies the new historical dynamics of European cities.

THE SPATIAL TRANSFORMATION OF MAJOR CITIES

From the trends we have described stem a number of spatial phenomena that characterize the current structure of major metropolitan centres in Western Europe. These centres are formed by the uneasy articulation of various sociospatial forms and processes that I find useful to specify in their singularity, although it is obvious that they cannot be understood without reference to each other.

First of all, the national-international business centre is the economic engine of the city in the informational global economy. Without it, there is no wealth to be appropriated in a given urban space, and the crisis overwhelms any other project in the city, as survival becomes the obvious priority. The business centre is made up of an infrastructure of telecommunications, communications, urban services and office space, based upon technology and educational institutions. It thrives through information processing and control functions. It is sometimes complemented by tourism and travel facilities. It is the node of the space of flows that characterizes the dominant space of informational societies. That is, the abstract space constituted in the networks of exchange of capital flows, information flows, and decisions that link directional centres among themselves throughout the planet. Because the space

of flows needs nodal points to organize its exchange, business centres and their ancillary functions constitute the localities of the space of flows. Such localities do not exist by themselves but by their connection to other similar localities organized in a network that forms the actual unit of management, innovation, and power.

Secondly, the informational society is not disincarnated. New elites make it work, although they do not necessarily base their power and wealth on majority ownership of the corporations. The new managerial technocratic political elite does however create exclusive spaces, as segregated and removed from the city at large as the bourgeois quarters of the industrial society. In European cities, unlike in America, the truly exclusive residential areas tend to appropriate urban culture and history, by locating in rehabilitated areas of the central city, emphasizing the basic fact that when domination is clearly established and enforced, the elite does not need to go into a suburban exile, as the weak and fearful American elite did to escape from the control of the urban population (with the significant exceptions of New York, San Francisco, and Boston).

Indeed, the suburban world of European cities is a socially diversified space, which is segmented in different peripheries around the central city. There are the traditional working class suburbs (either blue collar or white collar) of the well kept subsidized housing estates in home ownership. There are the new towns, inhabited by a young cohort of lower middle class, whose age made it difficult for them to penetrate the expensive housing market of the central city. And there are also the peripheral ghettoes of the older public housing estates where new immigrant populations and poor working families experience their exclusion from the city.

Suburbs are also the locus of industrial production in European cities, both for traditional manufacturing and for the new high technology industries that locate in new peripheries of the major metropolitan areas, close enough to the communication centres but removed from older industrial districts.

Central cities are still shaped by their history. Thus traditional working class neighbourhoods, increasingly populated by service workers rather than by industrial workers, constitute a distinctive space, a space that, because it is the most vulnerable, becomes the battleground between the redevelopment efforts of business and the upper middle class, and the invasion attempts of the countercultures trying to reappropriate the use value of the city. They often become defensive spaces for workers who have only their home to fight for, while at the same time meaningful popular neighbourhoods and likely bastions of xenophobia and localism.

The new professional middle class is torn between attraction to the peaceful comfort of the boring suburbs and the excitement of a hectic, and often too expensive, urban life. The structure of the household generally determines the spatial choice. The larger the role women play in the household, the more the proximity to jobs and urban services in the city makes central urban space attractive to the new middle class, triggering the process of gentrification of the central city. On the contrary, the more patriarchal is the middle class family, the more we are likely to observe the withdrawal to the suburb to raise children, all economic conditions being equal.

The central city is also the locus for the ghettoes of the new immigrants, linked to the underground economy, and to the networks of support and help needed to survive in a hostile society. Concentration of immigrants in some dilapidated urban areas in European cities is not the equivalent however to the experience of the American ghettoes, because the overwhelming majority of European ethnic minorities are workers, earning their living and raising their families, thus counting on a support structure that makes their ghettoes strong, family oriented communities, unlikely to be taken over by street crime.

It is in the core administrative and entertainment districts of European cities that urban marginality makes itself present. Its pervasive occupation of the busiest streets and public transport nodal points is a survival strategy deliberately designed so they can receive public attention or private business, be it welfare assistance, a drug transaction, a prostitution deal, or the customary police care.

Major European metropolitan centres present some variation around the structure of urban space we have outlined, depending upon their differential role in the European economy. The lower their position in the new informational network, the greater the difficulty of their transition from the industrial stage, and the more traditional will be their urban structure, with old established neighbourhoods and commercial quarters playing the determinant role in the dynamics of the city. On the other hand, the higher their position in the competitive structure of the new European economy, the greater the role of their advanced services in the business district, and the more intense will be the restructuring of the urban space. At the same time, in

those cities where the new European society reallocates functions and people throughout the space, immigration, marginality and countercultures will be the most present, fighting over control of the territory, as identities become increasingly defined by the appropriation of space.

The critical factor in the new urban processes is, however, the fact that urban space is increasingly differentiated in social terms, while being functionally interrelated beyond physical contiguity. There follows a separation between symbolic meaning, location of functions, and the social appropriation of space in the metropolitan area. The transformation of European cities is inseparable from a deeper, structural transformation that affects urban forms and processes in advanced societies: the coming of the Informational City.

THE INFORMATIONAL CITY

The spatial evolution of European cities is a historically specific expression of a broader structural transformation of urban forms and processes that expresses the major social trends that I have presented as characterizing our historical epoch: the rise of the Informational City. By this concept I do not refer to the urban form resulting from the direct impact of information technologies on space. The Informational City is the urban expression of the whole matrix of determinations of the Informational Society, as the Industrial City was the spatial expression of the Industrial Society. The processes constituting the form and dynamics of this new urban structure, the Informational City, will be better understood by referring to the actual social and economic trends that are restructuring the territory. Thus the new international and interregional division of labour ushered in by the informational society leads, at the world level, to three simultaneous processes:

> The reinforcement of the metropolitan hierarchy exercised throughout the world by the main existing nodal centres, which use their informational potential and the new communication technologies to extend and deepen their global reach.
>
> The decline of the old dominant industrial regions that were not able to make a successful transition to the informational economy. This does not imply however that all traditional manufacturing cities are forced to decline: the examples

of Dortmund or Barcelona show the possibility to rebound from the industrial past into an advanced producer services economy and high technology manufacturing.

The emergence of new regions (such as the French Midi or Andalusia) or of new countries (e.g. the Asian Pacific) as dynamic economic centres, attracting capital, people, and commodities, thus recreating a new economic geography.

In the new economy, the productivity and competitiveness of regions and cities is determined by their ability to combine informational capacity, quality of life, and connectivity to the network of major metropolitan centres at the national and international levels.

The new spatial logic, characteristic of the Informational City, is determined by the preeminence of the space of flows over the space of places. By space of flows I refer to the system of exchanges of information, capital, and power that structures the basic processes of societies, economies and states between different localities, regardless of localization. I call it 'space' because it does have a spatial materiality: the directional centres located in a few selected areas of a few selected localities; the telecommunication system, dependent upon telecommunication facilities and services that are unevenly distributed in the space, thus marking a telecommunicated space; the advanced transportation system, that makes such nodal points dependent on major airports and airlines services, on freeway systems, on high speed trains; the security systems necessary to the protection of such directional spaces, surrounded by a potentially hostile world; and the symbolic marking of such spaces by the new monumentality of abstraction, making the locales of the space of flows meaningfully meaningless, both in their internal arrangement and in their architectural form. The space of flows, superseding the space of places, epitomizes the increasing differentiation between power and experience, the separation between meaning and function.

The Informational City is at the same time the Global City, as it articulates the directional functions of the global economy in a network of decision-making and information processing centres. Such globalization of urban forms and processes goes beyond the functional and the political, to influence consumption patterns, lifestyles, and formal symbolism.

Finally, the Informational City is also the Dual City. This is because the informational economy has a structural tendency to generate a polarized occupational structure, according to the informational capabilities of different social groups. Informational productivity at the top may incite structural unemployment at the bottom or downgrading of the social conditions of manual labour, particularly if the control of labour unions is weakened in the process and if the institutions of the welfare state are undermined by the concerted assault of conservative politics and libertarian ideology.

The filling of downgraded jobs by immigrant workers tends to reinforce the dualization of the urban social structure. In a parallel movement the age differential between an increasingly older native population in European cities and a younger population of newcomers and immigrants forms two extreme segments of citizens polarized simultaneously along lines of education, ethnicity, and age. There follows a potential surge of social tensions.

The necessary mixing of functions in the same metropolitan area leads to the attempt to preserve social segregation and functional differentiation through planning of the spatial layout of activities and residence, sometimes by public agencies, sometimes by the influence of real estate prices. There follows a formation of cities made up of spatially coexisting, socially exclusive groups and functions, that live in an increasingly uneasy tension vis-à-vis each other. Defensive spaces emerge as a result of the tension.

This leads to the fundamental urban dualism of our time. It opposes the cosmopolitanism of the elite, living on a daily connection to the whole world (functionally, socially, culturally), to the tribalism of local communities, retrenched in their spaces that they try to control as their last stand against the macro forces that shape their lives out of their reach. The fundamental dividing line in our cities is the inclusion of the cosmopolitans in the making of the new history while excluding the locals from the control of the global city to which ultimately their neighbourhoods belong.

Thus the Informational City, the Global City, and the Dual City are closely interrelated, forming the background of urban processes in Europe's major metropolitan centres. The fundamental issue at stake is the increasing lack of communication between the directional functions of the economy and the informational elite that performs such functions, on the one hand, and the locally oriented population that experiences an ever deeper identity crisis, on the other. The separation between function and meaning, translated into the tension between the space of flows and the space of places, could become a major destabilizing force in European cities, potentially ushering in a new type of urban crisis.

BACK TO THE FUTURE?

The most important challenge to be met in European cities, as well as in major cities throughout the world, is the articulation of the globally oriented economic functions of the city with the locally rooted society and culture. The separation between these two levels of our new reality leads to a structural urban schizophrenia that threatens our social equilibrium and our quality of life. Furthermore, the process of European integration forces a dramatic restructuring of political institutions, as national states see their functions gradually voided of relevance, pulled from the top toward supranational institutions and from the bottom toward increasing regional and local autonomy. Paradoxically, in an increasingly global economy and with the rise of the supranational state, local governments appear to be at the forefront of the process of management of the new urban contradictions and conflicts. National states are increasingly powerless to control the global economy, and at the same time they are not flexible enough to deal specifically with the problems generated in a given local society. Local governments seem to be equally powerless vis-à-vis the global trends but much more adaptable to the changing social, economic, and functional environment of cities.

The effectiveness of the political institutions of the new Europe will depend more on their capacity for negotiation and adaptation than on the amount of power that they command, since such power will be fragmented and shared across a variety of decision-making processes and organizations. Thus instead of trying to master the whole complexity of the new European society, governments will have to deal with specific sets of problems and goals in specific local circumstances. This is why local governments, in spite of their limited power, could be in fact the most adequate instances of management of these cities, working in the world economy and living in the local cultures. The strengthening of local governments is thus a precondition for the management of European cities. But local governments can only exercise such management potential if they engage in at least three fundamental policies:

The fostering of citizen participation, on the basis of strong local communities that feed local government with information, present their demands, and lay the ground for the legitimacy of local government so that they can become respected partners of the global forces operating in their territory.

The interconnection and cooperation between local governments throughout Europe, making it difficult for the global economic forces to play one government against the other, thus forcing the cooperation of global economy and local societies in a fruitful new social contract.

New information technologies should make possible a qualitative upgrading of the cooperation between local governments. A European Municipal Data Bank, and a network of instant communication between local leaders, could allow the formation of a true association of interests of the democratic representatives of the local populations. An electronically connected federation of quasi-free communes could pave the way for restoring social and political control over global powers in the informational age.

Managing the new urban contradictions at the local level by acting on the social trends that underlie such contradictions requires a vision of the new city and new society we have entered into, including the establishment of cooperative mechanisms with national governments and European institutions, beyond natural and healthy competition of parties. The local governments of the new Europe will have to do their homework in understanding their cities, if they are to assume the historical role that the surprising evolution of society could offer them.

Thus the historical specificity of European cities may be a fundamental asset in creating the conditions for managing the contradictions between the global and the local in the new context of the informational society. Because European cities have strong civil societies, rooted in an old history and a rich, diversified culture, they could stimulate citizen participation as a fundamental antidote against tribalism and alienation. And because the tradition of European cities as city states leading the pace to the modern age in much of Europe is engraved in the collective memory of their people, the revival of the city state could be the necessary complement to the expansion of a global economy and the creation of a European state. The old urban tradition of Amsterdam as a centre of politics, trade, culture and innovation, suddenly becomes more strategically important for the next stage of urban civilization than the meaningless suburban sprawl of high technology complexes that characterize the informational space in other areas of the world.

European cities, because they are cities and not just locales, could manage the articulation between the space of flows and the space of places, between function and experience, between power and culture, thus recreating the city of the future by building on the foundations of their past.

"The Emergence of Mega-Urban Regions in Asia"

from *Beyond Metropolis: The Planning and Governance of Asia's Mega-Urban Regions* (2005)

Aprodicio A. Laquian

Editors' Introduction

Urban economies have always extended well beyond their borders into their "hinterlands." Even the earliest cities of the ancient New East engaged in long-distance trade with other regions and communities, some of them quite distant. And both mercantilist and colonial systems extended the reach of economic exchange to global proportions as early as the seventeenth century. But only in recent decades, with the end of the Cold War and the advent of modern telecommunications, have the world's economies become truly globalized and certain cities become "global cities."

In recent decades Asia has become a major focus of burgeoning globalized economic activity based on the dramatic modernization campaigns of central governments throughout the region and the availability of inexpensive labor for the manufacturing operations of both offshore and home-grown corporations. As a result, many Asian cities have experienced rapid population growth, both in the cities themselves and in their surrounding regions. In *Beyond Metropolis*, Philippine–Canadian scholar-activist Aprodicio Laquian notes that "Asia currently has twelve of the world's largest urban agglomerations. By 2015, this number may increase to fourteen." Most of these new Asian global cities are actually "mega-urban regions" extending out from the core of the major urban centers, and almost all of them have experienced extraordinary regional population growth over the course of the past twenty to forty years. In some cases, the growth has been haphazard and uncontrolled. But in other cases, central governments have created regional administrative structures to carry out the planning and development needs of the new "mega-urban" regions. The Tokyo metropolitan region, for example, includes the cities of Osaka, Kyoto, Kobe, and Nagoya. And Shanghai, a city of about thirteen million, is the "head of the dragon" of the entire Yangtze River delta region and is now "the dominant core of a megalapolitan region covering about 100,000 square kilometers with a population of 72.7 million."

Although some argue that the huge new mega-urban regions, playing out their roles as actors in the economics of globalism, have served to weaken or "hollow out" the nation-states in which they developed, Laquian argues that, in fact, the "city-regions lead their countries in the drive toward economic development." As cities have always been, the Asian mega-urban regions are "the centers of innovation and the generators of social change." As such, he concludes, "They play, and will most likely continue to play, a highly significant role in the development of the nation-states in which they are embedded." Laquian's statistics about Asian mega-urban growth invite comparison with Friedrich Engels and his description of Manchester (p. 50) and the other cities of the Industrial Revolution in England. And his analysis of how some cities succeed while others fail, and how even the most successful ones create vast slums for the underserved while great wealth is created for the fortunate, suggests the pattern of uneven development described by Saskia Sassen (p. 197) in her writings on global cities, technological development, and inequality.

Aprodicio Laquian is Professor Emeritus of Community and Regional Planning at the University of British Columbia, Vancouver, where he was the director of the Centre for Human Settlements from 1991 to 2000. Born in a tiny village in rural Luzon, Laquian received his BA from the University of the Philippines in 1959 and his Ph.D. from the Massachusetts Institute of Technology in 1965. Early on, he focused on housing issues, publishing *Slums Are for People* (Manila: East–West Center Press, 1971), *Slums and Squatters in Six Philippine Cities* (Ottowa: International Development Research Centre, 1972), *Housing Asia's Millions* (Ottawa: International Development Research Centre, 1979), and *Basic Housing* (Ottawa: International Development Research Centre, 1983). From 1982 to 1990, Professor Laquian worked for the United Nations and turned his interest to larger questions of urban development, working in East Africa, Oceania, and China and publishing *Population and the Urban Future* (State University of New York Press, 1982). The selection reprinted here is from *Beyond Metropolis: The Planning and Governance of Asia's Mega-Urban Regions* (Baltimore and Washington, DC: Johns Hopkins University Press/Woodrow Wilson Center Press, 2005). In 2000, Professor Laquian was briefly chief of staff for Philippine president Joseph Estrada.

The literature on recent urban growth in Asia is vast. Among the best books on Chinese cities since the market reforms of 1978 include Wenfang Tang and William L. Parish, *Chinese Urban Life Under Reform: The Changing Social Contract* (Cambridge: Cambridge University Press, 2000), John R. Logan (ed.), *The New Chinese City: Globalization and Market Reform* (Oxford: Blackwell, 2002), and Aimen Chen, Gordon G. Liu, and Kevin H. Zhang (eds), *Urbanization and Social Welfare in China* (Aldershot: Ashgate, 2004). For Japanese cities, consult Carl Mosk, *Japanese Industrial History: Technology, Urbanization, and Economic Growth* (New York: M.E. Sharpe, 2001), Carola Hein (ed.), *Rebuilding Urban Japan After 1945* (New York: Palgrave Macmillan, 2003), and André Sorensen, *The Making of Urban Japan: Cities and Planning from Edo to the Twenty-First Century* (London and New York: Routledge, 2002). For urbanization in India, consult Sujata Patel and Jim Masselos, *Bombay and Mumbai: The City in Transition* (Oxford: Oxford University Press, 2003) and Veronique Dupont, (ed.), *Delhi: Urban Space and Human Destinies* (New Delhi: Manohar Publications, 2000). For Kuala Lumpur, see Tim Bunnell, *Malaysia, Modernity, and the Multimedia Super Corridor: The Construction of a High-Tech City-Region* (London and New York: Routledge, 2003). For Singapore, see Ole Johan Dale, *Urban Planning in Singapore: The Transformation of a City* (Oxford: Oxford University Press, 1999) and Brenda S.A. Yeoh, *Contesting Space in Colonial Singapore: Power Relations and the Urban Built Environment* (Singapore: Singapore University Press, 2003). Peter Bialobrzeski's *Neon Tigers* (Ostfildern: Hatje Cantz Publishers, 2004) contains striking photographs of Asian mega-cities, including the image of Shanghai that appears on the cover of this book.

■ ■ ■ ■ ■ ■

THE EMERGENCE OF MEGA-URBAN REGIONS IN ASIA

Asia currently has twelve of the world's largest urban agglomerations. By 2015, this number may increase to fourteen. Economically and socially, these urban agglomerations already dominate life in many Asian countries. Physically, the growth of some of these agglomerations has been so rapid that the processes of planning and governance have not kept pace with urban expansion. With Asian mega-urban regions becoming so important in world urbanization trends, it is necessary to analyze and examine why some of these urban agglomerations continue to grow and others not, the problems they now have and will continue to face in the future, various approaches that

have been used or that may be used to deal with those problems, and some lessons that can be learned by urban agglomerations in other parts of the world faced with the same problems.

Even a cursory examination of the main features of Asian mega-urban regions reveals that they make up a complex mix of varied settlements due to the historical, economic, cultural, and technological factors that have influenced their development. Because of this complexity, it would be a mistake to lump together all the Asian cities in a homogeneous mass.

Indiscriminately mixing the dynamics of East Asian and South Asian cities, for example, would most likely obfuscate understanding of the specific conditions in each set of cities. Planning and governance interventions that may be successful in cities like Seoul and

Tokyo may not work at all in solving problems in Delhi, Dhaka, Karachi, Kolkata, or Mumbai. Planning mechanisms that have been used in countries in transition from centrally planned economies to more market-oriented systems (China and Vietnam) would probably not be very useful in dealing with problems in more market-dominated countries (India and the Philippines). Political traditions rooted in centralized control and bureaucratic hegemony would probably be ineffectual in managing cities created on the basis of decentralized authority and local autonomy.

Despite the differences among Asian mega-urban regions, however, they do share a number of characteristics, in much the same way that apples and oranges (or mangoes and mangosteens) can all be classified as fruits. The most notable commonality among these urban agglomerations is their population size, which ranges from less than 10 million to more than 26 million. A second important consideration is the fact that these mega-urban regions have spread outward into adjacent areas, enveloping villages, towns, and cities, and at times linking up with the urban fields of other city-regions. A third common characteristic of Asian mega-urban regions is the economic, social,

cultural, and political dominance that they exert on their national or regional hinterlands. Fourth and finally, on the basis of demographic, spatial, economic, political, and administrative features, it is possible to view all the Asian mega-urban regions as commonly following a pattern of organizing economic and social space so that they all have a densely developed mega-city core and an extended metropolitan region – and, some of them, a megalopolitan form as well.

Taking into consideration the commonalities and differences of the various Asian mega-urban regions included in this study, I have categorized them into four types. First are the *technologically advanced East Asian cities*. Osaka, Tokyo, and Seoul, like their counterparts in North America and Europe, have extremely low rates of population growth. After an explosive annual growth rate of 7.6 percent in 1950–75, Seoul is expected to grow at an almost stagnant 0.02 percent in 2000–2015. Similarly, Tokyo grew at 4.2 percent a year between 1950 and 1975 but is expected to grow at only 0.19 percent in 2000–2015. Osaka grew at 0.45 percent a year in 1975–2000 and will not grow at all in 2000–2015. Tokyo and Seoul have relatively well-defined administrative and political jurisdictions

Asia's largest urban agglomerations (2000 rankings)

Urban agglomeration	Area (sq km)	Population (millions)				Growth Rate	
		2000	2005	2010	2015	1995–05	05–2015
1 Tokyo	2187	26.4	26.4	26.4	27.2	0.3	0.1
2 Mumbai	4167	18.0	20.9	23.5	26.1	3.2	2.2
3 Kolkata	1785	12.9	14.1	15.6	17.2	1.7	2.0
4 Shanghai	6340	12.8	13.1	13.6	14.5	0.0	1.1
5 Dhaka	528	12.3	15.3	18.3	21.1	4.9	3.2
6 Karachi	1800	11.7	14.0	16.5	19.2	3.7	3.1
7 Delhi	3182	11.6	13.4	15.1	16.8	3.0	2.2
8 Beijing	16,807	11.0	11.0	11.5	12.2	0.2	1.1
9 Osaka	1890	11.0	11.0	11.0	11.0	0.0	0.0
10 Jakarta	654	11.0	13.1	15.3	17.2	3.6	2.7
11 Manila	636	10.8	12.4	13.8	14.8	2.9	1.7
12 Istanbul	1991	9.4	10.8	11.8	12.4	3.1	1.5
13 Tianjin	11,919	9.1	9.4	9.9	10.7	0.5	1.2
14 Seoul	11,718	10.0	9.8	9.8	10.0	−0.4	0.0
15 Bangkok	1568	7.2	8.0	9.0	10.1	2.1	2.3
16 Hyderabad	217	6.8	8.1	9.3	10.4	4.0	2.5

Source: United Nations (2001)

encompassing highly urbanized areas. They have homogeneous populations, mostly employed in the formal sector. Tokyo has succeeded in curbing the massive environmental pollution that plagued it in the past, and Seoul has managed to deal with its serious housing problem. Both cities have evolved planning and governance mechanisms that adequately meet their demand for basic urban services.

Second are the *mega-cities of China*. Large cities in China, such as Beijing, Guangzhou, Shanghai, and Tianjin, have had low population growth rates as a result of migration and population control policies strictly enforced in the past. Since China's shift from a centrally planned to a more market-oriented economic system in 1979, however, the populations of large Chinese cities have started to grow as a result of in-migration arising from a relaxation of the *hukou* household registration system (the adjusted expansion of metropolitan boundaries has also officially shown an "increase" in city-region populations). Though urban service levels in key Chinese cities are somewhat hard pressed, China's high level of economic growth (an average annual growth of gross domestic product of 8.5 percent since 1979) and a recent boom in infrastructure development are tending to keep services up to meet citizen demand. . . .

Third are the *primate cities of Southeast Asia*. Southeast Asian cities like Bangkok, Jakarta, and Metro Manila currently have moderate rates of population growth, although in 1950–75 they grew at annual rates of 4.8, 4.7, and 4.1 percent respectively. Annual growth rates in these cities are expected to decline in 2000–2015 to 3.0 percent in Jakarta, 2.8 percent in Metro Manila, and 1.9 percent in Bangkok. These cities are very important to national life because of their primacy and their political status as central government capitals. Though they are beset with many problems (inadequate housing, water shortage, congested traffic, environmental pollution), their recent investments in infrastructure (e.g., rapid transit) promise to improve life for city residents. They have also benefited from urban reforms attempting to bring about better urban and regional planning, as well as administrative coordination among local units within city-regions.

And fourth are the *South Asian cities*. Large cities in Bangladesh, India, and Pakistan, such as Dhaka, Delhi, Kolkata, Mumbai, and Karachi, continue to expand at relatively high rates, with population growth only expected to slacken in the 2000–2015 period. Between 1975 and 2000, Dhaka grew annually by 7.0 percent,

Delhi by 4.0 percent, Karachi by 3.7 percent, Mumbai by 3.1 percent, and Kolkata by 2.02 percent. . . .The urban agglomerations of South Asia are already plagued with problems such as slums and squatting, inadequate transport, high unemployment rates, and massive levels of environmental pollution. With their continued expansion, they will pose extreme challenges to urban planning and governance.

The vast differences among Asian urban agglomerations demand a more in-depth analysis of their growth patterns, with special emphasis given to the historical factors that have influenced their development. Some Asian cities, such as Beijing, Delhi, and Tokyo, have ancient origins and were the seats of powerful empires. Others, like Jakarta, Karachi, Kolkata, Metro Manila, Mumbai, and Shanghai, trace their roots to the colonial era, when they served mainly as trading and production outposts. The growth and development of Chinese cities was mainly controlled by central government policies focused on population and migration controls. Most Southeast Asian cities prospered because of their dominant roles as national capitals. All these factors are important considerations when attempting to explain the growth and development of each of the Asian mega-urban regions covered in this study.

EAST ASIAN CITIES

Big cities in East Asia are technologically advanced, having passed through the industrial and even post-industrial developmental stages. *Tokyo* has been the largest city in Japan since the middle of the nineteenth century, when, as Edo, it was the capital of the Shogunate. After the Meiji Restoration of 1868, Edo was renamed Tokyo and became the capital of Japan. Although Tokyo was greatly devastated by the earthquake and fire of 1923 and American carpet bombing during World War II (1941–45), the city was rebuilt and quickly became the dominant center in the country. Tokyo's population, which dropped to 3.5 million in 1945, increased at such a formidable rate that by 1969 it had reached 11.3 million.

Concerned about the expansion of Tokyo, the Japanese government pursued a policy of decentralizing industries and manufacturing to other urban centers such as Osaka and Nagoya. The structural adjustments of Japan's economy, however – particularly the shift from industrial production to information technology – meant the continued dominance of

Tokyo. Although Osaka and Nagoya initially attracted many rural–urban migrants from other parts of Japan, they eventually became out-migration areas for people flocking to the Tokyo metropolitan area. The concentration of people in the Tokyo region became even more marked with its emergence as a global center for finance and other services.

[. . .]

At present, the official Tokyo national capital region covers 2,187 square kilometers and is run as a self-governing unit encompassing twenty-three wards (ku). Aside from the twelve prefectures, it also covers an area containing twenty-six cities, five towns, and one village. This built-up area had a population of 26.5 million in 2001; this is expected to increase to 27.2 million by 2015. The most recent Tokyo Megalopolis Concept Plan seeks to expand this territory, resulting in boundary adjustments that may increase the population of the Tokyo urban agglomeration to 33 million by 2025.

Osaka is the economic and industrial center of the Kansai region, which also includes Kobe, Kyoto, and Nara. Osaka, like Kyoto, is not an ordinary prefecture (*ken*) but an urban prefecture (*fu*). Osaka prefecture covers approximately 1,890 square kilometers and had a population of 11 million in 2001 living in Osaka City, thirty-two other cities, ten towns, and one village. The whole Osaka–Kobe–Kyoto Metropolitan Area, however, had a population of 17 million.

[. . .]

Seoul, the capital of the Republic of Korea, does not qualify as a mega-city according to the city-size definition of the United Nations. But it is included in this study because of its important role in the development of the country. The city traces its origins as the capital of the Paekche Dynasty, one of the three ancient kingdoms in Korea. The Koryo Dynasty made Seoul its capital in 1067. With the ascendancy of the Chosun Dynasty in 1392, Seoul became Korea's national capital. The city flourished within and beyond its fortified walls until Korea was occupied by Japan between 1910 and 1945. Seoul was almost completely destroyed during the Korean War, but since 1953 it has been rebuilt and transformed into a manufacturing and industrial base. Starting in 1956, Seoul's population grew explosively, with the annual growth rate reaching 10.8 percent in 1957, 19.2 percent in 1959, 16.8 percent in 1961, and 15.8 percent in 1962. By 1968, Seoul's population had reached 4.3 million and was growing at 8.3 percent a year.

At present, the Seoul capital region is made up of the central city of Seoul, eleven suburban districts, and parts of Kyonggi province. It covers a territory of 11,718 square kilometers and had a population of 9.8 million in 2000. In 1995, the Korean government approved the Capital Region Management Plan, which provides the basic direction and guidelines for the location of economic activities as well as the distribution of population. As envisioned by this plan, by 2015 the Seoul national capital region will have a population of 20.2 million and will encompass Seoul, Inchon, Kyonggi province, nineteen cities, and seventeen counties.

CHINESE CITIES

China has had very large cities for centuries, with Chang'an (present-day Xi' an) said to have had more than a million inhabitants during the Tang Dynasty (A.D. 618–907), when it was the eastern terminus of the famed Silk Road, the trade route that linked Asia and Europe. China's current capital, *Beijing*, was founded as a military outpost in about 1045 B.C. under the Western Zhao Dynasty (1122–771 B.C.). It became Yanjing, the capital of the state of Yan, during the Spring and Autumn Period (722–481 B.C.) and the Warring States Period (403–221 B.C.). It was expanded by Kublai Khan as Dadu (the Great Capital) in A.D. 1267–1271. By 1949, therefore, when the victorious Communist regime declared Beijing the state capital, the city had been the imperial capital of China for more than 800 years.

Before 1949, the built-up area of Beijing was only 109 square kilometers. This was expanded to 707 square kilometers (with a population of 1.6 million) after the Communist takeover to make room for economic expansion as Chinese authorities pursued the socialist policy of transforming Beijing from a "consumptive" to a "productive" city. The Beijing master plans encouraged urban development outward – following a "palm and fingers" pattern, where transportation routes made up the fingers extending toward the southern and eastern parts of the region. Despite efforts to control development, however, the fingers around Beijing continued to extend outward, and urban development kept filling up the spaces between the fingers. In 1957, the Beijing master plan expanded the city's territory to 8,860 square kilometers. A master plan prepared in 1982 envisioned that

Beijing's population would be kept below 10 million by 2000. This figure was exceeded as early as 1992.

To keep abreast of development, the current master plan for Beijing, which was approved by the State Council in 1993, expanded the municipality's jurisdiction to include four inner-city districts, four nearby suburban districts, two outer suburban districts, and eight counties. The master plan has adopted the "scattered groups" or "urban clusters" strategy designed to turn the capital region into a polynucleated settlement. In this strategy, the central city area would have 6.5 million inhabitants and the rest of the inhabitants would be living in 14 satellite towns and 140 small and medium-sized towns.

[. . .]

China's largest city, *Shanghai*, owes its development to its coastal location, which has made it a natural center for overseas trade and commerce. After the defeat of China's Qin Dynasty in 1842 by Western powers (in the Opium War), Shanghai became a colonial enclave with separate concessions run by Britain, France, Japan, Germany, the United States, and other powers. Because of Shanghai's "colonial background," the Chinese authorities tended to neglect its development after 1949. Later, Shanghai was turned into an industrial base in accord with Soviet planning concepts favoring the development of heavy industries in big cities.

In 1958, the Shanghai municipality annexed the counties of Baoshan and Jiading, increasing its territory to 5,908 square kilometers. Seven satellite towns were also annexed in 1990, expanding Shanghai's jurisdiction to 6,340 square kilometers. At present, the Shanghai municipality encompasses ten urban districts within the city proper, four suburban districts, and six suburban counties. Economically, however, Shanghai dominates the Yangtze River Delta region . . . Shanghai, as the "head of the dragon" in the Yangtze River delta, is now the dominant core of a megalopolitan region covering about 100,000 square kilometers with a population of 72.7 million.

The Pearl River Delta cities of *Guangzhou, Hong Kong, Shenzhen, Zhuhai, and Macao* have functioned independently until recent times. Lately, however, they have tended to develop in an increasingly coordinated manner because of China's regional development policies. The Pearl River Delta, at present, has three levels of urban settlements. First, there are the 2 large cities of Guangzhou (6.7 million) and Hong Kong (6.9 million). At a second level are 9 medium-sized cities:

Shenzhen, Macao, Zhuhai, Foshan. Jiangmen, Zhongshan, Dongguan, Huizhau, and Zhaoqing. A third tier is made up of 22 small cities that have county status and nearly 300 towns. The Pearl River Delta region, therefore, is a polynucleated megalopolitan region where no one center dominates. It is projected to have a population of 51 million by 2022, 18 percent of which will be in Hong Kong.

SOUTHEAST ASIAN CITIES

Big cities in Southeast Asia are characterized by primacy – the very high concentration of a country's urban population in a single agglomeration. All the Southeast Asian cities included in this study have populations many times larger than the country's next-ranking urban settlement. They are also national capitals that have thrived by serving as the economic, commercial, administrative, and cultural centers of nation-states.

The city of *Manila* was founded in 1571 as the Spanish colonial capital of the Philippines. It thrived on the basis of centralized production of agriculture-based products as well as the main port of the galleon trade that linked the Philippines with Mexico and Spain. When the United States took over the Philippines in 1899, it kept Manila as the national capital. . . .

Manila was seriously devastated during World War II, but postwar re-construction began immediately without the benefit of careful planning. Internal migration – triggered by the push of widespread rural poverty and a communist-led insurgency in other parts of the country on the one hand and the pull of the "bright lights" of Manila on the other – expanded the city's population. By 1960, the four cities and four towns officially designated as parts of Metro Manila had reached a population of 2.1 million and occupied a territory of 362.2 square kilometers.

Since the mid-1970s, the City of Manila proper and the inner core municipality of San Juan have been losing population. The metropolitan population, however, has increased to the point that the jurisdiction of the national capital region has been expanded to cover thirteen cities and four towns. . . .

The rapid expansion of the Manila mega-urban region, in fact, has prompted the government to start formulating a development plan for the whole island of Luzon with Metro Manila as its core. Such a plan includes schemes for the national capital region, the

Calabarzon region . . ., other provinces in the central plains of Luzon, and the two special economic zones established on the former U.S. military bases at Subic Bay and Clark.

The capital city of Indonesia, *Jakarta*, was established as the capital of the Dutch colony in the East Indies about 465 years ago. As Batavia, it was planned to mimic a Dutch settlement, complete with a network of canals. The canals, unfortunately, became the breeding place for deadly tropical diseases, and the Dutch colonizers were eventually forced to move to higher grounds. The coastal settlement expanded southward, with the Dutch quarters in the center and the natives settling on the periphery. . . . The 1990 Indonesian census showed that the population of the Special Capital Region of Jakarta (Daerah Khusus Ibukota, or DKI Jakarta, also referred to as Jakarta Raya) had reached 8.2 million.

DKI Jakarta has the status of a province. Surrounding it is the metropolitan region referred to as Jabotabek, . . . Although Jabotabek is fragmented into local government units, a Joint Development Cooperation Board has been established to coordinate development activities in the region. Planning and governance in Jabotabek, however, is vastly complicated by the fact that many central government agencies exercise authority and power over activities in the region. The local units in the Botabek area are also within the jurisdiction of West Java province, and a number of large private development companies wield considerable clout in public affairs as well.

The United Nations set Jakarta Raya's population at 11.4 million in 2001, projected to increase to 17.3 million by 2015. However, the Jabotabek Metropolitan Development Plan . . . projects a population of 26 million as early as 2005, with more than 18 million people living in the built-up area of DKI Jakarta. There are even suggestions that for better comprehensive planning, a wider territory that includes the city of Bandung and its surrounding regencies and towns should be taken as the megalopolitan region that actually serves as Indonesia's national capital.

Since the signing of the Bowring Treaty between Britain and Thailand in 1855, *Bangkok* has grown rapidly because of its integration into the world economy. Bangkok's expansion has been fueled by lucrative rice production and exports, the growth of industry, manufacturing and tourism, and the economic boom following the Vietnam War. Late in 1960, the Greater Bangkok Plan 2533 was formulated, designed for a population of 4.5 million on 780 square kilometers of territory by 1990. This plan was later revised by the Greater Bangkok Plan 2000, which projected a population of 6.5 million people occupying 820 square kilometers by the end of the century.

In 1970, Bangkok was combined with Thonburi, across the Chao Phraya River, to form Greater Bangkok, increasing the combined city populations to 2.5 million. Two years later, the Government of Thailand decided to merge the two adjoining provinces of Phra Nakhon and Thonburi into a single city and together with Bangkok created the Bangkok Metropolitan Area (BMA). By 1980, the BMA's population reached 4.7 million. The continued outward expansion of Bangkok has prompted the creation of the Bangkok Metropolitan Region (BMR). . . .

Because of the outward adjustments of Bangkok's metropolitan boundaries, in 1988 the BMA's population was said to have "increased" to 5.7 million and the BMR's to 8.5 million. The United Nations, however, has set Bangkok's population at a smaller 7.3 million in 2000, projecting this to reach 9.8 million by 2015. This lower figure is obviously based on a much narrower definition of the metropolitan population and does not consider Bangkok's dominant role in the development of Thailand.

Taking a more expansionist view, the National Economic and Social Development Board (NESDB) estimated in 1990 that the BMR's population had reached 8.9 million and would increase to 12.6 million in twenty years. Furthermore, the NESDB noted the emergence of an "extended BMR" . . . The total population of this extended BMR is expected to grow from 12 million in 1990 to 17 million by 2010, increasing its share of the national population from 21.5 to 24.3 percent.

SOUTH ASIAN CITIES

Urbanization in South Asia can trace its ancient roots to the Indus Valley civilization that flourished in preindustrial settlements like Harappa in the Punjab and Mohenjo-daro in the valley of the Indus around 2500 to 1500 B.C. The origins of India's capital, *Delhi*, have been attributed to the legendary state of Indraphrasta, which is prominently mentioned in the Mahabharata epic. Indian historians have written about the seven cities of Delhi, claiming that the present city is the eighth capital of independent India. The medieval glory of Delhi can still be sensed in the city walls built by the Moghal emperor Shah Jahan during the seventeenth century or the monumental massiveness of the Red Fort.

The modern era in Delhi started with the transfer of India's colonial capital inland from coastal Calcutta (now Kolkata) in 1912. However, city historians indicate that the British colonizers started settlements in the so-called Civil Lines north of the walled city earlier than that date. A well-planned New Delhi was planned and built south of the old city in the 1920s. By 1951, India's capital had grown into a "town group" consisting of two cities, three major towns, and one minor town. All the local government units in the region were amalgamated into a single municipal corporation in 1958, although New Delhi and the Cantonment were allowed to keep their individual identities. In 1961, the Delhi metropolitan population reached 2.65 million for the entire "union territory" that constituted the federal capital of India. This territory was made up of the areas governed by the Municipal Corporation of Delhi (539 square miles with 2.3 million people), the New Delhi Municipal Committee (16.5 square miles with 261,545 people), and the Cantonment Board of Delhi (16.5 square miles with 36,105 people).

. . . In 2001, Delhi had a population of 13.3 million, occupying a territory of 1,397 square kilometers. At the projected growth rate of 3.4 percent a year (2000–2015), Delhi is expected to reach a population of 20.9 million by 2015. Some Indian planners acknowledge, however, that the current field of influence of Delhi extends way beyond its present territory and should encompass areas in the adjacent states of Uttar Pradesh, Haryana, and Rajasthan to form a larger megalopolitan region.

India's largest city, *Mumbai* (formerly Bombay), is the capital of Maharashtra state and the acknowledged commercial center of India. This port city thrived not only because of manufacturing (particularly cotton and textile) but also with international trade and commerce during the British colonial period. The Bombay Municipal Corporation Act of 1888 specified eight statutory authorities to manage specific activities of the city government – mainly public works, water and sewerage, health, and solid waste disposal. City authorities became so concerned about the heavy industrial pollution in the inner city that they sought to transfer many of the plants to the outskirts early in the city's history.

The first postindependence plans for Mumbai in 1948 proposed the creation of satellite towns north of the city. In 1958, another plan resulted in the establishment of a township across the Thane Creek to draw away population from the crowded inner city. This became New Mumbai in Thane District, which rapidly grew under the guidance and management of the City and Industrial Corporation of Maharashtra state. The New Mumbai Municipal Corporation was established in 1991 and assumed responsibility for nine of the twenty-five urban nodes in New Mumbai. . . .

At present, the Mumbai Metropolitan Region (MMR) includes Mumbai, New Mumbai, and Thane. Thane, a municipal corporation created in 1982, encompasses Thane District and thirty-two surrounding villages that all fall within the jurisdiction of the Thane Municipal Council. Thane District had a population of 1.2 million in 2001. Also within the MMR are eighteen urban centers located on the periphery of Greater Mumbai. Mumbai City had a population of 11.9 million in 2001, but the whole Mumbai Metropolitan Region had 16.5 million, projected to increase to 22.6 million by 2015.

Kolkata (formerly Calcutta) is the capital of West Bengal state and the second largest city of India. It started as a trading post near the end of the seventeenth century and became the capital of India under the British Empire. In 1912, India's capital was moved to Delhi, but Kolkata continued as an important commercial and industrial center for Eastern India. By 1966, the Kolkata Metropolitan District extended over nearly 490 square miles and contained a population of 7.5 million. Within the district, Kolkata City proper covered about 37 square miles and had 3 million inhabitants. Also in the district were thirty-four municipal towns, an almost equal number of nonmunicipal towns, and more than a hundred villages in a built-up area stretching on both sides of the Hooghly River over 20 miles north and 15 miles south of the twin cities of Kolkata and Howrah. By 2001, Kolkata City had reached a population of 4.5 million, and the metropolitan area had 13.3 million inhabitants. At an annual growth rate of 2.0 percent, Greater Kolkata is expected to reach a population of 16.7 million by 2015.

Administratively, the Kolkata metropolitan area is divided into three municipal corporations, thirty-four municipalities, and a number of urban localities. The Kolkata Metropolitan Development Authority is responsible for the overall planning of the region and is in charge of large-scale urban infrastructures. . . .

Dhaka, the capital of Bangladesh, became an important settlement during the seventeenth century, when it was the Mughal capital of Bengal province (1608–1704). The British took over Dhaka in 1765, and it became the capital of Bengal in 1905. After the partition of British India in 1947, Dhaka became the capital of the Pakistani province of East Bengal and, in 1956, the capital of East Pakistan . . . In 1950, Dhaka had a population of 417,000. But after independence in 1971, it grew at the high annual growth rate of 6.6 percent, so that by 1975 its population had reached 2.1 million. The city currently has a population of 12.5 million, which is projected to increase to 22.8 million by 2015, when it is expected to be the second largest urban agglomeration in the world, second only to Tokyo.

Karachi was settled as a small fishing village and became Pakistan's largest city, with a current population of more than 10 million. In 1843, when the British Army conquered the province of Sindh, the colonizers chose Karachi as a coastal base. When a cholera epidemic ravaged the settlement in 1846, a Conservancy Board was organized, which in turn was upgraded into a Municipal Committee in 1853. In 1933, the City of Karachi Municipal Act was passed, and the Municipality of Karachi was given the status of a municipal corporation. Karachi was made the capital of the new nation-state of Pakistan in 1947 after the partition of the country from India. By 1950, the population of Karachi had reached more than a million and was growing at 5.4 percent a year. The metropolitan population totaled almost 4 million in 1975. In 1976, the Karachi Municipal Corporation (KMC) was upgraded to a metropolitan corporation. A separate Zonal Municipal Committee was set up in Karachi in 1987, but in 1994 this committee was again merged with the KMC. Five district municipal corporations were also created in the Greater Karachi area in 1996 to adjust to the expansion of the metropolitan area.

The most far-reaching changes in the planning and governance of Karachi occurred on August 14, 2001, when the Government of Pakistan implemented the Devolution Plan creating the City District Government of Karachi (CDGK) as well as 18 town administrations and 178 union councils all over the country. By that date, the population of Karachi exceeded 10 million. Under the devolution scheme, the jurisdiction of the CDGK was extended to 18 towns. . . .

[. . .]

CITY-REGIONS IN AN ERA OF GLOBALIZATION

Globalization as manifested in its various forms is intimately linked to the growth and functioning of big cities and mega-urban regions. It is in city-regions where people congregate at high densities that social and economic changes are rapidly happening. At the same time, the rapid growth of mega-urban regions is also affecting globalization. As noted by the UN Center for Human Settlements:

It is in cities where global operations are centralized and where one can see most clearly the phenomena associated with their activities: changes in the structure of employment, the formation of powerful partnerships, the development of monumental real estate, the emergence of new forms of local governance, the effects of organized crime, the expansion of corruption, the fragmentation of informal networks, and the spatial isolation and social exclusion of certain population groups. The characteristics of cities and their surrounding regions, in turn, help shape globalization; for example, by providing a suitable labor force, making available the required physical and technological infrastructure, creating a stable and accommodating regulatory environment, offering the bundle of necessary support services, contributing financial incentives and possessing the institutional capacity without which globalization cannot occur.

Globalization forces are centered in cities, and they have profound effects on the physical form as well as the socioeconomic makeup of mega-urban regions. In theory, globalization is expected to be directly correlated with (1) more rapid rates of economic development and growth of very large "global cities," (2) reduced inequalities among social classes and various regions, and (3) improved changes for the development of a liberal democratic way of life. This study of Asian mega-urban regions suggests, however, that the positive effects of globalization are not always realized and apply to different places in different ways. Though tangible benefits from globalization have been achieved (e.g., the reduction in mortality rates due to global diffusion of medical knowledge, increased food production based on shared agricultural technologies, improved environmental protection and conservation), the benefits from these achievements have not been equally and equitably shared among all classes and groups within a city-region or a country. In fact, empirical data are tending to show that inequality in mega-urban regions is growing wider and spatial polarization is occurring, with rich people living in gated communities and poor residents in marginal settlements.

It has been suggested that the emergence of global cities linked together by economic, informational, political, and financial ties has served to weaken the nation-state. The dominant influence of global actors (multinational corporations, international financing institutions, networks of agglutinated business elites) is supposed to have resulted in the "hollowing out of the state" and its capture by particularistic local interests closely allied with international power holders. In the case of the Asian mega-urban regions covered in this study, however, it has been shown that though most of them aspire to global city status, some of their leaders are strongly committed to their national institution-building roles. The city-regions lead their countries in the drive toward economic development. They are the centers of innovation and the generators of social change. They play, and will most likely continue to play, a highly significant role in the development of nation-states in which they are embedded.

Most important of all, city-regions may play a very important role in the spread of norms of democratic governance, the value of environmental protection, and the ideals of social justice and basic human rights. As Aristotle said centuries ago, "City air makes man free." It is that commitment to *civitas*, the right to citizenship in a free community and assuming responsibility for actions that uphold the common good, that makes citizens' involvement the keystone for city-region development. Mega-urban regions and megalopolitan areas already are, and will continue to be, the dominant forms of settlements in the new millennium. Their planning and governance provide the main instruments for achieving their fullest development potentials.

"Planning Sustainable and Livable Cities"

(1998)

Stephen Wheeler

Editors' Introduction

The enormous and exponential growth of the world's population described by Kingsley Davis in the "Evolution of Cities" section of this anthology (p. 17) and the emergence of enormous mega-city regions described by Aprodicio Laquian (p. 489) have had profound and often catastrophic effects on the natural environment of the planet. Non-renewable energy sources have been consumed, forests cleared, buffalo slaughtered, species extinguished. Rain forests are disappearing. Today, many argue that global climate change caused by population growth, urbanization, and irresponsible fossil fuels consumption threatens to inundate low-lying cities and irreparably damage the climate of the entire world. To one degree or another, this is the message of the Brundtland Report (p. 337), Timothy Beatley's vision of "green urbanism" in Europe (p. 411), and the planning practices introduced by New Urbanists such as Peter Calthorpe (p. 342) and Andrés Duany (p. 192).

In the following selection on "Planning Sustainable and Livable Cities," Stephen Wheeler reviews the evolution of sustainable urban development thinking and ties it to the new concern with urban livability. He defines sustainable development as "development that improves long-term health of human and ecological systems." He dismisses as inadequate recent debates about "needs" (which are hard to distinguish from wants), the "carrying capacity" of areas (which are tough to pin down, particularly for people), and "sustainable end states" (since it is virtually impossible to decide on end states).

Wheeler lays out a helpful compendium of core themes in the sustainable urban development literature, a list of specific approaches that can guide planning practice, and numerous examples to learn from. Wheeler's theoretical framework provides real-world guidance to planners about how to carry out plans to foster sustainable development. Whatever their specifics, he argues, sustainable urban development strategies need to be *long term* – plans for twenty, fifty, one hundred years or longer, rather than year-by-year plans that optimize short-term gains at the expense of future welfare. Wheeler emphasizes that a core theme in urban sustainability planning must be attention to the natural environment, including wilderness areas. And, finally, urban sustainability planning requires holistic and interdisciplinary approaches connecting the insights of biologists and transit planners, agronomists, economists, and many other disciplines. Sustainable urban development planning requires that land use, transportation, housing, community development, economic development, and environmental planning all be woven together.

Wheeler outlines eight directions sustainable urban development should take to move toward his definition of sustainable urban development. He illustrates each suggested direction with specific, concrete examples of actual good things some cities have done. In contrast to unplanned sprawl that consumes land and leads to costly and inefficient infrastructure, for example, Wheeler urges planners to move in the direction of compact, efficient land use. Wheeler illustrates each of his other suggested directions – fewer automobiles, more efficient resource use, restoring natural systems, good housing, healthy social ecology, and sustainable economics – with specific suggestions and concrete examples of where sustainability has been successfully implemented.

Stephen Wheeler is an assistant professor in the Landscape Architecture Program at the University of California, Davis. Wheeler was a lobbyist for the Friends of the Earth organization, a board member of Urban Ecology, and, for eight years, the editor of *The Urban Ecologist*. He is the author of *Planning for Sustainability: Creating Livable, Equitable, and Ecological Communities* (London and New York: Routledge, 2004) and the co-editor, with Timothy Beatley, of *The Sustainable Urban Development Reader* (London and New York: Routledge, 2004).

Among the literature on the destructive effects of urbanization on the natural environment, see William Cronon, *Nature's Metropolis* (New York: W.W. Norton, 1992) and Mark Reisner, *Cadillac Desert* (New York: Penguin, 1993). Other important writings on sustainable urban development include Graham Haughton and Colin Hunter, *Sustainable Cities* (London: Regional Studies Association, 1994), *Sustainable America* by the President's Council on Sustainable Development (Washington, DC: Government Printing Office, 1996), and William Rees, *Our Ecological Footprint: Reducing Human Impact on Earth* (Philadelphia: New Society Publishers, 1996). Mike Davis's *Ecology of Fear: Los Angeles and the Imagination of Disaster* (New York: Vintage, 1999) is a highly rhetorical study that ties together ecological and cultural tendencies in contemporary cities, arguing that "market-driven urbanization has transgressed environmental common sense." Ian McHarg's classic *Design With Nature* (New York: Doubleday, 1969) and Patrick Geddes, *Cities in Evolution* (London: Williams & Norgate, 1915; reprinted in Richard LeGates and Frederic Stout (eds), *Early Urban Planning, 1870–1940*, London: Routledge/Thoemmes, 1998) are essential and foundational works that anticipate today's sustainable development, regional planning, and ecological design debates.

INTRODUCTION

The term "sustainable" is now widely used to describe a world in which both human and natural systems can continue to exist long into the future. The concept of "sustainable development" is used to refer to alternatives to traditional patterns of physical, social and economic development that can avoid problems such as exhaustion of natural resources, ecosystem destruction, pollution, overpopulation, growing inequality, and the degradation of human living conditions.

However, the notion of sustainability is still very recent, and understandings of what it means and how to apply it are still evolving. Many questions surround this concept. Does it in fact express a coherent and meaningful philosophy? Can it be defined in ways that the general public can relate to? Can it be used to generate consensus around specific urban planning directions? Can it avoid cooptation by existing political forces?

In the following pages I will propose a framework for thinking about sustainable development in the metropolitan context. After examining the origins of the sustainability concept, different definitions, and key themes, I will outline implications for urban and regional planning, and suggest processes through which sustainable development planning might be achieved. I will also consider the related concept of

"livability," another broad notion that is often used in the same breath as "sustainability." These terms are closely related, in that they both promote urban planning that enhances long-term community well-being. It is no accident that these concepts have come to the fore in urban planning discussions recently, because they address important unmet needs arising from the nature of twentieth-century urban development.

ORIGINS OF THE "SUSTAINABILITY" CONCEPT

The verb "sustain" has been used in English since the year 1290 or before, and comes from the Latin roots "sub" + "tenere," meaning "to uphold" or "to keep". The *Oxford English Dictionary* traces the adjective "sustenable" to around 1400 and the modern form "sustainable" to 1611. However, this term appears to have been used mainly in legal contexts until recently, as in "The Defendant has taken several technical objections to the order, none of which . . . are sustainable" (1884). Many other variants of "sustain" have existed for centuries, but only in the past several decades has the word "sustainable" emerged with its current meaning, perhaps most simply defined as "that which can be maintained into the future."

It is far from clear who was the first to use the term "sustainable development" in its current sense. Rather, it seems one of those inevitable expressions that so neatly encapsulates what many people are thinking that once the words are mentioned they quickly become ubiquitous. The birth of the sustainability concept in the 1970s can be seen as the logical outgrowth of a new consciousness about global problems related to environment and development, fueled in part by 1960s environmentalism, publications such as *The Limits to Growth*, and the first United Nations Conference on Environment and Development held in Stockholm in 1972. However, the notion of sustainability is also rooted in older environmental traditions, particularly "sustained yield" techniques of forest management developed by nineteenth-century German foresters. These concepts influenced American policy makers such as Gifford Pinchot, Theodore Roosevelt's chief forester, and natural resource scientists such as Aldo Leopold. Leopold's notion of a "land ethic" – a human responsibility to care for particular lands and ecosystems, discussed most fully in *A Sand County Almanac* (1948) – represents a fundamental shift from the view that natural resources should be seen in terms of their utility for human beings, toward the perspective that species and ecosystems have intrinsic value in their own right and should be stewarded and sustained indefinitely into the future.

In the post-World War II period a long line of environmentalist works such as William Vogt's best-selling *Road to Survival* (1948), Fairfield Osborn's *Our Plundered Planet* (1948), Rachel Carson's *Silent Spring* (1962), and Barry Commoner's *The Closing Circle* (1971) sounded a note of alarm about the global ecological situation, and helped tie the rise of ecological problems to ongoing patterns of industrial development. Particular events also helped change consciousness around development issues, such as the 1973 oil embargo during which millions of people suddenly realized that their fossil fuel use could not continue to expand forever. Social critics, futurists, feminists, and new age writers further prepared the way for discussions of sustainability by critiquing existing notions of development and proposing alternative paradigms that would emphasize the spiritual, the natural, and the human over values of profit and economic progress as traditionally conceived. At the same time humanistic and transpersonal psychologists pointed out ways in which human potential is shaped by the surrounding environment, and ways in which it can perhaps be shaped in healthier directions in the future. The implication of such work is that people and perhaps entire societies can evolve towards more conscious, compassionate, and sustainable modes of existence, given the right conditions.

The earliest specific reference to sustainability that I have been able to document occurs in the 1972 book *Limits to Growth*, in which Donella Meadows and other MIT researchers describe computer models showing a collapse of global systems in the mid-twenty-first century, but state optimistically that "It is possible to alter these growth trends and to establish a condition of ecological and economic stability that is sustainable far into the future." A 1974 conference of the World Council of Churches then issued a call for a "sustainable society," and the earliest book specifically discussing sustainability appeared in 1976, a volume entitled *The Sustainable Society: Ethics and Economic Growth* by Lutheran theologian Robert L. Stivers.

In the late 1970s the number of writings on sustainability grew rapidly. The sustainability literature got one of its strongest pushes from Lester Brown and others at the Worldwatch Institute, a group which began publishing an extensive series of papers and books related to global sustainability, including annual State of the World reports. The tide of literature expanded in the 1980s with the International Union for the Conservation of Nature's influential World Conservation Strategy (1980), the President's Council on Environmental Quality's Global 2000 Report (1981), and above all the 1987 report of the World Commission on Environment and Development, chaired by Norwegian Prime Minister Gro Harlem Brundtland. These documents warned about global environmental problems and critiqued notions of "development," although generally accepting the desirability of continued economic growth. The influence of the IUCN and Brundtland reports in particular stemmed from the broad participation of mainstream governmental officials and academics within these bodies, which gave their findings an air of authority going beyond the "alarmist" reports of the *Limits to Growth* researchers, Global 2000, or the Worldwatch Institute. The Brundtland Commission in particular received input from literally thousands of individuals and organizations from around the world. Initiated at the request of the United Nations Secretary-General, it followed in the footsteps of two other highly respected UN-sponsored commissions, the Brandt Commission on North–South Issues and the Palme Commission on

Security and Disarmament Issues. A more authoritative body to explore the topic would have been hard to find.

With the release of the Brundtland Commission report *Our Common Future* in 1987 and the United Nations "Earth Summit" conference in 1991, calls for sustainable development entered the mainstream internationally. "Sustainable city" programs emerged in many parts of the world, some resulting from grassroots activism, some based on municipal initiative, some benefiting from the support of national governments, and some facilitated by multilateral entities such as the European Community, the World Bank, and the UN. The 1996 UN Habitat II "City Summit" in Istanbul took slow but significant steps toward establishing global consensus on how the sustainability agenda can be applied to urban planning. National reports such as that of the President's Commission on Sustainable Development (PCSD) in 1996 attempted to establish sustainable development directions for particular countries. As the 1990s went on academics in fields such as urban planning began to delve into the subject. Although actual implementation of sustainability programs remains difficult, the persistence and spread of the concept over three decades indicates that sustainability is a notion of lasting importance.

DEFINITIONS

Unfortunately, no perfect definition of sustainable development has emerged. The most widely used is that of the Brundtland Commission: "development that meets the needs of the present without compromising the ability of future generations to meet their own needs." However, this formulation is open to criticism for being anthropocentric and for raising the difficult-to-define concept of needs. (Does every household really need two cars? A VCR? A 2,000-square-foot house on a 5,000-square-foot lot? What happens if every household worldwide has these things?)

Other definitions include that given by the World Conservation Union in 1991: "improving the quality of human life while living within the carrying capacity of supporting ecosystems." This version raises the problematic notion of "carrying capacity," which is useful to think about for educational purposes but extremely hard to pin down in practice. It is one thing to say that the carrying capacity of a given watershed is a certain number of white-tailed deer; it is far more

difficult to say that it is a certain number of human beings, when humans readily transport themselves and the resources they use over vast distances, and can substitute some resources for others if necessary. Although William Rees at the University of British Columbia argues that it is useful to try to calculate the "ecological footprint" of cities in terms of their resource use, my own belief is that attempts to actually define "carrying capacity" are best avoided.

Still other writers prefer to define sustainability in terms of preserving existing stocks of "ecological capital" and "social capital." This approach builds on the economic wisdom of living on the interest of an investment – in this case the earth's stock of natural resources – rather than the principal. For example, British economist David Pearce argues that "Sustainability requires at least a constant stock of natural capital, construed as the set of all environmental assets."

Most sustainability advocates throw up their hands when faced with the definition question, and fall back on Brundtland. However, my own preference is to move instead towards a relatively simple, process-oriented definition emphasizing long-term systemic welfare: "Sustainable development is development that improves the long-term health of human and ecological systems." This strategy avoids fruitless debates over "carrying capacity," "needs," or sustainable end states, while emphasizing the process of continually moving towards healthier human and natural communities. In theory the directions of this process can be agreed upon through participatory processes in which all relevant stakeholders are represented, and progress can be measured by means of various performance indicators.

CORE THEMES

The widespread use of the sustainability concept testifies to the strength and relevance of its underlying themes, both for urban planning and other fields. Foremost among these is a concern for the long-term perspective. Though it seems only common sense that planning and building should be for the long-term, this is manifestly often not the case in practice. In particular, long-term patterns of metropolitan growth, land use, resource use, and infrastructure development demand attention, giving new impetus to old quests such as halting suburban sprawl. From this perspective it is very important to think about expanding planning horizons from a year-by-year approach or even a 20-year horizon, to think instead about the effects of urban development over 50 years, 100 years, or longer.

A second main theme is concern about the earth's natural environment. As widespread as this sentiment is these days, it is remarkably recent within industrial society and should not be taken for granted. Also new is a recognition that current development patterns are leading to ecological and social problems on a global scale. Problems such as the greenhouse effect and damage to the earth's ozone layer were only taken seriously starting in the late 1980s. In its simplest formulation, concern about environmental problems often focuses on worries about global collapse or large-scale disaster. In more complex, systems-oriented formulations, environmental and social problems can be seen not so much as leading to specific disasters in specific time frames, but as contributing to an increasingly unstable and unhealthy global system, which could be plagued by any number of unpredictable catastrophes as well as by a generally increased level of suffering on the part of human beings and natural ecosystems. Either way, the environmental costs and risks of current development patterns are viewed as unacceptable by increasing numbers of observers. Hence the search for "sustainable" alternatives.

Lastly, the sustainable development discourse can be seen as based on a new recognition of the complex web of interconnections between different issues, fields, disciplines, and actors. This holistic and inter-disciplinary perspective, based on the ecological metaphor of the world as an organic system, has huge implications for planning. Among other things, it means that different specialties having to do with transportation, land use, housing, community development, economic development, and environmental protection should not be handled in isolation from one another, but should be integrated to the extent possible even while specific tasks are carried out. Coordination of economic, environmental, and social goals within planning is also necessary. Indeed, it is widely believed that social dimensions of sustainable development should be given equal weight to environmental goals.

In some ways the related concept of "livability" is a bit simpler, since it focuses less on abstract themes and more on specific human needs and people's subjective reactions to places. Dictionary definitions such as "fit to live in" or "conducive to comfortable living" actually work fairly well. Of course human needs are to a large extent culturally determined and are open to extensive debate. However, there is widespread agreement on basic elements that make cities and towns livable – a healthy environment, decent housing, safe public places, uncongested roads, parks and recreational opportunities, vibrant social interaction, and so on. Such elements obviously contribute to sustainability as well.

Livability themes are becoming more and more important to modern societies in which the basic problems of food, shelter, public health and sanitation that plagued nineteenth century cities have long since been solved, at least for most people. Instead, in the postindustrial world the emphasis is increasingly on "quality of life." It is no longer enough just to throw up cities and suburbs that are ugly, uncoordinated, automobile-dominated, and lacking in parks, sidewalks, local shops, community vitality, and sense of place. The question becomes How do we make these places green, safe, convenient, and human-oriented? How do we turn mass-produced urban landscapes into places that have character and nurture community? How do we make cities attractive and comfortable to all groups within society, including women, children, the elderly, and minorities? All of these concerns touch upon the task of making urban places "livable" in the long run.

IMPLICATIONS FOR URBAN DEVELOPMENT

Exactly what constitutes a "sustainable city" is impossible to determine, given the extent to which cities are embedded in the global context. To be absolutely self-sustaining, an urban region would need to wall itself off from the rest of the world and produce all food, energy, and materials locally. Such an autarkic model is generally infeasible and would be seen as undesirable by most residents.

Rather, it is more useful to speak of cities as moving *toward* sustainability. A metropolitan area might seek to move toward greater resource efficiency, environmental quality, social equity, and community vitality, while moving away from automobile dependency, non-renewable resource consumption, hazardous waste generation, and inequity. While local self-sufficiency may indeed offer many benefits for sustainability, this can be set forth as a value to be enhanced rather than as an absolute goal.

Until the early 1990s very little of the sustainable development literature focused on cities or patterns of urban development. Instead, writers addressed topics such as the global environmental crisis, ecological economics, critiques of prevailing models of international development, and the need for a transformation of values and mindsets. However, in recent years architects and planners have begun looking more specifically at what sustainability means for patterns of metropolitan development. Some authors have emphasized urban design and physical planning. Others have focused on environmental planning concerns having to do with the quality of air, water, and natural ecosystems. A number have stressed the need to address social problems and inequities within the urban community, and emphasize that environmental and social issues are inextricably linked. In all of these categories, urban sustainability advocates can be seen as building on the work of past planning visionaries such as Patrick Geddes, Ebenezer Howard, Lewis Mumford, Jane Jacobs, and Ian McHarg.

Although authors may have different emphases, there is substantial agreement on many dimensions of sustainable urban development. Such directions have been endorsed by documents such as UN's Agenda 21 and Habitat Agenda, professional manifestos such as the Charter of the Congress for the New Urbanism and the Local Government Commission's Ahwahnee Principles, and publications of the European Community and the President's Council on Sustainable Development.

"Sustainable urban development" might be defined as "development that improves the long-term social and ecological health of cities and towns". Based on this definition and sources such as those mentioned above, main directions for urban sustainability can be seen to include the following:

1. Compact, efficient land use. Land is perhaps our most important limited resource, and current urban development patterns are clearly consuming the landscape in unsustainable ways. Land is also often divided very inequitably, and in many parts of the world those inequities are increasing. A wide range of devices can help lead to more sustainable land. For example, urban growth boundaries (UGBs) have been adopted by Portland, Oregon, and eleven San Francisco Bay Area cities to restrain sprawl. To be effective in the long run, UGBs need to be coupled with policies to increase the efficiency of land use within already built-up areas, and to make these places more green, safe, attractive, and livable. Other types of land use controls can help preserve farm land, ecological habitat, and open space near cities. Meanwhile, urban park systems can be expanded and property tax systems changed to promote equity. Beyond specific land use changes, sustainable patterns of development are likely to involve alterations to the relationship between people and the land. In particular, a new balance between private property rights and human responsibilities toward the land is needed, as Leopold urged. The view of land as a commodity for human use and profit needs to shift towards a respect for the landscape as a thing of value in its own right, and a renewed sense of connection between human beings and the land they live on.

2. Less automobile use, better access. Current transportation systems contribute to a complex web of urban problems such as air pollution, congestion, blight, suburban sprawl, ecosystem destruction, and social fragmentation. Transport in more sustainable cities will most likely be based on several key principles: access by proximity, an inversion of the current transportation hierarchy, and demand reduction. Together these are likely to reduce the total amount that people need to travel, while allowing them to travel by far cleaner, more resource efficient, and more community-enhancing modes.

"Access by proximity" means solving transportation problems by bringing people closer to the places they need to go everyday. This is done primarily through land use changes, for example by promoting mixed-use development and the creation of neighborhood centers and "urban villages" which contain homes, workplaces, shops, and recreational facilities in close proximity to one another. Not coincidentally, fine-grained, mixed-use development also tends to create more interesting and livable places. "Inverting the transportation hierarchy" means placing the heaviest emphasis on the pedestrian, who represents the most energy efficient form of transportation and adds a much-needed human presence to the city. Bicycle planning should also be near the top of the priority list, followed by public transit. The automobile should be given lowest priority in the new hierarchy, and existing automobile subsidies reduced, although it is enormously difficult to do this politically. Finally, efforts to develop more sustainable transportation systems should also include looking at the "demand side" of the equation. By providing incentives to reduce the amount that people travel (the "demand side"), congestion problems can be solved and quality of life improved without building new roads or other infrastructure (the "supply side"). Pricing mechanisms are particularly useful to do this, such as higher parking charges, gas taxes, tolls, and vehicle registration fees.

3. Efficient resource use, less pollution and waste. Moving toward sustainability means paying greater attention to flows of energy and materials through human society, and planning for wiser use of resources. The overall challenge can be seen as one of moving from open-ended resource flows, in which nonrenewable resources are harvested, used once by human systems, and then discarded (often creating pollution and toxic waste problems in the process), toward closed-loop flows in which resources are reused and recycled.

A great number of mechanisms are available to urban and regional authorities to encourage this transition toward more sustainable resource flows. Energy conservation and materials recycling are two areas in which ordinary citizens can most directly take action through small daily initiatives, and so are good subjects for public involvement efforts. Municipal recycling programs are one of the most obvious areas in which cities can demonstrate their commitment to sustainable resource use. As in the transportation field, demand-side management programs offer great potential to reduce resource consumption. Since the late 1970s, for example, many utility companies around the US have offered consumers free or reduced-price compact fluorescent light bulbs, which typically use about one-fifth the electricity of incandescent models of similar wattage, as well as rebates on energy-efficient refrigerators, air conditioners, and water heaters. This saves both the consumer and the utility money, while reducing energy consumption. Unfortunately such programs were de-emphasized in the 1990s as many states shifted their attention to deregulating the utility industry.

Tougher energy conservation codes in building construction have also produced large energy savings in many cities and states. Pollution prevention programs are being developed to try to eliminate pollution before it is created rather than cleaning it up afterwards. "Industrial ecosystem" projects try to take a systematic look at manufacturing processes to see if the wastes from one industry can be used as inputs to another. And in the economic realm, attempts are being made to shift the costs of pollution from society as a whole onto the individual or group who produces it – what has become known as the "polluter pays principle."

4. Restoration of natural systems. Even though many urban areas may seem entirely artificial

– full of pavement and buildings, and often landscaped with non-native plants – still in almost every location there are many elements of the original ecosystem that can be reclaimed. Such restoration efforts add to the livability, ecological health, and overall sustainability of the urban region. Creek restoration, for example, is an idea that is catching on rapidly in many parts of the U.S. as well as overseas. Restoring a natural water-course provides corridors and habitat for wildlife as well as walkways and open space for people. It also helps reconnect urban dwellers to the bioregion, reminding them that they live in a natural world with cycles of rainfall and waterflow. Existing urban parks and areas of open space can benefit from restoration activities as well. Volunteer site restoration programs and stewardship approaches to watershed management can help this happen. Urban agriculture is another area in which nature is being brought back into the city and urban sustainability enhanced. Biointensive methods make it possible for urbanites to grow substantial amounts of food on very small areas of land. In both real and symbolic ways, urban gardening helps reconnect city dwellers with the earth. Finally, ecological restoration is urgently needed in many inner city areas which are often home to lower-income neighborhoods and communities of color. Abandoned or contaminated industrial land can be reclaimed and restored, while vacant lots can be recycled into parks, housing, and community gardens.

Cities are often seen as unlivable because they have lost any sense of connection between people and the natural world. People move to suburbs in search of trees and nature, and there is a widespread belief that densely settled areas cannot also be green. Restoring urban ecosystems can lead to healthier and more livable cities, while providing important amenities that can help entice residents back from suburbia.

5. Good housing and living environments. One of the main purposes of cities and towns is to create decent places for people to live, and if these do not exist or are not affordable, the urban system is bound to suffer. Housing affordability is a recurrent crisis in many North American cities and suburbs. Steps to address the affordability problem include active government construction of housing – which has a checkered history in the US, although a somewhat better record in European countries – support of nonprofit housing developers, tenant subsidies, and requirements that developers include a certain number of affordable units in any market-rate project. Also, many urban areas are characterized by ugly, homogenous, pedestrian-unfriendly development which writer James Howard Kunstler has called a "geography of nowhere." The design of housing and neighborhoods needs to be rethought in many cases to ensure that people have access to open space, meeting areas, shared facilities, shops, offices, public transportation, child care facilities, and other essentials which can make urban communities more livable.

6. A healthy social ecology. The health of human communities within an urban region is more difficult to grasp than the natural ecology. Certain social problems such as homelessness are quite obvious to anyone who walks down an urban street. Other deeply entrenched social problems, which help to decrease overall sustainability and livability in the long run, are often more hidden. Racism, for example, has been an enormous factor shaping American cities for many decades. Expressed in particular through the denial of housing, financing, insurance, or other necessities to persons of color, this factor has done as much as any other to contribute to the decline of many central city areas.

Promoting a healthy and sustainable social ecology means looking for every opportunity to enhance human community, opportunity and empowerment. It requires planners in particular to advocate on behalf of those groups who do not have access to power or expertise, and to fight for equity and justice. On a personal level, it requires an ability to put oneself in the shoes of any resident of an urban region and ask, What are the opportunities available to him or her? What is the environment like in which he or she lives? What public policies, design improvements, and social programs could help improve this environment?

7. A sustainable economics. Developing an economy that values the long-term health of human and natural systems is one of the biggest challenges related to sustainability. It would be a mistake to say that any one economic model holds all the answers, but in general a sustainable regional economy is likely be oriented around three principles. First, it is likely to be what Paul Hawken terms a "restoration economy" – one which helps restore the environmental and social damage done in the past, and that prevents new problems from occurring. Second, it is likely to be a "human-centered economy," one which meets real human needs and provides meaningful work to people at decent pay. Third, it is likely to be a locally-oriented economy, one which emphasizes local

ownership, local control, local investment, use of local resources, production for local markets. This does not mean that economic development policies should totally downplay export-oriented industries, but that they should encourage as much economic activity as possible to be rooted in particular communities and regions.

The sustainable economy is likely to meet these goals through a mixture of market mechanisms, government action, and incentives for social and environmental responsibility in economic decision-making. One important step toward a more sustainable economy will be to phase out industries or processes that consume large amounts of nonrenewable resources and produce large quantities of pollutants and toxics. In addition, economic sectors based on large-scale extraction of natural resources such as minerals and oil are unlikely to be sustainable in the long run, since these resources will eventually run out after enduring various price and supply shocks along the way. Likewise, sections of the regional economy based on government subsidy – such as military industries – are not particularly sustainable either, since these handouts may cease. Somewhat more arguably, sectors of the economy which support the automobile should not be considered sustainable, since cars and car-dependent patterns of suburban sprawl cause a great deal of ecological and social damage. In contrast businesses engaged in environmental cleanup, recycling, public transportation, affordable housing, organic food production and the like contribute to sustainability, in that these improve the social and environmental health of the region. But while some industries may dwindle in a sustainable economy and others grow, many will simply find ways of doing the same things better. This can have advantages for businesses as well as the environment.

Many writers have argued that a cooperative, locally-oriented economics, emphasizing worker, producer and consumer co-ops and small, locally-owned businesses, is healthiest for local communities. Such a system promotes economic democracy, local control, diversity of ownership, and social responsibility, and offers an alternative model to the global market economy envisioned under free trade agreements such as GATT. Such a free-market global economy led by large corporations has many negative effects on cities and towns, in that it tends to undermine local ownership and control, replace a diversity of small retailers with a few standardized chains, and export capital from local communities to financial centers in other parts of the world.

8. Community participation and involvement. One of the most important components of urban sustainability will be creation of more functional local and regional democracy, which in turn can bring about other positive changes. There is no single best way to promote this. But a package of policies aimed at opening up local political processes, insulating them from money and special interests, producing an educated and informed electorate, and promoting responsible local decision-making can help. Community participation in local planning and design is important, as is the broad-minded leadership of officials at local, state and federal levels of government, who must demonstrate that it is possible at every level to make decisions with global, regional, and local sustainability in mind.

9. Preservation of local culture and wisdom. Much of the strength of any particular urban region lies in its cultural traditions and the unique relationships that its residents develop with each other and with the land. This uniqueness gives a region vitality, helps it take advantage of particular local contexts, and makes it an interesting place to live. Local culture, history and wisdom can add to sustainability, and their best aspects should be preserved. Such preservation will often take conscious action by governments to encourage traditional crafts, languages, rituals, cultural practices, and building techniques; to protect important local products from mass-produced imports; to protect local farmland and resource stocks; and to integrate vernacular architecture and materials into local development.

HOW CAN SUSTAINABLE AND LIVABLE CITIES BE CREATED?

How can sustainability goals be put into practice for cities and towns? On the one hand, progress will depend on sustainability themes diffusing into all existing planning and development processes. On the other hand, specific planning efforts and changes to planning procedures may be necessary as well. Sustainability-oriented planning processes could focus on particular urban issues or problems, such as air quality or watershed management (as long as these are undertaken with an awareness of other related issues). Or city leaders and public participants could take a more comprehensive look at the sustainability of a city or region. Cities such as San Francisco, Seattle, Santa Monica, and Leicester, England, have taken this latter approach.

In this age of entrenched economic and political forces opposing sustainability, no single planning effort is going to set cities on a path towards a healthy long-term future. Rather, the need is for a long-term strategy emphasizing consensus processes, public education, political organizing, policy tools such as indicators and performance standards, development of vision documents and "best practices" examples, and the creation of institutions that can more effectively address physical planning and equity issues. Together, such efforts can develop the knowledge, political will, and institutional capacity to bring about change.

A systematic planning approach to promote urban sustainability might first seek to get a wide range of parties involved in efforts to improve the long-term health of a particular city or region. Participants would then attempt to reach agreement on particular values and goals that might move the community in more sustainable directions. Vision statements and review of "best practices" examples worldwide are often helpful in giving stakeholders ideas about the range of possible approaches. Performance standards and sustainability indicators could be developed to help measure whether or not the community is making progress toward long-term goals, and to allow policies and programs to be revised to better achieve their objectives.

The most difficult challenge comes in implementing sustainability visions, policies, and programs, and in modifying institutions so as to be able to do this. This process depends on effective political organizing, and on the development of a coalition of interests supporting common objectives around sustainability and quality of life. Participants should expect the process to be a long one. Most cities and towns contain entrenched political and economic forces with an interest in continuing unsustainable patterns of development, and so progress will be slow.

In the long run, sustainable development will require systemic cultural change that builds democracy and social capital (accumulations of trust and cooperation between people). The problems created by concentrations of economic power must be addressed, as well as the tendency of capitalist systems to reinforce values oriented toward short-term private profit rather than long-term social or ecological well-being.

Ultimately, moving towards sustainable cities will require a different mix of values than dominates urban development at present. There will still be opportunities for individuals to make a profit. However, the emphasis must be instead on caring deeply about the community, the region, and each other. Each development or planning decision must be evaluated in terms of its effects on the health of human and ecological communities. In *A New Theory of Urban Design* (1987), architect Christopher Alexander sets forth a single overriding rule of city development: "Every increment of construction must be made in such a way as to heal the city." This is not a bad guideline for sustainable development planning.

CONCLUSION

Planning for urban sustainability is still in the very early stages. As of now little progress has been made on turning today's huge, resource-consumptive metropolises and sprawling, even more resource-intensive suburbs into sustainable communities. But the seeds are being laid for future change, in terms of emerging consensus on what more sustainable and livable cities would be like, and on some processes and institutions that can help implement these planning directions.

Different approaches will have to be found for different cities. The Third World's rapidly growing megacities face the challenge of providing basic water, sewer, utility, and transportation services for their residents in ways that will be sustainable in the long run and that will avoid some of the problems that industrialized cities in the "developed" world have gotten into. First World cities face a different task: redeveloping urban areas that have plenty of infrastructure but fail to provide an ecologically or socially healthy urban environment. Sometimes the needs of cities in different parts of the world will be exactly opposite. For example, planners in the U.S. often seek higher residential densities to support transit, conserve open land, and promote community interaction. Yet in high-rise Hong Kong – the world's densest city – efficient land use is not an issue, and densities may fall without affecting sustainability.

It is important to remember that the current city is very recent. Its form and environment are heavily determined by technological innovations such as the automobile and the elevator which have only existed since the late nineteenth century. The creation of megacities of more than ten million people is an even more recent phenomenon, occurring only in the latter part of the twentieth century. Just as recent patterns of suburban development are now layered on top of nineteenth-century streetcar grids and eighteenth-century walking cities, so new and more ecological patterns of development may someday be layered on top of these, gradually bringing cities back into a better balance with the ecological limits of regions and the planet as a whole.

Twentieth-century suburbanization was in large part a reaction against the dirty, crowded, unhealthy cities of the industrial revolution, in which people were crammed into terrible housing virtually without amenities, to serve the needs of ruthless early forms of industrial capitalism. In a similar manner sustainable city initiatives of the next century may form a reaction against the excesses of twentieth-century culture, which is dominated by economic rather than environmental or social values. The transition toward more sustainable cities will not happen overnight. But through a growing ecological and social consciousness, the development of innovative models and examples, and better understandings of the policies, programs and designs appropriate to urban sustainability, new, more sustainable forms of urban development can come about.

RECOMMENDED READINGS

Beatley, Timothy and Kristy Manning, *The Ecology of Place: Planning for Environment, Economy, and Community*, Island Press, Washington, D.C., 1997.

Braidotti, Rosi *et al.*, *Women, the Environment and Sustainable Development: Towards a Theoretical Synthesis*, Zed Books, London, 1994.

Brown, Lester R., *Building a Sustainable Society*, Norton, New York, 1981.

Calthorpe, Peter, *The Next American Metropolis*, Princeton Architectural Press, Princeton, 1993.

Elkin, Tim and Duncan McLaren, with Mayer Hillman, *Reviving the City: Toward Sustainable Urban Development*, Friends of the Earth, London, 1990.

Girardet, Herbert, *The GAIA Atlas of Cities: New Directions in Sustainable Urban Living*, Anchor Books/Doubleday, New York, 1992.

Haughton, Graham and Colin Hunter, *Sustainable Cities*, Regional Studies Association, London, 1994.

Hawken, Paul, *The Ecology of Commerce*, HarperCollins, New York, 1993.

Holmberg, Johan, ed., *Making Development Sustainable: Redefining Institutions, Policy, and Economics*, Island Press, Washington, D.C., 1992.

Kelbaugh, Douglas, *Common Place: Toward Neighborhood and Regional Design*, University of Washington Press, Seattle, 1997.

Local Government Commission, *Land Use Strategies for More Livable Places*, Sacramento, 1992.

Lowe, Marcia, *Shaping Cities: The Environmental and Human Dimensions*, Worldwatch Paper 105, The Worldwatch Institute, Washington, DC, 1991.

Lyle, John Tillman, *Regenerative Design for Sustainable Development*, John Wiley & Sons, New York, 1994.

Maclaren, Virginia W., "Urban Sustainability Reporting," *Journal of the American Planning Association*, Vol. 62, No. 2, Spring 1996, 184–202.

Mitlin, Diana, "Sustainable Development: A Guide to the Literature," *Environment and Urbanization*, Vol. 4, No. 1, April 1992.

Norgaard, Richard B., *Development Betrayed: The End of Progress and a Coevolutionary Revisioning of the Future*, Routledge, New York, 1994.

Pearce, David, Edward Barbier, and Anil Markandya, *Blueprint for a Green Economy*, Earthscan Publications, London, 1989.

President's Council on Sustainable Development, *Sustainable America*, Government Printing Office, Washington, D.C. 1996.

Rees, William, *Our Ecological Footprint: Reducing Human Impact on Earth*, New Society Publishers, Philadelphia, 1996

Stren, Richard, Rodney White, and Joseph Whitney, eds., *Sustainable Cities: Urbanization and the Environment in International Perspective*, Westview Press, Boulder, CO, 1992.

United Nations, *Agenda 21: Program of Action for Sustainable Development*, New York, 1992.

Van der Ryn, Sim and Stuart Cowan, *Ecological Design*, Island Press, Washington, D.C., 1995.

World Commission on Environment and Development, *Our Common Future*, Oxford University Press, New York, 1987.

Zuckerman, Wolfgang, *End of the Road: The World Car Crisis and How We Can Solve It*, Chelsea Green Publishing, Post Mills, VT, 1991.

EIGHT

"The Teleserviced City"

from *E-topia: Urban Life, Jim –*
But Not as we Know it (2001)

William J. Mitchell

Editors' Introduction

Population growth, environmental constraints, and new communications technologies are widely seen as the key determinants of urban futures. But according to William J. Mitchell, the city of the future has already arrived. Many elements of the "e-topia" that he predicts will become the new urban paradigm based on new ways of organizing urban space, urban communities, and urban personalities that flow from the demands of digitalized electronic communications technologies have already become accepted parts of everyday life. ATMs, Mitchell points out, have largely replaced banks with human tellers, we order everything from books to clothing from online companies like Amazon.com and LandsEnd, and even going to the doctor has become intermixed with the delivery of digitalized information through electronic channels. Indeed, the worldwide digital network has become a "new form of urban infrastructure – one that will change the forms of our cities as dramatically as railroads, highways, electric power supply, and telephone networks did in the past."

William Mitchell studied at the University of Melbourne, Yale, and Cambridge and has taught at Yale, Cambridge, Carnegie-Mellon, the University of Virginia, the UCLA Graduate School of Architecture and Urban Planning, and the Harvard Graduate School of Design. He is currently Professor of Architecture and Media Arts and Head of the Program in Media Arts and Sciences at the Massachusetts Institute of Technology. For years he has studied the relationship between architecture and the built environment and the new technologies of mass communications, including photography, cinema, and the new telecommunications technologies.

Mitchell argues that virtual urban places have become as important as physical ones. Cities of the future, he writes, will be "characterized by live/work dwellings, twenty-four-hour pedestrian-scale neighborhoods rich in social relationships, and vigorous community life, complemented by far-flung configurations of electronic meeting places and decentralized production, marketing, and distribution systems."

In *E-topia*, Mitchell explores all aspects of communal virtuality in ways that illuminate both the present and the future. Dispassionately, without assuming an advocacy position either for or against the new technological realities, he details how computers have already revolutionized – and will continue to revolutionize – the way we live in our homes, entertain ourselves, work, consume, define ourselves, and interact both with neighbors and the natural environment. Personal identity, he writes, is destined to become "an electronically metered commodity." And although many may want to opt out of "the vast electronic Panoptikon" that the world of online space has become, the changes already made are changes that are here to stay.

Mitchell's e-topia invites immediate comparison with the other writings in this section, especially Manuel Castells's "informational cities" (p. 478) and Melvin Webber's 1968 predictions of technology-driven "post-urbanism" (p. 473). Mitchell's vision of virtual community may also be seen as an important ingredient to fulfilling the dream of Frank Lloyd Wright's Broadacre (p. 331) as a community of perfect individual autonomy and perfect social interconnectedness. Also of interest is Joel Kotkin, *The New Geography: How the Digital Revolution is*

Reshaping the American Landscape (New York: Random House, 2000). The definitive collection of writings on the social implications of digital telecommunications is Stephen Graham (ed.), *The Cybercities Reader* (London and New York: Routledge, 2004).

Mitchell is extraordinarily prolific, and his writings easily navigate between communications theory, philosophy, and social psychology. His earlier works include the seminal *City of Bits: Space, Place and the Infobahn* (Cambridge: MIT Press, 1996), *The Reconfigured Eye: Visual Truth in the Post-Photographic Era* (Cambridge: MIT Press, 1994), *The Logic of Architecture: Design, Computation, and Cognition* (Cambridge: MIT Press, 1990), and, with Donals Schon and Bish Sayal, *High Technology and Low-Income Communities* (Cambridge: MIT Press, 1999). More recently he has published *Me++: The Cyborg Self and the Networked City* (Cambridge: MIT Press, 2003) and *Placing Words: Symbols, Space, and the City* (Cambridge: MIT Press, 2005).

In ancient Rome you got better military protection and better spectacles than in the provinces. In Manhattan you get better medical care, restaurants, and haircuts than in a cow town. As everyone knows, availability of high-quality services is a major attraction of urban areas.

In the emergent computer-networked world, though, this reduces to a half-truism. Some services still depend on the local presence of the providers, but others can effectively be summoned and delivered from a distance. As a result, new patterns of service distribution are superimposing themselves on cities and rapidly displacing some older ones.

A TYPOLOGY OF SERVICE SYSTEMS

Service systems consist, in their barest and most obvious essentials, of service providers, service consumers, and effective means to connect the two. The various possible patterns of connection define an elementary typology of service systems. And the effects of digital telecommunications vary by type.

Before telecommunications, the privileged surrounded themselves by servants or slaves and summoned them verbally as required. Members of large service staffs were classified and named according to their roles – butlers, chambermaids, cooks, chauffeurs, gamekeepers, viziers, broom-wallahs, personal trainers, scribes, amahs, corporate lawyers, whatever. It mostly depended on maintaining close physical proximity, even when primitive systems of bells and buzzers began to augment direct verbal communication. And it reflected itself architecturally in provision of servants' quarters, janitors' cubbyholes, service stairs, outer offices, and so on.

As large modern cities grew, so did an alternative system of centralized service points – particularly for sophisticated, specialized services. These allowed economies of scale and could serve large populations at relatively low cost, but service consumers had to travel to them. Medical care, education, and many commercial services followed this pattern, and associated building types such as modern hospitals and schools emerged as a result.

One way to resolve the inconvenient contradiction between gaining economies of scale through centralization and remaining close to consumers through decentralization was to develop distributed systems of branches. Thus, for example, in the nineteenth and earlier twentieth centuries, large banking organizations had head offices in prominent downtown locations, back offices for centralized processing activities in low-rent suburban space, and huge numbers of branches to provide customer service in local communities. Retailing followed a similar pattern. The general result was that Main Streets, commercial strips, and shopping malls became collections of branches and franchises. And the downtown office towers in all but the largest global cities mostly contained branches of national and international organizations.

Yet another strategy was to service distributed populations by means of mobile providers. This has its roots in the ancient tradition of itinerant healers, teachers, peddlers, and cops on the beat. The disadvantages are that the mobile providers have to carry their tools of trade with them, and that it is hard to achieve economies of scale in this way.

Finally, in efforts to maximize the advantages and minimize the disadvantages of these basic patterns, all manner of hybrids developed. A large, central medical facility might be combined with a system of local clinics, mobile care units, and home care providers. A retailer might make use both of downtown showrooms and traveling salesmen.

SUMMONING ASSISTANCE

In the nineteenth century, early telecommunications technology was quickly adapted to the task of summoning mobile service providers from central locations as required. This speeded response times, and so made centralized services far more effective.

In 1852, for example, Boston began constructing a system of telegraphic call boxes connected to fire stations, and other cities soon followed. Together with the replacement of hand-drawn firefighting apparatus by horse-drawn fire engines, then by motorized fire trucks, this allowed stations to serve wider areas and larger populations.

Successive waves of telecommunication and transportation technology extended and elaborated the idea. By the 1880s, telephones were being installed in police stations, and police forces eventually began to make combined use of the telephone, the two-way radio, and patrol cars to service extensive areas. From 1928, the Royal Flying Doctor Service began providing medical care to the vast, sparsely populated Australian outback through use of light aircraft summoned by means of pedal-powered Morse code radio transmitter-receivers. Now, in the era of cellular telephones and pagers, service providers of every kind – from drain cleaners to brain surgeons – can continually be on call.

KEEPING TABS

In all of these systems, somebody still has to take the action of calling the cops, the medics, the firefighters, the plumbers, or the caterers. But by adding sensors to remote summoning capabilities, the tasks of monitoring need and summoning services as required can be automated.

Thus it is now routine to install fire and smoke detectors in buildings; these not only sound a local alarm, but in many cases they automatically call firefighters to the scene. Burglar alarms that sense opening of doors, breakage of windows, or motion within interior spaces function in much the same way. Continuous electronic monitoring, based on embedded

sensors, is beginning to revolutionize the maintenance of structures such as bridges and dams. In industry, it has long been commonplace to embed sensors in plant and machinery to signal malfunctions, and in a ubiquitously networked world this idea will increasingly be extended to automobiles and to domestic appliances of all kinds.

Consider automobile and truck tires, for example. Traditionally, it has been up to drivers and service mechanics to check the pressure manually, and to adjust it as required; neglect results in poor performance and excessive wear. A smart vehicle might perform these routine service tasks itself, making use of onboard pressure monitors, computers, and controllable pumps and valves to maintain constant pressure. But even smarter logging trucks in Alaska and British Columbia now link their onboard computers, via satellite, to geographic information systems and weather information systems and dynamically adjust tire pressures to current conditions. Overkill? Not if the performance payoff is there.

What works for structures and machines can also work for our own bodies. We are also likely to see proliferation of sophisticated medical monitoring devices hooked up to health care providers; once available only in hospital beds, they will increasingly take the form of unobtrusive wearable devices and of systems that provide continuous monitoring in the homes of those who require it.

[. . .]

SURVEILLANCE AND SECLUSION

All this, of course, superimposes yet another layer of electronically mediated social relationships on daily life. Wherever such electronic monitoring is carried out, it adds what are sometimes called quaternary social relationships – those that exist between observed and anonymous observer – to our primary, secondary, and tertiary relationships. And as vigilant civil libertarians have been quick to point out, we could well end up imprisoning ourselves in a vast electronic Panopticon.

More subtly, we will increasingly face tradeoffs between maintaining our privacy and getting better service by giving some of it up. If an online bookstore or CD store keeps track of your purchases, for example, it can automatically compare these with the purchases of other customers and use these comparisons to tell you what customers with similar interests to yours have been buying. It is a very effective collaborative filtering and recommendation mechanism, and it adds considerable value to the bookseller's service – but you might want to opt out of it if you discovered that the purchase profiles were also being sold to direct-mail marketeers. And you might be very upset if you discovered that nosy journalists were poking around in them.

What if you were checking into a fancy hotel? If the hotel could electronically access a detailed profile of your needs and preferences, it could organize the space and the menu to your liking. But is it worth it? Would you want to reveal so much in order to gain this benefit? Then, what about checking into a hospital? Does the greater criticality of the situation change things? Would you be prepared to reveal much more about yourself if it made a significant difference in the quality of your health care?

[. . .]

Under the most pessimistic scenarios, the mechanisms will inevitably fail, the powerful will always get whatever information they want, and we will end up with no privacy left. Under more optimistic ones, we will find effective ways to treat identity as an electronically metered commodity. We will turn it up or down, depending on the context.

DELIVERY AT A DISTANCE

Remote monitoring and summoning significantly change service systems – particularly medical care and emergency services – but it is remote delivery that really makes a difference. If you can pump a service out through a network, then you can extend the service area to wherever that network reaches – potentially globally. This creates large service markets, promises greater distributional equity, and is particularly good news for inhabitants of remote and underdeveloped areas, and for those who are immobilized by age or infirmity. Furthermore, the service agent on the end of the line can often become an unsleeping piece of software rather than a human operator.

In the most elementary case, as with audio and video entertainment, news, and some educational services, delivery reduces to transmission and display of an information stream. It may be synchronous, as with radio and television broadcasts, or it may be asynchronous, as with Web news servers. Either way, the network simply provides one-way pipes; the logic is little different from that of water supply systems.

With two-way telecommunication, remote delivery becomes an attractive option for service businesses that pursue the strategy of informing clients about their options, advising them on their choices, then eventually executing a purchase transaction of some kind. It works even in cases where the product or service purchased is eventually delivered in an utterly conventional way.

Travel is typical. Once, you had to go to a railway station, steamship office, or local travel agency to get information and advice and to purchase your tickets. Then, when the telephone came along, you could get much the same service by dialing in; airlines and other transportation companies began to rely on their call center operations, and travel agents began to spend most of their time on the phone. More recently, interactive Web sites have offered a third alternative; you can use them to consult comprehensive online databases of travel information, to conduct sophisticated searches to find flights and rates that meet your requirements, then immediately to make reservations and ticket purchases by executing online transactions. This has made it tougher for travel agents to survive (as they traditionally have) on commissions from ticket sales, and has forced them to compete on the quality of the information and advice that they can offer.

Some areas of retailing are going the same way. Online book and CD stores such as Amazon.com do not just offer convenience and twenty-four/seven service; they also compete with traditional bookstores by offering increasingly sophisticated information and advice. Their catalogs are extensive and detailed, contain summaries, reviews, and cross references, and can be searched in a variety of ways . . .

Banks and financial services vendors have also been affected dramatically. Deposits, withdrawals, and balance queries have become high-volume, low-price commodity transactions; they are increasingly handled remotely and automatically by ATM machines and electronic home banking systems, rather than by tellers behind counters as in the past. Bills get paid online rather than through the mail. And an increasing number of investors use inexpensive online trading sites instead of calling live brokers.

[. . .]

The scoffers who think these online services are little different from mail-supported and phone-supported services of the past, and that nothing can replace face-to-face interaction with a knowledgeable human specialist, are just not getting it. It is the ubiquity and speed of electronic delivery capabilities, combined with the possibility of effectively integrating electronic intelligence, that makes the crucial difference . . .

EXPANDING THE WEB OF INDIRECT RELATIONSHIPS

The general social effect of these new sorts of tele-service systems is to eliminate familiar middlemen. They are replaced by electronic systems and software.

Instead of going to a branch bank and meeting with a teller – perhaps someone you have come to know through regular contact – you interact with a faceless ATM machine or an electronic home banking system. Instead of buying a theater ticket at the box office (or from a scalper), you surf into a Web site, select a seat from an onscreen plan, and prepay by credit card. Rather than visit your friendly local retailer, you run searches on catalogs and click "order" buttons. Where you once had to line up at the Registry of Motor Vehicles to renew your driver's license, you now perform that task online.

So indirect, anonymous, electronically enabled relationships are proliferating in our daily lives, while certain kinds of face-to-face transactions (and the secondary social relationships with familiar interme-diaries that these have fostered) are correspondingly being reduced. Society as a whole is becoming more and more dependent on a vast, complex web of automated, electronic intermediation – our new, all-purpose go-between. The reductions in transaction costs, and the gains in market efficiency, are potentially enormous; not surprisingly, Bill Gates has written lip-smackingly of a coming era of "friction-free capitalism."

Many people, understandably enough, fear reduc-tion to a bitsucking subspecies of homo economicus, and the attendant loss of human contacts and rela-tionships. But let's ask the hard questions. What are these particular social relationships really worth? And what will replace them? For my part, I can well live without the human contact that I used to get from the bored and overworked clerks at the RMV, and I can put the time that I spent in line to much better use. And I doubt that I am alone in this.

[. . .]

TELEROBOTICS

All this applies to services that can successfully be disembodied. But what about those that traditionally have required not just exchange of information but also human hands, right there, on the spot? Can you get your car fixed remotely?

Well, you can certainly get your computer fixed – under some circumstances, at least. If you let a skilled technician log in remotely, you may be able to get software problems resolved without a site visit or taking the machine to a service center; large-scale networks would, in fact, be very difficult to maintain without this sort of remote servicing. As more and more machines get embedded software and network connections, they will also be serviced in this fashion. If they cannot be fixed directly, their problems will be diagnosed remotely, and technicians dispatched with appropriate parts and tools.

Where this strategy does not suffice, telerobots may, at least in principle, be brought in to do the job. Telerobots are remotely controlled machines that are capable of performing various physical tasks. They may be fixed in place, like industrial robots, or they may be mobile, like delivery vehicles. They may be wired up to telecommunications networks, or they may be linked wirelessly. Their every move may explicitly be controlled, or they may be equipped with some autonomous decision-making capability.

[. . .]

In general, telerobotics sounds complicated, expen-sive, and a bit perverse. Indeed it often is, but it can make practical sense under conditions where distances and travel costs are very high, where services must be delivered to dangerous localities, or where demand is widespread but skilled providers are confined to a few locations. Consider specialized surgery, for example. No doubt it normally makes eminent sense to have the surgeon in the room, with the patient. But is it necessarily better to transport a sick patient or busy surgeon over very long distances if a combination of telerobotics and digital imaging, delivered via a smart surgical suite, might substitute? What about battlefield or disaster situations, where surgeons may be too precious to put at risk on the front lines? And what if the demand for some specialized procedure is widely scattered around the world, but the necessary skills are only available in a couple of major centers? The need to provide service under these sorts of conditions has motivated intensive research into the possibility of telesurgery, and the development of some impressive prototype systems.

So, yes, you can sometimes use telerobotics to be in touch – literally in touch – with distant service providers. But don't get too excited about robotic limbs, tactile feedback devices, and shaking the booty electric. Not yet, anyway.

THE TELESERVICE PARADOX

The limitations of telerobotics are instructive. Whatever the successes of new teleservice systems, it remains true that some services – including many of the most humble – still depend on local presence of the providers. Teleworkers need to get their dry cleaning done, and don't want to go too far for it. Electronic traders may operate globally, but the janitors who empty their wastebaskets and vacuum their floors have to be right there on the spot . . . Chefs have to get the food to the table while it's hot. Telehairdressing and teledentistry seem very far off in the future. Add it all up and you can quickly see that the goods and services produced within a city for local consumption – as distinct from what regional economists term the "export base" – are likely to remain a very significant percentage of the total.

Thus concentrations of population and economic activity, once established, still have some powerful glue holding them together. Digital delivery of educational, entertainment, medical, retail, financial, and other services can be expected to produce new patterns of service availability within and among these concentrations, but it will certainly not dissolve them. Indeed, something of a paradox emerges; hot spots of electronically mediated activity – such as Manhattan's Financial District, the City of London, or the teleworking-millionaire enclave of Aspen – become magnets for the low-wage service workers who do the sorts of things that computers and electronically controlled machinery cannot. And, of course, these concentrations of service workers become part of the attraction of such localities for the more privileged. It is a dirty little semisecret, then, that all of these high-flying places have large, far less interesting and attractive, low-rent counterparts somewhere nearby.

On this one, though, the fat lady is still clearing her vocal chords. Paul Krugman has suggested – and he is probably right – that the wrong side of the tracks will eventually have its revenge. As networks spread,

smart places proliferate, and software becomes more and more capable, the prices of information-related services will be driven down. At the same time, the value of manually performed services that cannot readily be automated or remotely delivered will correspondingly rise. Cooks, gardeners, nannies, and plumbers will do increasingly well.

[. . .]

ELECTRONIC FRONTS, ARCHITECTURAL BACKS

Architecturally, the most striking consequence of teleservice is transformation of the traditional relationship between facade and back room. Many organizations are beginning to acquire electronic fronts and architectural backs.

Think, for example, of a retail store on an old-fashioned shopping street. The storefront presents the enterprise to the public, and the floor space immediately behind is where customers browse the stock, interact with sales staff, and make purchases. Behind that again is backroom stock and administration space which is not open to the public. Still further in the background, perhaps, is remote warehouse and head office space. Overall, there is a very clear hierarchy of visibility and public presence.

In the electronic equivalent of this store, though, the online interface takes over the functions of the street facade, the signage and display windows, and the retail floor; the software carries the full burden of mediating the organization's interactions with its customers. The backroom space remains; the need to store stock and accommodate administrative staff does not disappear. Locational constraints are loosened, however, and this backroom space can freely be distributed in whatever new patterns make practical sense. Furthermore, the buildings that provide this backroom space do not necessarily need to be in prominent, high-rent urban locations, nor do they need to perform representational roles. They can be remote and anonymous.

[. . .]

Not surprisingly, the character and distribution of electronically administered backroom space varies according to the natures of the products and services that organizations provide. Highly perishable items that require very quick delivery, like hot food, need backroom space distributed throughout service areas;

you cannot have a national pizza delivery center! Online supermarkets need warehouses located to support same-day delivery within metropolitan areas. Online bookstores, which rely on air and truck delivery, need large concentrations of backroom space at national and international transportation hubs. And financial services organizations, which do not deliver anything physically, can locate wherever rents and available labor forces make it attractive. Where back-room workers do not handle physical items, they can even be located in highly distributed telecommuter workspaces that have no spatial connection to customers whatsoever.

At the same time, an organization's electronic front changes the style and granularity of its public repre-sentation. Once, for example, banks were represented by branches located on Main Streets. Now, they are represented by much larger numbers of ATM machines, minibranches, and electronic home banking screens scattered in a very different, much more diffuse pattern.

SERVED AND SERVING SPACES REVISITED

Most importantly, though, teleservice demands a new way of thinking about the organization of architectural space at both building and urban scales.

Back in the 1960s, Louis Kahn drew an influential distinction between the served and serving spaces of a building. Served spaces were the sites of important human activities, while serving spaces accommodated the support activities and equipment that the served spaces required. Thus a laboratory floor might be a served space, with adjacent plant rooms and exhaust shafts as serving spaces.

Since then, network technologists have learned to think in similar ways – and even to reinvent similar terminology. The Web, and comparable network struc-tures, consist of client sites and server sites. Your home office might be a client site, for example, with your employer's intranet server supporting it.

Today, in the digital network era, the two traditions are beginning to converge. You might still relate smart served and serving spaces in the traditional way, by making them adjacent, but you might also establish their functional linkage through distant electronic connection. You can still partially read the functional organization of architectural space from floor plans and land use maps, but you must now look, as well, at the networking and the software.

"The Urban Future"

from *The City: A Global History* (2005)

Joel Kotkin

Editors' Introduction

Lewis Mumford's *The City in History* (1961) ran more than 600 pages, with over fifty pages of scholarly bibliography. And Sir Peter Hall's *Cities in Civilization* (1998) weighed in at over 1100 pages. So when Joel Kotkin published *The City: A Global History* in 2005, eyebrows were raised. How could anyone, urbanists wondered, cover the entire complex and multi-detailed history of the world's urban civilization in only 160 pages?

What Kotkin *can* do in 160 pages – and what he does very effectively – is give a dramatic overview of "the universality of the human experience" and identify the major phases and broad themes that have characterized urban life over the past five or six thousand years. One also gets a sense of the effectiveness of Kotkin's text as a "global" history. The book's brevity invites the reader to finish it in one or two sittings, and Kotkin is therefore able to juxtapose the major trends of urban development in Europe and Asia, through the classical and modern periods, without the feeling of abruptly jumping back and forth or carrying on unconnected parallel narratives. The result is remarkably cogent and level-headed.

Like Mumford, Kotkin recognizes that cities were spiritual, or sacred, places before they were centers for economic exchange or even political power, although these three urban characteristics – the spiritual, the "animating role of commerce," and the "ability to provide security and project power" – constitute the "critical factors" of the urban experience. As Kotkin surveys the history of cities in the Near East, in Asia, in Europe, in the "Islamic Archipelago" stretching from Morocco to India, and in the Americas, it is the health of all of these three factors working together that accounts for the rise of urban civilizations and the weakening of these functions that leads to decline and fall.

In his final chapter, "The Urban Future," Kotkin begins with what he calls "the crisis of the megacity," especially in the developing world, the globalization of urban functions, and the rise of "ephemeral cities" – Richard Florida would call them "cities of the creative class" (p. 129) – that thrive as providers of arts, ideas, and tourist experiences rather than as producers of material goods or even essential services. He argues that modern telecommunications will indeed contribute to decentralization and the "destruction of space" – taking a position sometimes similar, sometimes at odds with those of Melvin Webber (p. 473), Manuel Castells (p. 478) and William Mitchell (p. 510). There are dark clouds in Kotkin's view of the urban future: basic security, whether from local crime or international terrorism, has become tenuous for many urban residents; and environmental pollution leaves many others without access to clean air and water. But in the end, it is the sacred functions of the city that will prevail. Cities will always generate "a consciousness that unites their people in a shared identity" and be the places where "humanity's future will be shaped for centuries to come."

A former business journalist and broadcaster, Joel Kotkin is an Irvine Senior Fellow at the New America Foundation and a senior advisor to The Planning Center, a planning and environmental design consulting firm. He is the author of *California, Inc* (New York: Crown, 1982), *The Third Century: America's Resurgence in the Asian Era* (New York: Crown, 1988), *Tribes: How Race, Religion, and Identity Determine Success in the New Global*

Economy (New York: Random House, 1994), and *The New Geography: How the Digital Revolution Is Reshaping the American Landscape* (New York: Random House, 2000).

For a cultural explanation of the origins of contemporary terrorism, see Ian Buruma and Avishai Margalit, "The Occidental City" (p. 136). For more on the city as sacred space, see two books cited by Kotkin himself: Mircea Eliade, *The Myth of the Eternal Return* (Princeton: Princeton University Press, 1971) and Jacques Ellul, *The Meaning of the City* (New York: Vintage, 1967).

■ ■ ■ ■ ■ ■

The process of ascent and decline of cities is both rooted in history and changed by it. Successful urban areas today must still resonate with the ancient fundamentals – places sacred, safe, and busy. This was true five thousand years ago, when cities represented a tiny portion of humanity, and in this century, the first in which the majority live in cities.

The world's urban population, only 750 million in 1960, grew to 3 billion by 2002 and is expected to surpass 5 billion in 2030. These swelling ranks of city dwellers face a vastly changed environment in which even the most powerful urban area must compete not only with other large places, but with an ever wider array of smaller cities, suburbs, and towns.

THE CRISIS OF THE MEGACITY

These shifts will be felt most acutely among the sprawling megacities of the developing world. In the past, size allowed cities to dominate the economies of their hinterlands. Today, the very girth of the most populous megacities – Mexico City, Cairo, Lagos, Mumbai, Kolkata, São Paulo, Jakarta, Manila – is often more a burden than an advantage.

In some places, these urban giants have been losing out to smaller, better-managed, and less socially beleaguered settlements. In East Asia, the critical nursery of twenty-first-century urbanism, Singapore and, to a lesser extent, Kuala Lumpur have integrated themselves into the global economy more successfully than the far more populous Bangkok, Jakarta, and Manila.

Similarly, bloated size has, as one observer noted, "robbed Mexico City of its economic logic." Burdened by crime, congestion, and pollution, *La Capital* is often bypassed by entrepreneurs and ambitious workers for faster-growing, better-run cities such as Chilango, Guadalajara, and Monterrey, or across the border to urban areas of *el norte* itself.

In the Near East, megacities like Cairo and Tehran have suffered to keep pace with their exploding populations, while smaller, more compact centers such as Dubai and Abu Dhabi have flourished. Dubai, a dusty settlement of twenty-five thousand in 1948, saw its population approach 1 million fifty years later, while avoiding the economic stagnation that has haunted most of the Arab world.

As in Dubai, cosmopolitan attitudes and the accumulation of unique skills continue to have a major impact in determining successful cities. Openness to varied cultures and the clever employment of talent helped relatively small cities such as Tyre, Florence, and Amsterdam play out-size roles in their times. Similarly, in the twenty-first century, a small cosmopolitan city such as a Luxembourg, Singapore, or Tel Aviv often wields more economic influence than a sprawling megagiant of 10 or even 15 million.

THE LIMITS OF THE CONTEMPORARY URBAN RENAISSANCE

As the twentieth century drew to a close, megacities in the advanced countries seemed to be enjoying brighter economic prospects. There was a statistically small but notable increase in residential development even in some long-abandoned downtowns. Many people now predicted that the most cosmopolitan "world cities" – London, New York, Chicago, Tokyo, and San Francisco – had indeed irrevocably "turned the corner." "Neither Western civilization nor Western cities," suggested one keen observer, the historian Peter Hall, "show any sign of decay."

This new optimism rested largely on the impact of global integration and the worldwide shift from a manufacturing to an information-based economy. Cities like New York, London, and Tokyo, argued the theorist Saskia Sassen, occupy "new geographies of centrality" that provide "the strategic sites for management of the global economy." Behind these giants, she identified a secondary list of global centers, including variously Los Angeles, Chicago, Frankfurt, Toronto, Sydney, Paris, Miami, and Hong Kong.

These cities clearly enjoyed far better prospects than the rapidly shrinking great industrial cities such as Manchester, Liverpool, Leipzig, Osaka, Turin, and Detroit, which increasingly suffered from technological obsolescence and competition from developing countries. "The 'things' a global city makes," Sassen suggested, "are services and financial goods." These products, it was widely assumed, needed the unique skills and capabilities that existed only within the "global city."

THE "DESTRUCTION OF DISTANCE"

Such an assessment may be replacing the excessive pessimism of the 1960s with a magnified sense of optimism. Even the most evolved "global cities" now find the advantages of scale diminished by the rise of new technologies that, in the words of the anthropologist Robert McC. Adams, have accomplished "an awesome technological destruction of distance."

The ability to process and transmit information globally, and across great expanses, undermines many traditional advantages enjoyed by established urban centers. Throughout the last third of the twentieth century, secular trends, particularly in the United States, pointed to a continued shift even of corporate headquarters to the suburbs and smaller cities. In 1969, only 11 percent of America's largest companies were headquartered in the suburbs; a quarter century later, roughly half had migrated to the periphery.

These developments contradict the notion that a handful of mega-cities exercised the ultimate "command and control" centers of the global economy. Many elite service and financial firms remained in the established centers, such as Boston, New York, or San Francisco, but the clients "calling the plays" were just as likely to be operating in distant suburbs of Seattle, Houston, and Atlanta, or from abroad.

Even high-end services, the supposed linchpin of "global city" economies, have continued to disperse toward the periphery or smaller cities. This was even more marked among firms in the largest generator of new growth, the entrepreneurial sector. Improvements in telecommunications promise to further flatten economic space in the future, with choice jobs able to

shift to exurbs and even small cities such as Fargo, Des Moines, and Sioux Falls. One result has been the shift in the very landscape of growth, with suburban office parks widely favored over gleaming high-rise towers.

The global securities industry, for example, once overwhelmingly concentrated in the financial districts of London and New York, has gradually shifted an ever larger share of their operations to their respective suburban rings, other smaller cities, and overseas. The headquarters might remain in a midtown high-rise, but more and more the jobs are located elsewhere.

One particularly striking case comes from the retail industry. For much of the twentieth century, New York dominated a great deal of the retailing world; by 2000, not one of the sector's twenty largest firms was located there. New York-based fashion designers, advertising executives, trade show organizers, and investment bankers continued to play an important supporting role. The most critical power-shaping global retail lay elsewhere, in firms such as Wal-Mart, which operated effectively from Bentonville, Arkansas.

[. . .]

THE RISE OF THE EPHEMERAL CITY

Under these circumstances, even the best-positioned urban areas face severe demographic and economic challenges. Many of the young people lured to these cities in their twenties often depart when they start families and businesses. Increasingly, upwardly mobile immigrants, critical contributors to the urban resurgence, join the exodus. European, Japanese, and other East Asian urban centers face an even more extreme demographic crisis. Low birthrates are reducing the ranks of young people, the group most attracted to large cities, while choking off the traditional pool of immigrants from the countryside.

With economic growth, even in high-end services, shifting elsewhere, many leading cities in the advanced world increasingly rest their future prospects on their role as cultural and entertainment centers. These cities may now be morphing, as H. G. Wells predicted a century earlier, from commanding centers of economic life toward a more ephemeral role as a "bazaar, a great gallery of shops and places of concourse and rendezvous."

Cities have played this staging role since their origins. Central squares: the areas around temples, cathedrals, and mosques, long provided a place for merchants to sell their wares. Natural theaters, cities provided the overwhelmingly rural populations around them with a host of novel experiences unavailable in the hinterland. Rome, the first megacity, developed these functions to an unprecedented level. It boasted both the first giant shopping mall, the multistory *Mercatus Traini*, and the Colosseum, a place where urban entertainment grew monstrous in its size and nature.

In the industrial era, noted the French philosopher Jacques Ellul, "the techniques of amusement" became "more indispensable to make urban suffering bearable." By the twentieth century, industrialized mass entertainment – publishing, motion pictures, radio, and television – occupied an ever larger hold on the life of urban dwellers. These media-related businesses also accounted for a growing part of the economy in such key image-producing cities as Los Angeles, New York, Paris, London, Hong Kong, Tokyo, and Bombay.

By the early twenty-first century, this focus on the cultural industries began to inform economic policies in many urban areas. Instead of working to retain middle-class families and factory jobs or engage in economic competition with the periphery, urban regions placed increased focus on such ephemeral concepts as fashionability, "hipness," trend, and style as the keys to their survival.

In Rome, Paris, San Francisco, Miami, Montreal, and New York, tourism now stands as one of the largest and most promising industries. The economies of some of the fastest-growing centers, such as Las Vegas or Orlando, rely largely on the staging of "experiences," complete with uniquely eye-catching architecture and round-the-clock live entertainment.

Even in such unlikely places as Manchester, Montreal, and Detroit, political and business leaders hoped that by creating "cool cities," they might lure gays, bohemians, and young "creatives" to their towns. In some places, the accoutrements of this kind of growth – loft developments, good restaurants, clubs, museums, and a sizable, visible gay and single population – succeeded in reviving once desolate town centers, but hardly with anything remotely reminiscent of their past economic dynamism.

Cities in continental Europe – notably Paris, Vienna, and post-cold war Berlin – particularly embraced this reliance on a culturally based economy. Having largely failed to regain its status as a world business center, Berlin now celebrated its bohemian community as its primary economic asset. The city's relevance was increasingly defined not by the export of goods or services, but by its edgy galleries, unique shops, lively street life, and growing tourist trade.

THE FUTURE, AND LIMITS, OF GENTRIFICATION

In the twenty-first century, some cities or parts of cities may survive, and even thrive, on such an ephemeral basis and, with the support of their still dominant media industries, market that notion to the wider world. The brief but widely acclaimed rise of urban technology districts – such as New York's "Silicon Alley" or San Francisco's "Multimedia Gulch" during the dot-com boom of the late 1990s – even briefly led some to identify hipness and urban edginess as the primary catalyst for information-age growth.

Both of these districts ultimately shriveled as the Internet industry contracted and then matured, yet the market for new housing continued to grow. This demand came partly from younger professionals, but also from a growing population of older affluents, including those hoping to experience "a more pluralistic way of life." These modern-day nomads often reside part-time in cities, either to participate in its cultural life or to transact critical business. In some cities – Paris, for one – these occasional urban nomads constitute, by one estimate, one in ten residents.

The rush in many "global cities" to convert old warehouses, factories, and even office buildings into elegant residences suggests the gradual transformation of former urban economic centers into residential resorts. The declining old financial center of lower Manhattan seemed likely to revive not as a technology hub, noted the architectural historian Robert Bruegmann, but as a full- or part-time home for "wealthy cosmopolites wishing to enjoy urban amenities in the elegantly recycled shell of a former business center."

Over time, however, this form of culturally based growth may not be self-sustaining. In the past, achievement in the arts grew in the wake of economic or political dynamism. Athens first emerged as a bustling great mercantile center and military power before it astounded the world in other fields. The extraordinary cultural production of other great cities, from Alexandria and Kaifeng to Venice, Amsterdam, London, and, in the twentieth century, New York, rested upon a similar nexus between the aesthetic and the mundane.

Broader demographic trends also pose severe long-term questions for these cities. The decline in the urban middle-class family – a pattern seen in both the late Roman Empire and eighteenth-century Venice – deprives urban areas of a critical source for economic and social vitality. These problems will be particularly marked in Japan and Europe, where the numbers of young workers are already dropping. Superannuated Japanese cities face increasing difficulties competing with their Chinese counterparts, enriched by the migration of ambitious young families from their vast agricultural hinterlands.

Under these circumstances, it is difficult to imagine the continued dominance of the Italian fashion industry or Japan's preeminence in Asian popular culture as their populations of young people continue to decline. Over time, the economically ascendant cities around the world – Houston, Dallas, Phoenix, Shanghai, Beijing, Mumbai, or Bangalore – seem certain to generate their own aesthetically based industries.

Finally, the ephemeral city seems likely to face often profound social conflicts. An economy oriented to entertainment, tourism, and "creative" functions is ill suited to provide upward mobility for more than a small slice of its population. Focused largely on boosting culture and constructing spectacular buildings, urban governments may tend to neglect more mundane industries, basic education, or infrastructure. Following such a course, they are likely to evolve ever more into "dual cities," made up of a cosmopolitan elite and a large class of those, usually at low wages, who service their needs.

To avoid the pitfalls of an ephemeral future, cities must emphasize those basic elements long critical to the making of vital commercial places. A busy city must be more than a construct of diversions for essentially nomadic populations; it requires an engaged and committed citizenry with a long-term financial and familial stake in the metropolis. A successful city must be home not only to edgy clubs, museums, and restaurants, but also to specialized industries, small businesses, schools, and neighborhoods capable of regenerating themselves for the next generation.

[. . .]

THE SACRED PLACE

Far more important to the future of cities than constructing new buildings will be the value people place on the urban experience. Great structures or basic physical attributes – location along rivers, oceans, trade routes, attractive green space, or even freeway interchanges – can help start a great city, or aid in its growth, but cannot sustain its long-term success.

In the end, a great city relies on those things that engender for its citizens a peculiar and strong attachment, sentiments that separate one specific place from others. Urban areas, in the end, must be held together by a consciousness that unites their people in a shared identity. "The city is a state of mind," the great sociologist Robert Ezra Park observed, "a body of customs, and of unorganized attitudes and sentiments."

Whether in the traditional urban core or in the new pattern of development in the expanding periphery, such issues of identity and community still largely determine which places ultimately succeed. In this, city dwellers today struggle with many of the same issues faced by the originators of urbanity anywhere in the world.

Progenitors of a new kind of humanity, these earliest city dwellers found themselves confronting vastly different problems from those faced in prehistoric nomadic communities and agricultural villages. Urbanites had to learn how to coexist and interact with strangers from outside their clan or tribe. This required them to develop new ways to codify behavior, to determine what was commonly acceptable in family life, commerce, and social discourse.

In earliest times, the priesthood usually instructed on these matters. Deriving their authority from divinity, they were able to set the rules for the varied residents of a specific urban center. Rulers also gained stature by claiming their cities to be the special residences of the gods themselves; the sanctity of the city was tied to its role as the center for worship.

The great classical city almost everywhere was both suffused with religion and instructed by it. "Cities did not ask if the institutions which they adopted were useful," noted the classical historian Fustel de Coulanges. "These institutions were adopted because religion had wished it thus."

This sacred role has been too often ignored in contemporary discussions of the urban condition. It barely appears in many contemporary books about cities or in public discussions about their plight. This would have seemed odd not only to residents of the ancient, classical, or medieval cities, but also to many reformers in the late Victorian age.

"New urbanist" architects, planners, and developers, for example, often speak convincingly about the need for city green space, historical preservation, and environmental stewardship. Yet unlike the Victorian-era progressives, who shared similar concerns, they rarely refer to the need for a powerful moral vision to hold cities together.

Such shortcomings naturally reflect today's contemporary urban environment, with its emphasis on faddishness, stylistic issues, and the celebration of the individual over the family or stable community. The contemporary postmodernist perspective on cities, dominant in much of the academic literature, even more adamantly dismisses shared moral values as little more than the illusory aspects of what one German professor labeled "the Christian-bourgeois microcosmos."

Such nihilistic attitudes, if widely adopted, could prove as dangerous to the future of cities as the most hideous terroristic threats. Without a widely shared belief system, it would be exceedingly difficult to envision a viable urban future. Even in a postindustrial era, suggested Daniel Bell, the fate of cities still revolves around "a conception of public virtue" and the "classical questions of the polis."

Cities in the modern West, Bell understood, have depended on a broad adherence to classical and Enlightenment ideals – due process, freedom of belief, the basic rights of property – to incorporate diverse cultures and meet new economic challenges. Shattering these essential principles, whether in the name of the marketplace, multicultural separatism, or religious dogma, would render the contemporary city in the West helpless to meet the enormous challenges before it.

This is not to suggest that the West represents the only reasonable way to achieve an urban order. History abounds with models developed under explicit pagan, Muslim, Confucian, Buddhist, and Hindu auspices. The cosmopolitan city well predates the Enlightenment: It may have surfaced first in pagan Greek Alexandria, but it also flourished later in coastal China and India as well as throughout much of Dar al-Islam.

In our time, perhaps the most notable success in city building has occurred under neo-Confucianist belief systems, mixed with scientific rationalism imported from the West. This convergence, an amalgam of modernity and tradition, eventually overcame Maoism, which was intent on destroying all vestiges of China's cultural past. Today, it struggles both with the ill effects of unrestrained market capitalism and, particularly in China itself, with the self-interested corruption of a ruling authoritarian elite.

It is to be hoped that the Islamic world, having found Western values wanting, may find in its own glorious past – replete with cosmopolitan values and a belief in scientific progress – the means to salvage its troubled urban civilization. The ancient metropolis of Istanbul, with more than 9 million residents, has demonstrated at least the possibility of reconciling a fundamentally Muslim society with what one Turkish planner called "a culturally globalized face." The future success of such a cosmopolitan model, amid the assault from intolerant brands of Islam, could do much to preserve urban progress around the world in the new century.

Indeed, in an age of intense globalization, cities must manage to meld their moral orders with an ability to accommodate differing populations. In a successful city, even those who embrace other faiths, like *dhimmis* during the Islamic golden ages, must expect basic justice from authorities. Without such prospects, commerce inevitably declines, the pace of cultural and technological development slows, and cities devolve from dynamic places of human interaction into static, and ultimately doomed, congregations of future ruins.

Cities can thrive only by occupying a sacred place that both orders and inspires the complex natures of gathered masses of people. For five thousand years or more, the human attachment to cities has served as the primary forum for political and material progress. It is in the city, this ancient confluence of the sacred, safe, and busy, where humanity's future will be shaped for centuries to come.

Epilogue

Urban Studies and Planning (2006)

Richard T. LeGates

The prologue began by noting that studying cities is a vast and never-ending enterprise and that there is too much material for any one individual to master and always more to learn. *The City Reader*, fourth edition, only has space for fifty-seven selections. The book, section, and selection introductions place these readings in a wider context and the bibliographic material that accompanies each selection points readers to additional writings by the author of the selection and other writings related to the selection. Throughout this edition of *The City Reader* are references to additional readings in volumes in the ten-volume Routledge Urban Reader Series of which *The City Reader* is the anchor text. Many readers who have read the selections in *The City Reader* and the explanations of how these writings fit within the wider body of scholarship about cities will want to pursue some topics in greater detail. They are prepared to shift from the general to a more detailed roadmap.

Given how many different disciplines and professional fields contribute to understanding cities, this epilogue does not purport to recommend one right way to proceed further with the study of cities. It would be presumptuous for this epilogue to attempt to summarize all of the nearly 400 selections in the series. Rather, the epilogue provides a more detailed roadmap of fundamental issues and key writings related to cities. It contains references to many more important classic and contemporary writings on urban studies and planning than appear in *The City Reader* itself and in the bibliographic material that accompanies each selection. The selections in the Routledge Urban Reader Series were important sources in preparing this epilogue. Many of the selections cited are reprinted in one or another of the readers.

THE EVOLUTION OF CITIES

Part One of *The City Reader* deals with the evolution of cities. Historians, based in history departments within colleges of social science or humanities, have made the greatest contribution to understanding the evolution of cities. Archaeologists, geographers, political scientists, economists, architects, and other scholars have also contributed to an understanding of how cities evolved.

A fundamental question about the evolution of cities is what proportion of the population live in cities as opposed to rural areas or villages in different societies at different times in history and today. Demographer Kingsley Davis's selection on the urbanization of the human population (p. 17) provides an overview of this topic. Other demographers, archaeologists, and historians' studies of urbanization in different parts of the world at different times in human history provide a fuller (and sometimes conflicting) account of the course of urbanization in human history.

Tertius Chandler and Gerald Fox compiled a compendium of estimates of the population size of world cities from thousands of sources – reliable and not-so-reliable – and provide their own estimates of the population of cities worldwide for a period of three thousand years (Chandler and Fox, 1974). University of California, Berkeley historian Jan de Vries's study of European urbanization from 1500–1800 provides more credible estimates of the population of European cities at fifty-year intervals during these three centuries and an excellent account of the political, economic, and demographic changes that led to urbanization in early modern Europe (de Vries, 1984). French historian Fernand Braudel's history of the Mediterranean provides fascinating factual information on the rise of cities in the Mediterranean region from the Middle Ages through the early modern period (Braudel, 1996). Paul M. Hohenberg and Lynn Hollen Lees have written a succinct account of the rise of European cities from the Middle Ages through the middle of the twentieth century (Hohenberg and Lees, 1985). The United Nations provides annual estimates of the population size of 525 urban agglomerations with populations of 750,000 or more (United Nations, 2005). The World Bank provides annual estimates of what percentage of each country's population is urban (World Bank, 2005).

Beginning in 1790 in the United States and in 1800 in many European countries, national governments began compiling decennial censuses (US Census, 2002). Drawing on this census data, American social scientist Adna Ferrin Weber compiled a pioneering study of statistics on the growth of cities in the nineteenth century (Weber, 1899). Many other demographers and historians have used census data to enrich our understanding of the evolution of cities.

Enduring questions in urban studies are where, when, and why the first cities arose. Since the earliest cities in different regions of the world arose largely independently of other cultures and at widely different times, these questions are best framed as where, when, and why did the first cities *in different regions of the world* arise. Archaeologist V. Gordon Childe's account of the urban revolution in Mesopotamia (p. 27) was a pioneering contribution to the study of the origin of cities. Archaeologists have made the greatest contribution to our understanding of the origin of cities – often by excavations in the field digging up the remains of the most ancient cities themselves. But marshalling empirical evidence from the field to create theories about the evolution of cities, such as Childes' "urban revolution" theory, is also essential.

Most of the pioneering archaeological digs of ancient cities were undertaken by British archaeologists in the late nineteenth and early twentieth centuries. Today archaeolgists in China, India, Pakistan, Mexico, Guatemala, Peru, Zimbabwe, and elsewhere where early cities arose, as well as visiting archaeologists from England, the US and other countries, continue the study of the evolution of cities.

Based on excavations of the ancient Mesopotamian city of Ur in present day Iraq, Childe popularized a chronology of the first cities that has stood the test of time. His explanation of why the first cities arose based on deterministic, Marxist categories is largely discredited. Additional information on Mesopotamian cities is available from the man who actually dug up Ur – Leonard Woolley (Woolley, 1930).

British archaeologist Sir Mortimer Wheeler describes the ancient cities of Mohenjo-Daro and Harrapa in the Indus River Valley that arose about 4,500 years ago based on his excavations of these cities (Wheeler, 1966).

Another British archaeologist, James Mellaart, excavated the ancient settlement of Çatal Hüyük in Turkey (Mellaart, 1967). Mellaart concluded that trade in the hard, sharp obsidian found at Çatal Hüyük – not fertile soil – accounted for Çatal Hüyük's emergence. Meelaart believed that the residents of Çatal Hüyük raised crops from seeds they obtained in exchange for obsidian and diffused the most productive seeds to surrounding areas. That idea turns Childe's theory on its head. The idea that a very early proto-urban settlement led to innovation in agriculture appealed to Jane Jacobs, who popularized Mellaart's theory in *The Economy of Cities* (Jacobs, 1970).

Yet another British archaeologist, Dame Kathleen Kenyon, excavated the ancient settlement of Jericho in Israel. The earliest layers of Jericho proved older than Ur or Çatal Hüyük and enclosed a reasonably large settlement with massive stone walls. Accordingly Kenyon forcefully advocated for Jericho's role as the first city (Kenyon, 1979).

Each of these archaeologists prepared descriptions of what the earliest cities were like, devised their own definition of what population size, density, and other characteristics they considered sufficient to constitute a city, dated the earliest layer of the city and subsequent additions, and proposed a theory of why the city arose. Many current archaeological studies, cited in the introduction to the Childe selection, continue to report new findings on what early cities were like and to shed light (or confusion!) on continuing debates about when, where, and why the first cities arose.

Understanding what cities in different parts of the world were like at different times in history is enriched by first-person accounts of travelers who recorded what they saw on their visits. These accounts must be read with caution as the writers often exaggerated, interpreted what they saw through the perspective of their own culture, or simply made mistakes. Marco Polo's account of thirteenth-century Kin-Sai (p. 42), Ibn Battuta's description of thirteenth-century Constantinople (p. 45), Bernal Diaz's report on Aztec Tenochtitlán (p. 47), and Albrecht Dürer's description of early modern Antwerp (p. 49) are only a small sampling of the many first-hand accounts of what different world cities were like at different times that appear in travel books and literature. In addition to Friederick Engels's first-hand description of Manchester in 1844 (p. 50), there is a large literature describing the deplorable conditions of nineteenth-century industrial cities. British social reformer Charles Booth's accounts of conditions in nineteenth-century London contain an enormous amount of shocking detail (Booth, 1888, 1892).

Artists and novelists also enrich our understanding of cities in different parts of the world and at different times in history. Hartmann Schedel's woodcuts of cities (such as the image introducing Part One), Pieter Bruegel the Elder's scenes of sixteenth-century Dutch cities, and Canaletto's paintings of Venice provide a lively record of urban life that is difficult to capture in words. In America, New York City photographer Jacob Riis's photographs of poor, largely immigrant, areas in late nineteenth-century New York City provide graphic visual proof of poverty, crowding, and social disorder that complement Engels's written description of nineteenth-century Manchester (Riis, 1890).

Historians have provided the most important scholarly works on cities at different times in history. A number of texts and anthologies are commonly used in American urban history courses (Mohl, 1997; Chudakoff and Baldwin, 2005; Chudakoff and Smith, 2005). In a 1997 article titled "New Perspectives on American Urban History," historian Raymond Mohl proposed an eleven-category typology that provides a useful framework for organizing scholarship on American urban history (Mohl, 1997). Mohl's categories are urban political history, suburbanization, city and region, sunbelt cities, technology and the city, planning and housing, the urban working class, immigration and ethnicity, African–American urban history, urban policy history, and urban culture.

Kenneth Jackson's account of the drive-in culture of contemporary America (p. 59) is taken from a leading history of suburbanization. Samuel Bass Warner's study of Boston's Streetcar suburbs (Warner, 1962) is an excellent study of the impact of electric streetcars on suburbanization in the Boston area. The large literature on suburbia also includes works by political scientists like Robert Wood (Wood, 1958) and sociologists like Mark Baldassare (Baldassare, 1986).

Robert Fishman's description of technoburbia (p. 69) is an example of a study of "technology and the city." Sir Peter Hall has also written about cities and technology (Hall, 1998).

Urban Regional histories include Mellior Scott's history of the San Francisco Bay area (Scott, 1959) and Carl Abbot's study of the Portland, Oregon metropolitan region (Abbott, 2001).

Histories of sunbelt cities generally date from the 1980s when the distinction between sunbelt and frostbelt cities appeared clearer than it does today. They include books by Carl Abbott (Abbot, 1987), Richard Bernard and Bradley Rice (Bernard and Rice, 1984), William K. Tabb and Larry Sawers (Tabb and Sawers, 1984), and Raymond Mohl (Mohl, 1990).

The subfield of urban planning history has been developed by urban planners such as John Reps (Reps, 1965), Mellior Scott (Scott, 1985), Christine Boyer (Boyer, 1986), Donald Krueckeberg (Krueckeberg, 1994), and Dolores Hayden (Hayden, 2003) in the United States and Sir Peter Hall in England (Hall, 2002a). They have been joined by historians Carl Abbot (Abbot, 2001), Mary Corbin Sies and Christopher Silver (Sies and Silver, 1996), William H. Wilson (Wilson, 1994), Thomas Bender (Bender, 1982), David Schuyler (Schuyler, 1988), and Stanley Buder (Buder, 1990), by geographer James Vance (Vance, 1990), and by architectural historians A.E.G. Morris (Morris, 1996) and Spiro Kostoff (Kostoff, 1991, 1992). Many other historians and urban planners have written about urban planning history topics as varied as the building of Athens in the fifth century BCE, Michelangelo's work in renaissance Rome, Baron Hausmann's rebuild of Paris in the mid-nineteenth century, City Beautiful planning for early twentieth century Chicago, and the vision and reality of planned garden cities in early twentieth-century England.

British historian Edward Palmer Thompson pioneered working class urban history with his study of the British working class (Thompson, 1977). Historical studies of the urban working class in America include works by Herbert Gutman (Gutman, 1977), Gary Nash (Nash, 1986), and Daniel Rodgers (Rodgers, 1978).

Notable studies of the impact of immigration and ethnicity on cities include Oscar Handlin's studies of European immigration to New York, Boston, and other east coast cities (Handlin, 1951) and Ron Takaki's studies of Asian migration to west coast cities (Takaki, 1989).

Gary Nash's study of Blacks in colonial Philadelphia (Nash, 1991) is an example of African–American urban history. Key writings on African–American urban history are collected in an anthology edited by Raymond Mohl and Kenneth Goings (Mohl and Goings, 1996).

The history of urban policy includes studies of federal-city relations during the New Deal, World War II, the great society programs of the 1960s, and more recent urban policy (Gelfand, 1975; Funigiello, 1978; O'Connor, 1999). Some very recent scholarship focuses on how American cities are responding to the terrorist threat (Eisinger, 2004). Raymond Mohl and Arnold Hirsch have edited selections on the history of urban policy (Mohl and Hirsch, 1993).

Many historians have written accounts of urban culture. Lewis Mumford's pioneering *The Culture of Cities* (Mumford, 1938), Arnold Toynbee's *Cities of Destiny* (Toynbee, 1967), and Sir Peter Hall's magisterial *Cities in Civilization* (Hall, 1998) are noteworthy books on the history of urban culture.

The evolution of cities section concludes with Robert Fishman's perceptive description of the emergence of technoburbia (p. 69) – the latest phase in a long evolution and perhaps the precursor to new urban forms described more fully in Part Eight on the future of the city. Another important piece summarizing the status of use cities today was written by Norman Glickman and other scholars at Rutgers (Whylie, Glickman, and Lahr, 1998).

URBAN CULTURE AND SOCIETY

Scholars in sociology departments within colleges of social science have made the most notable contributions to the study of urban society. Jan Lin and Chris Mele's *Urban Sociology Reader* (Lin and Mele, 2004) in the Routledge Urban Reader Series contains many of the best classic and contemporary writings by urban sociologists.

Anthropologists, located in anthropology departments within colleges of social science, have made notable contributions to the study of human society from an anthropological perspective. Several recent anthologies contain writings by urban anthropologists (Low, 1999; Low and Lawrence-Zuniga, 2003; Zenner, Zenner, and Gmelch, 2004).

Questions that have fascinated urban sociologists are the way that living in an urban environment rather than a rural area or village affects relations among individuals and the personality of individuals. Louis Wirth's 1938 essay on urbanism as a way of life (p. 90) is the classic American article on this topic. Wirth's theories are grounded in early work by the first generation of European urban sociologists.

Sociology was the first social science discipline to turn its attention to the study of cities, as Europeans experienced the massive urban growth that accompanied the industrial revolution and confronted conditions like those discussed by Friedrich Engels in his study of the condition of the working class in Manchester, England in 1844 (p. 50). French sociologist Émile Durkheim and German sociologists Ferdinand Tönnies and Georg Simmel pioneered the study of urban society.

Tönnies distinguished between two basic types of social formation – *Gemeinschaft* (community) and *Gesellschaft* (society) (Tönnies, 1887). He saw the rural-and-small-town, homogeneous, pre-industrial social structure of German society governed by tradition and informal conventions being replaced by urban, impersonal, industrial society governed by contracts and formal legal rules. Economic relations were increasingly replacing relationships based on family, ethnicity, religion, and neighborhood. Tönnies saw society moving from a type of social organization characterized by a simple division of labor and a population bound by similar values and beliefs (what he called a *mechanically solid* type of society) to a type of society characterized by a complex division of labor and formal legal rules (what he called an *organically solid* type of society).

Georg Simmel, a contemporary of Tönnies, was concerned by the way in which the changing economic and social structure of large, late-nineteenth-century cities like his home city of Berlin was affecting the mental life of individuals (Simmel, 1903). Like Simmel, Tönnies saw urbanization as having essentially negative impacts on the human personality. Wirth drew on the work of Tönnies, Simmel, and Durkheim, synthesized them, and stimulated American scholarship on these important questions.

Wirth was a member of the Chicago School of sociology. In contrast to the detached, philosophical orientation of the first generation of European sociologists, the Chicago School of sociology encouraged faculty and students to use the city of Chicago as a laboratory. Researchers studied Blacks in Chicago's Black Belt (Drake and Cayton, 1945), Chicago's Gold Coast and nearby slums (Zorbaugh, 1929), hobos (Andersen, 1923), immigrant polish peasants (Thomas and Znaniecki, 1920) and many other social groups in Chicago.

The academic concern with the loss of community that urbanization may bring has prompted many city planners from Clarence Perry (Perry, 1929) to new urbanists Andrés Duany and Elizabeth Plater-Zyberk (p. 192) to design neighborhoods to increase human interaction. Robert Putnam's brilliant "Bowling Alone" essay (p. 120) has inspired a new generation of studies about the loss of community and declining social capital and a worldwide movement to engage young people in community service.

Not all social scientists agree with the nineteenth-century European sociologists, Wirth, and other Chicago School sociologists that urbanization is destructive of social relations and harmful to the human personality. Michael Young and Peter Willmott's classic study of family and kinship in East London found positive networks in a London slum and far less community in new housing estates built for the relocated slum dwellers (Young and Willmott, 1957). A decade later sociologist Herbert Gans characterized Boston's West end as an urban village where migrants from southern Italy retained many of the supportive relationships and folkways of their villages of origin (Gans, 1969). Gans lived in and studied the same working-class Italian neighborhood where Jane Jacobs once lived and where the street ballet she describes (p. 98) took place. Jacobs believed that positive social relationships in communities like the West End could combat crime and provide a good quality of life.

Sociologists continue to study the way in which urban society affects human behavior and the human personality. The rise of the Internet, cell phones, and instant messaging offers new avenues for research on cyber communities discussed in Stephen Graham's, *The Cybercities Reader* in the Routledge Urban Reader Series (Graham, 2003).

Race, class and gender are central concerns in urban sociology. Pioneering African–American sociologist W.E.B. Dubois' study of the Philadelphia Negro (p. 103) was one of the first, and remains one of the best, accounts of urban Black society. Anthropologist Oscar Lewis developed a controversial theory that a distinct culture of poverty exists that so shapes the aspirations of poor children that it is nearly impossible for them to break out of the culture of poverty (Lewis, 1966). Sociologist William Julius Wilson's study of unemployment and the Black underclass (p. 110) and his other studies of the urban underclass continue the debate (Wilson, 1978, 1987, 1996). Sociologists are not the only scholars concerned with issues of race, ethnicity, and discrimination. Urban designer Ali Madanipour has studied the emotionally charged issues of discrimination and social exclusion (p. 158).

Sociologists, historians, planners, urban designers and others are focusing increasing attention on issues of gender and sexuality as well as race and class. *The Urban Sociology Reader* (Lin and Mele, 2004) includes some of the best recent studies of inequality and social difference (Duncan, 1978; Portes and Manning, 1986; Wacquant and Wilson, 1989; Massey and Denton, 1993; Wacquant, 1993) and of gender and sexuality in cities (Markusen, 1981; Adler and Brenner, 1997; Donham 1998; Gilbert, 1998).

Urban ethnographies – detailed field research on urban subcultures usually conducted by anthropologists – continue the tradition of urban field study pioneered by the Chicago School sociologists. Anthropologist Elliot Liebow's poignant descriptions of poor, Black street-corner men (Liebow, 1967) and homeless women (Liebow, 1995) are exemplary examples of this tradition.

A number of sociologists, architects, and anthropologists study the relationship between humans and the built environment. Yale professor of architecture, urbanism, and American studies, Dolores Hayden (Hayden, 1997), sociologist Lynn Lofland (Lofland, 1994), and other scholars study the way in which humans use public space.

Anthropologists and sociologists have also studied the way in which human beings use the space around them. Robert Sommer has observed behavioral aspects of how people use space in such diverse settings as prisons, airports, and jury rooms (Sommer, 1969). These behavioral and social science studies provide guidance to architects and planners.

Each society produces a distinct culture, and scholars have studied urban culture. Lewis Mumford pioneered the culture studies approach to studying cities (Mumford, 1938). His answer to the question "What is a City?" (p. 85) is that a city is essentially a theater for the display of human culture. Historian Arnold Toynbee (Toynbee, 1967), geographer/planner Peter Hall (Hall, 1998), sociologist Sharon Zukin (Zukin, 1995), historians Gunther Barth (Barth, 1980) and Neil Harris (Harris, 1990), geographers David Harvey (p. 225) and Edward Soja (p. 166), social critic Mike Davis (p. 178), and many other scholars, have studied urban culture. Richard Sennett has collected classic essays on the culture of cities (Sennett, 1969). *The City Cultures Reader*, edited by Malcolm Miles and Tim Hall with Iain Borden (Miles, Hall, and Borden, 2003) – another volume in the Routledge Urban Reader Series – contains writing on urban culture.

URBAN SPACE

Geographers, generally located within geography departments in colleges of social science, have made the most notable contributions to our understanding of urban space. Geographers study spatial aspects of many of the topics sociologists study: for example how different income, ethnic, racial, and religious groups within a city or region are distributed in space. Essential writings on urban geography are collected by geographers Nick Fyfe and Elizabeth Kenny in *The Urban Geography Reader* (Fyfe and Kenny, 2005). Standard urban geography texts include books by Michael Pacione (Pacione, 2001), David Kaplan, James Wheeler, and Steven Holloway (Kaplan, Wheeler, and Holloway, 2003), and Paul Knox and Linda McCarthy (Knox and MacCarthy, 2005).

Burgess's seminal essay on the internal structure of the city (p. 150) sparked a debate on alternative models that would best describe the internal structure of cities. Descriptions of land economist Homer Hoyt's competing sector model of the internal structure of the city (Hoyt, 1939) and geographers Chauncey Harris and Edward Ullman's multiple nuclei model (Harris and Ullman, 1945) are reprinted in *The Urban Geography Reader* (Fyfe and Kenny, 2005). Writings about the internal structure of cities have been collected in an anthology edited by University of Toronto geographer Larry Bourne (Bourne, 1982).

German geographer Walter Christaller (Christaller, 1933) was interested not in the internal structure of cities but in how systems of cities related to each other. Based on his study of telephone connections in southern Germany during the 1930s, Christaller formulated central place theory – a unified explanation of the way in which cities of different sizes are related to each other in a distinct hierarchy or system. Christaller's work led to many empirical studies of systems of cities. Peter Taylor's study of the world city network (Taylor, 2004) and Sir Peter Hall and Kathy Pain's study of the polycentric urban network emerging in northwestern Europe (Hall and Pain, 2006) are recent studies in this tradition. University of Toronto geographers Larry S. Bourne and J.W. Simmons have collected studies of the internal structure of cities (Bourne and Simmons, 1978).

While much current scholarship focuses on spatial issues at the global scale, many scholars are concerned with spatial issues closer to home. J.B. Jackson's description of the almost perfect town (p. 184) is an example of writings on vernacular architecture pioneered by Jackson. Other writings on vernacular architecture inspired by Jackson's work are described in the introduction to the Jackson selection.

Ian Buruma and Avishai Margolit's selection on Occidentalism (p. 136) captures two current themes in urban scholarship. There has always been a strand of anti-urban sentiment in popular and scholarly opinion.

There is a growing interest in understanding the way in which the West is perceived by non-Western cultures. Buruma and Margolit argue that Western cities, long criticized by Western intellectuals, are increasingly seen as demonic places by Muslim extremists.

The increasingly integrated world economic system and the ever-closer connections among cities around the globe have captured the attention of many scholars. Scholarship on globalization is not confined to any one academic discipline or professional field. Professors of international relations, economics, business administration, urban planning, and sociology have written about the impact of globalization on cities. Recent writings on global issues are collected in Frank Lechner and John Boli's *Globalization Reader* (Lechner and Boli, 2003) and on global cities in Neil Brenner and Roger Keil's *The Global Cities Reader* (Brenner and Keil, 2005).

Related to interest in global systems of cities is attention to urbanized *regions* – what Scottish biologist Patrick Geddes (Geddes, 1915) termed "conurbations," French geographer Jean Gottman named "megalopolises" (Gottmann, 1964), and the United Nations calls "urban agglomerations" (United Nations, 2005). The largest of these areas – with populations in excess of 10 million people – are often called megacities, or more precisely megacity regions since they usually encompass more than one city and surrounding suburban areas. In addition to Aprodicio Laquian's description of Asian megacity regions (p. 489) notable studies of megacities include studies by Peter Hall (Hall, 1966), Alan Gilbert (Gilbert, 1996), Fu-chen Lo and Yue-man Yeung (Lo and Yeung, 1996), Janet Abu-Lughod (Abu-Lughod, 2000), and Saskia Sassen (Sassen, 2001, 2006). "New regionalists" like Peter Calthorpe and William Fulton (p. 342), Stephen Wheeler (Wheeler, 2002), and Andrés Duany and Elizabeth Plater-Zyberk (p. 192) believe that urban areas should be planned at different scales from the neighborhood level to the regional level.

The postmodern perspective is important in urban geography. University of California Los Angeles geographer Edward Soja (p. 166) and other members of the Los Angeles School have developed postmodernist theories of urbanization based on their observations of Los Angeles. Mike Davis (p. 178) writes in a similar vein.

Geographers have the main claim to a powerful technology – Geographical Information Systems (GIS) software (Greene and Pick, 2005; Longley, Goodchild, Maguire, and Rhind, 2005). GIS software allows users to create digital maps and – more importantly – to apply spatial statistics to the study of urban phenomena.

URBAN POLITICS, GOVERNANCE, AND ECONOMICS

Scholars located in political science departments within colleges of social science have made the most important contributions to our understanding of urban politics. Historians, sociologists, economists, geographers, lawyers, ethnic and women's studies scholars, and others have also contributed to the study of urban politics.

Essential writings about urban politics are collected in John Mollenkopf and Elizabeth Strom, *The Urban Politics Reader* (Mollenkopf and Strom, 2006) in the Routledge Urban Reader Series. Dennis Judd and Paul Kantor have edited another anthology of writings about urban politics (Judd and Kantor, 2005). John Mollenkopf has summarized the literature on urban politics since World War II (Mollenkopf, 1975, 1994). There are three main urban politics texts (Harrigan and Vogel, 2002; Judd and Swanstrom, 2005; Ross and Levine, 2005).

Philosophers and political operatives have long opined about how cities should be governed. As early as 360 BCE Plato philosophized about the ideal Republic (Plato, 2000). In sixteenth-century Florence Niccolò Machiavelli drew on the ruthless and tumultuous politics he observed in the feuding Italian city-states around him for a classic treatise on how a prince should govern (Machiavelli, 1513). Political science curricula incorporate these writings in political theory courses.

Just as the first-person accounts of cities in different parts of the world at different times in world history help make the evolution of cities come alive, accounts by first-hand observers of urban politics enliven our understanding of urban politics. The rapid growth of cities in the United States during the nineteenth century gave rise to a particular style of urban politics that has especially intrigued political scientists and historians: machine politics controlled by urban political bosses. The first-hand accounts of nineteenth-century political bosses and machines by James Bryce (p. 215), Jane Addams (p. 221), and William Riordan, paraphrasing George Washington Plunkitt (p. 219) are among the best first-hand accounts of nineteenth-century bosses and machines. Historians like Amy Bridges (Bridges, 1987), Leo Hershkowitz (Hershkowitz, 1977), and Adam Cohen and Elizabeth Taylor (Cohen and Taylor, 1992) have studied machines and bosses. They are joined by political scientists like Milton Rakove (Rakove, 1976) and journalists like Mike Royko (Royko, 1971). Studies of bosses and machines in New York City receive the most attention because of Civil War era Boss Tweed's notably corrupt ring and other colorful New York city bosses like Plunkitt and Richard Croker. There is also a great deal of scholarship about machines and bosses in Chicago, focusing on Mayor Richard Daley Senior's phenomenally efficient and long-lasting mid-twentieth-century machine. Richard Daley's son – also named Richard Daley – was the mayor of Chicago as this book went to press in 2006.

An important political movement at the end of the nineteenth and beginning of the twentieth century – the "progressive movement" – sought to reform local government and do away with bosses and machines. The progressive agenda included replacing strong mayors elected at large with weak mayors and professional city managers appointed by the city council (the council-manager form of government), civil service commissions empowered to hire and promote qualified local government employees through competitive examinations, term limits, and the initiative, referendum, and recall to make local government officials more accountable to citizens. Roy Lubove (Lubove, 1963), John Whiteclay Chambers II (Chambers, 2000), Roman Esperjo (Esperjo, 2003), Lewis L. Gould (Gould, 2000), Michael McGerr (McGerr, 2003), and others have written histories of the urban progressive movement. Public administrators David Osborne and Ted Gaebler (Osborne and Gaebler, 1993) continue the long tradition of writing on how cities are, and should be, managed. While the progressive agenda has largely been achieved, and machines in the form they took in the nineteenth century have nearly disappeared from the American landscape, machines and machine-like politics are still discernible – particularly in older east coast and Midwestern cities. Raymond Wolfinger describes the continuing presence of political machines and the reasons they continue to exist (Wolfinger, 1973).

Another question that has long intrigued students of urban politics is who really makes decisions in local government. Sociologist Floyd Hunter developed an elitist model of community power in a classic study of governance in Atlanta, Georgia (Hunter, 1953). Hunter concluded that power resided in the hands of a few wealthy individuals. Political scientist Robert Dahl developed a competing pluralist model of urban community power in his classic study of governance in New Haven, Connecticut (Dahl, 1961). Dahl concluded that a large number of very diverse people make local government decisions. Other pluralists like Herbert Kaufman and Wallace Sayre (Kaufman and Sayre, 1965), Edward Banfield (Banfield, 1970), Douglas Yates (Yates, 1978), and Andrew McFarland (McFarland, 2004) continue to debate competing models of community power with exponents of the elitist model of community power like Peter Bachrach and Morton Baratz (Bachrach and Baratz, 1970), William G. Domhoff (Domhoff, 2005), and John Manley (Manley, 1983).

Regime theory is an approach to studying community power developed by Steven Elkin (Elkin, 1987), Clarence Stone (Stone, 1989), Richard DeLeon (DeLeon, 1992) and other political scientists. Regime theory holds that like-minded organization of local elected officials, businessmen, church leaders, unions, the media, and other interest groups work together over time to achieve a common agenda. John Logan and Harvey Molotch describe business-oriented pro-growth regimes as growth machines (Logan and Molotch, 1987). Logan and Molotch's notion of growth machines has spawned a number of related studies. Regime theory has been extended to comparative studies of cities in Europe as well as the United States (Kantor, Savitch, and Vicari, 1997). There are a number of good summaries of regime theory (Mollenkopf, 1994; Judge, Stoker, and Wolman, 1995; Stone, 1995).

Race, ethnicity, and national origin are important in urban politics. The first nineteenth-century political machines were organized among Irish immigrants. Later, central and southern European immigrants formed important voting blocs in American cities. With the civil rights movement of the 1960s and today's Hispanic and Asian immigration, the political dimensions of race and ethnicity in urban politics are taking new forms. Political scientists Rufus Browning, Dale Rogers Marshall, and David Tabb wrote a classic study of racial politics in American cities (Browning, Marshall, and Tabb, 1984) and have recently edited a volume of more recent studies on this important topic (Browning, Marshall, and Tabb, 2003).

Urban-based political formations and urban issues play an important role in national politics. Political scientists like Richard Sauerkopf and Todd Swanstrom had studied the role of the urban electorate in national elections (Sauerkopf and Swanstrom, 1999).

Marxist theorists like economists James O'Connor (O'Connor, 1973), Larry Sawers and William Tabb (Tabb and Sawers, 1978), political scientist Ira Katznelson (Katznelson, 1981), geographers David Gordon (Gordon, 1978) and Henri Lefebvre (Lefebvre, 1991), posit structuralist theories of urban political power. Structuralist interpretations argue that underlying economic and class relationships predetermine most of what is important to decide at the city level. Structuralists see city governments as unable to control national and international capitalist forces. For them, urban politics is largely irrelevant.

Academic Urban Studies programs emerged in the 1960s and 1970s largely in response to the ghetto riots of the 1960s when urban problems were on the front burner of the national agenda. The urban problems perspective on urban studies organizes subject matter around problems cities face, such as race, poverty, crime, fiscal stress, pollution, and traffic congestion. This perspective tends to neglect urban assets and positive aspects of cities. While *The City Reader* is not organized around the urban problems approach, the selection by political scientist James Q. Wilson and criminologist George Kelling on the broken windows theory of crime (p. 256) is a good introduction to the urban problems perspective. Herbert Gans's anthology *People, Plans, and Policies* (Gans, 1991) contains additional readings on urban problems.

Just as geographers are focusing increasing attention on metropolitan regions as the appropriate unit for analyzing urban problems, increasingly political scientists and policy analysts propose regional solutions to metropolitan problems. Parallel to the new urbanist movement in city building is a new regionalist movement (Wheeler, 2002). Brookings Institution policy analyst Anthony Downs's new vision for metropolitan America (p. 245) proposes many new regionalist solutions to urban problems. Bruce Katz, also at the Brookings Institution, has edited an anthology of progressive thinking about regional issues (Katz, 2000) and developed his own progressive agenda for metropolitan America (Katz, 2004).

How can decision makers involve citizens and other stakeholders in local government decision making? This is a question many students of urban governance have asked. Sherry Arnstein's ladder of citizen participation (p. 233) provides a powerful metaphor to understanding levels of participation citizens can achieve in local decision making. British planner Patsey Healey (Healey, 2006) has developed an approach to collaborative planning that provides more ideas on how citizens might move to higher rungs on Arnstein's ladder.

Politics and governance is inextricably entwined with issues of economics and finance. Studies of the economies of cities are primarily the province of economists, located within economic departments in colleges of social science (and sometimes business schools, urban planning departments, or elsewhere). The selection by Wilbur Thompson (p. 266) provides an introduction to key concepts in urban economics. Thompson's classic textbook *A Preface to Urban Economics* (Thompson, 1965) introduced the last generation of urban economic students to these ideas. Arthur O'Sullivan's text on urban economics (O'Sullivan, 2002) is widely used in urban economics courses today and is a worthy successor to Thompson's original text in this field.

Some social scientists blend urban economics and other disciplines. Economic geographers study the production and exchange of goods within and between cities. Walter Christaller, for example, considered his central place theory to be a form of economic geography (Christaller, 1933). Historians interested in both economics and cities – like Belgian historian Henri Pirenne (Pirenne, 1969), French historian Fernand Braudel (Braudel, 1996), and American sociologist Janet Abu-Lughod (Abu-Lughod, 1991) – have developed the subfield of urban economic history and have illuminated the relationship of capitalist institutions and trade to the rise of cities. Paul E. Peterson describes the limits of the ability of individual cities to solve problems within their boundaries, given the limitations on their ability to generate tax revenue and the existence of problems beyond their borders that they do not control but which affect them (Peterson, 1981). Urban political economists like William Tabb and Larry Sawers (Tabb and Sawers, 1984) blend politics and economics – usually from a critical perspective. Myron Orfield combines interests in urban politics, geography, and law. His book *American Metropolitics* (Orfield, 2002) uses GIS to map the characteristics of local governments, develops a progressive theory for regional metropolitan reform, and explains how it can be implemented in law. Political scientists Peter Dreier, John Mollenkopf and Todd Swanstrom have written about the spatial dimensions of urban problems and policy (Dreier, Mollenkopf, and Swanstrom, 2001).

Two interdisciplinary subfields closely related to economics help inform urban studies. Urban public finance is often taught in economics departments, and sometimes in business schools. It contributes to an understanding of how local governments raise revenue and what they spend it on (Rosen, 2004). The subfield of regional science includes economists, geographers, and others who study regions rather than cities (Florax and Plane, 2004). Regional scientists generally employ quantitative methodologies. They have their own professional associations and scholarly journal.

URBAN PLANNING HISTORY AND VISIONS

Urban planning is usually taught at the graduate level in departments of city and regional planning located within colleges of architecture and urban planning or environmental design that also include architecture, landscape architecture, and sometimes other programs related to the built environment. Essential writings about urban and regional planning are collected in Eugenie Birch, *The Urban and Regional Planning Reader* (Birch, 2007) in the Routledge Urban Reader Series. Urban design is often included as part of urban planning education.

How cities should be planned has excited the interest of scholars since the earliest times. In *The Republic*, the Greek philosopher Plato (Plato, 2000) envisioned the ideal polis (city-state). In 27 BCE, the Roman architect Vitruvius formulated principles for planning cities, and in 1485, during the Renaissance, Florentine architect Leon Battista Alberti reminded his peers of how cities could contribute to the rebirth of humanism.

During the Industrial Revolution unplanned, congested, polluted, slums jammed with a newly impoverished urban proletariat dominated nineteenth-century industrial cities like Manchester, England (Engels, 1845). Landscape architect Frederick Law Olmsted (Olmsted, 1870) – who created New York's Central Park – led a national urban parks movement in the United States. Civil engineers like Baron Haussman in Paris rebuilt the infrastructure of cities, and settlement house workers like Jane Addams in Chicago studied, theorized about, and agitated for the improvement of nineteenth-century industrial cities. Ebenezer Howard (Howard, 1898) envisioned humane, social, garden cities surrounded by greenbelts in place of cities of this time – and lived to see several garden cities built. But it was not until the twentieth century that academic courses and degree programs in urban planning were created.

The University of Liverpool in England and the Massachusetts Institute of Technology in the United States began the first formal courses devoted to city planning. Liverpool established a degree program in town and country planning in 1909 and MIT instituted a course in urban planning the same year.

Professors of urban planning study cities in order to help plan them. Twentieth-century land use planning books by British garden cities advocate and regional planner Thomas Adams (Adams, 1932), lawyer and zoning expert Edward Bassett (Bassett, 1938), British architect/planner Patrick Abercrombie (Abercrombie, 1933), political scientist Alan Altschuler (Altschuler, 1966), New York city planning commissioner Alexander Garvin (Garvin, 2002) and urban planning professors T.J. Kent (Kent, 1964), F. Stuart Chapin (Chapin, 1965), and Anthony Catanese and James C. Snyder (Catanese and Snyder, 1988) helped educate planners and shaped the city planning profession in the United States and the United Kingdom.

Sustainable urban development is one of the most important visions shaping urban planning today. The selection from the Brundtland Commission Report (p. 337), and writings by Timothy Beatley (p. 411) and Stephen Wheeler (p. 510) introduce this important topic.

URBAN PLANNING THEORY AND PRACTICE

The professional field of urban and regional planning is informed by planning theory and teaches planners the practice of city planning. While theory and practice are related, it is helpful to treat each separately.

Sir Peter Hall's overview of the city of theory (p. 354) is among the best short overviews of planning theory. Nigel Taylor, Susan Fainstein and Michael Teitz have written other summaries of urban planning theory (Taylor, 1999; Fainstein, 2000; Teitz, 2000). A number of anthologies collect seminal works in urban planning theory (Taylor, 1998; Allmendinger, 2002; Campbell and Fainstein, 2003).

Paul Davidoff's selection on advocacy and pluralism in planning (p. 400) is an example of a writing on planning theory that illustrates what planning theory writings are like and how they can inform planning practice. Davidoff criticized the entire way in which urban planning in the United States (and elsewhere in the world) is conducted and called for a more pluralistic approach. Davidoff's theoretical writing inspired a practical advocacy planning movement that changed urban planning practice. As cities become increasingly multicultural, effective representation of diverse interests remains an issue of fundamental importance (Qadeer, 1997).

One organization with a long history of attempting to systematize the field of urban planning practice is the International City Management Association. The ICMA has produced a leading practitioner's text on urban planning practice since the early 1940s. Examining the changing themes represented in this seminal book is one good way to understand changes in urban planning practice. Examining the contents of the most recent ICMA green book – now titled *The Practice of Local Government Planning* (ICMA, 2000) – is one important way to understand major themes in urban planning practice. Planning professor Eugenie Birch has traced changes in the ICMA greenbook over the last sixty years (Birch, 2001). Her conclusions are reprinted in *The Urban and Regional Planning Reader* (Birch, 2007).

The most recent edition of the greenbook (ICMA, 2000) contains an introduction titled "Planning for People and Places," and chapters titled "Making Plans," "Planning in the Information Age," "Population Analysis," "Environmental Analysis," "Economic Analysis," "Development Planning," "Environmental Policy," "Transportation Planning," "Housing Planning and Policy," "Community Development," "Economic Development," "Urban Design," "Zoning and Subdivision Regulations," "Growth Management," "Budgeting and Finance," "Building Consensus," "Communities, Organizations, Politics, and Ethics."

In addition to the ICMA greenbook, current texts on urban and regional planning in the United States include books by Jonathan Barnett (Barnett, 2003), John M. Levy (Levy, 2005), Edward J. Kaiser and David Godschalk (Kaiser and Godschalk, 2006). In the United Kingdom, the leading texts on urban planning practice were written by Sir Peter Hall (Hall, 2002b) and Barry Cullingworth and Vincent Nadin (Cullingworth and Nadin, 2006). These works continue the tradition of earlier urban planning texts described in the section "Urban Planning History and Visions" (p. 349).

There are many areas of urban planning practice. Among the most important are land use planning, transportation planning, environmental planning, housing and community development planning, and economic development planning.

The selection on land use planning by Edward J. Kaiser and David R. Godschalk (p. 366) summarizes urban planning practice in the important area of land use planning. Kaiser and Godschalk trace the development of land use planning practice in the twentieth century and describe how the urban general plan stalwart family tree has branched into a number of sophisticated types of plans today. Kaiser and Godschalk, with Philip Berke, have written the standard text on land use planning. (Kaiser, Godschalk, and Berke, 2006).

Transportation planning relies on quantitative methods to forecast future travel demand and to allocate it between different transportation modes, such as automobile, bus, and light rail. Transportation planning is closely allied with civil engineering. Transportation planners look at land use and estimate the number of trips a particular development pattern will likely generate, and the origins and destinations of the trips. They forecast the modal split in the future trips – how many people are likely to drive, bus, bike, or take some other form of transportation to work and other destinations. Transportation planning and policy involves much more than number crunching. It also involves normative and political judgments about what form of transportation should be used and political judgment about what is feasible. Important recent transportation planning books by Robert Cervero (Cervero, 1998) and Anthony Downs (Downs, 2004) summarize transportation policy and planning issues.

Environmental planning in the United States (and much of the rest of the world) involves the preparation of environmental impact analyses. These analyses describe the probable impacts a proposed development will have on the natural environment and propose alternatives to minimize and mitigate environmental damage. Earth Island Press is a leading publisher of books on environmental issues and environmental planning. Earth Island Press Publishers works on sustainable urban development, green urbanism, design with nature, the new regionalism, transit policy and other topics cited in this epilogue (Van der Ryn and Cowan, 1995; Cervero, 1998; Calthorpe and Fulton, 2001; Hack, 2001; Hopkins, 2001; Gillam, 2002).

In addition to substantive planning topics, urban planning involves an understanding of urban planning processes. Alan Altschuler's study titled *The City Planning Process* (Altschuler, 1966) describes the failure of planners in the mid-1960s to pay much attention to citizens and other stakeholders (topics that were not included in the ICMA greenbook at that time). John Forester's selection on planning in the face of conflict (p. 387) describes many different approaches planners take to the politics of planning, conflict resolution, and collaborative planning. Journalist Robert Caro won a Pulitzer Prize for his monumental study of Robert Moses and the politics of planning in the twentieth-century New York region (Caro, 1975). Recently Patsy Healey and others have written books advising urban planners on collaborative planning (Healey, 2006).

PERSPECTIVES ON URBAN DESIGN

The question of how to design cities people will use and enjoy has long intrigued scholars associated with the design professions and, more recently, social scientists. Roman architectural theorist Vitruvius (27 BCE), Italian renaissance architect Leon Battista Alberti (1485) and a host of other early theorists and practitioners developed theories of urban design, and architectural practitioners implemented designs for cities long before urban planning or urban design existed as recognized academic fields. Early urban design is well described by A.E.G. Morris (Morris, 1996) and Spiro Kostoff (Kostoff, 1991, 1992). Kostoff's books contain magnificent photographs illustrating urban design. These early urban designers were concerned both with practical matters such as how to create cities that functioned efficiently and with matters of aesthetics – creating attractive cities that people enjoy.

Urban planning is strongly influenced by urban design. As Peter Hall notes (p. 354), academic urban planning departments were originally located within schools of architecture, staffed mostly by faculty trained as architects and oriented toward teaching their students to produce physical designs for cities. Urban planning education emphasized urban design until the 1960s. As planning departments recognized the limitations of urban design and adopted other approaches that Hall describes, such as systems thinking and Marxist analysis, urban design moved into the background in urban planning education. Fortunately some planners, social scientists, and design professionals continued to develop the theory and practice of urban design in the 1960s, 1970s, and 1980s, and recently there has been a major resurgence of interest in urban design.

Important contributions to understanding urban design come from architects like Christopher Alexander (Alexander, 1977, 1979), sociologists like William Whyte (p. 448) and Ray Oldenburg (Oldenburg, 1997), geographers like Edward Relph (Relph, 1976), urban planners like Allan Jacobs (Jacobs, 1985, 1995), and landscape architects like Claire Cooper Marcus and Carolyn Francis (Marcus and Francis, 1997). In his magnum opus, MIT urban design professor Kevin Lynch worked out a theory of good city form (Lynch, 1988). Professors of urban design like Jon Lang (Lang, 1994) and Jan Gehl (Gehl, 2001) continue this tradition.

Jon Lang's *Urban Design: The American Experience* (Lang, 1994) contains a good description of the urban design profession. Essential readings in urban design are collected in Elizabeth Macdonald and Michael Larice's *Urban Design Reader* (Macdonald and Larice, 2006) in the Routledge Urban Reader Series.

An enduring issue in urban design is to understand how people actually use physical space. The selection by late nineteenth-century Austrian architect Camillo Sitte (p. 427) shows how careful observation (of public spaces in European cities) can help design theory. Sitte's principles are fully developed in the book from which this selection is taken (Sitte, 1889). Urban planner/urban designer Allan Jacobs's insightful little book, *Looking at Cities* (Jacobs, 1985), is an indispensable guide to observing cities.

While most of the land area of modern capitalist cities is privately owned, urban designers concentrate their attention on public space – such as streets, parks, plazas, and public buildings (Carmona, Heath, Oc, and Tiesdell, 2003). Urban design may be carried out at the level of an individual building – for example requiring an individual downtown office building to comply with urban design guidelines intended to assure that it fits into the urban fabric. Urban design can also take place at the neighborhood level (Hester, 1976) and at the level of an entire city (most notably new cities like Letchworth, England; Chandigarh, India; Brasilia, Brazil). New regionalists like Andrés Duany and Elizabeth Plater-Zyberk (p. 192), Oliver Gillam (Gillam, 2002), Stephen Wheeler (Wheeler, 2002), and Peter Calthorpe and William Fulton (p. 342) argue that entire regions should be designed.

The desire to beautify cities drove much early design theory and practice and remains an important goal of urban designers today. European monarchs and prelates commissioned architects to produce beautiful palaces, cathedrals, boulevards, and public spaces to please kings and glorify God. Great public spaces in London, Paris, Rome, and other European cities reflect the values of secular and religious elites who had the power to shape city space. In contrast, through the nineteenth century most US cities were utilitarian places dominated by business elites who valued economic efficiency over urban aesthetics. In the late nineteenth and early twentieth centuries, Chicago architect/planner Daniel Burnham led the City Beautiful movement – a movement to beautify American cities along the lines of monumental European capital cities.

Burnham's *Plan for Chicago* (Burnham, 1909), recently reprinted by Princeton Architectural Press (1996), beautifully illustrates City Beautiful ideas. A contemporary of Sitte and Burnham, Charles Mulford Robinson (Robinson, 1901) stimulated a popular civic arts movement to improve and embellish American cities.

Urban design today draws on insights from empirical research by psychologists, sociologists, and other social scientists about how people perceive space to inform urban design practice. Urban design professor Kevin Lynch blended history, sociology, psychology, and insights from the humanities in *The Image of the City* (p. 438). Sociologist William Whyte's studies of urban park and plaza design (p. 448), planner Allan Jacobs's individual and collaborative studies of streets and boulevards (Jacobs, 1995; Jacobs, Macdonald, and Rofé, 2003), and Edmund Bacon's influential writings on urban design (Bacon, 1967), all draw upon social science findings to inform urban design principles.

Urban designers often confront issues of how to relate the built environment of cities to the natural environment. Ever since the pioneering work of Frederick Law Olmsted in the nineteenth century (p. 307), landscape architects have contributed to understanding the relationship between the natural and built environments. Systematic study of how to plan human settlements in harmony with the natural environment was pioneered by Scottish biologist Patrick Geddes before World War I (Geddes, 1915) and revitalized by Scottish landscape architect Ian McHarg's approach to design with nature (McHarg, 1969). Harvard urban planning professor Anne Whiston Spirn's *Granite Garden* is a classic work on nature in cities (Spirn, 1985). Harmonizing nature and the built environment resonates with advocates of sustainable urban development such as members of the Congress for the New Urbanism (Congress for the New Urbanism, 2000).

Ultimately many urban designers consider their work to be part of the larger enterprise of designing physical spaces that reflect a unified vision of society and humankind's relation to the world. A remarkable example of one early vision of how the physical design for a community may reflect a vision of humankind's relation to the world is the Plan for St. Gall. This is a land use plan prepared by Benedictine monks in Switzerland 1,200 years ago. It has been beautifully recreated by architectural historians Walter Horn and Ernest Born (Horn and Born, 1979; Price, 1982). The creators of the Plan for St. Gall envisioned a harmonious community in which residents would perceive order from the tiniest herb garden to the farthest reaches of the universe. Architects like Frank Lloyd Wright (p. 331) in the US, Raymond Unwin in England

(Unwin, 1909), and Le Corbusier (p. 322) in France approached the design of cities in the same spirit – seeing cities as the ultimate expression of the human spirit.

THE FUTURE OF THE CITY

A final strand in the study of cities is reflection on probable, possible, and desirable urban futures. Virtually all demographers anticipate massive population growth and continuing urbanization – particularly in the Third World. The urbanization of the human population that Kingsley Davis describes (p. 17) is continuing. The twenty-first century will surely see more congestion, sprawl, pollution, exhaustion of natural resources, extinction of species, and proliferation of megacities and vast urban conurbations, even if governments intervene to plan and regulate city development far more than they have ever done in the past.

Studies of sprawl and ex-urban development raise the possibility that the growth rate in cities may slow or even that counter-urbanization may occur in the future – at least in the developed world. Some writers still subscribe to the view Melvin Webber (p. 473) advanced in 1968 that information technologies will make cities unnecessary. A few even predict a withering away of cities, though empirical research by Saskia Sassen (p. 197) and other scholars has found no evidence that this is occurring.

In addition to prediction, urban futurists have formulated different normative visions for alternative urban futures. Ebenezer Howard (Howard, 1898), Raymond Unwin (Unwin, 1909), Patrick Geddes (Geddes, 1915), and Lewis Mumford (p. 85) opposed the unfettered growth of large cities and proposed regional systems of human-scale cities. Frank Lloyd Wright (p. 331) went further and advocated Broadacre city – a neo-Jeffersonian future in which each family would live on an acre. Soviet-era anti-urbanists and Maoists during the Great Proletarian Cultural Revolution took Marx's ideas of eliminating distinctions between the city and the countryside as a call for de-urbanization. Megastructuralists like Le Corbusier (p. 322) and Paolo Soleri (Soleri, 1974) took the opposite view and proposed building huge, self-contained structures.

Technology, particularly information technology, has a profound effect on cities and systems of cities. Scholars who have written about the impact of information and other technologies on cities are not localized in any one academic discipline. Professors from information science, communications, urban planning, history, sociology, economics, and other disciplines and professional fields have contributed to the study of information technology and cities. Scholars are particularly interested in the way in which information technology will affect spatial and social relationships in the future – particularly the issues of whether global access to information will lead to larger and more powerful global cities or disperse population and power, and to what extent information technology may lead to a digital divide and widen the gap between rich and poor. The issues concerning information technology introduced by Manuel Castells (p. 478), Saskia Sassen (p. 197), and William Mitchell (p. 517) are developed in books by Stephen Graham (Graham, 1996, 2001). Graham's *The Cybercities Reader* (London and New York: Routledge, 2003) brings together many of the best writing.

There is a large literature on the impact of transportation technology on cities. Geographer James Vance has studied the relationship between transportation technologies and urban space (Vance, 1986). Historians like Sam Bass Warner (Warner, 1962), Charles W. Cheape (Cheape, 1980), and Clifton Hood (Hood, 1993) have documented the way in which the electric streetcar and urban subway systems affected the spatial structure of cities and metropolitan regions and, as a consequence, their economic and social structure. Mark S. Foster (Foster, 1981), Kenneth Jackson (Jackson, 1985), Clay McShane (McShane, 1994), and Mark Rose (Rose, 1990) have studied the impact of automobiles on cities and suburbs. They are interested in public policy regarding freeways and automobiles and its impact on urban and regional physical form, social structure, and culture. Policy analysts like Anthony Downs (Downs, 2004) and urban planners like Robert Cervero (Cervero, 1998) write about transportation policy and planning issues.

CONCLUSION

Cities are civilization and the study of cities involves the study of humankind. No one can master all there is to know about cities and there will always be more to study and to understand as cities continue to evolve. But a great deal is known about cities and how to study them. Academic scholarship about cities is necessarily diverse, but it is increasingly well developed. Hopefully this epilogue to *The City Reader* will serve as a roadmap to the further study of cities for readers who want to travel further into this fascinating realm.

Bibliography

Abbott, Carl (1987) *The New Urban America: Growth and Politics in Sunbelt Cities*, revised edition. Chapel Hill: University of North Carolina Press.

—— (2001) *Greater Portland: Urban Life and Landscape in the Pacific Northwest*. Philadelphia: University of Pennsylvania Press.

Abercrombie, Patrick (1933) *Town & Country Planning*. London: T. Butterworth Ltd.

Abu-Lughod, Janet (1991) *Before European Hegemony: The World System A.D. 1250–1350*. Oxford and New York: Oxford University Press.

—— (2000) *New York, Chicago, Los Angeles: America's Global Cities*. Minneapolis: University of Minnesota Press.

Adams, Thomas (1932) *Outline of Town and City Planning*. New York: Russell Sage Foundation.

Addams, Jane (1898) "Why the Ward Boss Rules." *The Outlook*.

Adler, Sy and Joanna Brenner (1997) "Gender and Space: Lesbians and Gay Men in The City." *International Journal of Urban and Regional Research* 16(1): 24–34.

Alberti, Leon Battista (1485) *De Re Aedificatoria*. (Translated 1991 as *On the Art of Building in Ten Books* by Joseph Rykwert, Neil Leach and Robert Tavernor. Cambridge: MIT Press.)

Alexander, Christopher (1979) *The Timeless Way of Building*. New York: Oxford University Press.

—— (1977) *A Pattern Language: Towns, Buildings, Construction*. Oxford and New York: Oxford University Press.

Allmendinger, Peter (2002) *Planning Theory*. New York: Palgrave Macmillan.

Altschuler, Alan (1966) *The City Planning Process: A Political Analysis*. Ithaca: Cornell University Press.

Andersen, Nels (1923) *The Hobo*. Chicago: University of Chicago Press.

Arnstein, Sherry (1969) "A Ladder of Citizen Participation." *Journal of the American Institute of Planners* 35(4): 216–224.

Bachrach, Peter and Morton S. Baratz (1970) *Power and Poverty: Theory and Practice*. Oxford and New York: Oxford University Press.

Bacon, Edmund (1967) *Design of Cities*. New York: Viking Press. (Revised edition 1976. New York: Penguin.)

Baldassare, Mark (1986) *Trouble in Paradise*. New York: Columbia University Press.

Banfield, Edward (1970) *The Unheavenly City*. Boston: Little Brown.

Barnett, Jonathan (2003) *Redesigning Cities: Principles, Practice and Implementation*. Chicago: American Planning Association.

Barth, Gunther (1980) *City People: The Rise of Modern City Culture in Nineteenth-century America*. Oxford and New York: Oxford University Press.

Bassett, Edward (1938) *The Master Plan, With A Discussion of the Theory of Community Land Planning Legislation*. New York: Russell Sage Foundation.

Battuta, Ibn (1962) *The Travels of Ibn Battuta 1325–1354*. Hamilton Alexander Rosskeen Gibb (ed.). London: Hakluyt Society.

Beatley, Timothy (2000) *Green Urbanism: Learning from European Cities*. Washington, DC: Island Press.

Beaverstock, Jonathan V., Richard G. Smith, and Peter J. Taylor (2000) "World-city Network: A New Metageography?" *Annals of the Association of American Geographers* 90(1): 123–134.

Bender, Thomas (1982) *Toward an Urban Vision: Ideas and Institutions in Nineteenth-century America*. Baltimore: Johns Hopkins University Press.

Bernard, Richard and Bradley Rice (1984) *Sunbelt Cities: Politics and Growth Since World War II*. Austin: University of Texas Press.

Birch, Eugenie. L. (2001) "Practitioners and the Art of Planning." *Journal of Planning Education and Research* 20: 407–422.

—— (2007) *The Urban and Regional Planning Reader*. London and New York: Routledge.

Booth, Charles (1888) "Condition and Occupations of the People in East London and Hackney." *Journal of the Royal Statistical Society* 51: 276–331.

—— (1892) *Life and Labor of the People in London, Vol. I: East, Central and South London*. London: Macmillan.

Bourne, Larry (1982) *Internal Structure of the City: Readings on Urban Form, Growth, and Policy*. Oxford and New York: Oxford.

Bourne, Larry S. and James Simmons (eds) (1978) *Systems of Cities*. Oxford and New York: Oxford University Press.

Boyer, Christine (1986) *Dreaming the Rational City: The Myth of American City Planning*. Cambridge: MIT Press.

Braudel, Fernand (1996) *The Mediterranean and the Mediterranean World in the Age of Philip II, Vol. 1*. Berkeley: University of California Press.

Brenner, Neil (1998) "Global Cities, 'Glocal' States: Global City Formation and State Territorial Restructuring in Contemporary Europe." *Review of International Political Economy* 5(1): 1–37.

Brenner, Neil and Roger Keil (eds) (2005) *The Global Cities Reader*. London and New York: Routledge.

Bridges, Amy (1987) *A City in the Republic: Antebellum New York and the Origins of Machine Politics*. Ithaca: Cornell University Press.

Browning, Rufus, Dale Rogers Marshall, and David H. Tabb (1984) *Protest Is Not Enough: The Struggle of Blacks and Hispanics for Equality In Urban Politics*. Berkeley: University of California Press.

—— (2003) *Racial Politics In American Cities*, third edition. Upper Saddle River: Addison-Wesley.

Bryce, James (1888) *The American Commonwealth*. New York: Macmillan. (Reprinted 1996. Indianapolis: Liberty Fund.)

Buder, Stanley (1990) *Visionaries and Planners: The Garden City Movement and the Modern Community*. Oxford and New York: Oxford University Press.

Burgess, Ernest W. (1925) "The Growth of a City: An Introduction to a Research Project," in Robert E. Park, Ernest W. Burgess, and Roderick McKenzie (eds), *The City*. Chicago: University of Chicago Press.

Burnhan, Daniel (1909) *Plan of Chicago*. Chicago. (Reprinted 1996. New York: Princeton Architectural Press.)

Buruma, Ian and Avishai Margolit (2004) *Occidentalism*. New York: Penguin.

Calthorpe, Peter and William Fulton (2001) *The Regional City*. Washington, DC: Island Press.

Campbell, Scott and Susan Fainstein (eds) (2003) *Readings in Planning Theory*, second revised edition. Oxford: Blackwell Publishers.

Carmona, Matthew, Tim Heath, Taner Oc, and Steven Tiesdell (2003) *Public Places – Urban Spaces: The Dimensions of Urban Design*. New York: Architectural Press.

Caro, Robert (1975) *The Power Broker: Robert Moses and the Fall of New York*. New York: Vintage.

Castells, Manuel (1999) "European Cities, the Informational Society, and the Global Economy." *New Left Review* 204 (March/April): 18–32. (Reprinted in Richard LeGates and Frederic Stout (eds), *The City Reader*, fourth edition. London and New York: Routledge.)

Catanese, Anthony James and James C. Snyder (eds) (1988) *Urban Planning*, second edition. New York: McGraw-Hill.

Cervero, Robert (1998) *The Transit Metropolis*. Washington, DC: Island Press.

Chambers, John Whiteclay, II (2000) *The Tyranny of Change: America in the Progressive Era, 1890–1920*. New Brunswick: Rutgers University Press.

Chandler, Tertius and Gerald Fox (1974) *3000 Years of Urban Growth*. New York: Academic Press.

Chapin, F. Stuart, Jr (1965) *Urban Land Use Planning*. Urbana: University of Illinois Press.

Cheape, Charles W. (1980) *Moving the Masses: Urban Public Transit in New York, Boston, and Philadelphia, 1880–1912*. Cambridge: Harvard University Press.

Childe, V. Gordon (1928) *The Most Ancient Near East*. New York: Grove Press.

—— (1951) *What Happened in History*. Oxford: Oxford University Press.

Christaller, Walter (1933) *Die Zentralen Örte in Suddeutschland*. (Reprinted 2000 as *Central Places in Southern Germany*. New York: Prentice-Hall.)

Chudakoff, Howard P. and Peter Baldwin (eds) (2005) *Major Problems in American Urban and Suburban History*, second edition. Boston: Houghton Mifflin.

Chudakoff, Howard P. and Judith E. Smith (2005) *The Evolution of American Urban Society*, sixth edition. Upper Saddle River: Pearson Education.

Cohen, Adam and Elizabeth Taylor (1992) *American Pharaoh: Mayor Richard J. Daley – His Battle for Chicago and the Nation*. Boston: Back Bay Books.

Congress for the New Urbanism (2000) "Charter of the New Urbanism." San Francisco: Congress for the New Urbanism.

Connolly, Peter, illustrated by Hazel Dodge (1998) *The Ancient City: Life in Classical Athens and Rome*. Oxford and New York: Oxford University Press.

Cronon, William (1992) *Nature's Metropolis: Chicago and the Great West*. New York: W.W. Norton.

Cullingworth, John Barry and Vincent Nadin (2006) *Town and Country Planning in England and Wales*, fourteenth edition. London: Taylor & Francis.

Dahl, Robert (1961) *Who Governs?: Democracy and Power in an American City*. New Haven: Yale University Press.

Davidoff, Paul (1965) "Advocacy and Pluralism in Planning." *American Institute of Planners Journal* 31(6) (November): 331–338.

Davis, Kingsley (1965) "The Urbanization of the Human Population." *Scientific American* 213: 40–53.

Davis, Mike (1990) *City of Quartz: Excavating the Future in Los Angeles*. London: Verso.

DeGrove, John, with the assistance of Deborah A. Miness (1992) *The New Frontier for Land Policy: Planning and Growth Management in the States*. Cambridge: Lincoln Institute of Land Policy.

DeLeon, Richard (1992) *Left Coast City: Progressive Politics in San Francisco, 1975–1991*. Lawrence: University Press of Kansas.

De Vries, Jan (1984) *European Urbanization 1500–1800*. Cambridge: Harvard University Press.

Diaz, Bernal (1521) *Verdadera Historia de la Conquista de Nueva España*. (Translated 1963 as *The Conquest of Mexico* by J. M. Cohen. New York: Penguin Books.)

Domhoff, G. William (2005) *Who Rules America, Power, Politics and Social Change*, fifth edition. New York: McGraw-Hill.

Donham, Donald L. (1998) "Freeing South Africa: The 'Modernization' of Male–Male Sexuality in Soweto." *Cultural Anthropology* 12(1): 3–21.

Downs, Anthony (1970) *Urban Problems and Prospects*. Chicago: Markham.

—— (1989) *The Need for a New Vision for the Development of Large U.S. Metropolitan Areas*. New York: Salomon, Smith, Barney Inc.

—— (1992) *New Visions for Metropolitan America*. Washington, DC and Cambridge: Brooking Institution Press and Lincoln Institute of Land Policy.

—— (1998) *The Selected Essays of Anthony Downs, Vol. 2: Urban Affairs and Urban Policy*. Cheltenham and Northampton: Edward Elgar.

—— (2004) *Still Stuck in Traffic: Coping With Peak-Hour Traffic Congestion*. Washington, DC: Brookings Institution Press.

Drake, St. Clair and Horace Cayton (1945) *Black Metropolis: A Study of Negro Life in a Northern City.* Chicago: University of Chicago Press. (Revised and enlarged edition 1993. Chicago: University of Chicago Press.)

Dreier, Peter, John Mollenkopf, and Todd Swanstrom (2001) *Place Matters: Metropolitics For The Twenty-First Century.* Lawrence: University Press of Kansas.

Duany, Andrés and Elizabeth Plater-Zyberk (1993) "The Neighborhood, the District and the Corridor," in Peter Katz (ed.), *The New Urbanism: Toward an Architecture of Community.* New York: McGraw-Hill.

Du Bois, W.E.B. (1899) *The Philadelphia Negro.* New York: Lippincott.

Duncan, James S. (1978) "Men Without Property: The Tramp's Classification and Use of Urban Space." *Antipode* 10: 24–34.

Dürer, Albrecht (1527) *Journal of a Voyage to Belgium.* Reprinted by the Philosophical Library of New York.

Eisinger, Peter (2004) "The American City in the Age of Terror." *Urban Affairs Review* 40(1): 115–130.

Elkin, Stephen L. (1987) *City and Regime in the American Republic.* Chicago: University of Chicago Press.

Engels, Friedrich (1845) *The Condition of the Working Class in England in 1844.* Leipzig. (Reprinted 1999. New York: Oxford University Press.)

Esperjo, Roman (2003) *The Age of Reform and Industrialization, 1896–1920.* San Diego: Greenhaven Press.

Fainstein, Susan S. (2000) "New Directions in Planning Theory." *Urban Affairs Review* 35(4): 451–478.

Fainstein, Susan S. and Lisa J. Servon (eds) (2005) *Gender And Planning: A Reader.* New Brunswick: Rutgers University Press.

Fishman, Robert (1987) *Bourgeois Utopias: The Rise and Fall of Suburbia.* New York: Basic Books.

—— (2000) *The American Planning Tradition, Culture and Policy.* Princeton: Woodrow Wilson Center Press.

Florax, J.G.M. and David A. Plane (2004) *Fifty Years of Regional Science.* New York: Springer.

Florida, Richard (2002) *The Rise of the Creative Class: And How It's Transforming Work, Leisure, Community and Everyday Life.* New York: Basic Books.

Forester, John (1987) "Planning in the Face of Conflict." *Journal of the American Planning Association* 53(3)(Summer): 303–314.

Foster, Mark S. (1981) *From Streetcar to Superhighway: American City Planners and Urban Transportation, 1900–1940.* Philadelphia: Temple University Press.

Friedman, John (1986) "The World City Hypothesis." *Development and Change* 17: 69–83.

Funingiello, Philip (1978) *The Challenge to Urban Liberalism: Federal-City Relations during World War II.* Knoxville: University of Tennessee Press.

Fyfe, Nicholas and Judith Kenny (2005) *The Urban Geography Reader.* London and New York: Routledge.

Gans, Herbert (1969) *The Urban Villagers.* New York: Free Press.

—— (1982) *The Levittowners.* New York: Columbia University Press.

—— (ed.) (1991) *People, Plans, and Policies: Essays on Poverty, Racism, and Other National Urban Problems.* New York: Columbia Univesity Press/Russell Sage Foundation.

Garvin, Alexander (2002) *The American City: What Works, What Doesn't,* second edition. New York: McGraw-Hill.

Geddes, Patrick (1915) *Cities in Evolution.* London: Williams & Norgate.

Gehl, Jan (2001) *Life Between Buildings: Using Public Space,* fifth edition. Copenhagen: Danish Architectural Press.

Gelfand, Mark I. (1975) *A Nation of Cities: The Federal Government and Urban America, 1933–1965.* Oxford: Oxford University Press.

Gilbert, Alan (ed.) (1996) *The Mega-City in Latin America.* New York: United Nations University Press.

Gilbert, Melissa R. (1998) " 'Race,' Space and Power: The Survival Strategies of Working Poor Women." *Annals of the Association of American Geographers* 88(4): 595–621.

Gillam, Oliver (2002) *The Limitless City: A Primer on the Urban Sprawl Debate.* Washington, DC: Island Press.

Gmelch, George and Walter P. Zenner (eds) (2001) *Urban Life: Readings in the Anthropology of the City*, fourth edition. Long Grove: Waveland Press.

Gordon, David (1978) "Capitalist Development and the History of American Cities," in William K.Tabb and Larry Sawers (eds), *Marxism and the Metropolis*. Oxford and New York: Oxford University Press.

Gottman, Jean (1964) *Megalopolis: The Urbanized Northeastern Seaboard of the United States*. Cambridge: MIT Press.

Gould, Lewis L. (2000) *America in the Progressive Era, 1890–1914*. New York: Longman.

Graham, Stephen (1996) *Telecommunications and the City*. London and New York: Routledge.

—— (2001) *Splintering Urbanism*. London and New York: Routledge.

—— (ed.) (2003) *The Cybercities Reader*. London and New York: Routledge.

Greene, Richard P. and James B. Pick (2005) *Exploring the Urban Community*. New York: Prentice-Hall.

Gutman, Herbert (1977) *Work Culture Society*. New York: Vintage.

Hack, Gary (2001) "Planning Metropolitan Regions," in Jonathan Barnett (ed.), *Planning for a New Century: The Regional Agenda*. Washington, DC: Island Press.

Hall, Peter (1966) *The World Cities*. New York: McGraw-Hill.

—— (1998) *Cities in Civilization*. New York: Pantheon.

—— (2002a) *Cities of Tomorrow: An Intellectual History of Urban Planning and Design in the Twentieth Century*, third edition. Oxford: Blackwell.

—— (2002b) *Urban and Regional Planning*, fourth edition. London and New York: Routledge.

Hall, Peter and Kathy Pain (2006) *The Polycentric Metropolis*. London: Earthscan.

Handlin, Oscar (1951) *The Uprooted: The Epic Story of the Great Migrations that Made the American People*. Boston: Little Brown, & Co. (Second edition 1996. New York: Time Warner Trade Publishing.)

Harrigan, John J. and Ronald K. Vogel (2002) *Political Change in the Metropolis*, seventh edition. New York: Longman.

Harris, Chauncey and Edward Ullman (1945) "The Nature of Cities." *Annals of the Association of Political and Social Science*: 1–17.

Harris, Neil (1990) *Cultural Excursions: Marketing Appetites and Cultural Tastes in Modern America*. Chicago: University of Chicago Press.

Harvey, David (1973) *Social Justice and the City*. Baltimore: Johns Hopkins University Press. (Reprinted 1992. Oxford: Basil Blackwell.)

—— (1997) "Contested Cities: Spatial Process and Spatial Form," in Nick Jewson and Susanne MacGregor (eds), *Transforming Cities*. London and New York: Routledge.

Hayden, Dolores (1982) *The Grand Domestic Revolution: A History of Feminist Designs for American Homes, Neighborhoods and Cities*. Cambridge: MIT Press.

—— (1997) *The Power of Place: Urban Landscapes as Public History*. Cambridge: MIT Press.

—— (2002) *Redesigning the American Dream: Gender, Housing, and Family Life*. New York: W.W. Norton.

—— (2003) *Green Fields and Urban Growth 1820–2000*. New York: Vintage.

Healey, Patsy (2006) *Collaborative Planning: Shaping Places in Fragmented Societies*, second edition. New York: Palgrave Macmillan.

Herschkowitz, Leo (1977) *Tweed's New York: Another Look*. Garden City: Anchor Press.

Hester, Randolph (1976) *Neighborhood Space*. Stroudsburg: Dowden, Hutchinson & Ross.

Hohenberg, Paul M. and Lynn Hollen Lees (1985) *The Making of Urban Europe 1000–1950*. Cambridge: Harvard University Press.

Hood, Clifton (1993) *722 Miles: The Building of The Subways and How They Transformed New York*. New York: Simon & Schuster.

Hopkins, Lewis D. (2001) *Urban Development, The Logic of Making Plans*. Washington, DC: Island Press.

Horn, Walter and Ernest Born (1979) *The Plan of St. Gall: A Study of The Architecture and Economy of and Life in a Paradigmatic Carolingian Monastery*. Berkeley: University of California Press.

Howard, Ebenezer (1898) *Garden Cities of Tomorrow*. London: Swan Sonneschein. (Original title *Tomorrow: The Peaceful Path to Real Reform*.)

Howe, Frederick G. (1915) *The Modern City and Its Problems*. New York: Charles Scribner's Sons.

Hoyt, Homer (1939) *The Structure and Growth of Residential Neighborhoods in American Cities*. Washington, DC: Federal Housing Administration. (Reprinted 1972. Washington, DC: Scholarly Press.)

Hunter, Floyd (1953) *Community Power Structure: A Study of Decision Makers*. Chapel Hill: University of North Carolina Press.

International City Management Association (2000) *The Practice of Local Government Planning*. Washington, DC: ICMA.

Jackson, John Brinkerhoff (1952) "The Almost Perfect Town." *Landscape* 2(1)(Autumn).

—— (1970) *Landscapes*. Amherst: University of Massachusetts Press.

—— (1972) *American Space*. New York: Norton.

—— (1980) *The Necessity for Ruins, and Other Topics*. Amherst: University of Massachusetts Press.

—— (1984) *Discovering the Vernacular Landscape*. New Haven: Yale University Press.

—— (1994) *A Sense of Place, A Sense of Time*. New Haven: Yale University Press.

Jackson, Kenneth (1985) *The Crabgrass Frontier: The Suburbanization of the United States*. Oxford and New York: Oxford University Press.

Jacobs, Allan (1985) *Looking at Cities*. Cambridge: Harvard University Press.

—— (1995) *Great Streets*. Cambridge: MIT Press.

Jacobs, Allan and Donald Appleyard (1987) "Toward an Urban Design Manifesto." *Journal of the American Planning Association* 53(1)(Winter): 112–120.

Jacobs, Allan, Elizabeth Macdonald, and Yodan Rofé (2003) *The Boulevard Book: History, Evolution, Design of Multiway Boulevards*. Cambridge: MIT Press.

Jacobs, Jane (1961) *The Death and Life of Great American Cities*. New York: Random House.

—— (1970) *The Economy of Cities*. New York: Vintage.

Jewson, Nick and Susanne MacGregor (eds) (1987) *Transforming Cities*. London and New York: Routledge.

Judd, Dennis R. and Paul Kantor (2005) *American Urban Politics: The Reader*, fourth edition. New York: Longmans.

Judd, Dennis R. and Todd Swanstrom (2005) *City Politics: The Political Economy of Urban America*, fifth edition. New York: Longmans.

Judge, David, Gerry Stoker, and Hal Wolman (eds) (1995) *Theories of Urban Politics*. Thousand Oaks: Sage.

Kaiser, Edward J. and David R. Godschalk (1995) "Twentieth Century Land Use Planning: A Stalwart Family Tree." *Journal of the American Planning Association* 61(3) (Summer): 365–385.

Kaiser, Edward J., David. R. Godschalk, and Philip Berke (2006) *Urban Land Use Planning*, fifth edition. Chicago: University of Illinois Press.

Kantor, Paul, Hank V. Savitch and Serena Vicari (1997) "The Political Economy of Urban Regimes: A Comparative Perspective." *Urban Affairs Review* 32: 348–377.

Kaplan, David H., James O. Wheeler, and Steven Holloway (2003) *Urban Geography*. New York: Wiley.

Katz, Bruce (ed.) (2000) *Reflections on Regionalism*. Washington, DC: Brookings Institution Press.

—— (2004) *A Progressive Agenda for Metropolitan America*. Washington, DC: Brookings Institution Center on Urban and Metropolitan Policy.

Katz, Bruce and Robert Puentes (eds) (2005) *Taking the High Road: A Metropolitan Agenda for Transportation Reform*. Washington, DC: Brookings Institution Press.

Katznelson, Ira (1981) *City Trenches: Urban Politics and the Patterning of Class in the United States*. Chicago: University of Chicago Press.

Kaufman, Herbert and Wallace S. Sayre (1965) *Governing New York City: Politics in the Metropolis*. New York: Norton.

Kent, T.J. (1964) *The Urban General Plan*. San Francisco: Chandler Co. (Reprinted 1990. Chicago: American Planning Association.)

Kenyon, Dame Kathleen Mary (1979) *Archaeology in the Holy Land*. London: Methuen.

Kitto, H.D.F. (1951) *The Greeks*. London: Penguin Books. (Revised edition 1957.)

Knox, Paul and Linda McCarthy (2005) *Urbanization: An Introduction to Urban Geography*, second edition. New York: Prentice-Hall.

Kostoff, Spiro (1991) *The City Shaped: Urban Patterns and Meanings Through History*. Boston: Bulfinch.

—— (1992) *The City Assembled: The Elements of Urban Form Through History*. Boston: Little Brown.

Kotkin, Joel. (2005) *The City: A Global History*. New York: Modern Library.

Kropotkin, Peter (1913) *Fields, Factories and Workshops: Or Industry Combined with Agriculture and Brain Work with Manual Work*. New York: G.P. Putnam's Sons.

Krueckeberg, Donald (1994) *The American Planner*, second edition. New York: Methuen.

Lang, John (1994) *Urban Design: The American Experience*. New York: Van Nostrand Reinhold.

Laquian, Aprodicio A. (2005) "The Emergence of Mega-Urban Regions in Asia," in *Beyond Metropolis: The Planning and Governance of Asia's Mega-Urban Regions*. Washington, DC and Baltimore: Woodrow Wilson Center Press and Johns Hopkins University Press.

Lechner, Frank J. and John Boli (eds) (2003) *The Globalization Reader*, second edition. Cambridge: Blackwell.

Le Corbusier (1929) *The City of Tomorrow and its Planning*. (Translated 1967 from the eighth French edition of *Urbanisme* by Frederick Etchells. Cambridge: MIT Press.)

—— (1967) *The Radiant City*. New York: Orion Press.

Lefebvre, Henri (1991) *The Production of Space*. London: Blackwell.

LeGates, Richard and Frederic Stout (1998) *Early Urban Planning 1870–1940*. London: Routledge and Thoemmes.

Levy, John M. (2005) *Contemporary Urban Planning*, seventh edition. New York: Prentice-Hall.

Lewis, Nelson P. (1916) *The Planning of the Modern City*. New York: Wiley.

Lewis, Oscar (1966) "The Culture of Poverty." *Scientific American* 215(5)(October): 19–25.

Liebow, Elliot (1967) *Tally's Corner: A Study of Negro Streetcorner Men*. New York: Little Brown.

—— (1995) *Tell Them Who I Am: The Lives of Homeless Women*. New York: Penguin.

Lin, Jan and Christopher Mele (2004) *The Urban Sociology Reader*. London and New York: Routledge.

Lo, Fu-chen and Yue-man Yeung (eds) (1996) *Emerging World Cities in Pacific Asia*. New York: United Nations Press.

Lofland, Lynn (1994) *Analyzing Social Settings: A Guide to Qualitative Observation and Analysis*, third edition. New York: Wadsworth Publishing Co.

Logan, John R. and Harvey L. Molotch (1987) *Urban Fortunes: The Political Economy of Place*. Berkeley: University of California Press.

Longley, Paul A., Michael F. Goodchild, David J. Maguire, and David W. Rhind (2005) *Geographic Information Systems and Science*, second edition. New York. John Wiley & Sons.

Low, Setha M. (1999) *Theorizing the City: The New Urban Anthropology Reader*. New Brunswick: Rutgers University Press.

Low, Setha M. and Denise Lawrence-Zuniga (2003) *The Anthropology of Space and Place: Locating Culture*. Oxford: Blackwell.

Lubove, Roy (1963) *The Progressives and the Slums: Tenement House Reform in New York City, 1890–1917*. Pittsburgh: University of Pittsburgh Press.

Lynch, Kevin (1961) *The Image of the City*. Cambridge: MIT Press.

—— (1988) *Good City Form*. Cambridge: MIT Press.

Lynch, Kevin and Gary Hack (1984) *Site Planning*, third edition. Cambridge: MIT Press.

Macdonald, Elizabeth and Michael Larice (2006) *The Urban Design Reader*. London and New York: Routledge.

McFarland, Andrew S. (2004) *Neopluralism: The Evolution of Political Process Theory*. Manhattan: University Press of Kansas.

McGerr, Michael A. (2003) *Fierce Discontent: The Rise and Fall of the Progressive Movement in America, 1870–1920*. New York: Free Press.

McHarg, Ian (1969) *Design With Nature*. New York: Natural History Press.

Machiavelli, Niccolò (1513) *Il Principe* (*The Prince*). (Reprinted 1988. Cambridge: Cambridge University Press.)

McShane, Clay (1994) *Down the Asphalt Path: The Automobile and the American City*. New York: Columbia University Press.

Madanipour, Ali, Goran Cars, and Judith Allen (eds) (1998) *Social Exclusion in European Cities: Processes, Experiences, and Responses*. London: Jessica Kingsley.

Manley, John (1983) "NeoPluralism: A Class Analysis of Pluralism I and Pluralism II." *American Political Science Review* 77(2)(June): 368–383.

Marcus, Clare Cooper and Carolyn Francis (eds) (1997) *People Places: Design Guidelines for Urban Space*. New Jersey: John Wiley & Sons.

Markusen, Ann R. (1981) "City Spatial Structure, Women's Household Work, and National Urban Policy." *Signs* (Spring): 169–190.

Marsh, Benjamin Clark (1909) *An Introduction to City Planning*. New York: B.C. Marsh.

Massey, Douglas S. and Nancy S. Denton (1993) *American Apartheid: Segregation and the Making of the Underclass*. Cambridge: Harvard University Press.

Mellaart, James (1967) *Çatal Hüyük: A Neolithic Town in Anatolia*. New York: McGraw-Hill.

Miles, Malcolm and Tim Hall, with Iain Borden (2003) *The City Cultures Reader*, second edition. London and New York: Routledge.

Mitchell, William J. (2001) *E-topia*. Cambridge: MIT Press.

Mohl, Raymond (1990) *Searching for the Sunbelt: Historical Perspectives on a Region*. Knoxville: University of Tennessee Press.

—— (1997) *The Making of Urban America*, second edition. Lanham: SR Books.

Mohl, Raymond and Kenneth W. Goings (1996) *The New African–American Urban History*. Thousand Oaks: Sage.

Mohl, Raymond and Arnold R. Hirsch (1993) *Urban Policy in Twentieth-Century America*. New Brunswick: Rutgers University Press.

Mollenkopf, John (1975) "The Postwar Politics Of Urban Development." *Politics and Society* 5: 247–296.

—— (1994) "How to Study Urban Political Power," in *A Phoenix in the Ashes: The Rise and Fall of the Koch Coalition in New York City Politics*. Princeton: Princeton University Press.

Mollenkopf, John and Elizabeth Strom (2006) *The Urban Politics Reader*. London and New York: Routledge.

Morris, Anthony Edwin James (1996) *History of Urban Form Before the Industrial Revolution*, third edition. New York: Prentice-Hall.

Mumford, Louis (1937) "What is a City?" *Architectural Record* LXXXII (November): 58–62.

—— (1938) *The Culture of Cities*. New York: Harcourt Brace.

—— (1961) *The City in History*. New York: Harcourt Brace.

Nash, Gary (1986) *The Urban Crucible: The Northern Seaports and the Origins of the American Revolution*. Cambridge: Harvard University Press.

—— (1991) *Forging Freedom: The Formation of Philadelphia's Black Community, 1720–1840*. Cambridge: Harvard University Press.

Nolen, John (1927) *New Towns for Old*. Boston: Marshall Jones.

O'Connor, Alice (1999) "Swimming Against the Tide: A Brief History of Federal Policy in Poor Communities," in Ronald Ferguson and William Dickens (eds), *The Future of Community Development*. Washington, DC: Brookings Institution Press.

O'Connor, James (1973) *The Fiscal Crisis of the State*. New York: St. Martins.

O'Sullivan, Arthur (2002) *Urban Economics*, fifth edition. New York: McGraw-Hill.

Oldenburg, Ray (1997) *The Great Good Place: Cafés, Coffee Shops, Bookstores, Bars, Hair Salons and Other Hangouts at the Heart of a Community*. New York: Marlowe & Company.

Olmsted, Frederick Law (1870) *Public Parks and the Enlargement of Towns*. American Social Science Association.

Orfield, Myron (2002) *American Metropolitics: The New Suburban Reality*. Washington, DC: Brookings Institution Press.

Osborne, David and Ted Graebler (1993) *Reinventing Government: How the Entrepreneurial Spirit is Transforming the Public Sector*. Reading: Addison-Wesley.

Owen, Robert (1816) *The New View of Society and Other Writings*. (Reprinted 1977. London: J.M. Dent & Sons.)

Pacione, Michael (2001) *Urban Geography*. London and New York: Routledge.

Park, Robert E., Ernest W. Burgess, and Roderick D. McKenzie (1925) *The City*. Chicago: University of Chicago Press.

Perry, Clarence (1929) "The Neighborhood Unit," in *Neighborhood and Community Planning: Regional Survey, Vol. VII: Regional Plan of New York and Its Environs*. New York: Russell Sage Foundation.

Peterson, Paul E. (1981) *City Limits*. Chicago: University of Chicago Press.

Pirenne, Henri (1969) *Medieval Cities: Their Origins and the Revival of Trade*. Princeton: Princeton University Press.

Plato (360 BCE) *The Republic*. (Reprinted 2000. Minneola: Dover.)

Polo, Marco (1299) *The Travels of Marco Polo*. (Translated 1984 by Teresa Waugh from the Italian by Maria Bellonci. London: Sidgwick & Jackson.)

Porter, Michael (1995) "The Competitive Advantage of the Inner City." *Harvard Business Review* (May–June): 55–71.

Portes, Alejandro and Robert D. Manning (1986) "The Immigrant Enclave: Theory and Empirical Examples," in Susan Olzak and Joanne Nagel (eds), *Competitive Ethnic Relations*. New York: Academic Press.

Price, Lorna (1982) *The Plan of St. Gall: An Overview Based on the Three-Volume Work by Walter Horn*. Berkeley: University of California Press.

Putnam, Robert (2000) *Bowling Alone*. New York: Simon & Schuster.

Qadeer, Mohammed A. (1997) "Pluralistic Planning for Multicultural Cities: The Canadian Practice." *Journal of the American Planning Association* 63(4): 481–494.

Rakodi, Carole (ed.) (1997) *The Urban Challenge in Africa*. New York: United Nations Press.

Rakove, Milton L. (1976) *Don't Make Any Waves – Don't Back Any Losers: An Insider's Analysis of the Daley Machine*. Bloomington: Indiana University Press.

Relph, Edward (1976) *Place and Placelessness*. London: Pion.

Reps, John W. (1965) *The Making Of Urban America: A History of City Planning in the United States*. Princeton: Princeton University Press.

Riis, Jacob (1890) *How the Other Half Lives*. New York: Charles Scribner's Sons.

Riordan, William (1905) *Plunkitt of Tammany Hall: A Series of Very Plain Talks on Very Practical Politics*. New York: Dutton.

Robinson, Charles Mulford (1901a) *The Improvement of Towns and Cities: Or, The Practical Basis of Civic Aesthetics*. New York: G.V. Putnam's Sons.

—— (1901b) *Modern Civic Art, Or The City Made Beautiful*. New York: G.V. Putnam's Sons.

Rodgers, Daniel T. (1978) *The Work Ethic In Industrial America, 1850–1920*. Chicago: University of Chicago Press.

Rose, Mark (1990) *Interstate: Express Highway Politics, 1939–1989*. Knoxville: University of Tennessee Press.

Rosen, Harvey S. (2004) *Public Finance*, seventh edition. New York: McGraw-Hill/Irwin.

Ross, Bernard H. and Myron A. Levine (2005) *Urban Politics: Power in Metropolitan America*, seventh edition. New York: Wadsworth.

Royal Commission on the Distribution of the Industrial Population (1940) *Report*. [The Barlow Commission Report.] London: Her Majesty's Stationery Office.

Royko, Mike (1971) *Boss: Richard J. Daley of Chicago*. New York: Dutton.

Sandercock, Leonie and Ann Forsythe (1992) "A Gender Agenda: New Directions for Planning Theory." *Journal of the American Planning Association* 58(1)(Winter): 49–59.

Sassen, Saskia (2001a) *The Global City: New York, London, Tokyo*, second edition. Princeton: Princeton University Press.

—— (2001b) "The Impact of New Information Technologies and Globalization on Cities," in Arie Graafland and Deborah Hauptmann (eds), *Cities in Transition*. Rotterdam: 010 Publishers.

—— (2006) *Cities in a Global Economy*, third edition. Thousand Oaks: Pine Forge Press.

Sauerkopf, Richard and Todd Swanstrom (1999) "The Urban Electorate in Presidential Elections." *Urban Affairs Review* 35(1) (September): 72–91.

Sayre, Wallace Stanley (1965) *Governing New York City*. New York: W.W. Norton.

Schuyler, David (1988) *The New Urban Landscape: The Redefinition of City Form in Nineteenth-century America*. Baltimore: Johns Hopkins University Press.

Scott, Mellior (1959) *The San Francisco Bay Area: A Metropolis in Perspective*. Berkeley: University of California Press.

—— (1969) *American City Planning Since 1890*. Berkeley: University of California Press.

Sennett, Richard (1969) *Classic Essays on the Culture of Cities*. New York: Appleton-Century-Crofts.

Sies, Mary Corbin and Christopher Silver (1996) *Planning the Twentieth-century American City*. Baltimore: Johns Hopkins University Press.

Simmel, Georg (1903) "Die Grossstädte und das Geistesleben." (Reprinted 1950 as "The Metropolis and Mental Life," in Hans Gerth (translator) and Kurt H. Wolff (ed.), *The Sociology of Georg Simmel*. Glencoe: Free Press.)

Sitte, Camillo (1889) *Der Städtebau Nach Seinen Künstlerischen Grundsätzen*. (Reprinted 1979 as *The Art of Building Cities: City Building According to its Artistic Fundamentals*. Westport: Hyperion Press.)

Soja, Edward (1989) *Postmodern Geographies: The Reassertion of Space in Critical Social Theory*. London and New York: Verso.

—— (2000) *Postmetropolis: Critical Studies of Cities and Regions*. London: Blackwell.

Soleri, Paolo (1974) *Arcology: The City in the Image of Man*. Cambridge: MIT Press.

Sommer, Robert (1969) *Personal Space: The Behavioral Basis of Design*. New York: Prentice-Hall.

Soria y Mata, Arturo (1997) *The Linear City*. (Translated 1998 by Marcos Diaz-Gonzales in Richard LeGates and Frederic Stout (eds), *Early Urban Planning 1890–1950, Vol. I*. London: Routledge and Thoemmes.)

Spain, Daphne (1992) *Gendered Spaces*. Chapel Hill: University of North Carolina Press.

Spirn, Anne Whiston (1985) *The Granite Garden: Urban Nature and Human Design*. New York: Basic Books.

Stone, Clarence (1989) *Regime Politics: Governing Atlanta, 1946–1988*. Lawrence: University Press of Kansas.

—— (1995) "Looking Back to Look Forward: Reflections on Urban Regime Analysis." *Urban Affairs Review* 40: 309–341. (Updated version 2005 at <http://www.bsos.umd.edu/gvpt/stone/power2.html>.)

Tabb, William K. and Larry Sawers (eds) (1984) *Marxism and the Metropolis: New Perspectives in Urban Political Economy*, second edition. London and New York: Oxford University Press.

Takaki, Ron (1989) *Strangers from a Different Shore: A History of Asian Americans*. Boston: Back Bay Books. (Revised and updated edition 1998.)

Taylor, Nigel (1998) *Urban Planning Theory Since 1945*. Thousand Oaks: Sage.

—— (1999) "Anglo-American Town Planning Theory Since 1945: Three Significant Developments But No Paradigm Shifts." *Planning Perspectives* 14(4)(October): 327–345.

Taylor, Peter (2004) *World City Network: A Global Urban Analysis*. London and New York: Routledge.

Teitz, Michael B. (2000) "Reflections and Research on the U.S. Experience," in Lloyd Rodwin (ed.), *The Profession of City Planning, Changes, Images and Challenges: 1950–2000*. Cambridge: Center for Urban Policy Research.

Thomas, William Isaac and Florian Witold Znaniecki (1920) *The Polish Peasant in Europe and America. Volume 5* of Richard G. Badger (ed.), *Organization and Disorganization in America*. Boston: Gorham Press. (Reprinted 1958. Mineola: Dover.)

Thompson, Edward Palmer (1977) *Making of the English Working Class*. New York: Vintage.

Thompson, Wilbur (1965) *A Preface to Urban Economics*. Baltimore: Johns Hopkins University Press.

—— (1968) "The City As a Distorted Price System." *Psychology Today* 2(3)(August): 28–33.

Tönnies, Ferdinand (1887) *Gemeinschaft and Gesellschaft*. (Reprinted 2002 as *Community and Society*. Mineola: Dover.)

Toynbee, Arnold Joseph (ed.) (1967) *Cities of Destiny*. New York: McGraw-Hill.

Tugwell, Rexford G. (1939) "The Fourth Power," in *Planning and Civic Comment*. Harrisburg: J. Horace McFarland.

United Nations (2005) *Urban Agglomerations 2004*. New York: United Nations.

UN-Habitat (2006) *State of the World's Cities 2006/7: The Millennium Development Goals and Urban Sustainability: 30 Years of Shaping the Habitat Agenda*. London: Earthscan.

Unwin, Raymond (1909) *Town Planning in Practice*. (Reprinted 1993. Princeton: Princeton University Press.)

US Census (2002) *Measuring America: The Decennial Censuses from 1790 to 2000*. Washington, DC: US Government Printing Office.

Vance, James (1986) *Capturing The Horizon: The Historical Geography of Transportation Since the Sixteenth Century*. Baltimore: Johns Hopkins University Press.

—— (1990) *The Continuing City: Urban Morphology in Western Civilization*. Baltimore: Johns Hopkins University Press.

Van der Ryn, Sym and Stuart Cowan (1995) *Ecological Design*. Washington, DC: Island Press.

Vitruvius (Marcus Vitruvius Pollio) (27 BCE) *De architectura*. (Reprinted 2001 as *The Ten Books of Architecture*. Cambridge: Cambridge University Press.)

Wacquant, Loïc J.D. (1993) "Urban Outcasts: Stigma and Division in the Black American Ghetto and the French Urban Periphery." *International Journal of Urban and Regional Research* 17(3): 366–383.

Wacquant Loïc J.D. and William Julius Wilson (1989) "The Cost of Racial and Class Exclusion in the Inner City." *Annals of the American Academy of Political and Social Science* 501: 8–26.

Warner, Samuel Bass (1962) *Streetcar Suburbs*. Cambridge: Harvard University Press.

Webber, Melvin (1968) "The Post-City Age." *Daedalus* 97(4)(Fall): 1106–1107.

Weber, Adna Ferrin (1899) *The Growth of Cities in the Nineteenth Century*. New York: Macmillan.

Wheeler, Sir Robert Eric Mortimer (1966) *Civilizations of the Indus Valley and Beyond*. New York: McGraw-Hill.

Wheeler, Stephen (2002) "The New Regionalism." *Journal of the American Planning Association* 68(3): 267–278.

—— (2007) "Planning Sustainable and Livable Cities," in Richard LeGates and Frederic Stout (eds), *The City Reader*, fourth edition. London and New York: Routledge.

Wheeler, Stephen and Timothy Beatley (2004) *The Sustainable Urban Development Reader*. London and New York. Routledge.

Whyte, William H. (1989) *City*. New York: Doubleday.

Williams, Raymond (1999) "Metropolitan Perceptions and the Emergence of Modernism," in *The Politics of Modernism*. London: Verso.

Wilson, James Q. and George L. Kelling (1982) "Broken Windows." *Atlantic Monthly* 249(3)(March): 29–38.

Wilson, William H. (1994) *The City Beautiful Movement*. Baltimore: Johns Hopkins University Press.

Wilson, William Julius (1978) *The Declining Significance of Race*. Chicago: University of Chicago Press.

—— (1987) *The Truly Disadvantaged*. Chicago: University of Chicago Press.

—— (1996) *When Work Disappears*. New York: Knopf.

Wirth, Louis (1938) "Urbanism as a Way of Life." *American Journal of Sociology* 44: 3–24.

Wolfinger, Raymond (1973) "Why Political Machines Have Not Withered Away and Other Revisionist Thoughts." *Journal of Politics* 35(1)(February): 204–207.

Wood, Robert (1958) *Suburbia: Its People and Their Politics*. Boston: Houghton Mifflin.

Woolley, Leonard (1928) *The Sumerians*. Oxford: Oxford University Press.

—— (1930) *Ur of the Chaldees: A Record of Seven Years of Excavation*. New York: Scribner.

World Bank (2005) *World Development Indicators 2004*. Washington, DC: World Bank.

World Commission on Environment and Development (1987) *Our Common Future*. Oxford and New York: Oxford University Press.

Wright, Frank Lloyd (2002) *Broadacre City: A New Community Plan*. Scottsdale: Frank Lloyd Wright Foundation.

Wyly, Elvin K., Norman J Glickman, and Michael L. Lahr (1998) "A Top 10 List of Things to Know About American Cities." *Cityscape* 3(3): 7–32.

Yates, Douglas (1978) *The Ungovernable City: The Politics of Urban Problems and Policy Making*. Cambridge: MIT Press.

Young, Michael and Peter Willmott (1957) *Family and Kinship in East London*. London: Routledge & Kegan Paul.

Zorbaugh, Harvey Warren (1929) *Gold Coast and the Slum*. Chicago: University of Chicago Press.

Zukin, Sharon (1995) *The Cultures of Cities*. Oxford: Blackwell.

Illustration Credits

Every effort has been made to contact copyright holders for their permission to reprint plates in this book. The publishers would be grateful to hear from any copyright holder who is not acknowledged and will undertake to rectify any errors or omissions in future editions of this book. Following is copyright information for the plates that appear.

19 "Ford Plant, Detroit." Charles Sheeler (1927). © Museum of Modern Art, New York. Used by permission of the Museum of Modern Art, New York.

20 "The Critic." Weegee (Arthur Fellig) (1943). © 1994, International Center of Photography, New York. Bequest of Wilma Wilcox. Used by permission of the International Center of Photography.

21 Still from *The Crowd*. King Vidor (1926). Museum of Modern Art, New York, Film Stills Archive. © Museum of Modern Art, New York. Used by permission of the Museum of Modern Art, New York.

22 Still from *Metropolis*. Fritz Lang (1929). Museum of Modern Art, New York, Film Stills Archive. © Museum of Modern Art, New York. Used by permission of the Museum of Modern Art, New York.

23 Street in Seaside, Florida. © Duany–Plater-Zyberk. Used by permission of Duany–Plater-Zyberk.

24 North Garden, Mall of America, Minneapolis. © Mall of America/Simon MOA Management Co., Inc. Used by permission of Mall of America.

25 The Petronas Twin Towers, Kuala Lumpur. © Roger Mellor. Used by permission of the photographer.

26 New office buildings in Shanghai, China. © Peter Bialobrzeski. From Peter Bialobrzeski, *Neon Tigers: Photos of Asian Megacities* (Ostfildern, Germany: Hatje Cantz Verlag). Used by permission of Redux Pictures 116 E, 16th Street, 12th Floor, New York, NY 10003.

27 Neighborhood market in Rome. © Patti Walters. Used by permission of the photographer.

28 Sumner Street, San Francisco. © Derek Metzger. Used by permission of the photographer.

29 Walled plaza in Fez. © Paul Turner. Used by permission of the photographer.

30 Skateboarder along the San Francisco waterfront. © StoutFoto, San Francisco. Used by permission of the photographer.

31 Anti-globalization demonstration, Seattle, Washington. © Jason Morrison, Pacific Institute. Used by permission of the photographer.

32 World Trade Center under attack. © Wanda McCormick, Readio.com. Used by permission of the photographer.

33 Central Park, New York, 1863. © Museum of the City of New York. Used by permission of the Museum of the City of New York.

34 Arturo Soria y Mata's plan for a linear city around Madrid, 1894. From Arturo Soria y Mata, "The Linear City" in Richard LeGates and Frederic Stout (eds), *Early Urban Planning 1870–1940* (London: Routledge/Thoemmes, 1998).

35 Ebenezer Howard's plan for a Garden City, 1898. From Ebenezer Howard, "Garden Cities of To-morrow" in Richard LeGates and Frederic Stout (eds), *Early Urban Planning 1870–1940* (London: Routledge/ Thoemmes, 1998).

36 Plan for Welwyn Garden City, 1909. Royal Town Planning Institute. Public domain.

37 Le Corbusier's "Plan Voisin" for a hypothetical city of three million people, 1925. From Le Corbusier, *Urbanisme* (London: John Rodher, 1929). Public domain.

38 Plan for Radburn, New Jersey, 1929. Clarence Stein and Henry Wright architects. Public domain.

39 Frank Lloyd Wright's plan for Broadacre City, 1935. Frank Lloyd Wright, *When Democracy Builds* (Chicago: University of Chicago Press, 1945).

40 Paseo del Rio, San Antonio Texas. © Alexander Garvin. Used by permission of Alexander Garvin.

41 Quincy Market, Boston, Massachusetts © Alexander Garvin. Used by permission of Alexander Garvin.

42 Peter Calthorpe's plan for "The Crossings," Mountain View, California © Peter Calthorpe Associates. Used by permission of Peter Calthorpe.

PART PAGE ILLUSTRATIONS

PROLOGUE Lisa Ryan, 1998. © Lisa Ryan. Used by permission of Lisa Ryan.

PART 1 Hartman *Schedel*, WeltChronik (Nuremberg, 1493). Public domain.

PART 2 Gustav Doré, "London Traffic," 1872. Public domain.

Copyright Information

PROLOGUE

1 THE EVOLUTION OF CITIES

4 URBAN POLITICS, GOVERNANCE, AND ECONOMICS

5 URBAN PLANNING HISTORY AND VISIONS

6 URBAN PLANNING THEORY AND PRACTICE

7 PERSPECTIVES ON URBAN DESIGN

Index